Motor Control Technology for Industrial Maintenance

◀ Thomas E. Kissell ▶

Terra Community College
Fremont, Ohio

Upper Saddle River, New Jersey
Columbus, Ohio

Library of Congress Cataloging-in-Publication Data

Kissell, Thomas E.
Motor control technology for industrial maintenance /
Thomas E. Kissell.
p. cm.
Includes index.
ISBN 0-13-032291-1
1. Electric motors—Electronic control. 2. Electric motors—Repairing.
3. Electric machinery—Maintenance and repair. I. Title.
TK2851 T54 2002
621.46—dc21 2001034590

Editor in Chief: Stephen Helba
Executive Editor: Ed Francis
Production Editor: Christine M. Buckendahl
Production Coordinator: Lisa Garboski, bookworks
Design Coordinator: Robin G. Chukes
Cover Art: Neal Moss
Production Manager: Brian Fox
Marketing Manager: Jamie Van Voorhis

This book was set in Century Book by Carlisle Communications, Ltd., and was printed and bound by Courier/Kendallville, Inc. The cover was printed by The Lehigh Press, Inc.

Pearson Education Ltd., *London*
Pearson Education Australia Pty. Limited, *Sydney*
Pearson Education Singapore Pte. Ltd.
Pearson Education North Asia Ltd., *Hong Kong*
Pearson Education Canada, Ltd., *Toronto*
Pearson Educación de Mexico, S.A. de C.V.
Pearson Education—Japan, *Tokyo*
Pearson Education Malaysia, Pte. Ltd.
Pearson Education, *Upper Saddle River, New Jersey*

10 9 8 7 6 5 4 3 2 1
ISBN 0-13-032291-1

To Kathy
Because she is so special

◀ Preface ▶

This book was written to respond to the needs of modern industry to increase the knowledge base of every person who must install, troubleshoot, and maintain equipment with electrical controls. In the past, this job was delegated directly to only electrical technicians. Today, every company expects a wide variety of personnel to troubleshoot and repair these systems, including maintenance personnel, electricians, electronics technicians, automation technicians, and engineers.

Most of the motor control textbooks on the market today were written for advanced technicians and engineers. This book is designed to be used by anyone who must understand the operation of motor control components in motor control circuits and know how to troubleshoot large complex systems as well as individual components. The book is written so someone who has completed electrical classes in AC and DC fundamentals can easily understand how motor control components operate, and how to install or troubleshoot them. Basic AC and DC theory is integrated directly into each chapter at the point where the theory is used to help reinforce the basic theory students learned in their earlier courses. The author has found that this technique is also very useful for students who learned AC and DC electrical theory several months prior to taking the motor controls class, or for students who really did not understand how the AC and DC electrical theory actually works in application. Objectives are provided at the front of the chapter, and a comprehensive set of questions is provided at the end of each chapter. The questions can be assigned as homework or be used as a final test for each chapter.

The book is designed to be used for a one- or two-course sequence in motor controls. The author intentionally included more information than most classes have time to cover. The large amount of detail is provided so students can refer to the book as a reference when they are on the job and need the details to troubleshoot a component or circuit. The book is also designed so that each chapter is complete and can stand alone. This allows instructors to use only the chapters in their course syllabus, and use the remainder of the book for reference. Another option is to use the book for other electrical courses. For example, Chapter 13 (DC Motors), Chapter 14 (AC Motors), and Chapter 17 (DC and AC Drives) provide enough detail that they can be used in a motors class. Chapter 18 (Programmable Controllers) is complete enough to be used in a PLC class. This would allow students to pay for the book once and use it for multiple classes, which will save them money.

In the past 20 years the technology involved with motor controls has changed drastically, to the point of overwhelming some maintenance personnel and technicians today. Early motor control circuits consisted primarily of a motor starter and several pilot switches, such as pushbuttons and limit switches. When the pilot switches were closed, the motor would run, and if any of the switches were open, the motor would be deenergized.

Today, the technician is responsible for installing, programming, troubleshooting, and repairing programmable controllers and other automated systems that are considered part of the motor control system. A maintenance person or electrical technician in any factory today will probably be expected to work on a traditional motor control circuit with motor starters and pilot devices, process control systems for process heating, motion control systems that include servo systems or robotic applications, and sophisticated electronic and microprocessor-controlled devices.

Since these control systems represent a wide variety of technologies and theories, it is important that a book be available to explain all these systems at an introductory level, so that they are easy to understand. More complex material is added to each chapter so that each chapter contains sufficient information to allow an engineering student to design and install complex motor control systems. The book also provides extensive reference information that is usable when students get jobs and begin to work on the factory floor or for students who are currently working on the job. The book also shows students the value of hands-on learning. Each chapter provides detailed diagrams that can be wired and tested to develop wiring skills for installation and troubleshooting skills for making repairs.

Most students entering the electrical field do not understand that most equipment found in factories today has been designed to operate for 20 years or more. This means that numerous motor control systems that they will be expected to maintain will have been in existence for many years. Many motor control texts do not include these older circuits and components, and maintenance personnel do not have a resource to find out how these older components and circuits operate and how they should troubleshoot them.

The author has taken great pains to ensure that theory of operation, installation, and troubleshooting procedures for these traditional circuits are included in each chapter, so that new technicians can be introduced to them and understand their value. This text also provides information about the newest control circuits being brought into factories that are integrated with solid-state electronics, microprocessor controls, and programmable controllers. The theory of operation and troubleshooting techniques are presented for these types of equipment, too. The book goes into great detail to explain the implementation of old and new controls to programmable controller systems.

Troubleshooting is included in each chapter at the point where each motor control concept is introduced. The final chapter, Chapter 20, also provides a comprehensive in-depth study of how students should approach troubleshooting when they are on the job. This chapter is also unique in that it provides detailed instructions on how to convert a ladder diagram to a wiring diagram, and how to convert a wiring diagram to a ladder diagram. Students must learn to do these diagram conversions when they are troubleshooting so that they fully understand the importance of each.

The author provides information in this text that can be obtained only from years of experience on the factory floor, installing, troubleshooting, and maintaining motor controls and working with the best troubleshooting minds and technicians in the industry today. The massive amount of information is presented in such a way that new technicians can utilize it immediately and continually refer to it as they gain more experience. The book is designed to be the source of knowledge on electrical controls that a student will need to succeed.

Acknowledgments

I would like to thank my wife, Kathy, for all her help as graphic artist and project manager of this book in keeping all the figures and diagrams straight. I also want to thank her for her support and for putting up with all the activities associated with writing this book. I also thank my daughter, Kelly Starkey, who provided so much help in reviewing this book and helping with the questions and answers in the Instructor's Manual.

I want to thank Ed Francis for his encouragement and help at all stages of this book. I also thank Lisa Garboski, bookworks, for performing the magic that makes this book come to life.

Thanks also to the reviewers of this text for their helpful comments and suggestions: George Kracke, Ivy Tech State College; Jholam Massiha, University of Louisiana; and James E. Temple, Pennsylvania College of Technology.

Finally, I thank all of the following people and their companies for helping me obtain the pictures and diagrams for this book. Without their help, it would not be possible to write this book.

Kathy McCoin, Acme Electric Corp.;
Steve Rosenthal, Danaher Controls;
Paul Handle, Eaton Cuttler Hammer;
Julie Cohen, Square D/Schneider Electric;
Patricia Pruis, GE Motors and Industrial Systems;
Ted Suever, Triplett Corporation;
Rich Kirwin, Snap-on Tools;
Kate Wagner, Rockwell Automation;
Dennis Bush, A.O. Smith Electrical Products Company;
Jenifer Turski, Ridgid Tools;
Russ Stroback, Parker Hannifin Corporation;
Nancy Kelly and Dennis Berry, National Fire Protection Association;
Karen Bosco, Motorola Semiconductor;
Carl Fields, Electrical Apparatus Service Association Inc.;
Mark Lovell, Pepperl + Fuchs, Inc.;
Beverly Summers, Fluke Corporation;
Ron Powell, Klein Tools;
Harold Ray, Leonard Safety Equipment;
Peter Haywood, Ridge Tool Company;
Richard Kirwin, Snap-on Tools;
Sandy Turner, Greenlee Textron;
Cy Pfeifer, AMETEK National Controls;
Christian Benson, Banner Engineering Corp.;
Mike Delantoni, IDEC Corporation;
Dennis Kennedy, Cooper Bussmann;
Mark Lewis, Omron Electronics.

Thomas E. Kissell

◀ Contents ▶

Chapter 13

DC Motors 227

Chapter 14

AC Motors 245

Chapter 15

Motor Control Circuits 273

Chapter 16

Advanced Motor Control Circuits: Accelerating and Decelerating Circuits 283

Chapter 17

AC and DC Drives 305

Chapter 18

Programmable Controllers 321

◀ About the Author ▶

Thomas Kissell has been a technical instructor and consultant in the field of industrial automation at Terra Community College in Fremont, Ohio for over twenty years. He has spent the majority of his career working directly with area industries, where he taught electricians and technicians directly on the factory floor. This training included installation, troubleshooting, and repair of major automation and motor control systems. The author has integrated this practical approach to troubleshooting and repair into this text, which includes the theory of operation of all traditional motor control components and circuits as well as state-of-the-art controls and control systems used with programmable controllers and robots.

Over the years Thomas Kissell has worked directly with the following companies: Ford Motor Company, General Motors, Texas Instruments, Union Carbide, Libbey Owens Ford, American National Can Company, TRW, Whirlpool Inc., General Mills, Campbell Soup Co., and Heinz USA. The author taught courses to technicians and engineers in these companies, and he was able to work side by side with some of the best electricians and electrical technicians in the world, troubleshooting their most complex problems. At these times, the author became their student, and they taught him the tricks and troubleshooting procedures that consistently located and repaired electrical problems. These procedures have been documented and are included in every chapter of this text.

Thomas Kissell has written numerous other books, including *Understanding and Using Programmable Controllers; Motors, Controls and Circuits for Air Conditioning and Refrigeration Systems; Modern Industrial/Electrical Motor Controls; Electricity Fluid Power and Mechanical Systems for Industrial Maintenance; Electricity, Electronics and Control Systems for HVAC; Industrial Electronics*, 2nd ed.; and a lab manual for industrial electronics.

◀ Chapter 1 ▶

Shop Safety and Shop Practices

Objectives

After reading this chapter you will be able to:

1. List typical safety clothing and equipment you should use on the job.
2. Develop a plan and implement it for fire safety.
3. Inspect machines to ensure safety guards are in place and operating.
4. Identify problems with housekeeping to avoid accidents.

The most important part of working on the factory floor is personal safety. You are responsible for your own safety and the safety of others working with you. In this chapter you will learn about safety in regards to protecting yourself, protecting others, and basic machine safety that will ensure a safe workplace. You will also learn about basic shop practices that are created to help maintain a safe environment in the workplace. Your instructor will use many of the same safety rules and practices in your lab at school to ensure your safety in that work environment as well as to help you learn them.

1.1 Safety Glasses, Protective Clothing, and Equipment

When you are working on the factory floor you will be required to wear a variety of safety gear, such as safety glasses and earplugs, to protect yourself. Safety glasses with shatterproof glass lenses and side shields should be worn in your lab and the factory at all times to protect your eyes against foreign debris. Problems can also occur with electrical equipment if there is an electrical short; the metal from electrical contacts can turn into molten metal and spray into your face and eyes. Safety glasses with side shields will protect your eyes against these problems. Make sure your safety glasses are comfortable to wear over extended periods. You can get safety glasses with prescription lenses to match your regular glasses.

Figure 1–1 shows examples of several types of eye-protection equipment. Safety glasses, safety goggles, and safety shields are shown. Safety goggles with colored lenses are required when you are using torches or welding equipment. Safety shields must be used when you are mixing or pouring dangerous liquids that might splatter and get in your eyes or on your face.

In most factory settings you will also need to wear ear protection, which may include earplugs or ear muff protectors or both. Figure 1–2 shows typical earplugs and ear muffs. The conditions in your factory can be tested to see the decibel level of noise and the type of ear protection that will be best over a long period of time. Ear damage

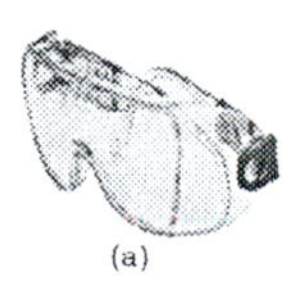
(a)

(b)

(c)

Figure 1–1 (a) Safety glasses; (b) safety goggles; (c) face shield.
(Courtesy of Leonard Safety Equipment)

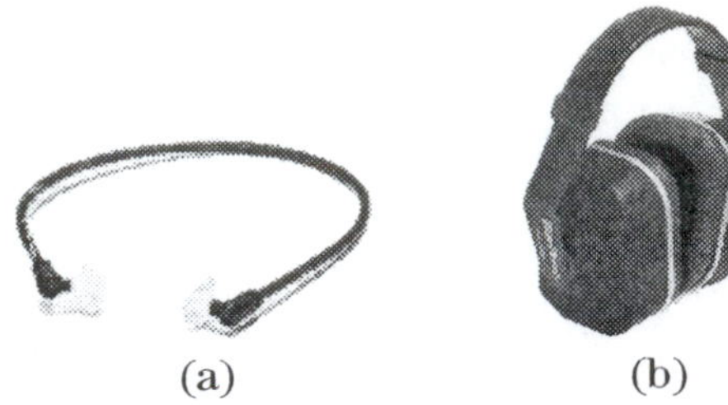

(a) (b)

Figure 1–2 (a) Earplugs; (b) ear muffs.
(Courtesy of Leonard Safety Equipment)

occurs very slowly over time and it is almost impossible to detect without sophisticated testing equipment. On any given day, you will not realize the noise is causing any damage, but after a number of years you may find that you cannot understand certain conversations around you or that you may need to turn up the volume on the television or radio that you use. All excessive noise in a factory is dangerous and will cause permanent damage to your hearing over time. Certain frequencies of noise will cause the deterioration to occur more quickly. You should begin to wear hearing protection as soon as you enter areas where you are exposed to noise, and continue to wear it at all times as long as you are in the exposed area.

You should also be aware of loose clothing and jewelry when you are working around equipment that has moving parts or if you work in electrical cabinets. Jewelry such as watches, finger rings, and other body rings needs to be removed when you are working where these items can be snagged on equipment or if they come into contact with exposed electrical terminals. Many accidents occur each year when jewelry becomes snagged on equipment and causes a laceration to the skin or broken fingers. In some cases, the jewelry may become entangled in equipment such as gears or belts and personnel may be pulled deeper into the machine where extreme injury or death may result.

Protective clothing may be required in applications where welding or cutting occurs. Other protective clothing and respirators may be required where spray painting or other similar activities occur. You will receive a briefing before beginning any job where protective clothing is required, and you should always follow the instructions provided.

1.1.1 Back Brace for Lifting and Carpel Tunnel Braces

Certain activities in industry require repeated activity that aggravates specific muscle groups. For example, if you must lift loads over ten pounds repeatedly, you should wear a back brace that is specifically designed to help prevent back muscle strains. Other problems may occur if your job requires repetitive motions, such as tightening nuts and bolts continually: The muscles in your hands and wrist will begin to weaken and they may begin to cramp. This condition can be corrected or prevented by special braces that provide support for the muscles. In some instances, the muscles are so deteriorated that the person must be removed from the job and given a different job. Some employers recognize the activities that cause these types of injuries and either provide protection or replace the job with some type of automation. You can help prevent these types of injuries by wearing the appropriate braces. You should also request help in lifting larger weights or determine the maximum weight a person should lift without equipment.

1.1.2 Steel-Toed Shoes and Hard Hats

Another safety concern in most factories involves protecting your head against falling objects or bumping into solid objects with your head. A hard hat will protect you against these hazards. Figure 1–3 shows an example of a typical hard hat.

In some factories, you will be required to wear steel-toed shoes to prevent injuries due to dropping heavy objects on your feet. Steel toes are available in a wide variety of shoes and boots. Figure 1–4 shows steel-toed boots and steel-toed tennis shoes. This al-

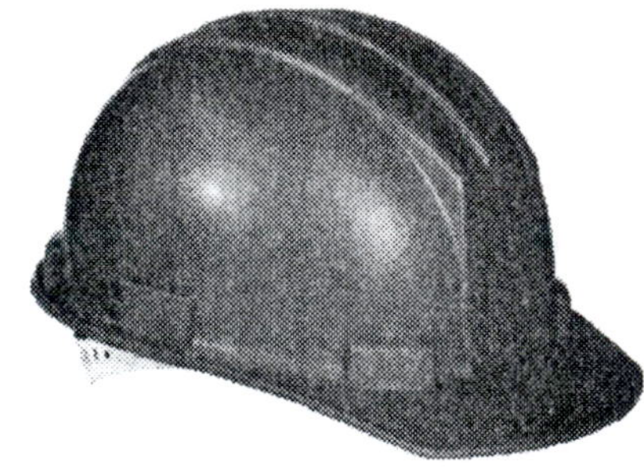

Figure 1–3 Hard hat.
(Courtesy of Leonard Safety Equipment)

(a) (b)

Figure 1–4 (a) Steel-toed boots; (b) steel-toed tennis shoes.
(Courtesy of Leonard Safety Equipment)

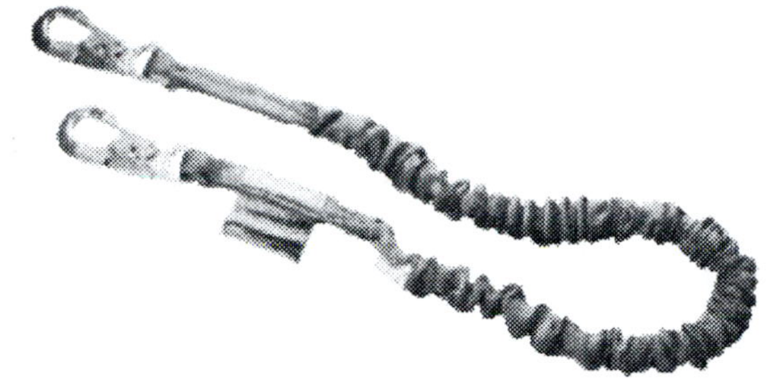

Figure 1–5 Safety harness.
(Courtesy of Leonard Safety Equipment)

lows you to wear shoes or boots that are both comfortable and safe. It is important to wear steel-toed shoes at all times in areas where they are required and there is a danger of foot injuries. Some shoes are also specially fitted with steel soles to prevent punctures from walking on sharp objects.

1.1.3 Safety Harness

At times you may need to work on areas above the floor. In most factories, electrical technicians may need to check fuses in an overhead bus duct or to change light bulbs in overhead fixtures. Any time you are working above the floor you should wear a safety harness. Figure 1–5 shows a typical safety harness that will protect you from falling. You must attach one end of the safety harness to your safety belt and the other ends must be connected to the equipment you are working on. If you slip and fall, the safety harness will prevent you from falling to the floor.

1.2 Safety Mats and Equipment to Prevent Fatigue

Some jobs in the factory require standing on concrete floors for long periods of time. In these conditions a cushion safety mat may be used to help absorb some of the shock and limit fatigue. Other mats are provided to keep personnel from coming into contact with wet or damp surfaces. Some automation applications that have robots or presses with moving parts provide mats that have special sensors in them to ensure that personnel are not in the area of the machine's moving parts when the machine starts. These mats may include safety light curtains that will automatically stop the machine if you move off the mat or break the light curtain.

1.3 Safety Light Curtains and Two-Hand Start Buttons

Safety light curtains are used to ensure that personnel are not in the work area of automation and machines. These safety devices are specifically designed to prevent machine operation where personnel may be injured. These are called active safety devices. They include machine start buttons that require two hands so that the operator's hands are not near any moving parts of the machine.

Many machines have two pushbuttons that are located approximately 2 feet apart. Both pushbuttons must be depressed at the same time to start the machine. Since the pushbuttons are mounted 2 feet apart, the operator must have one hand on each switch to start it. This ensures that both of the operator's hands are clear of the machine, so the operator will not be injured. These starting controls are added to machines that have moving parts that may injure hands if they are not clear when the machine is in operation. The two-handed pushbuttons are designed so they cannot be held down or bypassed in some manner. Since there are two switches, operators may try to hold one pushbutton down with a screwdriver or other tool to bypass the safety feature so they can start the machine with one hand. This is usually done to speed up loading and unloading the machine. The two-handed pushbutton circuit is specifically designed so that if one button is held down, the machine will not restart. Do not try to disable these types of safety circuits under any circumstances since the operator can be seriously injured.

1.4 Safety Gates and Shields

Another type of safety device that is added to machines in a factory is a safety gate and safety fence around large-scale automation such as robot cells. In some factories one or more robots are used to stack boxes onto pallets or other jobs where personnel may be in close proximity. In these applications the equipment, robots, and machines are enclosed behind fencing. When personnel must work on the equipment inside the fence, they must enter through a safety gate that is wired so that all motion inside the fence is stopped anytime the gate has been opened. When personnel leave the fenced-in-area, the gate is closed and the machines are reset.

1.4.1 Machine Guards

Machine guards are provided to keep hands and clothing out of rotating parts on machines such as shafts, gears, and belts. These guards are designed to be removed easily for maintenance and repairs, but they should remain in place any time the machine is operating. If a guard is removed during maintenance, do not try to restart the machine until the guard is back in place.

1.4.2 Glass Machine Guards

Some machines have guards that are see-through, such as glass guards on grinders and glass doors. Glass allows you to see into the moving part of the machine without removing the guard completely. For example, the glass guard on grinding machines is intended to be in place when the grinding operation is occurring, yet allow the operator to view through the guard.

Over time the glass may become dirty or get scratched to the extent that you cannot see through it. If the safety glass is dirty or damaged, it should be cleaned or replaced so that employees can utilize it as intended.

1.5 Checking Guards and Active Safety Devices

One of the most important activities that needs to be completed every day is a walk-around check to ensure that safety guards are in place and that all active safety devices are operational. This type of check is called a daily PM (preventative maintenance) check. This check should be made each day to ensure the safety of all personnel as they work around equipment or in dangerous areas of the factory. A more complete test and cleaning of these safety devices should be scheduled approximately twice a year. During this more in-depth test, the safety guards should be inspected for wear or deterioration of mounting mechanisms to ensure that the guards will withstand normal use and wear over the next 6 months.

A formal procedure should be provided that allows employees to write up any piece of equipment that has a missing guard or a piece of safety equipment that is not operating correctly. The procedure should include a means to inspect and repair the problem as quickly as possible. A complete record should be kept on each piece of equipment to track these types of safety problems so that inspections and repairs can be made on similar equipment in the factory. This is very important where a factory has more than one of any machine, such as multiple plastic presses. If a safety guard fails on one machine, it is a good bet that the same problem will occur on the other similar machines, especially if they are identical in make and model.

1.6 Cleanliness and Shop Housekeeping

One of the major conditions that affect shop safety is the cleanliness of the shop. Many accidents are caused by a cluttered shop where things are not put away or cleaned up. A routine should be developed to clean the shop on a regular basis. It is also important to have proper locations to store everything that is used around machines on the shop floor. When you locate oil spills or water spills you should make sure that they are cleaned up thoroughly and as quickly as possible. Water can allow conditions where someone may slip and fall, and it also may allow an electrical shock to occur if the power cords from hand tools, such as drills, are allowed to lie in the water.

Another problem with cleanliness is proper disposal of oily rags. It is important to have proper fire-rated storage cans for all oily rags. These containers have a fire-rated lid so that a fire cannot start in the container. It is also important to empty these containers on a regular basis.

1.7 Leaks

Oil leaks and water leaks should be cleaned up and repaired as soon as possible to limit the amount of liquid that is leaked. The leaks can cause exposure to hazardous material as well as provide a potential for slips and falls. In most factories, periodic maintenance schedules identify specific parts of each machine that should be checked for leaks on a daily basis. When leaks are first identified, they should be fixed immediately before the leak becomes excessive.

1.8 MSDS (Material Safety Data Sheets)

In a factory or industry you may work with any number of chemicals and other types of materials that are hazardous. In 1983, OSHA (Occupational Safety and Health Administration) required manufacturing companies to maintain and distribute data sheets on any material used in the factory. MSDS information may be stored in notebooks or on computer programs somewhere that is totally accessible to all personnel. Additional information is available by contacting the manufacturer of any material. In 1987, OSHA expanded this to all employers. When you are working where you may be exposed to a hazardous material you should wear the proper safety equipment and clothing and locate the MSDS information and follow the instructions provided therein.

1.9 First Aid and CPR

As many people as possible at every work site should have training for first aid and CPR (cardiopulmonary resuscitation). Local Red Cross agencies usually provide first aid and CPR courses for workers. If someone at the plant receives a severe electrical shock or collapses and stops breathing, he or she will need the aid of someone who is certified in CPR.

First aid safety includes first response to problems such as severe cuts and other lacerations or broken bones. If people are injured by machines, it is important to turn off power to the machine as quickly as possible, determine if you can safely move the injured person, and call for help immediately. First aid courses will show you how to apply pressure to deep wounds to stop bleeding. These courses will also show when it is unsafe to move someone with a head or neck injury.

1.10 Fire Safety

Fire safety is very important in industrial settings and it consists of prevention and fighting fires with fire extinguishers. The most important part about fire safety is to understand the basic principles of fire. Fire requires fuel, oxygen, heat, and a chain reaction (source of ignition) to start and sustain it. When you are trying to prevent or fight a fire, you must separate these. Fuels may include paper, rubber, oils, gases, or any other material that can burn.

The National Electric Code (NEC) classifies the locations and types of materials that have the possibility of burning as Class I, Class II, and Class III:

Class I Petroleum-Refining Facilities

- Petroleum distribution points
- Petrochemical plants
- Dip tanks containing flammable or combustible liquids
- Dry-cleaning plants
- Plant manufacturing organic coatings
- Spray-finishing areas (residue must be considered)
- Solvent extract plants
- Locations where inhalation anesthetics are used
- Utility gas plants, operations involving storage and handling of liquified petroleum and natural gas
- Aircraft hangers and fuel servicing areas

Class II Grain Elevators and Bulk Handling Facilities

- Manufacturing and storage of magnesium
- Manufacturing and storage of starch
- Fireworks manufacturing and storage
- Flour and feed mills
- Areas for packaging and handling of pulverized sugar and cocoa
- Facilities for the manufacture of magnesium and aluminum powder
- Coal preparation plants
- Spice-grinding plants
- Confectionary manufacturing plants

Class III Rayon, Cotton, and Other Textile Mills

- Combustible fiber manufacturing and processing plants
- Cotton gins and cottonseed mills
- Flax-processing plants
- Clothing manufacturing plants
- Sawmills and other woodworking locations

1.10.1 Sources of Ignition

One way to prevent a fire is to remove the source of ignition from the fuel. Some sources of ignition that you should be aware of include open flames from torches and heaters, smoking and matches, electrical equipment and heating elements, friction, high-intensity lighting, combustion sparks from grinding and welding, hot surfaces, spontaneous ignition, and static electricity. You must also store flammable liquids in proper storage containers and keep them away from open flames and sources of ignition at all times.

Many companies have proactive programs when any of these sources are exposed. For example, when you must use welding equipment or torches, safety covers and fire extinguishers are moved into place and additional personnel are used to watch for fire problems while the welding is completed. Other conditions are prevented by safety guards to keep fuels away from hot surfaces.

1.10.2 Fire Extinguishers

Fire extinguishers are rated by classification. Class A fire extinguishers can be used on ordinary combustible material like paper, wood, or clothing. This type of fire can easily be extinguished by water. A Class B fire extinguisher is designed for fires in flammable liquids, grease, and other material that can be extinguished by smothering or removing air (oxygen) from the chain reaction. Class C fire extinguishers are used on fires from live electrical equipment. The extinguishers for these types of fires must use material that is nonconducting, so that the person using the extinguisher is not exposed to additional hazard from electrical shock when fighting the fire. Class D fire extinguishers are used on fires in combustible metals such as sodium, magnesium, or lithium. Special extinguishers must be used to lower the temperature and remove oxygen from these fires.

Color codes for these fire extinguishers have been developed and it is important that you recognize the type of fire and use the proper extinguisher. Class A fire extinguishers are green, Class B fire extinguishers are red, Class C fire extinguishers are blue, and Class D fire extinguishers are yellow. It is also important that you recognize the type of material in your shop area and have the proper fire extinguisher for each material.

1.10.3 Fire Extinguisher Safety

Fire extinguishers should be inspected on an annual or semiannual basis. During the inspection, each fire extinguisher should be checked to ensure that it is fully charged and ready for use. All fire extinguishers should be clearly marked for the type of fire they are specified for and clearly displayed where all employees can locate them when they are needed.

1.10.4 Personnel Safety During a Fire

When a fire occurs in your building it is important that several things occur. The most important point is that the proper authorities are notified. This should include setting off all fire alarms in the building to warn the other personnel so that they can evacuate the area. The next thing that should occur is to notify the plant fire brigade if you have one, and notify the local fire company so that it can begin its response. If the fire is small, the local plant fire brigade may begin to fight the fire with fire extinguishers. If the fire is larger, the most important point is to get everyone safely evacuated from the building. Be sure you learn the proper evacuation routes for all areas of the factory that you work in. It is also important that procedure is in place to account for all personnel once they have evacuated.

1.11 Electrical Safety

When you are working in a factory it is important to completely understand how to work safely around electrical cabinets and circuits. This includes the inspection of electrical safety grounds to ensure that they are connected correctly and operational, and working safely around circuits that are under power. The safest way to work around electrical circuits is to turn power off and lockout and tag-out the electrical circuit. Lockout and tag-out procedures are required by OSHA and are explained in detail in Chapter 2.

In some cases you may be required to test voltage and current in circuits that have electrical power applied. When you must work on circuits that are under power be sure to wear heavy rubber sole shoes and try to work with only one hand exposed. If you work with both hands in an electrical panel, you may conduct electric current through one arm, through your body (near your heart), and out the other arm. This is the most dangerous type of shock; it may cause severe damage to you or stop your heart. If you are working with only one hand in the electrical panel at a time, the chances of conducting electrical current directly through your body during an electrical shock is minimal.

An important new innovation in electrical safety is shown in Figure 1–6. The Triplett Corporation has designed a new type of voltmeter display called a "heads-up display" that consists of a glass lens that is attached to a visor that is worn on your head. The heads-up display attaches to a digital voltmeter and displays the voltmeter reading directly on the glass lens that is located directly in front of your eye. The safety feature in the heads-up display allows the electrical technician to safely make voltage measurements in cabinets that have a large number of exposed electrical terminals. When you are working in an electrical cabinet there is always a danger that you will accidentally come into contact with an exposed electrical connection with your skin and receive a severe electrical shock, or that you will place the meter probes across two terminals and cause an arc. This is extremely dangerous when you have your hands inside the cabinet on voltmeter leads, and you are

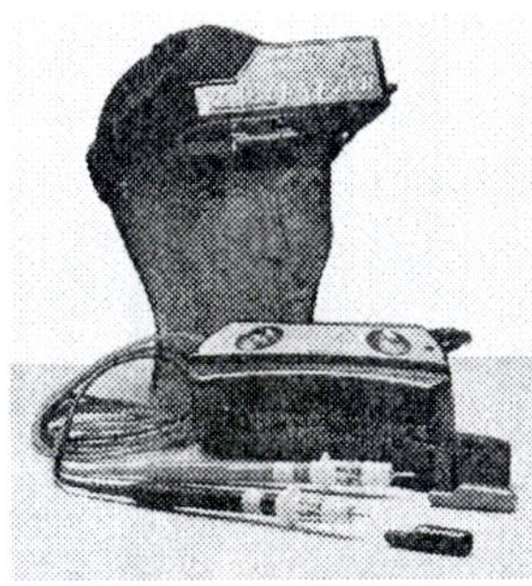

(a)

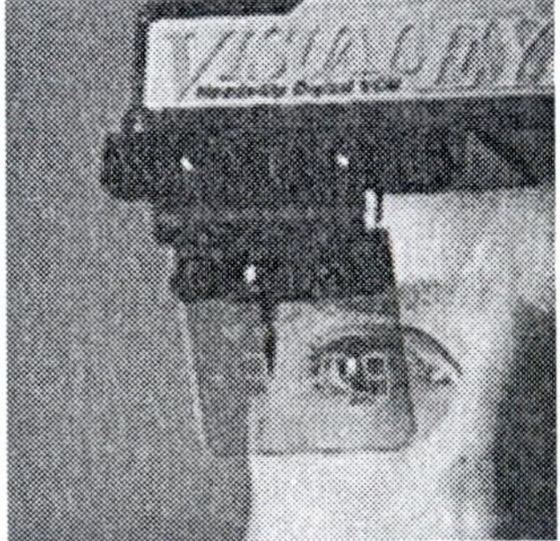

(b)

Figure 1–6 (a) Voltmeter with heads-up display so technician can work safely while making voltage measurements; (b) close-up of voltage displayed on heads-up display. (Courtesy of Triplett Corporation)

looking away from the meter leads and exposed electrical terminals to read the meter face. The heads-up display allows you to look at your meter leads and concentrate on the placement of the leads in the cabinet, while you see the meter reading on the display.

If you do not have a heads-up display on your voltmeter you should be careful to place the meter where you can easily read the face while you have your hands in the cabinet. In some instances, you may need a helper to read the meter as you concentrate on placing the meter leads on exposed electrical terminals.

1.11.1 Ground Fault Interruption Circuit Breakers

If you are using electrical hand tools or if you must work in areas that are damp or have standing water, it is important that all circuits for these conditions have ground fault interruption circuit breakers (GFIs). Figure 1–7 shows single-pole, two-pole, and three-pole GFIs. The GFI-type breaker measures all of the electrical current that is supplied to a circuit or power tool and compares it to the amount of current returning from the tool. If there is no electrical short circuit to ground and the power tool is working correctly, the amount of current returning from the circuit will be the same. If less current is returning, the GFI assumes that a short circuit is occurring and a dangerous condition will result so it will trip to the open position. This means you could be using a power tool that has a faulty power cord and as soon as a short circuit occurs and you start to get an electrical shock, the GFI will interrupt all current in the circuit and prevent you from being shocked. GFI protection is also available as a built-in feature for some extension or power cords for portable electrical power tools.

Another type of GFI has an adjustable detection setting. Figure 1–8 shows this type of ground fault circuit breaker. Figure 1–9 shows an electrical diagram for the adjustable GFI. When you are working with GFIs, you should never increase their trip point or tamper with their mechanism if they are tripping. If you tamper with the GFI, someone may receive a severe electrical shock. When GFIs are properly installed and maintained, they will prevent injury and death due to electrical shocks. It is important that you test all GFI circuit breakers at their predetermined test intervals.

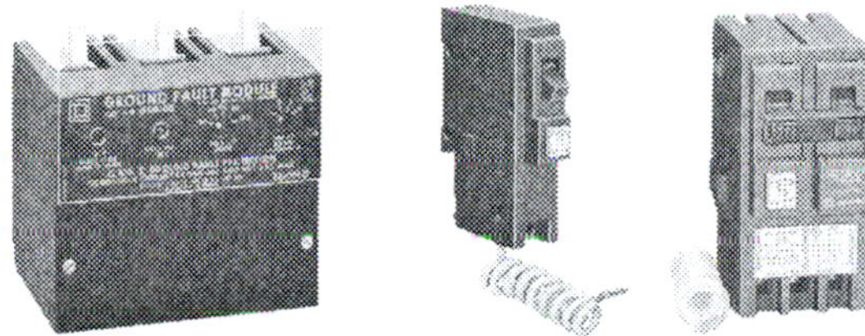

Figure 1–7 Examples of 3-pole, 2-pole, and single-pole ground fault interruption (GFI) circuit breakers.
(Courtesy of Square D/Schneider Electric)

Figure 1–8 Adjustable ground fault detection system.
(Courtesy of Rockwell Automation)

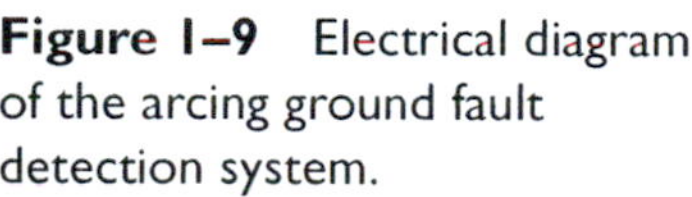

Figure 1–9 Electrical diagram of the arcing ground fault detection system.
(Courtesy of Rockwell Automation)

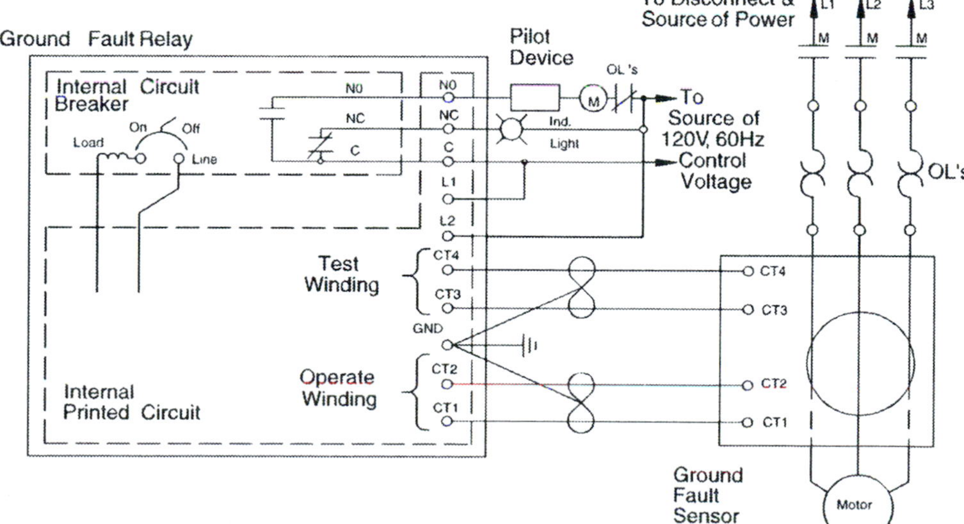

1.11.2
Ground Wires

All electrical circuits and metal power tools require an electrical ground wire. The ground wire is identified by its green color and is used to provide a direct path to ground, which will cause excessive electrical current to flow through fuses or circuit breakers and will cause them to trip.

If a power tool requires a ground circuit, it will have a third prong on its plug end. You should use these tools only in an electrical outlet receptacle that also has a ground terminal. If you have a power tool with a three-prong plug, you should never cut the ground prong off the plug so that it will fit a two-prong (nongrounded) receptacle. You should also never use a two-prong to three-prong adapter with a power tool that requires a three-prong plug. Some power tools have a plastic case; these tools will have a two-prong plug that does not need to be grounded when it is used. You will learn more about short circuits, grounding circuits, fuses, and circuit breakers in a later chapter in this text. It is important that you inspect the power cord of any electrical power tool that has a metal case to ensure that it has a ground wire.

1.11.3
Ground Indicators on Electrical Panels

Some electrical panels have a set of two white lights on the front that are used to indicate that the cabinet or equipment is properly grounded. If the equipment ground wire is faulty, the lights will not be illuminated. If you work on an electrical cabinet where the ground-indicating lights are not illuminated, it may indicate that the ground wire is now faulty and you should not touch the cabinet until you can determine if the ground wire is effective.

1.11.4
Electrical Enclosures That Provide Electrical Safety

At times you will encounter a variety of electrical enclosures that are specifically designed for safety. Figure 1–10 shows examples of the different types of enclosures, and Figure 1–11 shows a list of enclosures and explanations of where they should be used. It is important that you use the proper enclosure for the application that you are working with during installation

Figure 1–10 Enclosures used to provide safety in a variety of industrial applications.
(Courtesy of Rockwell Automation)

Figure 1–11 List of enclosures and explanations of their applications.
(Translated and reproduced with permission of the National Electrical Manufacturers Association. This standard may not be further reproduced. Only the original English version is valid for any legal or judicial consequences arising out of the application of the standard.)

ENCLOSURE RATINGS

NEMA 1	Enclosures are intended for indoor use primarily to provide a degree of protection against limited amounts of falling dirt.
NEMA 3	Enclosures are intended for outdoor use primarily to provide a degree of protection against rain, sleet, windblown dust, and damage from external ice formation.
NEMA 4	Enclosures are intended for indoor or outdoor use primarily to provide a degree of protection against windblown dust and rain, splashing water, hose-directed water, and damage from external ice formation.
NEMA 4X	Enclosures are intended for indoor or outdoor use primarily to provide a degree of protection against corrosion, windblown dust and rain, splashing water, hose-directed water and damage from external ice formation.
NEMA 6	Enclosures are intended for indoor or outdoor use primarily to provide a degree of protection against hose-directed water, the entry of water during occasional temporary submersion at a limited depth, and damage from external ice formation.
NEMA 6P	Enclosures are intended for indoor or outdoor use primarily to provide a degree of protection against hose-directed water, the entry of water during prolonged submersion at a limited depth, and damage from external ice formation.
NEMA 13	Enclosures are intended for indoor use primarily to provide a degree of protection against dust, spraying of water, oil and non-corrosive coolant.

Class	Locations
Class I	Petroleum-refining facilities
	Petroleum distribution points
	Petrochemical plants
	Dip tanks containing flammable or combustible liquid
	Dry-cleaning plants
	Plants manufacturing organic coatings
	Spray-finishing areas (residue must be considered
	Solvent-extract plants
	Locations where inhalation anesthetics are used
	Utility gas plants; operations involving storage and handling of liquefied petroleum and natural gas
	Aircraft hangers and fuel-servicing areas
Class II	Grain elevators and bulk-handling facilities
	Plants manufacturing and storing magnesium
	Plants manufacturing and storing starch
	Fireworks manufacturing and storage
	Flour and feed mills
	Areas for packaging and handling of pulverized sugar and cocoa
	Facilities for the manufacture of magnesium and aluminum powder
	Coal preparation plants
	Spice-grinding plants
	Confectionary-manufacturing plants
Class III	Rayon, cotton, and other textile mills
	Combustible -fiber manufacturing and processing plant
	Cotton gins and cottonseed mills
	Flax-processing plants
	Clothing-Manufacturing plants
	Sawmills and other woodworking locations

Figure 1–12 Typical hazardous locations for Class I, Class II, and Class III applications.
(Courtesy of Rockwell Automation)

or if you are replacing an enclosure. You must always ensure that the integrity of the seals on the different types of enclosures are protected when you open the enclosure and then replace the cover. If the enclosure has a seal, it is important that you inspect it and replace it carefully when you replace the cover.

1.11.5
Hazardous Locations

You should become familiar with the different types of hazardous locations that exist where you work. Figure 1–12 shows a list of hazardous locations. When you arrive at your new work site, you should ask about any hazardous locations and any safety equipment you must use when working in these locations.

1.12
Working Safely with Hand Tools and Power Tools

At times you will be required to work with hand tools and power tools, which increase the chances of having an accident. You should learn the safe operation of these tools and practice safe working habits when you use them. Always read and follow the directions for safe operation prior to using tools. It is also important to have proper safety equipment, such as safety glasses, when you are operating hand tools. Do not use tools if you have not received the proper instructions to use them safely.

1.13
Safety Briefings and Safety Meetings

Many companies use safety meetings to provide basic safety information on a weekly basis. This provides a forum to identify and discuss safety problems throughout the plant. Other plants use a safety committee to identify safety areas of concern and to post safety warnings and posters throughout the plant to keep everyone safety conscious.

Some companies have designated personnel who inspect the various shop areas for safety compliance, while other companies empower all employees to keep an eye on safety conditions throughout the plant. Remember that you are responsible for your own safety at all times. Make working safely a habit.

1.14
Safety Standards Organizations in the United States, Canada, and Europe

Figure 1–13 shows the different safety standards organizations found throughout the United States, Canada, and

EEMAC
Electrical Equipment Manufacturers Advisory Council (Canada)

UL
Underwriters Laboratories Inc. (US)

IEC
International Electrotechnical Commission (Europe)

CSA
Canadian Standards Associations

ANSI
American National Standards Institute (US)

DIN
Deutsches Institut fur Normung (Germany)

NEMA
National Electrical Manufacturers Association (US)

ISO
International Organization for Standardization (Europe)

VDE
Verband Deutscher Elektrotechnicker (Germany)

NEC
National Electric Code (US)

NFPA
National Fire Protecion Agency (US)

CENELEC
European Committee for Electotechnical Standardization

Figure 1–13 Standards organizations in the United States, Canada, and Europe.

Europe. The most familiar organizations are the National Electrical Manufacturers Association (NEMA), Underwriters Laboratories (UL), the National Fire Protection Agency (NFPA), and the National Electric Code (NEC). You will see the logos of these safety organizations stamped on the electrical equipment that you use on the job. If you want to learn more about the standards for each of these organizations, you should use the internet to check out the web site of each organization.

Questions

1. Explain why you should wear safety glasses with side shields at all times while you are working in a factory.
2. Explain why you should always wear earplugs or other types of ear protection while you are working in a factory.
3. List two safety items you can wear to protect yourself against falling debris in a factory.
4. Explain what MSDS is.
5. Identify a type of fire for each of the three classifications.

True and False

1. Class I fires include petroleum and solvent type fires.
2. The reason a two-hand pushbutton station is used on a stamping press is so that you can operate it with either switch.
3. You can use a Class A fire extinguisher on an electrical fire.
4. A Class B fire extinguisher is painted red.
5. A GFI (ground fault interruption) circuit breaker should be used with an extension cord if it is used in a damp or wet location.

Multiple Choice

1. The Class A type fire extinguisher is painted ________ and it is used for paper or wood fires.
   - **a.** red
   - **b.** green
   - **c.** yellow
2. A GFI (ground fault interruption) receptacle is ________
   - **a.** a special receptacle that has a third terminal for a ground wire.
   - **b.** a special receptacle for power tools that allows more than one power tool to be plugged into it at the same time.
   - **c.** a special circuit breaker that can determine if a short circuit occurs and automatically opens the circuit breaker to protect the circuit.
3. A two-handed pushbutton station is ________
   - **a.** a safety circuit that requires an operator of a stamping press to have both hands on a button for the press to begin its cycle.
   - **b.** a convenience circuit that allows an operator to operate the press from two different locations.
   - **c.** a single pushbutton that has an oversize spring that requires the operator to use both hands to depress it.
4. Ear protection should be used ________
   - **a.** at all times even in locations with moderate or small amounts of noise because this noise is damaging over a long period of time.
   - **b.** only in high noise areas.
   - **c.** only if a person has been tested and found to have some hearing loss.
5. If an electrical power tool has a power cord with a three-prong plug, and the only electrical outlet has two prongs, you should ________
   - **a.** cut the third prong (ground lug) off the power cord so that you can use the outlet.
   - **b.** find an adapter that allows you to plug a three-prong plug into a two-prong outlet.
   - **c.** not use this outlet because the power tool needs to be grounded for safe operation, and the two-prong outlet is not grounded.

Problems

1. Make a fire safety plan for your lab area and classroom that includes an evacuation plan.
2. Make a safety plan that includes a method of accounting for all personnel if evacuation must occur.
3. Make a safety plan that lists all the potential fire safety problems in your lab.
4. Make an electrical safety plan that includes a weekly inspection of all the power cords used for power tools to ensure that they are properly grounded.
5. Make a safety plan that includes the proper periodic inspection of all safety guards for all equipment in your lab.

◀ Chapter 2 ▶

Lockout, Tag-out

Objectives

After reading this chapter you will be able to:

1. Identify the lockout, tag-out procedure.
2. Identify sources of electrical energy.
3. Identify sources of other types of energy.
4. Identify sources of potential energy and kinetic energy on a machine.

A wide variety of safety precautions are used in factories today to ensure that workers are not endangered or injured while they are working. One program in use is called lockout, tag-out. This program is designed to ensure that all power to the machine is deenergized and locked in the off position prior to working on the system, and tags are attached to the disconnect switches to indicate who is working on the machine.

The lockout, tag-out program is mandated to all factories by the Occupational Safety and Health Administration (OSHA). The exact standard is OSHA Standard 1910.147. This program explains techniques that are to be used to ensure that uncontrolled release of energy is locked in the off position prior to working on the machine. The program is designed to prevent injury that may be caused by uncontrolled machine motion, electrical shock, electrical burns, or injuries from coming into contact with hazardous materials.

The National Institute for Occupational Safety and Health (NIOSH) has maintained data that show a wide number of factory accidents that resulted in serious injury or death. These accidents include personnel being crushed between moving parts on a machine, such as the clamps on a molding machine, having arms and legs caught in moving conveyor equipment, and amputation of arms or legs from blades. These accidents occurred because co-workers did not pay attention when workers were inside equipment and machines, or because workers did not understand the dangers of working around machinery that still has power applied to it.

The lockout, tag-out procedure is devised to allow maintenance and other personnel to safely work on and around machinery where unexpected start-up of the equipment or the unexpected release of stored hydraulic, pneumatic, or other type of energy may cause injury or death. The procedures should be used any time when equipment is being set up, inspected, or adjusted or any time when maintenance is being conducted.

These procedures are particularly important when guards must be removed to set up the machine or change molds and dies. During this time workers are especially vulnerable to conditions where body parts such as arms and legs are exposed to moving parts. In these conditions workers may sustain serious injury from a crushing accident or they may have their clothing snagged, which can pull them deeper into moving parts in the machinery.

The procedure should also cover the disconnection and lockout of electrical power when necessary. There are times when machines need to be lubricated or cleaned by personnel and it is important that all power is turned off. This will prevent personnel from being shocked if they come into contact with electrical terminals, or if the cleaning fluids and equipment they are using come into contact with electrical terminals. All hydraulic or pneumatic power should also be disabled at this time by bleeding their pressure to zero. This will prevent accidental activation of a hydraulic or pneumatic actuator, which could cause a cylinder to extend and crush someone.

Fuels must also be controlled. This includes exposure to fuel intakes and outlets and exposure to extreme heat such as from steam pipes or other exposed heating elements.

2.1 A Typical Lockout, Tag-out Policy

Each factory must create a policy for lockout, tag-out and ensure that the policy is being enforced. The following illustration may be used as an example, but it does not necessarily include all of the facets of your shop.

The policy ensures that electrical disconnect for all equipment shall be locked out to the deenergized state where possible while maintenance work is being completed on said equipment. If it is not possible to lock out the electrical disconnect, equipment must be protected against possible operation by unauthorized personnel during maintenance procedures. In all cases, all of the power sources for the equipment, such as electrical, hydraulic, pneumatic, and gravity, must be disabled and made inoperable so as to prevent injury to personnel. Employees must never try to operate or reapply any of the power sources that have been locked out or tagged out by another person. The person who applies the lockout or tag-out equipment is the only authorized person to remove these safety devices, unless the procedure continues past the work shift of the person. If the work continues past the work shift of the person applying the lockout device, the person should remove his or her lockout and allow the new shift workers to place their device on the equipment. If the device must be removed by personnel on a different shift, they must follow the procedure for such work.

2.2 Determining the Lockout, Tag-out Policy and Procedure

The lockout and tag-out procedures should be determined and enforced by each individual company. There are many sources for examples of what should be included in these procedures. This chapter provides a basic version of a typical procedure. It is limited to a few of the things that you should do during the lockout, tag-out procedure, and is not intended to be a complete procedure. Each work site has specific machines and equipment that may require additional procedures.

When a company is designing lockout and tag-out procedures, it must keep in mind the different types of energy that the system uses and the way the machine is engineered to operate, such as using gravity for part of the machine operation. While writing the procedure, the company must also have a plan to evaluate how well the procedure is working to prevent accidents, and finally it must also design a means to enforce its procedures to ensure they are being used at all times by all personnel. The evaluation and enforcement part of the procedure should also include continuing education. This will ensure that all personnel remain aware of the potential dangers that are ever present and that they will learn to perform the procedure automatically at all times.

2.3 Identifying Sources of Kinetic Energy and Potential Energy

Energy in and around a machine may take the form of potential or kinetic energy. *Kinetic energy* is energy that is in motion. This is the energy that is easy to see and understand when you are around a machine. For example, the energy that a plastic injection molding machine uses to move its mold (clamp) open and closed is easily observed when the clamp is moving. *Potential energy* is energy that is stored and is available for use after the power source is removed. Its dangers are not always so apparent. For example, a hydraulic system may be designed to store hydraulic fluid in a pressurized accumulator where it is waiting to be released when a hydraulic valve is activated and the stored pressure moves a hydraulic cylinder. What is dangerous about this form of energy is that it continues to be present even after the hydraulic pump is turned off and locked out.

Other forms of potential energy include parts of the machine that utilize gravity for part of its operation. For example, during each cycle, a metal stamping press may raise a die several feet above the parts that are being stamped from the die. If the machine stops with the die in the top-dead-center position above the die, it has the potential of falling from gravity when power is turned off. In this type of application, any part of the machine that could move from gravity must be supported in such a way that it cannot move when power is removed. Another problem arises when the machine or the die must be checked or tested while it is running.

Another example of potential energy is the energy that is stored in rotating parts of a machine, such as a rotary saw blade or rotary grinding wheel. In these applications, the machine uses an electric motor or similar power source to provide large amounts of energy to get the saw blade or grinding wheel up to top speed. When the blade or wheel is spinning at top speed, it has a large amount of energy stored in its mass, which will cause it to continue to spin even after the power to the motor is turned off. In most conditions, the heavier the blade or wheel, and the higher the speed, the more energy it will store and the longer it will continue to spin when power is turned off. In these applications, the company must evaluate the potential for danger and provide braking to stop the blades or wheels as quickly as possible. Another method of protection is to provide sufficient guards so that personnel cannot come into contact with the moving parts. In some cases the guards are interlocked with the moving parts to cause them to stop moving immediately any time the guard is moved or removed.

In some machines, springs are used to provide an additional source of energy. Any time a spring is under compression or extension it has the potential of causing physical harm. This type of energy is not always apparent, since some springs are not visible inside spring-return-type pneumatic cylinders. When the ends of the cylinder are removed during maintenance, the spring may release its energy and cause harm.

2.4 Designing Safety into the Machine

The personnel who are developing the lockout, tag-out procedure must be aware of all of these types of energy sources and the potential problems they present. They should evaluate the machine and provide guards and stops where necessary and build in other safety equipment to minimize the dangers to personnel.

2.5 Documentation and Training

Documentation and training are essential parts of the lockout, tag-out program. Documentation will ensure that each procedure is explained in detail so that personnel are continually reminded of using lockout and tag-out procedures at the appropriate times when they are on the job. Training is essential to ensure that the employees understand the proper procedure and perform it correctly. Training helps new employees to understand the dangers of working around equipment and reminds them that they each have a personal responsibility in ensuring that lockout and tag-out procedures are used at all times.

2.6 OSHA Inspections

The Occupational Safety and Health Administration may become involved if accidents occur that are caused by not following procedures or if incorrect lockout and tag-out procedures have been used. The problem becomes extreme if OSHA finds that the employer does not have lockout and tag-out procedures or if these procedures are not being enforced.

2.7 What Is Lockout?

Lockout procedures include manually disconnecting the electrical and hydraulic power and placing a locking device (usually a padlock) on the disconnect so that it cannot be reconnected without removing the padlock. Figure 2–1 shows a typical lockout safety center that

Figure 2–1 A typical lockout safety center and safety kit that provides all of the safety devices needed for electrical lockout. (Courtesy of Leonard Safety Equipment)

contains all of the safety equipment needed for a lockout procedure. If more than one employee is working on the equipment, each should place his or her padlock on the disconnect. A special device that allows multiple padlocks to be connected on the same disconnect is shown in Figure 2–2.

When the clasp is closed, up to 6 padlocks can be installed at one time. The clasp is secured around the disconnect and it cannot be opened until the last padlock is removed.

The lockout, tag-out procedure also requires that each padlock and lockout device be identified with a picture and the name of the employee who placed the lockout on the disconnect. Lockout devices must be supplied by the employer and are to be maintained by the employee.

In some cases, the work on a machine involves multiple personnel, and multiple padlocks will be applied to the disconnect. When all work is completed and it is time to return power to the machine, all the locks must be removed and all personnel must be accounted for so that they are not in danger.

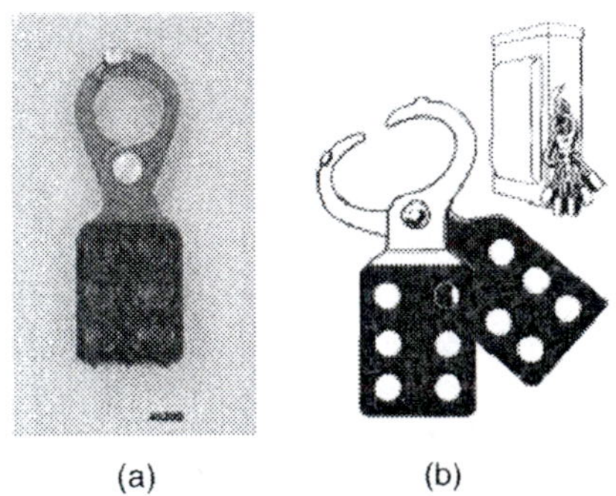

Figure 2–2 A typical lockout device that can accept multiple padlocks.
(Courtesy of Klein Tools)

2.8 What Is Tag-out?

Figure 2–3 shows a typical tag that is used for tag-out procedures. The tag has a written warning that explains why the machine has been tagged. This warning also explains that the power should not be applied to the machine until the person working on the machine has completed the work and is ready for power to be reapplied.

All tags used in the lockout, tag-out procedure should be uniform and standardized so that they are easily recognized by all employees. Figure 2–4 shows additional types of lockout tags.

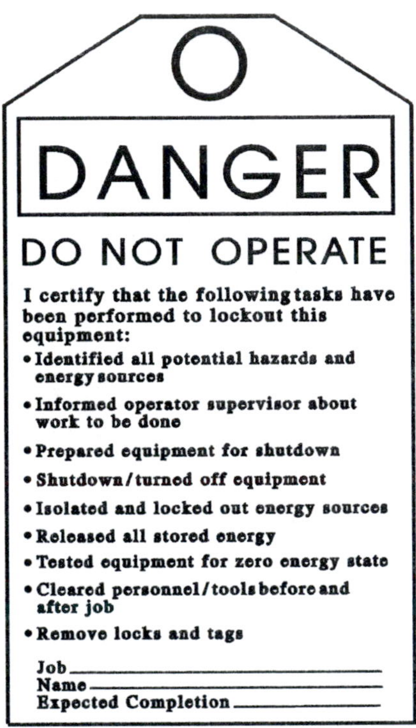

Figure 2–3 Typical lockout tag.

Figure 2–4 Additional types of lockout tags.
(Courtesy of Leonard Safety Equipment)

2.9 Designing a Lockout, Tag-out Procedure for Your Machines

The first step in a typical lockout, tag-out procedure is to evaluate each machine and determine all dangerous conditions that are present when the machine is running and when it is locked out for maintenance. This includes identifying all possible sources of power to the machine and listing them by type, such as electrical, hydraulic, and pneumatic. It also includes identifying any hazardous material, chemicals, sources of radiation, or exposure to sources of thermal (heat) energy.

The next step includes identifying all employees who may work on or around the machine (operators, inspectors, supervisors, and others). These employees should always be notified when the machine is going to be turned off and locked out and tagged out. There should also be a means of notifying affected personnel who work on other shifts. The notification should include who shut the machine down and what is being worked on. Other information could include the expected time when the machine will be in operation again.

The next step is the actual shutdown of the equipment. This includes pressing the stop button, moving any other switches to the off position, and making sure that all motion on the machine is stopped. All sources of energy should be secured in the off condition and all forms of potential energy should be neutralized so that they do not present a hazard. After you have identified all sources of energy and have ensured they are in the off position, you can apply lockout and tag-out hardware to ensure that they remain in the off position until you are ready to turn them back on.

It is important at this time to identify several methods of disrupting power that are unacceptable because they can easily be bypassed and pose a danger to workers. For example, if you simply remove the fuses to a machine and do not lock out the fuse box, someone could replace the fuses and apply power to the machine while you are still working on it. If you turn off a pneumatic hand valve but do not lock the valve in the off position by a chain or other blocking device, it may be turned on again and cause serious injury. Some equipment is controlled by a circuit breaker rather than a disconnect-type switch. Figure 2–5 shows lockout devices for circuit breakers. You can attach the lockout device to ensure that power to the circuit is turned off and remains off. If the equipment is powered by an extension cord, you must secure the end of the extension cord to ensure that the power is turned off to the equipment. Figure 2–6 shows typical lockout equipment for an electrical extension cord.

If the equipment has air pressure or steam pressure that must be turned off and secured, you will need to use a safety donut for locking out these valves. Figure 2–7 shows a lockout donut for a globe valve, and Figure 2–8 shows the lockout for a ball valve.

It is also important to identify any interlocking power sources or auxiliary power sources to each machine. For example, if a machine uses a robot to load and unload it, it is possible for some electrical signals from the robot cabinet to be sent to the electrical cabinet of the machine. If the machine is locked out, but the robot is not, you could have problems with the robot moving into the machine space, or you could have problems with electrical wires from the robot that terminate in the machine cabinet still being powered.

Figure 2–5 Circuit breaker lockouts.
(Courtesy of Leonard Safety Equipment)

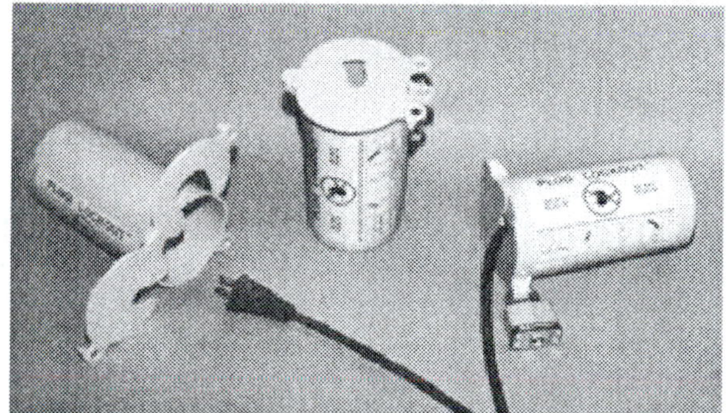

Figure 2–6 Lockout for electrical plugs and extension cords.
(Courtesy of Leonard Safety Equipment)

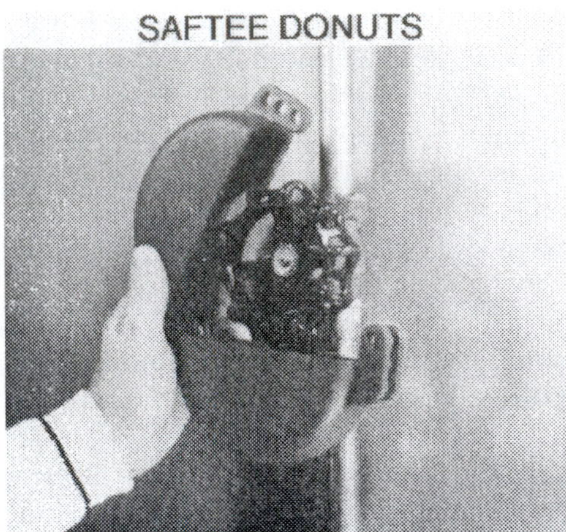

Figure 2–7 Safety donut for locking out air and steam valves.
(Courtesy of Leonard Safety Equipment)

Figure 2–8 Ball valve lockout.
(Courtesy of Leonard Safety Equipment)

Since many of these signals are 110 VAC, they can pose an electrical shock hazard.

After all power is shut down and locked in the off position, you should make a number of tests to ensure that all of the power is in fact turned off. You should also test all blocks that are in place to ensure that they are secure and holding the machine in a safe condition. After all tests are made you can begin your work.

2.10 Removing the Lockout, Tag-out Devices and Returning Power to the Machine

After all work has been completed on the machine, you are ready to remove the lockout and tag-out devices and return power to the machine. You must make several checks prior to removing the lockout and tag-out devices. First, ensure that all work is complete and that all tools and any test equipment have been removed. All guards and safety doors should be returned to the normal condition and tested to ensure that they operate correctly. Be sure that all parts of the machine that have been disassembled are reassembled properly.

When you have completed these tasks, you must physically account for all personnel who are in the area to ensure that they know and understand that you are ready to turn on power to the machine again. This is probably the most important part of the procedure, especially if more than one person is working on the machine, or if several employees routinely work on the machine when it is in normal operation.

When you are ready to apply power again, be sure you follow the operational checklist so that power is returned to the machine in the proper order. For example, some machines must have circulating pumps running before steam valves are returned to their open position.

2.10.1 Removing a Co-worker's Lockout and Tag-out Equipment

The general rule is that the person who placed the lockout on the machine must be the one who removes it. But there will be times when work is started on the day shift and completed 14 hours later on another shift. In this case the person who placed the lockout, tag-out equipment on the machine is not present to remove it. When this occurs, it is permissible for a supervisor or person of responsibility to remove the lock. The personnel on each shift who may remove a lock should be identified when the lockout, tag-out procedure is written. The supervisor may have a master key or permission to cut the lock off with bolt cutters. The person who removes the lock should always file a report to indicate why the lock was removed and who performed the action.

2.11 Review of Lockout, Tag-out Procedures

The lockout and tag-out procedure is designed to ensure that all sources of energy are locked out and tagged out whenever maintenance procedures require that power must be removed from the machine. The lock provides security to ensure that the disconnect switch remains in the off position, and the tag provides a means to identify who placed the lock on the disconnect. When the procedure is created, a comprehensive review of all safety aspects of a machine should be accounted for. The lockout, tag-out practices should include training and education to ensure that all employees understand the safety hazards that are present with each machine and that they continually use correct safety procedures when working on and around the machine. The procedures should identify all sources of kinetic and potential power as well as sources of hazardous material, thermal energy (steam), and radiation.

Lockout and tag-out procedures are mandated and enforced by OSHA and all procedures, equipment, and training are to be provided by the employer. Each employee is responsible for utilizing proper lockout and tag-out procedures and for maintaining the equipment that has been issued to him or her. Be sure to use the correct lockout and tag-out procedure any time you are working around equipment.

Questions

1. Explain what a lockout, tag-out procedure is.
2. What do the initials OSHA stand for?
3. What is kinetic energy?
4. What is potential energy?
5. Who is allowed to remove a padlock from a lockout?

True and False

1. A vertical press that is in the open position so that its platen could fall is an example of potential energy.
2. A fluid accumulator that is under pressure for a hydraulic press is an example of kinetic energy.
3. Lockout devices are available that allow more than one padlock to be connected to the same lockout.
4. The lockout device should have a picture of the owner on it in case the lockout is lost.
5. It is permissible to have electrical power applied to a machine if you need to make voltage readings as part of your troubleshooting procedure.

Multiple Choice

1. Who is responsible for creating the lockout, tag-out procedure in a factory?
   a. OSHA
   b. Each worker should create his or her own procedure.
   c. Each factory should create its own procedure.
2. The forms of energy that you should be aware of during a lockout procedure include _____
   a. hydraulic and electrical power.
   b. hydraulic, electrical, and potential energy sources such as gravity.
   c. only electrical power sources.
3. If an employee on the day shift places a padlock on a lockout device, and a second-shift employee completes the repair, _____
   a. the second-shift supervisor can remove the first-shift padlock.
   b. the second-shift employee can remove the first-shift padlock.
   c. no one on the second shift can remove the padlock; it cannot be removed until the employee that placed the padlock on the lockout comes back to work.
4. Prior to removing a padlock from a lockout device and restarting a machine you should ______
   a. replace all guards and turn on the power.
   b. alert all employees in the area that work on this machine that you are turning on the power and starting the machine.
   c. call the operator to start the machine.
5. The major problem with potential energy sources during a lockout procedure is that ______
   a. the hazard they present may not be readily apparent, such as a machine part that is supported by hydraulic pressure that will drop when the pressure is released.
   b. potential energy is much more dangerous than electricity.
   c. potential energy is not a real threat, like kinetic energy, so it can be ignored.

Problems

1. You are asked to create a lockout, tag-out procedure for a plastic injection molding machine that has both electrical power and hydraulic power. List all of the things you should be aware of for the lockout procedure.
2. List 4 sources of potential energy you should be aware of when you devise a lockout procedure.
3. List the things you should do prior to restarting a machine when you are ready to remove the padlock from the lockout device.
4. You are asked to help implement a lockout, tag-out procedure for your factory. Explain what you would consider as part of the education and training of all employees.
5. You are asked to help develop an inspection procedure for the lockout, tag-out procedures in your factory. Explain what you would include in creating the inspection part of the lockout, tag-out program.

Chapter 3

Tools for Electrical Technicians

Objectives

After reading this chapter you will be able to:

1. Identify the proper electrical meters to use for troubleshooting.
2. Select the proper electrical tools for punching holes or cutting holes in an electrical cabinet.
3. Select the proper electrical tools for cutting and bending electrical conduit.
4. Select the proper pliers for cutting and stripping wire.
5. Select the proper screwdriver for installing wiring in an electrical cabinet.
6. Select the proper wrenches to use in electrical maintenance.
7. Select the proper tool pouch for electrical tools.

An electrical technician must be able to properly use a wide variety of meters, construction tools, wire cutters, screwdrivers, wrenches, and other tools. This chapter introduces the tools you will be using on the job in installing, troubleshooting, and repairing electrical equipment.

3.1 Electrical Meters

As an electrical technician or person who will work on electrical systems in maintenance, you will need to be able to select the proper type of meter and use it correctly. Figure 3–1 shows a variety of types of meters you will use. Figure 3–1a shows a Klein Wiggy voltage tester, which is also called a solenoid-type voltage tester. This type of tester is available from a number of meter manufacturers and it operates much like a voltmeter in that it is able to determine the presence of voltage in a circuit. The advantage of a solenoid-type tester over a traditional digital meter is that the solenoid in the Wiggy requires a substantial amount of voltage and current to cause the meter to activate. When the meter leads are placed across a 110-VAC power source, the solenoid will activate and the meter indicator will "jump" to indicate the amount of voltage the meter is measuring. If the voltage is a "stray voltage" or a voltage that would cause a digital voltmeter to create a "phantom voltage" and it does not have current with it, it will not cause the solenoid to energize on the Wiggy. The other feature that makes the Wiggy different from a traditional analog or digital voltmeter is that it indicates the presence of voltage in three distinct ways. First, the meter has a needle that moves to indicate the amount of voltage numerically; second, it vibrates when its solenoid is energized by the presence of voltage; and, third, it will emit an audible hum when voltage is present.

The Wiggy and all voltmeters must have their leads placed across a voltage differential, such as the voltage supply lines, Line 1 to Neutral or Line 1 to Line 2, to indicate the presence of voltage. It is possible to put the meter leads on two separate wires that are connected to terminal L1 in the circuit. Since both wires come from the same source, there is no difference of potential and the meter will not indicate a voltage. This may give you a false sense of security, because you will assume that these wires are not energized. But if you cut through one of these wires and let your wire cutters touch metal conduit that is grounded, you could receive a severe electrical shock. This is true of all digital and analog

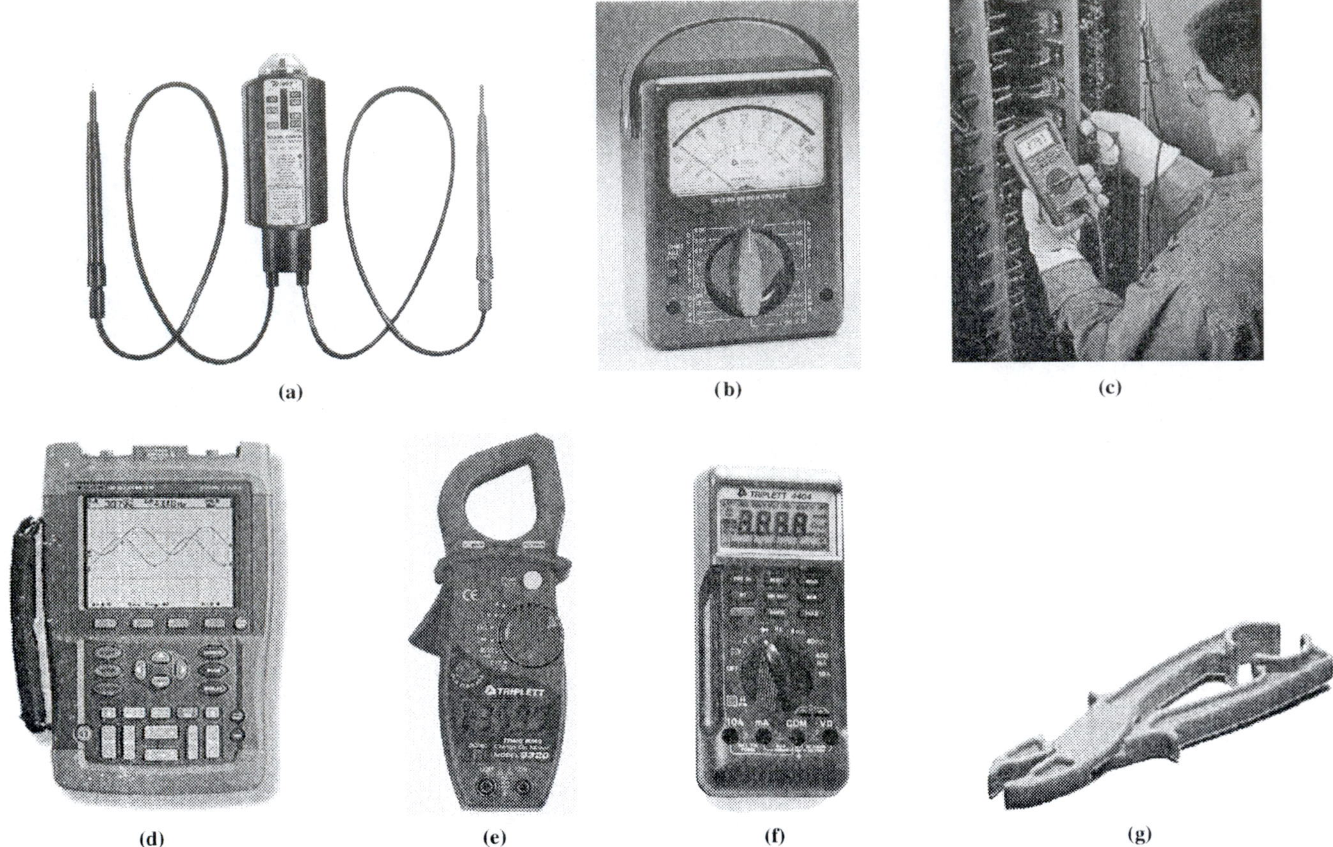

Figure 3–1 (a) Klein Wiggy solenoid tester (courtesy of Klein Tools); (b) Triplett analog volt, ohm, and milliamp (VOM) (courtesy of Triplett Corporation); (c) technician using a Fluke digital voltmeter (reproduced with permission of Fluke Corporation); (d) Fluke scopemeter (reproduced with permission of Fluke Corporation); (e) Fluke clampmeter (reproduced with permission of Fluke Corporation); (f) Triplett digital voltmeter (DVM); (g) Snap-on nylon fuse puller (courtesy of Snap-on Tools); (f) Triplett digital voltmeter (DVM) (courtesy of Triplett Corporation).

voltmeters. You should always test across the two lines and if you do not measure voltage, you should then test for voltage between each line and a known good ground wire or ground terminals.

The solenoid-type tester does not have high impedance like an analog or digital voltmeter. This means that a small amount of current will flow through the meter when you are making a voltage measurement. For this reason you must be careful using this type of meter around proximity switches and other electronic circuits where the small amount of current may cause erroneous circuit or component operation. You should always use an analog or digital voltmeter that has high impedance when you are making measurements in electronic components or electronic circuits.

There is an electronic tester available that will emit a buzz or tone when it is placed near any single wire that is energized with voltage, regardless of the presence of a neutral or ground wire. This tester will allow you to determine if a wire is energized with voltage or not.

Figure 3–1b shows a meter that indicates the amount of voltage and resistance with a meter movement over a numerical face. The amount of voltage is determined by the location of the needle over the face of the meter. A range switch allows you to select a range that is higher than the voltage you are trying to measure. You must be aware of the polarity of the leads if you are using this type of meter to measure DC voltage. If you have the polarity wrong (the positive lead on the negative voltage terminal), the needle on the meter will try to move backward. If this occurs, you can simply exchange the positive and negative leads or change the polarity switch and the meter needle will read up-scale. If you select a range that is too low, the needle will try to move too far and the meter could be damaged. You should always set the meter on the highest range and switch it to a lower range if the reading is too low. If you use the meter to measure AC voltage, the polarity does not matter, but you will need to set the meter to the highest reading and set the switch lower if needed.

Figure 3–1c shows a technician making a measurement with a digital voltmeter. Notice that the technician places the meter in a location where he can put the meter leads on the electrical terminals and look at the meter display to read the amount of voltage the meter is indicating. Experienced technicians tend to place the meter so that they can comfortably read the meter display and keep an eye on the terminals where the leads are touching. This helps to prevent accidental electrical shocks that may occur if you place the meter so that you must look away from the cabinet to read the meter.

The electrical meter in Figure 3–1d is called a Fluke scopemeter. The scopemeter can read voltage in a digital readout like a digital voltmeter, and it can produce a copy of the waveform of the voltage like an oscilloscope and can display the frequency as a number. The scopemeter is battery powered and has extremely high impedance, so it can be placed across any terminals in any electronic or electrical circuit and safely provide a measurement of voltage and frequency. Some models of the scopemeter have the ability to measure two separate signals and display the phase shift or time difference between the two. It is also possible to measure extremely small amounts of voltage and display the waveform it creates to determine if any "noise" or unwanted signals are present. This is useful in measuring computer level signals that are present in an electrical cabinet.

Figure 3–1e shows a clamp-on-type ammeter. This type of meter also has the ability to be used to measure voltage and resistance with probes, just like a digital voltmeter. The clamp-on meter is used to measure AC voltage. Some models have special circuits that allow DC current to be measured too. The clamp-on ammeter operates on the principle of electromagnetic induction, which is similar to the operation of a transformer. You can read more about transformers in Chapter 6.

Figure 3–1f shows a typical digital voltmeter, which displays the amount of voltage the meter is measuring. The digital voltmeter is sometimes called a DVM. The DVM can be designed to measure resistance and small currents up to 1000 milliamps (1 amp) and are called digital multimeters (DMM). The digital meter has a range selection button that allows you to set the range for the voltage or current you are measuring. Some of these meters are also autoranging, which means they will automatically adjust to the proper range. The newest models of these meters also have automatic mode selection so that the meter and circuit are protected if you set the meter for resistance and mistakenly place the probes across a voltage. When any meter is set to measure resistance, its internal circuitry creates a low impedance circuit that will act like a short circuit if the meter leads are placed across a voltage. If this were to occur, the meter would be severely damaged, and a large arc could be produced where the meter leads touch the electrical voltage source. Always be sure to set the meter for voltage if you are not absolutely sure the circuit does not have voltage present.

Figure 3–1g shows a fiberglass fuse puller. You may encounter a fuse mounted in a fuse socket that you suspect is faulty. If you want to remove the fuse from the socket, you will need to use a fuse puller that is made of insulating material so that you do not receive an electrical shock. The fuse puller allows you to grasp the fuse securely and pull it from its socket without exposing yourself to additional shock hazards in the electrical panel. You can also use the fuse puller to put a new fuse back into the socket.

3.2 Hole Punches and Hole-Cutting Tools

If you are requested to install electrical conduit or switches into an electrical panel, you will need a hole punch or hole saw to safely cut a hole in the metal cabinet. Figure 3–2 shows a number of tools specifically designed to punch or cut holes safely and efficiently in metal cabinets. The hole punch will make an exact-size hole in the cabinet and it is generally used for making holes that are larger than 1/2 inch in diameter. The hole punch is preferred over a drill to make larger holes in the cabinet, because it would require a very large drill motor to turn a larger drill bit to make holes larger than 1/2 inch. The hole saw is used when the metal is thinner. The hole punch shown in Figure 3–2a is manufactured by the Greenlee Company and is sometimes referred to as a Greenlee punch. Figure 3–2b shows how the hole punch operates.

If you are using the hole punch to cut a 3/4-inch hole in a metal electrical panel, you will need to drill a small 1/4-inch hole in the metal cabinet where you want to locate the larger hole for a switch. The 1/4-inch hole is used to allow the bolt for the hole punch to be inserted through the metal plate of the panel. The hole punch is disassembled so that the two parts of the punch are mounted on each side of the metal plate you are piercing. The bolt will be used to pull the two sides of the punch together to pierce the metal plate. Figure 3–2e shows a hydraulic version of the hole punch. The hydraulic-operated hole punch al-

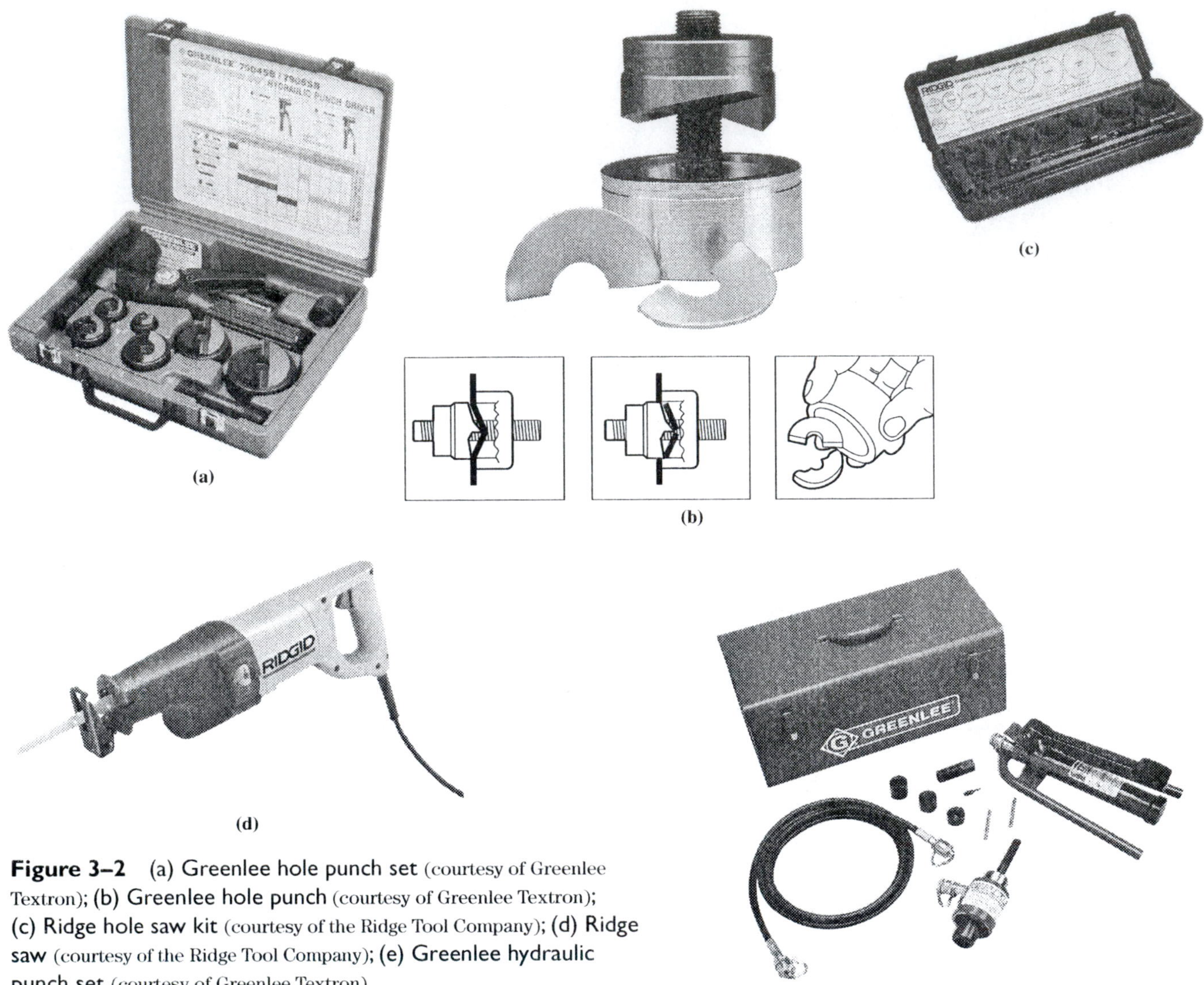

Figure 3–2 (a) Greenlee hole punch set (courtesy of Greenlee Textron); (b) Greenlee hole punch (courtesy of Greenlee Textron); (c) Ridge hole saw kit (courtesy of the Ridge Tool Company); (d) Ridge saw (courtesy of the Ridge Tool Company); (e) Greenlee hydraulic punch set (courtesy of Greenlee Textron).

lows larger punches to be used to make larger holes or when material is thicker. The hole punch die is assembled on the two sides of the metal plate that is to be pierced, and the hydraulic hand pump is used to supply hydraulic pressure through a hose to the die. Some larger hydraulic hole punches have a foot pedal for easier use.

Figure 3–2c shows a set of hole saws. These circular metal saws are mounted in a drill and are rotated at high speeds. The teeth in the saw cuts the metal just like a wood saw.

Figure 3–2d shows a reciprocating saw, which can use a metal cutting blade to cut conduit or other metal. This saw can be used as a cutoff saw or to cut free-hand holes into panels. It can also be used to cut off bolts or other metal shafts.

3.3 Electrical Conduit Bending, Cutting, and Threading Tools

Figure 3–3a shows a conduit bender that is used to make measured bends into metal conduit. This bender is used by placing your foot on the base plate of the bender and pulling on the handle. The bender has a set of indicators that show the amount of the bend in degrees. A hydraulic bender is available for bending larger conduits.

Figure 3–3b shows a flexible metal fish tape that is used to pull wire through a conduit. The fish tape is stored in a plastic case when it is wound up, and it is extended into the empty conduit until the hook on its end extends out of the conduit. You can attach the wire that

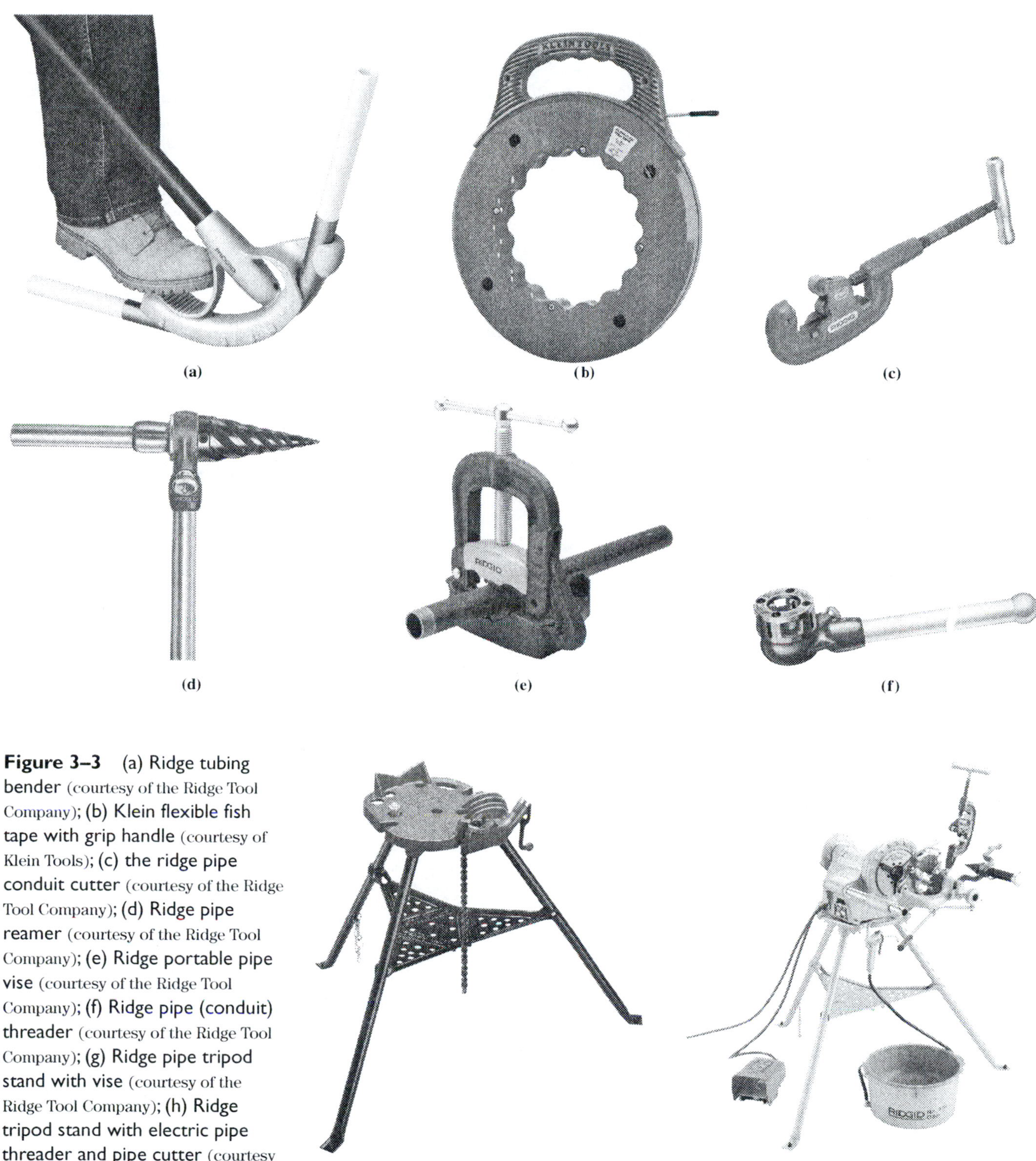

Figure 3–3 (a) Ridge tubing bender (courtesy of the Ridge Tool Company); (b) Klein flexible fish tape with grip handle (courtesy of Klein Tools); (c) the ridge pipe conduit cutter (courtesy of the Ridge Tool Company); (d) Ridge pipe reamer (courtesy of the Ridge Tool Company); (e) Ridge portable pipe vise (courtesy of the Ridge Tool Company); (f) Ridge pipe (conduit) threader (courtesy of the Ridge Tool Company); (g) Ridge pipe tripod stand with vise (courtesy of the Ridge Tool Company); (h) Ridge tripod stand with electric pipe threader and pipe cutter (courtesy of the Ridge Tool Company).

you are pulling into the conduit onto the hook of the fish tape. The fish tape is strong enough to pull multiple wires into the conduit. You must be very careful that you do not allow the metal fish tape to come into contact with exposed electrical wires or terminals, because the metal tape will conduct current that can cause severe electrical shock.

Figure 3–2c shows a cutting tool that is used to cut metal conduit. Figure 3–2d shows a tool that is used to ream the pipe and remove rough edges after the pipe cutter is used to cut it. This is an important step, since the rough edges of the pipe can cut or damage the insulation on wire that is pulled into the pipe or conduit.

Figure 3–2e shows a portable pipe vise and Figures 3–2g and 3–2h show models of vice clamps that are mounted on a tripod. The tripod provides a stable work surface to hold conduit or pipe while you are cutting or threading it. Figure 3–2f shows a pipe-threading tool. This tool is used to put pipe threads on metal conduit so that it can be threaded together with threaded fixtures.

3.4 Electrical Pliers, Wire Cutters, and Wire Strippers

Electrical technicians and electricians must use a wide variety of pliers, wire cutters, and wire strippers specifically designed for the electrical trade. Figure 3–4 shows a variety of wire cutters and strippers you may use on the job. The wire stripper is designed so that only the insulation on the wire is cut, leaving the wire undamaged when the jaws are closed tightly. The stripper has a variety of holes that fit wire gauges from size 8 to size 22. It is also possible to purchase wire strippers designed for smaller holes, and the larger cutters are used to cut large wires.

Figure 3–5 shows a variety of slip-joint, side-cutting, needle-nose, and clamping pliers that electricians will use on the job. The reason a wide variety of pliers are provided is to give you access in places that are hard to reach.

3.5 Screwdrivers for Electrical Installation and Maintenance

Figure 3–6 shows a variety of regular and Phillips screwdrivers. Some of the screwdrivers in this figure are called self-starting screwdrivers because they hold a screw securely so that it can be started into a threaded hole with one hand. Other screwdrivers are specifically offset to provide access to very tight spaces so that you can turn screws slightly to tighten or loosen them.

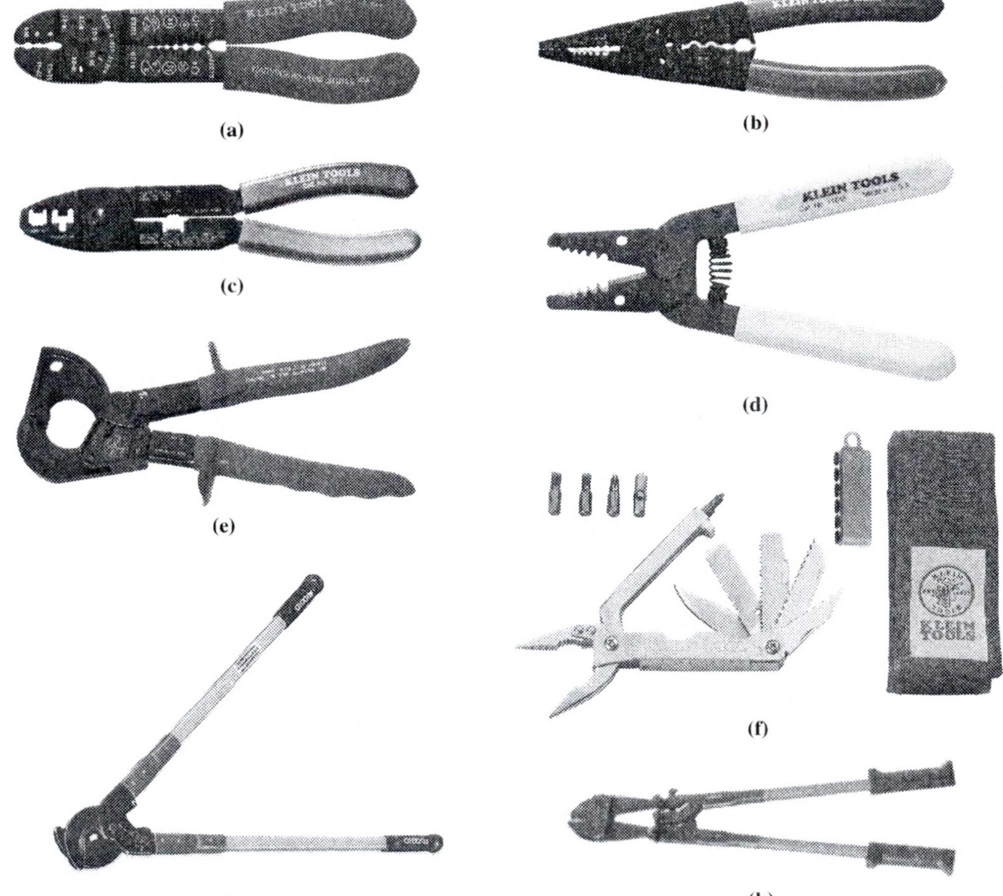

Figure 3–4 (a) Klein five-in-one tool that cuts bolts, cuts and strips wire, measures stud sizes, and crimps insulated terminals (courtesy of Klein Tools); (b) Klein combination wire-stripping and wire-cutting tool (courtesy of Klein Tools); (c) Klein modular plug crimping tool (courtesy of Klein Tools); (d) Klein wire stripper and cutter (courtesy of Klein Tools); (e) Klein ratcheting cable cutter (courtesy of Klein Tools); (f) Klein electrical maintenance multitool (courtesy of Klein Tools); (g) Ridge cable cutter (courtesy of the Ridge Tool Company); (h) Klein bolt cutter (courtesy of Klein Tools).

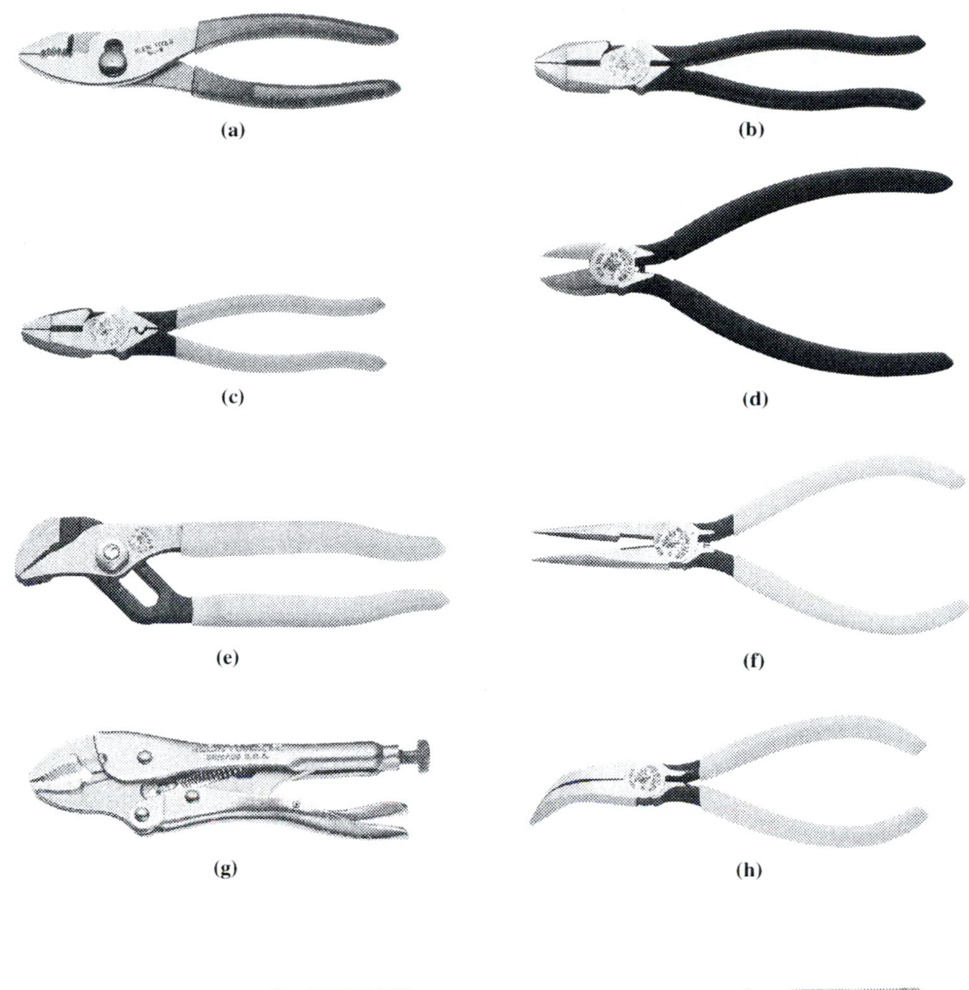

Figure 3–5 (a) Klein slip joint pliers; (b) Klein side-cutting pliers with smooth streamlined nose; (c) Klein side-cutting pliers with crimping die; (d) Klein electronic diagonal-cutting pliers; (e) Klein pump pliers; (f) Klein long-nose, side-cutting pliers; (g) Klein locking pliers with curved jaw and cutting knives; (h) Klein curved-nose pliers.
(Courtesy of Klein Tools)

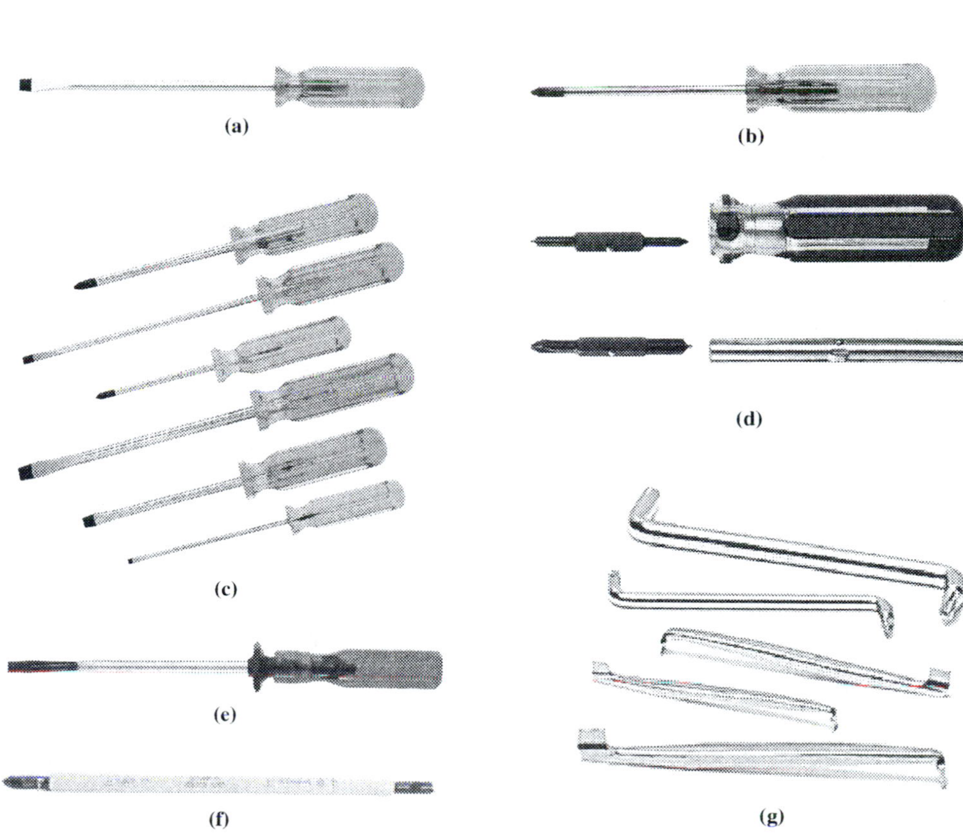

Figure 3–6 (a) Vaco 6-inch, flat-blade screwdriver; (b) Vaco #2 Phillips screwdriver; (c) Vaco 6-piece screwdriver set; (d) Vaco 4-in-1 screwdriver; (e) Klein starting screwdriver (courtesy of Klein Tools); (f) Klein starting screwdriver (courtesy of Klein Tools); (g) Snap-on offset screwdrivers (courtesy of Snap-on Tools).

3.6 Wrenches for Electrical Maintenance

Figure 3–7 shows a variety of wrenches electricians may use on the job. Figure 3–7a shows combination box- and open-end wrenches that are used to tighten or loosen nuts and bolts. These types of wrenches are available in metric and standard sizes. Figure 3–7b shows an example of a set of socket wrenches. The sockets snap onto a ratchet drive, which allows nuts and bolts to be tightened or loosened quickly. The ratchet is also very useful in tight spaces.

Figure 3–7c shows a box-end wrench that has a ratchet feature. This allows the box-end wrench to be used

Figure 3–7 (a) Klein combination wrench set (courtesy of Klein Tools); (b) Klein socket wrench set (courtesy of Klein Tools); (c) Klein box reversing ratchet set (courtesy of Klein Tools); (d) Klein nut driver set (courtesy of Klein Tools); (e) Klein hex key Allen wrench set (courtesy of Klein tools); (f) Klein T-handle hex key set (courtesy of Klein Tools); (g) Klein adjustable wrench (courtesy of Klein Tools); (h) Ridge pipe wrench (courtesy of the Ridge Tool Company); (i) Ridge chain wrench (courtesy of the Ridge Tool Company); (j) Ridge strap wrench (courtesy of the Ridge Tool Company).

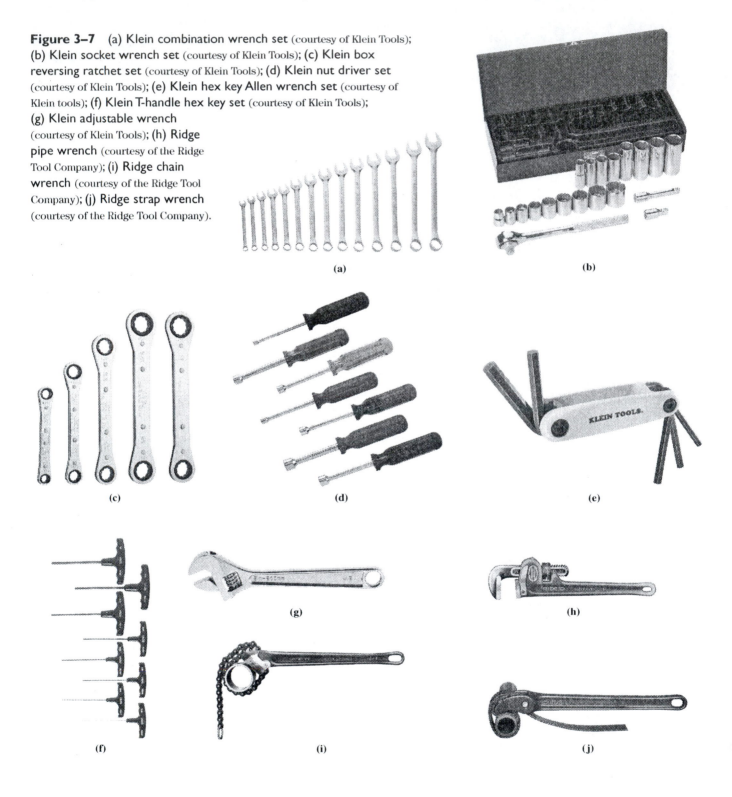

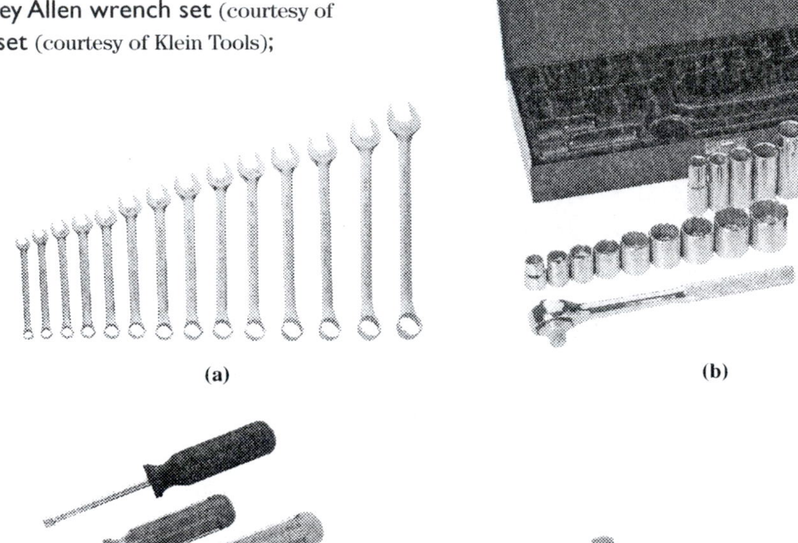

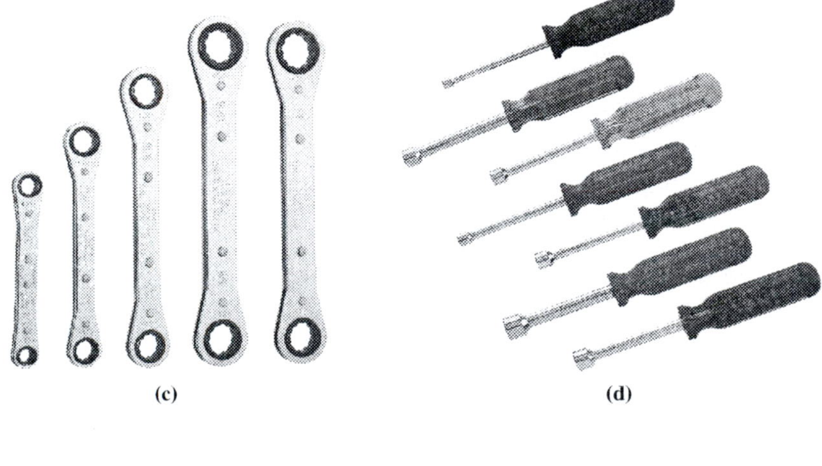

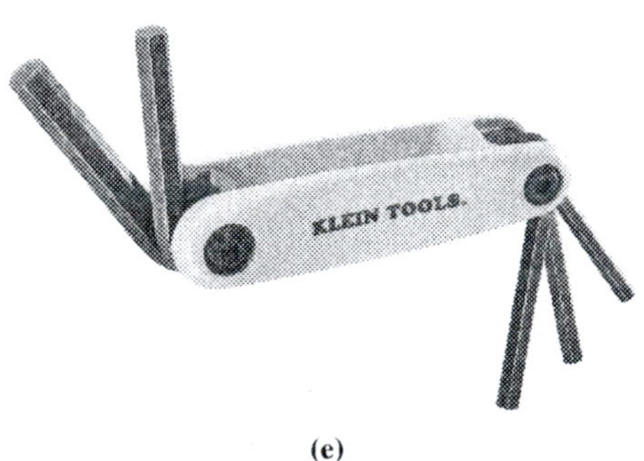

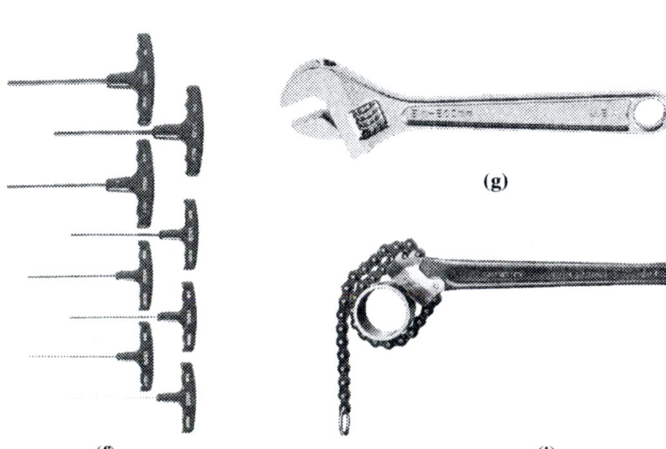

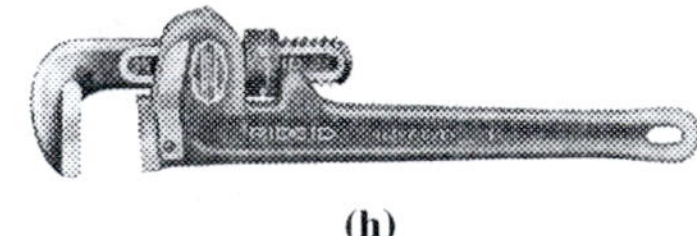

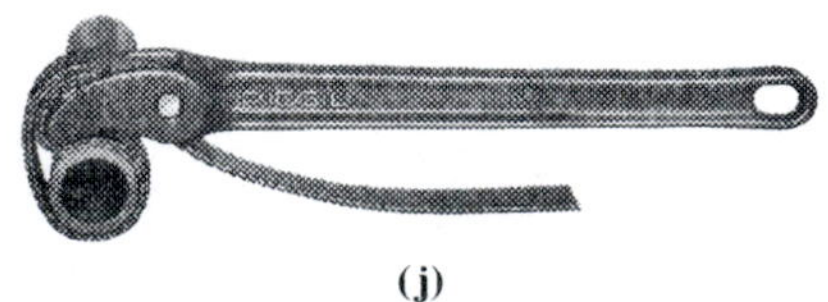

in tight spaces with a minimum of travel to tighten and loosen nuts and bolts. Figure 3–7d shows a set of hex key wrenches. These wrenches are sometimes called Allen wrenches. Figure 3–7e shows a set of nut drivers. Nut drivers are special wrenches that have an end that looks like a socket wrench and a handle that is similar to a screwdriver. The nut driver is very useful for removing and replacing small sizes of screws or bolts that have hex heads. Figure 3–7f shows hex key wrenches with a T-handle. The T-handle allows more torque to be placed on the wrench so that larger hex head bolts can be tightened or loosened.

Figure 3–7g shows an adjustable wrench. One type of adjustable wrench is called a crescent wrench. This wrench is extremely useful when you can carry only one wrench instead of a complete set of wrenches. The adjustable wrench can be used for a wide variety of bolt and nut sizes by simply changing the opening of its jaws.

Figure 3–7h shows a pipe wrench. The pipe wrench has an adjustable jaw that grips the smooth surface of a pipe or conduit so that it can be rotated to tighten or loosen its threads. Figure 3–7i shows a chain wrench and Figure 3–7j shows a strap wrench; both are used like a pipe wrench. These wrenches are adjustable wrenches used to loosen or tighten pipe and conduit.

3.7 Tool Sets and Tool Pouches

Figure 3–8a shows a general tool set that has wrenches and screwdrivers in it. The tool set has the most common tools that an electrician may use for maintenance. Figure 3–8b shows a tap and die set. You may need to use a tap to make interior threads inside a hole, or a die that makes

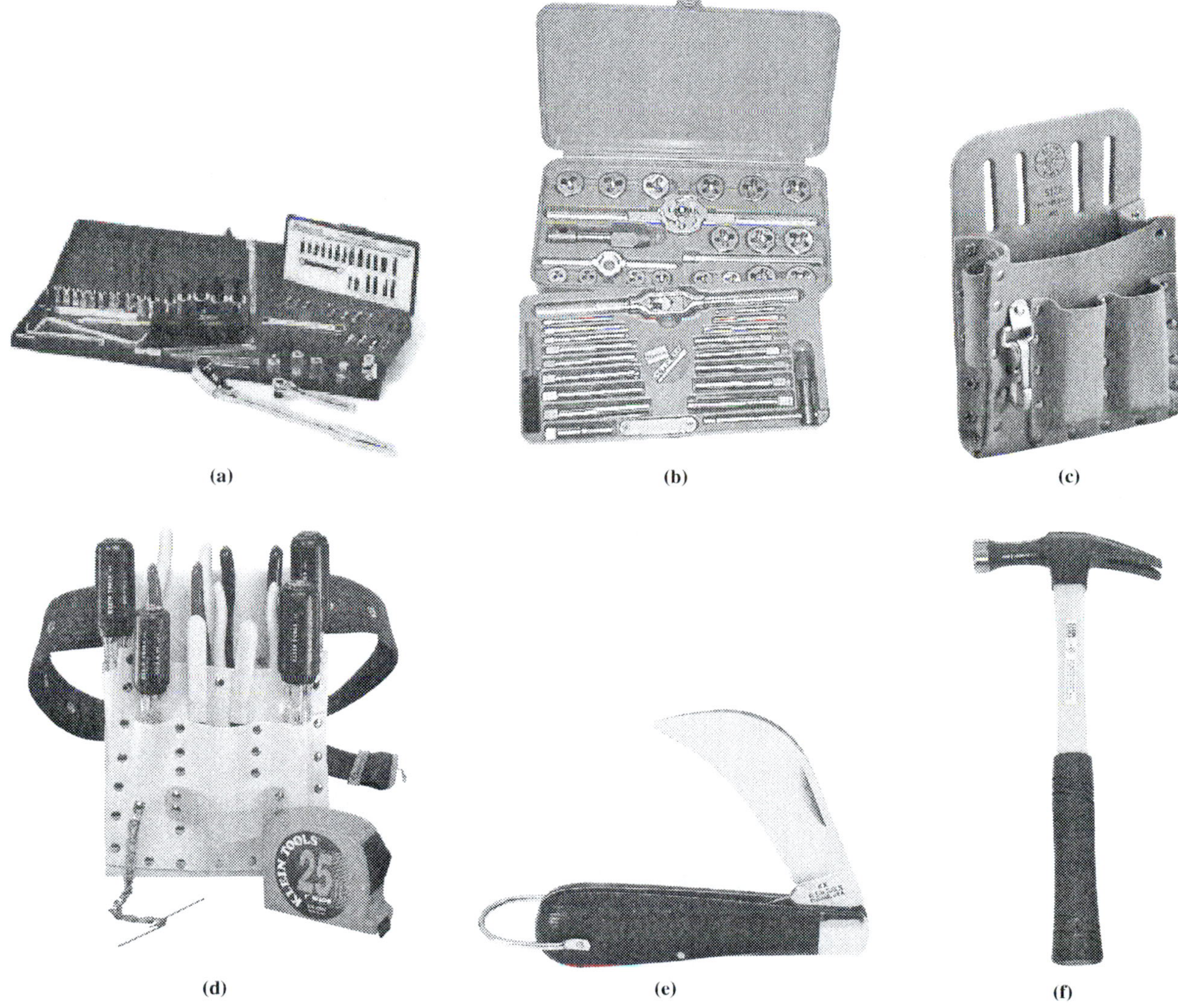

Figure 3–8 (a) Klein general tool set (courtesy of Klein Tools); (b) Snap-on tap and die set (courtesy of Snap-on Tools); (c) Klein tool pouch (courtesy of Klein Tools); (d) Klein electrician's tool pouch (courtesy of Klein Tools); (e) Klein electrician's pocket knife (courtesy of Klein Tools); (f) Klein hammer (courtesy of Klein Tools).

exterior threads on a bolt. The die is used to "dress" or clean up threads on a bolt, and the tap can be used to make initial threads or to dress up threads that are worn or damaged. You may need to drill a hole and tap threads into it to allow you to mount electrical components into an electrical panel with a screw or small bolt. The tapped hole takes the place of a nut, so you do not need to have a wrench on the back side of a panel to hold the nut.

Figure 3–8c and 3–8d show two different types of tool pouches. The tool pouch allows you to carry tools easily and safely so that they are always within reach. Figure 3–8e shows an electrician's pocket knife that is used to remove or "skin" the insulation off of a wire. Figure 3–8f shows a hammer that is used for maintenance. The hammer has a claw on one end to help remove nails or other fasteners.

3.8 Proper Use of Tools

You will be exposed to a wide variety of tools when you are on the job. It is important to understand that tools are designed for a specific job and that you should always use the proper tool for the right job. Many on-the-job injuries are caused by improperly using tools. If you do not have the correct tool to accomplish a job safely, you should request the proper tool.

Questions

1. Explain how a Wiggy-type voltmeter is different from a traditional analog-type voltmeter.
2. Identify four types of pliers.
3. How is a clamp on an ammeter used for troubleshooting a petition?
4. Explain how a hole punch is used to make a hole in a metal cabinet for a pushbutton switch.
5. Name three different types of screwdrivers.

True and False

1. A digital voltmeter indicates the amount of voltage it is measuring, with numbers on its display.
2. A Wiggy-type voltmeter indicates the presence of voltage with a needle on a scale and with a solenoid that bounces or buzzes.
3. You can safely remove a fuse with a common pair of pliers instead of a recommended fiberglass fuse puller.
4. A hole punch is a better tool to use than a drill to put a large hole in an electrical cabinet.
5. Wire cutters are available in a variety of sizes to allow all sizes of wire to be easily cut.

Multiple Choice

1. Which tool would you use to strip a wire to remove its insulation?
   a. Wire cutters
   b. Common pliers
   c. Wire strippers
   d. Side-cut pliers
2. A scopemeter can _____
   a. display voltage as a number.
   b. display the waveform of the voltage being measured.
   c. display the frequency of the voltage being measured.
   d. All of the above
3. Offset screwdrivers are used to _____
   a. provide more torque than a regular screwdriver.
   b. tighten or loosen hard to reach screws that have small clearance or are in tight spaces.
   c. reach screws in deep recesses.
   d. All of the above
4. A nut driver is _____
   a. a type of pliers.
   b. a type of wrench.
   c. a tool used to strip wires.
   d. a special offset screwdriver.
5. A tap and die set are used in this way:
   a. The tap makes interior threads and the die makes exterior threads.
   b. The tap makes exterior threads and the die makes interior threads.
   c. The tap can make both interior and exterior threads.
   d. The die can make both interior and exterior threads.
   e. Both c and d.

Problems

1. You are requested to mount three pushbutton switches in a new metal electrical panel. Select the tools that you would use to do this job.
2. You are requested to cut three pieces of wire and pull them through a new section of conduit. Identify the tools that would help you with this job.
3. You are requested to mount a panel meter that is 2 inches in diameter in the front of the metal electrical panel and secure it with four 1/4-20 bolts without nuts. Identify the tools that would help you complete this project.
4. You are requested to cut a piece of conduit to length and thread it on each end. Identify the tools you would select to do this job.
5. You are requested to loosen several screws with Phillips heads that are in a very tight location. Identify any tools that would help you with this process.

Chapter 4

Symbols and Diagrams

Objectives

After reading this chapter you will be able to:

1. Identify the symbol for each electrical component.
2. Understand the difference between a wiring diagram and a ladder diagram.
3. Identify the common abbreviation for electrical components.
4. Understand how to read the sequence of operation of a ladder diagram.
5. Understand the cross-reference functions in a ladder diagram.

Electrical diagrams and other drawings are used to communicate a large amount of information about the location and operation of equipment on which you will be working. This information includes the sequence of operation, operating instructions, and the location of electrical components that control a system. To include as much information as possible on a diagram, symbols are used to represent all the electrical components.

The diagrams that you will use to install, troubleshoot, and repair electrical equipment are typically drawn on a computer-aided drafting (CAD) system and converted to a drawing. The drawings can be used directly as a printout, or they can be converted to blueprints. The complete set of blueprints is required to understand fully how the machine has been designed to operate. The set will include a plant layout diagram that shows the location of the equipment in the factory and its relationship to other equipment, such as conveyors and material-handling equipment. A machine drawing shows the location of all electrical cabinets on the equipment.

The wiring diagram shows the location of all electrical components in the cabinet and the route of all wires between them. Wire numbers and cable numbers will also be included in this print. An elementary diagram shows the sequence of operation. This diagram is also called a *ladder diagram* because it shows the sequence of operation in separate lines that look like the rungs of a wooden ladder. This diagram is invaluable in troubleshooting a system when it becomes inoperative. The blueprints may also contain a page called the parts list, which includes a complete list of all electrical parts in the system with part numbers.

If the machinery is controlled by a programmable controller, its program will be shown on a print and will look very similar to the ladder diagram. A diagram that shows the type and number of input and output modules will be included in this set of prints, which will help you troubleshoot the system. At the end of each set of blueprints will be a parts list. The parts list will detail all components used in the equipment and provide part numbers and other important information, such as voltage and current specifications. When you determine that a part is inoperative and needs to be replaced, you will use the parts list to find the manufacturer's part number so that you can locate a new part in the parts crib, or provide information that will be needed to order a new one.

In this chapter we provide examples of all of these types of blueprints and explain how to read and use them in installing, troubleshooting, and repairing equipment. Additional information will provide pictures of some of the typical devices that are used in electrical systems and show their electrical symbols. This will help you relate the symbol to the actual component that you will be working with when testing and troubleshooting.

You will also see how the motor control process is displayed on paper so that electrical technicians will have available the vast amount of information that was used to design the machinery and motor control circuit. You will learn that the motor control process is a three-step process where inputs are used to sense the conditions in the system. This information is presented to another part of the control system, which must decide what action is required to be taken in terms of a signal sent to an output device such as a motor starter or solenoid. The diagrams and controls presented throughout this book will be related to the sequence of events (sense, decide, act) that are used to provide motor control. In the final part of this chapter we provide vital information about interfaces between electrical control systems and other important systems in the factory. These include electrical mechanical systems, such as robots, material-handling equipment, and hydraulic systems, that are used in virtually all types of manufacturing and process equipment. Other interfaces between the electrical systems and computer-controlled devices are also explained. The most common of these interfaces is between the programmable controller and the electrical–mechanical controls in the motor control system. When you have completed the chapter you will be able, properly and completely, to use all information that is provided in the blueprints and other diagrams that you will use when working with motor controls.

4.1 Electrical Symbols

Electrical components are identified by symbols on electrical wiring diagrams and ladder diagrams. These symbols were originally standardized by the Joint Industrial Council (JIC). In 1981 the JIC Board of Directors requested that the National Fire Protection Association (NFPA) incorporate the JIC standards into its standards. The new standards are called NFPA 79. The NFPA 79 standards have been updated continually through the years; its latest update was July 1997. This standard has also been adopted as an American National Standard and is used by the metalworking machine tool builders.

Figure 4–1 shows several typical electrical symbols that are used in various electrical diagrams. A picture of the component that is being represented is also shown with these symbols to give you an idea of what each symbol represents. The most difficult part of understanding symbols is to get a visual image of the component that is being represented by the symbol. This visual image is called a *referent*. After you have the image in your memory, you will be able to look at a symbol on a diagram and instantly visualize what component that symbol is representing and what it looks like when it is installed on the machinery. The link between the symbol and picture of the component is vital if you are to receive the full amount of information that is provided on

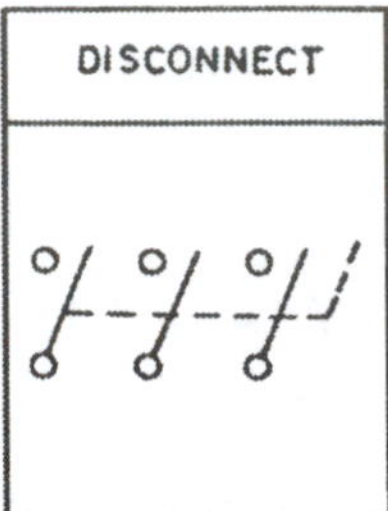

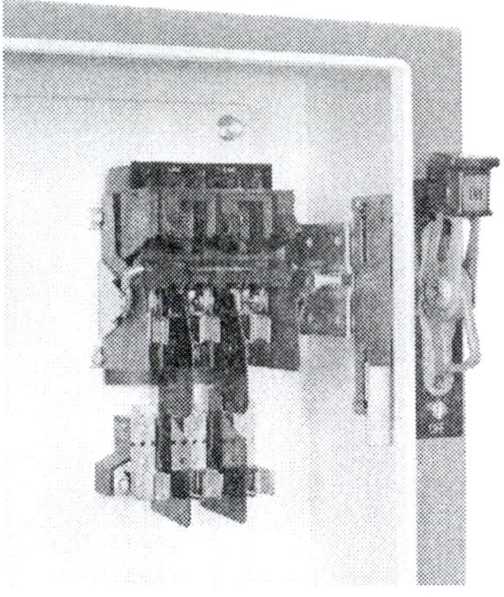

LIMIT			
NORMALLY OPEN	NORMALLY CLOSED	NEUTRAL POSITION	
			ACTUATED
HELD CLOSED	HELD OPEN	NP	NP

Figure 4–1 Electrical symbol and picture of electrical components: (a) disconnect switch; (b) limit switch; (c) pushbutton; (d) proximity switch.
(Courtesy of Eaton/Cutler Hammer)

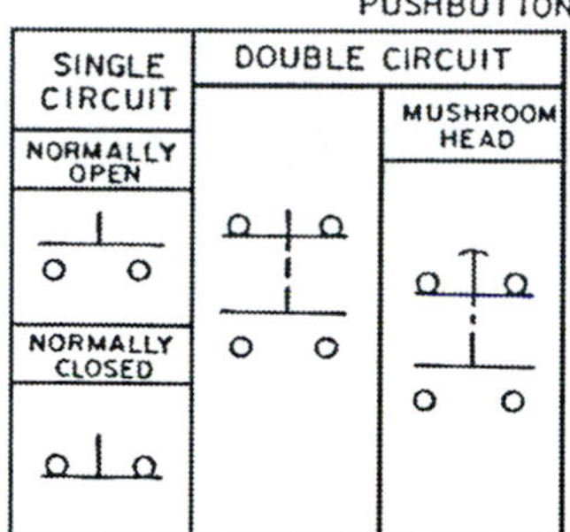

PROXIMITY SWITCH
CLOSED
OPEN

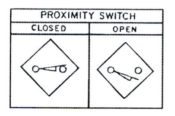

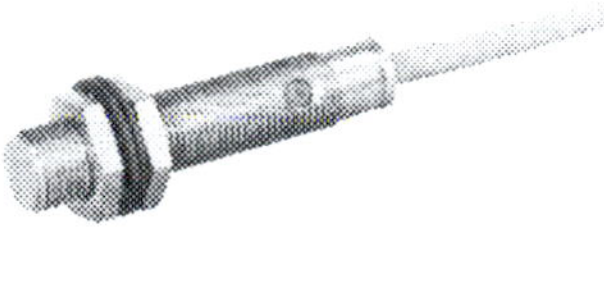

Figure 4–1 (continued)

any electrical diagram. If you find a symbol in an electrical diagram, you should try to find a picture of the component or an actual component to look at so that you will have an image of the component in your memory for future use.

The symbols that are shown in this figure are used in wiring diagrams and elementary (ladder) diagrams. They will show the number of terminals the component has and the identification that is used to mark each terminal. The diagram will also show the number of electrical parts that the component has, such as the number of contacts or poles.

A complete set of electrical symbols is provided in Figure 4–2. All the electrical components that are commonly used in motor control circuits are included in this set. These symbols include several standard methods of representing the operation of components. For example, a dashed line between sets of contacts indicates that each of the sets of contacts is operated by the same actuator. This is used to indicate that multiple-pole switches, such as disconnects and circuit breakers, will have all of their sets of contacts activated by the same condition. Another method of indication operation is shown in the symbol of plugging switches. A curved line with an arrowhead on each end indicates that this switch moves in a rotary motion when it is in operation.

Each switch has a different method of activating its contacts. The symbol of the operator normally looks like the device it is representing. For example, the symbol of the float switch (liquid-level switch) uses a ball to represent the float, and the symbol of the pressure switch uses a pressure element to represent the pressure operator of that switch. The location of the switch contact in the symbol of a switch is also used to indicate whether the contacts will open or close when power or the means of activation is removed. For example, the symbol for the normally open pressure switch shows the switch arm opening below the contact. This indicates that when pressure increases (goes up) the switch will move up and close. In the instance of the normally closed pressure switch, the switch arm is shown above the contact, which means that the switch will open when pressure increases. You can assume that when the force (pressure, temperature, flow, etc.) increases, the switch arm shown in the symbol is moved upward, and when it decreases, it moves downward.

Another important feature of the symbols is the symbol for conductors. In a diagram, conductors are represented by a line. If the conductor carries high voltage and high current, the line will be heavier than the lines for conductors that carry lower voltage and lower current. This means that the wires used to represent load circuit conductors will be darker and heavier than lines used to represent conductors in the control circuit.

The symbol for conductors will also indicate the presence of a junction of two or more conductors. If conductors cross each other and form an electrical junction, this point will be indicated by a dot with a small circle that is filled in. If the wires merely cross each other but are not connected electrically, no terminal point will be shown. As diagrams become increasingly complex, many wires must be shown crossing other wires to make connections at the proper terminal on each component. If a connection is being indicated, the terminal point will be shown.

STANDARD ELEMENTARY DIAGRAM SYMBOLS

The diagram symbols shown below have been adopted by the Square D Company and conform where applicable to standards established by the National Electrical Manufacturers Association (NEMA).

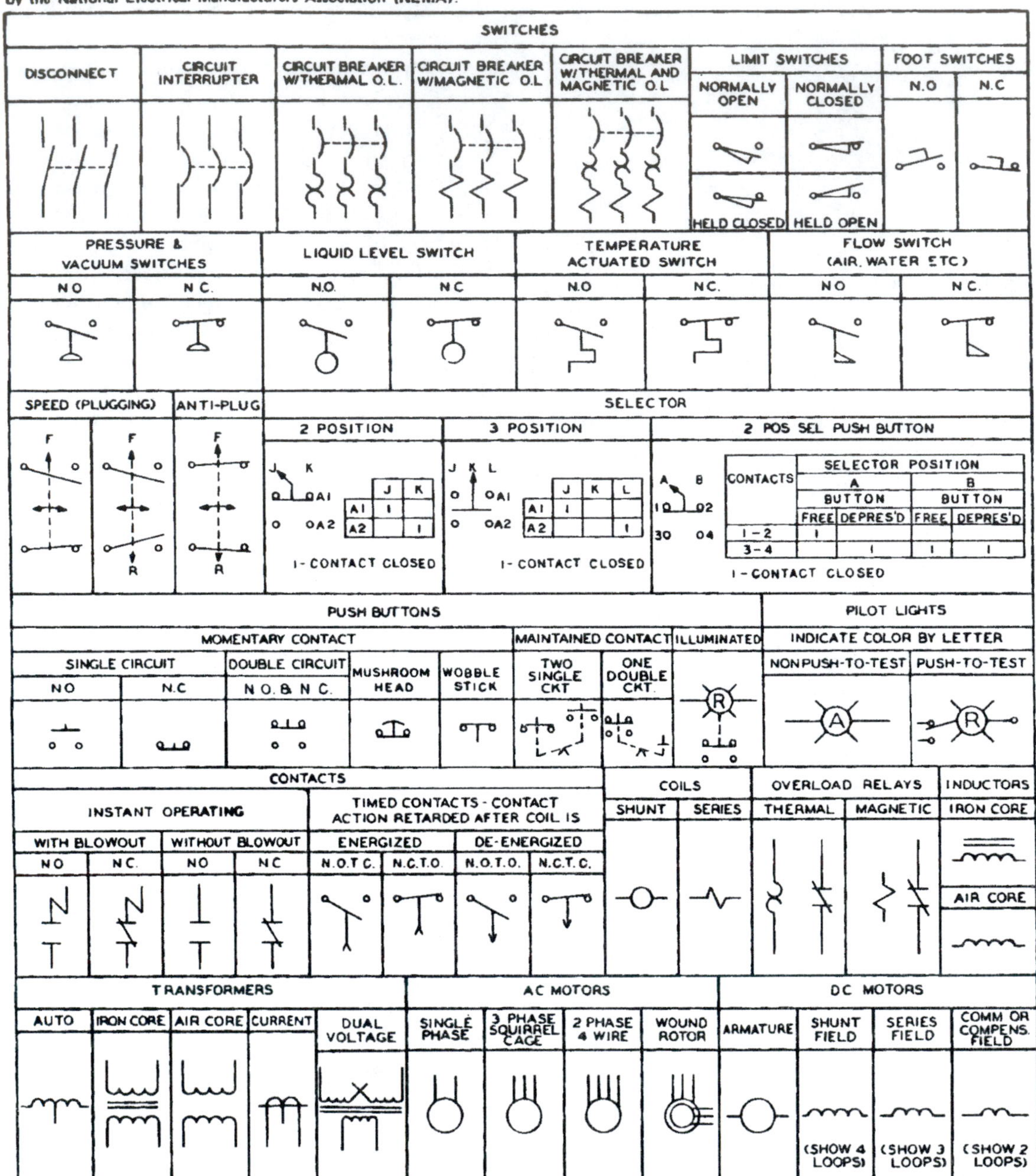

SWITCHES
DISCONNECT
CIRCUIT INTERRUPTER
CIRCUIT BREAKER W/THERMAL O.L.
CIRCUIT BREAKER W/MAGNETIC O.L.
CIRCUIT BREAKER W/THERMAL AND MAGNETIC O.L.
LIMIT SWITCHES
NORMALLY OPEN
NORMALLY CLOSED
HELD CLOSED
HELD OPEN
FOOT SWITCHES
N.O
N.C
PRESSURE & VACUUM SWITCHES
LIQUID LEVEL SWITCH
TEMPERATURE ACTUATED SWITCH
FLOW SWITCH (AIR, WATER ETC)
SPEED (PLUGGING)
ANTI-PLUG
SELECTOR
2 POSITION
3 POSITION
2 POS SEL PUSH BUTTON
1 - CONTACT CLOSED
CONTACTS
SELECTOR POSITION
BUTTON
FREE
DEPRES'D
PUSH BUTTONS
MOMENTARY CONTACT
MAINTAINED CONTACT
ILLUMINATED
SINGLE CIRCUIT
DOUBLE CIRCUIT
N.O. & N.C.
MUSHROOM HEAD
WOBBLE STICK
TWO SINGLE CKT
ONE DOUBLE CKT.
PILOT LIGHTS
INDICATE COLOR BY LETTER
NONPUSH-TO-TEST
PUSH-TO-TEST
CONTACTS
INSTANT OPERATING
WITH BLOWOUT
WITHOUT BLOWOUT
TIMED CONTACTS - CONTACT ACTION RETARDED AFTER COIL IS
ENERGIZED
DE-ENERGIZED
N.O.T.C.
N.C.T.O.
N.O.T.O.
N.C.T.C.
COILS
SHUNT
SERIES
OVERLOAD RELAYS
THERMAL
MAGNETIC
INDUCTORS
IRON CORE
AIR CORE
TRANSFORMERS
AUTO
IRON CORE
AIR CORE
CURRENT
DUAL VOLTAGE
AC MOTORS
SINGLE PHASE
3 PHASE SQUIRREL CAGE
2 PHASE 4 WIRE
WOUND ROTOR
DC MOTORS
ARMATURE
SHUNT FIELD
SERIES FIELD
COMM OR COMPENS. FIELD
(SHOW 4 LOOPS)
(SHOW 3 LOOPS)
(SHOW 2 LOOPS)

SUPPLEMENTARY CONTACT SYMBOLS

SPST, N O		SPST N.C.		SPDT		TERMS
SINGLE BREAK	DOUBLE BREAK	SINGLE BREAK	DOUBLE BREAK	SINGLE BREAK	DOUBLE BREAK	SPST - SINGLE POLE SINGLE THROW
						SPDT - SINGLE POLE DOUBLE THROW
DPST, 2 N O		**DPST, 2 N.C**		**DPDT**		DPST - DOUBLE POLE SINGLE THROW
SINGLE BREAK	DOUBLE BREAK	SINGLE BREAK	DOUBLE BREAK	SINGLE BREAK	DOUBLE BREAK	DPDT - DOUBLE POLE DOUBLE THROW
						N O - NORMALLY OPEN N.C - NORMALLY CLOSED

Figure 4–2 Standard set of elementary diagram symbols.
(Courtesy of Square D/Schneider Electric)

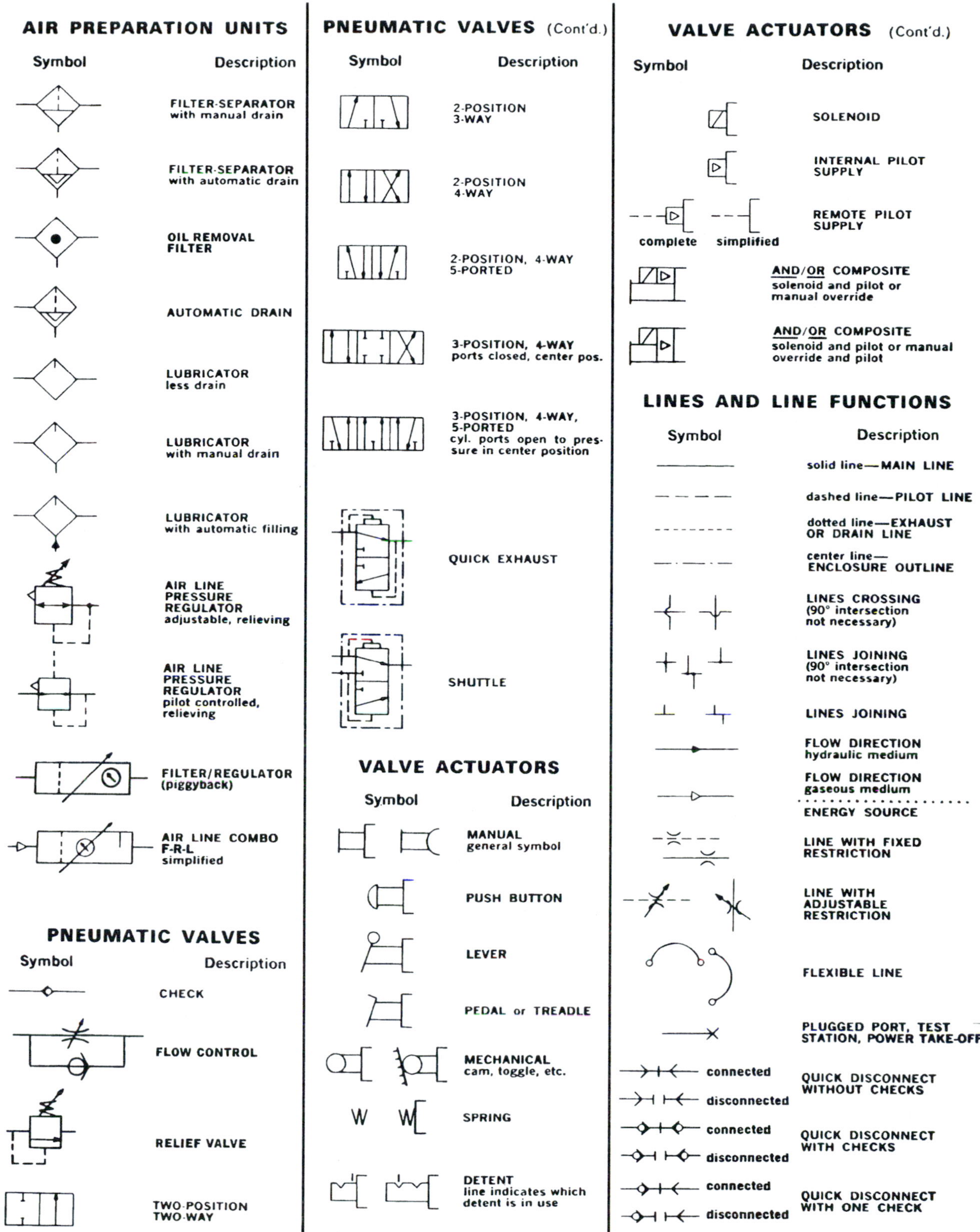

Figure 4–3 Fluid power symbols.
(Courtesy of Parker Hannifin)

4.1.1 Hydraulic and Pneumatic Fluid Power Symbols

Another set of symbols that are important to the troubleshooting technician is the set of fluid power symbols. The fluid power symbols consist of hydraulic and pneumatic components. Figure 4–3 shows a set of common fluid power symbols. When you are working on a machine such as a plastic press or metal processing machine, it may have hydraulic or pneumatic components that are controlled by the electrical control system. You can use these symbols to help identify components in the fluid power system.

4.1.2 Abbreviations Used with Electrical Symbols

Components in the diagram may also be identified by abbreviations. Figure 4–4 shows the common abbreviations for electrical components. An abbreviation provides a method of identifying multiple components that are used in the system. For example, every electrical technician can easily recognize the symbol for a limit switch. If more than one limit switch were used in a circuit, it would be difficult to tell them apart if they were not identified by an abbreviation that included a number. For instance, the first limit switch would carry the abbreviation LS1 and the second would be identified as LS2. The abbreviation LS helps identify the component symbol as a limit switch, and the numbers help identify which limit switch is being indicated. Abbreviations are also used to identify components that do not have a distinct symbol. For example, several electrical components use the symbol of a coil of wire. This symbol could indicate a motor winding, a transformer coil, a relay coil, or a solenoid. The abbreviation helps to identify these components positively.

Another common symbol used for electrical diagrams is a square or rectangular box. An abbreviation is used in conjunction with the box to indicate what type of component or circuitry is housed in the box. Each association uses its own standards for abbreviations. It is acceptable practice to indicate the full name with abbreviations that are not easily identified.

4.2 Electrical Diagrams

A large variety of electrical diagrams are required to present all the information about the location, installation, operation, and replacement of components that are connected together to provide a motor control system. These diagrams will be used to manufacture the machinery before it is brought into the plant. They will also be used to install the machinery correctly and to start it up. After the machinery has been installed and is operating, the diagrams will be used to troubleshoot faults that occur and to identify parts that must be replaced. If the machine is controlled by a programmable logic controller (PLC), a diagram will be used to show the interface between the PLC and the hardware components used as inputs and outputs. A copy of the PLC program will also be provided on a set of diagrams. This allows you to use the PLC to help locate the problem.

When a faulty part is located, another diagram will be used to determine wire numbers and terminal locations. These diagrams will also be used to determine which terminals should be used during troubleshooting tests, and the part number for the replacement part. After the replacement part has been located, the diagrams will be used to remove and replace the part and to make any preliminary setup or calibration adjustments. In this section of the chapter we present several different types of diagrams and explain how to utilize all the information that is provided on them.

Two of the most useful diagrams that are used in installation, troubleshooting, and repair are the wiring diagram and the elementary diagram. Figure 4–5 shows the wiring diagram for a typical motor starter. Figure 4–6 shows the ladder diagram for the same motor starter. In this chapter we will use a metal forming machine as our application. We will examine wiring diagrams and elementary (ladder) diagrams from this machine. The wiring diagram for this machine is provided in Figure 4–7 and an elementary (ladder) diagram for the machine is provided in Figure 4–8. These two diagrams represent the same set of components in two distinct different ways. The *wiring diagram* shows the location of the components in relation to each other in the electrical cabinet and also shows the exact location of terminals on the components and the location of wire bundles within the cabinet. Since the larger wires that provide voltage to the load are interspersed with the smaller control wires, the wiring diagram becomes very confusing to use in troubleshooting. Another version of this diagram is called the panel layout diagram. Figure 4–9 shows a panel layout diagram. This diagram looks very similar to the wiring diagram, except the wires are not shown, just the placement of each component. Another way to think about the panel layout diagram is that it will look exactly like a picture of the components in the electrical panel. The panel layout diagram and the wiring diagram are useful when used

COMMON ELECTRICAL ABBREVIATIONS

Abbreviation	Device	Abbreviation	Device	Abbreviation	Device
ABE	Alarm Bell	FLS	Flow Switch	QFE	Transistor, Field-Effect
ABU	Alarm Buzzer	FS	Float Switch	QSB	Transistor, Surface-Barrie
AH	Alarm Horn	FTB	Fusible Terminal Block	QT	Transistor, Triode
AM	Ammeter	FTS	Foot Switch	QTM	Thermistor
AT	Autotransformer	FU	Fuse	QU	Transistor, Unijunction
B	Brake Relay	GRD	Ground	QVR	Varistor
CAP	Capacitor	HTR	Heating Element	R	Reverse
CB	Circuit Breaker	INST	Instrument	REC	Rectifier
CI	Circuit Interrupter	IOL	Instantaneous Overload	RECP	Receptacle
CON	Contactor	LS	Limit Switch	RES	Resistor
CR	Control Relay	LT	Pilot Light	RH	Rheostat
CRA	Control Relay, Automatic	M	Motor Starter	RSS	Rotary Selector Switch
CRH	Control Relay, Manual	MAX	Magnetic Amplifier	S	Switch
CRL	Control Relay, Latch	MB	Magnetic Brake	SCR	Silicon Controlled Rectifie
CRM	Control Relay, Master	MC	Magnetic Clutch	SOC	Socket
CRU	Control Relay. Unlatch	MCS	Motor Circuit Switch	SOL	Solenoid
CS	Cam Switch	MF	Motor Starter - Forward	SS	Selector Switch
CT	Current Transformer	MR	Motor Starter - Reverse	ST	Saturable Transformer
CTR	Counter	MSH	Meter Shunt	SX	Saturable Core Rector
D	Diode	MTR	Motor	SYN	Synchro or Resolver
DAS	Diode Arc Suppressor	NLT	Neon Light	T	Transformer
DB	Dynamic Braking Contactor	OL	Overload Relay	TACH	Tachometer Generator
DISC	Disconnect Switch	PB	Pushbutton	TB	Terminal Block
DT	Tunnel Diode	PC	Printed Circuit	T/C	Thermocouple
DZ	Zener Diode	PL	Plug	TGS	Toggle Switch
F	Forward	PLS	Plugging Switch	TR	Time Delay Relay
FA	Field Accelerating Contactor	POT	Potentiometer	V	Electronic Tube
FB	Fuse Block	PRS	Proximity Switch	VAT	Variable Autotransformer
FF	Full Field Contactor or Relay	PS	Pressure Switch	VM	Voltmeter
FL	Field Contactor or Relay	NC	Photosensitive Cell	VS	Vacuum Switch
FLD	Field	Q	Transistor	W	Wattmeter

Figure 4–4 Common electrical abbreviations.

in conjunction with the ladder diagram for troubleshooting. As a technician, you will use the ladder diagram to determine the sequence of operation and flow of voltage, and use the panel layout diagram and wiring diagram to locate the component you want to test.

4.2.1 General Wiring Diagram

In this section we go into a little more depth about the wiring diagram and ladder diagram. If you examine the wiring diagram shown in Figure 4–5 you can see that

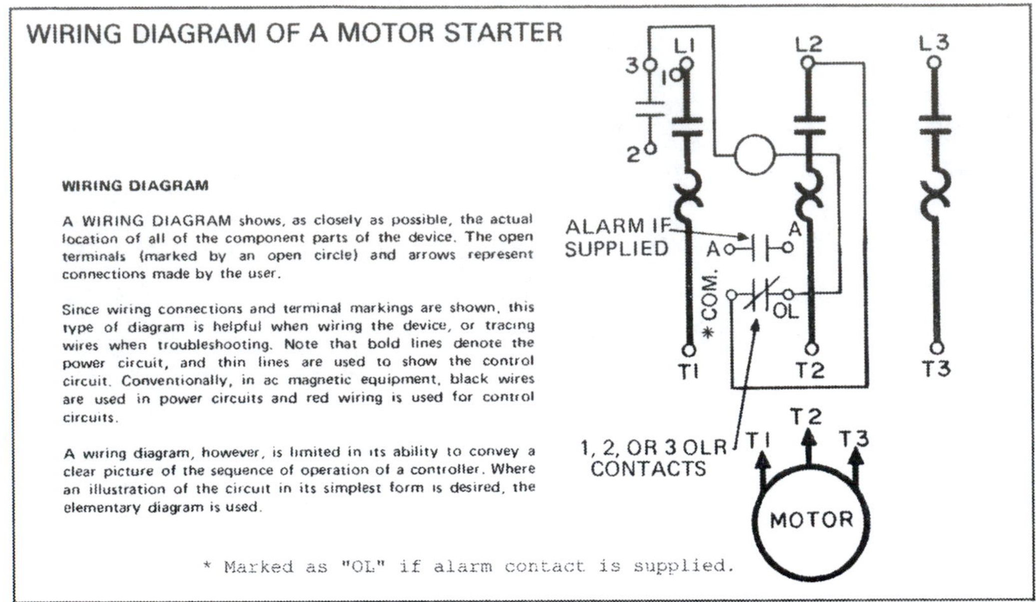

WIRING DIAGRAM OF A MOTOR STARTER
WIRING DIAGRAM
A WIRING DIAGRAM shows, as closely as possible, the actual location of all of the component parts of the device. The open terminals (marked by an open circle) and arrows represent connections made by the user.
Since wiring connections and terminal markings are shown, this type of diagram is helpful when wiring the device, or tracing wires when troubleshooting. Note that bold lines denote the power circuit, and thin lines are used to show the control circuit. Conventionally, in ac magnetic equipment, black wires are used in power circuits and red wiring is used for control circuits.
A wiring diagram, however, is limited in its ability to convey a clear picture of the sequence of operation of a controller. Where an illustration of the circuit in its simplest form is desired, the elementary diagram is used.
L1
L2
L3
3
1
2
ALARM IF SUPPLIED
A
A
*COM.
OL
T1
T2
T3
1, 2, OR 3 OLR CONTACTS
T1
T2
T3
MOTOR
* Marked as "OL" if alarm contact is supplied.

Figure 4–5 Wiring diagram of a typical motor starter circuit.

each component is represented by a figure that looks like the outline of the part. This means that the component will generally be represented by a box with terminal locations indicated in the proper locations. The main function of the wiring diagram is to show the location of components in relation to each other, and the location of specific terminals on each component. Wire numbers and terminal numbers are used to identify all conductors and terminals within the diagram.

The wiring numbers and terminal identification can also be used in conjunction with the ladder diagram shown in Figure 4–6. The ladder diagram shows the sequence of operation, and the wiring diagram provides the location of components and their specific terminals that are used when voltage and current tests are made. Generally, the ladder diagram is used to locate possible faults in the system, and the wiring diagram is used to indicate where the meter leads should be placed for the test.

The wiring diagram is provided by the machine builder, since it is also used when the machine is being built. If any wiring changes are made, or if a new component is added after the machine is installed, this diagram will be used to locate terminal points and spaces in the panel where the component can be located. Some wiring diagrams also provide grid numbers that are used to locate components or terminals in a large diagram. A cross-reference sheet is provided with the grid that indicates the location of each component or conductor on the diagram. The grid is necessary when a very large diagram is shown on a blueprint, and in cases where the diagram requires two or three pages of blueprints to display the entire system.

The electrical symbols shown in Figure 4–2 are not used in the wiring diagram. Some parts of the symbol may be used, but generally the wiring diagram uses the outline of the component rather than its symbol. The internal contacts of the device are generally shown using the electrical symbol for switches or contacts. The terminals of each component are shown in close proximity to where they will be found on the actual component. These terminals are generally shown as a small circle on the diagram. Other crucial information about the direction in which a component should be mounted is also listed on the wiring diagram. You can assume that the top of the diagram is the reference for the diagram, and you can determine the direction of mounting from the location of terminals or other identifying marks on the component.

4.2.1.1 WIRING DIAGRAMS OF COMPONENTS

Wiring diagrams of individual components are also provided for some components, such as motor starters. Figure 4–5 shows the wiring diagram of a full-voltage motor starter. The auxiliary contacts for the contactor are shown on the left side of the diagram between terminal 2 and 3, and other auxiliary contacts are shown in the diagram as terminal A and A. The coil for the motor

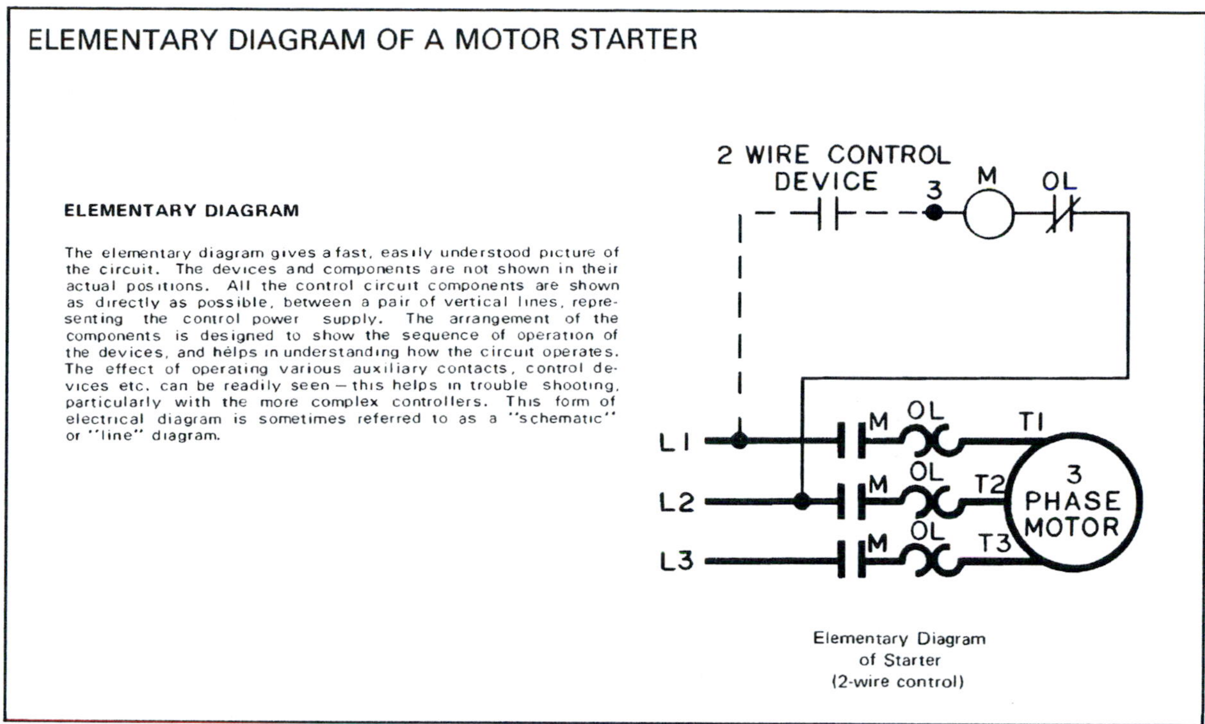

ELEMENTARY DIAGRAM OF A MOTOR STARTER
ELEMENTARY DIAGRAM
The elementary diagram gives a fast, easily understood picture of the circuit. The devices and components are not shown in their actual positions. All the control circuit components are shown as directly as possible, between a pair of vertical lines, representing the control power supply. The arrangement of the components is designed to show the sequence of operation of the devices, and hèlps in understanding how the circuit operates. The effect of operating various auxiliary contacts, control devices etc. can be readily seen — this helps in trouble shooting, particularly with the more complex controllers. This form of electrical diagram is sometimes referred to as a "schematic" or "line" diagram.
2 WIRE CONTROL
DEVICE
3
M
OL
L1
L2
L3
M
OL
M
OL
M
OL
T1
T2
T3
3
PHASE
MOTOR
Elementary Diagram
of Starter
(2-wire control)

Figure 4–6 Ladder diagram of a typical motor starter circuit.

starter is shown as a circle. The main terminals of all the contacts are identified with letters and numbers, such as T1 or L1. Many of the conductors cross each other but are not connected. Terminal points are indicated by a black dot, which indicates that two conductors are connected at that point. The actual location of the connection will be at the terminal connection on the motor starter contacts rather than as a splice in the middle of the conductor.

Some of the components in the diagram could be represented by their electrical symbol. These include the pushbutton switch, indicator lamps, and the heater and overload contacts in the overload device. In the diagrams provided by some manufacturers, symbols are used that do not belong to any standards that have been listed. These symbols may have originated when new electronic products were developed from traditional products or as a combination of several technologies, such as hydraulics and pneumatics with electronics.

The elementary diagram is shown in Figure 4–6. The elementary diagram is also called the ladder diagram and it is used to show the sequence of operation in the control circuit. Notice that the lines that represent the conductors in the control circuit are smaller than the lines used to represent the load circuit conductors. The smaller lines indicate that smaller-size wire is used in the control circuit. Figure 4–7 shows a larger example of a wiring diagram.

4.2.2 Ladder Diagrams

The ladder diagram shown in Figure 4–8 is the most widely used electrical diagram because it provides the sequence of operation for a system, which explains what should happen when the machine is started. The diagram is divided into two distinct sections. The top part of the diagram shows all components and conductors of the load circuit, and the lower part of the diagram shows the control part of the circuit.

The load circuit in this diagram is powered by 460-V, three-phase, 60-Hz AC, while the control circuit uses 115 V. The three-phase voltage lines in the load circuit are identified as L1, L2, and L3, which indicates where these lines should be connected during installation. The small circles on the left side of the disconnect switch indicate that the three high-voltage lines should be connected in the disconnect switch. Since the symbol shows a set of three fuses with the disconnect symbol, it indicates that the disconnect is a fusible type. The size of the fuses (15 A) is listed directly below the bottom fuse, which will help you check the actual size of the fuse in the panel. If any of the fuses should fail, they would be replaced with a 15-A fuse.

Three conductors are shown connected between the fused disconnect and the incoming terminals of the motor

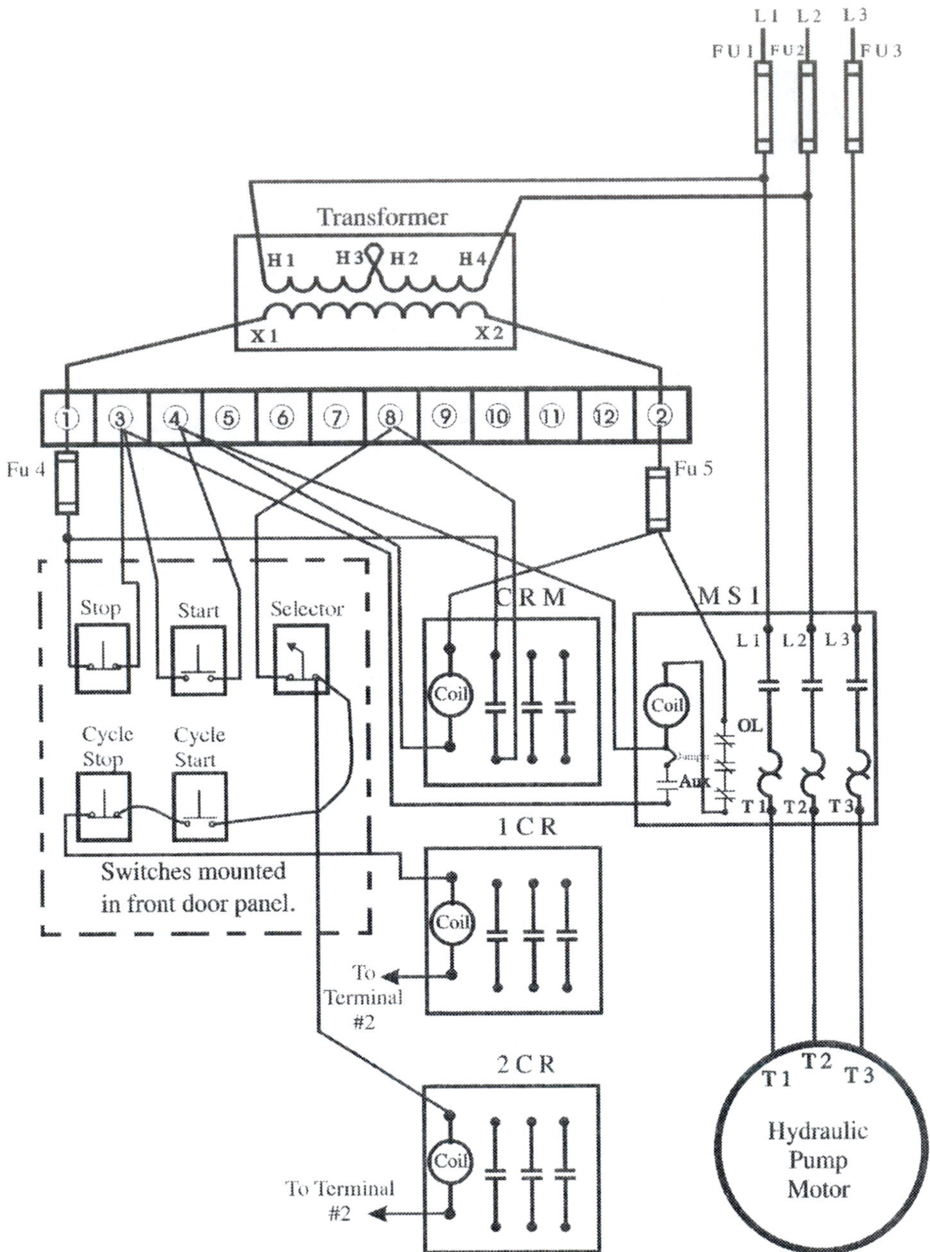

Figure 4–7 Electrical wiring diagram showing the location of electrical components in the electrical cabinet and the position terminals on each component.

starter contacts. The contacts are identified as 1M. You will also notice that the primary side of the control transformer is connected to L1 and L2. This is indicated by the small dark dot placed on the point where the connection is located. This diagram does not indicate whether the transformer should be connected at the fuses in the disconnect or at the contacts of the motor starter.

Notice that two motors are connected to the motor starter contacts in this diagram. The motors are identified by name, horsepower rating, value, and frame type. They also have an abbreviation (1MTR or 2MTR) that will be used with other components associated with the motor. Each motor has its own set of overload heaters, marked 1OL and 2OL. The contacts that the heaters activate will be shown where they provide the safety function, in line 1 of the control circuit diagram. This is a good example of the organization of the ladder diagram compared to the wiring diagram. In the wiring diagram, the heaters and overload contacts are shown beside each other because that is where they are physically located,

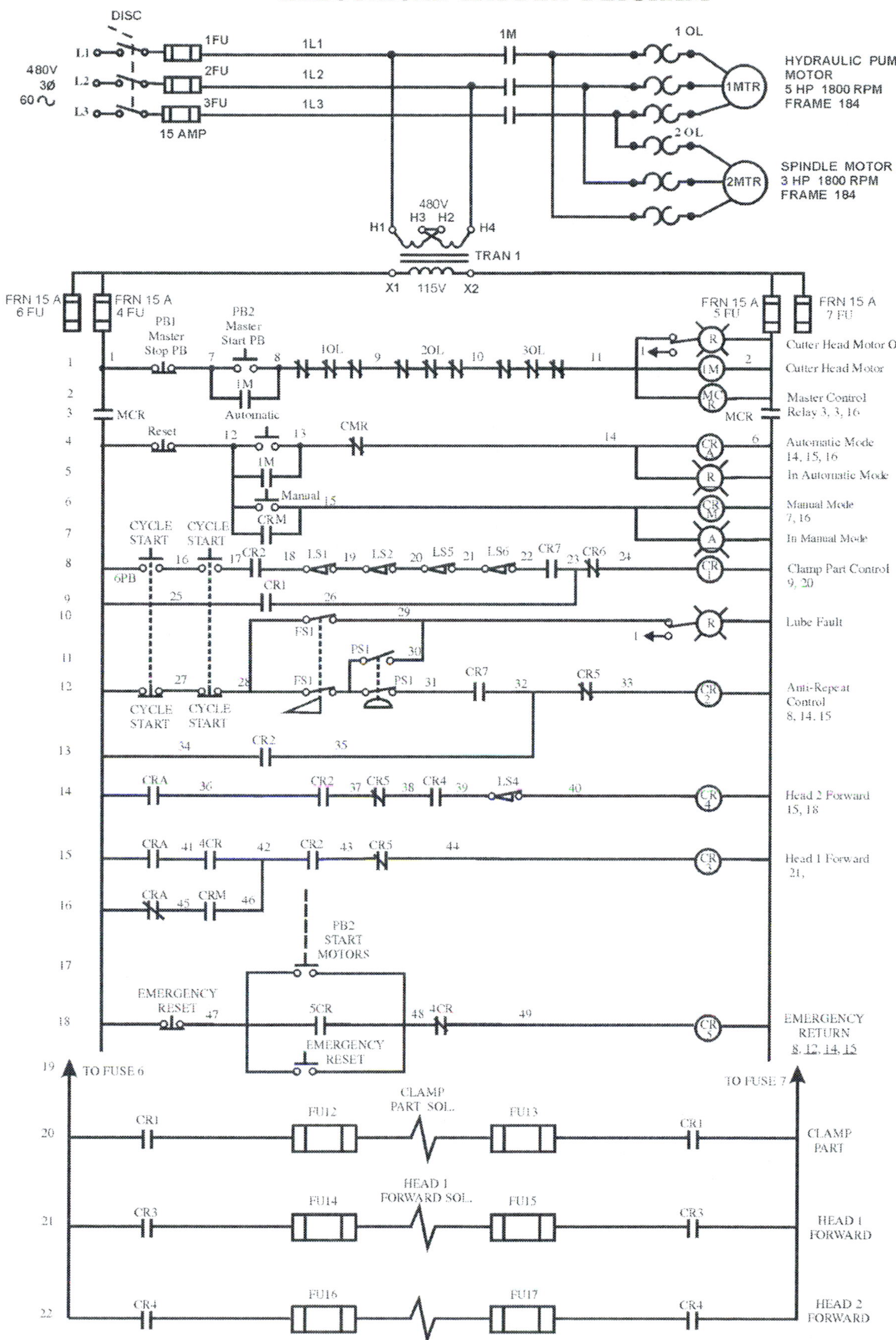

Figure 4–8 Ladder diagram of a metal-forming machine.

and the wiring diagram is trying to show the physical location of all parts. In the ladder diagram, the overload heaters are shown in the load diagram near the motor because that is where they are functionally located, and the overload contacts are shown some distance away in the control diagram, where they will provide the function of opening the circuit to the motor starter coil if the overloads sense that the motor is receiving too much current. These two components are shown in separate circuits because they provide two separate functions, even though they are physically located near each other.

The control portion of this diagram receives its power from the secondary of the control transformer. The lines are connected to terminals X1 and X2 of the transformer, and they are fused with 15-A fuses. You would need to look at the parts list for fuses 4FU and 5FU to determine that these fuses are an FRN-type fuse. Since the primary side of the control transformer is connected to the fuse disconnect, the control circuit will be powered with 115 V any time the disconnect is closed. The vertical line on each side of the control circuit that represents the conductors connected to the secondary of the transformer is called the *power rail* since it supplies power to the control circuit.

As you know, the ladder diagram has been designed to provide you with information concerning the sequence of operation for the system. This means that you should begin reading the diagram from top to bottom, and each line is read from left to right. Each line of the diagram has a line number. In this diagram the line numbers on the print run from 1 to 22. Other methods of indicating line numbers of larger diagrams include listing the page number in conjunction with the line number. For example, the fourth line on page 1 of the diagram would be numbered 1/4 and the fifth line on page 8 would be numbered 8/5 (similar line-numbering schemes may indicate the fifth line on page 8 in this manner). You should look for line numbers in the diagram because they will be used to indicate the source of coils for contacts and interconnecting wires from other pages in the print.

The numbers shown beside each wire in the diagram indicate a wire number. The same wire number will be attached to the end of the actual wire that is used to make the connection shown in the diagram. For example, in the diagram the wire that connects the start button to the overload contacts is identified with the number 8. The wire that makes that physical connection will also have the number 8 marked on it with wire number tapes. The wire numbers will help you locate the proper wire when you are troubleshooting the system. They will also be useful if any alterations must be made to the circuit.

Each switch in the circuit is identified in several ways. First each switch will be identified by its type, such as PB for pushbutton and LS for limit switch. Second, each switch will be identified by its function in the circuit, such as start, stop, or reset. A third way that each switch is identified is by its electrical symbol. The start pushbutton is shown as a normally open pushbutton, and the stop pushbutton is shown as a normal closed pushbutton. Several sets of contacts are also shown on line 14 and other lines. The contact symbol indicates that their operation is controlled by a relay coil rather than by some other physical operation, such as a limit switch or pushbutton.

4.2.2.1 INDICATING COILS AND CONTACTS

Contacts are controlled by the coils of relays and motor starters. The coils are shown connected against the right power rail. The conductor on this side of the circuit is identified by the number 6, and it is used as the return side of the control circuit. Each coil listed in this diagram has control over one or more sets of contacts. The exact number of contacts and their location in the program are shown at the right of the power rail on the right side of the diagram. For example, the coil for CR1 is shown at the right end of line 8. The numbers at the right side of the coil indicate that it has control of CR1 contacts in diagram lines 9 and 20. In some diagrams the type of contact (normally open or normally closed) is also indicated. If the contact is normally closed, the number is listed with an underline; the normally closed contacts for CR5 shown in lines 8, 12, 14, and 15 would be shown as $\underline{8}$, $\underline{12}$, $\underline{14}$, and $\underline{15}$. The normally open contacts would be shown as just a number. These numbers help you identify the lines where all the contacts that each coil activates are located.

In some diagrams the contacts will have a number above or below in addition to the coil identification. This number indicates on which line the coil that controls these contacts is located. For instance, the number 14 would be added to the normally open CR4 contacts that are shown on line 15 or line 18 of the diagram. The number 14 would indicate that the coil that controls CR4 is located on line 14 in this diagram. This number is called a *source number* and helps you identify the location of the coil.

You should also notice the presence of relay contacts indicated as MCR. One set of each of these contacts are located in each of the power rails of the diagram. The coil that controls these contacts is located on line 2 of the diagram and is called the *master control relay*. If these contacts open, all power to lines 4 through 18 will be disabled.

On several lines of the diagram, such as lines 1 and 10, you should notice a switch with an arrow that is near the lamps. The arrow also has a number connected to another conductor, as indicated by the number near the arrow. In this case the number 1 is shown near the arrow, which indicates that the wire with the arrow on it is actually connected to the left power rail, which is identified with the number 1. The arrow is used so that multiple lines are not shown crossing all over the diagram.

4.2.2.2 INDICATING JACKS AND PINS

Some automated systems found in factories today utilize multiconductor cables that have jack and pin connections. These cables plug into a socket in the control panel and at the equipment, to make connections simple. The connection symbol for the jack and pin (J-P4 12 12) will also indicate the cable number and the pin number so that you can test individual conductors within the cable. In this example, J-P4 indicates that this is cable 4 and the symbol shows the connection is for pin 12. The symbol will also be used to indicate which side of the connection has the jack and which side has the pins. In the ladder diagram, the jack and pins will be listed individually throughout the diagram on the lines where they occur, and in the wiring diagram the pin outline of the cable head will be used to show the location of the pins within the cable head and its physical shape.

4.2.2.3 SEQUENCE OF OPERATION

In some diagrams the actual sequence of operations is listed at the end of the diagram in English words. The main steps to the sequence of operation are listed by letters such as A through F. Substeps within each main step are listed by numbers as required. The sequence of operation also lists the function and operation of each limit switch. Other information within the set of diagrams is listed in this area of the diagram. It lists the last wire number used in the system and the highest number used to identify the control relays.

A title block is provided in the diagram to indicate the machine for which the diagram is prepared and the company and plant where the machine is installed. This block also indicates the sheet or page number of the diagram. This will be useful when a line shows a connection or a set of contacts that are referenced to a line on another sheet in the diagram. You can quickly locate it by looking for the sheet number in the lower right corner of the sheets.

4.2.3 Electrical Panel and Control Station Layout Diagrams

The third diagram in this series of electrical diagrams is the panel layout diagram (Figure 4–9). It is very similar to the wiring diagram in that it shows the location of the components and terminal boards within the panel and the location of all switches on the operator's console. The major difference between the layout diagram and the wiring diagram is that the layout diagram does not show any terminal connections on the components and it does not show the location of any wires or cables.

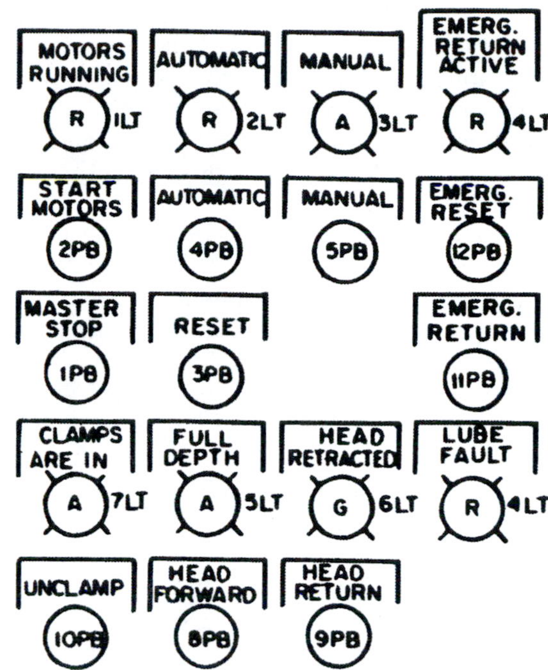

Figure 4–9 Panel layout diagram of operator panel for the metal-working machine.

In some cases the layout diagram is substituted for the wiring diagram, especially if sufficient detail is provided on the ladder diagram. In either case, the information provided in this type of diagram is used to locate components in the panel. This may not sound significant, but when you are troubleshooting, this diagram is invaluable. For instance, you may read the ladder diagram and determine that there is a good possibility that fuse 4FU is open. If you did not have a wiring diagram or a panel layout diagram, you would have to look all over the machine for the location of that fuse, and when you did locate a fuse, you would have to check wire numbers to determine if it is the correct fuse. In some units, you may find up to five separate electrical panels, and the components in each one will look similar to the others. This presents a real problem in locating the correct component and conductors to make the appropriate troubleshooting tests. In this diagram all pushbutton switches are identified by a circle that has the switch name above it, and all lamps are identified by a traditional lamp symbol, with the color of lens being indicated by a letter in the center of the symbol. All electrical components such as control relays, motor starters, and transformers are identified by a rectangular box with the abbreviation of the component name and number indicated in the box.

Terminal boards are identified by name, and each terminal on the strip is identified by number. The amount of voltage that is used on each terminal strip is also identified beside each strip. This is quite useful when you must make repeated voltage readings within the panel, because it will indicate at what range to set the voltmeter. You will notice that this diagram may list dimensions of blank space within the panel, or overall dimensions of the panel. These dimensions are provided in case you need space in the cabinet to mount additional components, such as motor speed drives or other solid-state, add-on equipment. The panel diagram may be used in conjunction with the ladder diagram during troubleshooting and installation procedures, or on its own to make modifications to the system.

4.3 Parts Lists

The parts list for this system is shown in Figure 4–10. This diagram shows a partial list of the components used to make the motor control system operate. The entire list of components may require several sheets of the diagram. The parts are grouped by component type. The manufacturer's part number is used in conjunction with other product specifications, such as voltage and current. If this system is used in a large factory, the inventory number may also be provided so that the availability of parts may be checked immediately by typing the inventory number into the computer terminal. The computer will indicate the number of parts with that number that are on hand and the location of the bin in which they are stored. Some systems also provide a purchase order that will be used to order the part from the storeroom or parts crib. This saves time for the technician, who must ordinarily fill out these forms by hand before the storeroom attendant can locate the part. This form is also used to order new parts when the inventory becomes too low.

The parts list diagram is useful for checking the size of components, such as fuses or motor starter heaters. In most factories, when a fuse is blown, it will be replaced with the same type of fuse as that found in the fuse socket. If someone used the wrong-size fuse at some time while troubleshooting, it is likely that the same size of fuse will be used to replace it when it goes bad again. This may cause problems if the fuse is of the wrong type (slow blow instead of fast blow, for example) or is the wrong size. From the information provided so far from the various types of diagrams, you can see that it is vitally important that all these diagrams be maintained as changes are made. It is also important that all changes to the diagrams be made by the engineering staff, and that all copies of the diagram that are located with the equipment and in files are then updated.

4.4 Diagrams of Systems That Use Programmable Logic Controllers

In some systems a programmable logic controller (PLC) is used for control. In this case several additional diagrams will be provided with the system. These include the program, which will look like a ladder diagram, and a wiring diagram that shows the location and address numbers of all input and output modules. Other pages will be provided that list the cross-reference name and number of all switches and outputs in the system and all variables that are used in the program.

Figure 4–11 shows a PLC program that would be used to provide the same operational sequence as that of the ladder diagram shown in Figure 4–8. In this diagram all switches are identified with the contact symbol. The PLC does not provide a set of symbols for each type of switch, such as limit switch or pressure switch, because it would require too much memory. Since each switch is shown in the program as a set of contacts, a comment and an address will be used to identify different switches. This means that you will need the PLC input/output (I/O) diagram and the cross-reference sheet when you are troubleshooting the system.

Figure 4–10 Partial parts list for the electrical system of the metal-forming machine.

Parts List

Part Name	*Part No.*	*Company*	*Number Used*
Fused disconnect, three-phase, 30 A 60 Hz	DG22INGB	Cutler-Hammer	1
15-A fuse	FRN 15	Bussman Fuse	3
3-hp size 1 motor starter	SCG-1	Square D Company	2
10-A fuse	FRN 10	Bussman Fuse	4
Pushbutton, 110 V	E22MPBCO	Cutler-Hammer	11
Indicator lamp 110 V	10250T -34R—red -34A—amber	Cutler-Hammer	8

The programmable logic controller I/O diagram is shown in Figure 4–12. This PLC program and the addresses used on it are for the Allen–Bradley MicroLogix programmable controller. The diagram for other brands of PLC systems will be very similar, with only the address numbering systems being substantially different. The PLC program is often called *relay ladder logic* (RLL) because it looks like a typical ladder diagram and the microprocessor executes the instructions as a logic program. In a logic program, the series and parallel contacts in a diagram are given names that refer to their operation. Figure 4–13 shows an example of each of the common logic circuits, with a truth table to indicate the condition of the output as the switches are opened and closed through all possible variations. In the truth table, a 1 indicates that a switch or output is on, and a 0 indicates that it is off. From the diagram you can see that when two switches are in series with each other and one output, the logic circuit is called an AND gate. In this case the word "gate" is used to indicate that components have been put together in a circuit for a specific set of output conditions. The gate is called an AND gate because the only time the output will be on (1) will be when switch A "and" switch B are on (1). You can see that the output will be off (0) for all other conditions of the switches. The AND gate for computer logic uses a bullet-shaped symbol with the inputs connected to the back edge and the output connected to the nose. In relay ladder logic programs found in the PLC, the AND function is represented by contacts connected in series.

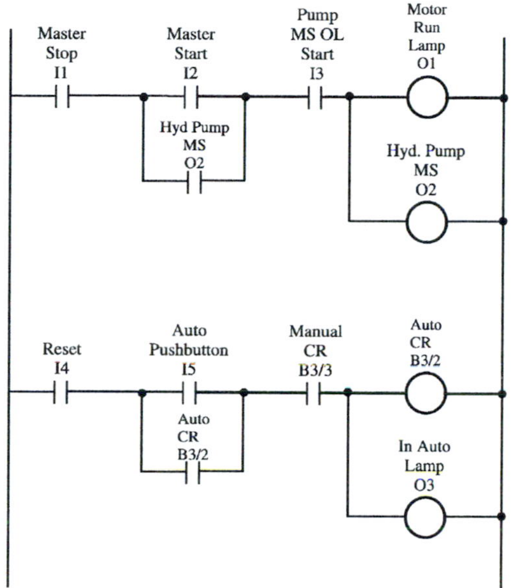

Figure 4–11 Partial programmable controller diagram of the motor control system for the metal-forming machine.

The OR gate is also shown in this figure, and you can see that in this circuit the two contacts are in parallel with each other. In this condition the output will be on (1) when either switch A "or" switch B is closed. This means that the output will be on for three of the switch conditions and off only when both of the switches are open. The computer logic symbol for the OR gate looks similar to that of the AND gate except that the back is concave rather than flat. The OR function is identified by parallel contacts in the RLL program for PLCs.

The NOT gate is similar to a normally closed set of contacts on a switch. You can refer this operation to a switch that has two sets of contacts, one set that is normally open and one set that is normally closed. The NOT gate is concerned only with the output from the normally closed set of contacts. This means that when the switch

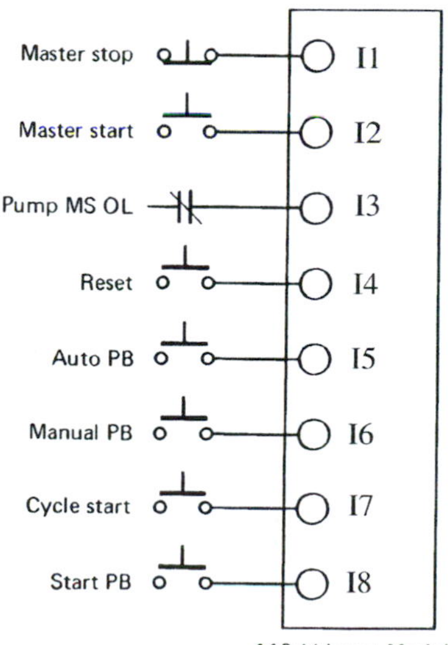

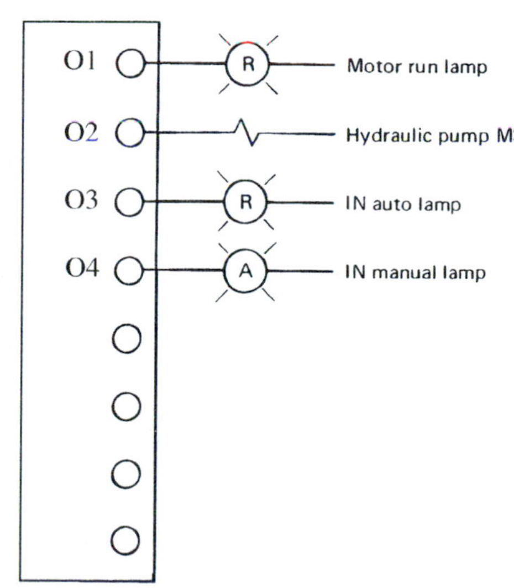

Figure 4–12 Partial input and output wiring diagram for the programmable controller program for the metal-forming machine.

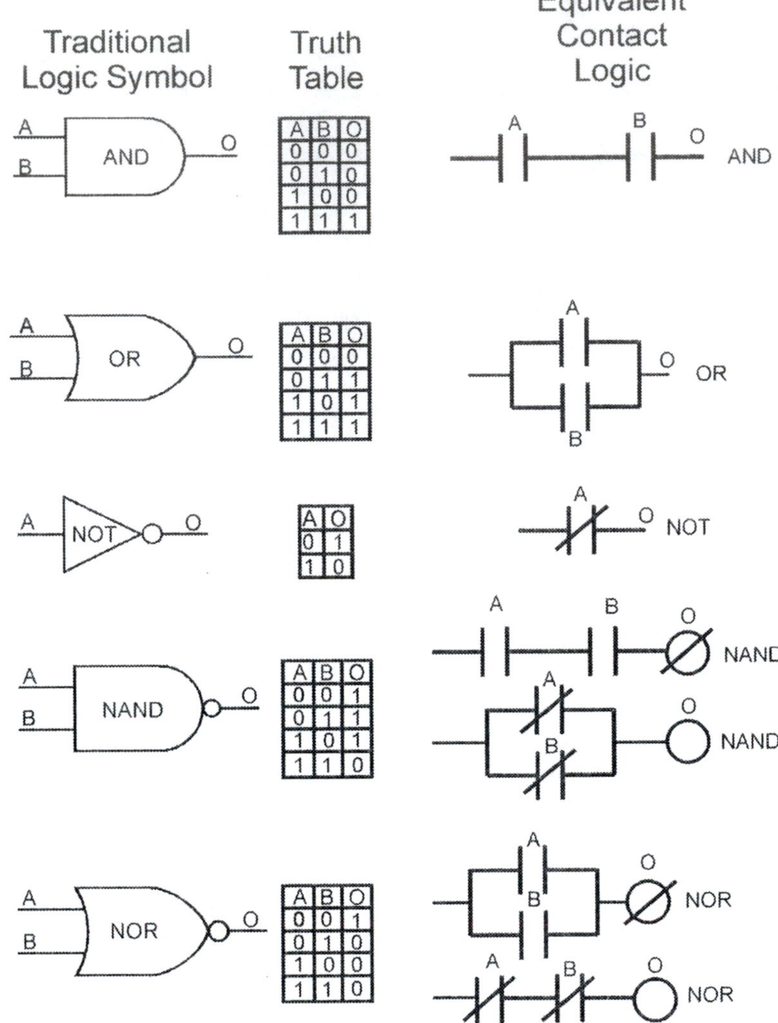

Figure 4–13 Truth tables, symbols, and equivalent ladder logic circuits for AND, OR, NOT, NAND, and NOR logic functions.

is "not" on, power will flow through the normally closed set of contacts to the output, and when the switch is on, the normally closed set of contacts will be open and no power will flow to the output. The symbol that is used in computer logic programs to represent the NOT function is a triangle on its side with a small circle drawn at the output. In the relay ladder logic diagram in a PLC, the NOT function is identified as a normally closed set of contacts.

The NOT function can be used alone or in conjunction with other logic functions. Two other symbols and truth tables are provided in this figure to represent the combination of the AND function and the NOT function to make a NAND gate, and the combination of the OR function and the NOT function to make a NOR gate. Notice that only the small circle at the output end of the gate is used to represent the NOT function. More information about the operation of these logic functions is provided in Chapter 18 with in-depth information about programmable logic controllers.

If the logic functions are used in a computer, the logic gate symbols are used, and when the logic is preformed in the programmable controller, the ladder diagram symbols (coils and contacts) are used. Coils and contacts are used for symbols in the ladder diagram because that is the language that electricians, technicians, and electrical engineers are used to. This means that the PLC will present a program that is readable by the personnel who must work with it daily.

The PLC program used for troubleshooting the control system is like the ladder diagram. Since the PLC program is in memory, and executes control of the circuit from a microprocessor, it is called a *soft-wired system;* the traditional motor control circuit is called a *hard-wired system.* The PLC system also provides a variety of status indicators that allow the technician to diagnose the condition of input switches and outputs such as solenoids and relays. The status indicators on the input modules will be shown on their diagram located on the module. There will be one indicator for each input, which will glow when voltage is received at that input. The status indicator for each output is shown located on the output module in the wiring diagram. Each indicator will glow when its output circuit is energized. These indicators allow the technician to look for the ones that are glowing to indicate which switches or outputs are energized, without testing each terminal for voltage.

The input and output module wiring diagram will also provide information that is necessary for locating the modules in the PLC rack and for identifying wire numbers and voltages. This information is particularly useful for installation and troubleshooting.

4.5 Machine Diagrams and Plant Layout Diagrams

Some diagrams that you will use on the factory floor are designed to show you where components are located on the machine or where the machines are located in relation to the entire factory. Machine diagrams are quite useful when you have not worked on a machine before or if you are not completely familiar with the machine design. In fact, most machine diagrams are drawn to scale, so the representation should be nearly exact. The machine diagram will indicate the location of key components, such as hydraulic manifolds, transformers, electrical panels, and disconnects. These are all devices that will be shown in their relative location on the wiring diagram, but you must remember that the wiring diagram is a two-dimensional diagram that can show only directional relationships (right and left, and above and below). In the case of solenoids located on a hydraulic manifold, they will be shown to the far left side of the wiring diagram to indicate they are located to the left of the electrical panel, but the wiring diagram does not show how far to the left. When you see the machine diagram, you can see that the hydraulic manifold is located inside the last door on the left end of the machine. Many other details, such as height and depth, may also be shown with the machine diagram. In fact, some machine diagrams show the machine as a multiview diagram that shows top, side, and front views, with several auxiliary views that indicate specific sections of the machine.

The plant layout diagram is basically a top-view diagram that provides a floor plan of the entire factory. From this type of diagram you will be able to see that all the machines in the plant are numbered, which provides an easy way to locate them and a method of identification for record keeping. Each aisle in the factory is also identified, like streets in a city. This provides reference for all personnel in the plant when they must move about. The identification will also be useful for dispatching technicians when problems exist on automated systems, robots, or conveyors that may be installed across large portions of the plant. For example, a problem may occur in a large automation line, but you would need to know in what section of the plant the actual problem exists. This could be narrowed down to one or two grid locations on the layout diagram, which would allow all maintenance personnel to arrive at the correct area. The plant layout diagram will also identify locations of power distribution switch gear and transformers. Large sections of bus ducts will also be shown on this diagram, which will be useful for additional installation of new equipment or other times when plant expansion is taking place.

4.6 Review

You have seen in this chapter that a variety of diagrams are provided and used on the factory floor for the installation, troubleshooting, and operation of equipment. Each diagram provides a different type of information presented in the manner that makes it most useful. For example, you learned that some diagrams, such as wiring diagrams, plant layout diagrams, and machine diagrams, are presented to show the location of equipment and components. Some of these diagrams will also show wire routes and wire numbers so that specific conductors can be located.

Other diagrams, such as the ladder diagram and the ladder logic program in PLCs, are provided to show the sequence of operation of equipment. The sequence of operation is provided by the machine designer so that a technician can determine the exact problem when a fault occurs. A diagram is also provided that lists all the components used in the electrical system. In this type of diagram, part numbers, specification of voltage and current, and location of crib storage are specified so that replacement parts can be identified and quickly located.

A large number of diagrams are needed to present all the information that is required to perform all work on the factory floor. If the diagram that you are using does not provide enough information, you should ask for associated diagrams and programs that may be helpful to you.

Questions

1. Explain how an NFPA 79 electrical symbol indicates what activates a switch.
2. Explain how the electrical symbol of a pressure switch indicates that an increase or decrease of air pressure causes the switch to activate.
3. You will find the terms *normally open* and *normally closed* used with switches. What is considered to be the normal condition?
4. Explain why an abbreviation may be used with a symbol.
5. Explain why a number such as LS1 or LS2 must be included with the abbreviation when identifying multiple limit switches in a diagram.

True and False

1. The elementary diagram is used to indicate the location of parts inside the electrical cabinet.
2. The wiring diagram is used to indicate the location of parts inside the electrical cabinet.
3. A PLC diagram includes drawings that show how inputs and outputs are connected.
4. A machine diagram or plant layout diagram indicates the location of each machine in the factory.
5. An elementary diagram is also called a ladder diagram.

Multiple Choice

1. The wiring diagram _____
   a. shows the location of each part in the electrical cabinet.
   b. indicates the sequence of operation for the machine.
   c. is also called an elementary diagram.
   d. All of the above
2. The numbers along the side of a ladder diagram _____
   a. indicate the rung number for the diagram.
   b. can be used for cross-reference purposes.
   c. indicate the number of parts in that line of the diagram.
   d. All of the above
   e. Only a and b
3. The symbol for a switch that is operated by pressure_____
   a. can be used as the symbol for a pressure switch.
   b. can be used as the symbol for a vacuum switch.
   c. can be used to indicate a temperature-operated switch.
   d. All of the above
   e. Only a and b
4. The parts list in a set of blueprints _____
   a. indicates the part number of each component in the system.
   b indicates the number of each part used in the system.
   c. indicates the size or rating for the part.
   d. All of the above
   e. Only a and c
5. The dashed line that connects two switch contacts that are in different lines of a ladder diagram indicates that _____
   a. the sets of contacts are actually mounted in the same switch and are activated by the same switch operator.
   b. the two sets of contacts are mounted in the same switch, but are activated by separate switch operators.
   c. the sets of contacts are controlled by the same type of switch, but are not connected in any other way.
   d. All of the above

Problems

1. You are asked to locate a motor starter inside an electrical panel. Which part of the blueprints would you use to find the location of the part in the panel?
2. You are asked to determine the sequence of operation of a machine. Explain which part of the blueprint would be useful for this job.
3. Explain why an elementary diagram is sometimes called a ladder diagram.
4. You are requested to find the part number of a motor starter in the system so you can purchase a new one. What part of the blueprints would you use for this job?
5. Explain why the programmable controller diagram must include documentation comments and an I/O cross-reference list.
6. You are asked to locate a specific machine in the factory so an electrician can be sent to work on it. Explain what parts of the blueprint you would use to show the electrician the location of the machine.

◀ Chapter 5 ▶

Overview of Motor Controls

Objectives

After reading this chapter you will be able to:

1. Explain the function of machine operational controls.
2. Explain the function of machine safety controls.
3. Identify components in the load circuit and in the control circuit.
4. Explain the difference between motion control and process control.
5. Explain how programmable controllers have changed motor controls.

Modern motor control has become a complex technology that includes simple control of motors and other loads as well as intricate motion control for precise positioning. It includes simple safety controls such as fuses and other controls as complex as programmable controllers. Some of these controls are designed to stand alone, while others must be interfaced to robots and automated manufacturing cells. In this book you will learn the basic theories that are used to design and implement modern complex motor controls. You will also learn of the changes that have evolved to bring motor control systems into the modern electronic and computer age. You will see how new knowledge of microprocessors, electronics, physics, and mechanics has been blended to provide the most sophisticated controls known to humankind.

Human beings have tried to control their environment and machines ever since the discovery of the wheel. As electricity was developed, the focus changed to controlling the operation of the electric motor. This early control consisted of ways to turn the motor on and off and methods of protecting the motor from damage. As transducers evolved it became possible to automate the controls so that something other than people could switch the motor on and off.

Some of the early controls included mechanical methods of motor control. For example, the switch that was used to turn the motor on and off could be customized to complement the action the motor provided. If the motor was used to move material like a winch, a limit switch would be used to detect this motion. If the motor was used to pump water, a float device could be mounted on the end of a rod that would move against the switch handle and cause it to open and close to control the motor. An example of this type control is shown in Figure 5–1, where you can see that the controls are related to the function the motor provides. These controls began to evolve at a time when human labor was

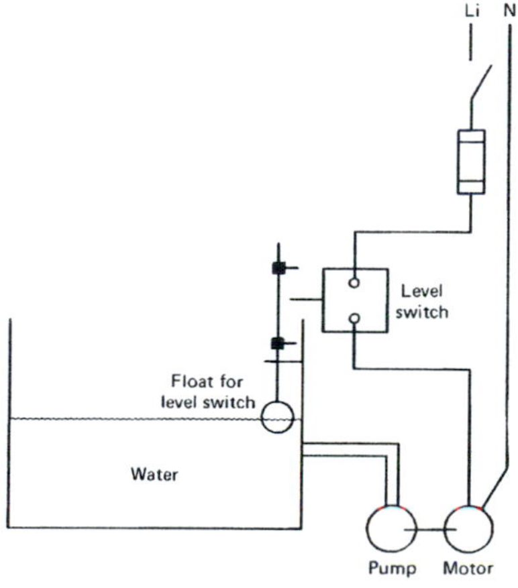

Figure 5–1 A manual liquid level control used to control the water level in the tank.

both cheap and plentiful. In fact, the cost of early controls was related directly to the expense of providing people to turn the switches on and off manually.

The motor control circuits that you will encounter on the job will fall into one of several classifications, according to the function they are designed to provide. These functions include operation of the equipment, safety of the equipment, and safety of personnel who must operate the equipment or work on it.

The circuits that you find in a control system operate in a similar manner regardless of their function. This means that all motor control circuits operate in essentially the same way. They must first sense conditions in their environment. The sensing signal is then sent to a device or circuit that must decide what to do when this condition arises, and the third part of the operation is the action that is prescribed based on the decision. This means that each circuit goes through these three distinct steps to turn a motor on or an output signal on or off. Figure 5–2 shows an example of a simple control application that uses these three steps.

This circuit uses a limit switch to determine if a box is in place on a conveyor line. If the box is in place, a push bar is extended to push the box off the conveyor. The sensor in this circuit is the limit switch, and the decision is determined on the basis of whether a box is or is not in place. The action part of the circuit is the solenoid valve that will cause the push bar to extend. If a box is not in place, the circuit will decide not to energize the solenoid, and if the box is in place, the solenoid will be activated.

A safety circuit could be designed in the same manner. The safety circuit could use a photoelectric device to indicate if anyone was near the operating equipment. If someone entered the danger area around the equipment, all machine operation would be terminated until the area was clear again and the system was restarted. A second type of safety circuit could also be designed to detect if any boxes were jammed in the machinery, which could cause damage to equipment.

Each of these circuits would operate in the same manner, in that they would use sensors to detect the conditions that exist at any time. The decision part of the circuit would be determined by the types of contacts used in the circuit (normally open or normally closed), and the action part of the system would be the solenoid turning on or off, or a relay shutting the equipment down when a dangerous situation was detected. These circuits also show the three main types of circuits used in motor controls: operational control circuits, equipment safety circuits, and personnel safety circuits.

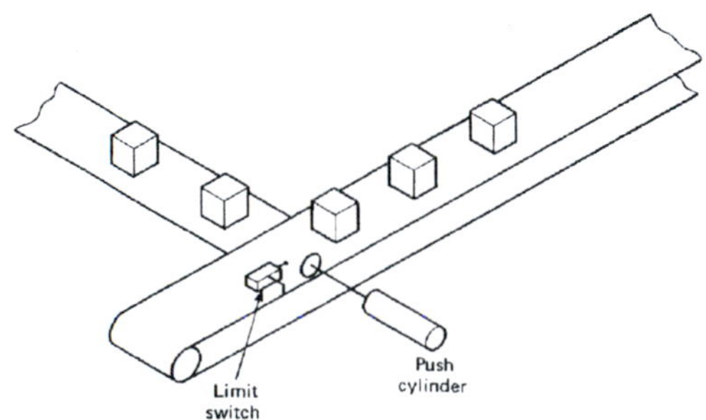

Figure 5–2 Automatic control circuit that utilizes a limit switch to control a sorting function for boxes.

5.1 The Need for Motor Controls

From the beginning, motor controls were developed for a variety of reasons. The primary reason was to control an operation or process, such as determining the number of parts a machine would manufacture, or to limit the time of a machine cycle. In process control, the temperature of a material or the level of product in a container was controlled.

Other controls were soon required to protect the motor and other electrical and mechanical equipment. These controls included fuses, circuit breakers, and clutches. The main function of the fuse and circuit breaker was to ensure that a motor would not receive too much current or be allowed to become damaged by heat. These devices have been evolved to a point where the newest fuses are programmable in both the amount of current they will allow to flow and the amount of time the condition is allowed to continue. This control is provided by a combination of sensors and microprocessors that are used as stand-alone protection or in combination with motor drives, programmable controllers, and robots.

The final type of control was designed for personal safety. These controls were needed to protect the people who had to work around the equipment and processes. As more knowledge was gained about the danger factors, more controls were necessary to provide adequate protection against them in both the short and the long term.

5.1.1 Machine Operational Controls

As you know, operational controls were first designed to control simple machine operations, such as start and stop functions. The simplest of these is a toggle switch, which is similar to the light switch in your home. More complex controls involve other types of switches that are incorporated into controls that can sense movement, level, pressure, flow, temperature, and force.

Some of the controls in use today are still manual or semiautomatic controls. They are used because they are the best type of control for the job. In some cases, control systems installed in the late 1970s are still in operation. Controls that were installed in the 1980s and 1990s may also seem archaic in comparison to today's controls, but as long as they are controlling systems in a satisfactory manner, they will continue to be used.

Some operational controls are used to provide speed control of the motor. These include a variety of belts, clutches, and transmissions to control the speed mechanically, and variable-frequency and variable-voltage controls that control the speed electronically. The new generation of solid-state devices and microprocessors have been actively incorporated to provide more complex control. It is now possible to change AC voltage to DC voltage in one simple circuit and to control the amount of voltage and current that is sent to a DC motor. This provides speed control and torque control that is needed to meet the variety of changing conditions that loads place on the motor. AC motors can have their speed, torque, and efficiency controlled by the same types of electronic devices that are used to alter the voltage, current, and frequency of the power the motor uses. These speed controls can be used as stand-alone devices or they can be incorporated into complex control systems such as programmable controllers or robotic work cells. The choice of operational control will depend on the complexity of the system and the amount of money that is to be invested.

The new trend in motor control is to provide data as well as control for a system. It is now as important to know the quality and quantity of the parts being manufactured as it is to produce the parts. In years past, as long as the machines were operational and parts were being produced, the job of the control system was considered complete. Today, it is important to test the parts that are being manufactured, and controls must be provided to make subtle changes in the system to make corrections when defects are found. In some systems the correction is made to machines manually; in more complex systems, the control is closed loop. Controls such as measuring devices and vision systems send analog data instead of the more simplified on/off signals to indicate that the product is good or bad. The analog values from the testing sensors can determine not only that the part is bad, but how much correction is needed in the machine to make good parts again.

Part of the control system that you will be involved with will be used to send the data that have been gathered to several different offices in the plant. In today's modern control systems the data must be analyzed by production control specialists who determine the number of parts that the system must make, and to quality control specialists who determine if the parts that are being produced meet the required standards. Data must be sent to product specialists such as engineers, technicians, and machine operators, who understand all facets of the production system. They must be consulted when parts are not meeting standards, so that they may suggest minor changes that can be implemented with a simple modification, or a complex modification that requires intricate changes to the system controls.

If the parts being made are manufactured from metal products, the people receiving the data may include engineers and technicians who understand the chemical and physical makeup of the raw material being used, as well as production specialists who understand molds and dies and metal forming. You will be required to work with all of these people in the plant to interpret their needs and to convert these needs to changes in the electrical control of the system.

5.1.2 Safety for Equipment and Personnel

Most safety devices were first built into equipment to protect the motor and working parts. As designs evolved, more safety features were required to protect the people who must work on or around the equipment. These safety controls included sensors to indicate overtemperature, overpressure, and spillage of dangerous or volatile substances.

Modern control systems provide safety circuits to protect the equipment, as well as the people who must work around the equipment. One common example of a safety circuit on a machine is the two-handed palm start buttons that require an operator to have one hand on each button to start the machine. This type of safety control circuit ensures that the operator's hands are nowhere near any moving parts or where they may become injured when the machine starts to move. If both of the operator's hands are not on the pushbuttons, the machine will not start. Other safety circuits are provided to protect people from the product that is being produced. In modern technology, many products, such as chemicals, pesticides, and pharmaceuticals, must be controlled during manufacturing and handling to avoid excess contact with human beings, and the production of other products, such as electroplating, that produce dangerous vapors, fumes, and by-products that must be controlled and disposed of safely.

The manufacture of these products requires extensive safety devices and circuits to detect the presence of dangerous conditions and to execute predetermined procedures to control the condition. These procedures can be designed into the control circuitry or programs that are used to provide motor control. The safety components will reside in the same electrical cabinet as the

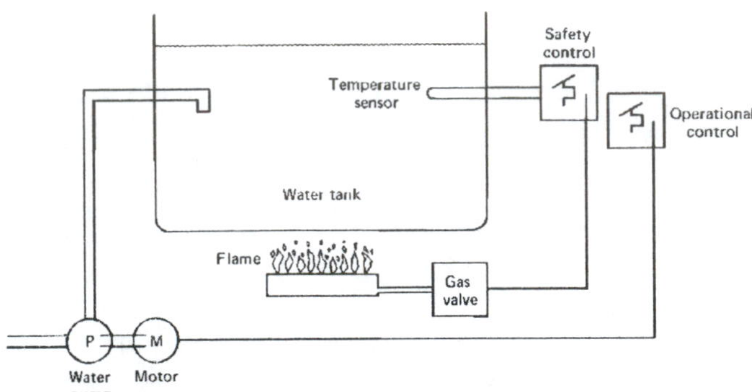

Figure 5–3 Example of a temperature sensor being used as an operational control and a safety control.

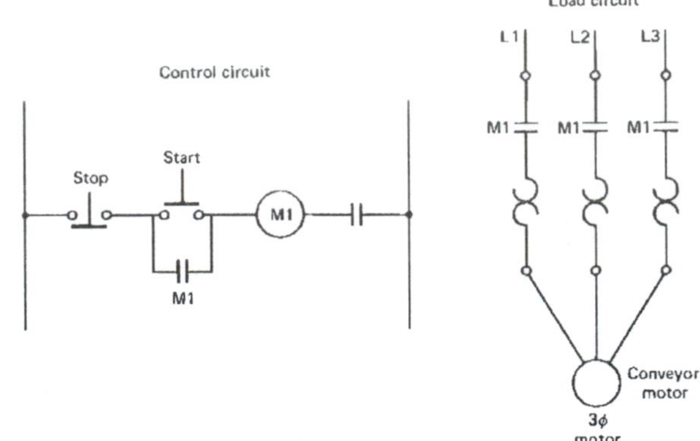

Figure 5–4 An electrical diagram showing an example of a control circuit and a load circuit.

operational components and production controls that are used to make the product. In fact, a particular type of sensor may be used in one part of the control system as an operational control and in another part of the system for a safety circuit. The control will provide a slightly different function and be connected to a different part of the control circuit, but when it comes to installation, calibration, and troubleshooting, you will use the same operational theory to determine if it is operating correctly.

In other systems it is possible to use the same sensor for production and safety control. For instance, a temperature control can be used to control the production temperature of a product and protect against overtemperature. Figure 5–3 shows an example of this type of control circuit. The high safety temperature level is set at 500F, the high operational temperature is set at 450F, and the low operational temperature is set for 350F. In this application, one sensor is used to sense the temperature, but different control strategies are used to determine what should be done when the temperature is in a specific range. If the temperature of the product is between 350 and 450F, nothing is changed in the system, which means that heat is maintained at the present level. If the product's temperature drops below 350F, more heat is added to the system. If the temperature increases above 450F, heat is gradually removed from the product, and if the temperature exceeds 500F, an unsafe condition is sensed, all heat is removed, and an alarm is sounded. From this circuit you can see that one type of sensor can be used to provide information to be used as operational control and also as safety control.

5.2 Control Circuits and Load Circuits

We have just learned that controls can be classified as operational and safety controls. Another way to classify the components so you can learn their operation and function is to divide the motor control circuit into two distinct parts. One part of the circuit has the function of providing *control* of the circuit. This is where sensing, decision, and action are initiated. The second part of the circuit is called the *load* circuit. This part consists of the motor and any devices used to provide power to the machinery. The load circuit usually uses higher voltage and more current than the control circuit. Figure 5–4 shows an example of a simple circuit that uses a start switch to energize a motor starter coil. The motor starter contacts will close to provide high voltage and sufficient current to the motor to allow it to turn the load. The load in this case is a conveyor. When the pushbutton is depressed, the motor starter coil is energized, which causes the contacts in the circuit to close. When the contacts close, they allow voltage to flow to the motor. When the stop button is depressed, the motor starter coil is deenergized and the contacts are returned to their open position, which deenergizes the motor.

The start button, stop button, and motor starter coil are all part of the control circuit. This is where the sense, decide, and act functions are executed. The part of the circuit where the motor starter contacts close and open to energize and deenergize the motor is part of the load circuit. In this case the motor is the load for the circuit.

In most circuits it is possible to determine which part of the circuit is functioning as the control and which part is operating as the load. In the circuit just described, the start and stop buttons are the control for the circuit, and they are located at some distance from the motor. This type of control is called *remote control* or *automatic control*. If a manual motor starter was used, the start switch would be located very close to the motor and the control circuit would be called a *manual control* circuit. In this book we explain how each component used in these circuits functions and methods for testing them. Other information will be provided for installation. When you understand the function of a component, it will be easier to troubleshoot it since you know what the device is supposed to do.

5.3 Motion Control

In some industrial systems, the movement of the machine must be controlled more precisely. As you know, motor control can be used to turn motors and other loads on and off, depending on the status of the input signal, but in some applications, the speed, position, and motion of the load must be controlled. This type of control, called *motion control*, is associated with robotic systems and other automated positioning systems. You will learn more about motion control in later chapters.

Specialized devices are used as controls in these systems to provide the intricate type of positioning. In a motion control system the speed with which a load is moved and the positions to which it is moved can be controlled by servomotors and stepper motors. A servomotor is a motor that has sensors to tell the motor controller where the motor shaft is and at what speed the shaft is turning. A stepper motor is a motor whose shaft's position can be precisely controlled by providing the motor with a specified number of pulses; the pulses cause the motor shaft to turn incrementally so that its position is always known. Both of these motion control systems are used for precision positioning. This type of control can also be provided by hydraulic servo systems. In the past, a robotic technician or a servo specialist was required to install, calibrate, troubleshoot, and repair these systems, but today, every electrical technician or electrician is expected to be able to understand their operation. The information in this book will help you understand the theory of operation used to design and operate motion control systems, and will provide you with diagnostic tests that can be used to determine the status of each component in a system.

Other types of motion control systems provide speed and torque control of motors through the use of motor drives. Such drives use variable frequency, variable voltage, and variable current to change conditions of the motor's speed and load capability. Solid-state devices called silicon-controlled rectifiers (SCRs) are used in DC speed controllers to provide voltage and current control for DC motors. A wide variety of magnetic and mechanical clutches are also used to allow the motor to operate at full speed and torque while adjusting the speed and torque characteristics of the load.

5.4 Process Control

Another specialized type of control system is known as *process control*. A process control system is a closed-loop system that has a set point, and the controller turns the output on and off to ensure that the process remains near the set point. An example of a process control system on an injection molding machine is the temperature control system. The machine operator puts a set-point temperature into the controller, such as 600F, and the process control system controls the heating element (output) on and off to ensure that the temperature remains near the set point. For many years these control systems were installed, calibrated, troubleshooted, and maintained by instrumentation technicians. In fact, most factories had only motor control systems for manufacturing products, or process control systems that were designed for the processing of products in batch or continuous form. This meant that you could study one type of control system, dedicate your career to specialized work on that system, and never have to work on other types of systems.

In modern industry you will find a mixture of motor control and process control built into each machine in both the process industry and the manufacturing industry. In process control the sensors usually transmit analog signals to an amplifier or analog-to-digital converter for use in a computer or controller, and in motor control the signal tends to be on/off in nature. An analog signal is similar to the volume control on a radio or a light dimmer switch where the signal can be set at any value between full on and full off. Other differences between the process control and motor control systems are now blended into complex machine control, and generally no distinction is made between them when you are assigned to work on them.

5.5 A Major Change

A major change has occurred in the controls industry that has affected what you will be doing when you work as an electrical technician or as an electrician. This change is the removal of barriers between motor controls, motion controls, and process controls. The advent of the programmable controller has helped to bring about this change.

When you go to work in industry today, you will be expected to work on any of the control systems listed above. Industry can no longer afford to separate the duties and hire three different people to work on these three types of systems. You will also find all three systems included in the same piece of machinery because they allow designers to produce machines that perform more functions and maintain more constant quality in the product that is manufactured.

For example, it is common today to work on a system such as a plastic injection molding machine that will have normal motor control circuits to turn it on or off, have process control circuitry to provide closed-loop temperature control, and have a hydraulic servo system to provide intricate motion control to control the position of the clamp by adjusting hydraulic pressure and flow. With this mixture of components and systems, you can see that you can no longer afford to learn only one of the systems and feel that you are ready to design, install, and service motor control systems. In addition to these controls, many factories are adding programmable controllers to manage the large control circuits, and robots are used to provide material-handling capabilities.

In this text we show you how this blend of controls functions. Even though motor controls, motion controls, and process controls have slightly different purposes, their operations are similar enough that you will easily understand them all. Some instructors teaching this course will have taught motor control courses for years, and the main topic of those courses will have been limit switches, pushbuttons, timers, and motor starters. If you could take a trip through any factory, whether it is brand new or 30 years old, you would notice that such components and controls are just the basic part of the control system. The controls that are found today include programmable controllers, robots, automated manufacturing, and motion and process control systems. If you cannot go to a factory, open any trade journal and look closely at the machinery advertisements. These ads will show how designers have incorporated the finest advanced controls to make their machines best at what they do. What the ads do not say is that this practice has required a drastic change in the content of what electrical technicians or electricians must study to become competent and remain so throughout their careerseven with changes that will not occur until years from now.

5.6 Innovations in Control Systems

A variety of innovations have been incorporated in motor control devices over the last few years. They include the use of solid-state devices in all areas of the detection and amplification of signals that were once too small to detect. Another innovation that has been integrated into motor control systems is the *programmable logic controller* (PLC). The PLC was originally introduced for use in motor control circuits as a relay replacement system. It has since grown into a complex, industrial, computer-controlled system that can provide motor control, motion control, and process control. In fact, multiple programmable controllers can be connected in a local area network so that data can be transmitted and received between each PLC and the network host. This allows the implementation of complex automated systems that extend beyond the confines of factory walls.

Another innovation that has developed from solid-state components is the *microprocessor.* The microprocessor is a complete computer on a chip that has become less expensive and more usable, so that it can now be interfaced directly with many motor controls. It is the heart of the PLC system, and it has been customized for use in many other controls. The timing and memory functions that it can provide make it a logical replacement for complex electromechanical devices such as timers, counters, and other sensors. Since microprocessors have been added to motor control devices, the devices now have the ability to be programmed and execute very complex operations.

5.6.1 Solid-State Devices

The advent of the solid-state diode and transistor has brought on another large change in the motor control industry. Solid-state devices include transistors, SCRs, and operational amplifiers (op amps). You will learn more about these devices in Chapter 19. These devices allow the newer solid-state controls to be much smaller and react much faster than could traditional mechanical and electromechanical devices. The transistor allows very small physical changes to be detected and sensed. For example, transistors added to photoelectric and proximity controls allow them to sense changes in motion as small as 0.01 inch. Other control devices, such as resolvers and encoders, have incorporated the transistor together with other solid-state components, such as the operational amplifier, to provide control to within 0.001 inch.

The transistor has been designed to be used in power switching circuits for controlling voltage and current to motors and other devices. Transistors are also quite useful in providing positive and negative bias to controls that must detect bipolar signals. The SCR and triac have been designed for work in power control circuits called thyristors. These devices provide control of exceedingly large voltages and currents at speeds that are compatible with microprocessors.

The op-amp has been developed from work with traditional transistor amplifiers to provide circuits on a solid-state chip that can detect changes in input sensors that were once thought impossible. These devices are also used to control analog outputs to motor speed controls and control of servo valves.

The discrete (individual) solid-state components that are presently used in motor controls are described throughout this text. The functions they provide are also explained with the theory of their operation. Newer developments have included some of these components in integrated circuits (ICs). The integrated circuit provides complete circuits on a single chip that can easily be tested and quickly removed and replaced when mounted in IC sockets on printed circuit boards. This allows motor control devices to be troubleshooted and maintained with ease. The research and development of these devices has led to the discovery of new applications for the microprocessor chip used in programmable controllers and other motor control devices.

5.6.2 Programmable Logic Controllers

Programmable logic controllers (PLCs) have altered the application of motor control as much as any other device or component that has been introduced into motor control circuits. The PLC was introduced in a General Motors Hydramatic Plant in 1969 as a reprogrammable control that could replace the vast network of control relays that were used to provide motor control logic. Since that time the PLC has evolved into a series of controllers that may be as simple as the first sequence controller designed in 1969, or as complex as any mainframe computer with multiple microprocessors that operate asynchronously.

The PLC provides the ability to make changes quickly and easily in motor control systems. The logic control that is executed by the PLC is called a *program* and it can be stored on a variety of memory devices, such as the floppy disk, hard drives, and other sites through the internet. This means that the program can be developed in an office and downloaded into the PLC by a technician working near the system, or from the network, which allows the programmer to be hundreds of miles away in another city. The PLC provides a method of controlling multiple analog and digital circuits that may be some distance from each other. It also provides a method of diagnosing problems that occur in the complex system by using its logic program for troubleshooting as well as control. The PLC provides a very complex system that can easily be interfaced to a wide variety of electronic sensors and controls. This is important since the variety of sensors and controls covers the complete spectrum of electronic signals, which range from TTL (transistor–transistor logic) through 220 VAC. This means that modules are available that can be plugged into a PLC's rack that can quickly allow the system to be programmed to control such complex operations as servo positioning, data communications, and report generation, and to receive traditional signals from limit switches and to send voltage to energize motor starter coils.

5.6.3 Microprocessor Control

The microprocessor has also been incorporated into a wide variety of controls used in motor control systems. The microprocessor allows a device to be programmed for operation that may occur many days in advance. The program may be permanently entered into the control, or it may be reprogrammable, such as timers and counters and motor speed drives. The microprocessor has also been added to many devices that are interfaced with PLCs, since they allow the device to execute sensing and control without requiring constant updating.

In some devices the microprocessor allows specific functions to be selected on a "use as needed" basis. This means that a device may be capable of being programmed as a timer or a counter, which allows fewer components to be maintained in inventory. If a timer becomes faulty, the control can be removed from inventory and programmed as a timer, and if a counter control is needed, the same device could be reprogrammed as a counter.

The microprocessor also provides other selections in motor controls, including choice of voltage, frequency, and range of input and output signals. The maximum value that the signal can reach is also programmable, which allows the microprocessor to provide set-point and dead-band control on process control instruments. All of these choices mean that you must be able to understand the ramifications of all the selections that can or must be made on microprocessor-controlled devices and systems so that the system can be troubleshooted.

5.7 Review

Control systems can be classified as motor control systems, motion control systems, or process control systems. All controls in these systems can be classified as operational controls or safety controls. The controls can be integrated into complete circuits called safety circuits or operational circuits. The safety control circuits can be designed to protect personnel or equipment. Motor control circuits can be broken into two distinct parts: the load circuit, where the motors and other loads are located, and the control circuit, where all of the controls are located. The load circuit generally uses higher voltage than the control circuit. As a technician you will learn to work on all of the controls in all of the circuits. This overview of modern motor controls is intended to provide you with an understanding of where motor controls started many years ago and how quickly they have progressed to the present state. You will need to understand the operation of many electrical, electronic, electromechanical, and mechanical devices to install, calibrate, troubleshoot, and maintain modern control systems. You should also understand how fast these controls continue to change. This means that devices you read about today will by next week be installed in your factory on the equipment you are responsible for maintaining. You will be expected to read a small amount of documentation or attend a short course or seminar and be able to maintain the equipment. You will also be expected to provide this information to your fellow workers and to teach them the principles that you have learned. The state of the art as we know it today will not stand still; so you must be prepared to continue to learn about these changes for the rest of your career.

For this reason it is vitally important that you fully understand the concepts and principles of operation for every device that is used in motor control devices and systems. In this way you will be able to understand any changes quickly, as they occur. This also ensures that you will have a job in the years to come. The better you understand these concepts, the more you can contribute to the success of your employer. This will aid in the employer's profitability, which will help to keep the company in business. The knowledge that you gain and possess can never be taken from you. The key to your career success is based on the amount of knowledge that you have. Even if the company that you are working for moves or goes out of business, you will have the necessary skills to be quickly reemployed.

Questions

1. Name three classifications of motor controls by their function.
2. Give two examples of motor controls provided to protect personnel who must operate equipment.
3. Give two examples of motor controls provided to protect a machine from damaging itself.
4. Provide an example of a motor control sensor that could be used for operation and for safety.
5. Explain the difference between the control part of a circuit and the load part of a circuit.

True and False

1. Operational controls are used to make the machine perform its function.
2. Safety controls are designed into the control system of a machine to protect the personnel who are operating the machine and to protect the machine from damaging itself.
3. Motion control uses servomotors and stepper motors to control the position and speed of moving parts in a machine.
4. The programmable controller allows motor controls to be fully integrated with motion control or process controls.
5. Microprocessor-controlled devices are seldom used in modern motor controls.

Multiple Choice

1. Two-handed palm start buttons that require an operator to have one hand on each button to start the machine are an example of ________
   a. operational controls.
   b. safety controls.
   c. process controls.
   d. motion controls.
2. A limit switch that indicates midtravel location on a machine is an example of ________
   a. operational controls.
   b. safety controls.
   c. process controls.
   d. All of the above
3. The coil of a motor starter is usually found in the ________
   a. control circuit.
   b. load circuit.
   c. safety circuit.
   d. All of the above
4. A pump motor that is connected to the contacts of a motor starter is usually found in the ________
   a. control circuit.
   b. load circuit.
   c. safety circuit.
   d. All of the above
5. Solid-state controls may include ________
   a. transistors.
   b. SCRs.
   c. op-amps.
   d. All of the above

Problems

1. Your supervisor thinks a robot could be used to help unload parts from a plastic injection press. Explain some of the devices that could be used as part of the motion control system.

2. Provide an example of a process control system that could be used on an injection molding machine.

3. Why have programmable controllers brought about so much change in motor controls?

4. Your supervisor would like to bring the machines in your area up to date and add some newer types of controls. Explain the types of electronic components and microprocessor controls that may be available.

5. Explain some typical safety components you might use in a control system for your machine.

◀ Chapter 6 ▶

Power Distribution and Transformers

Objectives

After reading this chapter you will be able to:

1. Explain how AC electricity is generated.
2. Explain how three-phase voltage is generated.
3. Identify the equipment used to distribute all power from the generating station to the factory.
4. Explain the operation of a step-up and step-down transformer.
5. Explain the operation of fuses and circuit breakers.

A large amount of power is controlled and consumed in modern industry. This power must be produced at a utility power plant and distributed to the industrial site. It must also be distributed all around the factory at appropriate levels to power motors and control circuits. This causes the power distribution system in modern factories to be as large as that of some small towns or cities. In fact, in many of the processing factories where metal and plastic parts are produced or machined, the power consumption will be higher than that of most small towns or suburbs.

This causes the power distribution system, from the utility where the power is produced to the point on the factory floor where it is used in individual motors or heating elements, to be an important link in the motor control function. As a technician you will be responsible for installing new distribution systems or altering existing ones to accommodate new machines or processes. This involves selecting proper equipment to meet codes and standards, and determining proper sizes for this equipment. It will include the installation of switchgear and conductors, providing adequate safety devices (fuses and circuit breakers), and troubleshooting the system when it fails to operate correctly.

In this chapter we explain how electrical power is produced and where the production stations will be located. We also explain how this power is transmitted from the generating station to the substation at the factory. The distribution of the power after it has entered the building is also described, including the identification of vital pieces of equipment involved in the distribution. Methods for selecting proper sizes of factory distribution equipment and installation procedures are also covered. You will be presented with typical faults that occur and troubleshooting procedures that will help you locate the fault quickly and make repairs or replace equipment.

In this chapter we also provide information about power distribution so that you will understand safety considerations. You will learn the possible dangers that high voltage represents and procedures for working safely around it. Extensive information is provided about proper selection and sizing of safety devices so that the system, equipment, and personnel will be protected. You will also learn to measure the capacity of the power distribution system so that you can determine if it meets the current needs of the plant and if extra capacity is available for additional motors and controls. This will help you understand the capacity of the utility company's distribution system.

6.1 Generating Electricity with an Alternator

The electrical energy that is used in industry today does not all occur naturally; it must be converted from some other energy source. The most common form of conversion is the electrical generator, which will produce a voltage when its shaft is rotated. It uses the principle of operation that a magnet passing a coil of wire will cause a current to flow in the coil of wire. The coil of wire in the alternator is called the *armature* and it is the stationary part of the machine. The magnetic field is produced by passing current through a small coil of wire called the *field*. Since the flux lines of the magnetic field must move past the coils of wire in the armature to produce a current, the field is mounted on the rotating shaft of the alternator.

The amount of voltage that the alternator produces can be controlled in several ways. One way is to spin the field faster, another way is to increase the number of turns of wire in the armature coil, and a third way is to increase the strength of the magnet that increases the flux lines in the field. Since the speed of the rotating field will determine the frequency of the AC voltage, the main method of increasing the amount of voltage that the alternator produces is to increase the current flowing in the field, which controls the strength of the magnetic field and controls the number of flux lines.

Figure 6–1a shows a diagram of a simple alternator. You can see that the armature is the stationary coil and the field is the rotating coil. As you know, the shaft is rotated to make the field's magnetic flux lines pass (cut through) the coils of wire in the armature. When this occurs, a strong current is induced (generated) into the armature coils.

Figure 6–1b shows the waveform of the induced current. From this diagram you can see that the waveform is sinusoidal. This occurs because the DC voltage causes a north and a south magnetic pole to be developed in the field coil. When the north magnetic pole passes an armature coil, the induced voltage will be positive. From the diagram you can see that when the magnetic field begins to pass the armature coil, only a few lines of force are cut and the amount of voltage produced is small. This voltage will increase proportionately as the rotating field coil causes more of the field's magnetic flux lines to be cut, and the voltage will peak in the positive direction as the maximum number of flux lines cut across the armature coil. As the field continues to rotate, fewer flux lines are cut, until the positive magnetic field moves beyond the armature coil. On the diagram you can see that the voltage peaked when the maximum number of coils were cut by the flux lines, and it decreases to zero as the magnetic field rotates away from the armature coil and cuts fewer lines. When the flux field is completely away from the coil and no lines are being cut, the voltage will return to zero. Since the field magnet is rotated on a shaft, the flux lines cut through the armature coils in an arc, which causes the voltage to increase and decrease smoothly, which produces the sinusoidal waveform.

After the north (positive) magnetic field passes the armature coil, the south (negative) magnetic field begins to cut across the armature. When this occurs, a negative voltage begins to be induced into the armature. From the diagram you can see that the negative voltage will also increase, until it peaks in the negative direction, and again decreases to zero volts as the negative

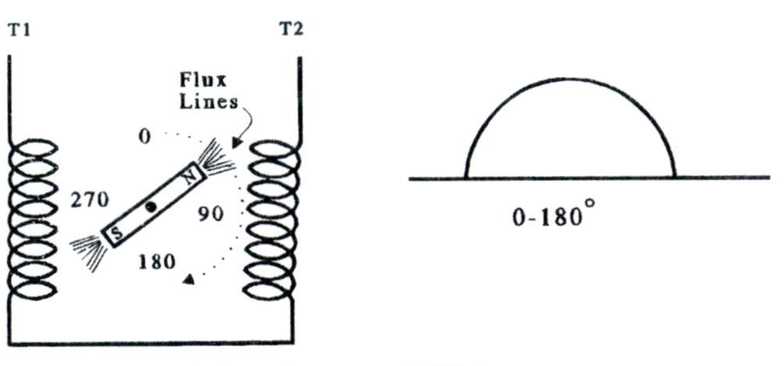

(a)The positive half of the AC sine wave 0-180 degrees.

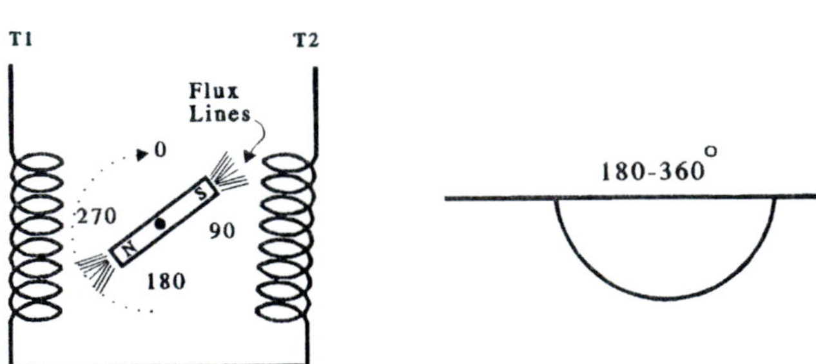

(b)The negative half of the AC sine wave180-360 degrees.

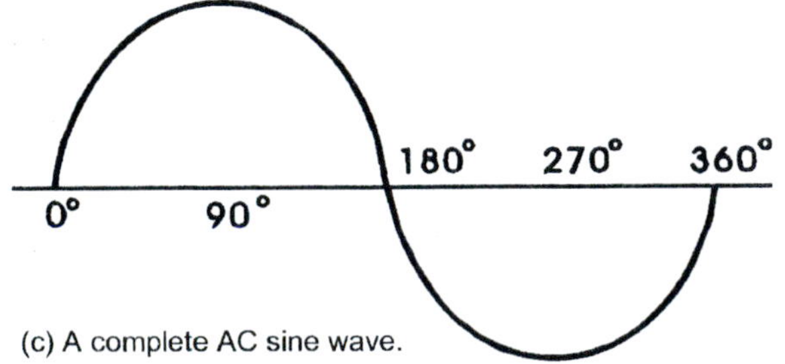

(c) A complete AC sine wave.

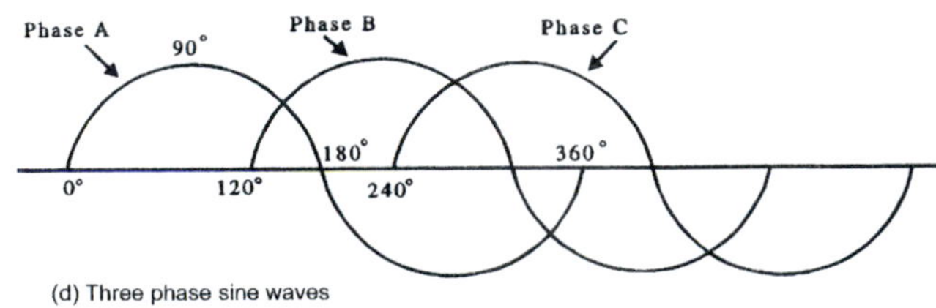

(d) Three phase sine waves

Figure 6–1 Diagram of (a) an alternator creating a positive half-wave; (b) an alternator creating a negative half-wave; (c) a complete single-phase AC sine wave; and (d) three-phase sine waves.

field moves past the armature. This positive and negative oscillation occurs at the speed of the rotating field. This speed is regulated to 60 cycles per second (60 Hz) for all utility-generated voltages in the United States. The maximum amount of voltage that is generated is determined by the amount of DC current applied to the field. The actual amount of voltage generated will vary from alternator to alternator, and will be stepped up by a set of transformers before it is sent out for distribution. Figure 6–1c shows an example of the complete single-phase AC voltage waveform. Industrial alternators produce three duplicate sets of AC voltages called three-phase voltage. Figure 6–1d shows an example of the three-phase AC voltage waveform. From this diagram you can see that the three sine waves are generated 120 degrees apart.

6.2 Three-Phase Voltage

The alternator is designed to produce as much power from the armature as possible. For this reason, three separate sets of armature coils are mounted in the generator, which will produce three separate voltages as the field rotates past them. These voltages are produced out of phase of each other by 120 electrical degrees. The three voltages are called *phases* and are identified as phase A, phase B, and phase C. They will carry phase shift all the way through the transmission system to the point in the factory where it is used.

To produce enough power for use in industrial and residential areas, the utility company must parallel several generators and generating stations together to increase the amount of current that is available. When these generators are connected together to make power available, it is called a *grid*. The grid allows extra power to be shifted from one area of the United States to another as the demand shifts. For instance, excess power can be generated with hydroelectric dams in areas where water is plentiful and the demand is not large. This power can be transmitted up to 1000 miles through the grid to areas of the country that could use the extra power. It is also possible to shift power across the grid to match load demands as they occur. For example, the peak demand period is early in the morning, when people are getting ready for work and commercial and industrial buildings are starting up for the day, and again in the evening when people come home from work and the sun is setting, which brings on a large number of appliances and lights. Since these peaks occur at different times in each time zone, power can be shifted eastward for the morning peak and westward for the evening peak. This type of shifting is also possible when one area of the country has an extra-high demand due to extremely hot or cold weather, which causes air conditioners and heating equipment to use more than normal amounts of power. If another part of the country is not experiencing this weather, the extra power can be produced at other generating stations and put into the grid for use.

6.3 Meeting Peak Demands

When the peak demands for power are experienced on site at individual factories, it may be too expensive to draw the extra current from the power company. The power company bases its rate on the amount of power that is used and the peak demand. The peak demand is measured as the highest amount of power consumed over any 20-minute period. The peak demand is then converted into a penalty charge that is multiplied by the total usage for the entire monthly billing period. This means that if the factory uses 25 percent more power for 1 hour once during the entire month, a penalty rate is multiplied against the entire bill. Some utility companies use a 15-minute period, while others use a slightly longer time to base the peak measurement, but as you can see, the penalty is assessed against the total usage, so the penalty may result in an increase of 15 to 40 percent.

Factories use several methods to combat this penalty, which include controlling load shedding with a computer to turn off nonvital loads, such as extra lighting or ventilation, when people are not using the area. Other controls limit the number of motors that can be started at any one time, to keep the peak inrush to a minimum. Other companies use *peaking generators* that they have at their site. These generators use natural gas or fuel oil to produce the extra amount of power that is required for a short period. Since the factory owns and maintains the generating equipment and pays for its own fuel, the penalty rate for the extra demand can be avoided. The electric bills for many factories reach several hundred thousand dollars per month. As the high-demand penalty will increase this by thousands of dollars, peaking generators are easily seen to be cost-effective.

Utilities also use smaller generators, called peaking generators, at peak demand periods. Some hydroelectric generating stations use their excess power to pump extra water into peaking lakes to be released during peak demand periods. These lakes are rather small and do not provide the extra capacity naturally, so the extra water that is pumped into them allows the generating station to meet the peak demand without having to build more generating capacity.

6.4 Distributing Generated Power

To be useful, the power that is produced at a generating station must be distributed to all locations in a region. Some generating stations are located where water is plentiful, because cooling water is essential in the process to return the steam that has passed the turbine wheel back to liquid form. The cooling water is used to condense the steam so that it can be pumped back to the boiler. Other power-generating plants are located where large amounts of coal can be shipped into the site easily, or, in the case of nuclear power plants, they may be located where populations are not so dense.

In each of these cases, the power is produced in large amounts where it is not being used, which means that it must be distributed. Figure 6–2 shows the path the power will take to reach a factory where it will be consumed. The first stage of the distribution system is the transformer substation at the generating site. This substation steps the voltage up to transmission levels. Some of the power from this substation (up to 35 percent) is returned to the generating plant for its operation. The rest of the power is transmitted out through wires connected to the large transmission towers that support the high-voltage wires. The power is usually transmitted at levels above 100,000 V. Some long-distance transmission lines use voltage levels up to 900,000 V.

A second substation is provided near the factory to step this large voltage down to levels around 2000 to 4000 V for transmission around the local commercial and residential areas. The voltage for commercial areas is set at a higher level than that for residential use.

A third substation may be provided at the factory site, where the voltage is reduced for transmission within the factory. If the factory does not use a large amount of power, one set of transformers is used instead of the substation. These transformers are part of the factory's power distribution and they provide the proper amount of voltage to the various areas of the factory. At various points the power can be distributed through service entrance equipment, lighting panels, motor control centers, and bus ducts. The substation at the factory must provide switchgear, transformers, and switchboards to receive the utility high voltage safely and to step it down to the levels required for factory distribution. Figure 6–3 is a power distribution diagram that shows how voltage is distributed throughout a factory. As you can see, there are several ways to receive and distribute the high voltage from the utility lines. This voltage may enter the factory substation at levels from 2.4 to 200 kV. The first devices inside the substation are the circuit breakers. They provide overcurrent protection, short-circuit protection, and a means of disconnecting the system when it must be worked on. Circuit breakers are available to protect circuits up to 3000 A, and they will provide an interruption capacity up to 8000 A rms (root mean square) symmetrical. The *interruption capacity* is the amount of current the circuit breaker can safely handle during a short circuit without damaging itself.

The next piece of equipment in the substation is the high-voltage switchgear. From the diagram you can see that once the main circuit is brought into the substation, it is divided into sections by the switchgear. The switchgear allows each circuit to be isolated and disconnected from all the other circuits for installation and maintenance work. This also allows for several different levels of voltage to be distributed throughout the plant. The high-voltage switchgear has voltage capacities up to 150 kV and 1200 A. In some systems it is

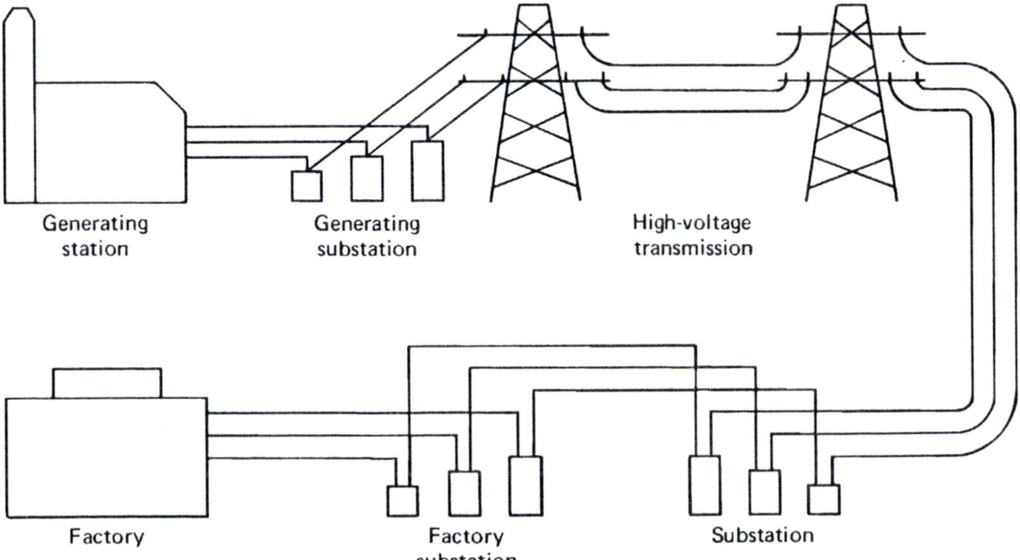

Figure 6–2 Diagram of a power distribution system showing electricity flow from the power plant to the factory.

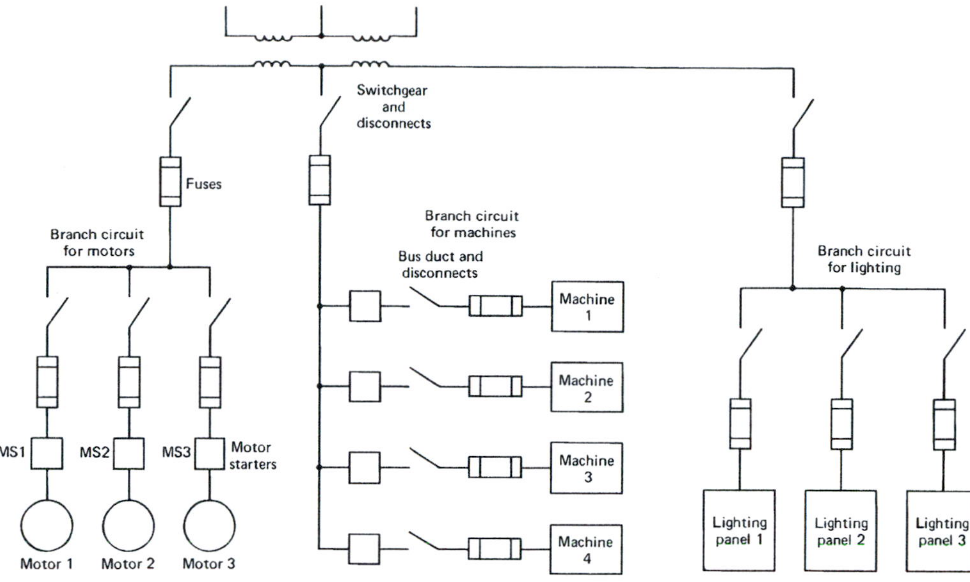

Figure 6–3 Diagram of the power distribution inside a factory.

preferable to use a step-down transformer prior to the circuit breaker and main switchgear so that the voltage rating of these devices can be reduced.

After the power comes through the high-voltage switchgear, it will be routed through step-down transformers, which will drop the voltage from the transmission level (up to 200 kV) down to in-plant transmission levels, which are usually less than 5 kV. The exact voltage level used for transmission in each factory will be different depending on the transmission distance and the amount of current that is needed. Higher current levels require larger wire sizes, which weigh more. This also causes the switchgear, circuit breakers, fuses, and terminal connectors to be larger than when higher voltages are used. Another consideration in the level of transmission voltage is the location of transformers. Some factories try to locate all their transformers in the substation area of the plant or in the powerhouse, so that they can be maintained more easily. This means that voltage is reduced at this point and the feeder circuits will be a little longer than if the transformers were located nearer the equipment where the power was being consumed.

After the voltage is stepped down by the transformers, it is routed through switchboards, load centers, motor control centers, and lighting panels. This equipment allows the power to be broken into feeder circuits and individual branch circuits. The feeder circuits group common types of loads with respect to their voltage specifications, current requirements, or applications. This strategy allows for selective protection with a combination of fuses and circuit breakers, and balancing the amperage load from phase to phase and circuit to circuit within the distribution system.

6.5 Transformers

Transformers are needed to step up or step down voltage levels. For example, when voltage is generated it needs to be stepped up to several hundred thousand volts so its current will be lower when it is transmitted. Since the current is lower, the size of wire used to transmit the power can be smaller so the power can be transmitted over longer distances. When the voltage arrives at a city, it needs to be stepped down to approximately 40,000 V so that it is less dangerous. This level of voltage is sufficient to transmit power throughout a city. When the voltage arrives at a factory it must be further stepped down to 480, 240, or 208 V and 120 V for all of the power circuits in the factory, including the office. Transformers provide a means to step up or step down this voltage. Step-up transformers are used in some industrial applications, such as air cleaners, to increase the line voltage (120 VAC) to several thousand volts. Different parts of the country will use slightly different amounts of voltage, such as 440, 220, 208, and 110 V. These values are determined by the amount of the supply voltage by the power company and the different variations of transformers used. For this text we will try to limit our discussion to 480, 240, 208, and 120 V so that you do not become confused.

Another type of transformer is also used in factory electrical systems to step down 240 VAC to 120 V. The 120 VAC is used as control voltage for all the controls in the machine systems so that higher voltages are not used in start and stop buttons or other controls that people touch when they are setting them. The lower

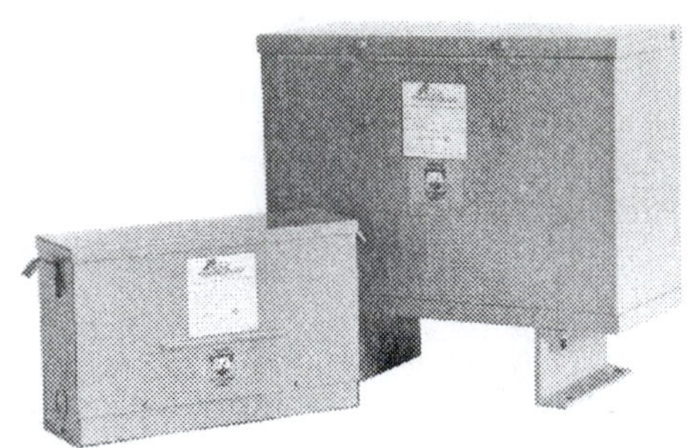

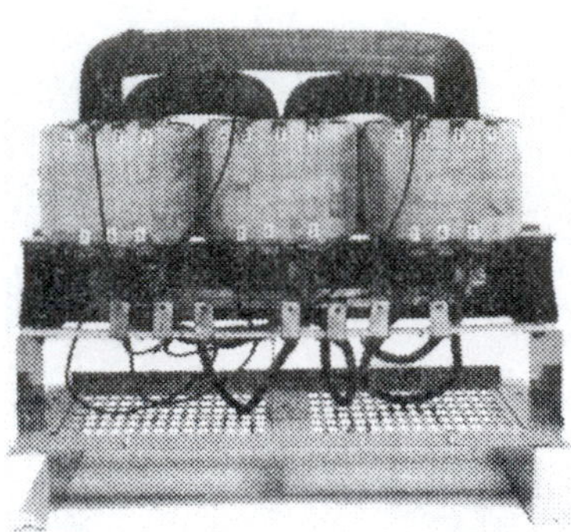

Figure 6–4 (a) A typical control-type transformer (Courtesy of Eaton/Cutler Hammer); (b) a typical three-phase transformer; (c) a three-phase transformer with its cover removed (Courtesy of Acme Electric).

voltage provides a degree of safety against electrical shock and it also allows these controls to last longer. This type of transformer is called a control transformer and it is shown in Figure 6–4a.

As a technician you will encounter different voltages, such as 440, 220, 208, and 120 VAC, while you are working on electrical systems in factories. The transformers that convert this type of voltage are shown in Figure 6–4b. A three-phase transformer is shown in Figure 6–4c with its cover removed. As a maintenance person in the factory you will be responsible for connecting transformers and ensuring that the proper amount of voltage is provided to each machine. At times you will also need to check the connections at a specific transformer in the factory and check the voltage at its input and output terminals. When the proper amount of voltage is not present, you will need to review the main power distribution print to determine which transformers are required to step the voltage up or down to the proper level. This chapter explains how transformers operate and how the various levels of voltage are derived from connecting transformers to provide the proper amount of three-phase and single-phase voltage. The chapter also explains how to take voltage readings at the transformer and disconnect to troubleshoot the loss of a phase.

6.5.1 Operation of a Transformer and Basic Magnetic Theory

The transformer consists of two windings (coils of wire) that are wrapped around a laminated steel core. The winding where voltage is supplied to a winding transformer is called the primary winding. The terminals on the primary side of the transformer are identified as H1 and H2. The winding where voltages come out of the transformer is called the secondary winding.

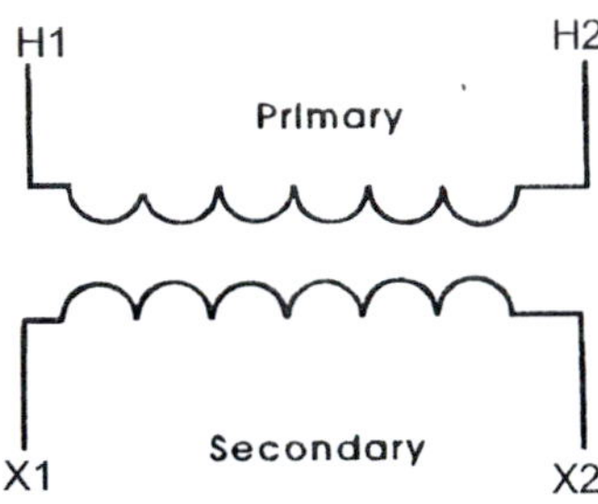

Figure 6–5 Electrical diagram of a basic transformer showing the primary and secondary windings.

The terminals for the secondary winding are identified by the letters X1 and X2. Figure 6–5 shows a diagram of a typical transformer with the primary and secondary coils identified.

A transformer works on a principle called *induction,* which occurs when current flows through the primary winding and creates a magnetic field. The magnetic field produces flux lines that emanate from the wire in the coil as current is flowing through it. When this current is interrupted or stopped, the flux lines will collapse and the action of the collapsing flux lines will cause them to pass through the winding of the secondary coil that is placed adjacent to the primary coil. The AC sine wave continually starts at 0 V and increases to a peak value and then returns to 0 V and repeats the waveform in the negative direction. The process of increasing to peak and returning to zero provides a means to create flux lines in the wire as current passes through it, and then allows the flux lines to collapse when the voltage returns to zero. Since the AC voltage follows this pattern naturally 60 times a second, it makes the perfect type of voltage to operate the transformer.

Figure 6–6 shows the 4 stages of voltage building up in the transformer and the flux lines building and collapsing at each point as the sine wave flows through the

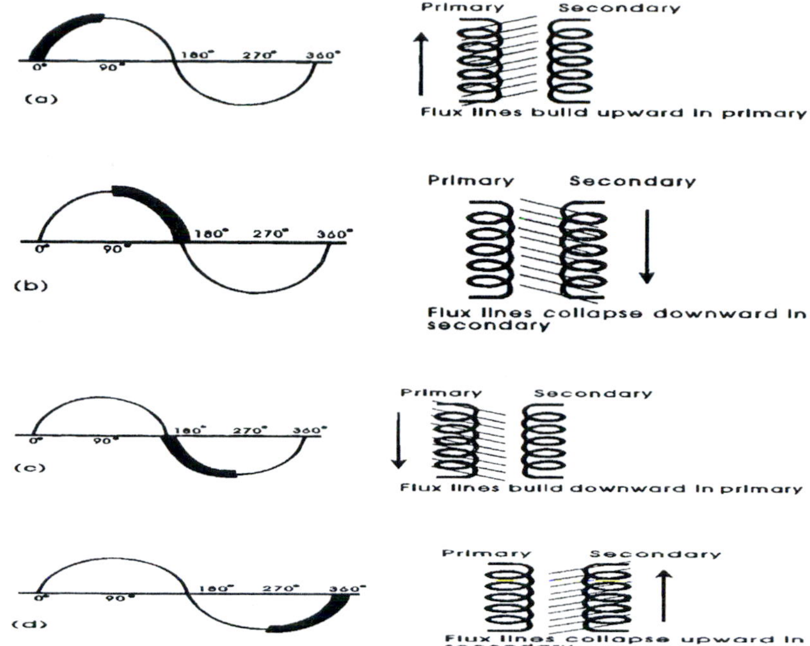

Figure 6–6 (a) AC voltage builds to a positive peak as the sine wave moves from zero to 90 degrees. (b) AC voltage drops off from peak at 90 degrees to 0 V at 180 degrees. During this time the flux lines in the primary coil begin to collapse and cut across the coils of the secondary winding. (c) AC voltage builds to a negative peak as the sine wave moves from 180 to 270 degrees and the flux lines build in the primary coil again. (d) When voltage decreases from the negative peak back to zero as the sine wave moves from 270 to 360 degrees, the flux lines collapse.

primary coil. When the AC voltage returns to zero during each half-cycle, the flux lines that were created in the primary winding begin to collapse and start to cross the wire that forms the secondary coil of the transformer. When the flux lines from the primary collapse and move across the wire in the coil of the secondary winding, the electrons in the secondary winding begin to move, which creates a current. Since the current in the secondary winding begins to flow without any physical connection to the primary winding, the current in the secondary winding is called an *induced current.*

The following provides a more technical explanation of the relationship between the AC voltage and the windings of the transformer. The magnetic field in the primary winding of the transformer builds up when AC voltage is applied during the first half-cycle of the sine wave (0 to 180 degrees). When the sine voltage reaches its peak at 90 degrees, the voltage has peaked in the positive direction and it begins to return to 0 V by moving from the 90-degree point to the 180-degree point. When the voltage reaches the 180-degree point the voltage is at 0 V and creates the interruption of current flow. When the sine wave voltage is 0 V, the flux lines that have been built up in the primary winding collapse and cross the secondary coils, which creates a current flow in the secondary winding.

The sine wave continues from 180 to 360 degrees and the transformer winding is energized with the negative half-cycle of the sine wave. When the sine wave is between 180 and 270 degrees the flux lines are building again, and when the sine wave reaches 360 degrees the sine wave returns to 0 V, which again interrupts current and causes flux lines to cross the secondary winding. This means the transformer primary energizes a magnetic field and collapses it once in the positive direction and again in the negative direction during each sine wave. Since the sine wave in the secondary is not created until the voltage in the primary moves from 0 to 180 degrees, the sine wave in the secondary is *out of phase by 180 degrees* to the sine wave in the primary that created it. The voltage in the secondary winding is called *induced voltage* because it is created by induction. The induced voltage is developed even though the primary coil of the transformer does not make electrical contact with the secondary coil in any way.

6.5.2 Connecting a Transformer to a Disconnect for Testing

An easy way to test a control transformer is to connect it directly to a disconnect and apply power to the transformer primary circuit. When voltage is applied to the primary circuit, voltage should be available at the secondary. The amount of voltage at the secondary will depend on the rating of the transformer. Figure 6–7 shows a transformer connected to 240 V at the disconnect. The secondary voltage available at this transformer at terminals will be approximately 120 V. This figure also shows a turns ratio of 2:1. Figure 6–8 shows a similar transformer connected to a disconnect to provide 480 V to the primary circuit. Since this transformer is rated as

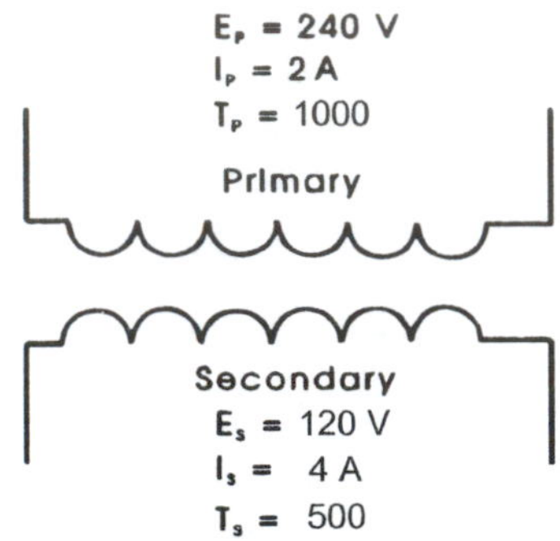

Figure 6–7 Diagram of a transformer that shows the primary voltage (V_p), primary current (I_p), and the number of turns of wire in the primary (T_p).

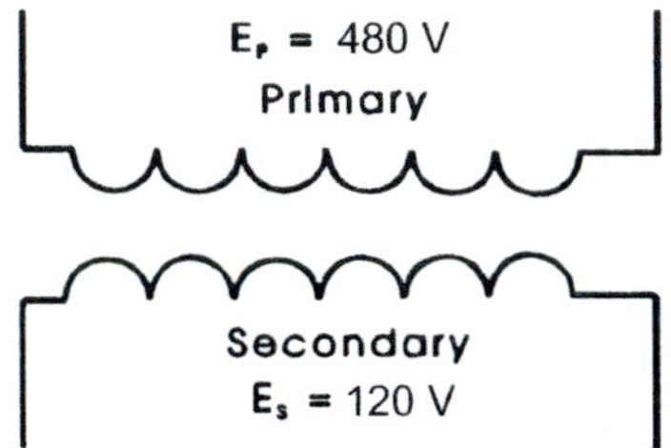

Figure 6–8 Diagram for transformer problem.

480/120, the secondary voltage will also be 120 V. The turns ratio for this transformer is 4:1.

If the equipment is used in the office part of the factory application, such as heating or air conditioning, the supply voltage may be 230 or 208 V, depending on the transformer connections that are used by the utility company that supplies the power. The secondary voltage for the thermostat will be stepped down to 24 VAC. Since the equipment manufacturer does not know where the equipment will be used, it may provide a control transformer whose primary side can be wired to either 230 or 208 V and its secondary side will produce 24 V. This is accomplished by providing a second connection that is *tapped* in the primary winding.

6.5.3 Transformer Voltage, Current, and Turns Ratios

The amount of voltage a transformer will produce at its secondary winding for a given amount of voltage supplied to its primary is determined by the ratio of the number of turns in the primary winding compared to the number of turns in the secondary. This ratio is called the *turns ratio*. The amount of primary current and secondary current in a transformer is also dependent on the turns ratio. Figure 6–7 shows a transformer and indicates the primary voltage (E_p), the primary current (I_p), the number of turns in the primary winding (T_p), the secondary voltage (E_s), the secondary current (I_s), and the number of turns in the secondary winding (T_s).

The primary and secondary voltage, primary and secondary current, and the turns ratio can all be calculated with the following formulas. The turns ratio for a transformer is calculated from any of the following three formulas:

$$\text{Turns ratio} = \frac{T_s}{T_p}$$

$$\text{Turns ratio} = \frac{V_s}{V_p}$$

$$\text{Turns ratio} = \frac{I_p}{I_s}$$

Another calculation you can make is the ratio of primary voltage, secondary voltage, primary turns, and secondary turns:

$$\boxed{\frac{E_p}{E_s} = \frac{T_p}{T_s}}$$

From this ratio you can calculate the secondary voltage:

$$\boxed{E_s = \frac{E_p \times T_s}{T_p}}$$

The ratio of primary voltage, secondary voltage, primary current, and secondary current is

$$\boxed{\frac{E_p}{E_s} = \frac{I_s}{I_p}}$$

Notice that the ratio of voltage to current is an inverse ratio. From this ratio you can calculate the secondary current:

$$\boxed{I_s = \frac{E_p \times I_p}{E_s}}$$

If you were provided the information that the primary voltage is 110 V, the primary current is 2 A, the primary turns are 458, and the secondary turns are 100, you could easily use these formulas to calculate the secondary voltage and the secondary current:

$$E_s = \frac{E_p \times T_s}{T_p}$$

$$\frac{110\text{ V} \times 110\text{ turns}}{458\text{ turns}} = 24\text{ V}$$

$$I_s = \frac{E_p \times I_p}{E_s}$$

$$\frac{110\text{ V} \times 2\text{ A}}{24\text{ V}} = 9.16\text{ A}$$

Example

Calculate the secondary current and secondary turns of the transformer shown in Figure 6–8.

Solution

Find the turns ratio:

$$\frac{E_p}{E_s} = \frac{240\text{ V}}{24\text{ V}} = 10{:}1$$

Since the number of turns in the primary is 458, the number of turns in the secondary will be 45.8.

The secondary current can be calculated from the formula

$$I_s = \frac{E_p \times I_p}{E_s}$$

$$\frac{240\text{ V} \times 2\text{ A}}{24\text{ V}} = 20\text{ A}$$

6.5.4 Step-Up and Step-Down Transformers

If the secondary voltage is larger than the primary voltage, the transformer is known as a *step-up transformer*. If the secondary voltage is smaller than the primary voltage, the transformer is known as a *step-down transformer*. The control transformer that was presented in the previous sections is an example of a step-down transformer because the secondary voltage is smaller than the primary voltage. Step-up transformers are used to boost the secondary voltage, which also has the effect of lowering the secondary current so that more power can be transferred on smaller size wire. Some applications where a step-up transformer is used in the industry include electrostatic air cleaners and the ignition system of an oil furnace. The oil furnace uses an ignition transformer that is a step-up transformer to increase the 120 VAC that is used to power the furnace to approximately 16,000 V. The high voltage is supplied to the igniter points so that the high voltage can jump the gap between the ignition points to create a spark that will ignite the fuel oil as it comes out of the nozzle.

6.5.5 VA Ratings for Transformers

The VA (volt ampere) rating for a transformer is calculated by multiplying the primary voltage and the primary current, or the secondary voltage and secondary current. For example if the primary voltage is 110 V and the primary current is 2 A, the VA rating for the transformer would be 220 VA. The VA rating indicates how much power the transformer can provide. As a technician you can say that the primary VA is equal to the secondary VA. If you were a circuit designer, you would need to be more precise and you would find that the secondary VA is slightly less than the primary VA because of transformer losses. It is important that the VA rating of the transformer is large enough for the application. If the VA rating is too small, the transformer will be damaged and fail prematurely. Any time you need to replace a transformer you must be sure that the voltage ratings match and that the VA size of the replacement transformer is equal to or larger than the original transformer.

6.5.6 The 120-VAC Control Transformer

The control transformer provides 120 VAC in the secondary for use in factory control systems. Utility companies in different areas of the country provide voltage that is between 110 and 120 V. This means that one part of the country may refer to its lower voltage as 110, while another part calls its voltage 115 or 120 V. For this section the voltage will be referred to as 110 VAC. The 110-VAC secondary voltage is necessary because some of the control devices require 110 VAC and the larger loads in the system, such as pumps or other motors, require 240 VAC. The control transformer allows the larger loads in a system like pump motors and conveyor motors to be supplied with 240 V, while it provides the control circuit with 110 VAC.

The 240/110-VAC-type of control transformer operates on a similar principle as the 110/24-VAC transformer. A picture of the control transformer was shown in Figure 6–4a. The diagram for this transformer is provided in Figure 6–9. You can see that the primary windings of this transformer are identified with the letter "H" and the secondary windings are identified with the letter "X." The primary winding for this transformer is constructed in two equal sections (windings). The first winding has its terminals identified as H1 and H2; the second winding has its terminals identified as H3 and H4. The primary winding is divided into two separate

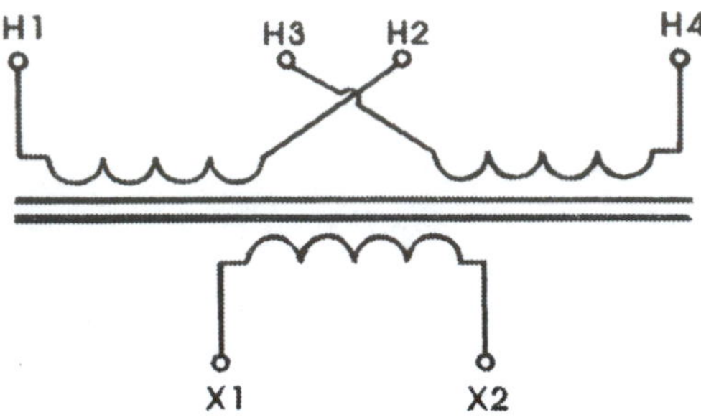

Figure 6–9 Diagram of a control transformer whose secondary voltage is 110 V. The primary windings of the transformer are identified as H1, H2, H3, and H4. The secondary windings are identified as X1 and X2. The primary voltage can be 480 or 240 V.

windings so that the transformer primary side can be powered with 480 or 240 V. If these two sections are connected in series with each other, the primary side of the transformer will be powered with 480 V, and if the two sections are connected in parallel with each other, the transformer will be powered with 240 V.

6.5.7 Wiring the Control Transformer for 480-VAC Primary Volts

The control transformer can also be powered with 480 VAC by connecting its two primary windings in series. Figure 6–10 shows three diagrams of the control transformer. Figure 6–10a shows the transformer as you would see it on a wiring diagram without any jumpers connected. The H2 and H3 terminals are positioned so that they can be connected either in series or in parallel. In this application the jumper is connected between terminals H2 and H3 to connect them in series. Figure 6–10b shows the physical location of the jumper when it is connected across these terminals. Figure 6–10c shows the equivalent electrical diagram. The reason the two windings are connected in series is because each winding must be rated for 240 V, and the primary voltage is 480 VAC. Since the windings are connected in series, each of them will receive 240 V of the total 480 V.

6.5.8 Wiring the Control Transformer for 240-VAC Primary Volts

The control transformer can also be connected to operate with 240-VAC primary volts and produce 110 VAC on the secondary terminals. Figure 6–11 shows a set of three diagrams for the same control transformer that is connected for 240 VAC primary. Figure 6–11a shows the electrical diagram of the control transformer as you would see it in a circuit diagram without any jumpers. The reason this diagram is shown is that some equipment manufacturers will show the control transformer in the electrical diagram for the system since they are not sure whether the system voltage will be 480 or 240. This diagram indicates that the control transformer is to be connected by the technician when the unit is installed and the supply voltage can be verified. Figure 6–11b shows the diagram with two jumpers in place to connect the two primary windings in parallel with each other. This diagram will help you connect the jumpers to the correct terminals so that the windings are connected in parallel. Figure 6–11c shows the equivalent circuit of the two windings in parallel as you would see them in the system electrical diagram provided by the equipment manufacturer. Sometimes it is difficult to see that the connections of the two jumpers actually connect the two windings in parallel, so the diagram in Figure 6–11c shows how the windings look when they are in parallel.

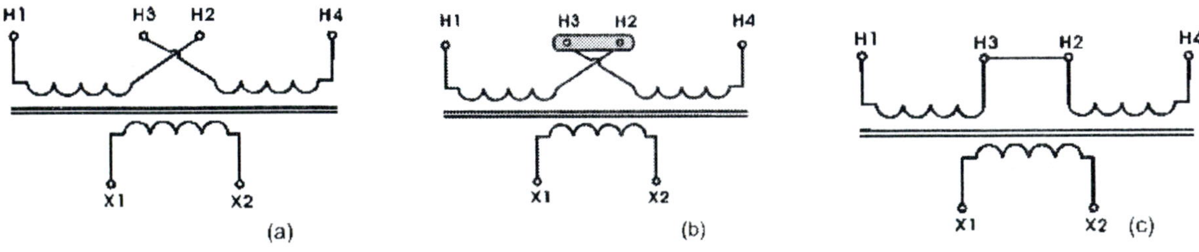

Figure 6–10 (a) Electrical diagram of the control transformer. Notice that the H2 and H3 terminals are positioned so that a jumper can be placed on them to connect them in series or parallel. (b) Diagram showing the jumper in place connecting the two primary windings in series so the primary winding can accept 480 V. (c) The equivalent electrical diagram showing the two primary windings connected in series.

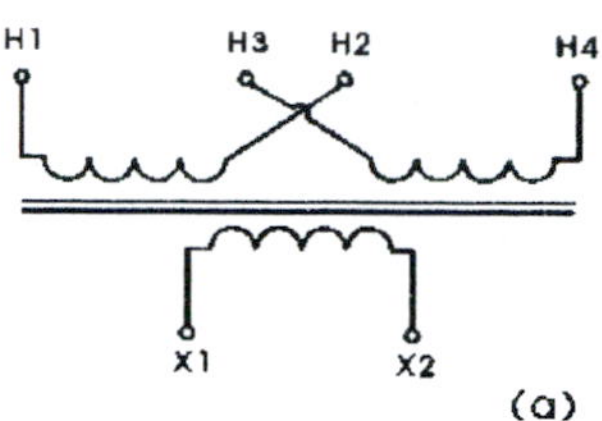

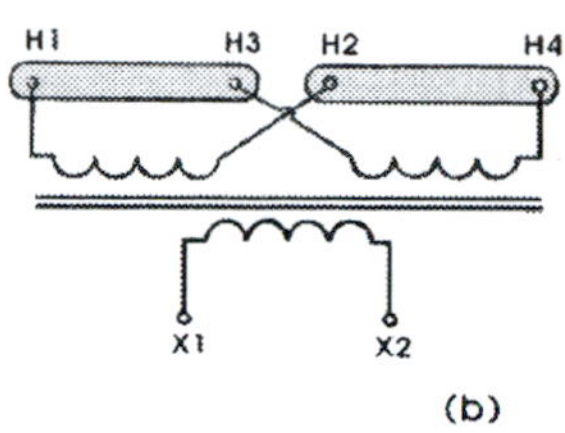

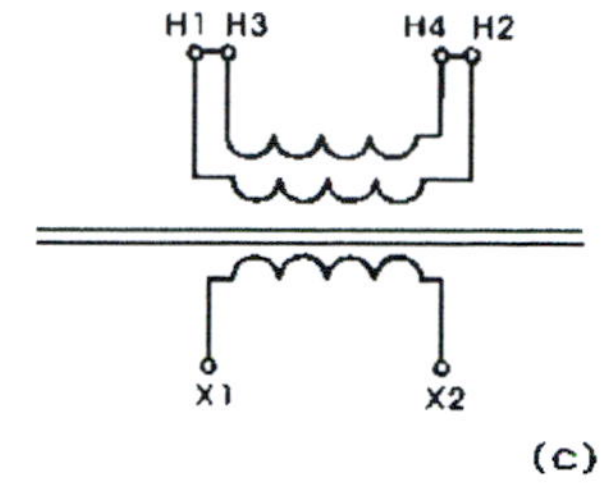

Figure 6–11 (a) Electrical diagram of a control transformer. (b) Diagram that shows the jumpers in place to connect the primary windings in parallel to connect the transformer for 240 V. (c) The equivalent diagram that shows the two primary windings connected in parallel.

As a technician you will be responsible for making the proper jumper connections based on the primary voltage that the system will have. If the jumpers on the transformer are connected for 480 VAC primary when the unit is shipped, and the unit is connected to 240 V, you will need to make the changes in the jumper so that the secondary winding of the transformer provides exactly 110 VAC.

6.5.9 Troubleshooting a Transformer

As a technician you will need to troubleshoot a variety of transformers. If the transformer you are testing is not connected to power you can test each of the windings with an ohmmeter for continuity. Each of the coils should have some amount of resistance that indicates the amount of resistance in the wire that is in each coil. A measure of infinite resistance (∞) indicates that the winding has an open, and a measure of 0 Ω indicates that one of the windings is shorted.

Another way that you can test a transformer is by applying power to the primary winding. It is important that you provide the continuity test first to detect any shorts before you apply power. If the transformer windings are not shorted, you can apply any amount of AC voltage that is equal to or less than the primary rating of the transformer. If the transformer is operational, a voltage will be present at the secondary terminals of the transformer. If no voltage is present at the secondary, be sure to check all of the connections to ensure that they are correct. If no voltage is present at the secondary, the transformer is defective and should be replaced. Remember that the transformer is a primary winding and a secondary winding that are placed in close proximity to each other and when AC voltage is applied to one winding, induction will cause voltage to be available at the other winding.

The transformer may have a problem where its secondary voltage is higher or lower than its rating. If this occurs, it is possible that the amount of primary voltage is incorrect or that the transformer jumpers are not connected properly. For example, if the primary voltage is 208 V instead of 240 V, the secondary voltage will be less than 110 V, which will cause a problem. Be sure that the secondary voltage is the proper amount before you use the transformer in a circuit.

6.6 Three-Phase Voltage

Industrial equipment installed in factories may require three-phase AC voltage or single-phase voltage. Three-phase voltage is usually indicated on equipment data plates as by the symbol (3ϕ) voltage. Three-phase voltage is generated as three separate AC sine waves. The diagram in Figure 6–1c showed an example of the three sine waves, identified as phase A, phase B, and phase C. The phases are separated from each other by 120 degrees. The 120-degree phase separation is caused by the generator windings being 120 degrees out of phase with each other. Three-phase voltage is generated and transmitted because it is more efficient to produce than single-phase voltage, since the generator can have three separate windings.

6.6.1 Why Three-Phase Voltage Is Generated

Three-phase voltage is more useful than single-phase voltage for larger equipment and systems because it provides more power than single-phase voltage for an equal size system. If a 10-horsepower (hp) motor is required as the hydraulic pump motor, a single-phase motor would be physically larger than the 10-hp, three-phase motor. The three-phase motor will be smaller because it has three sets of windings and receives three equal sources of voltage (L1, L2, and L3), whereas a single-phase motor would receive only two sources of power (L1 and L2). Since the power is shared by three circuits instead of two, the wire sizes, fuse sizes, and switch sizes are also smaller for a three-phase system.

For example, if a 10-hp motor is connected to a single-phase, 230-V power source, each line (L1 and L2) would need to be capable of providing 50 A. A three-phase, 10-hp motor connected to a three-phase power distribution would need only 28 A from each wire. This means that the three-phase system could use much smaller wire, which would be lighter and less expensive.

6.6.2 Three-Phase Transformers

The three-phase transformer you will encounter on the job was shown in Figure 6–4b. The transformer may be mounted near the equipment, or it may be located in a transformer vault (a special room where transformers are mounted). In some cases the transformers are mounted on a utility pole just outside of the commercial site. Figure 6–4c showed the three-phase transformer with its cover removed so you can see three separate transformer windings. The three-phase transformer operates exactly like a single-phase transformer in that AC voltage is applied to the primary side of the transformer and induction will cause voltage to be created in the secondary winding. In the case of the three-phase transformer, the 120-degree phase shift of the three-phase voltage applied to the primary windings will be maintained in the secondary side of the transformer.

When you are working on a commercial system, you may need to make connections between the transformer and the disconnect box or between the transformer and a load center. You may also need to make voltage tests at the terminals of the transformer. These connections will be made at the terminal strip of the transformer as shown in Figures 6–12 and 6–13. The three-phase transformer is essentially three single-phase transformers that are connected together in a *wye configuration* or *delta configuration*. In most cases, you will not be requested to make the physical connections at the transformer, but you must understand the amount of voltage that will be available when a transformer is connected in a wye or delta configuration. Since the primary voltage may be rather high (1300–1700 VAC) the electrical technicians for the electric utility company or technicians from a high-voltage service company will make the primary voltage connections for the transformer and connect the secondary to a disconnect box or a load center. You will be expected to make the connections for the equipment at the disconnect or at the load center.

6.6.3 The Wye-Connected, Three-Phase Transformer

Figure 6–12a shows a diagram of a wye-connected, three-phase transformer. The shape of the transformer windings looks like the letter Y, which gives this type of configuration its name. This diagram shows only the secondary coils of the transformer. It is traditional to show only the primary side connections or only the secondary side connections when discussing a transformer power distribution system, since showing both the primary and secondary connections in the same diagram tends to become confusing.

The amount of voltage measured at the L1–L2, L2–L3, and L3–L1 terminals on the secondary winding will be 208 V for the wye-connected transformer if its turns ratio is set for low voltage. If the turns ratio of the transformers is set for high voltage, the amount of voltage between each of the terminals is 480 V. If 480 V is needed, a transformer with a higher turns ration would be used. The primary and secondary voltage for each transformer is provided on its data plate and can be specified when the transformer is purchased and installed.

6.6.4 The Delta-Connected, Three-Phase Transformer

Figure 6–12b shows a delta-connected, three-phase transformer. The shape of the transformer windings in this diagram looks like a triangle, Δ. This shape is the Greek letter D, which is named *delta*. This diagram also shows only the secondary side of the three-phase transformers.

The amount of voltage measured at L1–L2, L2–L3, and L3–L1 is 230 V for the delta-connected transformer if it is wired for its lower voltage. If the delta-connected transformer is wired for its higher voltage, the voltage between

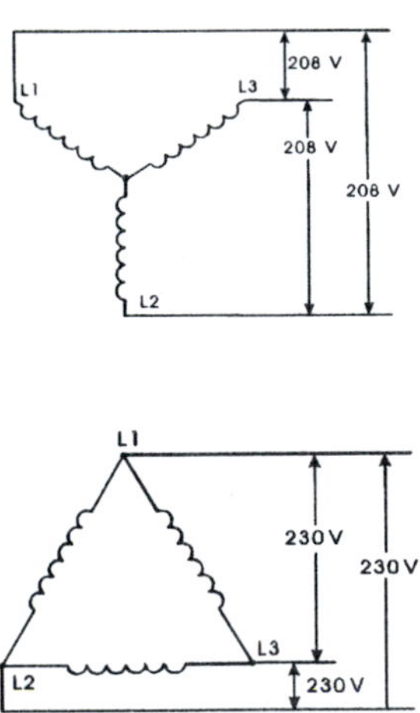

Figure 6–12 A diagram that shows the secondary winding of a three-phase transformer connected in (a) a wye configuration and (b) a delta configuration.

L1–L2, L2–L3, and L3–L1 is 480 V. If the delta-connected transformer is wired for its lower voltage, it is very easy to differentiate it from a wye-connected transformer. If the delta- and wye-connected transformers are connected for their higher voltage, you cannot tell them apart, since both will provide 480 V between their terminals.

You will usually purchase equipment so that its voltage requirements match the voltage supplied by the transformers at the commercial or industrial site. This means that you do not need to determine if the transformer is connected in a delta configuration or a wye configuration. You will need only to measure the amount of voltage at the secondary of the transformer and if it is 480 V, you will need to ensure that the system is rated for 480 V. If the secondary voltage is 230 V, the system must be rated for 230 V, and if it is 208 V, the equipment must be rated for 208 V.

If you need to know if the transformer windings are connected as delta or wye, you can check the physical connections or you can make an additional voltage measurement between each line and the neutral terminal of the transformer if one is provided. The next section explains these measurements.

6.6.5 Delta- and Wye-Connected Transformers with a Neutral Terminal

Figure 6–13 shows the diagrams of a wye- and a delta-connected transformer, each with a neutral terminal (N). The neutral terminal on the wye-connected transformer is at the point where each of the three ends of each individual winding are connected together. This point is called the *wye point*. The neutral point for the delta-connected transformer is actually the midpoint of the secondary winding that is connected between L1 and L2. This point is essentially the center tap of one of the transformer windings. Traditionally, it will be the winding that is connected between L1 and L2 if a neutral connection is used with the three-phase transformer.

Figure 6–14a shows the amount of voltage that you would measure between terminals L1 and L2 of a wye-connected transformer and between terminals L1 and N and L2 and N. Notice that the L1–L2 voltage is 208 V, so this transformer is wired for the lower voltage. The L1–N and L2–N voltage is shown as 120 V. Figure 6–14b shows the amount of voltage that you would measure between terminals L1 and L2 and between L1 and N and L2 and N for a delta-connected transformer. The L1–L2 voltage is 230 V, so this transformer is connected for its lower voltage. The L1–N and L2–N voltage is shown as 115 V. Since the neutral point for the delta-connected transformer is exactly halfway on the transformer winding, the L1–N and L2–N voltage will always be exactly half of the L1–L2 voltage.

This is the main difference between a wye- and a delta-connected transformer. The L1–N and L2–N voltage for any delta-connected transformer is always exactly half, and the L1–N or L2–N voltage for a wye-connected transformer will always be more than half the voltage of L1–L2. The exact amount of the L1–N voltage can be calculated by dividing the L1–L2 voltage by 1.73, which is the square root of 3 ($\sqrt{3} = 1.73$). The square root of 3 is used because of the relationship between the phase shift of the three phases. Notice that 208 divided by 1.73 is 120 V.

The same relationship between L1–L2 and L1–N voltage exists when the transformers have a turns ratio for their higher voltage. Figure 6–15a shows a wye-

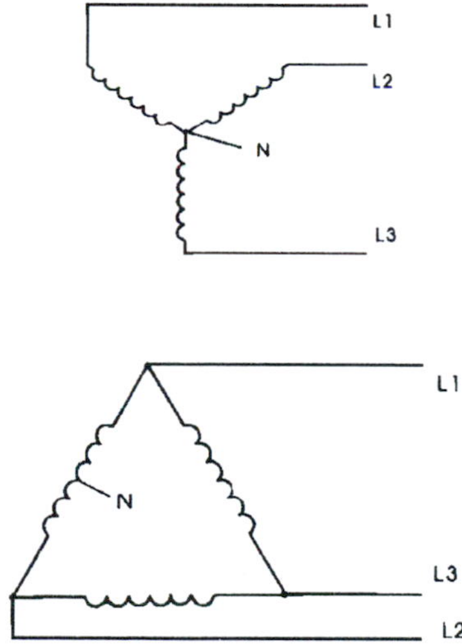

Figure 6–13 (a) Electrical diagram of a wye-connected transformer with a neutral point. (b) Diagram of a delta-connected transformer with a neutral tap.

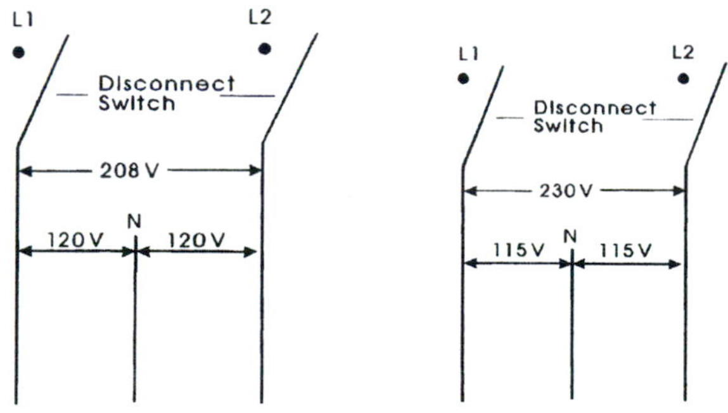

Figure 6–14 (a) Voltage from L1–L2 for a wye-connected transformer is 208 V and from L1–N or L2–N is 120 V. (b) Voltage from L1–L2 for a delta-connected transformer is 230 V, and from L1–N or L2–N is 115 V.

connected voltage between L1 and L2 of 480 V, and between L1 and N of 277 V. The 277 V can be calculated by dividing 480 by 1.73. The 277 V that comes from L1–N or L2–N is generally used for fluorescent lighting systems in commercial buildings. Since the supply voltage originates from a three-phase transformer, the lighting system in a commercial or industrial building will also use L3–N so that voltage is used from all three legs of the transformer. This voltage is referred to as single-phase voltage, since only 1 line of the three-phase transformer is used for each circuit. For example, L1–N is a single-phase circuit.

L1–L2, L2–L3, or L3–L1 could each be used to supply complete power for a machine's electrical system. Even though this type of power supply would use two legs of the transformer, it is still called single-phase power supply because only one phase of voltage is used at a time. For example, if the system is powered with voltage from L1–L2, during any given half-cycle of AC voltage, the power source for the system would come from L1, and then during the next half-cycle it would come from L2. The source of power would continue to oscillate between L1 and L2, but only one phase is in use at any time.

Figure 6–15b shows that the higher voltage for the delta system between L1 and L2 is also 480 V, but this time the L1–N or L2–N voltage is 240 V. Again the L1–N or L2–N voltage is exactly half of the supply voltage, since the neutral point on the transformer is the center tap of one of the transformers. It is very easy to distinguish between a wye-connected power source and a delta-connected power source by measuring the L1–L2 and L1–N voltage. If the L1–N voltage is half of the L1–L2 voltage, the system is a delta-connected system. If the L1–N voltage is more than half, it is a wye-connected system. Remember that the L1–N wye voltage can always be calculated by dividing the L1–L2 voltage by 1.73.

6.6.6 The High–Leg Delta System

When a three-phase transformer system is used for the power source for an electrical system it may have a neutral tap. A three-phase system does not need to have a neutral to operate correctly. The neutral is added only if the lower voltage is needed for some part of the system. You have just learned in the previous section that if the three-phase voltage is rated for 230-V and has a neutral tap, the transformer can also provide 115-V between one of the lines and the neutral tap to provide power for a single-phase 115-V pump motor. Figure 6–16 shows a diagram of this connection.

You must be aware of a problem that occurs if the transformer with a neutral tap is connected as a delta transformer. At first it appears that you could connect the 115-V motor to any line (L1, L2, or L3) and the neutral tap and the motor would receive 115-V. In the diagram shown in Figure 6-16 you can see the neutral tap is connected exactly halfway along the transformer winding that is connected to L1 and L2. The voltage between L1 and neutral is 115-V, and the voltage between L2 and neutral is 115-V, which is always exactly half of the 230-V supplied to L1 and L2. The problem occurs when you use L3 and neutral for the connection. The voltage between L3 and neutral is not exactly half.

Since the center tap is not between L3 and L1 the voltage for L3–N must come from one complete phase (230 V) and half of the next phase (115 V). Since this voltage uses two phases, the 230 V and the 115 V are out of phase, and the resultant voltage from them is 208. This voltage is not the same as the 208 voltage that occurs between L1 and L2, L2 and L3, or L3 and L1 of the wye-connected transformer because the neutral point is not at a midpoint.

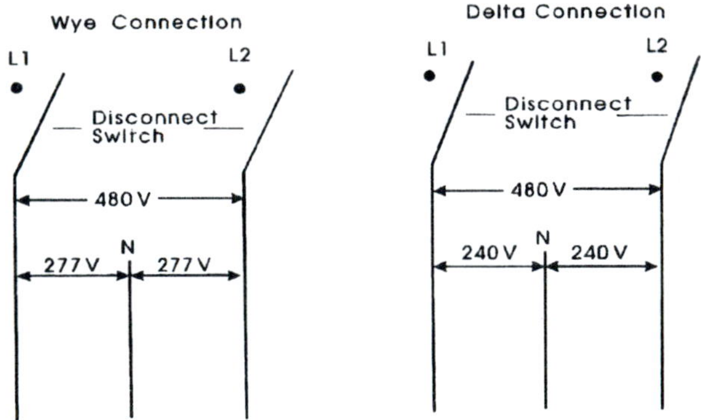

Figure 6–15 (a) Voltage from L1–L2 for a wye-connected transformer is 480 V and from L1–N or L2–N is 277 V. (b) Voltage from L1–L2 for a delta-connected transformer is 480 V, and from L1–N or L2–N is 240 V.

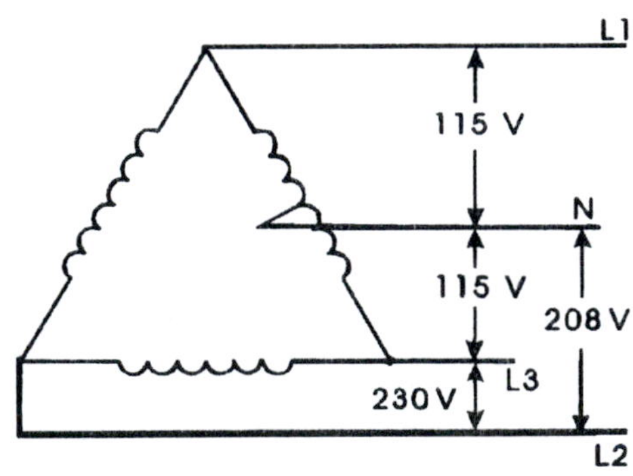

Figure 6–16 A high-leg delta voltage occurs between L3 and N because the center tap that produces the neutral point for the transformer is midpoint in the L1–L2 winding.

Since the 208 V between L3 and N comes from two phases, it will cause the transformer to overheat if it is used to power any components that require 208 V. For this reason the voltage is called the *high-leg delta voltage*, to indicate that it is derived from L3–N of a delta-connected transformer, and that it should not be used to power 208-V components. The L3 leg of the transformer is very usable when it is used with L1–L3 or L2–L3 as part of the three-phase system or 230-V single-phase system. The only problem occurs when the L3 terminal is used in conjunction with the neutral, which creates an L3–N voltage of 208 V.

The L3 terminal in any power distribution box for a delta-wired system should always be marked with an orange wire to identify it as the high-leg delta. In some areas of the country, the high-leg delta is also called the *wild leg*. The high-leg delta voltage of 208 V may occur between L1 and N if the neutral point is produced by a center tap of the L2–L3 winding, or it may occur between L2 and N if the neutral point is produced by a center tap of the L1–L3 winding.

6.6.7 Autotransformers

Another type of transformer is often needed in machine power circuits to provide a small increase or decrease in secondary branch circuit voltage. For example, most factories try to standardize one or two voltage levels throughout the plant for all motor and heating loads. This voltage is generally 480 and 208 V. At times a new piece of equipment, such as a robot or older metal machining equipment, will be installed that requires a slightly higher voltage (such as 220 or 230 V). In this case, the equipment may become damaged if it is operated at 208 V, so an autotransformer will be provided to produce the extra voltage. The autotransformer is very useful because it allows the voltage to be increased or decreased the small amount to make the supply voltage match the equipment requirements. When the autotransformer is used to step up voltage from 208 to 220, it allows the 208 V to flow directly through the conductors without using the transformer. The power going through the transformer is the small amount to make up the 12-V difference. This means that the size of the wire in the autotransformer can be much smaller than that in a traditional step-up transformer because only the wattage (power) due to the 12 V needs to be accounted for in sizing the transformer wires.

The autotransformer shown in Figure 6–17a may have one or two primary and one or two secondary windings. Each winding can produce 12, 16, or 24 V, which can also be combined to produce 32 and 48 V. Notice that the primary windings are marked H1, H2, H3, and H4, while the secondary windings are marked X1, X2, X3, and X4. This means that the transformer can be connected as shown in Figure 6–17b and provide 220 or 244 V between terminals H1 and X2 or X4. As you can see, 208 V are provided from the primary side, and the other 12, 24, or 36 V are provided from the secondary. This also means that the load rating of an autotransformer is calculated only on the part of the voltage and current that is transformed through its secondary.

The autotransformer can be connected to provide slightly lower voltage in applications where the supply voltage is 240 and the load needs 208 V. Figure 6–17c

Figure 6–17 Autotransformer (buck–boost) connections that are used to increase or decrease voltages a small amount.

H1 H2 H3 H4

X1 X2 X3 X4

Formula for calculating kVA for buck-boost transformers:

$$\text{kVA} = \frac{\text{output volts} \times \text{secondary amps}}{1000}$$

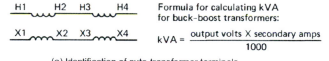

(a) Identification of auto-transformer terminals

H1 208 V 220 V Input H4 AU

H2 H3 X2 X2 X3 X4

230 V 240 V Output

Formula for calculating maximum secondary amperes:

$$\text{Max. amp.} = \frac{\text{nameplate kVA} \times 1000}{\text{secondary volts}}$$

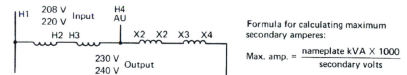

(b) In the "boost" configuration the transformer will increase 208 V to 230 V, or increase 220 V to 240 V.

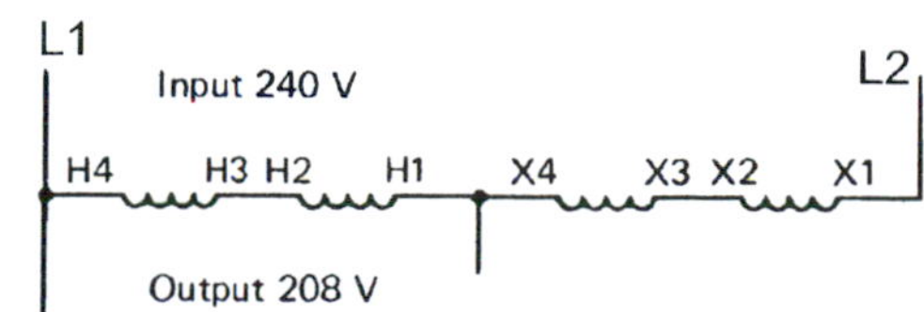

(c) In the "buck" configuration the transformer will lower 240 V to 208 V.

shows this type of connection. The autotransformer can be connected so that the secondary voltage is added to the primary, or so that the secondary voltage is subtracted from the primary. For this reason the transformer is also called a *buck–boost transformer*. The secondary of the autotransformer can also be used as a step-down transformer to provide 12, 16, 24, 32, or 48 V for control circuits.

6.6.8 Three-Phase Voltage on Site

The source for three-phase voltage will be an electrical panel or a load center. The voltage actually comes from the three-phase transformers, but as a maintenance technician you will not be expected to make connections right at the transformers. Instead, technicians from the utility company will connect the wires between the transformers and the load centers or main disconnects. Figure 6–18 shows an example of a vacuum substation breaker. The voltage on the primary side of these transformers is very large, so you must not enter the area where these transformers are located unless you have special training and special clothing and gear that will protect you against the high voltage.

After voltage leaves the substation breakers, it goes to the switchboard (Figure 6–19) for power distribution. From the switchboard, power is sent to the meter panel (Figure 6–20). The meter panel allows the factory to measure the voltage, current, and power factor of the voltage that it is bringing in. After voltage is brought to the meter panel, it is distributed throughout the factory through a series of overhead bus ducts. Figure 6–21 shows an example of a plug-in-style and feeder-style bus duct. Figure 6–22 shows types of bus plugs and connections for the bus duct. Figure 6–23 shows a wireway, trench duct, and a cable tray, which are used to support cable in overhead

Figure 6–19 Switchboard for power distribution.
(Courtesy of Square D/Schneider Electric)

Figure 6–20 Meter panel for power distribution.
(Courtesy of Square D/Schneider Electric)

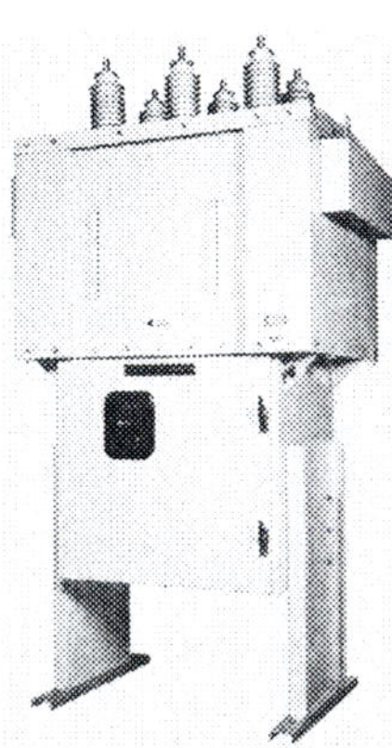

Figure 6–18 Example of vacuum substation breaker.
(Courtesy of Square D/Schneider Electric)

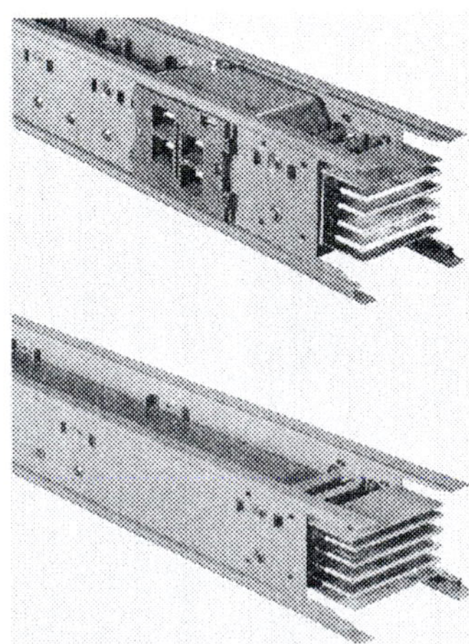

Figure 6–21 Plug-in-style and feeder-style busway.
(Courtesy of Square D/Schneider Electric)

distribution so that it does not sag or stretch. These products are also used to ensure that the cable stays well above the floor where it may get damaged by forklifts or other equipment moving about the factory. Bus plugs are used to make connections at any point on the bus duct where equipment may need it. The bus plugs also provide a place to mount fuses for the circuit, and a means to turn off the power during lockout and tag-out procedures. You must take great care any time you are working around the power distribution system or bus duct, because these locations can cause serious injury to personnel who come into contact with the bare conductors inside the bus or at other points in the system.

The final point in the power distribution system is the electrical cabinet that is located on the machines. The electrical cabinet provides a place to mount the relays and other controls and it also provides another means to disconnect (turn off) all power to the machine. Figure 6–24 shows several types of disconnect switches that you will find on the door of an electrical cabinet.

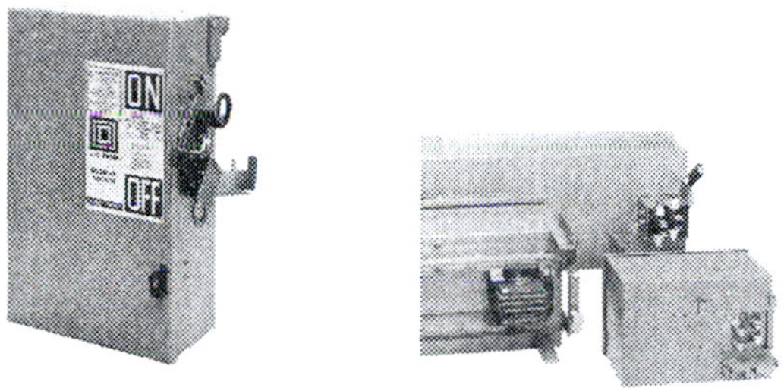

Figure 6–22 Bus plugs and plug-in unit connections. (Courtesy of Square D/Schneider Electric)

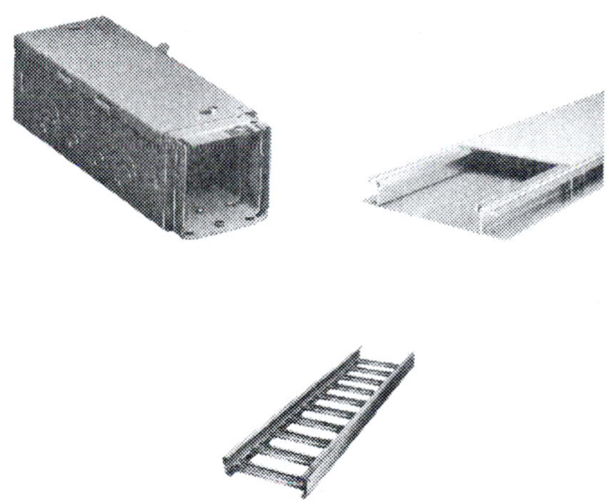

Figure 6–23 Various ways to support cables for power distribution: (a) wireway; (b) trench duct; (c) cable tray. (Courtesy of Square D/Schneider Electric)

These disconnect switches are designed so a technician can open the panel while power remains on. This is important when the system must have routine maintenance while it is under power. The handles also have a mechanism that allows the disconnect to be locked in the off position during lockout and tag-out procedures.

In some parts of the factory, power is needed for outlets and lighting. A load center is used to mount circuit breakers to control each individual circuit for outlets and lighting. The load center may also be called a circuit breaker panel, and if the panel is used primarily for lighting, it may be called a lighting panel. Figure 6–25a shows an example of a load center and Figure 6–25b shows the load center with its cover removed. Figure 6–26a–c shows examples of single-pole, two-pole, and three-pole circuit breakers, which you will

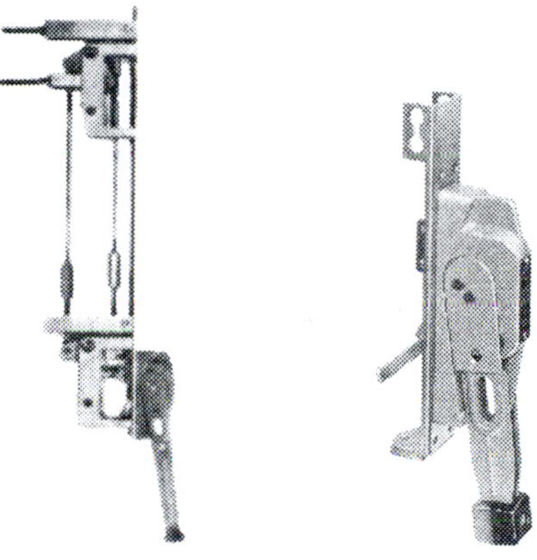

Figure 6–24 Disconnect switches for control panels. (Courtesy of Square D/Schneider Electric)

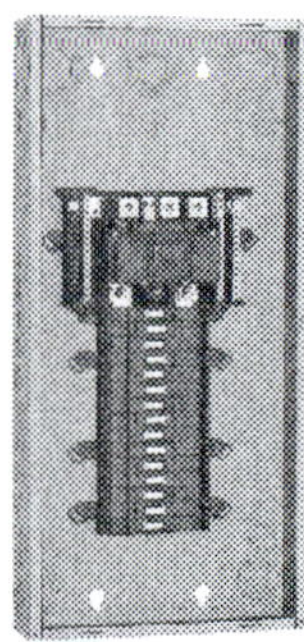

Figure 6–25 (a) Load center circuit breaker panel used for lighting loads and other circuits on the shop floor; (b) internal view of load center. (Courtesy of Square D/Schneider Electric)

Figure 6–26 (a) Single-pole circuit breaker for load center; (b) two-pole circuit breaker for load center; (c) three-pole circuit breaker for load center; (d) stand-alone molded case circuit breaker for control panels. (Courtesy of Square D/Schneider Electric)

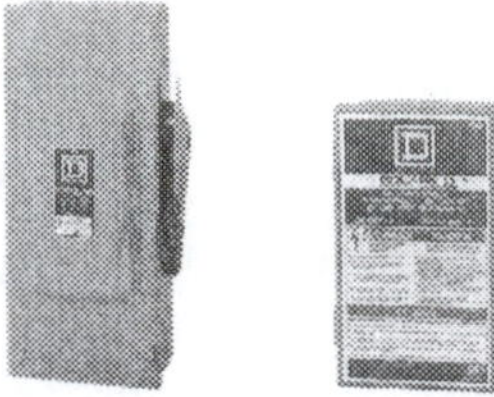

Figure 6–27 General-duty, light-duty, and heavy-duty safety switches. (Courtesy of Square D/Schneider Electric)

find mounted in the circuit breaker panel. The two-pole and three-pole circuit breakers are designed so that all circuits are tripped when any one leg of the breaker detects an overcurrent. The important point about circuit breakers is that they can be reset if the overcurrent condition is cleared. Figure 6–26d shows an example of a stand-alone molded case circuit breaker, which you may find controlling current for stand-alone equipment, such as drill presses, grinders, and other shop floor equipment. These larger circuit breakers are usually mounted in the power panel for a machine or in a smaller enclosure specifically designed for it. You should never mount a stand-alone circuit breaker in an exposed area such as on a wall without an enclosure.

Another component you will find in the power distribution system in a factory is a fused disconnect. Figure 6–27 shows several types of fused disconnects, and Figure 6–28 shows a fused disconnect with its door open. You will need to open the door to test the fuses and replace them if they are faulty. As a general rule, power is supplied to the terminals at the top of the fused disconnect prior to the switch. When the switch is open, power to the fuses is disconnected, so that you can safely remove or replace them. Always check across the fuses for power before you try to remove the fuses and always use a plastic or fiberglass fuse puller tool instead of metal pliers when you are removing and replacing the fuses. The plastic fuse puller will provide a large degree of insulation so that you can safely work around the disconnect. Remember that even though the switch disconnects power to the fuses, the top terminals of the fused disconnect will still have power supplied to them.

On closer inspection of the picture of the three-phase disconnect in Figure 6–28, you can see the box uses three fuses. As you know, incoming voltage is connected to the line side terminals at the top of the disconnect and the line side terminals are identified as L1, L2, and L3. The disconnect has a handle on the right side that is used to turn its switch on and off. If the disconnect switch arm is in the up position the switch is closed and the disconnect will have voltage at the load side terminals at the bottom of the disconnect. The load side terminals are identified as T1, T2, and T3.

Figure 6–28 A three-phase fusable disconnect switch used to provide fuse protection, lockout, and disconnect for equipment. (Courtesy of Square D/Schneider Electric)

The three-phase load center is similar to the circuit breaker panel for a single-phase system, except it will have three individual circuits. In smaller shops that have only a few pieces of equipment, the equipment may be connected directly to the circuit breakers in the load center. This is also generally the case for loads such as exhaust fans and lighting, since they need to be individually protected. The load center is specifically designed to mount three-phase circuit breakers or each individual circuit that will be used. This means that if you have 4 exhaust fans mounted in the ceiling and they are connected to the load center, each one will be connected to its own individual circuit breaker. When you are installing a new fan or servicing existing equipment, you will need to locate the load center and circuit breakers for the circuit you intend to use, or you must locate the fusible disconnect if one is installed near the unit location.

6.6.9 Installing Wiring in a Three-Phase Disconnect

When you are ready to connect the wires between the disconnect and the equipment you are installing, you will need to ensure that the source of three-phase voltage is turned off and a padlock is used to lock out the system. You may even want to go back to the original source of the voltage circuit that supplies voltage to the disconnect (such as the bus plug) and lock out power there also. As you learned in the chapter on lockout and tag-out, the padlock should have a tag attached to it that has your picture and name on it to identify you as the technician who has locked out the system. The lockout and tag-out procedure ensures that you can safely work on the system without anyone accidentally turning the power on when you are still working in the panel.

When power was originally distributed to this part of the factory, the supply voltage wires were installed at the three terminals at the top of the disconnect box. These terminals are called the *line side terminals* and they are identified as L1, L2, and L3. You will make your connections at the bottom terminals, identified as T1, T2, and T3. These terminals are called the *load side terminals*. The other ends of the wires will be connected in the power panel for the equipment. This process is called *connecting the field wiring*.

If you are extending power from a disconnect to a circuit breaker panel, you will need to make connections for field wiring in a circuit breaker panel and connect the wires to its *main breaker* at the top of the panel. The individual circuit breakers will receive power any time the main breaker is in the "on" position. You would need to add a three-phase circuit breaker for each piece of equipment that you were providing power for. The wires that leave the circuit breaker panel are routed through conduit to the machine power panel where they are connected at the panel disconnect in the equipment.

6.6.10 Testing for a Bad Fuse in a Disconnect

At times you will be working on a system that receives its power from a three-phase disconnect. The system will not run and you may suspect that one or more of the fuses are bad. You can check for a blown fuse by measuring the voltage or by removing the fuses and testing them for continuity with an ohmmeter. If you select to check the fuses with a voltage test, you will need to test the incoming voltage at L1, L2, and L3 to determine that the supply voltage is good. Check the voltage between terminals L1 and L2, L2 and L3, and L3 and L1. The voltage at each of these sets of terminals should be the rated voltage for the system, such as 480 V. If you do not get the same voltage reading at each set of terminals, the incoming power has a problem and you will need to determine where the voltage has been interrupted.

The next test should be at the bottom terminals of the disconnect, at T1–T2, T2–T3, and T3–T1. Each of these readings should be the same as the voltage readings at the line side of the disconnect. If any of the voltage readings are lower than the line side readings or 0 V, you have an open fuse. At this point the simplest way to find the bad fuse is to turn the disconnect off and remove all of the fuses and test them for continuity. Be sure to use a fuse puller when you are removing or installing fuses in the disconnect. Also remember that the line side terminals of the disconnect are "hot" (powered) even though the disconnect switch is open.

The reason the load side voltage may be lower than the line side voltage rather than 0 V is that voltage may feed back through two of the windings in any three-phase motor or three-phase transformer that is connected to the load side of the disconnect. It is difficult to determine the exact amount of voltage that is dropped when voltage feeds back through other windings, so you must test for an open fuse if the line side voltage is not the same as the load side voltage at each set of terminals where you take the reading. If you find an open fuse, you will need to replace it with one that has the same current rating and same voltage rating.

You can also measure for voltage across each fuse and if you measure full voltage across any fuse, the fuse is open and the voltage is actually "backfeed voltage."

6.6.11 Single-Phase Voltage from a Three-Phase Supply

At times you will need to provide a single-phase source of voltage for a small motor or pump or for shop tools, such as grinders and drill presses. The single-phase voltage can be either 480, 208, or 230 V L1–L2. In Figure 6–23 you can see the connections between the load side terminals of a three-phase disconnect and the line side terminals of a single-phase disconnect. If the single-phase circuit requires a neutral, a neutral wire can be connected between the neutral terminal of the three-phase disconnect and single-phase disconnect, and between the three-phase disconnect and the transformer.

This type of circuit may also be used in applications inside a factory or commercial location where three-phase voltage is supplied to the system and you need to provide a single-phase voltage source to power smaller shop equipment. Once you understand that you can get single-phase voltage from the three-phase system, you will begin to see how the AC power distribution system works in industrial, commercial, and residential applications.

The remaining chapters of this book will continue to refer to the supply voltage as you would find it connected to the equipment at the line side terminals. As a technician you will need to be aware of where this voltage comes from and how it is distributed so that you can trace the circuits if no voltage is present at the unit.

6.7 Fuses and Circuit Breakers

The motors, switchgear, transformers, and power distribution conductors must be protected against short-circuit current. The motors and other loads must also be protected against slow overcurrents or sustained overload that will allow them to overheat and become damaged. *Short-circuit current* is defined as any current that exceeds the normal full-load current of a circuit by 10 times. When short-circuit current occurs unprotected, any conductor or switchgear that is involved will be severely damaged by excessive magnetic forces and by extremely high heat levels that will melt most metal objects. Fuses, circuit breakers, and magnetic overloads can provide protection for motors and other loads against slow overcurrents, and fuses and circuit breakers can provide short-circuit protection.

6.7.1 Slow Overcurrents

The slow overcurrent develops from overloading devices or from malfunctions in motors and other loads. For example, if a motor-driven conveyor is overloaded, the motor will be required to draw extra current to try and move the load. If the motor is not protected, it will draw the extra current and begin to overheat. If the transformer that supplies voltage to the motor is also fully loaded, it will become overloaded when the motor draws the extra current. If the overload condition continues for 10 to 20 minutes, enough heat will be built up in the motor and transformer to cause the insulation in both of these devices to break down and deteriorate.

The same problem will occur if the bearings in a motor become dry and begin to wear. After the bearing has operated without any lubrication, it will begin to heat up and seize on the motor shaft, which will in turn cause the motor to draw excessive current. This condition will cause the motor and transformer to overheat to a point where they are completely damaged because the overcurrent will continue as long as the motor is running.

6.7.2 Protecting Against Slow Overcurrent

In all these examples the problem is caused by an increase in normal operating current to the point where damage can occur. Circuit breakers can sense either the heat or magnetic forces as they increase beyond the maximum safe level, and fuses and overloads can sense increased heat that the overload creates. Separation of these devices will cause a set of contacts or a conducting element to open any time the current increases above the safe level. This presents a problem with some loads, such as motors, that have a very large inrush current when they start. A 5-hp motor operating on 208 V will draw 16.5 A at full load. This motor will draw up to 99 A, which is six times the amount of full-load current, when it starts. This presents a problem in protecting against overcurrents, because a circuit breaker or fuse that is sized to protect the motor during full-load current (16.5 A) would trip when the motor is started, and if it is sized to allow the motor to start (99 A), it will not provide adequate protection when the motor is running at full-load current.

Several solutions to this problem are available. One of them is a motor starter with heaters and overloads, and another is inverse-time circuit breakers. Each of these devices provides several minutes of time delay before it trips and takes the motor off line. The operation of these mechanisms is explained in detail in Chapter 7. The theory of their operation involves allowing small overcurrents to exist for up to 4 or 5 minutes, and allowing larger overcurrents for less than 10 seconds. These times are based on the amount of time a specific overload can exist before a motor begins to sustain damage. The devices must be sized properly to provide adequate protection, and may be adjusted slightly once they are installed. The only problem that remains with the inverse-time circuit breaker and overloads is that they cannot sense a short circuit and open the circuit fast enough to provide interruption capacity.

6.7.3 Short-Circuit Currents

When a conductor from one potential comes in contact with the system ground or a conductor from another potential, a short circuit can occur. The short circuit provides little or no impedance to the power source, so that the current can rise to 100 times the full-load current levels in one or two cycles. This means that currents of 50,000 A or larger are possible if the current is allowed to build. In a short-circuit condition, the current continues to increase with each cycle of the AC voltage. As this current increases, it will build up powerful magnetic forces and tremendous amounts of heat energy that will cause the metal conductors and terminals to melt and explode.

A circuit breaker is an electromechanical device that requires approximately one-half an AC cycle to sense the short circuit and another half-cycle to trip its mechanical contacts. Since the circuit breaker requires one-and-a-half cycles to open, the short-circuit current may still be able to reach 40,000 to 45,000 A before the circuit is fully opened. In some cases, the heat is so immense that it will actually weld the contacts of the circuit breaker together so that they cannot open even though the circuit breaker's trip mechanism has activated.

A fuse can sense the overcurrent as it begins to build and its link will melt before the current increases to a dangerous level. The fuse element will open as soon as the current reaches the overcurrent level. This means that the short-circuit fault will be sensed and opened in less than a half-cycle. The only problem with the single-element fuse is that it must be sized up to six times the full-load current level to allow the motor to start. This leaves two alternatives when providing short-circuit and overload current: fuses in combination with motor starters or circuit breakers, or the use of a dual-element fuse.

6.7.4 Types of Fuses

Several types of fuses are available for various applications in motor control systems. The two main types of fuses are cartridge and plug fuses. Figure 6–29 shows several cartridge and a plug fuse in different sizes. Some of these fuses provide blades to make connections in the fuse holders, and other fuses provide connections at the neck of the fuse. The plug fuse comes in several sizes, based on the current rating of the fuse. A rejection feature is provided on cartridge, blade, and plug fuses. Examples of the rejection features are also shown in this figure. The rejection feature prevents a fuse with a larger current rating from being substituted for the fuse of a smaller size that was originally specified. This is accomplished by providing a matching rejection feature in the fuse holder with the feature that is provided for the fuse size. This means that each size of fuse has a different type of rejection feature, or it is located at a different point on the fuse body for each amperage rating.

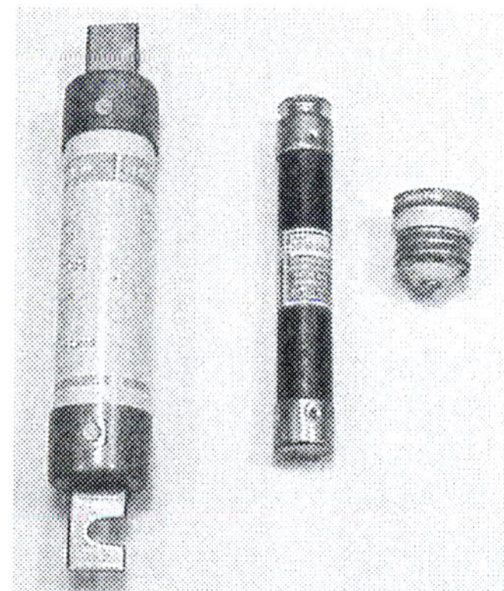

Figure 6–29 (a) A blade-type fuse; (b) a cartridge-type fuse; (c) a screw-base-type plug fuse.

6.7.5 Single-Element and Dual-Element Fuses

The dual-element fuse combines the best of overcurrent protection with short-circuit protection. Figure 6–30 shows cutaway diagrams of single-element fuses and Figure 6–31 shows diagrams of dual-element fuses. You can see that the single-element fuse is made of a single

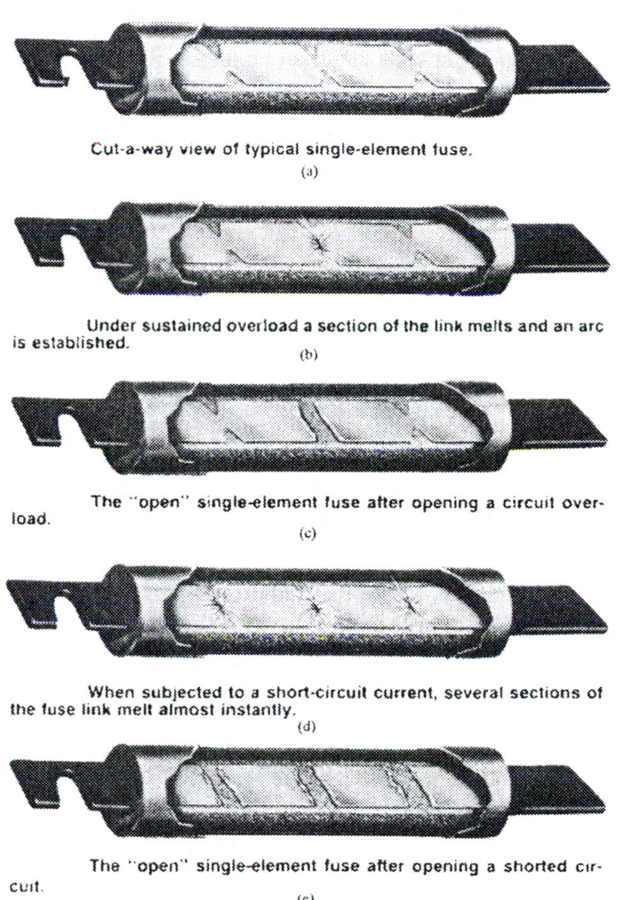

Cut-a-way view of typical single-element fuse.
(a)
Under sustained overload a section of the link melts and an arc is established.
(b)
The "open" single-element fuse after opening a circuit overload.
(c)
When subjected to a short-circuit current, several sections of the fuse link melt almost instantly.
(d)
The "open" single-element fuse after opening a shorted circuit.
(e)

Figure 6–30 A series of pictures of single-element fuses. The top picture shows a good fuse, and the lower pictures show the fuse in the process of opening.
(Courtesy of Cooper Bussmann)

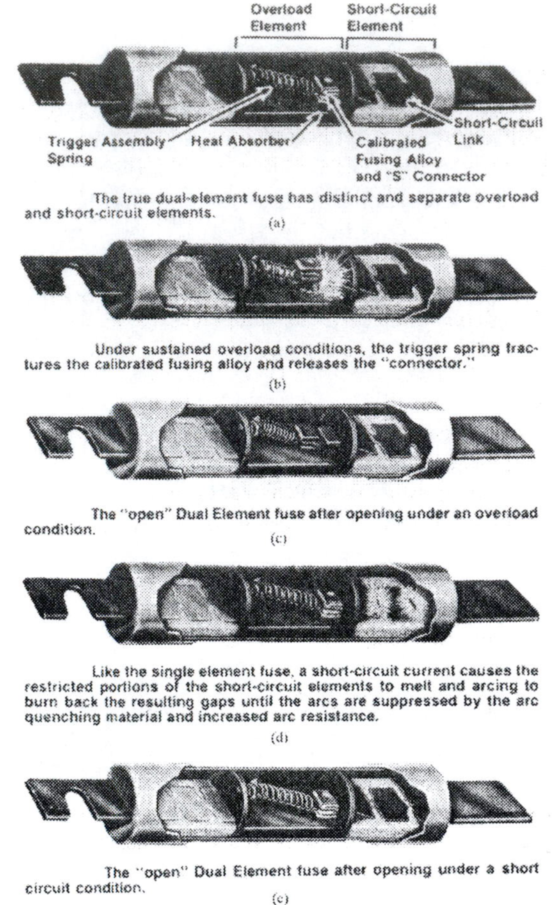

Figure 6–31 A series of pictures of dual-element fuses. The top picture shows a good fuse, and the lower pictures show the fuse in the process of opening.
(Courtesy of Cooper Bussmann)

conducting element with several neck-down sections. The neck-down sections provide a point where the short-circuit current will be concentrated and cause the metal to melt. Since the element is made of thin metal, the larger the short-circuit current becomes, the quicker the element will melt.

In the cutaway diagram of a dual-element fuse, you can see that the dual-element fuse provides a short-circuit element and an overload element. A short-circuit element is located on each end of the fuse, while the overcurrent element is located between them. The overcurrent element is made of a spring-loaded section with a heat absorber. The heat absorber has a small solder pot that anchors one end of the spring link. When the full-load current exceeds the specified limit, the solder will become increasingly warm. If the overload continues, the heat will increase sufficiently to allow the solder to melt. When the solder melts, the trigger spring will fracture the calibrated fusing alloy and release the connector.

The overload element is sized to allow the fuse to sustain a 500-percent overload for approximately 10 seconds. If the overload is smaller, such as 300 percent, it will sustain it for approximately 60 seconds, and if it is larger, such as 800 percent, it will be cleared in approximately 1.5 seconds. This means that the small overloads will be allowed to continue for several minutes since the small heat buildup will not damage the motor, and large overloads will be cleared quickly since they will cause heat buildup that will permanently damage the motor. This design allows the motor to draw high inrush current, up to 500 percent (5 times) of the full-load current for 10 seconds, which is adequate time to allow the motor to start, and it will also allow the motor to develop extra horsepower to meet an increased load demand for several minutes, since the extra current that is drawn will not cause the fuse to open.

The dual-element fuse provides another safety feature, which involves encasing the short-circuit element in silica sand. When the short-circuit current is applied to the short-circuit element, tremendous amounts of heat are built up while the element is melting. If this heat is allowed to build up, gases can be released when the metal is melted and cause the casing of the fuse to rupture. When silica is placed around the short-circuit element, it will absorb the extra heat and use it to melt the sand into a semiliquid state. Since the silica is forced to change state, it will absorb more heat than the reaction can produce, which results in the excessive energy being controlled without damaging the fuse or the hardware and enclosures that are used to mount the fuse.

The ability of the fuse to clear short-circuit currents safely is called its *interruption capacity*. The interruption capacity is listed as the maximum number of amperes that the fuse can safely clear. The interrupting capacity of modern current limiting fuses may be as large as 200,000 A. Fuses can also be used in the power distribution system for the expressed purpose of providing interruption capacity for protecting the system equipment and switchgear against large short-circuit currents.

6.7.6 Sizing Fuses According to the National Electric Code

When a power distribution system is designed, the fuses must be correctly sized and specified to provide protection for switchgear and conductors against short-circuit current and overload protection for motors. Figure 6–32 shows the number and types of fuses that are required in a power distribution system. The electrical specification is the same regardless of the brand name of the fuse that is being used.

Fuses are selected on the basis of the application they are protecting. These applications include circuits

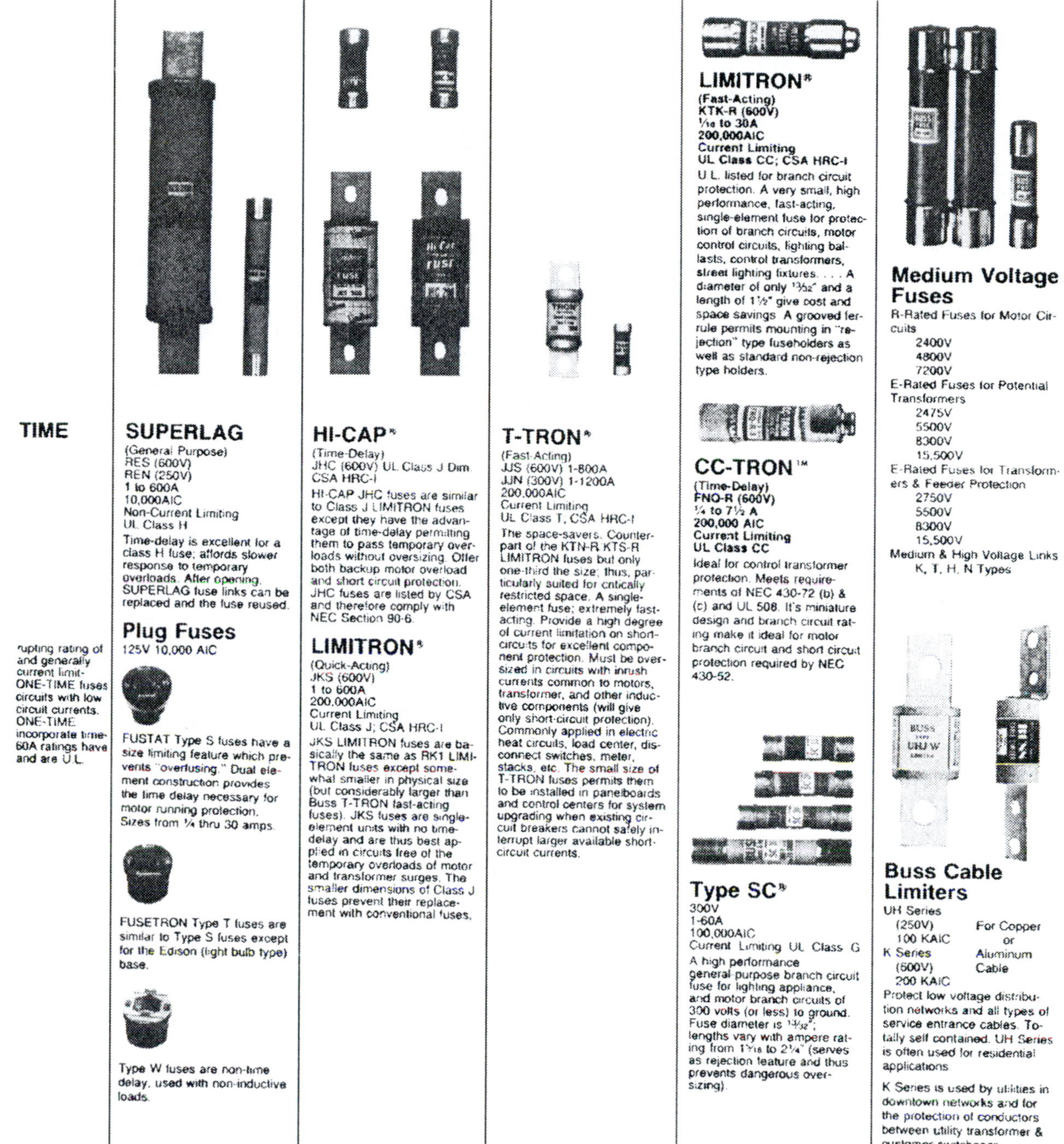

Figure 6–32 Examples of the wide variety of fuses available for protection of all types of circuits and equipment. (Courtesy of Cooper Bussmann)

with motors only, circuits with lighting loads only, circuits with a mixture of motors and lighting loads, and circuits with motors and power factor correction. Fuses are also specified for feeder and branch circuit applications. When fuses are used in conjunction with circuit breakers and motor starters they should be sized so that the fuse does not open before these resetting components.

From Figure 6–32 you can see that a wide variety of fuses must be available to meet the various loads that may be encountered. The fuse sizes have been standardized somewhat to ensure that loads of all sizes are adequately protected. Plug fuses are available in sizes that increase 1.15 percent starting at $\frac{3}{10}$ of an ampere. Cartridge fuses are available in sizes from $\frac{1}{10}$ to 600 A.

Buss Power Distribution Fuses

HI-CAP®
(Time-Delay)
KRP-C (600V)
601 to 6000A
200,000AIC
Current Limiting
UL Class L; CSA-HRC-L

The all-purpose silver linked fuse for both overload and short-circuit protection of high capacity systems (mains and large feeders). Time-delay (minimum of four seconds at five times amp rating for close sizing. Unlike fast acting fuses, pass harmless surge currents of motors, transformers, etc., without overfusing and any sacrifice of short-circuit current limitation (component protection). The combination use of $\frac{1}{10}$ to 600 ampere LOW-PEAK dual-element time-delay fuses and 601 to 6000A KRP-C HI-CAP fuses is recommended as a total system specification. Easily selectively coordinated for blackout protection. Size of upstream fuse need only be twice that of downstream HI-CAP or LOW-PEAK fuses (2:1 ratio). HI-CAP fuses can reduce bus bracing; protect circuit breakers with low interrupting rating as well as provide excellent overall protection of circuits and loads.

LIMITRON®
(Fast-Acting)
KTU (600V)
601 to 6000A
Current Limiting
200,000AIC
UL Class L; CSA-HRC-L

Silver-linked fuse. Single-element units with no time-delay. Very fast-acting with a high degree of current limitation; provide excellent component protection. Particularly suited for protection of circuit breakers with lower interrupting ratings, and non-inductive loads such as lighting and heating circuits. Can be used for short-circuit protection only in circuits with the inrush currents. Must be oversized to prevent opening by the temporary harmless overloads with some sacrifice of current limitation. In motor circuits, must be sized at approximately 300% of motor full-load current and thus will not provide the overload protection of HI-CAP KRP-C fuses.

LIMITRON®
(Time-Delay)
KLU (600V)
601 to 4000A
Current Limiting
200,000AIC
UL Class L; CSA-HRC-L

10 seconds delay (minimum) at 500% of amp rating. Not as current limiting as KRP-C or KTU fuses.

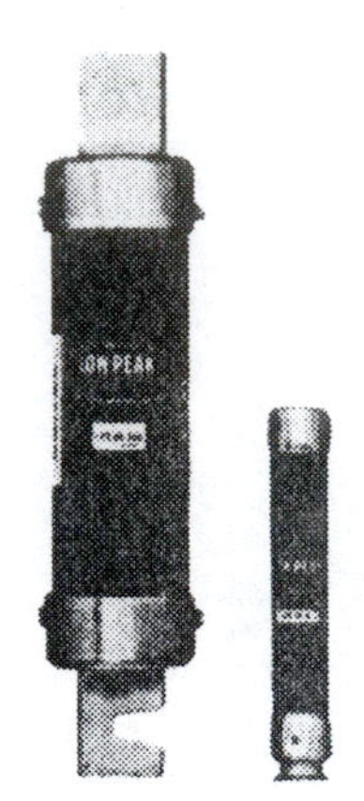

LOW-PEAK®
(Dual-Element, Time-Delay)
LPS-RK (600V)
LPN-RK (250V)
$\frac{1}{10}$ to 600A
200,000AIC
Current Limiting
UL Class RK1; CAS HRC-I ("D")

High performance, all-purpose fuses. Provide the very high degree of short-circuit limitation of LIMITRON fuses plus the overload protection of FUSETRON fuses in all types of circuits and loads. Can be closely sized to full-load motor currents for reliable motor overload protection as well as backup protection. Close sizing permits the use of smaller and more economical switches (and fuses); better selective coordination against blackouts; and a greater degree of current limitation (component protection). LOW-PEAK fuses are rejection type but fit non-rejection type fuseholders. Thus, can be used to replace Class H, K1, K5, RK5 or other RK1 fuses.

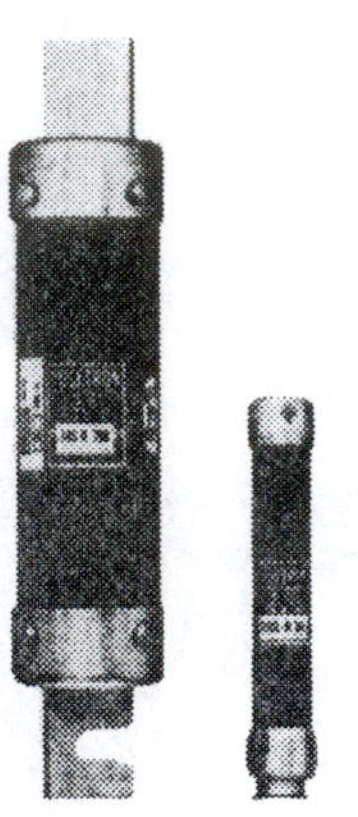

FUSETRON®
(Dual-Element, Time-Delay)
FRS-R (600V)
FRN-R (250V)
$\frac{1}{10}$ to 600A
200,000AIC
Current Limiting
UL Class RK5; CSA-HRC-I ("D")

Time-delay affords the same excellent overload protection of LOW-PEAK fuses of motors and other type loads and circuits having temporary inrush currents such as caused by transformers and solenoids. (In such circuits, LIMITRON fuses can only provide short-circuit protection). FUSETRON fuses are not as fast-acting as LOW-PEAK fuses and therefore cannot give as high a degree of component short-circuit protection. Like LOW-PEAK fuses, FUSETRON fuses permit the use of smaller size and less costly switches. FUSETRON fuses fit rejection type fuseholders and can also be installed in holders for Class H fuses. They can physically and electrically replace Class H, K5, and other Class RK5 fuses.

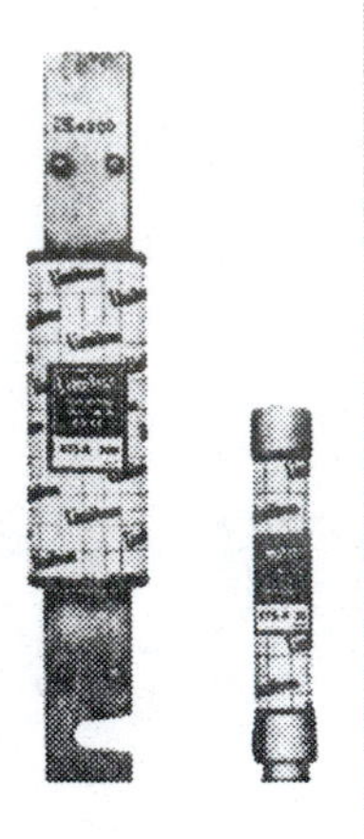

LIMITRON®
(Fast-Acting)
KTS-R (600V)
KTN-R (250V)
$\frac{1}{10}$ to 600A
200,000AIC
Current Limiting
UL Class RK1, CSA HRC-I

Single-element, fast-acting fuses with no time-delay. The same basic performance of the 601-6000A KTU fast-acting LIMITRON fuses. Provides a high degree of short-circuit current limitation (component protection). Particularly suited for circuits and loads with no heavy surge currents of motors, transformers, solenoids, and welders. LIMITRON fuses are commonly used to protect circuit breakers with lower interrupting ratings. If used in circuits with surge currents (motors, etc.) must be oversized to prevent opening and thus only provide short-circuit protection. Incorporate Class R rejection feature. Can be inserted in non-rejection type fuseholders. Thus, can physically and electrically replace fast acting Class H, K1, K5, RK5, and other RK1 fuses.

ONE-
(General Purpose)
NOS (600V)
NON (250V)

$\frac{1}{10}$ to 600A
10,000AIC
Non Current Limiting
UL Class H
(K-5 in Sizes 1-60A)

With an inter
10,000 amp
not considered
ing, Class H
are used in
available short-
Single-element
fuses do not
delay. The 1-
a 50,000 AIC,
Class K-5.

Figure 6–32 *Continued*
(Courtesy of Cooper Bussmann)

6.7.7 Fuses for Solid-State Applications

Solid-state devices, such as rectifier diodes, SCRs, inverters, and motor drives, require current-limiting protection that can respond faster than traditional fuses. These devices are very sensitive to short-circuit current, so the fuse that is used to protect these devices should be able to sense the fault and clear it rapidly. Specialized fuses have been developed that have a high degree of current limitation for semiconductor applications.

It is important to locate fuse protection on both sides of solid-state devices, since they can be damaged by short circuits that occur on the AC or the DC side of the circuit. The fuses used in these circuits must be sized according to the rms current rating of the circuit rather than the average current. This is required because the fuse must also protect against excessive heat buildup in the solid-state components caused by problems in the circuit. The nature of the solid-state circuit is to shift all current to the remaining components when one component fails. This means that if the circuit is overloaded and one component starts to fail before the others, it will ultimately shift its entire load to the remaining devices that are already overloaded. This tends to cause two or more solid-state components to fail when a fault occurs, if proper protection is not used. The solid-state fuses are designed to sense the overload and clear them before any components are affected. They will also afford proper protection to all remaining devices in a circuit if a solid-state device fails from internal breakdown. It is very important to size these fuses according to measured voltages and currents as well as specified values so that they can protect the circuit properly.

6.7.8 Cable-Limiting Fuses

Cable-limiting fuses are specialized fuses that are intended to protect cables from damage caused by short circuits. As you know, when a short circuit is allowed to occur, the current will cause both heat damage and magnetic distortion of all conductors that are in the faulty circuit. The cable limiter is specially designed to protect the cable against short-circuit current. The physical design of the fuse provides a termination point for the conductor and a hardware fitting so that it can easily be bolted to power distribution terminal strips. These specialized fuses should be located to protect the cables used in the power distribution system.

6.7.9 Troubleshooting and Testing Fuses

A fuse is the only component in a motor control system that is intended to be destroyed when it is doing its job properly. The main tendency for the troubleshooting technician is to test the fuses to find the one that has opened and then quickly replace it to put the equipment into production again. The fuse is in the circuit to protect the equipment from short circuits and overcurrents, but it is important to understand that the fuse is the only witness to the fault. This means that after a fuse is found to be faulty, you should check it carefully to determine whether it opened because of a short circuit or a slow overcurrent. The first part of the troubleshooting sequence is determining which fuse is the cause of the open circuit. As you have seen in the power distribution circuit diagrams, fuses are located throughout the system to protect against problems. When you are called to test the circuit, you will be called because a machine or group of machines has stopped operating. This could be caused by several problems that will be explained in other chapters, but in this case we will assume that one of the fuses has detected a fault and has opened to clear the fault.

You will need to take several voltage tests at various points in the circuit to determine which fuse has opened. The best way to test the fuses with voltage applied is to place the voltmeter probes on the top of fuses (where voltage comes in) from two different potential sources. The neutral or ground point in the system could also be used as a reference for one of the probes while the other probe is used to touch each fuse where power is incoming to determine if voltage is present. If voltage is present at the top of the fuse, it indicates that power is being supplied to the point. Next, the bottom of the fuse should be checked, and if voltage is not present at the bottom, the fuse is opened. If voltage is present at the bottom, one other test should be made to ensure that voltage is not backfeeding. This test requires that you place both terminals across each fuse in the box. If a fuse is open, it will read full voltage across it, and if the fuse is still in good condition, the voltage reading will be zero across it. This may confuse you at first, but Figure 6–15 shows how backfeeding can occur. If the circuit contains a motor starter or other switch that opens and isolates the load, the feedback problem will not occur.

From the diagram you can see that feedback can occur through the top set of fuses since the load that they are protecting is a transformer. In this case you should use both tests to find the open fuse. If two fuses are blown, you will find them with the test that checks for voltage at the top of the fuse in reference to ground or between phases. If you test across each fuse when two or more fuses are blown, the test that uses ground or another phase as reference may confuse you because of the backfeeding problem. In this case the bad fuse will be found by reading voltage directly across each fuse. As you can see, you must use a combination of each of these techniques to find all fuses that are blown. It is also best to check each fuse for continuity with an ohmmeter or test lamp after they have been removed from the circuit. *Do not try to test the fuse for continuity while power is applied, since the ohmmeter is a low-impedance meter and will cause a short circuit in any circuit that has power applied where you touch the probes.* You can also test a blown fuse to a motor circuit with a clamp-on ammeter. If only one fuse is blown, the motor will try to start by pulling current through the two good fuses. The ammeter will indicate which of the three fuses is not conducting any current and that is the bad fuse.

Another problem that may confuse you with a blown fuse is caused when the fuse is subject to a slow overcurrent. In these cases, the link in the fuse is trying to pull away from the melting solder, and the solder in each fuse may not have melted enough to allow the spring to break away. If this occurs, the fuse will test good with voltage tests while it is in circuit and with the continuity test while it is out of circuit, yet the motors and loads will not be able to start when power is applied because sufficient current cannot pass through the partially melted solder to power the circuit. If you have tested the fuses and they seem to be good, yet the motor is not receiving sufficient power, you can recheck each suspected fuse. This time when you have the fuse out of its holder, tap one end of the fuse on the side of the machine or on the floor. This will cause the spring to break away if the solder is partially melted.

Figure 6–33 The CUBEFuse is a new type of fuse package that offers world-class protection, increased safety, reduced size, and improved electrical performance. (Courtesy of Cooper Bussmann)

Some experienced technicians have learned to automatically change all the fuses in a circuit that has been involved in a slow overcurrent because the remaining fuses have been overheated and are probably stressed, which means that they may trip prematurely when the motor is started again. It is always important to check each fuse that you find opened to see if it has blown because of a short circuit or a slow overcurrent. If the fuse is a plug type, its window will be black from the high current. If the window is clear but the fuse is open, shake the fuse and listen for solder to roll around, which indicates that the fuse has cleared a slow overcurrent.

If the fuse is a cartridge type, you may need to cut it in half with a band saw to determine if it opened from a short circuit or slow overcurrent. Once you have determined why the fuse has opened, you can look for the source of the problem. If the fuse has cleared a slow overcurrent, you should look for overload conditions on the motor. You should also have an ammeter in place when the motor is started again. This will help you determine the extent of the overload and indicate when you have cleared the problem that is causing it.

If the fuse has opened because of a short circuit, you will have to begin disconnecting sections of the circuit until your test equipment indicates that the short circuit has been removed. At that point you will need to check more closely the circuit that has been disconnected for indications of a short circuit.

A new type of fuse package, called a cube fuse, is now available that offers world-class protection and reduces the physical size of the fuses. Figure 6–33 shows an example of these new fuses. The package has indicators that show if the fuse is good or bad, and the package also allows the fuse to be handled more safely when it is removed and replaced. These fuse packages allow fuses to be mounted directly to DIN rail and conductors can be connected directly to the terminals on the fuses.

6.8 Conductors

One part of the power distribution system that is involved at all points in the system is the conductor. Conductors can be made of aluminum or copper wire. Aluminum is generally used for long-distance distribution because of its lighter weight. Once the power is inside the plant, copper conductors are generally used for distribution. Copper conductors will be solid in bus bar and bus way applications, and they will be stranded conductors for all other applications.

The conductors are sized by the amount of amperage they can carry. A table of typical conductor sizes and their ampacities is provided in Figure 6–34. The standard used to determine the size of each conductor is called the *American wire gauge* (AWG). These tables have been established by the National Electric Code, and they are used to select the proper size of conductor to carry the load. Conductors are classified by the type of insulation that is used as a cover. The cover also provides the voltage rating of the wire. Typical voltage ratings for conductors used in motor control applications include 300, 600, and 1000 V. The voltage and current ratings of the conductor should not be exceeded under any circumstances.

The type of covering is also listed in Figure 6–34. Abbreviations for each of these coverings are listed in the NEC tables and help determine the type of wire that should be selected for each application. The outer covering of the wire serves several purposes, including protecting the conductors from coming in contact with metal in the cabinet or other conductors. The cover also provides a location to stamp all specification data regarding the wire, including the voltage rating, AWG size, temperature specification, and type of covering.

Figure 6–35 shows examples of terminal blocks mounted on DIN rail. Terminal blocks are used to provide a safe method of connecting two or more wires together. Figure 6–36 shows several diagrams of terminal blocks with their covers removed so you can see how the wires are held in place. Figure 6–37 shows several other types of termination connectors used to secure wires when they are connected to electrical components. It is important that the correct type of connector is used to provide the termination connection for wires when they are connected to electrical equipment, or the wire will overheat at the point of connection and cause damage to the wire and the electrical equipment.

Table 310-17. Allowable Ampacities of Single Insulated Conductors, Rated 0 through 2000 Volts, in Free Air Based on Ambient Air Temperature of 30°C (86°F)

Size	Temperature Rating of Conductor. See Table 310-13.						Size
	60°C (140°F)	75°C (167°F)	90°C (194°F)	60°C (140°F)	75°C (167°F)	90°C (194°F)	
AWG kcmil	TYPES TW†, UF†	TYPES FEPW†, RH†, RHW†, THHW†, THW†, THWN†, XHHW† ZW†	TYPES TBS, SA, SIS, FEP†, FEPB†, MI, RHH†, RHW-2, THHN†, THHW†, THW-2†, THWN-2†, USE-2, XHH, XHHW†, XHHW-2, ZW-2	TYPES TW†, UF†	TYPES RH†, RHW†, THHW†, THW†, THWN†, XHHW†	TYPES TBS, SA, SIS, THHN†, THHW†, THW-2, THWN-2, RHH†, RHW-2, USE-2, XHH, XHHW†, XHHW-2, ZW-2	AWG kcmil
	COPPER			ALUMINUM OR COPPER-CLAD ALUMINUM			
18	....	....	18	....	....	....	....
16	....	....	24	....	....	....	....
14	25†	30†	35†	....	....	....	....
12	30†	35†	40†	25†	30†	35†	12
10	40†	50†	55†	35†	40†	40†	10
8	60	70	80	45	55	60	8
6	80	95	105	60	75	80	6
4	105	125	140	80	100	110	4
3	120	145	165	95	115	130	3
2	140	170	190	110	135	150	2
1	165	195	220	130	155	175	1
1/0	195	230	260	150	180	205	1/0
2/0	225	265	300	175	210	235	2/0
3/0	260	310	350	200	240	275	3/0
4/0	300	360	405	235	280	315	4/0
250	340	405	455	265	315	355	250
300	375	445	505	290	350	395	300
350	420	505	570	330	395	445	350
400	455	545	615	355	425	480	400
500	515	620	700	405	485	545	500
600	575	690	780	455	540	615	600
700	630	755	855	500	595	675	700
750	655	785	885	515	620	700	750
800	680	815	920	535	645	725	800
900	730	870	985	580	700	785	900
1000	780	935	1055	625	750	845	1000
1250	890	1065	1200	710	855	960	1250
1500	980	1175	1325	795	950	1075	1500
1750	1070	1280	1445	875	1050	1185	1750
2000	1155	1385	1560	960	1150	1335	2000

CORRECTION FACTORS

Ambient Temp. °C	For ambient temperatures other than 30°C (86°F), multiply the allowable ampacities shown above by the appropriate factor shown below.						Ambient Temp. °F
21-25	1.08	1.05	1.04	1.08	1.05	1.04	70-77
26-30	1.00	1.00	1.00	1.00	1.00	1.00	78-86
31-35	.91	.94	.96	.91	.94	.96	87-95
36-40	.82	.88	.91	.82	.88	.91	96-104
41-45	.71	.82	.87	.71	.82	.87	105-113
46-50	.58	.75	.82	.58	.75	.82	114-122
51-55	.41	.67	.76	.41	.67	.76	123-131
56-60	....	.58	.71	....	.58	.71	132-140
61-70	....	.33	.58	....	.33	.58	141-158
71-80	....	....	.41	....	....	.41	159-176

†Unless otherwise specifically permitted elsewhere in this *Code*, the overcurrent protection for conductor types marked with an obelisk (†) shall not exceed 15 amperes for No. 14, 20 amperes for No. 12, and 30 amperes for No. 10 copper; or 15 amperes for No. 12 and 25 amperes for No. 10 aluminum and copper-clad aluminum.

Figure 6–34 Table for conductor sizing for the National Electric Code. (Reprinted with permission from NFPA 70-2000, the *National Electric Code*®, Copyright © 1999, National Fire Protection Association, Quincy, MA 02269. This reprinted material is not the referenced subject which is represented only by the standard in its entirety.)

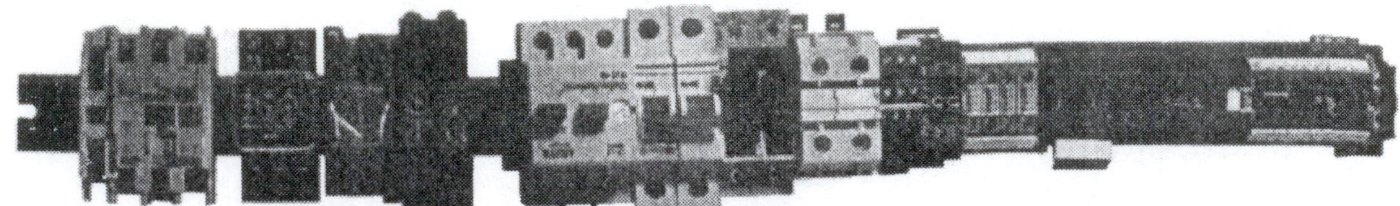

Figure 6–35 Examples of terminal blocks mounted on DIN rail.
(Courtesy of Eaton/Cutler Hammer)

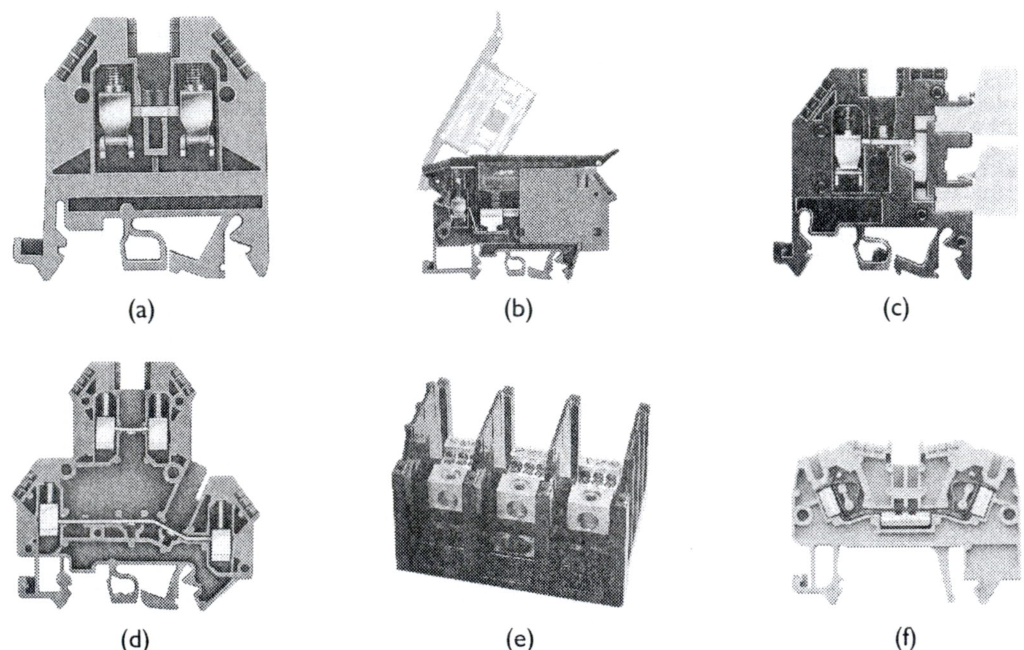

Figure 6–36 Examples of individual types of terminal blocks available to terminate wiring inside electrical panels: (a) single level feed through; (b) series fuse blocks; (c) feed through pluggable block; (d) series dual-level block; (e) power distribution terminal block; (f) spring clip block.
(Courtesy of Eaton/Cutler Hammer)

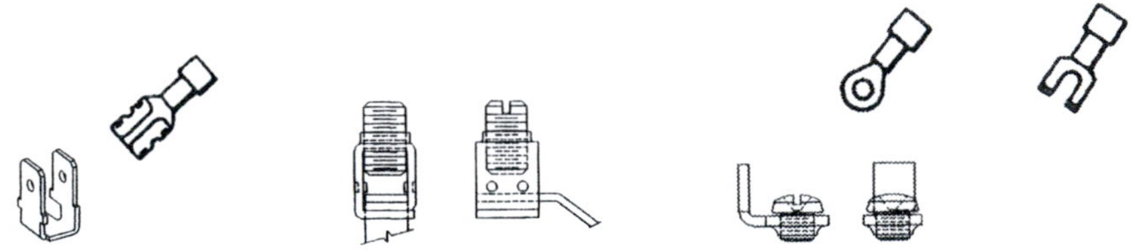

Figure 6–37 Types of wire termination connectors.
(Courtesy of Eaton/Cutler Hammer)

Questions

1. Explain the term *power distribution.*
2. Why must you understand the power distribution system if you are only going to install, troubleshoot, or repair industrial control systems?
3. Explain where the source for all electrical power is for a power distribution system.
4. Identify the location of the electrical generating plant in your area where the power for your building originates.
5. Name four sources of energy used to generate electricity.

True and False

1. A wire tie (cable tie) is used to secure wires in electrical panels.
2. The primary voltage of a step-up transformer is larger than the secondary voltage.
3. A control transformer is used to provide 110 V to the control components in a circuit.
4. A slow overcurrent is a very large current that occurs when resistance drops to near zero during a short circuit.
5. A dual-element fuse can protect a circuit against a short circuit and a slow overcurrent.

Multiple Choice

1. A delta-wired transformer will have ________ at its secondary terminals.
   a. 100 V
   b. 208 V
   c. 240 V
   d. 330 V
2. A wye-wired transformer will have ________ at its secondary terminals.
   a. 100 V
   b. 208 V
   c. 240 V
   d. 330 V
3. A single-element fuse ____________
   a. has a single element that protects against slow overcurrents.
   b. has a single element that protects against short-circuit current.
   c. has two elements that protect against both short-circuit current and slow overcurrents.
   d. is rated for a single voltage.
4. A dual-element fuse ____________
   a. has a single element that protects against slow overcurrents.
   b. has a single element that protects against short-circuit current.
   c. has two elements that protect against both short-circuit current and slow overcurrents.
   d. is rated for two voltages.
5. The rejection feature in a fuse ________
   a. ensures that the fuse cannot be used in a circuit that has a larger current rating than the fuse.
   b. ensures that the fuse cannot be used in a circuit that has a smaller current rating than the fuse.
   c. ensures that a dual-element fuse cannot be replaced with a single-element fuse.
   d. All of the above

Problems

1. The power supply for your factory is 208 V (three phase). You receive a new machine that requires 230 V three phase. Explain what you would do to get the proper voltage to the machine.
2. You are requested to select a step-up transformer that has a 2:1 ratio. Determine the secondary voltage if the primary voltage is 220 V.
3. You are requested to calculate the efficiency of the transformer you are working with. You have measured the input voltage at 480 V and 200 A and the secondary voltage is 240 V and 380 A.
4. Your boss asks you to test the three fuses that provide protection for a three-phase motor. You suspect that one of the fuses is blown, because the motor will not start and only hums when it tries to restart. Your boss reminds you to use both a voltmeter and a clamp-on ammeter to be sure you locate the faulty fuse. Explain what problems you might encounter using the voltmeter, and why the clamp-on ammeter is useful.
5. Use the diagram in this chapter to draw a sketch to show how power is distributed from the generating facility to a factory where it is used.

6. Draw a simple transformer and explain how voltage is produced in the secondary winding.

7. You have just installed an autotransformer to increase 208 voltage supply to 240 V. Your boss asks you what an autotransformer is and how it works. Provide a sketch of autotransformer windings and identify the terminals. Then explain how the autotransformer works and how it is different from a step-up transformer.

8. Your boss is concerned about summer brown-out at the factory where you work. Explain two ways the electrical utility company is able to meet peak demands on their generating system in the event of a brown-out.

9. Identify 5 of the NEMA enclosure types and what they are used for.

◀ Chapter 7 ▶

Manual Control Devices

Objectives

After reading this chapter you will be able to:

1. Explain the operation of a disconnect switch.
2. Explain the operation of a manual motor starter.
3. Identify the parts of a manual motor starter.
4. Explain the operation of undervoltage devices.
5. Explain the operation of a manual reduced voltage starter.
6. Troubleshoot a manual motor starter.

Manual controllers are widely used in a large variety of industrial applications in the form of manual motor starters, drum switches, and disconnects. These applications include on/off controls with job and reverse capabilities for pumps, compressors, fans, conveyors, grinders, drill presses, mixers, choppers, meat cutters, textile looms, and woodworking and metalworking machines. Manual controls are also useful for noninductive loads, such as resistive heat and lighting. Sometimes disconnects are used for both on/off control and protection of motors. The disconnect switch was introduced in Chapter 6, where it was used primarily for power distribution applications. Some additional information about the disconnect being used as a manual control is provided in this chapter.

Earliest manual controls were simple knife switches, which had a movable part and a stationary part. Figure 7–1 shows a disconnect switch. The stationary part of the switch provides two isolated terminals where wires are connected. In addition to providing a location to mount the field wires, the terminal also provides a V slot where the movable part of the switch will seat when it is closed. The movable part of the switch is called the knife blade and it is hinged at one end. The knife blade will seat tightly into the V slot at each terminal and it will make electrical contact with each terminal and provide a low-resistance path for electrical current. The knife blade is made from copper and has a nonconductive handle applied so that the switch can be safely operated manually.

The knife switch has evolved from the simple switch that was used for earliest motor control to the point where it is used in modern disconnect switches. The knowledge gained from problems with early

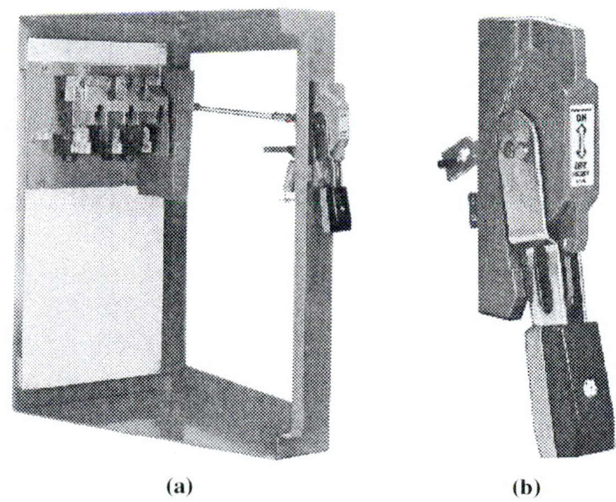

(a) (b)

Figure 7–1 (a) A simple manual control used to operate the disconnect switch and open the door of an electrical cabinet. Notice the switch can be adjusted any depth of cabinet. (b) A close-up of the disconnect switch handle. (Courtesy of Eaton/Cutler Hammer)

switches has been used to design contacts that are used in modern motor starters. This chapter will help you understand the operation of various manual controls as well as applications for which they are well suited. Wiring diagrams are provided to show distinct differences between various starters, and as an aid during field wiring and troubleshooting. Procedures for selecting the proper type of control and enclosure are also detailed. This includes methods of sizing the controls to provide adequate protection for the motor loads and noninductive loads being controlled. Installation of manual controls is also detailed. These procedures will be useful during initial installation or when the control is replaced. The final part of this chapter includes troubleshooting and diagnostic procedures for these controls and shows these controls wired directly to motors for speed control and reversing. It is assumed that you understand the principles of DC and AC motors. If you need to review information pertaining to methods of reversing motors and changing motor speeds, see Chapters 13 and 14.

7.1 Disconnect Switches

The simple type of knife switch has evolved into the modern disconnect for motor control circuits. The disconnect switch was introduced in Chapter 6, where it was used as a disconnect for power distribution systems. The information in this chapter shows applications in which the disconnect can be used as a motor control to switch equipment where the motor may be operated for long periods without being turned off. A ventilation fan is an example of this type of application. The disconnect switch will meet most electric code requirements, since it provides a means to disconnect power from the motor, and it can also be fused to provide motor overload protection. Figure 7–2 shows a picture of a three-pole disconnect switch. The electrical symbol for the disconnect is also shown. The disconnect is available as a single-pole, two-pole, or three-pole switch. The single-pole type is used for single-phase 110-V circuits, the two-pole type is used for single-phase 208/220-V circuits and DC circuits, and the three-pole switch is used for three-phase AC circuits.

The disconnect is rated for the maximum voltage and maximum current to which the contacts will be subjected. If the disconnect is a fusible type, the fuses and the disconnect must be sized to match the load. Since the disconnect will be used to disconnect and protect a motor without any other controls, the fuses must be a dual-element type and capable of allowing the motor to start while it pulls locked-rotor amperage while protecting it at its full-load amperage rating. Another useful feature that the disconnect provides is lockout protection. The disconnect can be locked in the off position with a padlock for safety or security reasons.

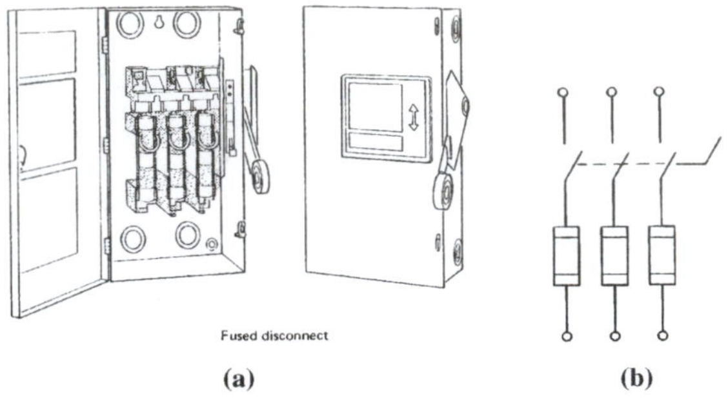

Figure 7–2 (a) A fused disconnect. (b) Electrical diagram of a fused disconnect.
(Courtesy of Square D/Schneider Electric)

In most applications the disconnect will be used primarily to satisfy the National Electric Code® (NEC) article pertaining to motor protection. This article requires that for safety, all motor-driven equipment have a disconnecting means within easy access, which provides the operator with a way to safely disconnect power to motorized equipment. The disconnect also provides a method to lock out the switch when it is turned off. This enables personnel to work on the machinery safely.

7.2 Manual Motor Starters

Several types of manual motor starters are used in a variety of applications for starting, jogging, reversing, and stopping motors. The starters can be used for control of single-phase and three-phase AC and DC motors. Motor starters also give the motor protection against overcurrent, which provides a means of safely cycling a motor on and off numerous times per hour without overheating or damaging the switching mechanism. The manual motor starter can also provide a feature that allows the motor to restart automatically after a loss of power. A picture and electrical diagram for a typical manual motor starter are shown in Figure 7–3. The motor starter is usually installed in an enclosure where its pushbutton or toggle switch is accessible to an operator.

Manual motor starters are generally used for controlling and protecting smaller motor loads or noninductive loads. Like the disconnect switch, the starter is available with a single-pole switch for single-phase,

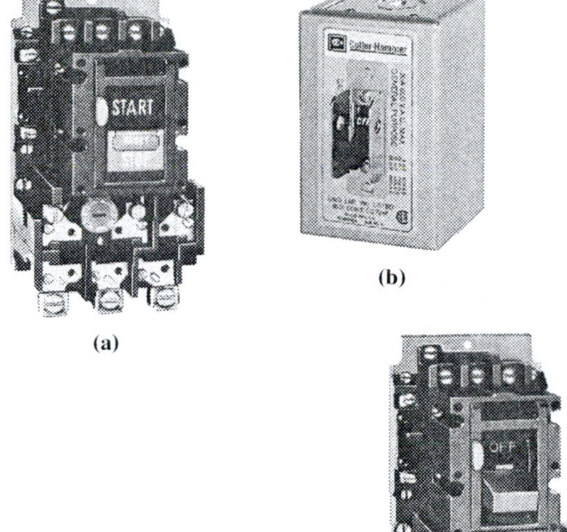

Figure 7–3 Types of switch operators for manual motor starter (a) pushbutton switch, (b) selector switch, (c) toggle switch.
(Courtesy of Eaton/Cutler Hammer)

110-V loads and a two-pole switch for 230-V loads up to 3 hp. Three-pole starters are available for loads up to 10 hp. Other types of manual starters are available for applications that require motors to be reversed and for two-speed motors. An autotransformer type is also available for reduced-voltage starting applications.

Figure 7–4 shows a typical manual motor starter and its diagram. The starter has three major parts: the base assembly, the movable contact carrier, and the operator assembly. The base assembly provides the mounting holes and plastic assembly to mount the stationary part of the contact assembly. The stationary part of the contact assembly provides a terminal point, which is a means of securing the field wiring for the line and load wires. The movable contact carrier contains the movable contacts and spring assembly as well as any undervoltage protection, and the operator assembly provides the means to activate the manual starter. The operator assembly can use pushbutton switches, selector switches, or toggle switches to activate the switch. The function of each of these parts and their operation are explained so that you will be able to install, troubleshoot, and replace major parts of the starter, or the complete device.

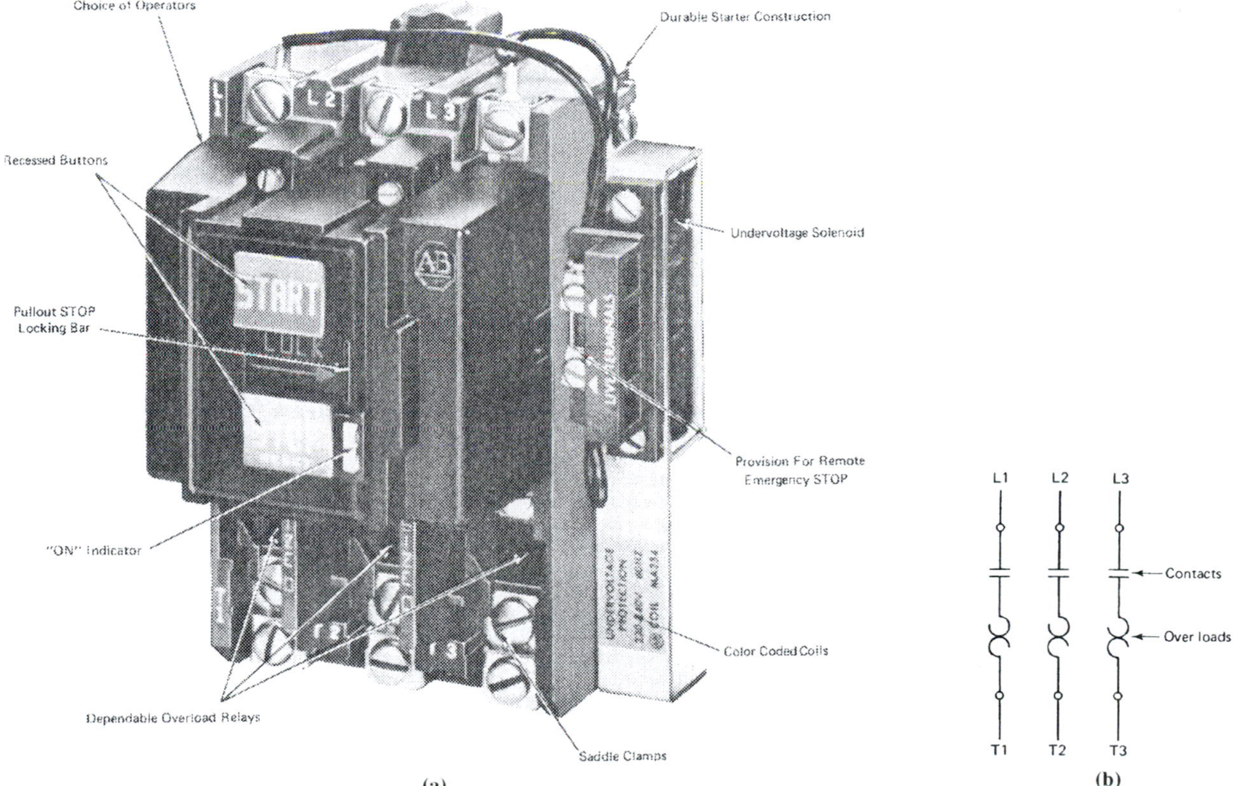

Figure 7–4 (a) Picture of manual motor starter. (b) Electrical diagram of manual motor starter.
(Courtesy of Rockwell Automation)

7.2.1 Operator Assembly, Base Assembly, and Contact Carrier

The base assembly for the manual motor starter is the largest part of the starter. It is made of a durable nonconductive plastic material that will house the stationary part of the contact assembly for the starter. It also provides a base on which to mount the operator assembly. The operator assembly is the part of the manual starter that activates the contact assembly. The operator assembly is available with a variety of switches. A pushbutton switch is available with the buttons flush mounted or recessed. Recessed pushbuttons are for applications where the buttons may be activated accidentally. Flush-mounted pushbuttons are available for applications where the switches must easily be activated by operators who must wear gloves or have other problems depressing the switches. Both the pushbuttons and contacts in the starter are of the maintained type, since the starter is activated manually. This means that when the start pushbutton is depressed, the starter will remain in the energized mode until the stop pushbutton is depressed. When the motor is deenergized by the stop button, the starter will remain in that mode until the start button is depressed again.

The toggle switch operates similarly to a single-pole switch used to control residential lighting. This type of switch is usually used for smaller starters. When the toggle switch is switched to the on position, the starter contacts will close and energize the motor. The toggle switch is also a maintained type of switch and will remain in the position to which it has been switched.

The selector switch operator is generally used for applications where the control has two or more positions. This type of switch is useful for applications where three functions are needed, such as hand, auto, and off; or forward, reverse, and off. Some manufacturers may use different names for these operators, but the function is the same.

The base assembly also provides a plastic barrier between each set of contacts. This barrier prevents arcing between sets of contacts that are of different potential, such as between L1 and L2. A cover called an *arc hood* is placed above the contacts to prevent arcing to anything outside the starter. Sometimes the potential at a set of contacts may become very large if a motor does not start or if a short circuit occurs. In these cases, a large arc may be drawn to any metal, such as an enclosure, and the arc hood prevents the arc from damaging any part of the starter assembly.

The contacts in the starter can be designed to operate in one of several ways. Figure 7–5 shows several of these designs. One type of design uses contacts that are flat (Figure 7–5a). Both the stationary and movable parts of the contacts are shown in this diagram. The stationary part of the contacts consists of two isolated contacts that have wiring terminals to connect field wiring to their terminals. The movable contacts are mounted so that they will align perfectly with stationary contacts when the movable contact carrier is depressed. The movable contact carrier is designed to prevent the contacts from twisting as they seat with the stationary contact. The contacts in the movable carrier actually float in the carrier to allow exact alignment when they seat with the stationary contact. A spring is used to provide pressure on the contacts as they close so that they will not bounce, rebound, or vibrate.

Notice that two contact points are used for each set of contacts. This arrangement is called *double-break contacts*. Double-break contacts allow larger currents and voltages to be switched on and off without causing severe arcing or overheating. The movable contact assembly also provides a snap action mechanism that causes the contacts to make (close) and break (open) quickly, which limits arcing. The contacts are manufactured from a silver alloy. Cadmium and cadmium oxide are mixed with the silver to make the best material for conduction and to prevent contact pitting. The parts of the contacts that come together to provide a circuit for current flow are made to withstand arcing from voltage and current surges that occur when the contacts are closing or opening. The metal on the tip of the contacts tends to melt and trans-

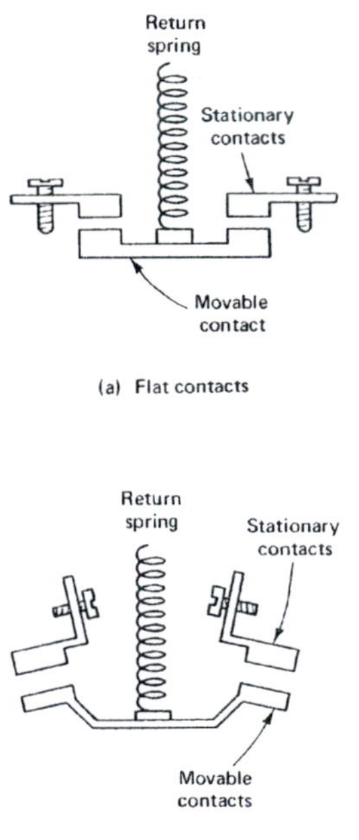

Figure 7–5 Typical contact arrangements used in manual motor starters: (a) flat contacts, (b) angle contacts.

fer from one of the contacts to the other during arcing. This reaction, very similar to arc welding, will leave a pit in one of the contacts and a deposit in the shape of a hill on the other contact. On earlier contact sets, pitting was so common that procedures for dressing or cleaning the contacts were provided.

Modern motor starters use new metal alloys that resist pitting, and the small amount of arcing that takes place will actually help clean the contact surface. The newer contacts are made from silver alloys and do not require cleaning or dressing. In fact, any time the contacts are filed or dressed with sandpaper, they tend to lose more of their plating and begin to deteriorate faster. The coatings on these modern contacts resist oxidation and can withstand continual operation. If the contacts must be cleaned, use a burnishing tool.

Another contact design used in motor starter contacts allows contacts to be mounted on a slight angle. This type of contact arrangement is shown in Figure 7–5b. The contacts are mounted at an angle to provide tension on them when they are closed. This arrangement allows the contacts to mate more easily during the travel as they come together. The contacts will operate very similarly to the flat contacts. The contacts are rated according to the amount of current that will pass through them safely. The maximum amount of voltage that can be applied to the contacts is determined by the contact rating as well as the rating of the base assembly. Figure 7–6 shows how wires are connected to the contacts with a saddle clamp or lug.

Typical voltage ratings of manual starters are 115, 200–230, and 460–570 V AC and 125 and 250 V DC. The manual starter's current rating may be identified in amperes, wattage, or horsepower. Typical current ratings are usually shown as a horsepower rating. Figure 7–7 shows typical AC and DC horsepower ratings for manual starters. You may need to review the horsepower-to-amperage conversion formula to determine exact current ratings.

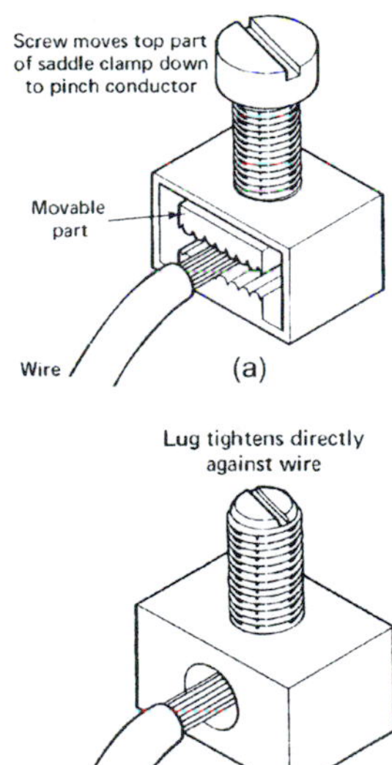

Figure 7–6 (a) Saddle clamp. (b) Box lug.

Make sure that both the voltage and current ratings for the starter exceed the actual voltage and current that is controlled to the load. The manual motor starters are also identified by size. The sizes are shown in Figure 7–7. Notice that size 00, called "double aught," is the smallest. The sizes for manual starters include 00, 0, and 1. The amount of current that each size of starter can control is dependent on the load voltage.

For example, you can see from Figure 7–7 that a size 1 starter can safely control a 1-hp single-phase motor at 115 V, or a 7.5-hp three-phase motor at 200 to 230 V, and 10 hp at 460–575 V. The reason for this difference is that motors draw less amperage at higher voltages and the contacts can withstand a specific maximum amount of current safely as long as the voltage rating is not exceeded. For example, a 3-hp motor at 115 V will draw 34 A, a 7.7-hp motor at 230 V will draw 40 A, and a 10-hp motor at 460 V will draw 14 A. Each of these loads can be controlled safely by a size 1 starter.

The contacts in manual motor starters are "normally open" contacts. The number of sets of contacts is determined by the application for the switch. Figure 7–8 shows the diagram for several types of manual motor starters. The motor starter in Figure 7–8a with a single set of normally open contacts is used for single-phase 110-V circuits. The motor starter is required only to break (open and close) the line side of the circuit. The neutral side of the motor circuit should not be broken. The motor starter in Figure 7–8b that has two sets of normally open contacts is used for 208/220- or 480-V AC single-phase applications or DC voltage applications. In single-phase AC voltages, both of the hot lines should be broken by the contacts. The motor starter in Figure 7–8c uses three contacts to break each of the three-phase lines.

7.2.2 Overload Device

The manual motor starter provides a device to detect and prevent motor overloads. Any motor is capable of drawing current in excess of its rated load. This condition is called an *overload*. It can occur because of bad bearings or from trying to turn a load that has become too large, such as a conveyor or a mixer being overloaded. When the motor draws too much current, its insulation will become overheated and begins to break down and deteriorate to the point of damaging the motor. Another damaging problem may occur even though the motor current is controlled within tolerance when a

Figure 7–7 Maximum horsepower ratings for NEMA size starters.
(Translated and reproduced with permission of the National Electrical Manufacturers Association. This standard may not be further reproduced. Only the original English version is valid for any legal or judicial consequences arising out of the application of the standard.)

Single-phase motors

NEMA Size	Full-Voltage Starting 115 V	230 V
00	1/3	1
0	1	2
1	1	3
$1\frac{1}{2}$	3	5
2	—	$7\frac{1}{2}$
3	—	15

Polyphase motors

NEMA Size	Full-Voltage Starting 200 V	230 V	460 V, 575 V
00	$1\frac{1}{2}$	$1\frac{1}{2}$	2
0	3	3	5
1	$7\frac{1}{2}$	$7\frac{1}{2}$	10
2	10	15	25
3	25	30	50
4	40	50	100
5	75	100	200
6	150	200	400
7	—	300	600
8	—	450	900
9	—	800	1600

motor begins to overheat from starting and stopping too frequently.

The overload device in the manual motor starter is designed to check for too much current and open the motor starter circuit before the motor windings become damaged. Several types of overload devices are used in modern motor starters. Figure 7–9a shows a picture of a single-pole overload device.

There are two basic types of overloads. One type uses a solder pot to sense the heat that is built up from excessive motor current. This type of control is called *solder pot overload* or *melting alloy thermal overload*, and it has several main parts: the solder pot, the heater (a wire that produces heat when current flows through it), the ratchet shaft with rewind spring, the overload contacts with return spring, and the reset button. The ratchet shaft is mounted in the solder pot, and it has spring tension applied to it that tries to make it rotate. Since the shaft is secured in solder and the solder is solidified, it cannot turn. The teeth of the ratchet are beveled so that they catch the end of the contact actuator lever. The rewind spring on the ratchet shaft keeps pressure on the contact actuator lever so that it cannot move. When the contact actuator lever is secured in place by the ratchet, it keeps pressure on the contacts, which keeps them closed. When the contacts are closed, they compress a very strong spring. This spring is constantly exerting pressure to try and open the contacts. Any time the solder becomes soft, the ratchet shaft will become free from the solder and the rewind spring will cause the ratchet to spin. When the ratchet spins, the contact actuator lever is released and the contact return spring forces the contacts open.

The overload contacts are connected in series with the motor. A heating element called a heater is also connected in this series circuit. This means that any current that flows to the motor must also pass through the heater. The heater is physically mounted so the heat that it generates is conducted directly to the solder pot. When the motor draws normal current the heater does not generate enough heat to melt the solder in the solder pot. But when the motor draws excessive current, the heater generates enough heat to cause the solder to begin to melt and release the ratchet shaft. When the ratchet shaft spins, the contact lever is released and the contact return spring forces the overload contacts open. Since the motor is connected in series with the overload contacts, the motor is deenergized. Figure 7–9b

L1 N

T1

1ϕ
Motor
115 V

(a)

L1 L2

T1 T2

1ϕ
Motor
208 V

(b)

L1 L2 L3

T1 T2 T3

3ϕ
Motor
208 V

(c)

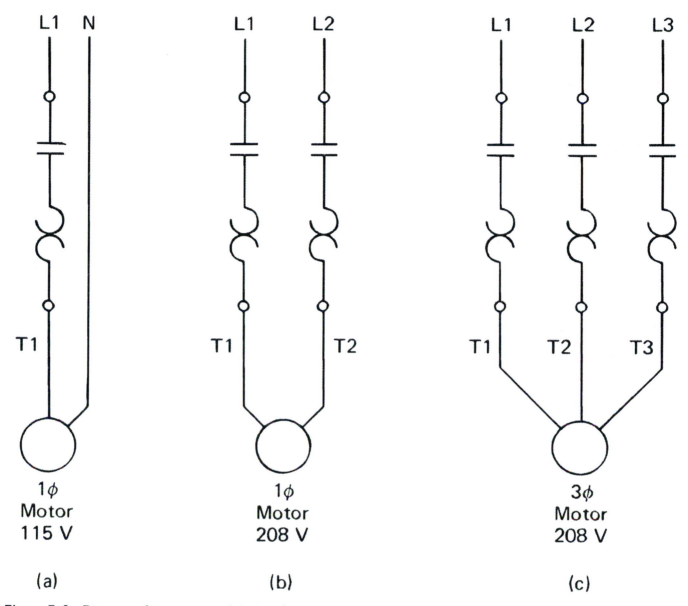

Figure 7–8 Diagrams of one-, two-, and three-pole motor starters.

shows type W heaters, and Figure 7–9c shows type J heaters.

The motor starter overload can be reset by pressing on the reset button. The reset button rotates the ratchet shaft so that it rewinds the shaft spring. The contact actuator lever is allowed to return to its reset condition, where it is again resting against the ratchet teeth. If the reset button is pressed too soon, the solder will not be solidified enough to hold the ratchet shaft, and the shaft spring will again spin the ratchet and allow the contacts to open. Sometimes the overload must be allowed to cool down for several minutes before it can be reset.

Since the overload contacts should open only when the motor is actually drawing too much current, the

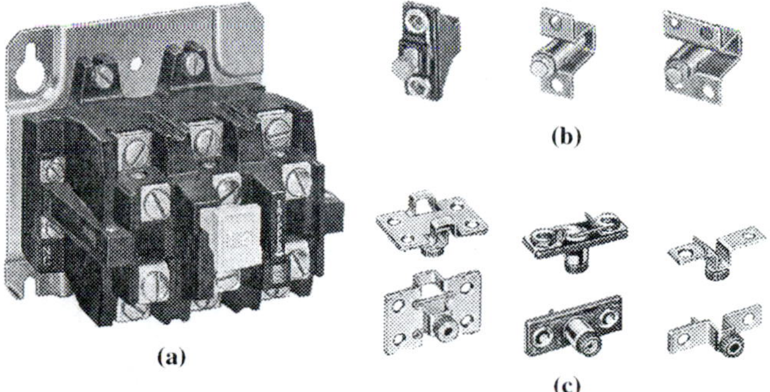

Figure 7–9 (a) An overload block for a manual motor starter. (b) Type W heaters. (c) Type J heaters for manual motor starters.
(Courtesy of Rockwell Automation)

TABLE 144

Heater Element Cat. No.	Full Load Amperes				
	Size 0	Size 1	Size 2	Size 3	Size 4
W10	0.18	0.18	—	—	—
W11	0.20	0.20	—	—	—
W12	0.22	0.22	—	—	—
W13	0.24	0.24	—	—	—
W14	0.26	0.26	—	—	—
W15	0.29	0.29	—	—	—
W16	0.32	0.32	—	—	—
W17	0.35	0.35	—	—	—
W18	0.38	0.38	—	—	—
W19	0.42	0.42	—	—	—
W20	0.46	0.46	—	—	—
W21	0.51	0.51	—	—	—
W22	0.56	0.56	—	—	—
W23	0.62	0.62	—	—	—
W24	0.68	0.68	—	—	—
W25	0.75	0.75	—	—	—
W26	0.82	0.82	—	—	—
W27	0.90	0.90	—	—	—
W28	0.99	0.99	—	—	—
W29	1.09	1.09	—	—	—
W30	1.20	1.20	—	—	—
W31	1.32	1.32	—	—	—
W32	1.45	1.45	—	—	—
W33	1.59	1.59	—	—	—
W34	1.75	1.75	—	—	—
W35	1.93	1.93	—	—	—
W36	2.12	2.12	—	—	—
W37	2.33	2.33	—	—	—
W38	2.56	2.56	—	—	—
W39	2.81	2.81	—	—	—
W40	3.09	3.09	—	—	—
W41	3.40	3.40	—	—	—
W42	3.74	3.74	—	—	—
W43	4.11	4.11	—	—	—
W44	4.52	4.52	—	—	—
W45	4.97	4.97	—	—	—
W46	5.46	5.46	5.60	—	—
W47	6.01	6.01	6.15	—	—
W48	6.60	6.60	6.76	—	—
W49	7.26	7.26	7.43	—	—
W50	7.98	7.98	8.17	—	—
W51	8.78	8.78	8.98	—	—
W52	9.65	9.65	9.87	—	—
W53	10.6	10.6	10.8	—	—
W54	11.7	11.7	11.9	—	—
W55	12.8	12.8	13.1	—	—
W56	14.1	14.1	14.4	—	—
W57	15.4	15.4	15.7	—	—
W58	16.8	16.8	17.1	—	—
W59	18.3	18.3	18.6	—	—
W60	—	19.8	20.1	—	—
W61	—	21.3	21.7	25.5	—
W62	—	22.7	23.1	28.1	—
W63	—	24.4	24.8	31.0	32.0
W64	—	26.2	28.6	34.0	35.0
W65	—	28.2	30.5	37.0	38.5
W66	—	—	33.0	40.0	42.5
W67	—	—	35.5	43.5	46.5
W68	—	—	38.0	47	51
W69	—	—	40.5	51	55
W70	—	—	43.5	55	59
W71	—	—	47.0	59	64
W72	—	—	—	63	69
W73	—	—	—	67	74
W74	—	—	—	71	79
W75	—	—	—	76	84
W76	—	—	—	80	90
W77	—	—	—	85	96
W78	—	—	—	90	102
W79	—	—	—	—	107
W80	—	—	—	—	113
W81	—	—	—	—	118
W82	—	—	—	—	124
W83	—	—	—	—	130
W84	—	—	—	—	135
W85	—	—	—	—	—

TABLE 55

Heater Element Cat. No.	Full Load Amperes					
	Size 00	Size 0	Size 1	Size 1P	Size 2	Size 3
JJ14	0.22	0.22	0.22	—	—	—
JJ13	0.24	0.24	0.24	—	—	—
JJ12	0.27	0.27	0.27	—	—	—
JJ11	0.30	0.30	0.30	—	—	—
JJ10	0.33	0.33	0.33	—	—	—
JJ9	0.36	0.36	0.36	—	—	—
JJ8	0.40	0.40	0.40	—	—	—
JJ7	0.44	0.44	0.44	—	—	—
JJ6	0.48	0.48	0.48	—	—	—
JJ5	0.53	0.53	0.53	—	—	—
JJ4	0.58	0.58	0.58	—	—	—
JJ3	0.65	0.65	0.65	—	—	—
JJ2	0.71	0.71	0.71	—	—	—
JJ1	0.78	0.78	0.78	—	—	—
J1	0.87	0.87	0.87	—	—	—
J2	0.95	0.95	0.95	—	—	—
J3	1.05	1.05	1.05	—	—	—
J4	1.16	1.16	1.16	—	—	—
J5	1.28	1.28	1.28	—	—	—
J6	1.41	1.41	1.41	—	—	—
J7	1.55	1.55	1.55	—	—	—
J8	1.70	1.70	1.70	—	—	—
J9	1.87	1.87	1.87	—	—	—
J10	2.06	2.06	2.06	—	—	—
J11	2.27	2.27	2.27	—	—	—
J12	2.51	2.51	2.51	—	—	—
J13	2.78	2.78	2.78	—	—	—
J14	3.07	3.07	3.07	—	—	—
J15	3.38	3.38	3.38	—	—	—
J16	3.72	3.72	3.72	—	—	—
J17	4.10	4.10	4.10	—	—	—
J18	4.52	4.52	4.52	—	—	—
J19	4.98	4.98	4.98	—	—	—
J20	5.49	5.49	5.49	—	—	—
J21	6.04	6.04	6.04	—	—	—
J22	6.66	6.66	6.66	—	—	—
J23	7.35	7.35	7.35	—	—	—
J24	8.13	8.13	8.13	—	—	—
J25	8.96	8.96	8.96	—	—	—
J26	9.90	9.90	9.90	—	—	—
J27	—	10.9	10.9	11.0	—	—
J28	—	12.0	12.0	12.2	—	—
J29	—	13.2	13.2	13.4	—	—
J30	—	14.6	14.6	14.8	—	—
J31	—	16.1	16.1	16.3	—	—
J32	—	17.7	17.7	17.9	18.3	—
J33	—	—	19.5	19.8	20.2	—
J34	—	—	21.4	21.8	22.2	—
J35	—	—	23.6	24.0	24.4	—
J36	—	—	26.0	26.4	26.9	—
J37	—	—	28.5	29.0	29.8	—
J38	—	—	—	32.0	33.0	—
J39	—	—	—	35.0	36.5	40.5
J40	—	—	—	38.5	40.5	45.5
J41	—	—	—	—	45.5	51
J42	—	—	—	—	—	56
J43	—	—	—	—	—	62
J44	—	—	—	—	—	68
J45	—	—	—	—	—	74
J46	—	—	—	—	—	82
J70	—	—	—	—	—	90
J71	—	—	—	—	—	—
J72	—	—	—	—	—	—
J73	—	—	—	—	—	—
J74	—	—	—	—	—	—

Figure 7–10 (a) Selection tables for type W heaters for overloads. (b) Selection tables for type J heaters for overloads.
(Courtesy of Rockwell Automation) (Courtesy of Rockwell Automation)

overload heater must be sized to the exact amount of current the motor is drawing during full-load conditions. The heater and overload unit are manufactured as a single unit so that they are easy to size and install. This also makes the overload tamperproof.

The overload and heater are widely used for motor protection because they can be reset without changing any parts. This allows the electrician to determine the cause of the overload and reset the motor starter after the condition has been cleared. Sometimes the condition that is causing the motor to overheat is not readily apparent, so the electrician will reset the motor starter and observe the motor when it is energized again. This may include taking a full-load motor amperage reading with an ammeter. If the motor continues to trip the overload device, the electrician must determine if the motor is faulty or if the load has changed in some way. This test procedure requires the load to be disconnected from the motor to see if it is still drawing overload current. If the overload disappears, the load is at fault. If the overload continues, the motor is at fault.

The procedure for correctly sizing motor starter heaters to specific motor and load applications is very important. The heater cannot be expected to protect the motor if it is not sized properly. The sizing procedure involves determining the motor horsepower, the full-load motor amperage, and the temperature conditions around the motor and motor starter. A table for sizing the heaters is shown in Figure 7–10.

The second type of overload device is called a *bimetallic thermal overload relay.* This type of overload relay is not used as often as the solder pot type of overload, but you should understand its operation. The bimetallic thermal overload uses a U-shaped bimetal strip as the heater for the thermal overload. The bimetal strip has an additional variable-resistance element that will allow the heater to be precisely adjusted to the full-load motor current. The bimetal heater is wired in series with the motor. When the motor is drawing normal current the heater will become slightly warm, but the heat is not enough to cause the U-shaped element to change position. If the motor begins to draw excessive heat, the U-shaped element begins to move and the contacts begin to open. When the element has moved enough, the contacts will snap open crisply and deenergize the motor. The overload is designed to snap the contacts open so that they do not arc excessively and wear out prematurely.

The bimetal type of overload can be converted in the field for manual reset or automatic reset. If the overload is converted for automatic reset, the contacts of the U-shaped element are allowed to return to the closed state when the element cools. Automatic reset allows motors and controls that are mounted in a remote or inaccessible location to be restarted without being serviced. A set of alarm contacts are provided to energize an indicator lamp to alert maintenance personnel that an overload condition is occurring. The alarm contacts may also be used where continuous processes are controlled. In an application such as continuous pouring of glass, the process should not be interrupted. In this application, if an overload is sensed, the circuit is wired so that the motor contacts cannot deenergize the motors. Instead, the alarm contacts will energize an indicator lamp, which will warn the operator that an overload exists. If the overload continues for several minutes, maintenance personnel are called. The motors are deenergized at the end of the pouring cycle and checked for faults. In some applications the process is so expensive that some damage to the motor may be acceptable so that the pouring process can be completed.

Some applications using manual starters can also use dual-element fuses to protect the motor against short-circuit current and slow overcurrents. The inverse time characteristics of the dual-element fuse were explained in Chapter 6. You may need to review this information to better understand the methods of motor protection and control that are available.

7.2.3 Undervoltage Protection

Motor starters also are available that can provide undervoltage protection.This type of control actually provides loss of voltage protection as well as detecting small undervoltages. Figure 7-11 shows an example of a manual motor starter with under voltage protection. This type of protection is required by OSHA and ANSI for many applications where a motor must be reset manually if the motor stops for any reason. Since the motor starter uses a manual switch operator such as a pushbutton or toggle switch, the contacts are maintained and they will stay in the condition (open or closed) where they were last actuated. For example, in drill-press or woodcutting applications, severe damage could occur if the motor stops for any reason and is suddenly restarted. This could occur if power was lost for a few seconds during a power outage or lightning storm. The loss-of-voltage device will trip the motor starter contacts to their open position even though the start pushbutton is depressed or if the toggle switch is in the on position. This trip action requires someone to come to the motor starter and manually reset the start button to allow the motor to restart. When the operator comes to restart the motor, a check can be made of the condition of the work in the cutter blade so that the work can be repositioned prior to restarting the motor, to prevent damage to the machine. If the machine were allowed to restart on its own, the material in the cutter

Figure 7–11 Mechanical diagrams of a manual motor starter with undervoltage protection.
(Courtesy of Rockwell Automation)

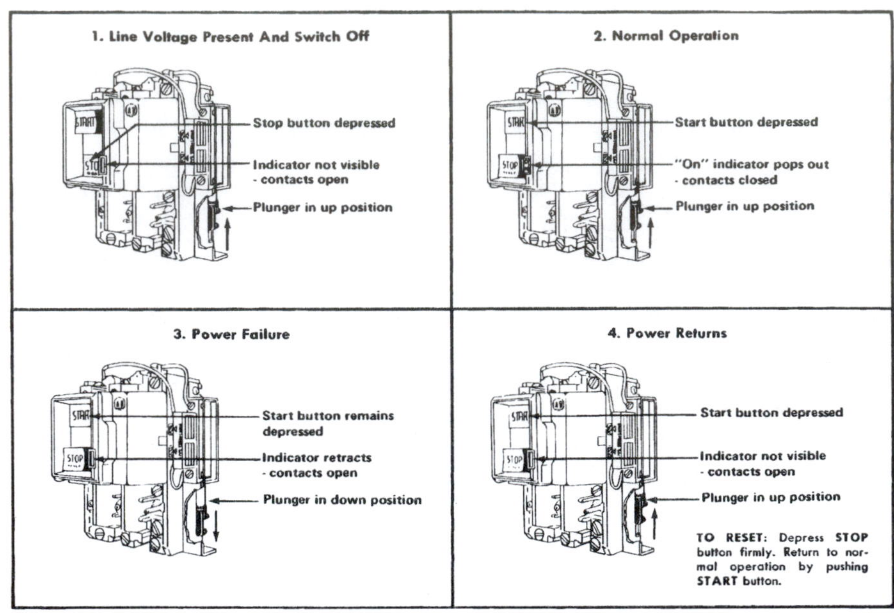

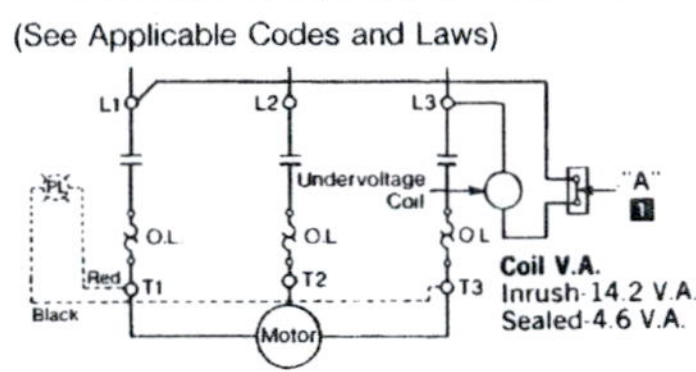

blade might be out of position and become damaged, or the cutter head might become jammed and damage the machine or malfunction and cause an accident.

The undervoltage sensor can also detect substantial drops in voltage caused by large motors and spot-welding loads cycling on and off. The lowering of voltage may also occur during a brown-out condition. This may occur during peak usage periods in large industrial and metropolitan areas, such as during lengthy warm periods in the summer. During this time, residential usage may cause utility supplies to drop 10 to 15 percent. This lowering of voltage will cause motors to overheat. If the motor is susceptible to overheating, the motor should be protected by an undervoltage monitor. In these cases the motor starter would be deenergized by the undervoltage sensor even though power was not lost completely. The motor starter would remain deenergized until manually reset and the undervoltage condition cleared.

A continuous-duty solenoid has been added to the manual motor starter to provide undervoltage protection. The solenoid is wired directly across the line-side terminals, such as L1 and L2. This allows the solenoid to be energized any time that power is applied to the motor starter. If voltage becomes low, the solenoid allows the motor starter to drop out.

The start and stop buttons on the manual starter are both maintained switches. This means that whichever switch is depressed, it will stay in that position until the other switch is depressed. The two switches are mounted on a toggle actuator that is similar to a seesaw. When the start button is depressed, the mechanism pushes the stop button to the off position. When the stop button is depressed, it will automatically cause the start button to move to the off position. This action maintains the switch that is depressed until the other switch is activated.

When power is applied and the start button is depressed the "on" indicator on the front of the motor starter is showing, which indicates that the contacts are closed and the motor is energized and running. The undervoltage solenoid's plunger is in the up position because power is still applied to the line side of the motor starter.

When the undervoltage device has detected a loss of voltage, it drops out the contacts even though the start button is still depressed. The undervoltage solenoid's plunger drops to the down position because it has detected a loss of voltage and it will remain in the down position even if supply voltage returns, until the starter is manually reset. You would be able to determine that the undervoltage device has tripped because the starter is deenergized and the motor is no longer running even though the start button remains depressed. The "on" indicator would also be retracted so that it was not visible.

To reset the starter after the undervoltage device has detected the loss of voltage, the stop button must be firmly depressed. The motor can then be restarted at any time by again depressing the start button. Remember that the undervoltage solenoid will not be returned to the up position if power has not returned to the line side of the motor starter, and the starter cannot be reset.

An emergency stop switch can be added to the undervoltage monitor circuit. Since the undervoltage solenoid is activated electrically, a remote stop switch with normally closed contacts can be connected in series with the undervoltage solenoid. Any time the stop switch is activated, the contacts will open and deenergize the undervoltage solenoid. Once the solenoid is deenergized, the motor starter must be manually reset, just as if a voltage loss has been detected. Additional stop switches can also be added to this series circuit as needed. If any of the stop switches are depressed, current to the solenoid will be interrupted and the motor starter would need to be reset manually.

7.2.4 Types of Manual Starters

Several types of manual starters are available for a variety of applications. These starters provide protection for single-phase and three-phase AC motors and DC motors. Figure 7–12 shows an electrical diagram for each of these types of starters. In Figure 7–12a a single-phase, single-pole starter is shown. This starter is used to control fractional-horsepower motors that operate on 115 V. Notice that only L1 has a set of contacts and an overload device. In accordance with the National Electric Code the neutral line is not allowed to have contacts or be interrupted.

The starter shown in Figure 7–12b is a two-pole starter used to control single-phase motors that operate on 208–220 V. Both lines L1 and L2 have a set of contacts to disconnect power from the motor, but only one overload is required because the motor current that flows to the motor through L1 must return through L2. Since this creates a series circuit, the single overload can sense any increased current and protect the motor. Figure 7–12c shows a two-pole starter with a pilot light

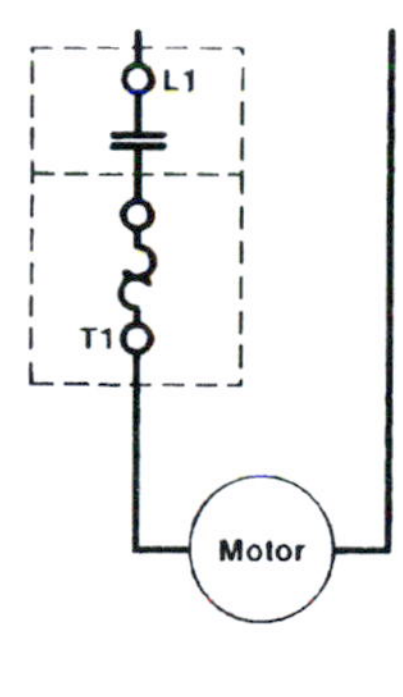

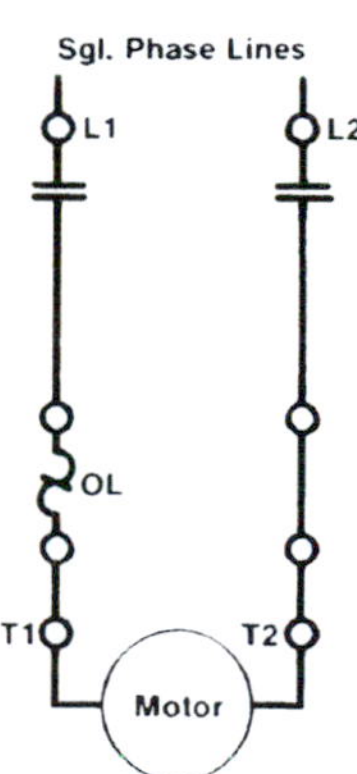

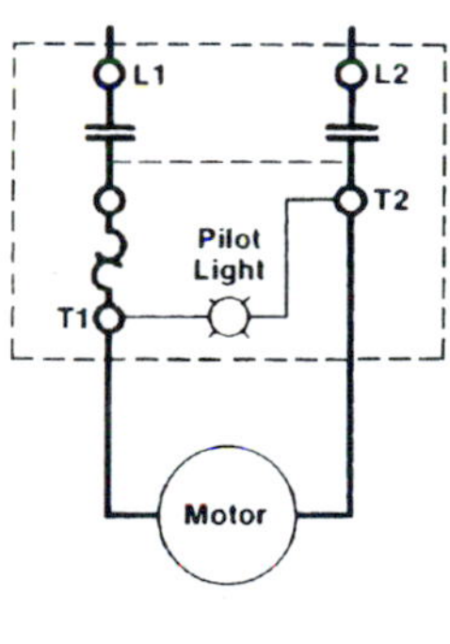

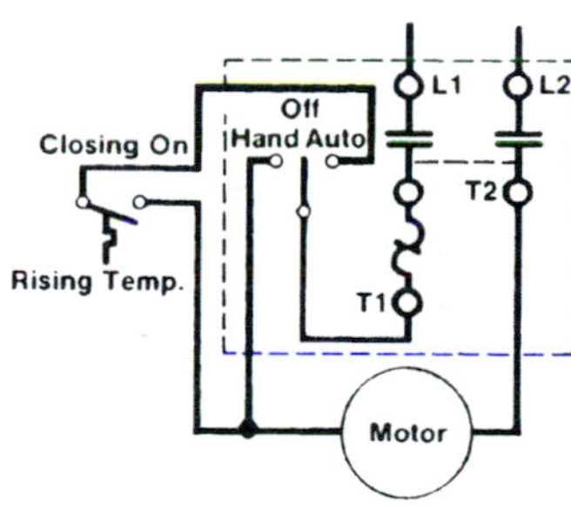

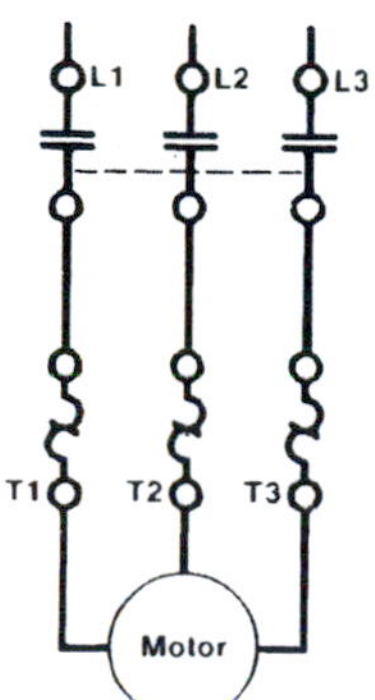

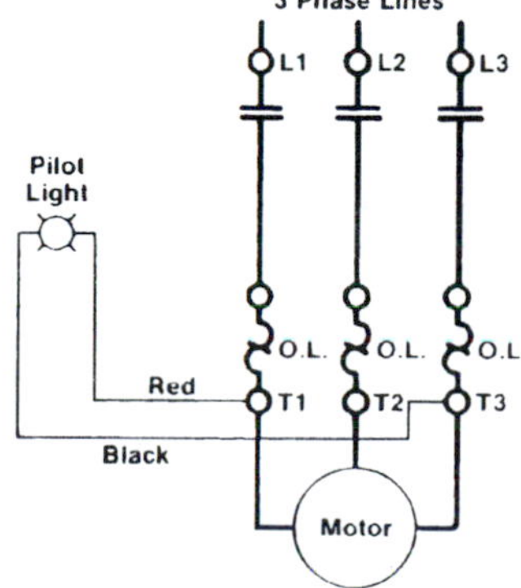

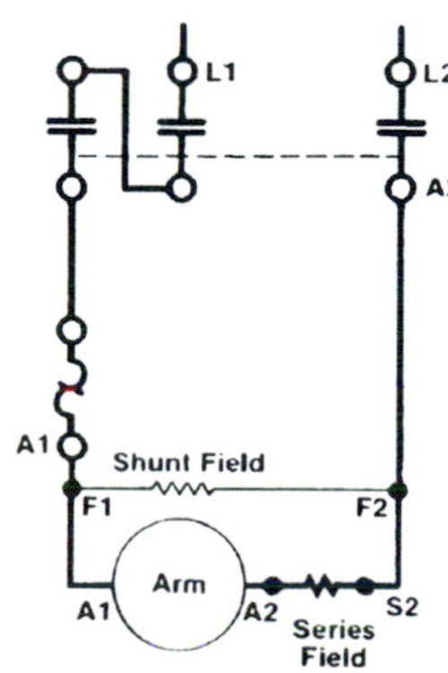

Figure 7–12 Diagrams of one, two, and three pole manual motor starters.
(Courtesy of Rockwell Automation)

connected across the load-side terminals. In this configuration, the pilot lamp will be illuminated any time the motor is energized. The pilot lamp is mounted on the cover of the motor starter enclosure so that it is visible at a distance. Another variation of the two-pole motor starter connects a selector switch and pilot device in series with L1. When the selector switch is in the hand position, the motor is connected directly to L1 and will run any time the starter is energized. When the selector switch is moved to the auto position, the pilot device is connected in series with L1. This means that the motor will run only when the starter is energized and the operating condition on the pilot device has been reached. In the case of the temperature switch shown in the diagram, the motor will run any time the temperature exceeds the temperature set point on the switch. This control is wired to control a cooling fan, which will be energized any time the temperature exceeds the set point. This type of application could utilize any type of pilot device as long as its contacts matched the current and voltage requirements of the motor. In another typical application for this type of control circuit, a level switch is used to control a sump pump. Any time the level in the sump reaches the set point, the pilot switch closes and energizes the pump motor. The motor starter provides overload protection for the pump motor in this application. The diagram in Figure 7–12d shows a single-pole switch utilized where only one line of the source voltage is broken.

The motor starter in Figure 7–12e shows a three-pole switch. This type of starter is used to control three-phase motors. Notice that an overload is provided for each of the three lines. The dashed lines below the three sets of contacts indicate that the three contacts operate as a set. This means that each of the three contacts will always be in the same condition, either open or closed. Figure 7–12f shows a three-pole starter with a pilot lamp connected across the load-side terminals T1 and T3. This lamp will be illuminated any time the motor is energized.

Figure 7–12g shows a DC motor starter. Only one overload is required since the two motor lines make a series circuit. The shunt field is wired directly across the two motor starter terminals marked A1 and A2. This configuration places the shunt field directly across the applied voltage so that the motor will see maximum starting torque.

Any starter may have a pilot light assembly added after it has been installed. The pilot light should always be wired to the load-side terminals to indicate the status of the motor starter. Any time the motor starter's contacts are energized, the pilot light will be illuminated, indicating that the switch is operating correctly and that the motor is energized. If the motor starter is deenergized for any reason, the indicator lamp will also be deenergized. This allows an operator to determine the status of a machine from some distance away. This is useful in a noisy environment when an operator is responsible for keeping several machines operating at the same time. The lamps are available with red or green lenses to indicate that the motor is energized or deenergized.

Several styles of manual motor starters are available with a locking guard that allows the handle to be locked in the off position. This provides security when the system must be deenergized for critical operations or maintenance. This also ensures that only qualified personnel may operate the system.

7.2.4.1 TWO-SPEED AND REVERSING MANUAL MOTOR STARTERS

Manual starters are also available for two-speed motor applications. Figure 7–13 shows the diagram for a two-speed motor controlled by a manual motor starter. Notice that the two-speed motor starter has two separate sets of contacts. One of these sets is for the low-speed winding and the other set is for the high-speed winding. These contacts are controlled by a toggle or selector switch that has three positions: high speed, low speed, and off. A mechanical interlock is used to prevent both high-speed and low-speed contacts from being closed or energized at the same time. This interlock automatically toggles the two sets of contacts to opposite conditions. This type of starter is useful for applications where two-speed fans are used.

Another specialized type of manual motor starter is a reversing starter. A diagram of an application using a reversing starter is also shown in Figure 7–13. This type of starter also utilizes two separate sets of contacts. One set of contacts allows the motor windings to receive voltage directly from the line-side terminals, which will cause the motor to run in the forward direction. The other set of contacts will reverse two phases of the three-phase voltage line voltage for a three-phase motor, or reverse the start winding for single-phase motors. If the motor starter is for reversing DC motors, the second set of contacts will reverse the polarity of the line-side voltage for DC series type and reverse the field windings for shunt and compound DC motors. In some DC motor applications resistors are added to protect the motor when it is reversed. If you need more review of reversing AC and DC motors, refer to Chapters 10 and 11.

7.2.4.2 MANUAL REDUCED-VOLTAGE STARTERS

Another specialty type of manual starter allows motors to be started by reducing voltage to decrease the amount of line current that is drawn when a motor is started. This type of control uses an autotransformer to

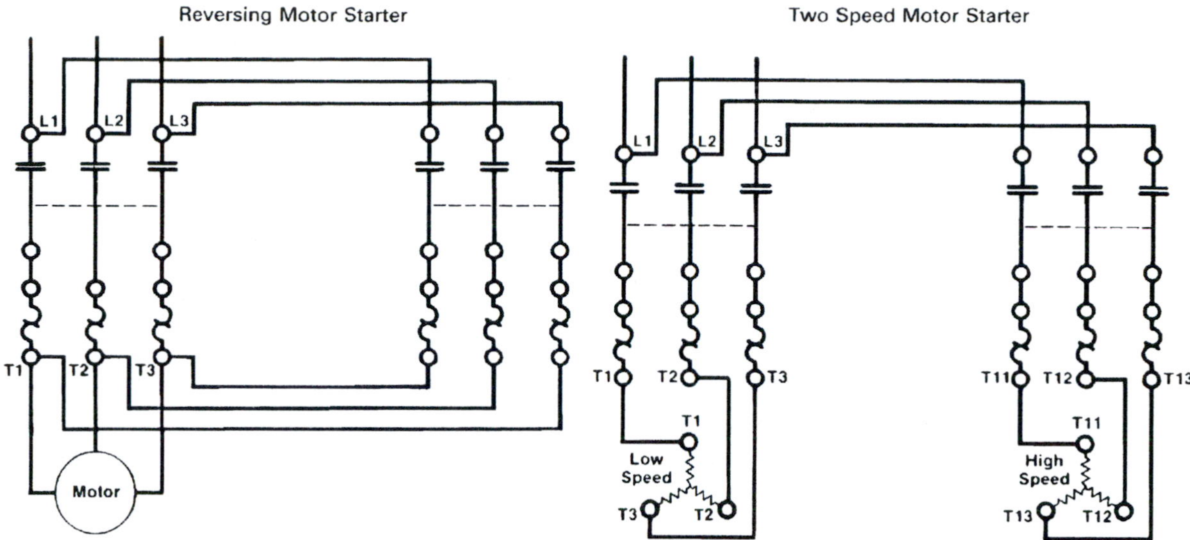

Figure 7–13 Diagrams of two-speed and reversing manual motor starter.
(Courtesy of Rockwell Automation)

allow larger current to flow in the secondary winding of the autotransformer and in the motor winding while current is reduced in the supply line. The autotransformer provides the function of a step-down transformer. By decreasing the secondary voltage to the motor, it will draw less current in the primary side of the circuit. The smaller primary current will keep the overall current demand lower so that the switchgear and other control components will last longer. It will also provide a savings in the electric bill by keeping the demand factor lower.

The autotransformer type of control provides better torque during motor starting than a resistive type of reduced voltage starter because resistors decrease the starting current as well as the starting voltage. When a motor loses current, it will also lose torque.

To understand fully the principles of motor starting and the need for reduced-voltage starting that is used in this type of control, you may need to review motor principles in Chapter 14. The autotransformer reduced-voltage starter uses a three-position switch for normal operation. When the operator is ready to start the motor, the switch is moved to the start position. In this position the switch connects the three-phase supply voltage lines to terminals L1, L2, and L3. Terminals L1 and L3 are each connected to their own transformer at the terminal marked 100. L2 is connected to the transformer terminal marked 0. The 0 terminal on the two transformer windings are jumpered and connected to terminal T2. T2 is connected directly to the motor without using any contacts. Terminals T1 and T3 are connected to the 65-percent tap on their respective winding. When the switch is in the start position, the contacts connect T1 and T3 to the motor windings.

In the starting configuration the motor receives reduced voltage. Since this voltage is reduced through a transformer, the current through the transformer is increased. The net effect is that the motor receives reduced voltage with increased current, whereas the three-phase supply actually uses less current. This allows the motor to be started without causing the demand meter to register a large current demand. After the motor begins to accelerate, the operator can turn the switch to the run position. This usually occurs when the motor reaches near full speed. When the switch is changed to the run position, the contacts disconnect the motor from the transformer and reconnect it directly across the line voltage. Notice that the overloads are in this circuit and will be used only when the starter is in the run position. This configuration keeps the overloads out of the starting circuit, where current will be large and tend to overheat them.

Any time the operator wants to stop the motor, the motor starter switch is moved to the stop position. This action disconnects the motor windings from voltage provided by either running or starting circuitry. An undervoltage solenoid is used to protect the motor from trying to restart if a power failure occurs or if the motor is deenergized at any time. One or more remote stop switches can be connected in series with this solenoid to provide safety for personnel who work near the motor.

An ammeter may also be installed on the motor starter to provide the operator with an indication of when the motor has reached sufficient speed to change the switch from start to run. A time-delay relay is also available to prevent the undervoltage solenoid from tripping if voltage dips for a few seconds while other motors are starting. If the brown-out or low-voltage condition exists longer than the setting on the time-delay relay, the undervoltage relay will trip and protect the motor.

This type of control uses air-break contacts or oil-immersed contacts. These contacts are rated for the high interruption capacity required for starting larger motors. Oil-immersed contacts are rated only for NEMA type 1, general-purpose application and are not usable where explosive conditions exist. The oil-immersed contacts utilize heavy copper contacts that work well in oil. The oil helps to keep the contacts cool over long periods of operation where motors are cycled frequently. The duty cycle for this type of control is critical. A duty cycle based on setting the autotransformer taps at 65 percent with an inductive load where the current is three times the full-load current would allow one start every 4 minutes, up to a total of four starts. Then the motor must be rested for 2 hours. The time required for the motor to start and reach running speed should not exceed 15 seconds at any time. Since these are maximum ratings, the motor can easily be started two or three times over a longer period without damage, or started several times and then allowed to remain running for a while. Remember, it is when the motor is being started that the large inrush current creates enormous amounts of heat in the motor and starter mechanism. In industrial applications, it is important to identify the number of starts that a motor is expected to have during normal use; the motor-starting equipment can be sized accordingly.

7.3 Manual Drum Switches

Manual drum switches are used for reversing single-phase and three-phase AC squirrel-cage motors and DC shunt and compound motors. The drum switch is also available for controlling multispeed motors with two to four speeds. This type of switch is capable of controlling 7-hp AC motors at 200–300 V, 7.7-hp AC motors at 460–575 V, and 2-hp DC motors at 115–230 V. If larger motors are used, a reversing magnetic starter should be used in place of the drum switch. One application of the drum switch is for reversing motors such as overhead hoists that are used to move heavy objects, such as press dies and molds. The hoists must have the ability to reverse a motor quickly to handle these loads safely. Other uses for the drum switch include controlling multispeed motors in applications such as blowers, mixers, compressors, pumps, tumblers, and special machine tools.

Handle end

Reverse	Off	Forward
1 o——o 2	1 o o 2	1 o o 2 (1 connected to 3, 2 connected to 4)
3 o——o 4	3 o o 4	3 o o 4
5 o——o 6	5 o o 6	5 o——o 6

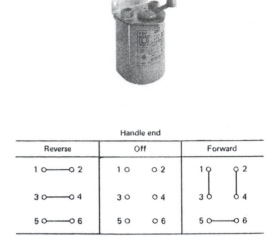

Figure 7–14 (a) Picture of drum switch (manual control). (b) Diagram of drum switch.
(Courtesy of Square D/Schneider Electric)

The drum switch can be used to control these types of motors directly, but it does not include any type of motor overcurrent protection or undervoltage protection. This means that the motor must be protected by a fused disconnect or magnetic starter. Several options are also available with most drum switches to provide interlocks with the magnetic starter to protect the motor against these perils.

Figure 7–14 shows a photograph and switching diagrams of a reversing drum switch. Notice that the switch has three sets of contacts with six separate terminals. The contacts are numbered 1 through 6 and are switched with a three-position selector. These positions include forward, reverse, and off. In the reverse position, each of the terminals on the left side makes contact with the terminal directly opposite to provide three separate circuits. This connects 1 to 2, 3 to 4, and 5 to 6. When the switch is in the forward position, contacts 1 and 2 are connected to the terminal directly below it in the diagram. This connects 1 to 3 and 2 to 4. Terminal 5 remains connected to terminal 6 just as when the switch was in the reverse direction. In the off position, the six terminals are isolated from each other.

Figure 7–15 shows several diagrams of a single-phase AC motor, a three-phase AC motor, and DC shunt and compound motors being reversed by a drum switch. These diagrams show the switch in its most commonly used applications. Notice in the diagram for a single-phase AC motor (Figure 7–15a) that the start winding must be reversed for the motor's rotation to be reversed. The start winding is connected to terminals 2 and 5, and the run winding is connected to terminals 1

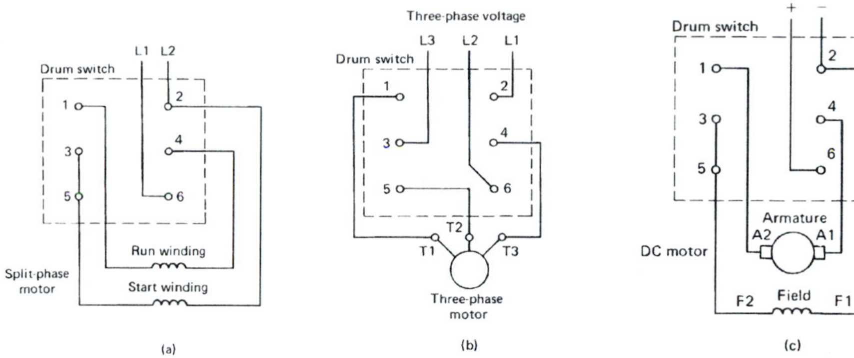

Figure 7–15 (a) Electrical diagram of split-phase motor connected to a drum switch. (b) Electrical diagram of three-phase motor connected to a drum switch. (c) Electrical diagram of a DC motor connected to a drum switch.
(Courtesy of Square D/Schneider Electric)

and 4. Single-phase voltage L1 and L2 are connected to terminals 6 and 2. Terminals 3 and 5 are jumpered. When the drum control is switched to the forward position, terminal 1 is connected to terminal 3, terminal 2 is connected to terminal 4, and terminal 5 is connected to terminal 6. In this configuration L1 voltage is applied to terminal 5 of the start winding and L2 voltage is applied to terminal 8 of the start winding. This allows the motor to run in the forward direction. When the switch is reversed, the drum switch contacts change so that terminal 1 is connected to terminal 2 and terminal 3 is connected to terminal 4. This action applies L1 voltage to terminal 8 of the start winding, and L2 is connected to terminal 5 of the start winding. By reversing the applied voltage to the start winding, the motor will now run in the reverse direction. When the switch is changed to the off position, L1 voltage is disconnected from the motor, which would deenergize it. If the motor were connected to 110 V, the neutral line would be connected where L2 is. This would allow the drum switch to reverse the motor, and when the switch is in the off position only the L1 voltage is disconnected. The neutral line remains connected to the motor at all times so that the switch will meet the National Electric Code, which states that the neutral line should never be opened by a switch or disconnect.

In Figure 7–15b the drum switch is connected to reverse rotation for a three-phase motor. Since the shaft rotation of a three-phase motor may be reversed by reversing the applied voltage to any two of the three motor terminals, the drum switch can accomplish this easily. When the switch is moved to the reverse position, L1 is connected to T1, L2 is connected to T2, and L3 is connected to T3. When the drum control is switched to the forward connection, L1 is connected to T3, L3 is connected to T1, and L2 remains connected to T2. By reversing the T1 and T3 connects, the motor will rotate in the opposite direction. When the control is switched to the off position, L1 and L2 are disconnected from the motor terminals, which will deenergize the motor. Two of the three supply voltage lines must be disconnected to deenergize a three-phase motor, and wherever possible, all three of the lines should be disconnected.

In Figure 7–15c the drum control is used to reverse the rotation of a DC shunt and a DC compound motor. The principle used to reverse the direction of rotation of a DC motor is to reverse the voltage polarity to the field winding. The field winding is wired to the control like the start winding of a split-phase AC motor. This means that when the switch is changed to the reverse position, the field winding is reversed to the power supply. In a compound motor, both the series field and the shunt field are reversed.

7.3.1 Changing Motor Speeds with a Drum Switch

The drum switch can also be used to change motor speeds. This function is useful where a motor may require several speeds, such as a multispeed grinding or other machine tool applications. The multispeed drum switch is available with 3, 5, 6, or 9 sets of contacts. The speed of a three-phase motor may be varied with the torque remaining constant or with the torque allowed to vary.

7.3.2 Installing and Troubleshooting the Drum Switch

Drum switches can be installed in surface-mounted enclosures or cavity enclosures. The cavity enclosure allows the switch to be flush mounted on an operator's panel. The drum switch also provides a choice of handle types, including the pistol grip, the tear drop, and the knob and pointer. The first part of the installation procedure involves the selection of the type of drum switch, based on the motor control application and the type of motor. These conditions will determine the number of sets of contacts the switch will require. The second part of the procedure is to connect the motor and line voltage wires to the drum switch according to the wiring diagram. Be sure that power to the line wires has been deenergized at the disconnect. This will allow the wires to be handled without any danger of electrical shock.

After the field wires are connected, the disconnect can be closed to energize the supply voltage lines. The switch should be operated through all the selections to ensure that the motor is operating correctly. If any of the switch modes do not operate correctly, the troubleshooting procedures should be used.

The drum switch should be troubleshooted in the same way as other switches or pilot devices. This means that since the switch opens or closes one or more sets of contacts, each of the sets can be tested. The reversing drum switch has three sets of contacts that are connected in two different configurations when the switch is operated. A switch diagram must be used to determine the terminals that are connected in each mode. A matrix provided at the bottom of this type of diagram indicates which contacts are closed when the handle is in the forward, reverse, or off position. An ohmmeter is used to test these contacts if the switch is removed from the circuit. Remember that the ohmmeter will show continuity through motor windings if the motors remain connected to the switch. These readings are confusing, because they can show that a set of contacts are closed even though they are open. This occurs because the motor winding provides a path for the ohmmeter current to backfeed instead of flowing through the contacts. This causes the technician to assume that the contacts are operating correctly and a fault in the switch contacts may be overlooked. In general, this means that a continuity test will be useful only if the switch has been disconnected from both the motor wires and the supply voltage wires.

Since you must disconnect the wires from the switch to make an ohmmeter test, it will be faster to make a voltage test with power applied to the circuit and switch. A voltmeter should be used to make these tests. In this case, one of the meter probes should be attached to one of the supply voltage lines while the other probe is moved from terminal to terminal to test for an open set of contacts. This test will show exactly where voltage is lost in the drum switch circuit. As with other electrical circuits, the motor to supply voltage may be at fault rather than the drum switch. Therefore, the complete circuit should be tested rather than selecting individual components and testing them at random until the problem is found. After the fault is found, the bad components should be replaced and the switch should be tested after the circuit is put back into commission.

7.4 Installing the Manual Motor Starter

One of the steps in the process of installing a manual motor starter is sizing the starter and overload to match the motor load. The selection process takes several conditions into consideration in choosing a proper heater for each application. One of these conditions is the ambient temperature at the location at which the motor and the starter will be located. Another condition is the service factor of the motor being controlled. The final condition is the motor's rated full-load amperage.

Typical heater tables show heaters by part number and the amount of full-load motor current that each will protect. If the motor has a service factor of 1.15, select the heater size nearest the full-load current rating listed on the motor name plate. This table assumes that the motor temperature rise will not be over 40°C (104°F). Sometimes the motor's full-load amperage will not be listed on the heater table, or special conditions will be encountered that will require you to adjust slightly the selection from the list. If the motor's full-load amperage does not exactly match one of the heaters on the list, select the next-higher-rated heater. If the temperature will be higher at the motor starter than where the motor is mounted, you should also select the next-higher-rated heater. If the temperature will be lower at the motor starter than where the motor is mounted, select the next-lower-rated heater. If the motor has a service factor of 1.0 or if it is rated for continuous duty, select a heater that is one rating smaller than is listed on the chart. After the motor is in operation, be sure to check the actual amperage that the motor is pulling to determine that the motor is not overloaded and that it is pulling slightly less than its rated nameplate value.

Another step in the installation procedure involves selecting a proper enclosure for the motor starter. Enclosures are rated according to the environment from which they are protecting the controller. Sometimes a consideration is to keep out elements such as acid fumes or moisture that may damage the control. Other times a consideration is to keep explosive fumes away from the contacts of the control where sparks or arcing occur. The National Electrical Manufacturers Association (NEMA) has classified each of these conditions and written specifications that rate the enclosures for all electrical equipment. Figure 7–16 shows the classification type for motor starter enclosures. This table applies to enclosures for both manual motor starters and full-voltage starters. Notice that some of the enclosure types are specified for outdoor use, while others are rated for indoor use. You must also understand that the enclosure will not meet the intended specifications if the proper type of

NEMA enclosures for starters

TYPE	ENCLOSURE
1	General purpose—indoor
2	Dripproff—indoor
3	Dusttight, raintight, sleet tight—outdoor
3R	Rainproof, sleet resistant—outdoor
3S	Dusttight, raintight, sleetproof—outdoor
4	Watertight, dusttight, sleet resistant—indoor & outdoor
4X	Watertight, dusttight, corrosion resistant—indoor & outdoor
5	Dusttight—indoor
6	Submersible, watertight, dusttight, sleet resistant—indoor & outdoor
7	Class I, group A, B, C or D hazardous locations, air-break—indoor
8	Class I, group A, B, C or D hazardous locations, oil-immersed—indoor
9	Class II, group E, F or G hazardous locations, air-break—indoor
10	Bureau of mines
11	Corrosion-resistant and dripproof, oil-immersed—indoor
12	Industrial use, dusttight and driptight—indoor
13	Oiltight and dusttight—indoor

Figure 7–16 NEMA classification for motor starter enclosures.
(Courtesy of Rockwell Automation)

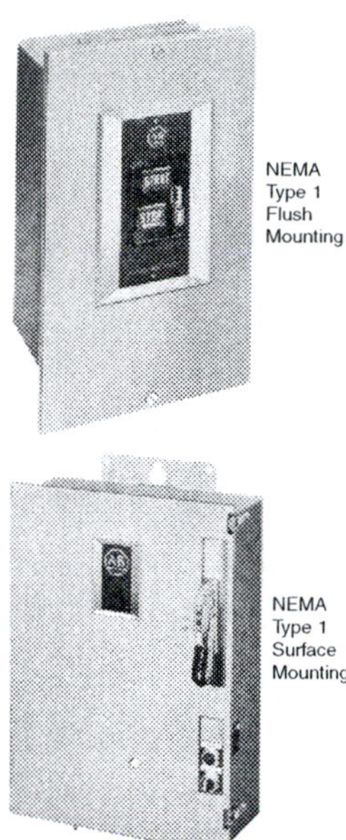

Figure 7–17 (a) Flush-mounted manual motor starter enclosure. (b) Surface-mounted manual motor starter.
(Courtesy of Rockwell Automation)

hardware is not used. This means that a type 4X enclosure will not be watertight if a regular thin-wall conduit is used instead of a watertight conduit. Be sure that the location where the starter will be mounted is specified on the installation blueprint so that the proper enclosure and hardware can be selected.

Most enclosures are available in both flush-mounted and surface-mounted enclosures. Figure 7–17 shows an example of each type. Each enclosure is available with pushbutton or toggle operators mounted on the cover. This provides the widest possible choice of equipment, to meet the requirements of any application.

After the proper heater and enclosure have been selected, the motor starter is ready to be installed. This procedure should be accomplished with the aid of a field wiring diagram. You should also use a blueprint that shows the exact location to mount the starter and enclosure. The enclosure should be mounted and all interconnecting conduits should be installed before you begin to wire the control. In some cases, the starter will need to be removed from the enclosure so that it can be mounted properly. After the enclosure is mounted correctly, the starter can be wired in accordance with the field wiring diagram. If the control is a single-pole, single-phase starter, it will be easier to connect the field wires if the starter is removed from its enclosure. After the field wires have been secured properly, the starter can be remounted into the enclosure and the cover can be replaced.

If the control is a larger starter, it will be easier to connect the field wiring with the starter mounted in the enclosure. This procedure will also allow the wires to be properly routed in the motor starter enclosure so that they will not rub or chafe against any metal parts or inhibit the motor starter parts from operating correctly. Be sure to apply field wiring markers to any wires that are connected to the motor starter. These markers should remain on the wires permanently for any diagnostics or troubleshooting procedures that may be required if the motor starter is suspected of being faulty at a later date. After you have completed the installation procedure and have all the field wires connected, the supply power should be applied and the motor should be tested. If the motor starter has an undervoltage solenoid, be sure to test it by deenergizing power at the nearest disconnect. The motor starter should be in the on position and the motor should be running for this test. The pilot-lamp indicator can also be tested during this procedure if it is being used. Remember that the quality of the installation will determine how well the motor starter will operate over its operational life.

7.4.1 Troubleshooting the Manual Motor Starter

At some time you will be required to troubleshoot a manual motor starter that is suspected of being faulty. It is important to remember what basic parts are used in the motor starter and to be able to identify them from the field wiring diagram. In most cases you will be called to troubleshoot a motor that has malfunctioned, and you will find that the motor is controlled by a manual motor starter. This means that you must troubleshoot the complete motor circuit and not pick on parts of the system to test randomly. Some electricians have found by experience that the motor starter overload will be the most likely problem in this circuit. Other typical problems will be the loss of supply voltage or a faulty motor. This means that these components can be tested first and the majority of the time the problem can be quickly found and repairs can be made.

At other times it will be necessary to make the diagnostic tests in an orderly fashion to isolate the problem. The test points are shown in Figure 7–18. This test procedure should be used if the circuit fault is not readily apparent or if the motor does not restart when the motor starter is reset. The first test in this procedure should be for supply voltage across the terminals marked L1 and L2. If the starter is for a three-phase application, the test should be across all three of the supply voltage lines, two at a time. If the supply voltage is DC, be sure to set the meter for DC voltage. If voltage is not present at these terminals, it means that the supply voltage has been lost and you should check the disconnects and fusing in the motor starter circuit.

If voltage is present across these terminals, proceed to the next test point. One of the voltmeter probes should be left on the L2 terminal during the next test. This will provide a point of reference for the test. The other probe should be moved to test point A (between the contacts and the overload). If voltage is not present at this point, it means that the motor starter contacts for line 1 are not passing power. Check the contacts and replace them if necessary. If voltage is present at this terminal, it means that the L1 contacts are passing voltage correctly. Continue the test by reversing the voltmeter probes. Leave one of the probes on terminal L1 and place the other probe at test point C. If voltage is not present, it indicates that the L2 contacts are faulty. Check these contacts and replace them if necessary. If voltage is present at this test point, it indicates that the L2 contacts are operating correctly. If the starter has three sets of contacts, leave one probe on the L1 terminal and duplicate this test at the point between L3 contacts and its overload. If the starter has an undervoltage or loss of voltage solenoid, be sure that it is energized. If the undervoltage solenoid is not energized, you will need to check for a remote stop button that may have been activated or that may be malfunctioning. Remember, the undervoltage solenoid may be tripped because of a previous power failure or low-voltage condition that has since cleared. Try to reset the motor starter by first pressing the off switch and then pressing the on switch. If voltage is present at test points A and B, continue the procedure. Leave one of the meter probes on terminal L2 and move the other probe to terminal T1. If voltage is not present at this point, it indicates that the line 1 heater is open. Allow the motor starter to cool down for several minutes and try to reset the overload mechanism. If voltage is present at the T1 terminal, reverse the voltmeter probes and level one of the probes at terminal L1. Touch the other voltmeter probe to terminal T2 and check for voltage at this point. If voltage is not present at this point, the L2 overload is open. If voltage is present at this point, you can assume that the starter is operating correctly and providing voltage to the motor. Additional tests should be made on the motor to determine the exact fault. You may need to repair or replace the motor to make the circuit operational again. You may need to review the information on troubleshooting motors that is listed in Chapters 13 and 14 to locate the fault that exists in the motor.

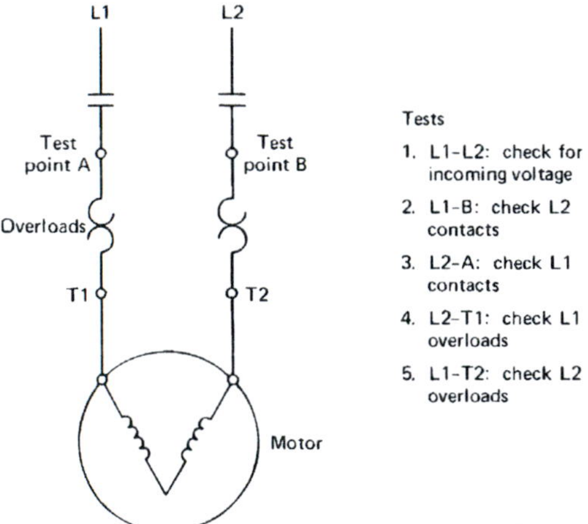

Tests

1. L1-L2: check for incoming voltage
2. L1-B: check L2 contacts
3. L2-A: check L1 contacts
4. L2-T1: check L1 overloads
5. L1-T2: check L2 overloads

Figure 7–18 Troubleshooting test points for a manual motor starter.

Some technicians prefer to start this test procedure at the motor terminals and work their way back to the motor starter and disconnect. Each of these methods of testing the circuit is acceptable. Also note that the contacts in most starters can be replaced rather than replacing the complete starter. You will need to check the parts catalog for your starter to see if replacement contacts are available. If they are available, they should be kept in stock so that they can be changed if your troubleshooting procedure proves they are faulty. Replacing faulty contacts is faster and less expensive than changing the complete starter. Instructions are provided with the contacts to help you remove the faulty one and install the new ones properly.

Questions

1. Explain the operation of a reduced-voltage manual motor starter. Use an electrical diagram to assist your explanation.
2. Explain how undervoltage protection operates on a manual motor starter.
3. Explain the operation of an automatically resetting overload.
4. Explain why the duty cycle of an autotransformer reduced-voltage starter is critical.
5. State two uses for manual motor starters.

True and False

1. The function of the arc hood on a manual starter is to contain arcs that occur when contacts are opened or closed.
2. Contacts on motor starters should never be filed when they are dirty or pitted.
3. Double-break contacts are used in starters to allow larger currents and voltages to be switched on and off without causing severe arcing or overheating.
4. The manual motor starter uses a fuse that is mounted directly on it to prevent motor overloads.
5. Manual drum switches are used for reversing single-phase and three-phase AC squirrel-cage motors and DC shunt and compound motors.

Multiple Choice

1. The function of the overload and heater on a manual starter is to _____
   a. ensure that the temperature of the motor starter remains constant.
   b. ensure that motors that are connected to the manual motor starter cannot receive excessive current.
   c. allow the motor to be reversed quickly.
   d. ensure that the temperature of the motor remains constant.
   e. Both a and d
2. The type 1 NEMA enclosure is rated for _____
   a. drip-proof indoor use.
   b. general-purpose indoor use.
   c. dust-tight indoor use.
   d. rainproof outdoor use.
3. When you test a drum switch for continuity, which of the following would you expect to see when the switch is in the forward position?
   a. Continuity between terminals 1 and 3
   b. Continuity between terminals 2 and 4
   c. Continuity between terminals 5 and 6
   d. All of the above
   e. Only a and b
4. If an overload cannot be reset immediately after it has tripped, the problem may be that _____
   a. the solder in the overload is not solidified yet.
   b. the motor has an open winding.
   c. the motor starter needs its fuses replaced.
   d. the start button is defective.
5. A size 1 motor starter is rated for _____ three-phase (poly phase) motor at 230 V.
   a. 1 hp
   b. 3 hp
   c. 7.5 hp
   d. 10 hp

Problems

1. Draw the symbol for a disconnect and a fused disconnect. Explain the operation of the disconnect switch.
2. Draw a diagram that shows the electrical and mechanical operation of a manual motor starter. Be sure to identify all parts on the starter.
3. If you are using a size 2 starter to control a machine at 208 V, what would the allowable amperage be? If this starter is used on a machine connected to 480 V, what would the allowable amperage be?
4. Explain when you would use a single-pole, a two-pole, and a three-pole motor starter. Draw an electrical diagram of a motor connected to each of these starters.
5. Draw an electrical diagram of a drum switch.

◀ Chapter 8 ▶

Magnets, Solenoids, and Relays

Objectives

After reading this chapter you will be able to:

1. Describe a magnet and explain how it works.
2. Explain the difference between a permanent magnet and an electromagnet.
3. Identify ways to increase the strength of an electromagnet.
4. Explain what flux lines are and where you would find them.
5. Explain electromagnetic induction.
6. Identify the contacts and coil in a relay and contactor.
7. Explain the operation of a relay and contactor.
8. Identify the components in a control circuit.
9. Select the proper size of contactor from a NEMA table.
10. Identify the basic parts of a solenoid and explain their functions.

8.1 Magnetic Theory

The theory of operation for all types of transformers, motors, and relay coils can be explained with several simple magnetic theories. As a technician or maintenance person who works on equipment on the factory floor, you will need to fully comprehend all magnetic theories so that you will understand how these components operate. You must understand how a component is supposed to operate before you can troubleshoot it and perform tests to determine if it has failed. Understanding magnetic theory will make this job easier. It is also important to understand that some of the magnetic theories rely on AC voltage. These theories will be introduced in this chapter, and more detail about AC voltage and magnetic theories that use AC voltage will follow in the next chapter. In this chapter you will learn about permanent magnets first and then you will learn about electromagnets.

Magnet is the name given to material that has an attraction to iron or steel. This material was first found naturally about 4000 years ago as a rock in a city called Magnesia. The rock was called magnetite and was not usable at the time it was discovered. Later it was found that pieces of this material could be suspended from a wire and it would always orient itself so that the same ends always pointed the same direction, which is toward the earth's north pole. Scientists soon learned from this phenomenon that the earth itself is magnetic. At first the only use for magnetic material was in compasses. It was many years later that the forces caused by two magnets attracting or repelling could be utilized as part of a control device or motor.

As scientists gained more knowledge and equipment became available to study magnets more closely, a set of principles and laws evolved. The first of these discoveries showed that every magnet has two poles, called the north pole and south pole. When two magnets are placed end to end so that similar poles are near each other, the magnets repel each other. It does not matter if the poles are both north or both south, the result is the same. When the two magnets are placed end to end so that the south pole of one magnet is near the north pole of the other magnet, the two magnets attract each other. These concepts are called the first and second laws of magnets.

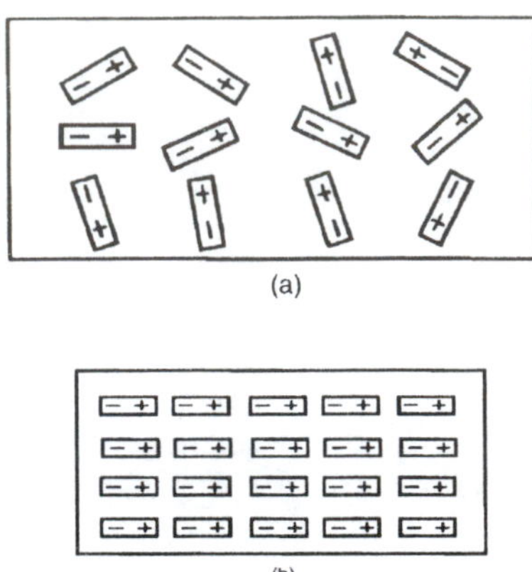

Figure 8–1 (a) Example of metal that has its dipoles randomly placed, which makes it a very weak magnet or is not magnetic at all. (b) Example of metal in which the dipoles are aligned to make a very strong magnet.

When sophisticated laboratory equipment became available it was found that this phenomenon is due to the basic atomic structure and electron alignments. When the atomic structure of a magnet was studied, it was found that its atoms were grouped in regions called *domains* or *dipoles*. In material that is not magnetic or that cannot be magnetized, the alignment of the electrons in the dipoles is random and usually follows the crystalline structure of the material. In material that is magnetic, the alignments in each dipole are along the lines of the magnetic field. Since each dipole is aligned exactly like the ones next to it, the magnetic forces are additive and are much stronger. In material where the magnetic forces are weak, it was found that the alignment of the dipole was random and not along the magnetic field lines. The more closely this alignment is to the magnetic field lines, the stronger the magnet is. Today we refer to a piece of soft iron that has all of its dipoles aligned as a *permanent magnet*. The name permanent magnet is used because the dipoles remain aligned for very long periods of time, which means the magnet will retain its magnetic properties for long periods of time. Figure 8–1a shows a diagram of nonmagnetic metal that has its dipoles randomly placed, and Figure 8–1b shows a piece of metal that is magnetic and has its dipoles aligned to make a strong magnet.

8.2 A Typical Bar Magnet and Flux Lines

Figure 8–2 shows the bar magnet that is made of soft iron that has been magnetized. The magnet is in the shape of a bar, and it has its north and south poles identified. Since the bar remains magnetized for a long period of time, it is called a permanent magnet. The magnet

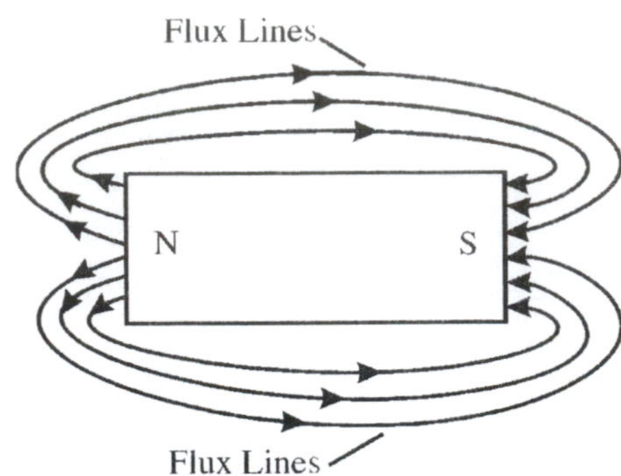

Figure 8–2 Example of a bar magnet. Notice that the poles are identified as north (N) and south (S). Flux lines are shown emanating from the south pole to the north pole.

produces a strong magnetic field because all its dipoles are aligned. The magnetic field produces invisible *flux lines* that move from the north pole to the south pole along the outside of the bar magnet. The diagram in Figure 8–2 shows that these flux lines are lines of force that are in a slight arc as they move from pole to pole.

Since the flux lines are invisible you will need to perform a simple experiment to allow you to see that the flux lines do exist and what they look like as they surround the bar magnet. For this experiment you will need a piece of clear plastic film, such as the plastic sheets used for overhead transparencies, and some iron filings. Place the plastic sheet over a bar magnet and make the plastic as flat as possible and sprinkle iron filings over it. The filings will be attracted by the invisible flux lines as they extend in an arc from the north pole to the south pole along the outside of the magnet. Since the flux lines begin at one pole and stretch to the other, the highest concentration of flux lines will be near the poles. The iron filings will also concentrate around the poles, but a definite pattern of flux lines can be seen along each side of the bar magnet. If an overhead projector is available, the image of the flux lines can be projected onto a screen or blackboard so that they can be seen more easily. The pattern of these fillings will look similar to the diagram in Figure 8–2. The number of flux lines around a magnet is directly related to the strength of the magnet. A stronger magnet will have more flux lines than a weaker magnet. The strength of a magnet's field can be measured by the number of flux lines per area. Since the strength of a magnet's field is based on the alignment of the magnetic dipoles, the number of flux lines will increase as the alignment of the magnetic dipoles increases.

Some material, such as alnico and Permalloy, make better permanent magnets than iron, since the alignment of their magnetic domains (dipoles) remains consistent even after repeated use. You may find these materials used in some expensive controls and motors, but normally the permanent magnet will be made of soft iron. The reason permanent magnets are useful in many types of controls, especially in motors and generators, is because the soft iron produces residual magnetism for long periods of time over many years. Permanent magnets have several drawbacks for use in some applications. One of these is that the magnetic force of a permanent magnet is constant and cannot be turned off when it is not needed. This means that if something is attracted to a magnet, it will remain attracted until it is physically removed from the force of the flux lines. Another problem with a permanent magnet's flux field being constant is that it cannot easily be made stronger or weaker if circumstances so require.

8.3 Electromagnets

An electromagnet is made by connecting a coil of wire to an electric cell (battery). The electromagnet has properties that are similar to those of a permanent magnet. When a wire conductor is connected to the terminals of the battery, current will begin to flow and magnetic flux lines will form around the wire like concentric circles. If the wire was placed near a pile of iron filings while current was flowing through it, the filings will be attracted to the wire just as if the coil were a permanent magnet. Figure 8–3 shows several diagrams indicating the location of magnetic flux lines around conductors. Figure 8–3a shows that flux lines will occur around any wire when current is flowing through it. You can set up several simple experiments to demonstrate these principles. In one experiment you can insert a current-carrying conductor through a piece of cardboard and place iron filings around the conductor on the cardboard. When current is flowing in the wire, the filings will settle around the conductor in concentric circles showing where the flux lines are located. As the amount of current is increased, the number of flux lines will also increase. The flux lines will also concentrate closer and closer to the wire until the current reaches *saturation.* When the flux lines reach the saturation point, any additional increase of current in the wire will not produce any more flux lines.

When the straight wire is coiled up, the flux lines will concentrate and become stronger. Figure 8–3b shows an example of flux lines around a coil of wire that has current flowing through it. Since the flux lines are much stronger in a coil of wire, most of the electromagnets that you will encounter will be in the form of coils. For example coils are used in transformers, relays, solenoids, and motors.

One advantage an electromagnet has over a permanent magnet is that the magnetic field can be energized and deenergized by interrupting the current flow through the wire. The strength of the magnetic field can also be varied by varying the strength of the current flow through the conductor that is used for the electromagnet. This is perhaps the most important theory of magnetism, since it can be used to change the strength of magnetic fields in motors, which causes the motor shaft to produce more torque so it can turn larger loads. Since the flux lines are invisible you may need to perform an experiment to prove that the magnetic field becomes stronger as current flow through the coil is increased. Figure 8–4 shows how to set up this experiment. Wrap a wire into a coil and connect the ends to a dry cell battery and place it near a pile of iron filings. Add a variable resistor to the circuit to increase or decrease the current. When the current is set to a minimum as in Figure 8–4a, the magnetic field around the coil of wire will attract only a few filings. As the current is increased as in Figure 8–4b, the number of filings the magnetic field will attract also increases until the current causes the magnetic field in the wire to reach saturation. When the saturation point is reached, any additional current flowing in the wire will not produce additional flux

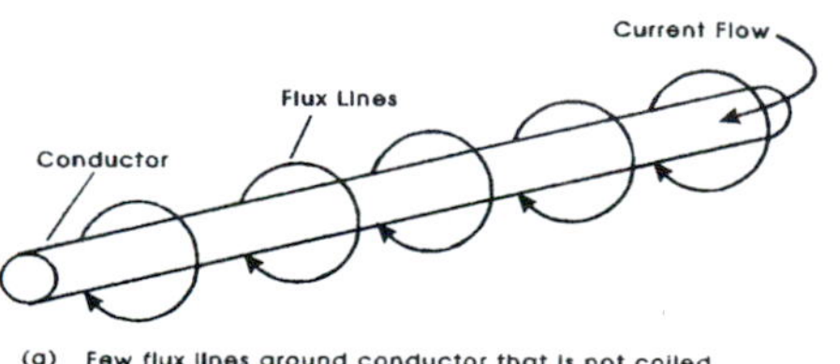

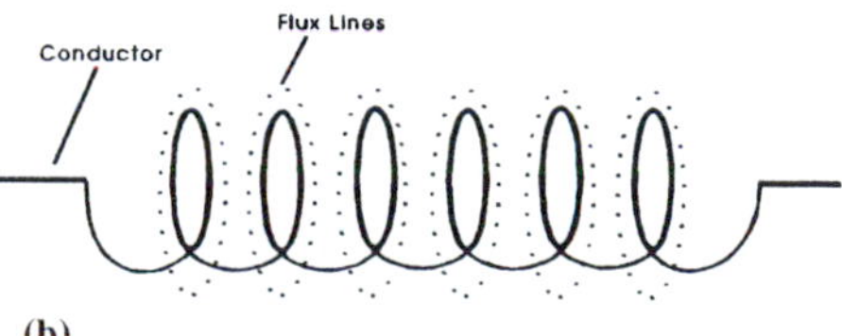

Figure 8–3 (a) Flux lines around a straight wire that is carrying current, and (b) a coil of wire that is carrying current. Notice that the number of flux lines increases when the wire is coiled.

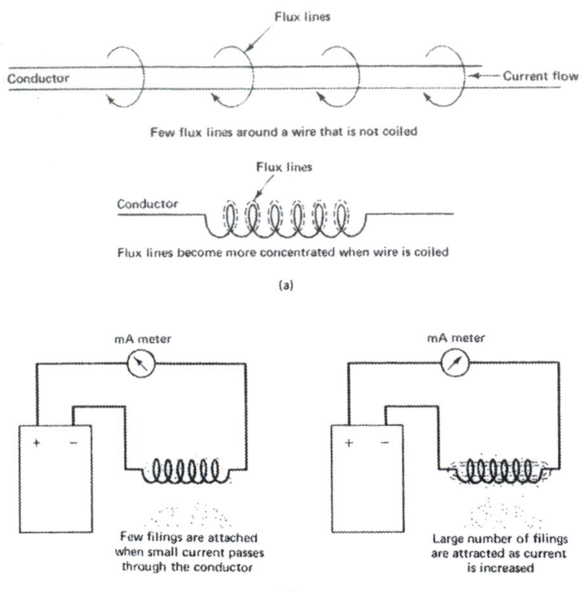

Figure 8–4 Small amount of current flowing in a coil of wire creates a small number of flux lines that attract iron filings.

lines. When a switch is added to this circuit, the magnetic field can be turned on and off by turning the switch on and off to interrupt the flow of current in the coil. When the switch is opened, current is interrupted and no flux lines are produced, so the magnetic field will not exist.

Components such as electromechanical relays and solenoids will use the principle of switching the magnetic field on and off. When current flows in the coil, the magnetic field will cause the contacts in the relay or the valve in the solenoid to close. When the current to the coil in these devices is interrupted and the magnetic field is turned off, springs will cause the relay contacts or valves to open More information about relays and solenoids is provided in later chapters.

This principle is also used to turn motors on and off. When current flows through the coils of a motor, its shaft will turn. When current is interrupted, the magnetic field will diminish and the motor will stop rotating. This point is important if the wire that is used in the coil develops an open. When current stops flowing in the coil due to the open circuit, the shaft of the motor will stop turning.

8.3.1 Adding Coils of Wire to Increase the Strength of an Electromagnet

Another advantage of the electromagnet is that its magnetic strength can be increased by adding coils of wire to the original single coil of wire. The increase of the magnetic field occurs because the additional coils of wire require a longer length of wire to be used, which provides additional flux lines. The magnetic field will be stronger when the coil is more tightly wound because the flux lines are more concentrated. This means that very fine wire is used in some electromagnets to maximize the number of coils. As smaller wire is used, the amount of current flowing through it must be reduced so that the wire is not burnt open.

Some motors will use this principle to increase their horsepower and torque rating. These motors will have more than one coil that can be connected in various ways to affect the torque and speed of the motor's shaft. *Torque* is defined as the amount of rotating force available at the shaft of a motor. Coils can also be connected in series or in parallel to affect the torque and speed of a motor.

8.3.2 Using a Core to Increase the Strength of the Magnetic Field of a Coil

The strength of a magnetic field can also be increased by placing material inside the helix of the coil to act as a core. The farther the core is inserted into the coil, the stronger the flux field becomes. When the core is removed completely from the coil, it is considered to be an *air coil magnet* and the magnetic field is at its weakest point. If a soft iron is used as the core, it will strengthen the magnetic field, but it also creates a problem because it has excessive residual magnetism, which is unwanted. Residual magnetism means the core will retain magnetic properties when current is interrupted in the coil, which will make it like a permanent magnet. This problem can be corrected by using laminated steel for the core. The laminated steel core is made by pressing sheets of steel together to form solid core (Figure 8–5). When current flows through the conductors in the coil, the laminated steel core enhances the magnetic field in much the same way as the soft iron, and when current flow is interrupted, the magnetic field collapses rapidly because each piece of the laminated steel will not retain sufficient magnetic field.

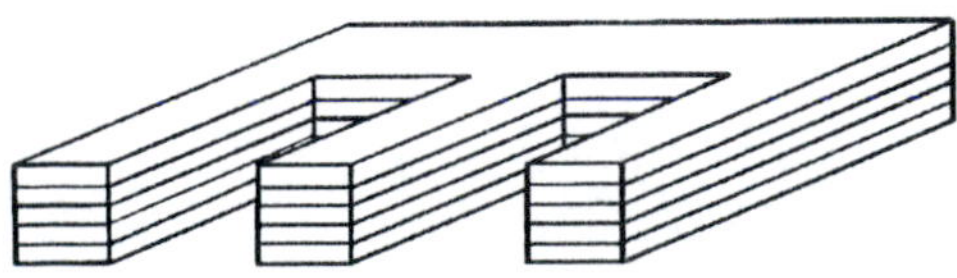

Figure 8–5 Pieces of steel pressed together to form a laminated steel core for use in electromagnets.

8.3.3
Reversing the Polarity of a Magnetic Field in an Electromagnet

When current flows through a coil of wire, the direction of the current flow through the coil will determine polarity of the magnetic field around the wire. The polarity of the magnetic field around the coil of wire is important because it determines the direction a motor shaft turns in an AC or DC motor. If the direction of current flow is reversed, the polarity of the magnetic field will be reversed, and the direction a motor shaft is turning will be reversed. The direction of rotation is very important in some motors such as fan motors and pump motors. In these applications you will be requested to make changes to the connections for the windings in the motor or to the supply voltage for a three-phase motor to make the motor rotate in the opposite direction. These changes take advantage of changing the direction of current flow through a coil or changing the polarity of the supply voltage with respect to the other phases so that the motor will reverse its rotation.

As you are leaning advanced theories about motors and other electromagnetic components, diagrams will be presented to explain more complex concepts. A universally accepted method of identifying the direction of current flow has been developed. Figure 8–6a shows a diagram that indicates the location of the flux lines in a coil of wire and shows the direction the flux lines travel around a straight current-carrying conductor. A dot or a cross (X) is used to mark the conductor to indicate the direction current is flowing. In Figure 8–6a a dot is used to indicate that current is flowing toward the observer. The flux lines in this diagram show that their flow is in a clockwise motion. In Figure 8–6b, a cross is used to indicate that current is flowing away from the observer and the flux lines are shown moving in the counterclockwise direction. If you know the direction of current flow, the advanced theories will allow you to determine the polarity of the magnetic field and predict the direction a motor's shaft will rotate.

Figure 8–6c shows another way to determine the directions of the flux lines, the direction of current flow, and which pole is the north pole. This method is called the *left-hand rule*. From this diagram you can see that you need to know either the direction of the current flow or the direction of the flux lines to determine the polarity of the coil. Normally the direction of the current flow is easy to determine with a voltmeter by determining which end of the wire is positive and which is negative. Current (electron flow) is from negative to positive.

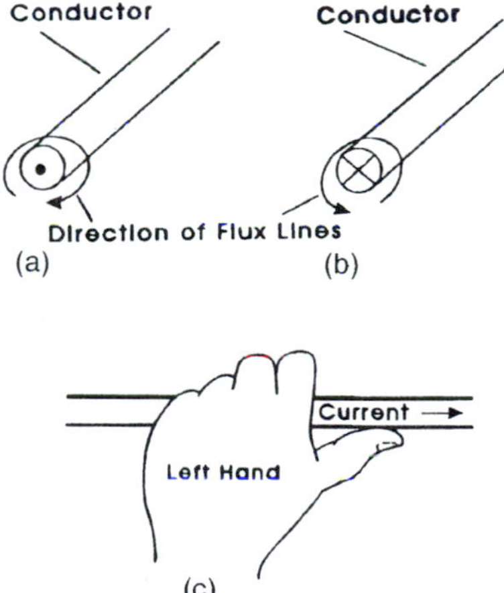

Figure 8–6 (a) The direction of flux lines around a wire when current is flowing in the wire toward you. The dot indicates that the direction of current flow is toward the observer. (b) the direction of the flux lines around a wire when current is flowing in the wire away from you. The X indicates current is flowing in the wire away from the observer. (c) an example of the left-hand rule. The thumb is pointing in the direction of the current flow in the wire, and fingers are pointing in the direction of the flux lines as they move around the wire when current is flowing.

8.4
An Overview of AC Voltage

Before you can understand more about magnetic theories you will need to know more about AC voltage and current. This section provides an overview of AC voltage. *AC stands for alternating current.* This name is derived from the fact that AC voltage alternates positive for half of a cycle and then negative for one half of a cycle. Figure 8–7a shows the characteristic AC waveform. The AC waveform shown in this figure is called a *sine wave*.

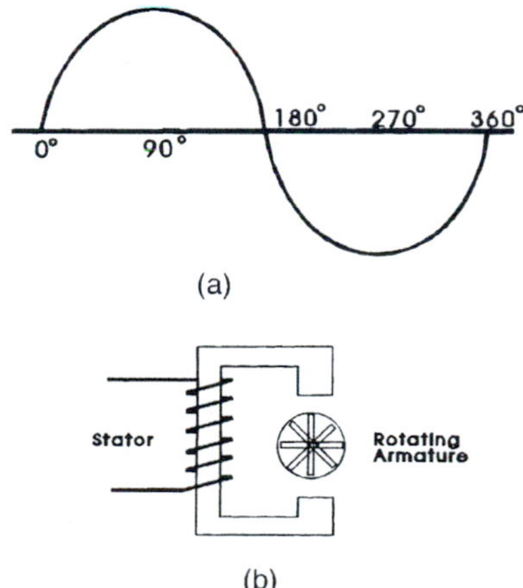

Figure 8–7 (a) A sine wave for AC voltage. (b) a simple alterator that has a moving coil and a stationary coil. The alternator produces the AC sine wave.

The AC waveform is created by an alternator that looks similar to a motor. Figure 8–7b shows a diagram of an alternator. The alternator has a rotating coil of wire that is mounted on a shaft and stationary coil that has current flowing through it so that it creates a magnetic field. The shaft of the alternator is rotated by an energy source such as a steam turbine. When the rotating coil of wire passes through the magnetic field, an electron flow is created in the coil of wire. Since the coil of wire rotates, it will pass the positive magnetic field and then the negative magnetic field during each complete rotation of 360 degrees. This action causes the voltage sine wave to be produced and the sine wave is also identified as having 360 degrees. In the United States, the speed of rotation of the alternator is maintained at a constant rate so that the sine wave will have a frequency of 60 cycles per second, which is also called 60 hertz (Hz).

8.5 Electromagnetic Induction

Electromagnetic induction is another magnetic principle that is used in transformer and motor theory. This theory states that if two individual coils of wire are placed in close proximity to each other, and a magnetic field is created in one of the coils so that flux lines are created around it, the flux lines will collapse across the second coil when the current is interrupted in the first coil. When these flux lines collapse across the coils of the second coil, they will cause a current to begin to flow in the second coil. This current flow will be 180 degrees out of phase with the current flow that created the magnetic field in the first coil.

Since AC voltage periodically builds voltage to a positive peak and interrupts it when the sine wave reaches 180 degrees, it is ideal to create the magnetic field that builds and collapses. When the current flows, it causes the flux lines to build in the coil and when the current flow is interrupted it causes the flux lines to collapse. Since AC voltage in America is generated at 60 Hz, the magnetic field in the first coil will build and collapse 60 times a second. The flux lines will also build and collapse across the windings of the second coil 60 times a second.

One component that uses two coils is called a transformer. The first coil in the transformer is called the *primary winding*, and the second coil in the transformer is called the *secondary winding*. The ratio of the number of turns of wire in the primary winding and the secondary winding will determine the ratio of the voltage in the primary winding and secondary winding. Another important point to remember is that since the two coils are electrically isolated, the voltage produced in the second coil by induction is totally isolated from the voltage in the first coil. The isolation allows the number of turns of wire in the two coils to be different so that a different amount of voltage can be created in the second coil. If the number of turns of wire in the first coil and second coil are identical, the amount of voltage induced in the second coil will be approximately the same as the voltage in the first coil.

In motors, electromagnetic induction is used to create a second magnetic field in a part of the motor that rotates. Since the shaft of the motor must become magnetic and rotate freely from the stationary part of the motor, it would not be practical to make a permanent connection with a wire to the rotating member so that it could receive current to create its magnetic field. Electromagnetic induction creates the second magnetic field without any connections. In Chapter 6, you learned about turns ratios and in future chapters you will learn more about transformer turns ratios and magnetic theories for motors.

8.6 Review of Magnetic Principles

When you are learning about components that utilize electromagnetic principles you can return to this section to review the basic concepts about them. The important electromagnetic principles are listed below:

1. In a magnet, the dipoles are aligned.
2. Every magnet has a north pole and a south pole.
3. Like magnetic poles repel, and unlike magnetic poles attract.
4. A permanent magnet is useful because its magnetic field is residual and can remain strong over a number of years.
5. When current passes through a conductor, magnetic flux lines will form around the conductor in concentric circles.
6. The strength of the magnetic field around a conductor will be proportional to the amount of current flowing through it until the amount of current reaches the saturation point of the electromagnet.
7. The magnetic field of an electromagnet can be turned on and off by interrupting the current flow through the coil of wire.
8. The polarity of an electromagnetic coil is determined by the direction of current flow through the conductor.
9. If the conductor is wrapped in coils, the strength of the magnetic field is increased.
10. When more coils of wire are added to an electromagnet, the magnetic field strength is increased.

11. A core will cause the field of an electromagnet to be strengthened.
12. The core must be made from laminated steel when AC voltage is used to power the coil so that the magnetic field can quickly build and collapse in the core as the AC voltage changes polarity every 1/60 of a second.

8.7 Solenoids Used in Industrial Systems

A solenoid is a magnetically controlled valve. You learned earlier in this chapter about the power of current flowing through a coil of wire and causing a strong magnetic field to develop. The strong magnetic field in the solenoid valve is used to make the valve plunger move from the open to the closed position or from the closed to the open position. The valve is part of the plumbing system, such as an inlet valve for the water system. Solenoid valves allow the plumbing system to be interfaced with the electrical system so the plumbing system can be automated.

8.7.1 Basic Parts of a Solenoid Valve

Figure 8–8 shows a typical solenoid valve in the closed position and in the open position. From these figures you can see that the solenoid valve is a plumbing-type valve that has a movable plunger that is used to close off the valve. The valve can be designed so that the plunger is held in either the open or the closed position by spring pressure. When electric current is applied to the coil in the valve, the strong magnetic field will cause the plunger to move upward. If the valve is the type where spring pressure is causing the plunger to stay in the closed position, the magnetic field will cause the plunger to open the valve passage. If the valve is the type where spring pressure is causing the plunger to stay in the open position, the magnetic field will cause the plunger to close the valve passage.

The coil for the solenoid is exactly like the coils that you learned about in the chapter for relays and contactors. Since the coil is basically a long wire that is coiled on a plastic spool, it will have two ends. You can test the coil for continuity and you should notice that each type of solenoid coil has some amount of resistance that will vary from approximately 50 to 1500. The exact amount of resistance does not matter. Because the coil is made from one long piece of wire, it is doubtful that the amount of resistance will change. The main problem a coil will have is an open somewhere along its wire.

8.7.2 Troubleshooting a Solenoid Valve

At times you will be working on a system that has one or more solenoid valves and you will suspect that a solenoid valve is not operating correctly. The first test you should make when troubleshooting a solenoid valve is to read its data plate and determine the amount of voltage the valve should have and then measure the amount of voltage that is available across the two wires that are connected to the coil. The coil of any solenoid valve must have the correct amount and type of voltage applied to its two wires if it is to conduct sufficient current to cause the plunger to move. This means that if the valve is rated for 120 VAC, you should measure 120 VAC across the two coil wires. If the voltage is missing or the incorrect amount, the coil will not operate correctly. If the proper amount of voltage is not present, you will need to use troubleshooting procedures to locate the open in the circuit that is causing the loss of voltage before it reaches the coil.

If the proper amount of voltage is present and the coil will not activate, you must make several additional tests. You can test to see if the coil has developed a magnetic field by placing a metal screwdriver near the coil to see if it is attracted by the magnetic field. If the screwdriver is attracted to the coil by the magnetic field but the valve is not opening or closing,

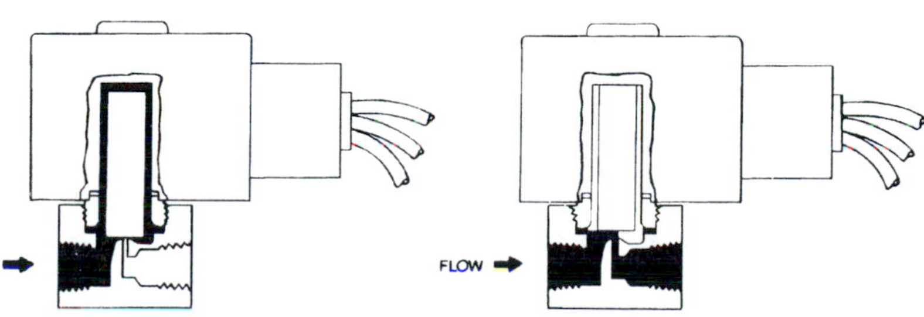

Figure 8–8 A solenoid valve in the closed position and the open position.

the valve seat may be frozen in place and the valve will need to be replaced.

If the correct amount of voltage is present at the coil wires, then the coil may have an open in its wire. You can test the coil wire with a continuity test. Be sure to turn off all voltage to the solenoid and disconnect the coil wires so that the coil is isolated from backfeeding to the remainder of the circuit. If the continuity test indicates infinite resistance at the highest resistance range, you can suspect the coil has an open and it should be replaced. The coil can be replaced without unsoldering the valve from the plumbing systems.

8.8 Relays and Contactors in the Control Circuit

The control circuit in any industrial electrical system is the heart of the system. It consists of the control devices such as start/stop switches, limit switches, or pressure switches to determine when to energize or deenergize each system. Each control circuit uses relays or contactors to control the various motors, such as pump motors and fan motors. This section introduces the main components that are used in the control circuits of these systems and their theory of operation. The components in the control circuit are energized and in turn they energize components such as motors, solenoids, or pumps in the load circuit. In most electrical diagrams the control circuit is shown at the bottom and the load circuit is shown at the top.

8.8.1 The Theory and Operation of a Relay

The relay is a magnetically controlled switch that is the main control component in a typical electrical system. Figure 8–9 shows a picture of a typical relay and Figure 8–10 shows an exploded view of a typical relay, which consists of a coil and a number of sets of contacts. The coil becomes an electromagnet when it is energized and

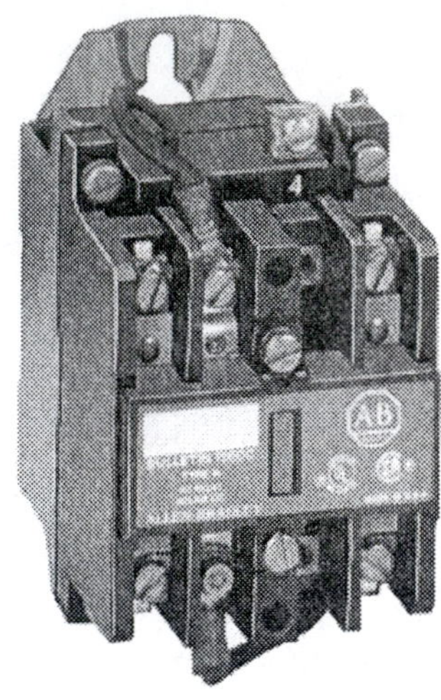

Figure 8–9 A typical relay.
(Courtesy of Rockwell Automation)

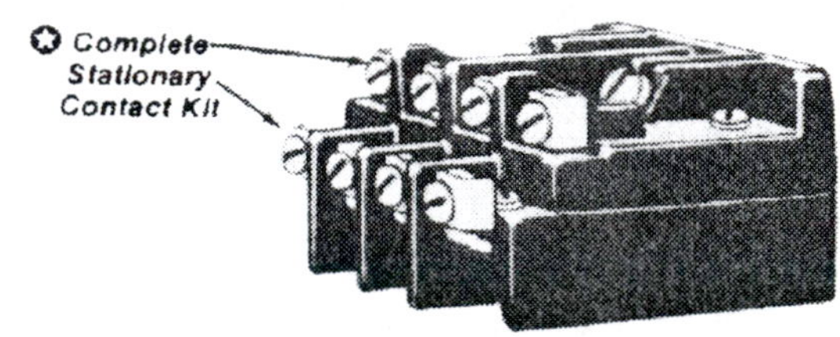

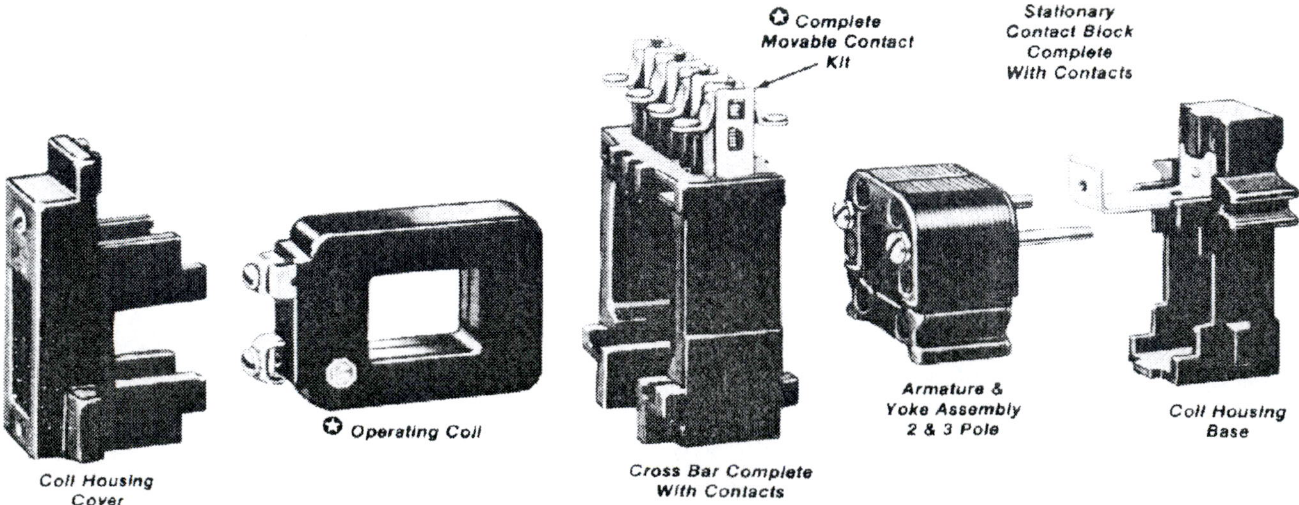

Figure 8–10 An exploded view of a relay. Notice that the coil and contacts are mounted so that the magnetic field of the coil will make the contacts move.
(Courtesy of Rockwell Automation)

its magnetic field causes each set of normally open contacts to close and each set of normally closed contacts to open. The contacts are basically a switch that is operated by magnetic force. The part of the relay that moves and causes the contacts to move is called the armature. Power is applied to the coil of the relay first, and the magnetic flux causes the armature to move and the contacts to change position. The coil is part of the control circuit, and the contacts are part of the load circuit.

Figure 8–11 shows the electrical symbol for the coil and contacts of a relay. The armature is not shown in the electrical symbols since it is considered part of the contacts. When you encounter a relay in the control panel of a system, you must relate the physical parts with the electrical symbols that are used to identify the components. You must also learn to envision its operation as two separate pieces, the coil and contacts, even though they are mounted near each other and operate almost simultaneously. It is important to understand that the coil must be energized first, and a split second later the magnetic field built up in the coil will cause the contacts to move.

8.8.2 Types of Armature Assemblies for Relays

You will encounter many different types of relays and different name brands of relays that look like they all function on different theories of operation. In reality the operation of all relays can be broken into four basic methods because one of four basic armature assemblies is used in all relays. Figure 8–12 shows an example of these four basic armature assemblies. A few relays will use hybrid armatures that use some features of more than one type of armature assembly.

Figure 8–12a shows an example of a horizontal-action-type armature for a relay. In this example, a set of stationary contacts are mounted in the horizontal position. A set of movable contacts are mounted on the armature assembly directly across from the stationary set. The armature is mounted inside the coil in such a way that when the coil is energized, its magnetic field will cause the mass of the armature to center in the middle of the coil. Since the armature is slightly offset when it is placed in the coil, it will move to the left to center itself directly in the middle of the coil when the magnetic field is developed in the coil. The movement of the armature to the left will shift the movable contacts to the left until they come into contact with the stationary set of contacts. When the movable contacts and stationary contacts touch each other, they will complete a circuit that allows large amounts of current and voltage to pass through them like a normal switch when it is in the closed position. When the current to the coil is deenergized, a small spring will cause the armature to shift back to the left to its original position.

Figure 8–12b shows an example of a Bell crank-type armature assembly. In this example, the coil is mounted above the armature, and when it is energized it produces a magnetic field that pulls the armature upward. The movable contacts are connected to an arm that is

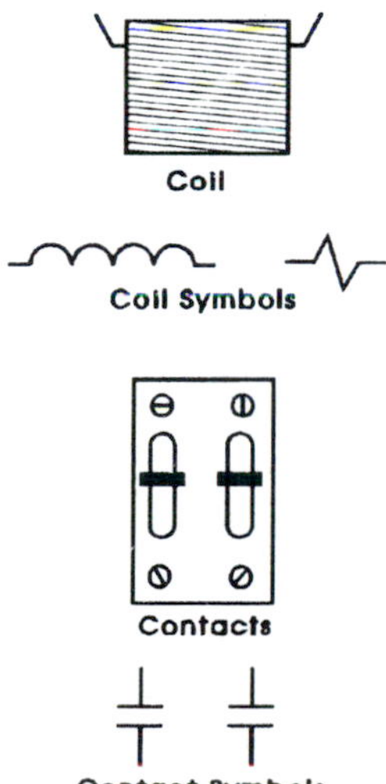

Figure 8–11 The electrical symbol for the coil and contacts of a relay. The armature is not shown in the electrical symbols since it is considered part of the contacts.

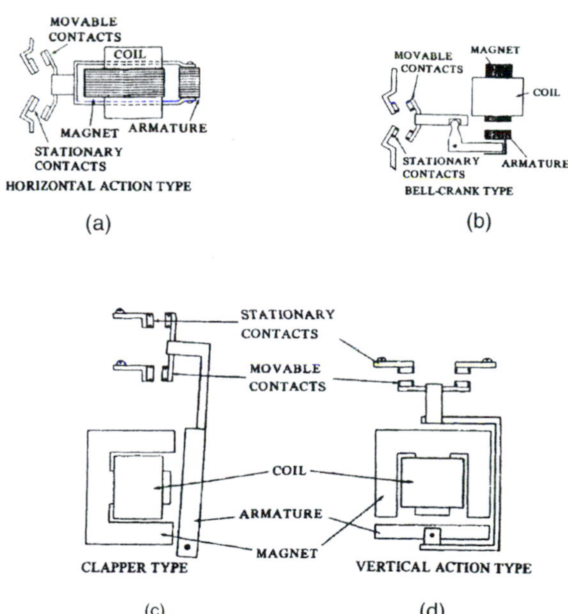

Figure 8–12 Four types of armature assemblies: (a) horizontal action type; (b) Bell crank; (c) clapper; (d) vertical action.
(Courtesy of Square D/Schneider Electric)

bent at a right angle. When the end of the arm is pulled upward by the magnet, the other end of the arm is moved to the left. This action shifts the movable contacts against the stationary contacts to complete the electrical circuit. When the coil is deenergized, gravity will cause the armature to drop downward away from the coil. This movement causes one end of the arm to also move downward, which causes the left end of the arm to shift back to the left, which causes the contacts to move to the open position again.

Figure 8–12c shows an example of a clapper-type armature assembly. The armature is a large arm on the right side of the coil. The armature (arm) has a pin through the bottom to act as an axis. The movable contacts are mounted to the top of the armature at the left side. When the coil is energized, it will create a magnetic field that pulls the armature to the left toward the coil. This action causes a small amount of travel at the bottom of the arm and a large amount of travel at the top of the arm since the bottom of the arm is held in place at the axis. Since the movable contacts are mounted near the top of the arm, they will move to the left a significant distance until they come into contact with stationary contacts. When the coil is deenergized, a small spring will pull the armature to the left, which will cause the contacts to return to their open position.

Figure 8–12d shows an example of the vertical action-type armature assembly. The armature is mounted to a bracket that is shaped like the letter C. The bottom part of the C-shaped bracket is mounted directly to the armature and a set of movable contacts are mounted directly to the top of the C-shaped bracket. When the coil is energized, it creates a magnetic field that pulls the entire C bracket upward until the armature is pulled tight against the coil. This movement causes the contacts that are mounted to the top of the bracket to shift upward until they touch the stationary contacts. When the coil is deenergized, gravity will cause the complete bracket to drop downward and move away from the coil, which causes the movable contacts to move away from the stationary contacts.

8.8.3 Pull-in and Hold-in Current

When voltage is first applied to a coil of a relay, it will draw excessive current. This occurs because the coil of wire presents only resistance to the circuit when current first starts to flow. As the flow of current increases in the coil, inductive reactance begins to build, which will cause current to become lower. When the current is at its maximum, it will create a strong magnetic field around the coil, which will cause the armature to move. When the armature has moved, it will cause the induction in the magnetic coil to change so that less current is required to maintain the position of the armature.

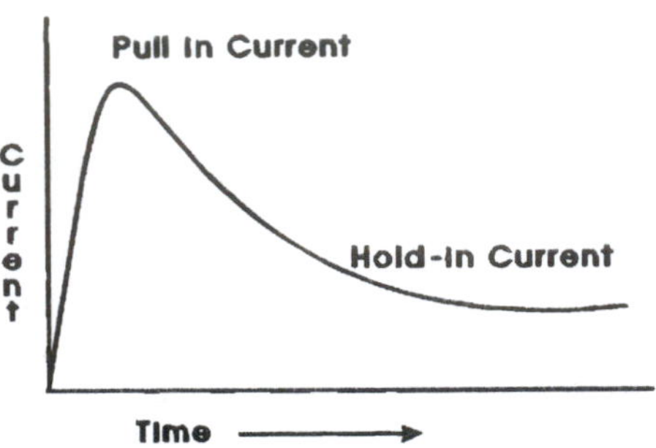

Figure 8–13 A diagram that shows pull-in and hold-in current for relay coils. Notice that the pull-in current is approximately 3–5 times larger than the hold-in current.

Figure 8–13 is a diagram of the pull-in current and the hold-in current. The pull-in current is also called the inrush current, and the hold-in current is also called seal-in current. The pull-in current is typically 3 to 5 times larger than the hold-in current. At times, the supply voltage will be slightly lower in the summer. This condition is called a brownout and it may cause a problem with activating relays. If the voltage is too low, it may not be able to supply sufficient current to pull the armature into place when the relay is first energized because the pull in current is too small.

If the system is running when the brown-out condition occurs, the relay coil will probably remain energized because the amount of hold-in current is sufficient to keep the armature in place, even though the voltage is low. Since the brown-out condition occurs when it is very hot outside, it will most likely affect air conditioners, which will typically remain running through this condition. If the air conditioner ever gets the space enough so it can cycle off during a brown-out condition, it will probably not energize again because of the low voltage and the large amount of pull-in current required.

8.8.4 Normally Open and Normally Closed Contacts

A relay can have normally open or normally closed contacts (Figure 8–14). The word "normal" for contacts indicates the position the contacts are in when no voltage is applied to the coil. The contacts can be held in their normal position by a spring or by gravity. The contacts will move from their normal position to their energized position when power is applied to the coil.

Some types of contacts can be changed or converted from normally open to normally closed in the field. Other types are manufactured in such a way that they cannot be changed. Figure 8–15 shows examples of converting normally open contacts to normally closed contacts while the relay is installed. The contacts can be converted in the field by the technician by simply removing them from the relay and turning them upside down. When a set of normally open contacts are inverted, they become normally closed contacts, and when normally closed contacts are inverted, they will become normally open contacts. This means that as a technician in the field, you can change the contacts in a relay to get the exact number of normally open or normally closed contacts needed for the application.

(a) Normally Open Contacts

(b) Normally Closed Contacts

Figure 8–14 Symbol for (a) normally open and (b) normally closed contacts.

8.8.5 Ratings for Relay Contacts and Relay Coils

When you change a relay that is worn or broken, you must ensure that the coil of the new relay matches the voltage of the control circuit exactly. This means that if the voltage for the control circuit is 110 VAC, the coil must be rated for 110 V. If the control voltage is 220 VAC, the coil must be rated for 220 VAC. The voltage rating for a relay coil is stamped directly on the coil. If the coil is rated for 24 VAC it will be color coded black. If the coil is rated for 110 VAC, it should be color coded red or have a red stamp on the coil. If the relay coil is rated for 208 or 230 VAC, it will be color coded green or identified with a green stamp or green printing on the coil. DC coils are color coded blue. The current rating for a relay coil is seldom listed on the component. If it is important to know the current

Figure 8–15 Example of changing normally open contacts to normally closed contacts in the field by turning them over. Normally closed contacts can also be converted to normally open contacts in the same manner. (Courtesy of Rockwell Automation)

ADDING or CONVERTING CONTACT CARTRIDGES HAVING "SWINGAROUND" TERMINALS

General Instructions (Specific cases below.)

1.1 **Adding a contact cartridge:**

As received, accessory cartridges are in the normally open mode with terminal screws adjacent to N.O. symbols. If normally closed mode is desired, convert contact as indicated in Step 1.2 below. When cartridges are inserted, the terminal screws must face the front. The clear cover may face either side. **Do not install more than 8 N.C. contacts per relay.** When installing one cartridge, locate it at an inner pole position. When installing 2 cartridges, locate both in inner or outer (balanced) positions.

1.2 **Converting a contact to its alternate mode (N.O. ⇄ N.C.):**

Withdraw an assembled cartridge for replacement or conversion by inserting the blade of a suitably-sized screwdriver under a terminal screw pressure plate. Slide cartridge out. See Figure 2. Back the terminal screws out of the cylindrical nuts a sufficient amount (approximately 2 turns for a fully-tightened screw) to permit rotation of each screw and nut assembly to its alternate position. See Figure 3.

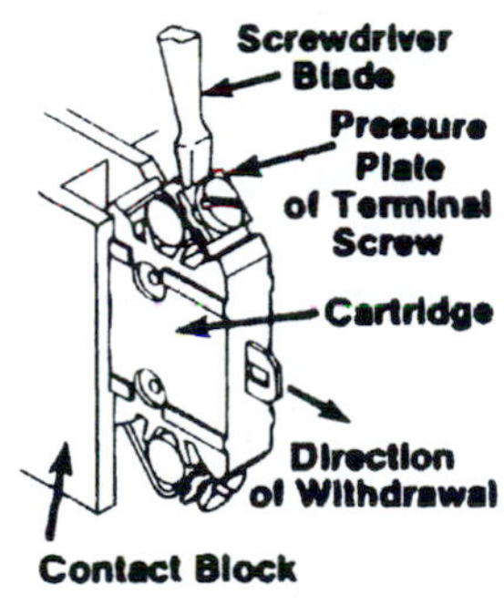

FIGURE 2

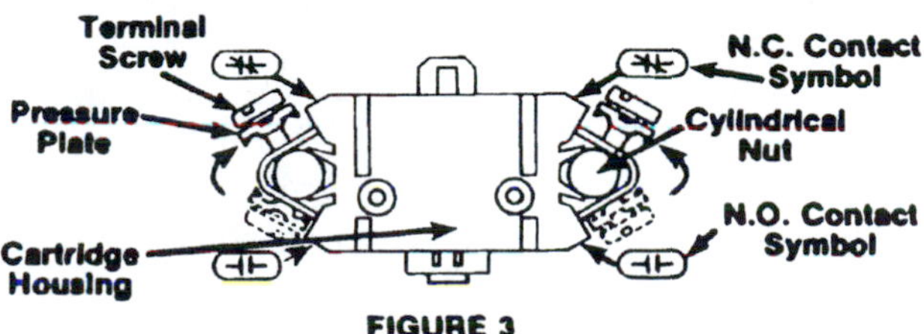

FIGURE 3

rating for the coil you can look for it in the catalog or on the specification sheet that is shipped with the new relay. If you change a relay you must also make sure that the rating for the contacts meets or exceeds the current rating and the voltage rating of the load that it will be connected to. For example, if the contacts of the relay are used to energize a 240-VAC fan motor that draws 3 A, the contacts must be rated for at least 240 V and 3 A. It is permissible to have the contacts in this example rated for more voltage and current such as 600 V and 10 A.

Contact ratings are grouped by voltage and by current. The voltage ratings are generally broken into two groups, 300 and 600 V. This means that if you are using the contacts to control 240 or 208 VAC, you would use contacts that have a 300-V rating. If you are using contacts to control a 480-VAC motor, you would need to use 600-V-rated contacts.

The current ratings of contacts are listed in amperes or horsepower. The current rating or horsepower rating must exceed the amount of current the relay is controlling. This means that if the relay is controlling 12 A, the contacts would need to be rated for over 12 A of current. The current rating of the contacts and the voltage rating for the contacts will be printed directly on the contacts, or on the side of the relay.

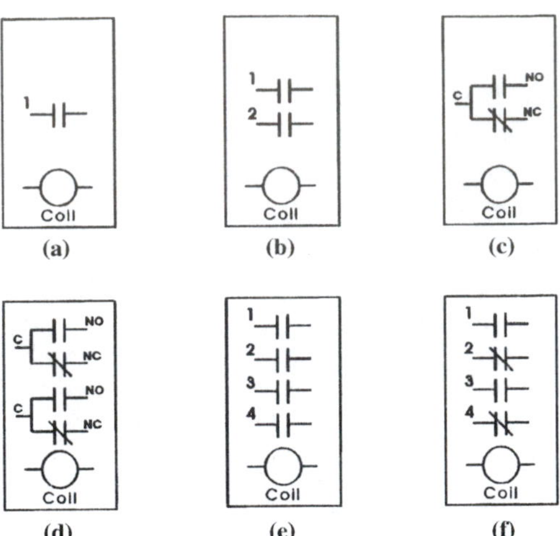

Figure 8–16 (a) A relay with a single set of normally open contacts. This type of relay is called a single-pole, single-throw (SPST) relay. (b) A relay with two individual sets of contacts. This relay is called a double-pole, single-throw (DPST) relay. (c) A relay with two sets of contacts that are connected at one side at a point called common (C). The output terminals are identified as normally open (NO) and normally closed (NC). This type of relay is called single-pole, double-throw (SPDT) relay. (d) A relay with two sets of SPDT contacts. This type of relay is called double-pole, double-throw (DPDT) relay. (e) A relay with multiple sets of normally open contacts. (f) A relay with a combination of normally open and normally closed sets of contacts.

8.8.6 Identifying Relays by the Arrangement of Their Contacts

Some types of contact arrangements for relays have become standardized so that they are easier to recognize when they are ordered for replacement or when you are trying to troubleshoot them. The diagrams in Figure 8–16 show some of the standard types of relay arrangements. Figure 8–16a shows a relay with a single set of normally open contacts. This type of relay could also have a single set of normally closed contacts instead of normally open contacts. Since this relay has only one contact and it can only close or open this type of relay, it is called a single-pole, single-throw (SPST) relay. The word pole in this identification refers to the number of contacts, and the word throw refers to the number of terminals the input contacts can be switched to. Since the contact in this relay has one input and it can only be switched to a single output terminal, it is said to have a single throw.

Figure 8–16b shows a relay with two sets of normally open contacts. Since this relay has two sets of single contacts it is called a double-pole, single-throw relay (DPST). The double-pole part of the name comes from the fact that the relay has two individual sets of normally open contacts, and the single throw part of the name comes from the fact that each contact has only one output terminal. When the coil is energized, both sets of contacts will move from their normally open position to the normally closed position. The two individual sets of contacts could be normally closed instead of normally open and the relay would still be called a DPST relay.

Figure 8–16c shows a relay with a set of normally open and a set of normally closed contacts that are connected on the left side. The point where this connection is made is called the *common point* and it is identified with the letter C. When the relay coil is energized the normally open part of the contacts will close, and the normally closed part of the contacts will open. Since these contacts basically have a common point as the input terminal, and two output terminals, one normally open (NO) and one normally closed (NC), it is called a single-pole, double-throw relay (SPDT). The most important part of this relay is that the contacts have two terminals on the output side so it is called a double-throw relay. The single-pole, double-throw relay is used where two exclusive conditions exist and you do not want them both to ever occur at the same time. For example, if this relay controls the air conditioning and the furnace for the offices of a factory, you would not want them to be on at the same time. By connecting the air conditioning system to the normally open terminal on the right side of the relay, and the furnace to the normally closed terminal, you have created the conditions so that the furnace and air conditioning system cannot be on at the same time.

Figure 8–16d has two sets of single-pole, double-throw contacts so it is called a double-pole, double-throw relay (DPDT). In this case the word double throw is used because two sets of normally open/normally closed contacts are provided. Each set has a common point on its left side (input side) and a terminal that is connected to the normally open set and a terminal that is connected to the normally closed set on its right side. This type of relay would be used where the exclusion is needed and the loads are 208 or 230 VAC where they need power from both L1 and L2. In this type of application, L1 would be connected to the common terminal (C) of one set of contacts, and L2 would be connected to the common terminal of the other set. This would cause L1 and L2 to be switched the same way in both conditions.

Figure 8–16e shows multiple sets of normally open contacts. This type of relay can have any number of sets of normally open contacts. The additional sets of contacts can be added to original contacts in some types of relays. If the original relay is manufactured with this provision, you can purchase the additional contacts and add them to the original relay by placing them on top of the original relay and tightening the mounting screws to make the additional contacts operate with the relay armature. The contacts for this type of relay could all be normally closed if the application required it. The main feature of this type of relay is that it can have any number of contacts.

Figure 8–16f shows a relay with multiple sets of individual normally open and normally closed contacts. The combination of normally open and normally closed contacts can be any mixture of the two. In most cases, the contacts in this type of relay are convertible in the field and as a technician, you could add sets of contacts and change them from normally open to normally closed or vice versa as needed. In most original installations, the relays are provided in the original equipment and you will only need to identify them for installation and troubleshooting purposes. Later, if the equipment is modified or if additional equipment, such as solenoids, motors, or pumps, are added, you may need to locate additional contacts on a relay to connect the add-on equipment so that it will operate correctly with the original system.

8.8.7 Examples of Relays Used in Industrial Electrical Systems

Relays are used in a variety of applications in industrial electrical systems. Many components, such as timer blocks and additional contact decks, can be added to a simple relay to allow it to provide a more complex function. Figure 8–17 shows examples of several types of relays that you will find in typical industrial systems. In some newer systems, the relays use plug-in terminals so that the relay can be removed and replaced without changing any wiring. The relay is simply plugged into a socket-type base. This allows the plug-in relay to be changed out on a machine very quickly and with very

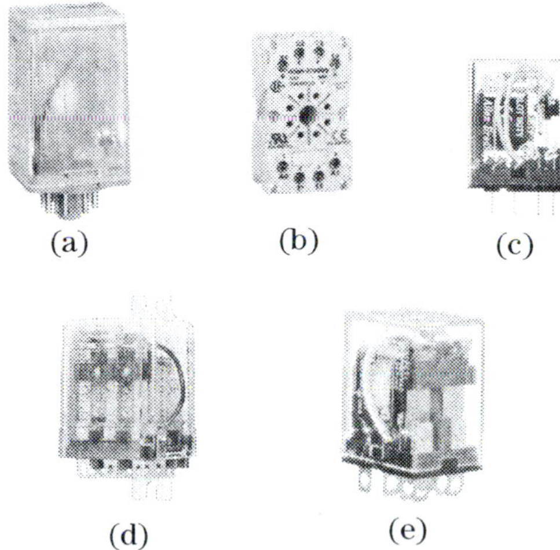

Figure 8–18 (a) General-purpose relay with 8 pin terminals; (b) screw terminal tube base sockets for 8 pin relays; (c) ice-cube-style relay with blade-style terminals; (d) flange-mounted relay with blade-style terminals; (e) miniature square base relay with solder-style terminals. (Courtesy of Rockwell Automation)

Figure 8–17 Various types of relays available for industrial applications: (a) add-on-type solid-state timer; (b) add-on-type pneumatic timer; (c) AC latching relay; (d) AC standard relay; (e) AC standard relay; (f) electrically held relay; (g) magnetic latching relay with sealed contacts. (Courtesy of Rockwell Automation)

little chance for wiring errors. Examples of plug-in relays are shown in Figure 8–18. You will begin to become familiar with relays and recognize them in the control panels of the equipment you work on as you spend more time troubleshooting and wiring electrical cabinets.

8.8.8 Adding Additional Contacts with a Contact Deck

Figure 8–19 shows a picture of a contact deck and an individual set of contacts. You can add additional contacts to most relays by bolting the deck onto the existing sets of contacts. The deck provides the mechanism to ensure all of the additional sets of contacts operate when the coil is energized. The individual contacts can be removed and replaced if you find they are faulty during troubleshooting. This is the acceptable process, rather than trying to repair any contacts. You can also determine the number of sets of normally open and normally closed contacts in any relay by changing the contact sets from open to closed or from closed to open by simply turning them over in the deck.

8.8.9 Miniature Relays

There will be times when the amount of current that a relay must switch will be very small. For these applications, you may want to use a smaller relay. Relay manufacturers produce a line of contactors that are designed specifically for small current applications. Figure 8–20 shows examples of several miniature relays. These relays were produced specifically to fit in locations where space is limited. Many of them have special mounting that allows them to be mounted on snap-on rails called DIN rails.

Figure 8–19 Example of a contact deck and extra contacts.
(Courtesy of Rockwell Automation)

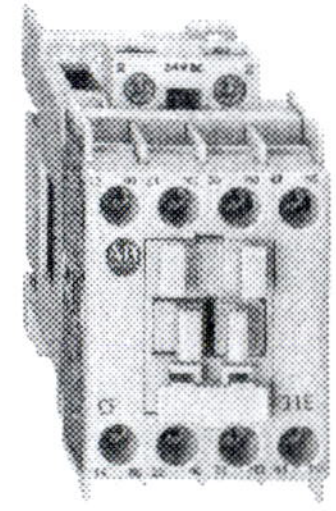
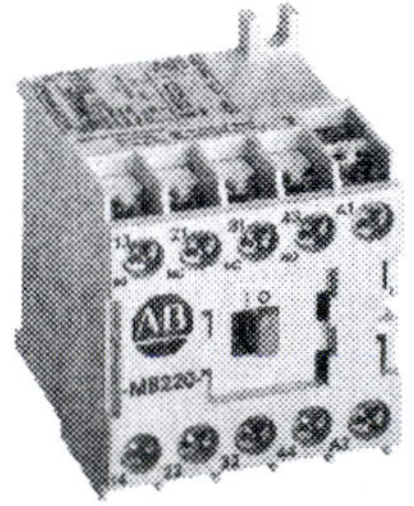

Figure 8–20 Miniature relays.
(Courtesy of Rockwell Automation)

8.8.10 Relays with Sealed Contacts

Some industrial applications require the contacts of the relay to be sealed or encapsulated so that the contacts are not exposed to the atmosphere. One type of relay with sealed contacts is called a *reed relay*. If the contacts become faulty, you can change individual sets of contacts just like you can with regular types of relay contacts. Figure 8–21 shows two types of relays that use sealed contacts.

8.8.11 Changing the Coil in a Relay

If you troubleshoot a relay and determine that its coil is faulty, you can disassemble the relay and remove and replace the relay coil. The coil is encapsulated and it has two terminals to mount the control wiring to. It is important to replace the coil with one that has the exact voltage rating. At times when you are trying to locate a replacement relay in your parts room, you will come across a similar relay, but the coil will have the wrong voltage. For example you may be looking for a relay with a 110-V coil, and you have located one with a 220-V coil. If you connected the 220-V coil in a control circuit that is powered by 110 V, there would not be sufficient voltage to pull the relay contact in. You could disassemble the relay on a work bench and remove the 220-V coil and replace it with a 120-V coil. If you connected the coil rated

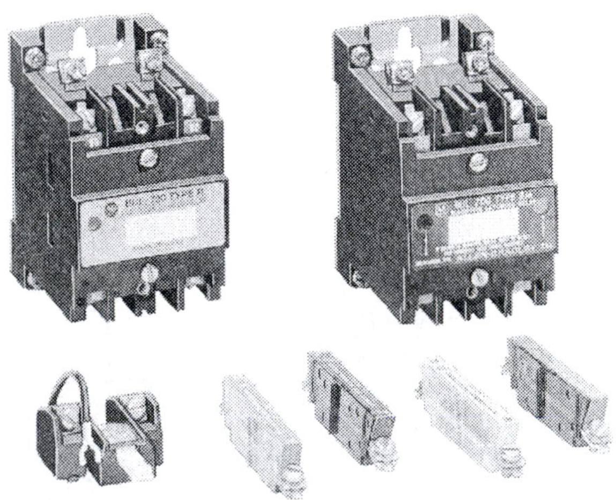

Figure 8–21 Sealed switch relays with sealed contacts.
(Courtesy of Rockwell Automation)

for 110 V to 220 V, the 220 V would damage the coil rated for 120 V when you applied power to the coil again. Most relay manufacturers make their relays so that the 120-V coil and the 220-V coil are the same size, which allows them to be interchanged, so you must be sure their voltage matches exactly. You can purchase additional coils to have on stock as replacements, in case you locate a faulty relay that has only a bad coil.

Figure 8–22 shows a typical coil. The coil is encapsulated in plastic and it has one or two holes designed into its center to allow the armature part of the relay to move through it. The armature must be free to move in and out of the center of the coil, or the coil will draw too much current. If the armature becomes stuck outside of the relay, the coil will draw excessive current that will damage the coil permanently by burning the wire in the coil and causing an open to occur in the coil wire. The coil will also be damaged if you apply power to it when it is removed from the relay or if you are bench testing it. When you are bench testing a coil, you should test it only for continuity. The coil is considered good if you have any amount of resistance that is less than infinite and more than zero. The reason it is difficult to specify the exact amount of resistance a coil will have is that the resistance will vary with the amount of wire that is used in the coil. As the amount of wire increases, the amount of resistance will increase. If you determine that the reading is infinite, you can predict there is an open somewhere in the coil and it should be replaced.

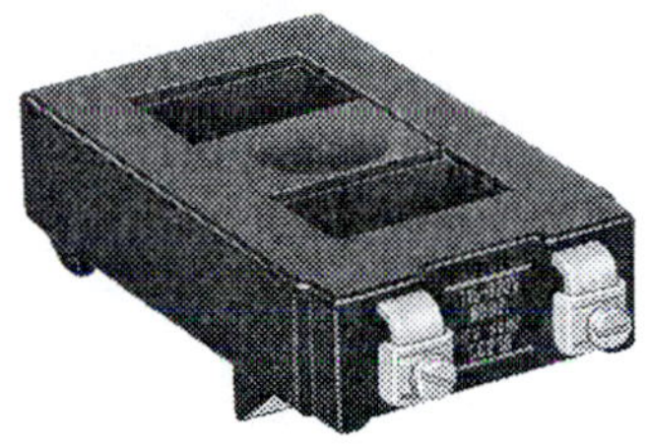

Figure 8–22 Typical encapsulated coil. (Courtesy of Rockwell Automation)

8.8.12 Typical Coil Voltage Ratings

The voltage rating for a relay coil will be stamped directly on the coil. As you learned earlier, the coil that is rated for 24 VAC will be color coded black. If the coil is rated for 110 VAC, it should be color coded red or have a red stamp on the coil. If the relay coil is rated for 208 or 230 VAC, it will be color coded green or identified with a green stamp or green printing on the coil. DC coils are color coded blue. You should remember that the amount of current a coil draws will depend on the voltage rating and the number of contacts the coil must move. The current rating is not listed on the specifications printed on the coil, and it is generally not important, since the relay coil is designed to handle the amount of current needed to make the contacts move. Figure 8–23 shows a table of voltages and the code letters for AC-powered relay coils.

Figure 8–23 Typical voltage ratings for AC relay coils. (Courtesy of Eaton/Cutler Hammer)

Coil Volts and Hertz	Code Suffix
120/60 or 110/50	A
240/60 or 220/50	B
480/60 or 440/50	C
600/60 or 550/50	D
208/60	E
277/60	H
208-240/60 ①	J
240/50	K
380-415/50	L
550/50	N
24/60, 24/50 ②	T
24/50	U
32/50	V
48/60	W
48/50	Y

① NEMA Sizes 00 and 0 only
② NEMA Sizes 00 and 0 only
Sizes 1-8 are 24/60 only

Questions

1. Explain the difference between a permanent magnet and a electromagnet.
2. Define the term saturation as it applies to an electromagnet.
3. Explain the principle of electromagnetic induction.
4. Identify two ways to increase the strength of an electromagnet.
5. Explain why it is important that the magnetic field of an electromagnet can be turned on and off.

True and False

1. When dipoles are aligned in a material, it is a good magnet.
2. Any time current flows through a conductor, magnetic flux lines are created around the conductor.
3. The polarity of the magnetic field of an electromagnet is determined by the direction of current flow through the conductor.
4. Like poles of magnets attract.
5. The strength of the magnetic field for a permanent magnet can be changed easily.

Multiple Choice

1. _____ can have the strength of its field changed easily.
   a. A permanent magnet
   b. An electromagnet
   c. A dipole
2. The polarity of an electromagnet can be changed by _____
   a. changing the direction of current flow through the coil of wire.
   b. changing the amount of current flow through the coil of wire.
   c. changing the frequency of the current flow through a coil of wire.
3. A laminated steel core is used in electromagnets that have AC voltage applied to them because _____
   a. laminated steel is more economical to use than soft iron.
   b. laminated steel is easily formed to any shape, which makes it more usable in complex components.
   c. laminated steel allows the magnetic field to build and collapse quickly.
4. A magnetic field is created in a coil of wire when _____
   a. current flows through the wire.
   b. current stops flowing through a wire.
   c. the coil has a core.
5. The left-hand rule is used to _____
   a. determine the amount of magnetic flux in a coil.
   b. determine the number of coils in an electromagnet.
   c. determine the polarity of the magnetic field in an electromagnet.

Problems

1. Draw a sketch of a permanent magnet and show where flux lines occur.
2. Draw a sketch of an electromagnetic coil and show where flux lines occur.
3. Draw a sketch of an electromagnetic coil with an open somewhere in the coil of wire. Explain what effect the open will have on the magnetic field.
4. Draw two permanent magnets with their poles placed so the two magnets will attract each other.
5. Perform the experiment presented in Section 8.1 and explain why the flux lines are concentrated around the ends of the permanent magnet.

Chapter 9

Contactors and Motor Starters

Objectives

At the end of this chapter you will be able to:

1. Identify the contacts and coil of a contactor and motor starter.
2. Explain the operation of a contactor and motor starter.
3. Identify the components in a control circuit.
4. Select the proper size of contactor from a NEMA table.
5. Explain the differences between NEMA specifications and IEC specifications.

The contactor is similar to the relay in that it has a coil and contacts, but it is more closely related to the magnetic motor starter because of its applications and size. In fact, some models of magnetic motor starters are made by adding a set of overloads to the contactor. At one time the distinction between relays and contactors was based strictly on size. The *relay* was classified as a set of contacts rated less than 20 A that are controlled by a coil. A *contactor* was classified as a set of contacts rated more than 20 A controlled by a coil. Today, this classification becomes less clear because power relays are available with contacts rated up to 30 A. Another reason this classification is no longer valid is that a relay and a contactor can utilize solid-state components instead of a coil. This caused the new classification to include any device that could energize and deenergize an electrical circuit repeatedly. The contactor and relay are now distinguished more by their applications than their size. The relay is often used for general-purpose switching applications and light-duty inductive and resistive loads. It is also used for control or logic purposes in electrical circuits. The contactor is used for motor-starting applications and to control lighting, heating, transformer, and capacitive loads.

The operation of the contactor is explained in this chapter. The basic design is discussed and the function of all major parts is provided. Additional information is provided regarding NEMA sizes, types of auxiliary contacts, and typical applications and circuits. Installation, preventive maintenance, and troubleshooting procedures are also detailed.

The last part of this chapter continues this discussion of the difference between a contactor and a magnetic motor starter. The motor starter has provisions for possible motor overload protection. Since the motor starter is the most widely used motor control, a detailed explanation of its basic parts and operation is provided. Additional information is given regarding typical applications for the basic motor starter, including wiring diagrams. A detailed comparison of traditional motor starter specifications from the National Electrical Manufacturers Association (NEMA) and the European standards from the International Electro-technical Commission (IEC) is provided. This chapter also provides some basic knowledge of solid-state motor controls, and more detailed information about solid-state devices is provided in Chapter 19.

9.1 Contactors

Contactors are being used to control motors and other types of noninductive loads, such as industrial heating and lighting loads. The contactor is designed similar to the relay in that it has a magnetic coil that controls the operation of one or more sets of contacts. The contactor can also have any number of auxiliary contacts rated slightly less than the main line contacts added to the main contactor frame. The contactor is available with mechanical or electrical holding and latching mechanisms.

9.1.1 Basic Parts of the Contactor

Figure 9–1 shows a contactor and an exploded view of a typical size 1 contactor. You can see that it has six basic parts. The largest part of the contactor is the base. It has several other pieces molded or fastened to it permanently, including the mounting plate, arc hood, and stationary contacts with field wiring terminals.

The mounting plate provides the method of mounting the contactor to an electrical panel. It also provides the mounting platform for the contactor base and keeps all of the moving and stationary parts properly aligned. The base is generally made of high-quality die-cast metal so that it can easily absorb and dissipate heat from the coil. The base also provides tapped threaded pads for all the additional parts of the contactor to be mounted to securely. Since the additional parts, such as contact carrier and coil, are mounted to the base with screws, they can easily be removed and replaced when they become faulty. The stationary contacts and field wiring terminals are mounted with screws to the top of the base. Since these contacts are also mounted with screws, they can be changed as part of the contact assembly in the event that they wear prematurely.

Arc quenchers are mounted on the stationary contacts, where they can absorb any arc drawn between the movable and stationary contacts as the contactor opens and closes them. This shield also prevents dangerous arcs from coming out of the contacts to other metal parts, such as the electrical cabinet. The contact carrier serves several vital purposes. The first is to provide a platform for the movable contacts. It also provides a housing for the coil and magnet assembly. When the coil is energized, the magnetic field pulls the contact carrier upward until the movable contacts seat against the stationary contacts. This assembly allows the contactor to have only one movable part

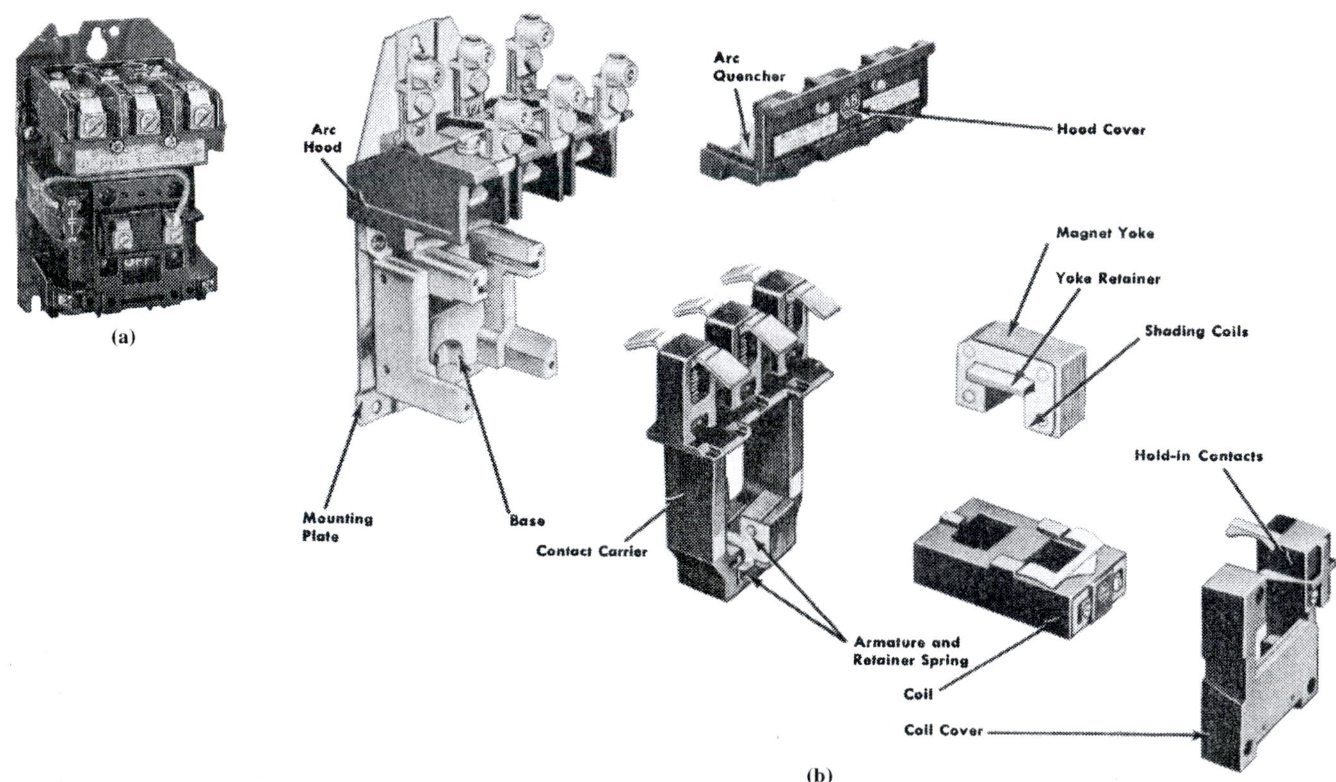

Figure 9–1 (a) A typical contactor; (b) exploded view of a typical contactor.
(Courtesy of Rockwell Automation)

when it is fully assembled. The contact carrier also provides a place to mount the armature and retainer spring. The molded coil is held in place in the middle of the contact carrier by the magnet yoke and clips. The coil cover is then mounted to the base with screws that prevent the coil from falling out of the contact carrier. The coil cover provides enough room for the contact carrier to be pulled up when the coil is energized. Since the contact carrier is the only moving part for the contactor, the operation is rather simple and reliable. The magnet is made with a permanent air gap that allows the contact carrier to drop out freely and prevent residual magnetism. The contacts are the double-break style and are made of solid cadmium oxide silver, which provides high conductivity and resistance to welding and pitting. The field wiring terminals provide a means of connecting field wiring to the two sides of the stationary contacts. When the double-break movable contact segment is seated to the stationary contacts, a conductive bridge is made across them, allowing current to travel from the line terminal to the load terminal.

9.1.2 Types of Contacts and Protection Features

The contactor is available with a variety of contacts for different applications. The contactor is also available with any number of poles up to five. When the contactor is purchased, the number of main contacts must be specified. If the contactor is used for DC circuits, a single-pole, normally open contactor is available. Up to five normally open main contacts are available for AC loads.

The contacts are made from cadmium oxide and silver to resist corrosion and wear due to arcing. This material also helps AC arcs extinguish themselves. Several types of protection are built into the contactor to prevent damage due to arcing. These features include arc chutes, arc traps, and blowout coils. The arc chutes and arc traps are provided on DC contactors to route the arc that is produced when the contacts are opened or closed, away from damaging the contact surface. DC loads produce larger arcs when contacts are opened or closed because DC current flows in one direction. AC arcs are not as excessive since the AC current is sinusoidal. The AC current changes direction during every cycle and returns to zero at the 180-degree point in the cycle. This action tends to keep to a minimum the arc caused by contacts opening and closing.

Since the DC current flows in only one direction, the magnetic field can become excessive and large arcs are caused when the contacts are opened or closed. The arc will begin when the contacts start to open and continue to grow as the distance between the contacts continues to grow. The arc will be at its largest when the contacts are fully opened. The arc that is developed when the contacts are opened tends to be much larger than the arc that is produced while the contacts are closing.

The arc produced when the contacts are closing is caused by inductive loads pulling locked-rotor amperage (LRA). The locked-rotor amperage, also called *inrush current*, tends to be very large while the motor is coming up to running speed. This current will be at its largest when the contacts are initially closed. The contactor provides arc chutes and arc traps to keep these arcs in check. The idea of these features is to limit the arc as much as possible and to shunt any residual arc to a position in the contactor away from the contact surface. The contactor also provides protective covers over the contacts to prevent the arc from coming out from the contacts and hurting someone. The arc chutes and arc traps provide a location in the contactor where the arc can be extinguished without damaging the contactor. These features are called *arc suppression* and are provided in one means or another in all contactors. You should not operate a contactor under load with any of these protective devices removed or damage may occur. When the contactor is inspected during preventive maintenance, it is important to see that the arc suppression features of the contactor are operational and undamaged.

9.1.3 Contact Sizes

Contactors are rated by NEMA according to the size of load that they can handle safely. The load rating for a contactor is based on the size of the contacts. Figure 9–2 shows a NEMA rating for contactors. The smallest contactor is a size 00 and the largest is a size 9. This table shows the load rating as a maximum continuous current or as a maximum horsepower. Notice that these ratings include the operating voltage and indicate whether the supply voltage is single phase or three phase.

Figure 9–3 shows two of these contactor sizes. These photographs provide an idea of the relative size of these contactors. The size 3 is about 10 inches tall and can carry a maximum of 90 A in a continuous load. The size 5 is about 13 inches tall and can carry a maximum load of 270 A. These amperage capabilities are also within the range of a power relay, which can cause confusion when relays and contactors are classified strictly by size. The ratings for each of the contactors include the largest horsepower load that can be controlled by plugging. The smallest contactors are not rated for plugging and should be limited to continuous operation. Plugging is a process where full reverse voltage is applied to a motor while it is running to cause it to stop quickly.

NEMA Size	Contin-uous Amp. Rating	600 VOLTS MAXIMUM				
		Maximum Horsepower Rating 2 Full load current must not exceed the "Continuous Ampere Rating"			Maximum Horsepower Rating For Plugging Service 1	
		Volts	Single Phase	3 or 2 Phase	Single Phase	Three Phase
00	9	120	⅓	¾	—	—
		208	—	1½	—	—
		240	1	1½	—	—
		480	—	2	—	—
		600	—	2	—	—
0	18	120	1	2	½	1
		208	—	3	—	1½
		240	2	3	1	1½
		480	—	5	—	2
		600	—	5	—	2
1	27	120	2	3	1	2
		208	—	7½	—	3
		240	3	7½	2	3
		480	—	10	—	5
		600	—	10	—	5
2	45	120	3	—	2	—
		208	—	15	—	10
		240	7½	15	5	10
		480	—	25	—	15
		600	—	25	—	15
3	90	120	—	—	—	—
		208	—	30	—	20
		240	—	30	—	20
		480	—	50	—	30
		600	—	50	—	30
4	135	120	—	—	—	—
		208	—	50	—	30
		240	—	50	—	30
		480	—	100	—	60
		600	—	100	—	60
5	270	120	—	—	—	—
		208	—	100	—	75
		240	—	100	—	75
		480	—	200	—	150
		600	—	200	—	150
6	540	208	—	200	—	150
		240	—	200	—	150
		480	—	400	—	300
		600	—	400	—	300
7	810	208	—	300	—	—
		240	—	300	—	—
		480	—	600	—	—
		600	—	600	—	—
8	1215	208	—	450	—	—
		240	—	450	—	—
		480	—	900	—	—
		600	—	900	—	—
9	2250	208	—	800	—	—
		240	—	800	—	—
		480	—	1600	—	—
		600	—	1600	—	—

Figure 9–2 Size ratings for contactors with horsepower and current ratings.

(Translated and reproduced with permission of the National Electrical Manufacturers Association. This standard may not be further reproduced. Only the original English version is valid for any legal or judicial consequences arising out of the application of the standard.)

Figure 9–3 Pictures of contactor sizes 3 and 5. The size 3 is approximately 5 inches tall and the size 5 is approximately 12 inches tall.

(Courtesy of Rockwell Automation)

(a)

Auxiliary Contacts

Bulletin 500 Line contactors and starters can accommodate a total of eight auxiliary contacts. Four of these contacts can be used without increasing panel space. Each contact easily snaps into place and is held firmly to guard against dislodging by vibration. Contacts are identified with a N.O. or N.C. symbol which is clearly visible from the front of the starter. The contacts are bifurcated and positioned in a vertical plane for reliability.

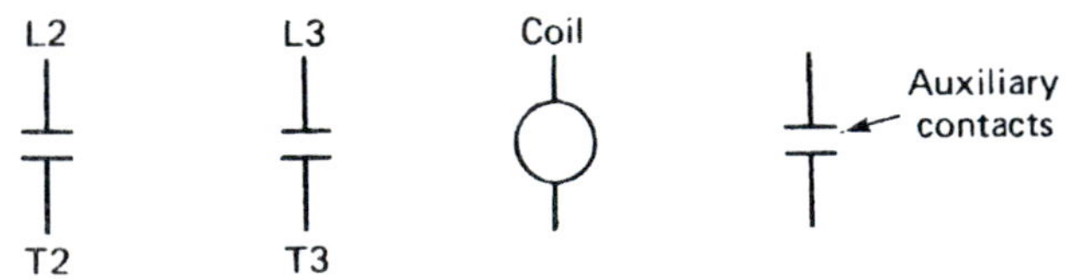

Electrical diagram of auxiliary contacts

(b)

Figure 9–4 (a) Auxiliary contacts mounted on a contactor; (b) diagram of main, coil, and auxiliary contacts.

(Courtesy of Rockwell Automation)

9.1.4 Auxiliary and Special-Purpose Contacts

The main contactor can have a variety of contacts added to it for special purposes. Auxiliary contacts can be added to the frame of the main contactor so that they can be operated by the contactor's coil. When the coil is energized, the contact carrier will also activate the auxiliary contacts. Figure 9–4 shows auxiliary contacts added to a contactor. A diagram is provided to show the location of these contacts and how they are operated. An electrical diagram is also shown in this figure, which will be used in all wiring diagrams and ladder diagrams to indicate circuits that utilize auxiliary contacts.

Auxiliary contacts can be normally open or normally closed. Some types are field convertible so that they can be converted from normally open to normally closed, or vice versa. In the diagram and on the contactor the types of contacts that are used will be identified clearly so that you will be able to troubleshoot the contactor. The auxiliary contacts are also rated by NEMA. The auxiliary contacts for a size 0 contactor are rated for 18 A. The auxiliary contacts for a size 1 are 27 A and for size 2 are 45 A. These contacts are heavy enough to control fairly large loads. Other types of auxiliary contacts are rated for pilot duty only. These types of contacts are usually used as sealing or maintaining contacts. They may also be used as control contacts to energize other lines of the circuit.

Other contacts are available to be used as alarm contacts. These contacts are usually wired to an indicator lamp. When the contactor is energized, the indicator lamp will be energized through the alarm contacts. When the contactor is deenergized the alarm contacts can be wired as a set of normally closed contacts. In this type of application, the indicator lamp would become energized when the contactor is deenergized. Mechanical holding or interlock contacts are available for use with most contactors. The mechanical interlock operates like a latch. When the coil is energized, the contact carrier is pulled up so that movable contacts seat to the stationary contacts. When the contact carrier is pulled up as far as its stroke will allow, a mechanical interlock is tripped and keeps the contact carrier in this position even when current is interrupted to the coil. The reset or unlatch coil must be energized for the interlock to be released.

This figure also shows NEMA ratings for the latch coil and unlatch coil. From this table you can see that the reset coil is smaller than the main coil, since it needs only to pull the interlock mechanism and allow the contact carrier to drop out. The coil current will be smaller as the voltage is increased.

9.1.5 Coils for Contactors

The coils for all the NEMA starters are available in a variety of voltages, which allows the contactor to be used in most any AC or DC control circuit. Figure 9–5 provides a table that shows the AC voltage on which these coils will operate. It also lists the pull-in and seal-in currents for each of these coils. Notice that AC coils are available for 120-, 208-, 240-, 480-, and 600-V circuits. Another table in this figure shows the typical operating time it takes for the coils to operate each of the NEMA

COIL CURRENTS

NEMA Size	No. of Poles	Inrush Current (Amps.) 60 Cycles					Sealed Current (Amps.) 60 Cycles				
		120V	208V	240V	480V	600V	120V	208V	240V	480V	600V
00	1-2-3	0.50	0.29	0.25	0.12	0.07	0.12	0.07	0.06	0.03	0.02
0	1-2-3-4	0.88	0.50	0.44	0.22	0.17	0.14	0.08	0.07	0.04	0.03
1	1-2-3-4	1.54	0.89	0.77	0.39	0.31	0.18	0.10	0.09	0.04	0.04
2	2-3-4	1.80	1.04	0.90	0.45	0.36	0.25	0.14	0.13	0.06	0.05
3	2-3	4.82	2.78	2.41	1.21	0.97	0.36	0.21	0.18	0.09	0.07
	4	5.34	3.08	2.67	1.33	1.07	0.39	0.23	0.20	0.10	0.08
4	2-3	8.30	4.80	4.15	2.08	1.66	0.54	0.31	0.27	0.14	0.11
	4	9.90	5.71	4.95	2.47	1.98	0.61	0.35	0.31	0.15	0.12
5	2-3	16.23	9.36	8.11	4.06	3.25	0.81	0.47	0.41	0.20	0.16
6	2-3	0.62	Current Shown is The Very Small AC Current Passing Through Coil of Control Relay. Any Standard Duty 120V Control Station May Be Used In This Circuit.				0.082	Current Shown is The Very Small AC Current Passing Through Coil of Control Relay. Any Standard Duty 120V Control Station May Be Used In This Circuit.			
7	2-3	0.62					0.082				
8	2-3	0.62					0.082				
9	2-3	1.2					0.16				

(a)

OPERATING TIME

Size	Approximate Operating Time in Milliseconds 3 Pole Contactors	
	Pick-up	Drop-out
00	28	13
0	29	14
1	26	17
2	32	14
3	35	18
4	41	18
5	43	18
6	88	40
7	88	45
8	118	94
9	118	84

(b)

Figure 9–5 (a) Typical inrush current and seal-in current for contactor coils; (b) typical operating times for coils. (Courtesy of Rockwell Automation)

contactors. These data are useful in selecting the proper contactor and coil for original installation or as a replacement part.

Figure 9–6 shows a typical encapsulated coil. These coils are made by tightly winding small wire on a core. The coil is then dipped in thermoset epoxy and terminals are provided for field wiring connections. Most manufacturers use color codes to identify the voltage rating for each coil. Some manufacturers use a different color thermoset epoxy for each voltage, while others use a color patch on the side of the coil to make the identification. All coils will be clearly marked with the voltage rating and the operating frequency.

The design of a DC coil assembly is slightly different from that of an AC coil assembly. The DC coil is built around a solid steel core, which can build extremely strong magnetic fields. The DC coil also has built-in shunt plates that retard the magnetic flux that builds in the coil until the applied voltage increases to the minimum coil pickup value. When the applied voltage reaches the coil pickup value, the flux will quickly build and the magnetic field will snap the contact carrier into place. This action causes the contacts to come together in a snap action, which limits the arc across the contacts. It also prevents the contacts from pitting and wearing when the contacts are energized.

The coils that are used in AC circuits are made from simple encapsulated coils of wire. The major difference between the AC and DC assembly is in the design of the laminated sections used in the AC coil assembly. The AC coil utilizes laminated sections instead of the solid steel coil used in DC coils. The laminated sections use a design that is similar to the type used in AC relay coils. The contactor coil assembly requires an air gap to allow the contactor to break loose cleanly when the coil is deenergized. The laminated sections provide protection against eddy current buildup and residual magnetism, just as in a relay coil.

Some coils provide an internal thermal cutout that is built into the coil when it is manufactured. The thermal cutout is located in the middle of the coil where heat initially builds up when the coil begins to overheat. This cutout will open the coil circuit and deenergize the coil, which protects it from overheating. When the coil cools down the cutout will reset and allow current to flow again. Since the cutout resets itself automatically, this type of coil should be used only with contactors that are wired for manual reset. This means that the contactor is controlled by a momentary-type start button. The start button is sealed with a set of auxiliary contacts, and when the thermal cutout deenergizes the coil circuit, the auxiliary contacts drop the circuit out until the pushbutton is pushed manually. The coil must be cool enough for the cutout to reenergize or the start button will not energize the coil again.

This type of protection will prevent the coil from burning out due to overheating. This could be caused by the contact carrier jamming and not allowing the armature to center itself in the coil. When this occurs, the coil's impedance remains low and the coil will draw excessively large currents and cause the coil to overheat. The coil could also overheat if the applied voltage became too large. This could occur if the utility voltage increased due to minimal loading on its generator. Coil voltage could also become too large if the wrong-size coil was used. This may occur when a coil or complete contactor is replaced by one of incorrect size.

Contactor coils can also be protected from excessive voltage transients with a surge suppression device. These devices are manufactured small enough to be mounted directly across the coil terminals. Like the coil, the surge suppression device is encapsulated in epoxy plastic to withstand the rugged factory floor environment. These devices must be matched with the coil voltage and may not be available for the larger coil voltages from all manufacturers.

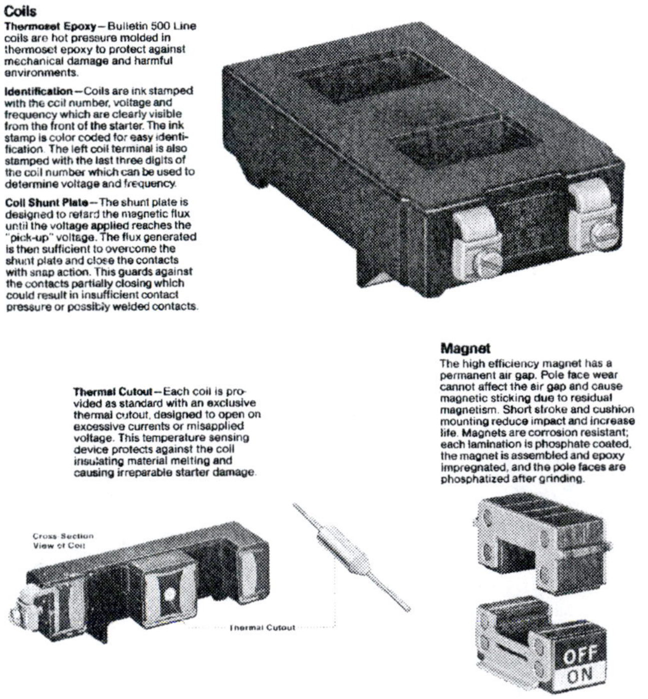

Figure 9–6 A typical encapsulated coil for a relay, contactor, or motor starter.
(Courtesy of Rockwell Automation)

9.1.6 Applications for Contactors

Contactors can be used for a variety of motor starting applications. These applications include starting and jogging single-phase and three-phase AC and DC motors of all sizes. Since the contactor is available in sizes up to a NEMA size 9, it can be used to start motors as large as 1600 hp. Some applications in motor control allow a contactor to be used instead of a motor starter, as long as some type of current protection, such as fuses, are used with the contactor to protect the motor against overcurrent conditions. The motor starter provides integral overcurrent protection in the form of heaters and overloads and is the control of choice for modern control circuits; however, you will encounter contactors that are used to control motor applications on some control systems and will need to know how to troubleshoot these circuits.

The contactor can be used in two-wire and three-wire motor starting circuits. Figure 9–7a shows an example of a two-wire circuit and Figure 9–7b shows a three-wire circuit. The control circuit in each circuit controls the coil of a motor starter or contactor. The motor that is connected to the contacts is not shown in these diagrams to keep them simple. The motor could be a three-phase motor or a single-phase motor. The two-wire circuit is so named because only two wires are connected to the coil. This means that the coil circuit is one large series circuit. Any pilot devices are located in this series circuit. The circuit in this diagram shows a float switch in series with the coil of the contactor used to control a large sump pump. The float switch will close its contacts any time the level of water in the sump in which it is mounted rises above the predetermined set point. When these contacts close, current is passed to the contactor's coil. When the contactor is energized, its contacts are energized and the motor will begin to run. Since the motor is used in a pumping application, it will be driving a water pump from its shaft. When the water level in the tank reaches the predetermined low set point, the float switch contacts will be opened and the coil will be deenergized and allow the contacts to drop out and stop the pump.

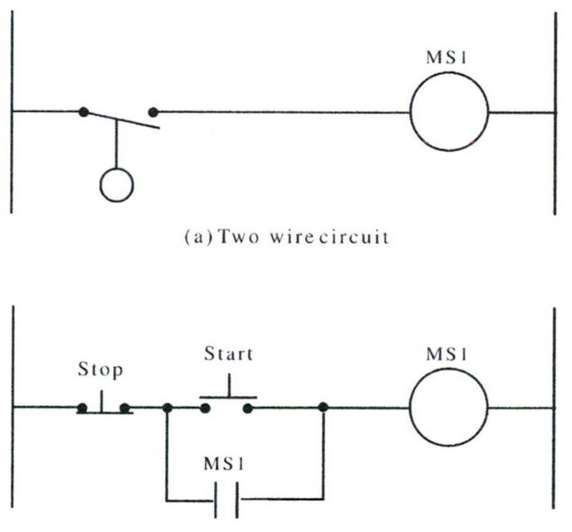

Figure 9–7 (a) A two-wire circuit; (b) a three-wire circuit. (Translated and reproduced with permission of the National Electrical Manufacturers Association. This standard may not be further reproduced. Only the original English version is valid for any legal or judicial consequences arising out of the application of the standard.)

Notice that the contactor does not provide overcurrent protection for the motor; overcurrent protection must be provided by another means. In a discussion later in this section we explain the method of adding overloads to the contactor. The motor in this application is protected with a fused disconnect. The fuses in the disconnect are the dual-element type, which allows the motor to start with large inrush current and yet protects the motor at the full-load current level under normal operating conditions. The fuses will blow and protect the motor any time the full-load current exceeds a safe limit. The fuses also provide excellent short-circuit protection for this application.

A three-wire control circuit can also be used to control a contactor. The diagram in Figure 9–7b shows the contactor in this application controlling a motor that is driving a conveyor. In the three-wire circuit a set of auxiliary contacts are used in conjunction with a momentary-type start/stop button. When the start button is pressed, its contacts close as long as the button is depressed. Current is passed through the start button to the stop button. As long as no one is pressing on the stop button, its contacts will remain closed and pass current to the contactor's coil. The contactor's coil will energize and pull its contacts closed. The normally open auxiliary contacts will also close when the coil is energized. These contacts are in parallel with the start button contacts and will seal them in when they are closed. The auxiliary contacts provide an alternative route for current to pass around the normally open start button contacts. This means that current will flow through the auxiliary contacts around the start button after it has been released and has returned its contacts to the normally open position. This allows adequate current to reach the coil through the auxiliary contacts and keep it energized.

When the stop button is depressed, all current to the contactor's coil will be interrupted and the coil will deenergize and allow all the contacts to drop out. When the auxiliary contacts drop open, they will keep the circuit deenergized until the start button is depressed manually.

A variety of safety devices can be added in series with the stop button contacts. If any of these safety devices sense an unsafe condition, their contacts will open and the circuit will be deenergized just as if the stop button had been depressed. Since the auxiliary contacts will drop out the seal-in circuit, the circuit will have to be restarted manually by pressing the start button.

9.1.7 Jogging a Motor with a Contactor

The contactor can also be used to control motors for jogging and plugging applications. The jogging and plugging circuits are explained in detail in Chapter 15. In this section we concentrate on the contactor. The basic principle of the jog circuit provides single-button control that allows the operator to use the start pushbutton to energize the motor for short periods of time without sealing in the contactor coil with auxiliary contacts. In Figure 9–8 you can see how a jog/run selector switch is added to the circuit. The jog/run button has two sets of starting contacts, marked jog and run. When the button is in the normal starting position, the switch allows current to flow to the "run" terminal so that when the start button is depressed, power is allowed to flow to auxiliary contacts, which allows the contacts to seal in the coil like a normal three-wire system. When the jog/run selector is switched to the jog position, the switch is moved to an open position, which disables the seal-in circuit. When the start pushbutton is depressed, the coil is energized, and the auxiliary contacts close, but the current used to seal the circuit is interrupted by the open that is caused by the jog/run switch. This means that the coil will remain energized only as long as the start pushbutton is depressed. Whenever the pushbutton is released, current to the coil is deenergized and the coil drops the main load contacts out, which stops the motor.

The contactor must be rated for jogging to be used in this type of application. The reason for the special rating is that the motor will draw locked-rotor amperage every time the motor is started. Since this current is 5 to 10 times the normal running current, the contactor will tend to overheat from starting too often. Contactors that are rated for jogging are manufactured to handle the extra heat. Some manufacturers derate the load capacity of their normal contactor when it is used for jogging.

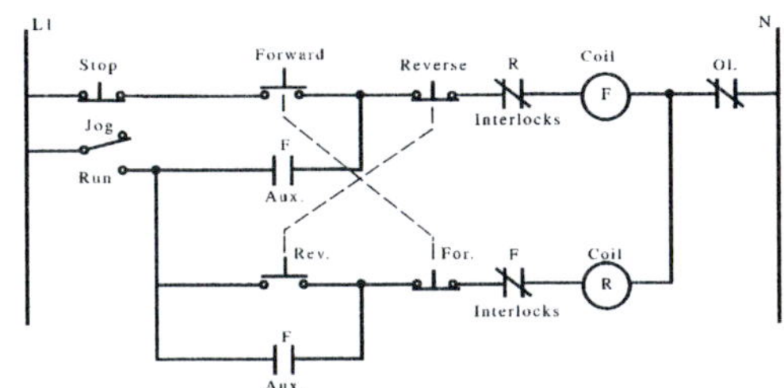

Figure 9–8 A typical jogging circuit.

9.1.8 Reversing a Motor with a Set of Contactors

A motor can also be reversed using two contactors. The contactors can be interlocked electrically or mechanically. Figure 9–8 which we used to show the job application also shows the forward and reversing operation using contactors. One of the contactors is used to control the motor in the forward direction and the second contactor is used to control the motor in the reverse direction. When the forward pushbutton is depressed, the forward contactor coil is energized, which pulls in the forward contacts and applies power to the motor in the forward direction. The forward contactor has one set of normally open auxiliary contacts connected in parallel with the forward pushbutton to seal it in and a set of normally closed auxiliary contacts wired in series with the reverse contactor's coil. Each of these sets of contacts is identified with the letter F to indicate that they will be energized by the forward contactor coil.

The second contactor is used to apply reverse voltage to the motor to cause it to operate in the reverse direction. When the reverse pushbutton is depressed, the reverse motor starter will not become energized because the normally closed set of contacts controlled by the forward contacts will be energized to the open position, which disables the reverse coil circuit.

This means that the stop pushbutton must be pressed first to deenergize the forward contactor. When the forward contactor is deenergized, the normally closed contacts controlled by the forward contactor will return to the closed position and allow the reverse contactor circuit to become operational if the reverse pushbutton is depressed. When the reverse pushbutton is depressed, the reverse contactor coil will become energized and the motor will begin to operate in the reverse direction. The normally closed set of auxiliary contacts that are controlled by the reverse contactor will disable the forward contactor's coil. This prevents the forward contactor and reverse contactors from coming on at the same time, which would cause a direct short circuit across the two contactors.

The reversing contactor can be used for applications where the motor must be capable of operating in the forward and reverse directions, such as elevators and lifts. This circuit could also be used for reversing conveyors or machine tool applications.

9.1.9 Using a Contactor to Control Lighting and Heating Loads

The contactor is well suited for controlling lighting loads and other noninductive loads. Lighting loads such as incandescent filaments used in tungsten lights can be controlled with contactors. The contactor allows the lights to be controlled from a remote location. It also allows multiple switches for controlling the lighting system, which is very useful in large industrial areas. The contactors can also be used for ballast lighting used in mercury arc lighting. These types of lighting systems present an inductive load during starting, which may cause some arcing when the contactor is energized. Contactors for these applications provide three, four, or five poles for specialized lighting circuits. The contactor can control individual lights or complete circuits. The contactor's coil is also available for 277-V applications. This allows the coil to use the same voltage that is used for the main power. Any type of pilot device can be used to activate the control circuit. This includes pushbutton or selector switches. In some applications, such as commercial lighting in stores or public buildings, key lock switches are used to energize the contactor. This prevents the lights from being turned off or on accidentally.

The contactor is also well suited for controlling electric heating loads. Unlike motors, resistive heaters do not need overload protection, so they can be controlled by a contactor and protected by a fused disconnect. The single-element fuse in a disconnect is sufficient to meet all NEC code requirements. A circuit breaker could also be used as branch circuit protection and provide protection for the heating elements.

Since the contactor does not require overload protection, it can use a two-wire or three-wire control circuit with thermostatic pilot switches to switch the coil off and on. The contactor is useful for controlling single-phase or three-phase heating systems, since they are available with a variety of poles.

The contactor is also rated for inductive, resistive, and radiant types of heating systems. These types of systems can be controlled using a thermocouple to provide accurate temperature control in an industrial furnace.

Another application for the contactor is switching large industrial capacitive circuits used for power factor correction. Since the contactor's coil can be energized by rather small voltages and currents, it can easily be interfaced with a microprocessor power factor control. The amount of leading or lagging phase shift can be programmed into the controller and when the power factor exceeds the set point, large banks of capacitors can be switched into the circuit by the contactors to correct it. This type of application requires only that sets of contacts be switched open or closed at the appropriate times as directed by the controller unit. If the power factor can be controlled within the limits specified, the utility bill could be reduced by thousands of dollars a month in some cases.

9.1.10 NEMA and IEC Standards for Contactors

There are a large number of manufacturers of contactors from North America and around the world. Several associations and commissions have agreed upon sets of tests to provide some standardization of electrical contactors. The major association for North America is the National Electrical Manufacturers Association (NEMA), and the commission for the rest of the world is the International Electro-technical Commission (IEC). Some manufacturers refer to the IEC as the European standards. The IEC makes recommendations for manufacturers to test products and publish technical data for product comparisons. NEMA has established standards for electrical products and test specifications for equipment. Many of these specifications and standards have been adopted by other agencies, such as the Underwriters Laboratories (UL).

The IEC and NEMA do not perform testing like UL does, but they do help establish test parameters. The net effect of these organizations is that they help engineers and technicians compare and select the correct-size product for the application. The IEC was established in England in 1906 and has been making recommendations with over 42 member nations since that time. NEMA was established in 1926 and presently has over 550 members, representing manufacturers of all types of electrical products. More information about IEC and NEMA is provided at the end of this chapter in Section 9.4 after you have learned more about magnetic motor starters.

9.2 Magnetic Motor Starters

The magnetic motor starter is one of the most widely used motor controls. The starter provides operational contacts such as the contactor and additional built-in overcurrent protection. The overcurrent protection is in the form of thermal or magnetic overloads. An overload can be added to a contactor for motor protection, or it can be considered an integral part of the motor starter. In reality, the same overload that is built into the motor starter is also available for use on a manual motor starter and can be added to a contactor to provide motor protection.

In this section we explain the operation of electrical and mechanical components of a magnetic motor starter. The operation of the overloads will also be explained. Typical applications and circuits will be presented as well as methods of selecting the proper size of motor starter and overloads. In the last part of this section we explain installation and troubleshooting procedures that you will be able to utilize in the classroom and later on the factory floor while you are on the job.

Figure 9–9 shows a photograph of a typical magnetic motor starter, also called a magnetic starter. A magnetic starter looks very similar to a contactor; in fact, it has many of the same components: contact assembly, contacts, coils, and auxiliary contacts. The main difference between a contactor and a motor starter is that the motor starter has overload protection built into it to protect motors from overcurrent. The contact assembly operates the same as the assembly in the contactor. When the coil is energized the contact assembly is pulled up so that the armature is centered in the coil. This upward action causes the movable contacts to press closed to the stationary contacts. When the armature assembly is pulled up into the coil, the pole faces on the armature are pulled tight to the upper pole face. A small air gap remains between the pole faces so that the magnetic field will be broken when the coil is deenergized.

Since the armature assembly with the contact carrier is the only movable part of the magnetic starter, it can be pulled straight up and has a very short vertical stroke. This means that the coil can snap the contacts closed, which causes a very small amount of contact wear.

9.2.1 Parts of the Magnetic Motor Starter

It is important that you be able to recognize the basic parts of a magnetic motor starter, since you will have to

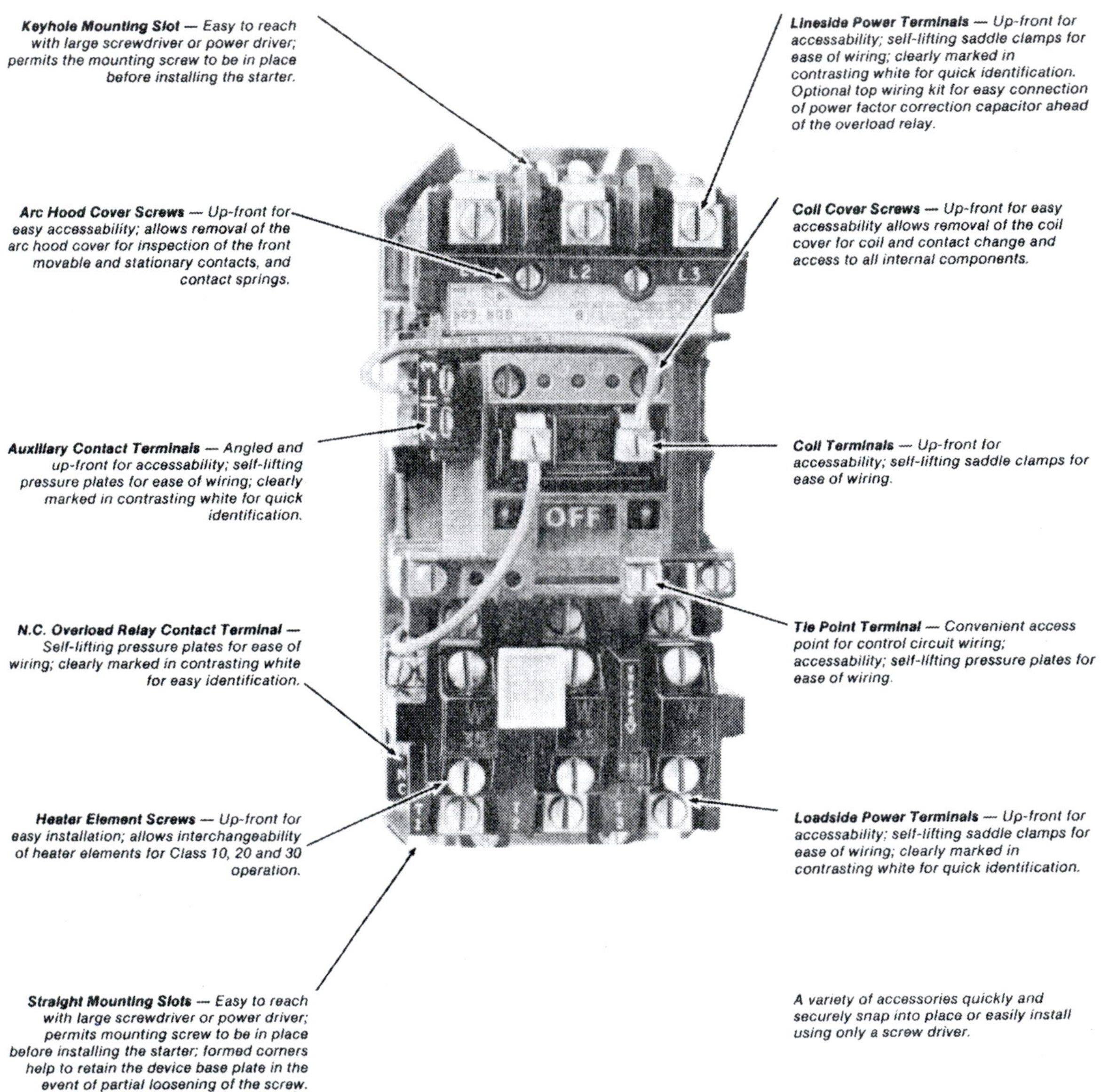

Figure 9–9 Identification of the main parts of a motor starter.
(Courtesy of Rockwell Automation)

install, troubleshoot, and repair the starter. For example, if you need to test the coil and later change it, you will need to know where the coil is and how the contactor comes apart. This is also true if you need to check the contacts and later change them.

Figure 9–10 shows a cutaway view of a motor starter, and Figure 9–11 shows an exploded view of the contacts and coil of a magnetic starter as they come together and operate as a starter. From these figures you can see that the base is attached to the mounting plate. The mounting plate is used to mount the starter to the electrical panel or enclosure. The stationary contacts are mounted in the top of the base. They have the field wiring terminal points for both the line-side terminals and the load-side terminals.

The overloads are also built into the base. This is the part of a magnetic starter that makes it different from a contactor. The overloads can be attached externally to the contactors discussed earlier in this chapter, or they are built into the molded plastic base assembly. The location of the overloads on the front of the starter allows them to be inspected and changed easily. This is necessary since you must size the heaters on the overload to the running current of the motor and sometimes they must be exchanged for the next larger or smaller size while the starter is installed in its enclosure or panel.

The other parts of the starter that are part of the base assembly are the built-in arc hoods specially molded into the plastic to prevent the arc from one set of contacts from getting near any other set of contacts or other parts of the starter. You can also see that the contacts are covered by a hood cover mounted on the front of the contacts with mounting screws. The hood cover contains the arc quencher, which helps extinguish an arc as soon as it begins to develop.

A set of mounting terminals are provided on the mounting plate for the auxiliary contacts to be attached. This location is immediately to the left of the base. It

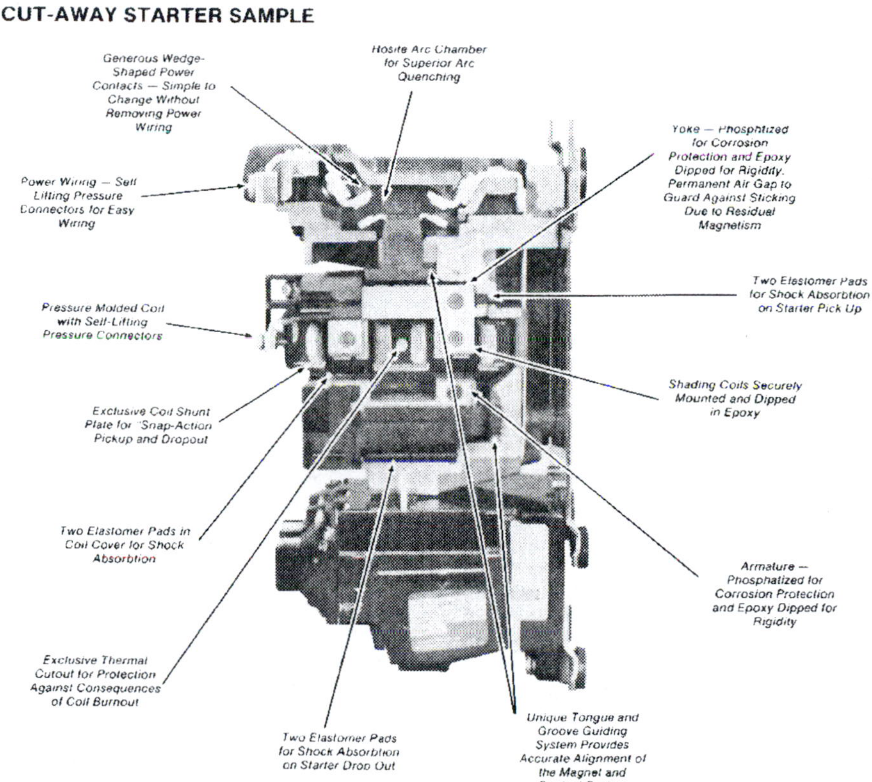

- Vertical lift gravity drop out design.
- Armature, contact carrier and contacts lift as a "single-piece" in a short vertical stroke.
- No pivot points, drop out springs or linkages.
- Elastomer pads reduce shock and contact bounce to extend mechanical and electrical life.
- Coil and yoke are keyed for proper reassembly.
- Domed surface on top of coil permits self alignment of the pole faces.

Figure 9–10 Cutaway diagram of a magnetic motor starter.
(Courtesy of Rockwell Automation)

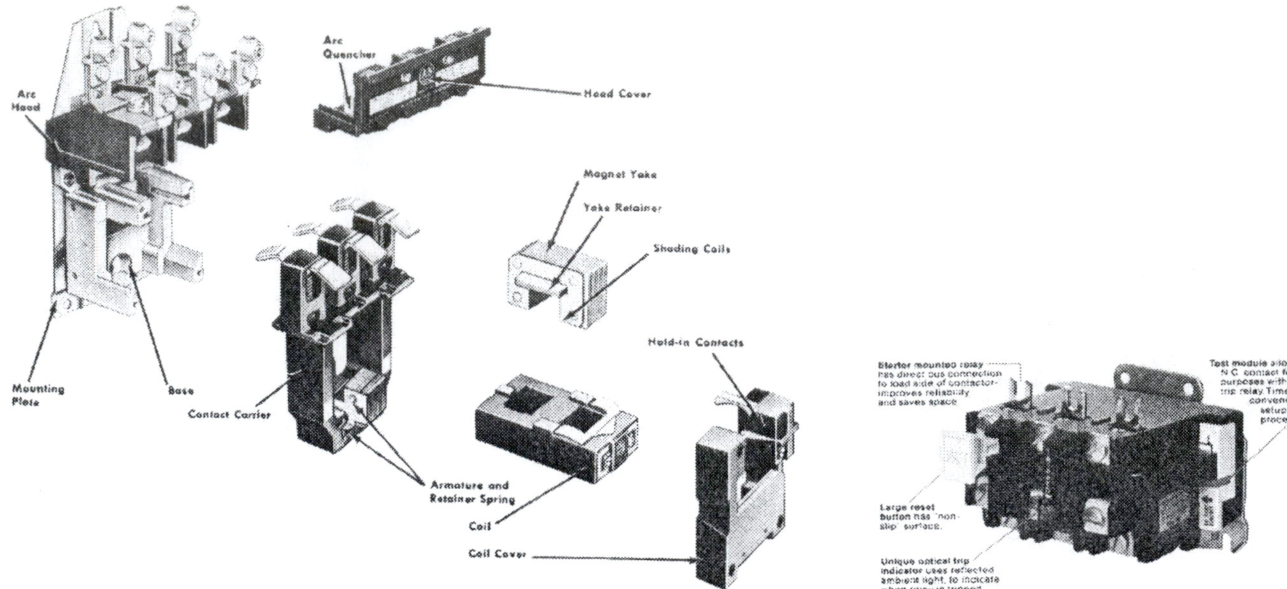

Figure 9–11 (a) Exploded view of a magnetic motor starter with its contacts, coil, and overloads identified. (Courtesy of Rockwell Automation)

allows the auxiliary contacts to be mounted in close proximity to other component parts of the starter so that interconnecting wires can be kept as short as possible.

Another component shown in this diagram is the contact carrier, which holds the movable set of contacts. Notice that each of the contacts has a spring mounted under it to help it remain tightly seated when it is closed against the stationary set of contacts. The contact carrier also houses the magnet assembly. The magnet assembly consists of the coil, which is molded in epoxy, and the laminated magnetic yoke, which has a shading coil attached. The yoke is placed through the coil and the coil is then snapped into the contact carrier. After all the parts are assembled into the contact carrier, the carrier is mounted in the base section of the starter, where it can move in the vertical direction to cause the contacts to close and open when the coil is energized. It pulls the magnet yoke upward, where it presses against the movable contacts and causes them to seat against the stationary contacts.

It is also very important that you be able to recognize the parts of the starter with the symbols used to represent them in the various types of diagrams. This is a very important concept to understand, since you will be using the wiring diagram and electrical ladder diagram to install and troubleshoot the magnetic starter. This means that you must be able to visualize the physical component when you see the symbol for it on an electrical diagram. You must also visualize the operation of these parts when you are trying to troubleshoot various components, such as the coil or contacts.

9.2.2 Wiring Diagram

Figure 9–12 shows several wiring diagrams used to represent the magnetic motor starter. It is important that you learn to recognize this group of symbols as a magnetic starter when they appear on an electrical diagram. The best technicians and troubleshooters are quickly able to recognize the symbols on a diagram as an operating component and its individual parts, and then transfer this concept to the physical parts when they must locate them for making electrical tests or change parts.

Each of the component parts of a starter is identified in the diagram. The symbols in the diagram are placed to show the actual location where you would find them if you were looking at the front of the diagram. This diagram is presented as a map of where you would expect to find the parts on the starter. Each electrical connection (terminal) is represented by a small circle. The three terminals at the top of the diagram are marked L1, L2, and L3. Since they are shown at the top of the diagram, that is where you would find them when you are looking at the starter.

The next symbol, which you will find directly under the line terminals, represents the load contacts. There are three sets of normally open load contacts and one

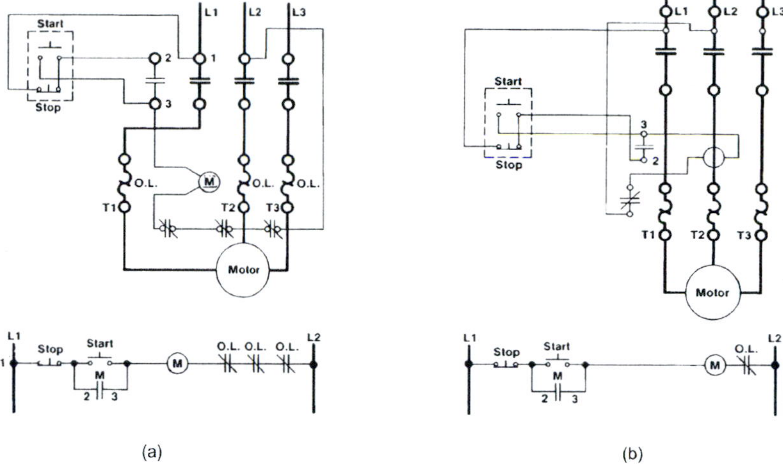

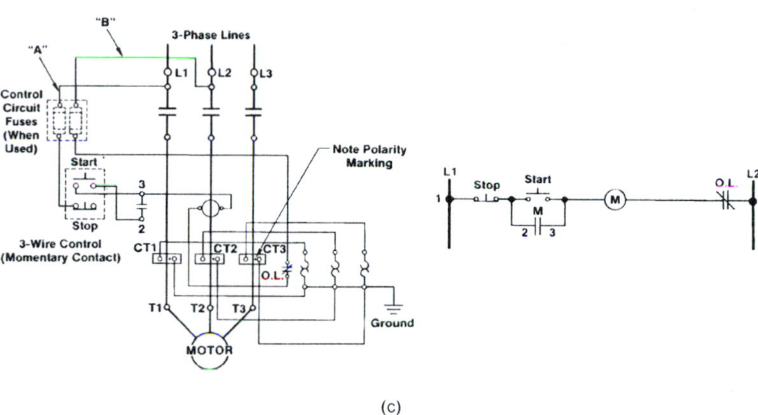

Figure 9–12 Three different ways to show a ladder diagram and wiring diagram of three-phase motor starter with three overloads.
(Courtesy of Rockwell Automation)

set each of normally open and normally closed auxiliary contacts.

The load contacts are drawn with a heavy dark line to represent that they carry the large motor or load current and the auxiliary contacts are drawn with a thinner line to show that they carry control current. In some diagrams the auxiliary contacts are drawn with a different color, but they will always be shown with thinner lines. This is another example of trying to show the physical relationship of components in a wiring diagram. Since the load contacts are physically much larger than the auxiliary contacts, the load contacts are always represented on the wiring diagram as being larger.

Since the load contacts are shown directly under the terminals marked L1, L2, and L3, this is exactly where you would expect to find them physically on the starter. The auxiliary contacts are shown on the left side of the starter, and the normally open pair are designated as being mounted on the rear pair. In Figure 9–9 you will notice that the auxiliary contacts are located on the left side of the starter. There are four terminals on the auxiliary contacts: Two are mounted on the top of the contact assembly and two are located at the bottom of the assembly. The front terminal on the top and bottom make up the normally closed set, and the rear terminals on the top and bottom make up the normally open set.

Three terminals are identified directly below the load contacts. The small circle on the diagram indicates that there is a terminal screw located at this point on the starter. The heaters for each overload are shown directly below these terminals. The heaters are part of the overload assembly, so they are identified by the letters OL.

In the diagram, three terminals marked T1, T2, and T3 are shown connected to the overload heaters. These terminals are the points where the wires from the load (motor) are connected. *Remember: The terminals marked L1, L2, and L3 identify the terminals where three-phase line voltage is connected and the terminals marked T1, T2, and T3 are where the load is connected.* This convention has been adopted for use on all North American motor starters.

The normally closed overload contacts are shown between the heater in line 1 and the heater in line 2. This diagram shows one set of overload contacts. The symbol for a starter may show one, two, or three sets of OL contacts.

Since the coil is physically located in the middle of the starter, in this diagram it is shown in the middle of the starter as a circle with two terminals. Three small wires are also shown in this diagram. Two of the wires are connected to the coil. One connects the coil to the overload contacts and the other is connected to the auxiliary contacts. The third small wire connects the other side of the normally closed overload contacts to the terminal point at L2. These wires are connected to the starter when it is manufactured. This allows for simple field wiring, since a pilot device such as a start/stop station can be connected between L1 and the top terminal on the normally open auxiliary contacts to complete the field wiring during installation.

9.2.3 Ladder Diagram

The magnetic starter can also be shown in an elementary (ladder) diagram. Remember that the ladder diagram shows the sequence of operation for the entire circuit, rather than the location of components, like the wiring diagram. Figure 9–12 shows both the wiring diagram and the ladder diagram of several motor starter circuits. The most important point to remember is that the ladder diagram is generally shown as two distinct sections. The diagrams in this figure show only the control circuit for the motor starter, and not the motor that is connected to the contacts of the motor starter. The wiring diagram does show the motor connected to the motor starter contacts. The control circuit shows a start/stop button connected in series with the starter coil. The coil is identified by the letter(s) M or MS for motor starter. If more than one motor starter is shown in the diagram, the coils will be identified as MS1, MS2, MS3, and so on. The load contacts and auxiliary contacts will be identified by the same letters used to identify the coil. The same mark is used to indicate which contacts the coil will activate when it is energized. This is extremely important if additional contacts are used in the circuit, because sometimes these contacts will be shown on a line of the diagram that is some distance away from the coil that activates them. In fact, sometimes the contacts will be shown closer to another coil in the circuit, which may cause some confusion if the contacts and coils are not clearly identified.

The ladder diagram should be read as a sequence of operation. This means that the start button must manually be pressed closed momentarily to allow control circuit voltage to flow to the starter coil and energize it. When the coil is energized, its magnetic field closes the three sets of load contacts, which allows three-phase voltage to pass to the motor. As you know, this voltage also flows through the overload in each line, so that the motor can be protected against overcurrents.

When the armature pulls up to close the line contacts, it also closes the auxiliary contacts. Since these contacts are wired in parallel around the start button, they will provide an alternative path for current when they are closed. This action seals the circuit in, so that the coil remains energized even when the start button is allowed to return to its normally open position. When the stop button is depressed, current to the coil is interrupted and the coil becomes deenergized. This allows the contacts to drop out and deenergize the motor. The auxiliary contacts that are used to seal in the coil circuit are also allowed to open, which breaks the seal circuit. This requires that the start button be depressed again to reenergize the circuit.

Another way the circuit can become deenergized is if one of the overloads senses an overcurrent condition. When this occurs, the overload contacts are opened and will remain open until they are manually reset. This causes the coil current to be deenergized and the line contacts will drop open to stop the motor. If the heater element in the overload cools down, the coil will not become energized again until the overload is reset and the start button is depressed again.

If additional pilot devices or control relays are used to design circuit logic, they will be placed in the control circuit. Examples of these logic circuits are shown throughout this book. Always remember when you are looking at a diagram that wiring diagrams are trying to show location of components and ladder diagrams are trying to show sequence of operation.

9.2.4 Overload Assembly and Bimetal Overload

The overload assembly is considered to be an integral part of the magnetic motor starter. The overload is very

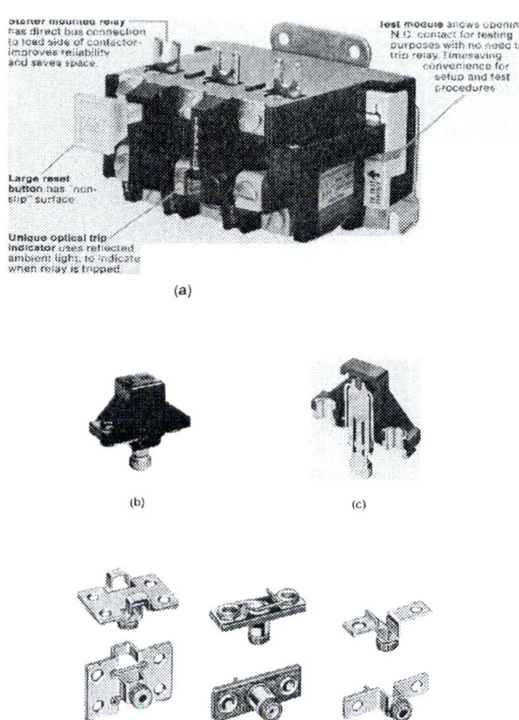

Figure 9–13 An overload assembly with the basic parts identified: (a) a type W heater; (b) cutaway picture of type W heater; (c) type J heater.
(Courtesy of Rockwell Automation)

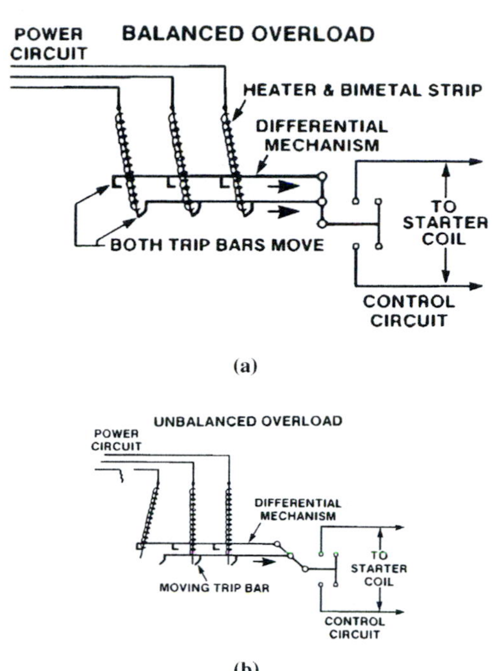

Figure 9–14 Diagram of a bimetal overload relay with (a) a balanced overload, and (b) an unbalanced overload.
(Courtesy of Rockwell Automation)

similar to the overloads used on manual starters. Some manufacturers use the same overload for both their manual and magnetic starters. In this section we review the operation of the overload assembly in terms of its use in magnetic starters and the interconnections between the overload contacts and the magnetic starter's coil.

Figure 9–13 shows an overload assembly with the basic parts identified. From this figure you can see that this overload is for a three-phase motor starter and the individual overload assemblies are connected side by side so that they will fit under the line contacts in the starter. Each individual overload has two terminals. The top terminal is where the overload is connected to the terminal for the line contacts and the bottom terminal is used to connect the field wires for the motor. From the picture of the overload assembly you can also see the location of each heater. A close-up picture and diagram of several heaters are also provided, in which you can see the ratchet gear at the end of the heater that is inserted into the heater assembly.

Figure 9–14 shows two diagrams that animate the action of the bimetal overload assembly as it heats up from an overcurrent. The bimetal-type overload is the oldest type of overload that is still in use. It is not as accurate as eutectic overloads or solid-state overloads but it is still widely used in existing motor control circuits.

The bimetal overload uses a bimetal strip that is formed so that it will move a specific distance when it is heated. The movement is sufficient to trip the overload contacts. The heat is generated by current flowing though a heating element that is mounted beside the bimetal strip. The three-phase overload has three individual elements. In Figure 9–14a you can see what happens when all three legs of the three-phase circuit have an equal overload and all three elements trip at the same time. In the diagram for Figure 9–14b, you can see the overload will also trip if only one leg has an overcurrent.

The accuracy of the bimetal overload is dependent on the proper size heater being installed on the overload block. The heaters are actually different sizes of resistors that create a specific amount of heat when current flows through them. The bimetal overload will trip with the same amount of heat every time. The heaters cause different amounts of current to produce the amount of heat required to trip the overload. If the incorrect size of heater is installed in the overload, the overload will not protect the motor against excessive current correctly. Some overloads have a fixed-size heater with an adjustment knob that allows the overload to be set for different

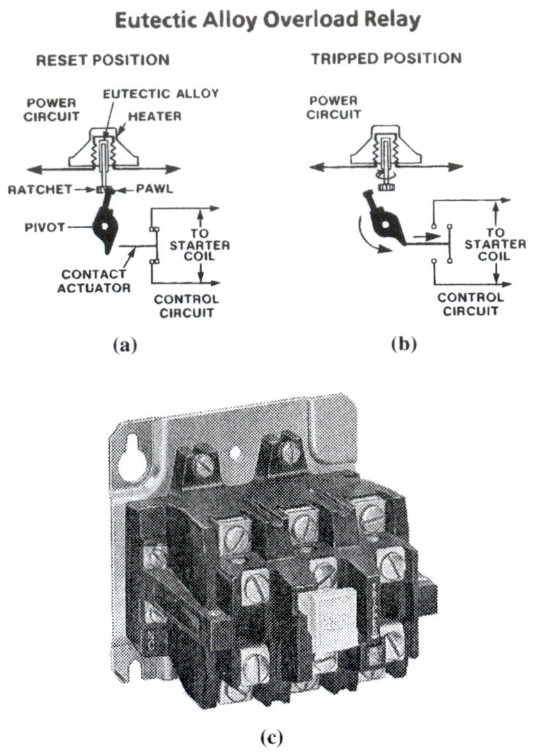

Eutectic Alloy Overload Relay

RESET POSITION — TRIPPED POSITION

POWER CIRCUIT, EUTECTIC ALLOY HEATER, RATCHET, PAWL, PIVOT, TO STARTER COIL, CONTACT ACTUATOR, CONTROL CIRCUIT

(a) (b)

(c)

Figure 9–15 Diagram of a eutectic alloy overload in (a) the reset position and (b) the tripped position; (c) a typical eutectic overload block.
(Courtesy of Rockwell Automation)

motor sizes. If the overload you are working on has this type of overload, you will need to measure the motor current with a clamp-on ammeter and set the overload adjustment for the amount of current the motor is drawing. Another way to set the overload is to use the motor ratings supplied on the motor data plate.

9.2.5 Eutectic Overload with Ratchet, Pawl, and Pivot

The eutectic overload operates on a different principle than the bimetal overload. Eutectic means low-melting-temperature solder. Figure 9–15 provides a diagram that shows how the eutectic-type overload operates. From this diagram you can see that the overload has four parts: the ratchet and shaft, the pawl that is connected to the end of the pivot, the overload contacts and actuator, and the heating element. The heater for the eutectic-type overload is actually a self-contained module that has a shaft with a tooth-gear on one end that extends from the heater. The operation of the overload depends on the design of the ratchet and pawl and how they actuate the pivot. The ratchet is built into the end of the overload shaft and has a series of teeth that are similar to the teeth in a common gear. These teeth are set on the shaft so that they provide a tight pocket for the pawl to rest in between any two consecutive teeth.

The other end of the shaft is surrounded by the eutectic alloy (solder) and is prevented from turning as long as the solder is solidified. A spring is connected to the pivot to try and turn it in a counterclockwise direction. The pawl is located on the end of the pivot in such a way that it will make contact with the teeth of the ratchet. When the overload is in the reset condition, the spring on the pivot is applying pressure against it so that the pawl is trying to spin or turn the shaft in the clockwise direction. Since the solder is solidified, it will prevent the shaft from turning.

When the overload's heater element senses an excess of current, extra heat is generated from the element to the eutectic alloy (solder) and begins to melt it. When the solder becomes soft enough, the spring pressure on the pivot is strong enough for the pawl to turn the ratchet and shaft. When the ratchet spins about one-fourth of a turn, it moves far enough to release the pivot and pawl. When the pivot and pawl move this slight distance, they will activate the overload contacts and open them.

The reset button on the overload assembly presses the pivot and pawl back into their original position. If the solder has solidified around the shaft again, the overload mechanism can be reset. If you try to reset the overload too quickly after it has tripped, the solder will still be soft and the spring pressure on the pivot will cause the trip action again and keep the overload contacts open.

9.2.6 Protecting Against Loss of One Phase

The three-phase voltage supplied to a motor through a motor starter may lose one phase. This condition will cause the motor to overload the other two phases. This means that two of the three overload assemblies will build up heat rapidly and cause the overload contacts to trip. A trip bar is used to ensure that all the sets of overload contacts trip when any one of the overload heaters senses an overcurrent. This will also protect a motor that becomes faulty and developed an open in one of its coils. When this occurs, the other two windings will draw excessive current and cause the overload to trip.

The drive bar is located at the end of the pivot for each overload. If any one of the heaters melts the solder around its shaft, it will activate its pivot and pawl assembly. When the pivot moves, it will also move the drive bar, which will cause the other two pivots to break loose from their ratchet teeth. This arrangement ensures that if any of the overloads senses excessive current and is activated, the drive bar will also be moved and cause the other two overloads to trip.

9.2.7 Sizing the Heater Element for Motor Protection

The heater is a coil of precisely measured resistive wire. The amount of resistance in the wire is calculated so that the element will give off an exact amount of heat for a given current. This ensures that the heater element will heat the eutectic alloy in the overload assembly to an exact temperature to melt its solder and release the shaft and pawl. Since each motor size will draw a different amount of current, a variety of different-sized heater elements must be available to provide the proper amount of heat to melt the eutectic alloy in the assembly when the motor draws too much current. If the heater is too small for the application, it will produce enough heat to melt the solder when the motor is drawing normal full-load current.

Heaters are rated by NEMA according to the amount of time it will take them to melt the solder in the overload mechanism when the motor draws 600 percent of the rated full-load current. This means that the motor is drawing an overload equal to six times the rated full-load current. The heaters are grouped into three classes: A class 10 heater will melt the solder in the assembly in 10 seconds, a class 20 heater will melt the solder in 20 seconds, and a class 30 heater will melt the solder in 30 seconds. If the amount of overload is less than 600 percent, it will take longer for the heaters to melt the solder. This is not a problem for the motor's protection since the motor can withstand smaller amounts of current for longer periods without causing damage to the windings. Class 10 heaters should be used in applications for protecting hermetic motors, such as air-conditioning compressors, submersible pumps, or other motors with short locked-rotor-time capabilities. Class 20 heaters are recommended for general-purpose applications, and class 30 heaters should be used in applications where motors require longer periods to come up to full speed, such as motors that are turning flywheels, grinding wheels, saw blades, and other high-inertia loads.

You can find sizing tables for selecting all classes of heaters. Allen–Bradley identifies their class 10 heaters as type J, their class 20 heaters as type W, and their class 30 heaters as type WL. To select the proper-size heater, you need to find the column in the heater table with the size of starter that you are using in your application and locate the amount of full-load amperage that is listed on the motor data plate. When you have found the motor's full-load current rating in the column for the starter that is being used, look for the part number of that particular. For example, if the motor in your application is a three-phase motor and you were selecting the heater for a size 2 starter, you would use the table for that starter. If the motor's data plate indicates that the motor draws 6.43 A, you would select a heater to protect the motor and install that size heater into the overload block. If the current listed on the motor's data plate does not exactly match the current listed in the table, you would select the heater with the amount of current listed that was nearest the motor's rating. If the temperature at the motor starter is higher than the temperature at the motor, the heater with the next-higher rating should be selected. If the temperature at the controller is lower than the temperature at the motor, the heater with the next-lower rating should be selected.

9.2.8 Solid-State Overloads

The newest type of overloads are the solid-state type. The solid-state overload is more accurate than mechanical styles of heaters, such as bimetal and eutectic types. The solid-state-type heater has electronic components that sense the amount of current flowing through the overload and cause it to open like a traditional mechanical-type overload. Figure 9–16 shows several types of solid-state overloads. The solid-state-type overload can be mounted directly onto a contactor, or it can be mounted in a panel some distance from the contactor.

9.2.9 Sizing Magnetic Motor Starters

Magnetic motor starters are sized according to their ability to carry motor current. This ability is based on the size of the load contacts. Figure 9–17 shows the NEMA ratings for magnetic starters, which are size 00 through size 9. The NEMA sizes for magnetic starters are the same as those used for manual starters. This table shows continuous amperage rating for the motor as well as the horsepower rating. Other ratings are shown in this table for tungsten and infrared lamp loads, resistance heating loads rated in kilowatts (kW), switching transformer primaries in kilowatt amperes (kWA), and kilovolt amperes reactive (kVAR) ratings for switching capacitive loads. From this table you can see that a NEMA size 2 starter can handle 3 hp for a single-phase motor at 115 V, and 7.5 hp for a single-phase motor at 230 V. If the starter is used in a three-phase application, it can handle 10 hp at 200 V, 15 hp at 230 V, and 25 hp at 380, 460, or 575 V.

If the motor is used for plugging or jogging it must be derated, which means the next larger starter on the table must be used. The starter must be derated for plugging and jogging applications because they cause the motor to draw locked-rotor current more frequently than a regular starting application and this continuous starting will cause the starter to overheat if it is not sized larger.

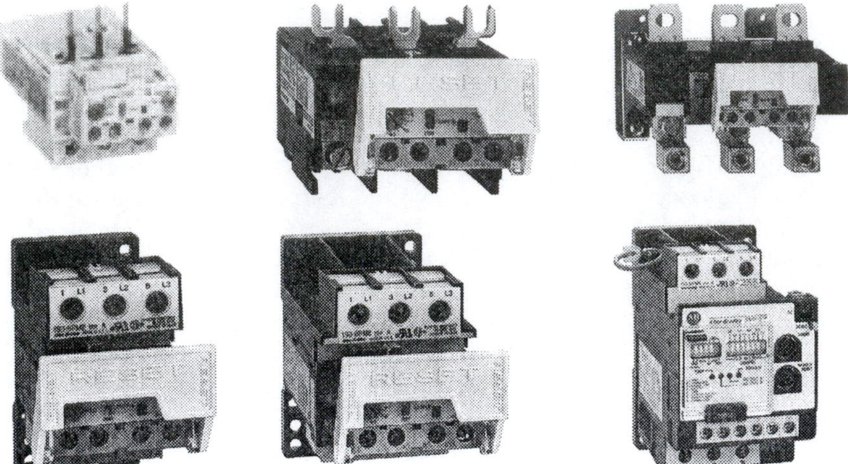

Figure 9–16 Examples of starter mount and panel mount solid-state overloads.
(Courtesy of Rockwell Automation)

Figure 9–17 Table showing electrical rating for motor starters.
(Courtesy of Rockwell-Automation)

Ratings for NEMA Full Voltage Starters

NEMA Size	Continuous Ampere Rating	Maximum Horsepower Rating Full Load Current Must Not Exceed "Continuous Ampere Rating" — Motor Voltage			
				50 Hz	
		200V	230V	380V-415V	460V-575V
00	9	1-1/2	1-1/2	2	2
0	18	3	3	5	5
1	27	7-1/2	7-1/2	10	10
2	45	10	15	25	25
3	90	25	30	50	50
4	135	40	50	75	100
5	270	75	100	150	200
6	540	150	200	300	400
7	810	–	300	600	600
8	1215	–	450	900	900
9	2250	–	800	1600	1600

9.2.10 Types of Auxiliary and Interlock Contacts

Several types of auxiliary contacts are available for use on magnetic starters. These contacts are available as normally open and normally closed contacts rated for pilot duty. This means that the contacts are sized for small control currents. They can be added to the motor starter after it has been installed for some time or during initial installation. The auxiliary contacts are mounted on the side of the starter and are activated by the movable contact carrier when the coil pulls it into place. These auxiliary contacts, shown in Figure 9–18, are similar to the ones used on the contactors.

Several types of interlocks are also available. The interlocks are a type of contacts that can be added directly to the side of the magnetic motor starter. These interlock contacts are activated in the same way as auxiliary contacts. The interlock contacts have an activating mechanism that is sticking out of the bottom of the contact assembly.

9.2.11 Coil Ratings

AC and DC coils are available for magnetic starters. AC coils are available for 24-, 110/115-, 120-, 208/220-, 240-, 279-, 380-, 440-, 480-, 550-, and 600-V circuits. DC coils are

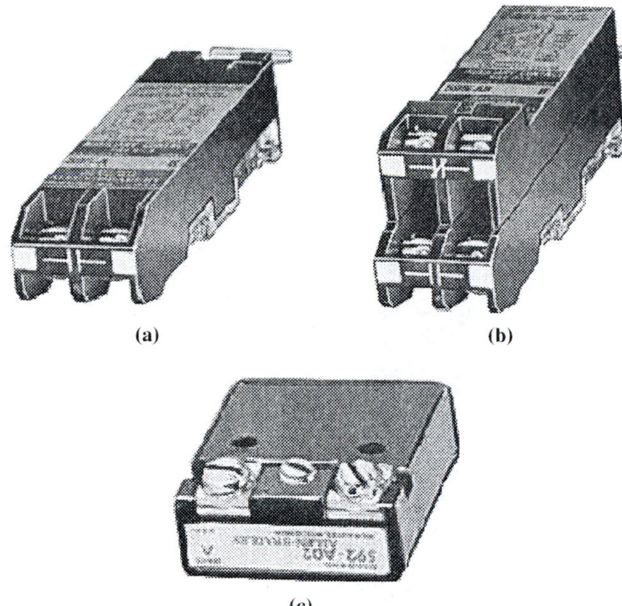

Figure 9–18 Example of auxiliary contacts for motor starters and overloads: (a) normally open auxiliary contacts; (b) contact block with both normally open and normally closed auxiliary contacts stacked in same block; (c) contact block available as normally open or normally closed. (Courtesy of Rockwell Automation)

available for AC and DC starters. Typical coil voltages include 24, 48, 120, and 240 V. DC coils are made to fit the smaller starters, such as NEMA sizes 0 to 5. When you are selecting a starter for your application, you can specify a coil that will operate on AC or DC control voltage. Be sure that the coil voltage specification matches the exact amount of voltage supplied by the control circuit.

Coils for magnetic starters have typical pickup and seal-in currents similar to those listed for contactors. In most cases, all you need to do is match the coil for the proper voltage that is used in the control circuit and match the coil for the size of magnetic motor starter it will be expected to operate. All manufacturers provide extensive data for coil sizes and voltages for various applications. This information is useful for initial installations and for replacing parts.

9.2.12 Types of Magnetic Motor Starters

Several different types of magnetic motor starters are available for different motor control applications. The most common type of control is called the full-voltage starter. This means that the starter applies full voltage to the motor for starting and control purposes. When the contacts of the magnetic starter are closed, full-line voltage is applied to the motor and it will draw maximum locked-rotor current. This causes the motor to start with the maximum amount of starting torque on its shaft so that it can start heavy loads.

This type of starter is also called an across-the-line starter, since it simply closes the line contacts that connect the motor across the incoming voltage wires (lines). This means that no control other than the contacts are placed in the circuit with the motor. In this type of configuration the motor is allowed to see full applied voltage as soon as the contacts are closed.

9.2.12.1 COMBINATION STARTERS

A combination starter combines a magnetic motor starter in a cabinet with a disconnect switch. The disconnect switch is the same type of switch that was shown in Chapter 7. Figure 9–19 shows two combination starters: a combination starter with eutectic overload is shown on the left, and a combination starter with solid-state-type overload is shown on the right. The disconnect switch can be the fused type or the nonfused type, or may have a circuit breaker instead of a fuse for circuit protection. If a fused disconnect is used, the fuse must be a dual-element type (inverse-trip-time characteristic) so that enough locked-rotor current during starting is allowed to pass through the fuse without tripping it. The inverse time characteristic will trip or blow the fuse if the motor draws excessive current for an extended period during its running cycle. If a circuit breaker is used, it should also have an inverse time characteristic so that the motor is allowed enough time to come up to speed when it is pulling locked-rotor current during the time that the motor is trying to start.

The combination starter allows a compact installation by including the motor starter in the same cabinet as the disconnect. The start/stop button for the starter is mounted on the outside of the cabinet door. This allows the starter and disconnect to meet the requirements of the National Electric Code, which requires the motor to have a disconnecting means when it is used to power machinery. This type of control also allows the motor starter to be mounted without requiring a separate enclosure. Time is also saved during installation since the starter is wired directly to the disconnect by the manufacturer.

9.2.12.2 MOTOR CONTROL CENTERS

In some applications, such as with multiple conveyors that pass through several floors in a factory, it is not practical to mount the motor starter near the motors. Instead, all the motor starters are located in a panel called a motor control center. This type of installation groups all the motor starters together for easy installation and troubleshooting. Figure 9–20 shows several examples of motor control centers.

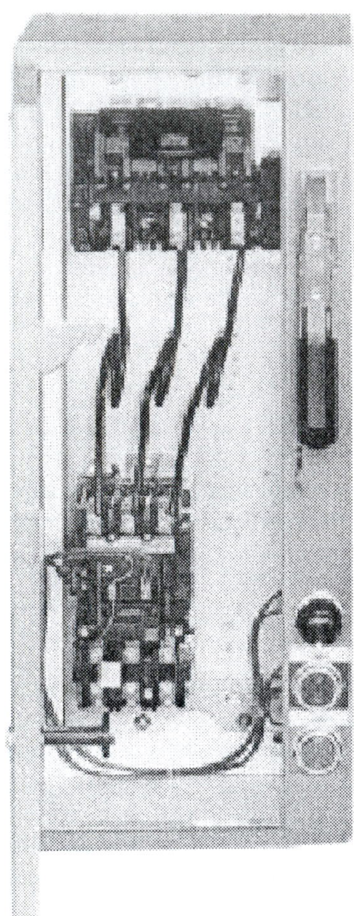

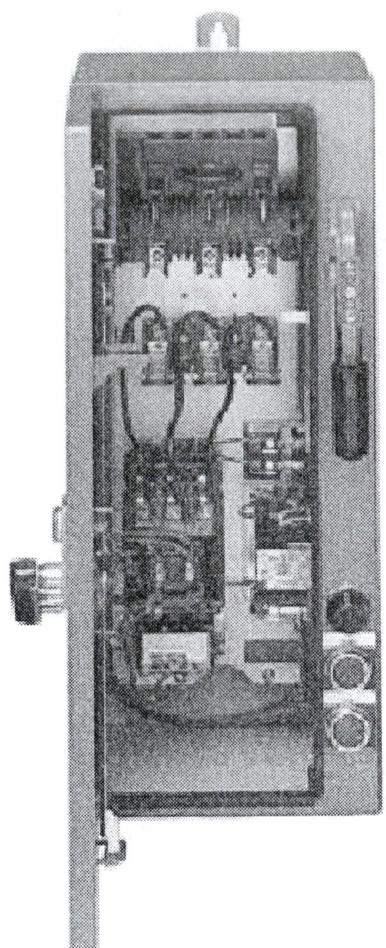

Figure 9–19 Combination motor starter in an enclosure. The starter on the left has eutectic-type overload and the one on the right has solid-state overload. (Courtesy of Rockwell Automation)

The motor control center is also used to control remotely located equipment, such as ventilating fans and equipment mounted in the ceiling of a building. A reset button is mounted on the door of each motor starter. If a motor starter's overload trips, the reset button can be pressed without opening the door.

The motor starters in the motor control center are available with circuit breakers or fuses for protection with their heaters and overloads. Starters are also available as reversing starters or two-speed starters. The motor control center provides an option that allows you to mount motor starters together, or you can mount them in individual enclosures near the motor, as explained previously.

9.2.12.3 CONTROL TRANSFORMER WITH THE STARTER

In some applications the supply voltage to the disconnect is over 400 V. In these applications it is not safe to use the high voltage as the control voltage since it is a safety hazard to apply high voltage to start/stop buttons. For these applications, a step-down control transformer is provided in the combination starter. The control transformer drops the voltage to 120 or 208 V, which can be used safely in the control circuits and pilot devices.

9.2.12.4 REVERSING STARTERS

Another specialty type of starter is the *reversing magnetic starter*. The reversing starter is actually a starter and contactor mounted in the same cabinet. The starter is used for forward operation of the motor and the contactor is used for reverse operation. The starter and the contactor each have their coil. The overload assemblies for both of the controls are mounted on the starter.

Figure 9–21 shows a reversing starter. The two controls are located side by side in the same enclosure. The wiring diagram in Figure 9–22 shows the location of the components for these controls. This is a good example of where a wiring diagram is not suited for explaining the operation of a circuit. The wiring diagram is important because it shows the location of all the components and the terminal points for all field wiring connects. The elementary (ladder) diagram does not show locations

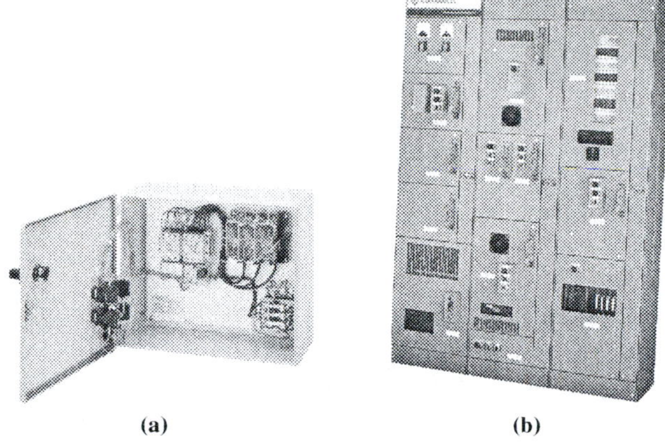

Figure 9–20 (a) Motor starter mounted in a motor control center. (b) Motor control center.
(Courtesy of Rockwell Automation)

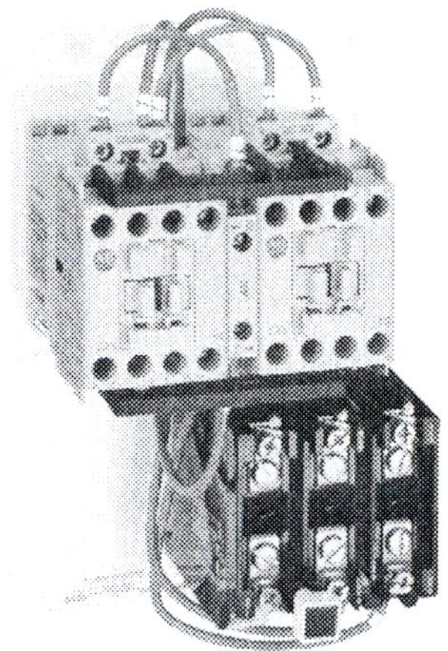

Figure 9–21 Reversing motor starter.
(Courtesy of Rockwell Automation)

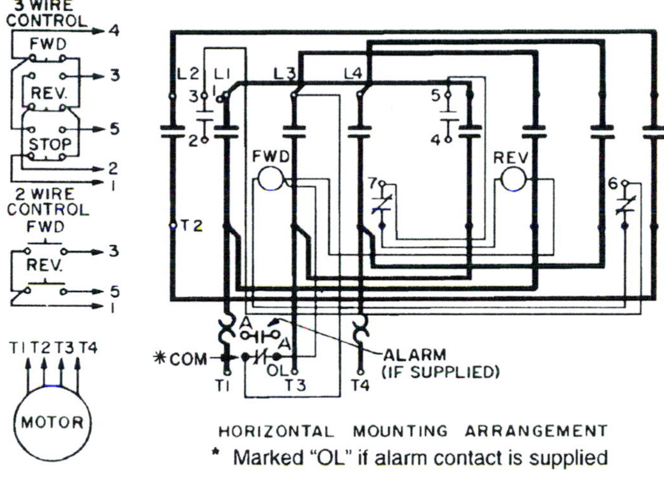

Wiring Diagram

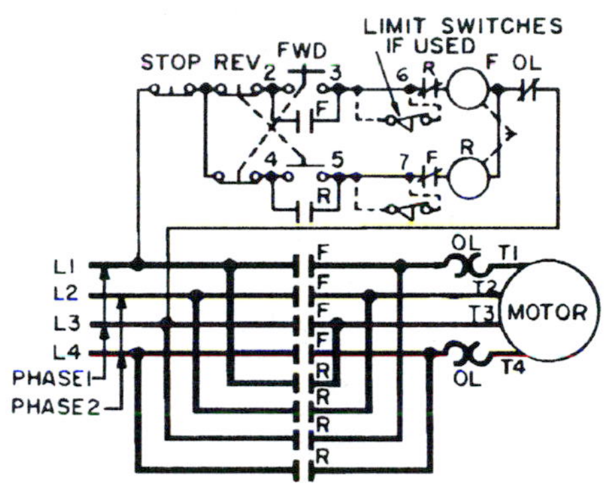

Elementary Diagram

Figure 9–22 Wiring diagram and elementary (ladder) diagram of reversing starter.
(Courtesy of Square D schneider Electric)

where the field wires should be connected, but it does show the sequence of operation of the reversing circuit.

From the ladder diagram you can see that the coil for the forward starter is identified with the letter F and the coil for the reversing contactor is identified with the letter R. All the contacts in this diagram that are controlled by the forward coil are also identified by the letter F, and all the contacts controlled by the reverse coil are identified with the letter R.

When the forward start button is depressed, the forward coil is energized, which activates all the sets of contacts marked with an F. Notice that the forward coil closes the normally open seal-in contacts that are wired in parallel with the forward start button, and opens the normally closed set of contacts that are wired in series with the reverse coil. The normally closed contacts act as an electrical lockout for the reverse coil circuit. The forward load contacts are also closed, so that the motor begins to operate in the forward direction.

Some reversing starters provide a mechanical interlock in addition to the electrical interlock. The mechanical interlock provides a mechanism that physically prevents the reversing contacts from closing any time the forward contacts have been energized. The interlock also prevents the forward contacts from closing if the reversing contacts have been closed first. The mechanical interlock is identified on the ladder diagram as a dashed line.

If you look at the load contacts of each control, you will notice that two legs of the three-phase voltage are exchanged at the reversing starter to make the motor run in the opposite direction. This also presents a potential problem if both sets of contacts would be closed

Figure 9–23 Two-speed combination starter.
(Courtesy of Rockwell Automation)

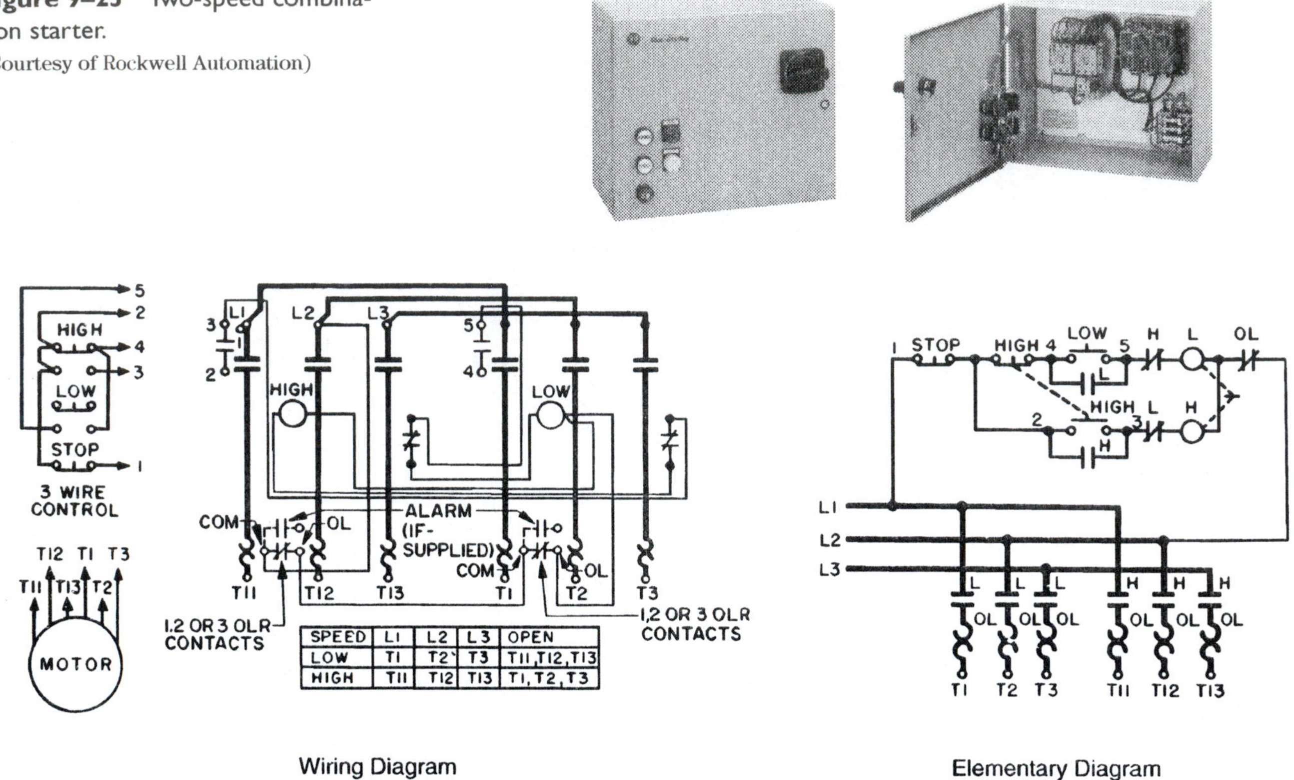

Figure 9–24 Wiring diagram and elementary (ladder) diagram of a two-speed motor starter.
(Courtesy of Square D)

at the same time, since the reverse contactor would provide a direct short circuit between L1 and L2. The electrical and mechanical interlocks are provided to prevent both contacts from being closed at the same time. *Never try to press the reverse contacts closed by hand with a screwdriver or other tool when the forward contacts are closed, because this will cause a severe arc to occur from a short circuit.*

When the motor is running in the forward direction, the stop button must be depressed to deenergize the forward contactor before the reverse contactor can be energized. When the forward contactor is deenergized, the normally closed contacts that make up the forward electrical interlock are allowed to return to their normally closed condition. This allows the reverse start button to be depressed so that the reverse coil can be energized. When the reverse coil becomes energized, all the contacts identified by the letter R are activated. The normally open set of reverse contacts that are connected in parallel across the reverse pushbutton act as a seal for it in the circuit. The normally closed set of reverse contacts that are connected in series with the forward pushbutton will be opened so that they act as an electrical interlock for the reverse contactor. The dashed line on this contactor indicates that a mechanical interlock will be engaged so that both sets of contacts cannot be energized. If the motor is required to run again in the forward direction, the stop button must be depressed first to disengage the electrical and mechanical interlocks. The heaters for the overloads are placed in lines so that they are in the circuit when the motor is running forward and reverse. The contacts for the overloads are in the control circuit next to the stop button. This location ensures that they are wired in series with both the reverse and forward control circuits so that they can protect the motor at all times.

Another type of reversing starter is also available, which allows the motor to be switched from the forward direction to the reverse direction without using the stop button. This type of control is used for jogging and similar applications that require the motor to be reversed quickly. It uses only electrical interlocks. The interlocks in this case are slightly different in that they allow control to be switched from the forward direction directly to the reverse direction without using the stop button. The interlock still prevents both sets of contacts from closing at the same time. Operation of the control contacts allows the reversing coil to become energized while it is deenergizing the forward contacts.

9.2.12.5 TWO-SPEED MAGNETIC MOTOR STARTERS

Two-speed motor starters use two magnetic starters to provide control for two-speed motors. Figure 9–23 shows a photograph and Figure 9–24 shows a wiring diagram and a ladder diagram of a typical two-speed motor starter. This type of control uses two separate magnetic motor starters housed on the same mounting plate, which allows the two starters to be mounted side by side in the same enclosure. Each starter has its own set of overloads, which allows the motor to be protected when the motor is operated on either high speed or low speed.

The ladder diagram is used to show the operation of the starters in this application. The wiring diagram shows the two starters in the location where you would find them in relation to each other. The dark lines in the diagram indicate the heavier load wires, and the thin lines in the diagram indicate the lighter control voltage wires. The motor in this application is a two-speed, two-winding, three-phase motor. When the motor is wired for low speed, L1, L2, and L3 are connected to T1, T2, and T3, respectively. Terminals T11, T12, and T13 remain open. When the motor is wired for high speed, L1, L2, and L3 are connected to T11, T12, and T13, respectively. T1, T2, and T3 remain open. Additional information on motor speed applications is provided in Chapter 14. The terminals for the high-speed winding, T11, T12, and T13, are connected to the three load-side terminals in the first starter. Terminals for the low-speed winding, T1, T2, and T3, are connected to the load-side terminals of the second starter. It is very important that you connect the motor leads to the load terminals on the starters in the exact order as listed in the diagram. If any of the leads are interchanged, the motor will try to operate in the opposite direction when it changes speed.

Operation of the two-speed circuit is indicated in the ladder diagram. Each motor starter has its own coil. The high-speed coil is identified by the letter H and the low-speed coil is identified by the letter L. All the contacts in the diagram identified by the letter H are activated by the high-speed coil and all the contacts identified by the letter L are activated by the low-speed coil.

This circuit is controlled by a set of pushbuttons. The top pushbutton is for high speed, the middle pushbutton is for low speed, and the bottom pushbutton is the stop button. The pushbutton for high speed consists of a set of normally open contacts and a set of normally closed contacts. The stop button and the low-speed button consist of only one set of normally open contacts.

When the high-speed pushbutton is depressed, the normally open set of contacts in the switch energize the high-speed coil. The normally closed set of contacts in the high-speed switch are opened when the high-speed pushbutton is depressed. This ensures that the low-speed coil cannot energize at the same time as the high-speed coil. When the coil to the high-speed motor starter is energized, the main load contacts are closed and the motor begins to run at high speed. The normally open set of auxiliary contacts from the high-speed starter are wired in parallel across the normally open set contacts on the high-speed pushbutton. These auxiliary contacts close and seal the high-speed pushbutton when it is depressed momentarily. Any time the stop button is depressed, the high-speed coil is deenergized. If the overload contacts open, the coil will also be deenergized.

Any time the low-speed button is depressed, the coil for the low-speed motor starter is energized. The low-speed coil can even be energized if the high-speed coil is energized, since the normally closed set of contacts on the high-speed pushbutton will be returned to their closed position whenever the high-speed pushbutton is released. Because these contacts are closed when the motor is running at high speed, the low-speed pushbutton can be depressed and the low-speed coil will become energized. The high-speed load contacts will be dropping out while the low-speed contacts are pulling in. The slight time difference will ensure that both sets of contacts will not be energized at the same time. The extra set of contacts in the high-speed pushbutton also ensures that only the high-speed pushbutton will be energized if both the high- and low-speed pushbuttons are depressed at the same time.

9.2.13 Installing the Magnetic Motor Starter

Motor starters are installed and field wire connections are made according to the wiring diagram. The ladder diagram is not much help in the installation procedure. Remember that some manufacturers install a wire between the starter's coil and overload contacts and between the coil and the auxiliary contacts at the factory. This prewiring allows the auxiliary contact to be used to seal in a momentary start pushbutton, and it ensures that the overload contacts are used in series with the coil. It also saves time and money for field wiring that would need to take place to make these connections at the time of installation. If the starter is being used in a two-wire control circuit, the wire from the motor starter coil and the auxiliary contacts will need to be removed.

The field wiring should be connected to the motor starter in accordance with the wiring diagram. The control voltage may not use the same source as the load voltage. In some circuits the control voltage will be supplied from a control transformer. Be sure to test the amount of voltage that will be used for the control circuit prior to applying power to the control circuit.

If a reversing motor starter or two-speed motor starter is being installed, be sure that you determine the correct phase sequence of the supply voltage lines for each motor starter. If the sequence is reversed on either of the starters, a rotational reversal may occur when the motor changes speed, or the motor may not change rotation when you expect it to with the reversing starter.

A variety of NEMA enclosures are available for the different types of applications that the motor starter may be used for. Some of the enclosures provide an airtight seal for explosion-proof applications. The complete list of enclosures was provided in Chapter 7, where installation of manual motor starters was explained. If you need to review the types of enclosures that are available, refer to the information in that chapter.

The enclosure type is specified on the wiring diagram for each application. Be sure that all connecting hardware, such as nipples and conduits, matches the enclosure type. Also be sure that the enclosure cover fits properly when you have completed the field wiring. After the installation is complete, the application should be thoroughly tested. If any problems are detected, use the troubleshooting procedure listed in the next section.

Figure 9–25 Solid-state, reduced-voltage motor starters.
(Courtesy of Rockwell Automation)

9.2.14 Troubleshooting the Magnetic Motor Starter

The magnetic motor starter is troubleshooted with a combination of procedures that are similar to those used to troubleshoot manual motor starters and relays. You should always start the troubleshooting procedure with a test to see if the coil is energized or not. *In a magnetic motor starter, the coil must be energized before the contacts will close.*

A diagram is provided for use during the installation procedure. From this diagram you can see that the line wires are connected to the terminals marked L1, L2, and L3. The wires to the motor will be connected to the terminals marked T1, T2, and T3. The control wires should be connected at the coil terminal and at the last overload. In some motor starters the interconnecting wires from the coil through the overloads and auxiliary contacts may be installed by the manufacturers to save you time. If this has not been completed, or if you must make minor changes in your circuit, be sure to check these connections carefully as they are completed.

Since the coil must energize before the contacts will close, it is important to determine if the coil is receiving the proper amount of voltage to energize. Most manufacturers provide some type of visual indicator on the face of the contactor's coil to indicate that the coil is energized. You can refer back to Figure 9–9 to see a typical indicator. The indicator is usually quite visible so that troubleshooting technicians can see it from a distance. Other starters also provide indicator lamps or pilot lamps to show when the coil is energized. If the starter is housed in an enclosure, the pilot lamp or indicator may be mounted on the cover of the enclosure.

Once you have determined the status of the coil, you can begin to form a troubleshooting strategy. If the coil is not energized, your test should focus on the control circuit. If the coil is energized, the tests should focus on the load circuit. First, test that the incoming voltage is the proper amount. The next tests should be to see that voltage is passing through each set of contacts. It is possible that one or more of the load contacts is defective from arcing. Another problem in the circuit is an improperly sized overload heater. If the heater is too small, the motor will continually trip the overload contacts when the motor is operating correctly.

If the contacts are passing voltage correctly, the next test should be performed on the motor. You may need to refer to Chapters 13 and 14 for troubleshooting procedures for DC and AC motors. If the motor is receiving the proper amount of voltage but does not operate correctly, it is defective and must be tested further or replaced.

9.2.15
Replacing Contacts and Coils

If the motor starter line contacts or auxiliary contacts are found to be defective, they can be changed without replacing the complete starter. Most manufacturers provide contact kits that contain new contacts and springs. The contacts can be installed when the motor starter is mounted in place. The front cover of the starter can be removed and the contacts can easily be replaced. You may need to refer to the exploded view diagram at the beginning of this section to get a better idea of where the contacts are located.

The coil of a motor starter can also be replaced while it is mounted in place. It is important to ensure that the coil voltage specification matches the coil that is mounted in the starter. For some coils color codes are used to indicate the amount of coil voltage, while for other coils the proper voltage and frequency for the coil are shown stamped on the front of the coil. The motor starter must be disassembled so that the coil can be removed and replaced. After the coil has been replaced it can be tested and the circuit can be put back into commission.

9.3
Solid-State Motor Starters

Today, solid-state motor starters are available. Figure 9–25 show examples of several solid-state motor starters. These starters are used in any application where a normal motor starter could be used. They provide a variety of advantages over the traditional across-the-line starter.

Since the solid-state motor starters use solid-state devices instead of contacts to switch current and voltage, they can provide a variety of voltages, such as less than full voltage that increases the life of the motor and controls overheating from excessive starts. The solid-state components used to switch the voltage on and off also means that there are no contacts to wear out as in a traditional motor starter. These same components are used in solid-state motor protectors. Figure 9–26 shows examples of solid-state motor protectors and the diagrams used to connect them in the circuit with a motor.

If you choose the reduced voltage mode, you can start the motor with 30 to 90 percent of full voltage to ensure less current draw during start-up. In Chapter 14 you will learn that when a motor receives full voltage when it is first started, it will draw 3–10 times the amount of its full-load current rating. This provides the motor with the starting torque it needs to get a large load moving, but it also tends to overheat the motor windings and cause the motor starter contacts to be susceptible to a large arc of current when the contacts are closing. This will cause the motor starter to wear out over time.

The solid-state motor starter can be set up to limit the maximum amount of current that is provided to the motor. This feature also allows the solid-state motor starter to act like a programmable fuse. The motor starter can be programmed to control the amount of

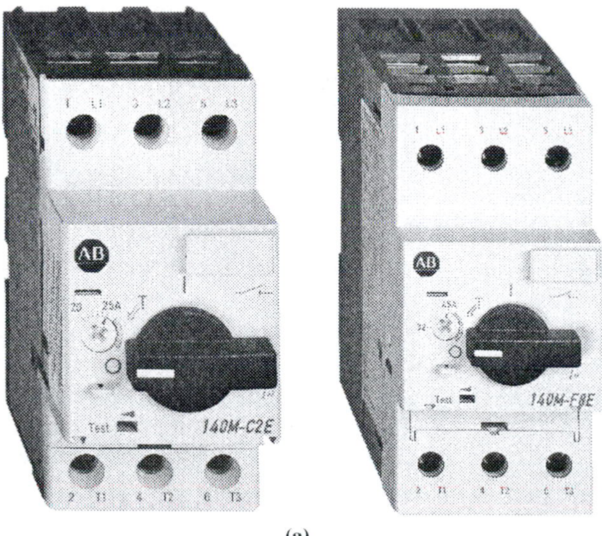

(a)

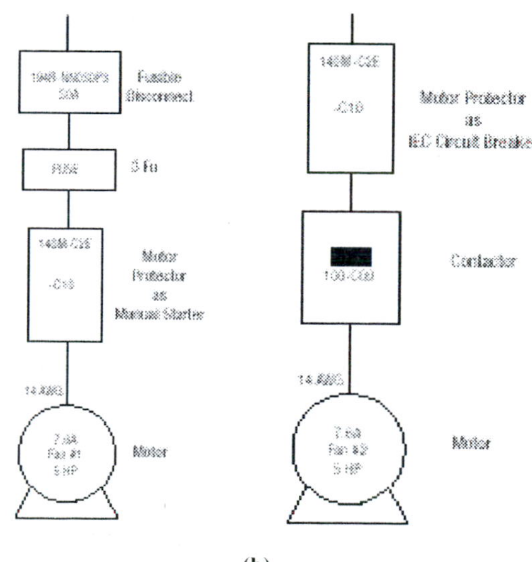

(b)

Figure 9–26 (a) Solid-state motor protectors. (b) Electrical diagrams of solid-state motor protectors used with contactors. (Courtesy of Rockwell Automation)

output voltage to the motor to maximize the power factor in the circuit and optimize the energy usage of the motor.

The solid-state motor starter can also provide several desirable features for stopping, such as a coasting stop or soft stop. It can also provide a braking stop or ramped stop that can be adjusted 1–999 seconds. The solid-state motor starter can also be set up to provide DC braking current to stop the motor.

The last feature that makes solid-state motor starters more usable than traditional starters is that they can be connected to networks to indicate their status. This means you can purchase a solid-state motor starter and connect it to a network like Device Net or Control Net, which will indicate to the system whether the contacts are open or closed, and if the starter has tripped due to excessive current or low voltage. This will make it easier to troubleshoot the motor starter when it has a problem.

9.4 NEMA and IEC Standards for Contactors and Motor Starters

One of the problems facing a new technician is that there are two types of relays, contactors, and motor starters: NEMA Standard products and IEC Standard products. The first thing you will notice is that NEMA products are generally larger than IEC devices that have similar ratings. When you open a panel of new equipment, you may find all of the components are smaller and carry the IEC standard, or the panel may be larger and all of the components are NEMA standard. This section provides some background so you will understand why each set of standards is used.

The National Electrical Manufacturers Association was established in 1926 and it created standards for a large number of manufacturers of all types of electrical equipment from North America and around the world. NEMA has established standards for electrical products and test specifications for equipment. The standards and test specifications allow comparison of similar products to meet the needs of engineers who specify new equipment in systems and for technicians who must replace parts on existing systems. In simple terms this means that if you purchase a motor starter that is manufactured to NEMA standards, you can be sure it will safely control the voltage and current that is specified on its rating sheet. Many of these specifications and standards have been adopted by other agencies, such as the Underwriters Laboratories (UL). NEMA does not perform testing like UL does, but it does help establish test parameters. The net effect of these organizations is that they help engineers and technicians compare and select the correct-size product for the application.

The International Electro-technical Commission was established in England in 1906 and has been making recommendations with over 42 member nations since that time for electrical products. Some manufacturers refer to the IEC as the European standards. The IEC makes recommendations for manufacturers to test products and publish technical data for product comparisons. Relays, contactors, and motor starters are rated by both of these groups in different ways. The IEC does not specify sizes of individual contactors, but they do provide some recommendations of test results for life expectancy of the electrical contacts and mechanical mechanism.

The most obvious difference between the products is the size difference for devices rated for the same voltage and current. The size difference came from conditions in the two regions of the world when the component standards were originally established. IEC standards were designed between the time of World War I and World War II when manufacturing material was in short supply and very expensive in Europe. For this reason, IEC contactors tend to be much smaller and more compact. By comparison, raw material was not a problem in North America, so their contactors were made for durability and life expectancy. In fact, at the time NEMA standards were developed, the trend for manufactures was to design electrical products for military specifications and the products were expected to last a very long time and cost was not an initial issue. Each NEMA product specification tried to ensure that the products would last a long time and not wear out prematurely. All of these factors tended to make the NEMA contactors and motor starters larger than was necessary for minimum operational standards.

Since original IEC contactors were smaller and more compact, they have maintained these basic sizes for current models. When North American manufacturers designed contactors to be used in European cabinets, they found that the larger NEMA-style contactors were not accepted since they caused cabinet sizes to become too large. On the other hand, many engineers were afraid that the smaller IEC-style device would not stand up to the operational standards that they were used to. Today, North American manufacturers like Allen–Bradley, Square D, and Cutler–Hammer produce two lines of products, one that meets the NEMA standard and another that meets the IEC standard. This is also true of equipment manufacturers from Europe and Asia. Figure 9–27 shows examples of IEC contactors and Figure 9–28 shows an example of an IEC motor starter.

One other major difference between IEC contactors and NEMA contactors relates to the way they interact

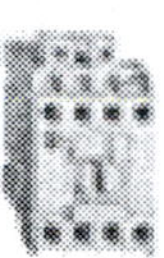

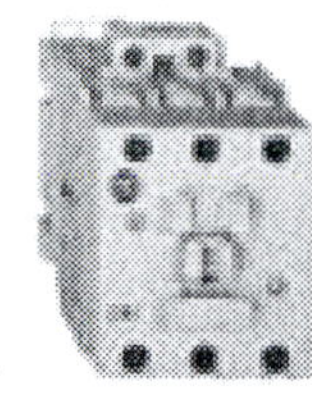
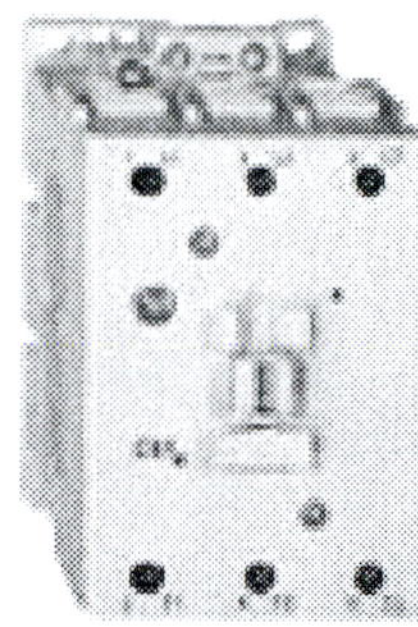

Figure 9–27 Examples of IEC contactors. (Courtesy of Rockwell Automation)

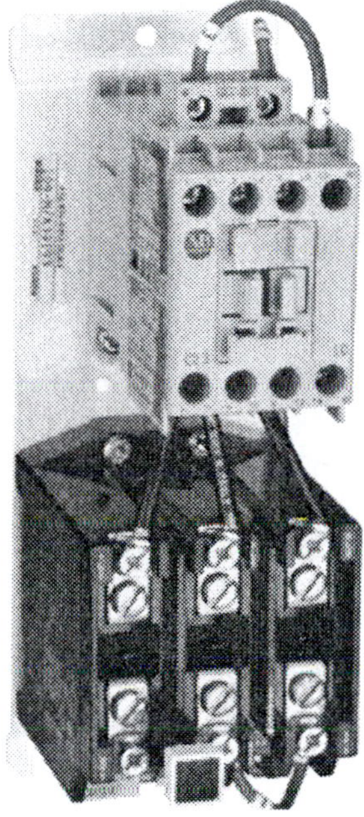

Figure 9–28 An IEC motor starter with eutectic alloy overload.
(Courtesy of Rockwell Automation)

with European and North American motors. North American motors are designed to withstand heat buildup caused by overloading and continual starting, stopping, or jogging. The European-style motors tend to overheat more rapidly, which causes excessive wear on the contacts. This means that overload components must be sized to protect the motors from rapid heat buildup.

The NEMA standards rate contactors and motor starters in 11 sizes (size 00 to size 9). Figure 9–29 shows the IEC utilization categories. From this table you can see that the contactors are rated according to the type of load application, such as AC-1 for noninductive or slightly inductive loads, or AC-3 for squirrel-cage motors for normal running operations. Contactors designed for DC applications are also identified on this chart. These devices are identified as DC-1 through DC-5.

Figure 9–30 shows a table with the IEC test load specifications. Each of the AC and DC categories of devices is shown. The notes at the bottom of this table help explain the ratings shown in the table. The following abbreviations are used in the table: *Ie* is rated operational current, *Ue* is rated operational voltage, *I* is for current when the contacts are closing (make), *U* is for voltage before make, *Ur* is for recovery voltage, and *Ic* is for current when the contacts open (break).

The important specification from this table is the amount of current the contactor can safely handle during testing. This value is listed in column *I/Ie*. *Note that these ratings are for testing purposes only; they are not intended as continuous-load conditions.* These ratings are used as guidelines for manufacturers to design contactors and for personnel who must select contactors for original equipment (OEM) or as replacement parts. For example, the AC-1 contactor can be expected to handle safely 100 percent of its rated current load. The AC-2 contactor can handle 2.5 times or 250 percent of its rated load, and the AC-3 and AC-4 can handle up to six times or 600 percent of its rated loads during the test cycle. The AC-3 and AC-4 must safely handle larger currents than their ratings for short periods, such as when motors are started, jogged, or reversed. Since the AC-1 contactor is not used for inductive loads, it should never be subjected to inrush currents since it is not rated for that type of application.

Notice that the operational voltage and the recovery voltage *Ur/Ue* is derated 17 percent for the AC-3 contactor. The reason for this is that when a motor is operating at full load, a back EMF (counter EMF) is generated by the motor windings and causes the net voltage to be approximately 17 percent less. This allows the contacts to carry the load when contacts are opening without design changes.

One last point about the differences between NEMA rated devices and IEC rated devices is the way their terminals are marked. Traditional terminal markings for NEMA devices has been L1, L2 and L3 for terminals where incoming power is attached, and T1, T2 and T3 have been used for identifying terminals where

Figure 9–29 Utilization categories for IEC contactors and motor starters.
(Courtesy of Rockwell Automation)

Category		Typical Applications
ac	AC-1	Non-inductive or slightly inductive loads, resistance furnaces.
	AC-2	Slip-ring motors: Starting, plugging ➊.
	AC-3	Squirrel-cage motors: Starting, switching off motors during running. Starting, switching off the motor only after it has come to its full speed.
	AC-4	Squirrel-cage motors: Starting, plugging ➊, inching ➋. Plugging or inching duty which is making and breaking the motor load at locked rotor current levels.
dc	DC-1	Non-inductive or slightly inductive loads, resistance furnaces.
	DC-2	Shunt-motors: Starting, switching off motors during running.
	DC-3	Shunt-motors: Starting, plugging ➊, inching ➋.
	DC-4	Series-motors: Starting, switching off motors during running.
	DC-5	Series-motors: Starting, plugging ➊, inching ➋.

Utilization Categories ➍

➊ By plugging, is understood stopping or reversing the motor rapidly by reversing motor primary connections while the motor is running.

➋ By inching (jogging), is understood energizing a motor once or repeatedly for short periods to obtain small movements of the driven mechanism.

Note — The application of contactors to the switching of rotor circuits, capacitors or tungsten filament lamps shall be subject to special agreement between manufacturer and user.

AC-3 Starting, switching off the motor only after it has come to its full speed.

AC-4 Relates to plugging or inching duty which is making and breaking the motor load at locked rotor current levels.

TABLE VI. PRODUCT COMPARISON IEC TYPE AND NEMA TYPE

	ISSUE	IEC TYPE	NEMA TYPE
1	Starter Size	Smaller / horsepower (rating)	Larger / horsepower (rating)
2	Starter Price	Lower price / horsepower (rating)	Higher price / horsepower (rating)
3	Contactor Performance	Electrical life — 1 Million AC-3 Operations 30,000 AC-4 Operations typical when tested per IEC 158-1	Electrical life typically 2.5 to 4 times higher than equivalently rated IEC device on the same test *
4	Contactor Application	Application Sensitive — greater knowledge and care necessary	Application easier — fewer parameters to consider
5	Overload Relay Trip Characteristics	Class 10 typical — designed for use with motors per IEC recommendations calibrated for 1.0 service factor motors	Class 20 typical — designed for use with motors per NEMA standards calibrated for 1.15 service factor motors
6	Overload Relay Adjustability	Fixed heaters. Adjustable to suit different motors at the same horsepower (heaters not field changeable)	Field changeable heaters allow adjustment to motors of different horsepowers
7	Overload Relay Reset Mechanism Characteristics	RESET/STOP dual function operating mechanism typical	RESET ONLY mechanism typical
		Hand/Auto Reset Typical	Hand Reset Only
8	Fault Withstandability	Typically designed for use with fast acting, current limiting European fuses	Designed for use with domestic time delay fuses and circuit breakers

Figure 9–30 Comparison of equal-size IEC and standard motor starters.
(Courtesy of Rockwell Automation)

power is taken out of the device. Some devices now use an identification scheme where the terminals are numbered and no letters are used. In these devices, numbers 1, 3, and 5 are used to identify terminals where voltage is coming in to the device, and numbers 2, 4, and 6 are used to identify terminals where voltage is taken out of a device. Another method of identifying terminals is by letters only. In this scheme the letters R, S, and T are used to identify the terminals where voltage comes into a device and the letters U, V, and W are used to identify the terminals where voltage is taken out of a device. Since these devices can be interchanged and used with each other, some newer devices have more than one set of terminal markings stamped on the terminals where power is brought in and taken out of a device.

9.5 Life and Load Curves for Contactors

Another important specification for contactors is the life expectancy of the electrical contacts and the mechanical assembly. One way of rating these conditions is by the number of expected cycles the component is expected to complete successfully during its lifetime. You can request data from the manufacturer that provides a graph that indicates the expected life of contactors. From these tables and graphs you will be able to obtain to determine the life expectancy of the product. For example, the contacts of an AC-3 contactor are expected to have 1.15 million cycles and 10 million mechanical cycles during its life. Remember that an AC-3 contactor is rated for general-purpose motor applications.

Contactors may be rated for severe applications, such as plugging, jogging, and inching, that cause the motor to see inrush or LRA current repeatedly. When the contactor is used in this type of application, its life expectancy is reduced to 30,000 cycles. Most equipment representatives will assist you in making equipment comparisons for selecting the proper components for your applications.

9.6 Miniature Contactors and Motor Starters

Electrical equipment manufacturers have produced another line of products to compete with the size and price of IEC products. These devices are similar in size to the IEC products but may have some life expectancy ratings that are closer to NEMA standards. These products are called miniature contactors and miniature motor starters. Figure 9–31 shows examples of these devices. Be sure to read the specifications for these products and observe their size limitations when specifying them as replacement parts or as new parts.

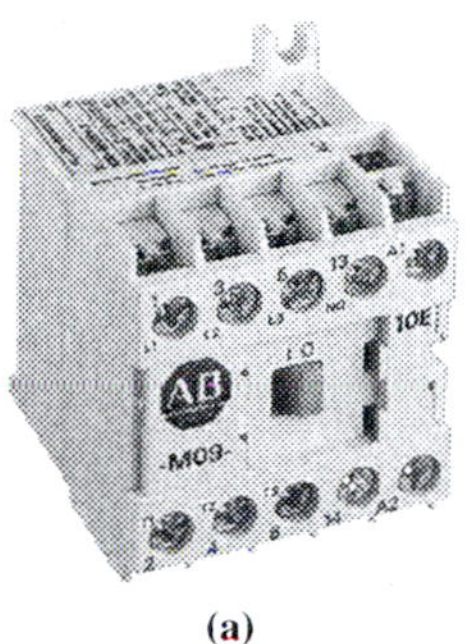

(a)

(b)

Figure 9–31 (a) Miniature contactors; (b) miniature motor starter.

(Courtesy of Rockwell Automation)

Questions

1. Explain how a contactor and motor starter are similar.
2. Explain how a motor starter is different from a contactor.
3. Explain the function of a thermal cutout on a coil and how it works.
4. Explain the methods that manufacturers use to identify different voltage ratings for their coils.
5. Explain how auxiliary contacts can be mounted on a contactor and what they could be used for.

True and False

1. The contacts on a contactor are made of solid cadmium oxide silver.
2. A mechanical interlock keeps the contact carrier in position even when current is interrupted to the coil.
3. The arc chutes and arc traps are provided on DC contactors to route the arc away from damaging the contact surface that is produced when the contacts are opened or closed.
4. NEMA stands for National Electrical Manufacturers Association.
5. The relay provides integral overcurrent protection in the form of heaters and overloads.

Multiple Choice

1. You will find the identifications ________ on a motor starter.
   a. L1 on the line side and T1 on the load side
   b. T1 on the line side and L1 on the load side
   c. L1 on the line side and the load side of the first set of contacts
   d. T1 on the line side and the load side of the first set of contacts
2. IEC stands for ________
   a. Inductive/Electrical-rated Contactor.
   b. International Electrical Contactor.
   c. International Electro-technical Commission.
   d. Inductive-type contactor.
3. The main difference between NEMA and IEC standards is that ________
   a. IEC devices are larger than NEMA devices with the same current rating.
   b. IEC devices are smaller than NEMA devices with the same current rating.
   c. there is no difference in size between NEMA and IEC components.
   d. IEC devices can be used only in foreign equipment.
4. An overload in a motor starter ________
   a. provides protection against overcurrent.
   b. provides protection against overvoltage.
   c. provides protection against both overvoltage and overcurrent.
   d. provides protection against motor overspeed.
   e. All of the above
5. A eutectic-type overload is different than a bimetal type overload because ______
   a. the eutectic type operates on the principle of low-temperature solder and the bimetal type operates on the principle of dissimilar metals.
   b. the bimetal type operates on the principle of low-temperature solder and the eutectic type operates on the principle of dissimilar metals.
   c. the eutectic type works better for three-phase motors.
   d. the bimetal type works better for three-phase motors.

Problems

1. Draw a diagram of a motor starter and identify all of its parts.
2. Explain how you would test a contactor that was controlling a motor if the motor would not run when you pressed the start button.
3. Draw a ladder diagram of start/stop buttons controlling a motor starter. Be sure to use a ladder diagram and show a three-phase motor connected as the load.
4. Explain how the ratchet, pawl, and pivot operate as an overload device. Draw a brief sketch of this device.
5. Explain how you would troubleshoot a motor starter. Do not forget to include checking the overload.

Chapter 10

Pilot Devices

Objectives

After reading this chapter you will be able to:

1. Explain the operation of a pushbutton switch, limit switch, flow switch, level switch, and pressure switch.
2. Identify the electrical symbol of each type of pilot device.
3. Explain how a reed switch operates.
4. Explain the difference between a momentary pushbutton and a maintained pushbutton.
5. Use the voltage-loss test method to locate opens in control circuits.

Pilot devices are switches used to control small amperages in motor control circuits. These switch contacts can be activated by motion, temperature, flow, liquid level, or other industrial actions. The main feature of a pilot device is the amount of current (amperage) it can handle. Since these switches are typically used in the control portion of a circuit, they are also called control devices. In some cases these devices can be connected directly to small motors, but generally they are used to control current for the coils of relays, motor starters, and solenoids. Small loads such as indicator lamps are also considered pilot devices. Information concerning indicator lamps used as pilot loads is provided later in this chapter. Figure 10–1 shows the typical current ratings of pilot switches. The type of current and the amount of current are both listed. Additional columns indicate the amount of current the switch will control on a continuous basis and during make or break. The term *make* indicates the time when the contacts close and allow current to begin flowing to the load. The term *break* indicates the time when contacts open and cause an interruption of current flow to the load. The characteristics of inductive and resistive loads vary considerably when current begins to flow and again when current is interrupted. For instance, an inductive load such as a coil will draw very large currents when current first begins to flow. This current is called inrush current. It will be two to five times larger then the continuous current the load will draw. The amount of inrush current will also vary between loads that use DC voltage and loads that use AC voltage. Inrush current for AC loads is typically much higher than that for loads in DC circuits. Since the inrush current exists for only a few AC cycles, the contacts may have a continuous current rating of 15 A and be able to withstand inrush currents up to 40 A.

When current to an inductive load is interrupted by switch contacts opening, the inductive load generates a moderate amount of back EMF (voltage), which will cause an increase in the current across the switch contacts. This increase of current looks like a spark or arc across the contacts and tends to continue for only a few AC cycles. If the load is pure resistive, such as an indicator lamp, inrush current will not occur. In this type of circuit the contact rating is based on the amount of continuous current to the load. If the contacts are used as part of multiple sets of contacts, such as double pole, double throw, the current rating must be derated. Derating is required because larger currents may arc between the different sets of contacts. The voltage rating

MAXIMUM CURRENT RATINGS FOR CONTROL CIRCUIT CONTACTS — TYPES C, XA, T, FT, and A

Switch Type	Contacts	AC — 50 or 60 Hz						DC			AC or DC
		Volts	Inductive 35% Power Factor				Resistive 75% Power Factor	Volts	Inductive and Resistive Make and Break Amperes		
			Make		Break		Make and Break Amperes		Single Throw	Double Throw	Continuous Carrying Amperes
			Amps.	VA	Amps.	VA					
C	SPDT	120	60	7200	6	720	6	125	0.55	0.22	10
		240	30	7200	3	720	3	250	0.27	0.11	10
		480	15	7200	1.5	720	1.5	600	0.1	....	10
		600	12	7200	1.2	720	1.2	..	..	....	..
C	DPDT	120	60	7200	6	720	6	125	0.22	0.22	10
		240	30	7200	3	720	3	250	0.11	0.11	10
		480	15	7200	1.5	720	1.5	600	...	..	10
		600	12	7200	1.2	720	1.2	..	..	...	...
C	**Reed*** SPST	120	15	1800	1.5	180	1.5	125	0.55	....	5
		240	7.5	1800	0.75	180	0.75	250	0.27	....	5
		480	3.75	1800	0.375	180	0.375	....	....	....	5
		600	3	1800	0.3	180	0.300	....	....	...	5
XA	Reed* SPST	120	2.0	240	0.2	24	0.2	125	0.2	...	0.5
		240	1.0	240	0.1	24	0.1	....	...	.	0.5
AW, AO-2 and AO-6 AB, AP	SPDT SPST•	120	40	4800	15	1800	15	125	2.0	0.5	15
		240	20	4800	10	2400	10	250	0.5	0.2	15
		480	10	4800	6	2880	6	600	0.1	0.02	15
		600	8	4800	5	3000	5	....	....	...	15
AW, CO-3 and CO-6 CB, CC, CP	DPDT DPST	120	30	3600	3	360	3	125	1.0	0.2	10
		240	15	3600	1.5	360	1.5	250	0.3	0.1	10
		480	7.5	3600	0.75	360	0.75	600	0.1	....	10
		600	6	3600	0.6	360	0.6	....	..	....	10
AO-1, AC	SPDT SPST•	120	40	4800	15	1800	15	125	0.5	0 25	15
		240	20	4800	10	2400	10	250	0.25	0.1	15
		480	10	4800	6	2880	6	600	0.05	...	15
		600	8	4800	5	3000	5	....	....	...	15
T and FT	SPDT Quick Make and Break	120	150	18,000	20	2400	20	125	5.0	....	20
		240	75	18,000	12.5	3000	12.5	250	1.0	....	20
		480	37.5	18,000	6.25	3000	6.25	600	0.2	....	20
		600	30	18,000	5	3000	5.0	....	....	...	20
	All Slow Make and Break	120	60	7200	6	720	6	....	...	...	20
		240	30	7200	3	720	3	....	....	....	20
		480	15	7200	1.5	720	1.5	....	...	....	20
		600	12	7200	1.2	720	1.2	...	...	...	20

• SPST versions are rated ½ HP at 110 and 200 VAC.
* Use of a transient suppressor will extend life of the switch when using on heavy electrical loads.
+ VITON is a Registered Trademark of DuPont.

Figure 10–1 Table for electrical contact ratings for pilot devices.
(Courtesy of Square D/Scheider Electric)

of contacts will also affect the amount of current the contacts can safely handle. As the voltage rating increases, the current rating decreases. This derating occurs because the VA rating of the contacts remains essentially the same. If the contacts are used in double-pole, double-throw arrangements, the VA rating remains exactly the same. This chart represents contact ratings for one manufacturer, but they will be nearly the same for pilot devices made by other companies.

The contacts used in pilot devices can be either normally open or normally closed. They will generally be made of a material that will withstand arcs without becoming damaged. Other types of contacts are also frequently used in pilot controls. One of these is the *reed switch*, which contains a set of contacts sealed in a glass or plastic capsule. Terminal leads that are connected to the contacts do not break this seal as they leave the capsule. This keeps the contacts hermetically sealed so that the environment will not cause them to corrode. The capsule also allows the reed switch to be used in explosive atmospheres without danger of starting fires.

Since the reed switch is fully encapsulated, the pilot operator cannot touch it to cause the switch contacts to activate. Instead, a magnet or magnetic strip is provided on the movable part of the reed switch. A stronger magnet is provided on the pilot switch's operator and is placed in close proximity to the reed switch. When the pilot switch's operator moves close to the reed, the contacts come directly under the magnetic force and are pulled closed. When the pilot switch's operator moves away from the reed switch, it moves the strong magnet far enough from the magnet on the reed switch to limit the magnetic force so that spring tension can return the contacts to their normal location. This type of operation allows the reed contacts to be used extensively with any type of pilot device. Specialized contacts provide for a variety of applications, such as fast response, microprecision movement, and in a logic circuit. Springs can be added to the contacts to provide rapid snapover center action to make the contacts close and open faster. Other contacts are built so that a very small amount of travel from the pilot device actuator will cause the contacts to change from open to closed.

In this chapter we explain the function of pilot devices used in modern motor controls. Each family of pilot devices is introduced with pictures and electrical symbols. Operation, installation, and troubleshooting are also explained. Typical applications and electrical diagrams are used to provide an understanding of their use. The contacts in each of these devices are very similar, if not exactly the same. The major difference between pilot devices is the type of operation used to make these contacts open and close. For example, in a pushbutton switch, the contacts are closed by manual motion (someone pressing on the switch). In a limit switch the contacts are closed or opened by machine motion, and in a pressure switch the contacts change position by an increase or decrease of pressure. The electrical symbol for each pilot device will be similar. The contact portion of the symbol will show if the contacts are normally open or normally closed. They will also show if the pilot device controls single or multiple sets of contacts. The lower portion of each symbol will show what type of operation is required to activate the switch contacts. These fundamentals will be explained in depth in the remainder of this chapter.

10.1 Pushbutton Switches

Pushbutton switches are very widely used in motor control circuits for modern industry. Perhaps the most familiar pushbutton switches are the start pushbutton and the stop pushbutton. Figure 10–2 shows pictures and Figure 10–3 shows the electrical symbols of various pushbuttons. The picture shows a square multifunction oiltight pushbutton operator, a heavy-duty metal pushbutton operator, a heavy-duty double-insulated pushbutton, and a push–pull operator. From this figure you can see that the pushbutton can operate normally open or normally closed contacts. It can also operate multiple sets of contacts, such as single-pole, double-throw, which is labeled a two-position pushbutton, and double-pole, double-throw, which is labeled a double-pole pushbutton. Another special type of pushbutton switch is called the maintained type. The first four pushbutton switch symbols represent momentary pushbuttons, which mean they will return to their open position after they are depressed and released. The fifth symbol shows a maintained pushbutton symbol, which indicates this switch will stay in the position that is activated. The fifth

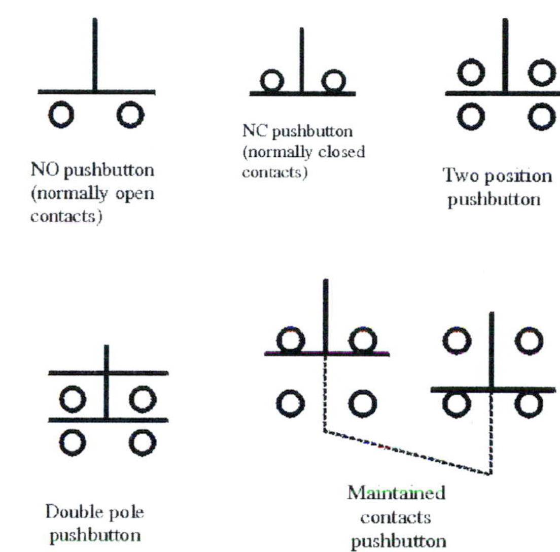

Figure 10–3 Pushbutton switch symbols.

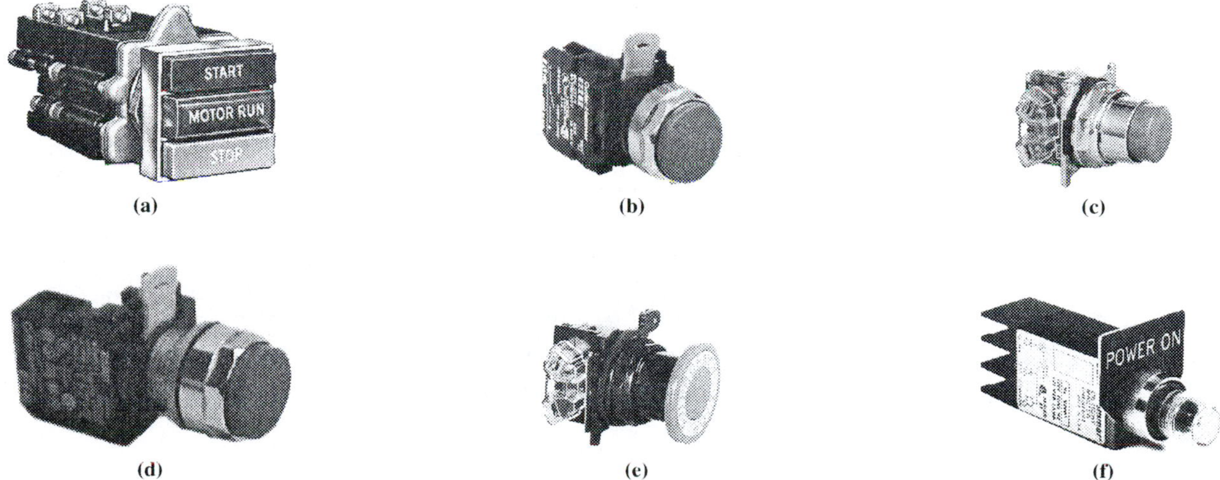

Figure 10–2 Pushbutton pilot devices: (a) square multifunction oiltight pushbutton operator; (b) heavy-duty metal pushbutton operator; (c) heavy-duty double-insulate pushbutton; (d) heavy-duty metal pushbutton switch operator; (e) push–pull stop pushbutton switch; (f) heavy-duty oiltight indicating master test light.

(Courtesy of Eaton/Cutler Hammer)

symbol that shows the maintained contact also shows another symbol convention where multiple sets of contacts that are operated by the same pushbutton are connected with a dashed line. Any time you see a dashed line connecting multiple sets of contacts you should understand that when the switch is activated, all of the sets of contacts will change condition at the same time.

Figure 10–4 shows the contact blocks that are used with pushbuttons and other types of switches. The contact blocks usually have one set of normally closed contacts and one set of normally open contacts. You can add two or three contact blocks on any switch and all of the sets of contacts will change state at the same time. You can use an ohmmeter or a diagram sheet to determine which terminals on the contact block act as a set of normally open or normally closed contacts.

Figure 10–5 shows an exploded view of a typical pushbutton. You can see from this picture how a pushbutton switch is taken apart so that it can be installed into an operator panel. This picture also shows how the contact block is mounted to the switch operator head. If the switch used additional contact blocks, they would be mounted directly to the first contact block and the operator head would activate the contacts in both blocks.

In some diagrams the abbreviation PB is included with the pushbutton switch symbol. If multiple pushbuttons are used in a system, they are identified as PB1, PB2, PB3, and so on. This is a standard method of identifying multiple sets of contacts. Figure 10–6 shows examples of a start and stop pushbutton in a traditional start/stop station. It is also possible to have more than one start or stop pushbutton in a system. This would be useful for conveyor systems that span over 100 feet. You could have a start and stop pushbutton on each end of the conveyor and a set in the middle so workers do not have to walk a long distance to start or stop the conveyor. Multiple start and stop pushbuttons may also be used where conveyors or other equipment extend between two different floors in a factory. An example of circuits using multiple pushbuttons is shown in Figure 10–6. From this diagram you can see that any time additional start pushbuttons are used, they must be added

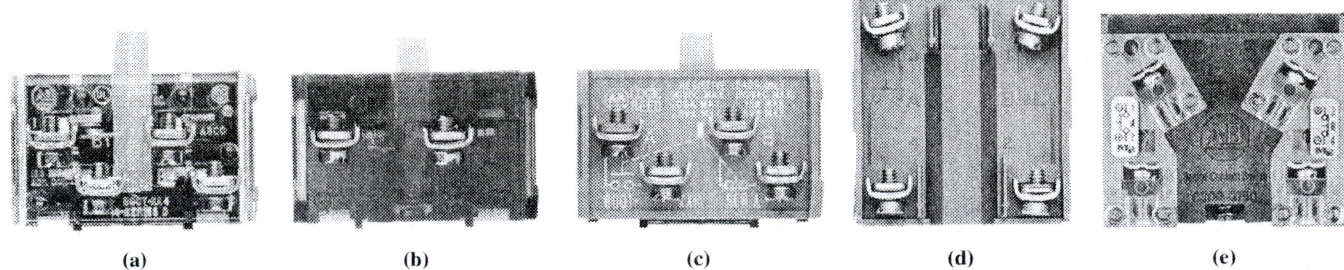

Figure 10–4 Types of contact blocks for pushbutton switches and other pilot switches: (a) shallow block; (b) low-voltage contact block; (c) logic reed block; (d) sealed switch block; (e) stackable sealed switch block.
(Courtesy of Rockwell Automation)

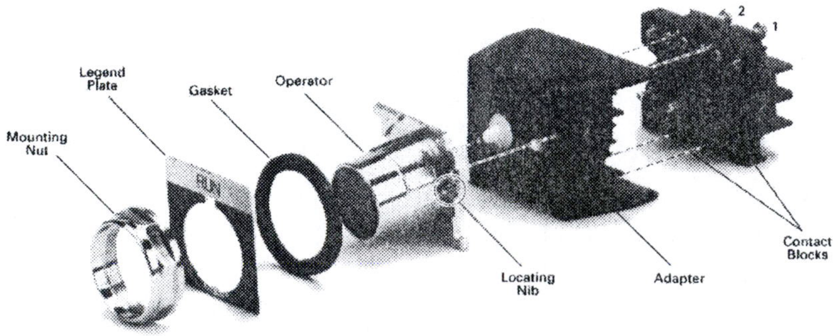

Figure 10–5 Exploded view of a pushbutton switch.
(Courtesy of Eaton/Cutler Hammer)

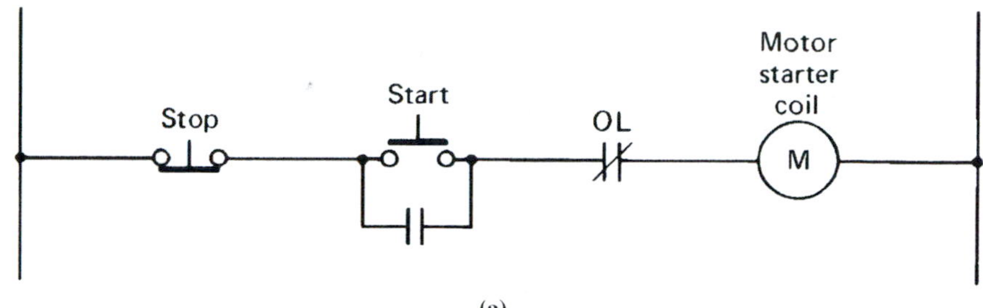

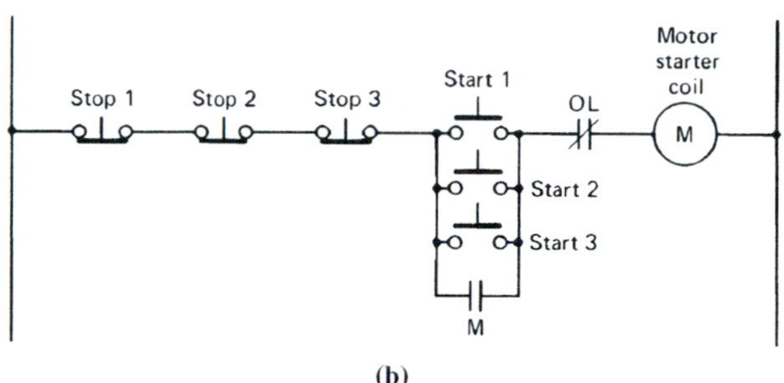

Figure 10–6 (a) Typical start/stop circuit. (b) Start/stop circuit with additional start and stop pushbuttons.

Figure 10–7 Multiple pushbutton switches. (Courtesy of Rockwell Automation)

in parallel with the original start pushbutton, and when multiple stop pushbuttons are used, they must be connected in series with the original stop pushbutton.

In some applications, multiple types of pushbuttons and other switches are mounted into an operator's panel. Figure 10–7 shows examples of typical operator panels. A number of switches may be mounted together for similar applications, such as the start–stop, run–jog, forward–reverse, or up–down function for an overhead lift. All the pushbuttons for a machine's operation may be mounted together in one panel, called the operator's panel.

If you look closely at the picture in Figure 10–7 you can see another type of pushbutton switch that is available to make a complex automated system operation safe so that only authorized personnel can operate the system. This type of pushbutton is called a security pushbutton. It has a lock with a key, and when it is locked and the key is removed it cannot be operated. This ensures that only qualified personnel operate the system. Another variation of this type of pushbutton has a lockable cover. Such a cover can be added to any pushbutton switch to ensure secure operation.

Another way to classify pushbutton switches is by the type of operator (button) that is used. For example, the operator can be flush mounted with the panel or it can be extended. It can also be fully shrouded or half-shrouded. The shroud protects the switch from being depressed accidentally by activity around the operator panel. Another type of button used on pushbutton switches is the *mushroom head*. The head is enlarged in the shape of a mushroom so that it is easy to depress. The mushroom switch is generally used as an emergency stop switch and may also be used as a start cycle switch. In these applications it is called a *palm button*. The size of the operator button is also used to classify pushbutton switches. In some applications the switch button is enlarged to a jumbo size so that it is easy to locate on the operator panel. This type of switch is generally used for emergency stop applications. The size of the pushbutton switch can also be reduced so that more switches can be fitted into the existing panel area. The reduced-size pushbutton switch is called a *miniature* or *compact switch*. Even though the size of the switch has been reduced, the amperage and voltage ratings are the same as for full-sized switches.

The shape of the pushbutton head is also used for classification. Pushbutton switches are available in round-head, rectangular-head, and square-head types. The heads are also available in different colors, such as red, black, green, yellow, white, blue, and orange. The pushbutton head may also contain an indicator lamp, which may be red, amber (yellow), green, or white.

Another classification of pushbutton switches is by the environment or location where they are installed. These switches are available in oiltight, watertight, weatherproof, corrosion-resistant, and explosion-proof assemblies. Another way to provide protection for the pushbutton switch is to add a protective boot. This rubber boot will keep the switch protected against dirt and larger particulates but will not make the switch moisture-proof or explosion-proof.

Another variation of the pushbutton switch that you will find is a pushbutton with a pilot light that is also called an indicator light. The pilot light is designed into the switch and it is wired so that it indicates the state of the circuit the switch is controlling.

10.1.1 Installing the Pushbutton Switch

You may be required to install or replace a pushbutton switch. To complete these tasks, you will need to understand both pushbutton symbols in the ladder diagram or wiring diagram like the ones shown in Figure 10–3 or 10–5. You may also need to look at the contact blocks shown in Figure 10–4 or the exploded view diagram in Figure 10–5 to understand how the switch is assembled in an operator panel. The symbols in the diagrams will be useful in the installation procedure because they show the wiring from component to component and whether contacts are normally open or normally closed. The wiring diagram is also used to show the location of terminals on the switch and the location of switches on a panel board. Terminal diagrams are usually available in the manufacturer's technical data or you can determine the contact sets with an ohmmeter. This test can be accomplished by placing the ohmmeter across a set of terminals and depressing the pushbutton. If the meter shows continuity when the button is depressed and infinity when it is released, the two terminals that the meter leads are touching are part of a set of normally open contacts. If the meter reads continuity when the button is not depressed and reads infinity when the button is depressed, the terminals belong to a set of normally closed contacts. If the meter shows infinity when the button is depressed and when it is released, the two terminals are not part of a set. If the meter shows continuity when the button is pressed and when it is released, the two terminals are connected together as a common point.

The second step of installation for the pushbutton switch is to disassemble the switch so that it will fit through the face of the panel. The wires are connected directly to the part of the switch called the *contact block*. The contact block is connected to the switch operator with screws. This allows the switch to be changed out later without removing any wiring. The front side of the switch includes a locking nut and legend plate. The locking nut is also called the *mounting ring*. When it is screwed down, it will pull the switch assembly tight to the panel. Grounding nibs are provided on the mounting surface of the switch to provide a positive ground circuit. The nibs will cut through the paint and into the metal of the panel surface as the mounting ring is tightened.

Legend plates are used to identify each switch. Since the pushbutton switch can be used for a variety of applications, it is important to identify each switch on the panel for safe operation. Typical legend plates available for use with pushbuttons include *start/stop*, *on/off*, *up/down*, *jog/run*, *forward/reverse*, *hand/auto*, *high/low*, *manual/auto*, and *open/closed*. Other legend plates are also available, but they may have to be special ordered.

10.1.2 Troubleshooting the Pushbutton Switch

The pushbutton switch is easy to test if you suspect it to be faulty. You should remember that if the circuit will not operate correctly, the problem could be any one of several things, such as the loss of the power supply, a loose or broken wire, a faulty pilot switch, or a faulty load (motor starter or relay coil). This means that you must troubleshoot the complete circuit instead of picking on individual components and testing them at random. The test that will be outlined in the following procedure is called the *voltage-loss* test and you should learn to use this test whenever you are trying to locate a faulty component or wire in an inoperable circuit.

Start the procedure by testing for voltage across the load terminals. If voltage is present at the power supply and through all the pilot switches and interconnecting wires, the remainder of the test should focus on the load. If no voltage is present at the load terminals, the test should focus on the source voltage and the loss of it through a pilot switch. If voltage is not present at the power supply, be sure to check for a blown fuse, open disconnect, or other power-supply problems. If voltage is present, it may be 24, 120, 240 V or higher, and it may also be AC or DC. At this point in the test, all that you are concerned with is that the supply voltage is the same voltage that is specified on the electric control diagram.

After you have determined that the proper amount of supply voltage is available, the remainder of the tests should focus on the pilot switches and interconnecting wires. The fastest and most accurate test for locating an open wire, loose terminal, or faulty pilot switch is to test for a voltage drop (loss). This test requires that power remain applied to the circuits while you touch the meter probes to each test point. *You must be aware of where you are placing your hands, tools, and meter probes at all times because severe electrical shock could result if you come in contact with the electrical circuit.* Even though leaving power applied to a circuit while you test it presents an electrical safety condition, it is necessary because control circuits tend to be complex and you may not be able to find the problem with the power off.

Figure 10–8 shows the test points where you should place the voltmeter probes to execute the test. Remember that this test can be used to locate problems in any circuit, regardless of the number and types of pilot devices. The test points are identified as A through J. The first test should be made with the voltmeter probes touching points A and J, which is across the power supply. Using this test is to make sure that your meter is on the right range. From this point, leave one of the voltmeter probes on point J and move the other probe to point B. If no voltage is available at point B, and you had voltage at point A, this test indicates the wire between the stop pushbutton and the power supply is open. You can turn off the power and test the wire for continuity and you should be able to verify that the wire has an open. You should remember that a continuity test uses the resistance range on the volt ohm milliamp meter. If the wire has an open, the meter will indicate infinite resistance, and if the wire is good it is said to have continuity and the meter will indicate low resistance. The wire must be isolated (one end removed from its terminals) so that you do not measure resistance from a parallel circuit that may be connected to wire you are testing. If your continuity test indicates that the wire has an open, you should remove and replace the wire to put the circuit back in service.

If you measured voltage when the meter probe is on point B and point J, the test indicates that the circuit is good to point B. You can move the meter terminal to point C. If you do not have voltage at point C and you had voltage at point B, the test indicates the stop pushbutton is open. Remember that the stop pushbutton is a normally closed pushbutton, so you should have voltage at point C. You can check to be sure that the stop button is not depressed. Remember that it may be a maintained type of switch and require someone to return it to its normal condition. If the switch is in its normal condition and no power is present at point C, the switch is faulty and must be replaced.

If you are testing for a stop switch that will not deenergize the relay coil, be sure to test point C and ensure that voltage is not present there when the stop switch is depressed. If voltage is still present at point C even when the stop button is depressed, you can assume that its contacts are welded or stuck together and that it must be replaced.

To continue the test for a loss of voltage in this circuit, we will resume from the last point where voltage is present at point C, and continue the test by moving the probe on to point D. If the voltmeter indicates that voltage is present at point C, and not at point D, your test indicates that there is an open in the wire that connects the stop button and the start button. Since the voltage test indicates a loss of voltage at point D, you can turn off power to the circuit and test this wire for continuity and if this test indicates an open, you should change the wire.

If you have voltage at point D, you have determined that the circuit is operating correctly to this point and you can move to point E. Your test should indicate that you do not have voltage at point E, because the start

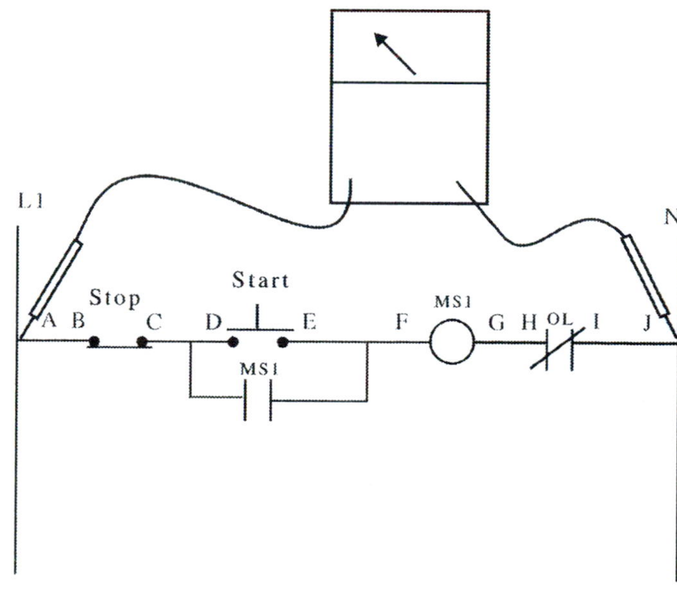

Figure 10–8 An electrical diagram used to show troubleshooting procedure.

pushbutton is a normally open switch. You will need to depress the start pushbutton and check for voltage at point E. If you do not have voltage at point E when the switch is depressed, you should turn off the power and test the switch for continuity. If the switch is faulty during the continuity tests, you should replace it.

If you have voltage at point E, your next test should be at point F. If you do not have voltage at point F, you can predict that the wire between the start pushbutton and the coil of the motor starter has an open in it. You can turn off the power and test this wire for continuity.

If you have voltage at point F, you should move the meter lead from point J to point G. This will have the meter probes touching both sides of the motor starter coil. Since the coil is a load, it should have full voltage applied to it if it is supposed to become energized. If you have voltage at points F and G, you can predict that the circuit wire and switches are all operating correctly and providing voltage to the coil. This is perhaps the most important point in troubleshooting. The main function of all of the wire and switches in every circuit is to provide the proper voltage to the load. In this circuit the load is the motor starter coil and if you measure voltage at its terminals and the contacts are not pulled in, you can predict the circuit wires and switches are operating correctly, and you should suspect the motor starter coil is faulty. You can turn off the power to the circuit and test the motor starter coil for continuity. If the coil is good, the test should indicate resistance in the range of 20 to 2000 Ω, depending on how much wire is in the coil. If the test indicates infinite resistance, the coil is open and it should be replaced. Remember, do not apply voltage to a coil that is not installed in a relay.

If you have voltage when the meter leads are on points F and J, but did not have voltage at terminals F to G, your test has indicated that the circuit has an open between point G and point J. The final two tests to locate the problem in the circuit are perhaps the most difficult to understand, because it appears as though the circuit is operating correctly when your test determined that voltage is present at point F to J. In reality, what has happened is that your tests have proven that the wire and switches on the left side of the motor starter coil are operating correctly. Since the coil will not energize, you can then predict that the problem is in the wire between points G to J or that overload contacts are open.

The best way to locate the fault in this part of the circuit is to move your left meter lead to point F and leave it there for the remainder of the tests, and to resume the test by moving the right meter lead to test points J, then to I, and then to H. The point where you lose voltage is the place where the open is in the circuit. For example, if you have voltage from point F to point J, but you do not when you move the right meter probe to point I, it indicates that the wire between J and I has an open. Turn off the power and test the wire for continuity. If your test indicates that you have voltage at terminals F to I, you should move the right meter lead to point H. If you do not have voltage at points H to F, it indicates that the overload contacts are open. You can verify this by turning off the power and testing the overload contacts for continuity. If you have voltage from points F to H, you can predict that the last wire to be tested, wire G to H, has an open. You can turn off the power and test the wire for continuity. Remove and replace the wire if it is faulty.

10.2 Why It Is Important to Test a Circuit with Power Applied

You may wonder why it is important to test a circuit with voltage applied to the circuit. You may even think that this is a dangerous practice. You will find that it is a very safe way to test a circuit and it is also the most reliable. You may think that it would be safer and easier to test the entire circuit for continuity with the power off. The problem with testing a circuit for continuity is that it is a very slow process because all of the wires must be disconnected before they can be reliably tested, and because it is very easy to have faulty readings from parallel circuits that you are not aware of. If you expect to become a qualified electrical technician, you must be able to repeat the voltage loss test explained in the previous section without making mistakes. You must also be able to make the test quickly and accurately.

10.3 Limit Switches

Limit switches are one of the most widely used pilot devices in modern motor control circuits. They are quite useful since they can convert machine motion into an electrical signal. Limit switches provide a variety of functions, such as limiting machine travel and sequence systems when machine motion has traveled a predetermined distance. Figure 10–9 shows an example application of a limit switch used to determine the depth to which a hole should be drilled. The drill will move down into the hole until the cam trips the limit switch and turns the motor off and disengages the drill clutch. An electrical diagram for this system is also shown in this figure. You can see that the limit switch cycles power to the drill and clutch when the drill travels to the predetermined depth. If a new hole depth is required, the cam on the arm can be adjusted to the new depth. Whenever the cam travels to a point where it activates the limit

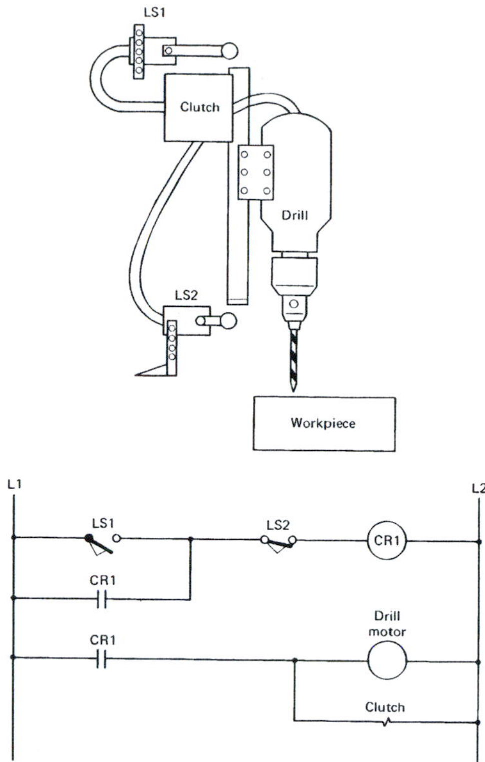

Figure 10–9 Electrical diagram and sketch of limit switch controlling the depth of a drilled hole.

switch, it will deenergize the clutch and drill and a spring will return the drill to the top of its stroke.

Figure 10–10 shows the electrical symbols for various types of limit switches. There are basically four types of limit switches: normally open held open, normally open held closed, normally closed held open, and normally closed held closed. The electrical switch symbol is essentially the same for all types of switches, while the type of operator varies slightly. The operator part of the symbol is shown as part of a triangle that is attached under the switch portion. The triangle signifies the limit switch cam or arm, which causes the limit switch contacts to activate. You can see from the electrical symbols that the location of the switch portion will be above the line for a limit switch that is held open, and below the line for a limit switch that is held closed. If the switch is shown in the closed position in the symbol, it is considered normally closed; if the switch is shown in the open position, it is considered normally open. Remember that the "normal" position is the position the switch would be in had no machine motion occurred. The "normal" position is also the position the contacts would be in had you removed the switch from the machine and bench tested it. The abbreviation for the limit switch is LS. If multiple limit switches are used in a circuit, they will be identified as LS1, LS2, LS3, and so on.

LIMIT SWITCHES	
Normally Open	Normally Closed

 Held Closed	 Held Open

Figure 10–10 Electrical symbols for limit switches.

The limit switch can also activate multiple contacts. These contacts can be groupings of normally open and normally closed contacts. They can be activated as multiple sets of single-pole contacts or as combinations, such as single-pole, double-throw or double-pole, double-throw. The switch contacts are generally mounted in the switch body so that they can be changed without having to unmount the switch. Some switch bodies also allow for plug-in contacts, which allow for quick replacement.

The basic difference between limit switches relates to the type of operator used to activate the contacts. Figure 10–11 shows the different types of limit switch operators. The operators are grouped as roller arm actuators, plunger actuators, fork lever or yoke actuators, and wobble head and cat whisker actuators. The *side rotary actuator* is activated by the motion of the arm as it is moved through an arc. The actuator arms shown in Figure 10–12 can be mounted to the shaft of the limit switch. When the actuator arm is moved, it rotates this shaft and a cam on it will depress the switch contacts and cause the normally open set to close and the normally closed set to open. In some applications, a roller is connected to the arm to provide smooth movement and easy travel as machine motion moves the arm. The roller also prevents the machine from causing wear on the switch. In some roller arm actuators, the roller is adjustable on the arm to provide precision limit control. The roller can be set for preliminary trials and then accurately adjusted to cause precise machine travel.

The *cross roller plunger* or *top roller* is a variation of the side and top roller plunger switches. It is very useful in locations where the top or side plunger types cannot be mounted to detect machine travel accurately. The top round plunger is activated by motion that presses directly on the top of the plunger, causing it to depress. This type of switch is very useful to indicate that a sliding door has moved to its proper location. It can also be used to indicate that a moving part of a machine has traveled the predetermined distance and activated the switch. The adjustable top round plunger provides a screw mechanism on the plunger for precision adjustment. The screw mechanism allows the length of the

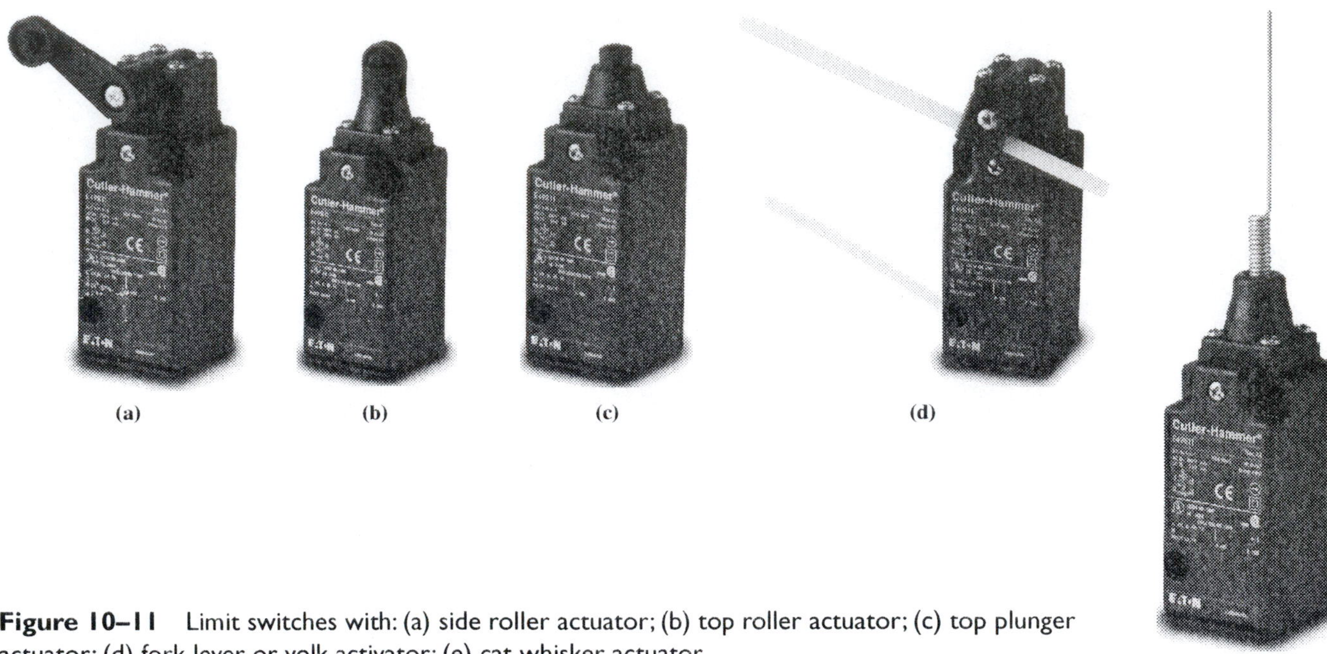

Figure 10–11 Limit switches with: (a) side roller actuator; (b) top roller actuator; (c) top plunger actuator; (d) fork lever or yolk activator; (e) cat whisker actuator.
(Courtesy of Eaton/Cutler Hammer)

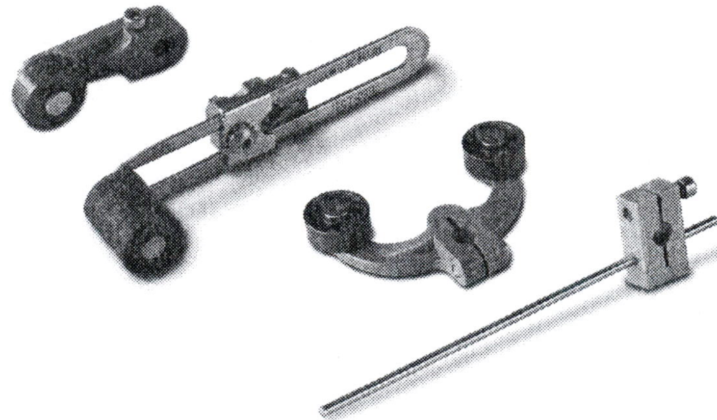

Figure 10–12 Typical actuator arms that are used with limit switches.
(Courtesy of Eaton/Cutler Hammer)

plunger to be increased or decreased slightly by rotating the screw mechanism. Since the thread on this mechanism is very fine, the amount of machine travel can be adjusted for precise repeatability. The side round plunger provides the same operation for horizontal machine travel. These two types of limit switches allow both horizontal and vertical machine travel to be detected without the need to design special brackets to make one type of switch design fit every application.

Another type of limit switch operator is the *plunger actuator*. This type of operator requires linear motion to depress the plunger. The side roller plunger is mounted in a similar manner. The cam that moves across the roller is tapered so that at one end of the travel it will cause the plunger to depress, and at the other end of the travel it will not touch the roller at all. In some cases the cam may have a double taper or multiple tapers. Each time the high point of one of the tapers passes directly over the roller, the plunger will be depressed and activate the switch. This type of switch is very useful in applications where midtravel points are being sensed. An example of this application would be sensing the near-zero or near-home position on liner-actuated robots, where the robot will travel at high speed until it reaches the near-home point. At this point the limit switch would be activated and cause the robot motor to change to low speed as it continued the short distance to the actual home position. The top roller plunger is mounted so that an inclined cam can move past the roller, causing the plunger to depress. This type of switch must be accurately mounted since overtravel on the plunger will damage the switch.

The *wobble head* and *cat whisker actuators* operate on a similar principle. These limit switches use an extended wobble stick or cat whisker to detect machine movement. The extended wobble stick allows the limit switch to be mounted in a safe location where machine travel will not damage it. In applications such as where a robotic arm is used to remove parts from a press, the arm must be clear of the press before it begins its travel. The wobble stick allows the arm to test to see if it is coming close to any fixed portion of the machinery where it could be damaged. Since the wobble stick can be extended, they are available in several lengths. It is also important to remember that the cat whisker actuator is much more sensitive than the wobble stick type. The cat whisker actuator uses a set of microprecision contacts that are activated by a very small amount of whisker travel.

The *fork lever actuator* is used where machine travel in two directions must be detected. This type of actuator is also called a *yoke*. When the machine travel passes the fork in one direction, the yoke is rotated forward so that the front actuator is pressed down, which in turn leaves the rear wheel up where it will be activated when the machine travel reverses. When the machine travel is reversed, it strikes the rear yoke and moves it down. When the rear yoke moves down, the front yoke moves up where it will be in the path of forward machine travel. Since the switch is a maintained type, each time the machine travel moves one side of the yoke down and the other side up, it will remain in that position until the machine travel is reversed and moves the yoke in the opposite direction. This type of switch is very useful in surface grinding applications, where the grinding wheel is moved over the surface in one direction and then reversed. When the machine bed travels forward to the end of its stroke, it will move the yoke and activate the limit switch contacts, which will reverse the motor that drives the bed travel. When the motor reverses, the bed will travel full stroke in the opposite direction until the cam on the other end of the bed trips the yoke in the opposite direction. This action causes the motor to switch to the forward rotation again. As this action continues, the bed cross travel is indexed by screw action. When the cross travel has moved the bed the full distance, it will strike another limit switch, which will turn the motor off.

Another aspect of limit switches is their method of returning to their original positions. Some limit switches are returned by spring action. When the switch lever is activated, the return spring is wound up so that when the machine travel is not pressing on the limit switch lever, the spring will return it to its original position. Other types of return action include gravity and yoke or fork lever return. The gravity return requires the switch to be mounted in such a way that gravity will return the switch lever to its original position. It is important to mount this type of switch in such a way that the machine travel in one direction activates the switch by moving the forward lever of the yoke down. This motion causes the rear lever part of the yoke to move up so that it is in the path of machine travel in the opposite direction. On the return travel, the machine presses the rear yoke down, which returns the yoke to its original position.

10.3.1 Installing and Adjusting the Limit Switch

When you are planning the installation of a limit switch, you must consider that the switch will need to be tested from time to time and its internal contacts may need to be removed and replaced. Figure 10–13 shows an exploded view of a typical limit switch. The limit switch is designed to be repaired in the field with minimal disturbance to the system. This means that parts of the system are installed as permanent components, and the switch should be positioned so that its cover can be removed and the switch can be tested.

Figure 10–14 shows several other environmental conditions that may affect the operation of the switch. Be sure to watch for mounting locations where heat may build up. These locations may include surfaces where heaters are used, such as injection molding machines or foundry application. This also means that you should not mount limit switches near the opening of furnace doors. When the door of the furnace opens, excessive heat may alter the plastic parts of the cam rollers or parts to the limit switch lever. If you must mount a limit switch in areas where high temperatures occur, be sure to mount it away from direct heat or use reflective shields. Sometimes a rod can be attached to the part of the machine that is being sensed. This rod will act as an extension of the machine motion and activate the limit switch even though it is mounted some distance from the heat source. A reflector can be used to deflect heat away from the switch. A rod can also be used in this application. The rod could be mounted so that it can travel through the reflector shield. The switch is mounted behind the shield and the rod passes through the shield to activate the limit switch.

You should be careful not to mount limit switches in the lower parts of metal-cutting machines such as mills and lathes. In these applications, machine-cutting oils or moisture from coolants may drip down into the switch and cause it to corrode and malfunction. You should look for locations that are above the level where the cutting oil or coolant may fall. It is usually possible to design a cam or operator that can trip the limit switch when it is mounted away from these conditions.

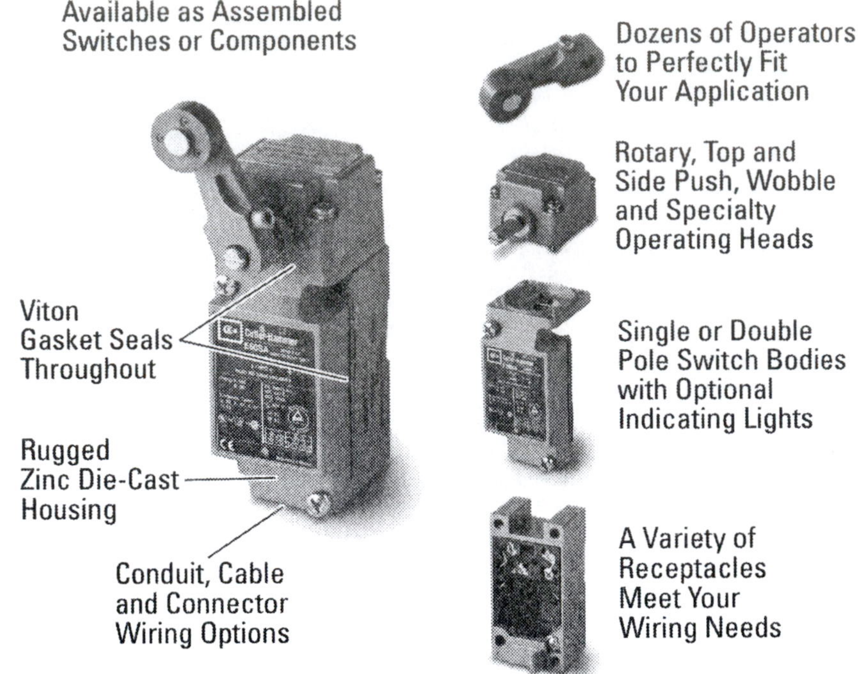

Figure 10–13 Basic parts of a limit switch shown ready for installation. (Courtesy of Eaton/Cutler Hammer)

Another possible solution to this type of application is an oiltight or watertight switch.

Some locations on stamping presses may be unsuitable for mounting limit switches because the excessive vibrations from the stamping operation may cause the switch to become loose in its mounting. In this type of application look for locations where you can mount the limit switch so that it is not attached to the press. This means that the machine motion that is being detected is transferred from the press by means of a rod or cam to the lever of the limit switch. The limit switch can be mounted nearby on a separate mounting stand or pole so that it is isolated from direct machine contact and vibration.

In all limit switch installations you must test the machine travel so that you are sure it will activate the switch each time motion occurs. In some installations the motion at the very end of a machine part may not be uniform during each machine cycle. In these applications you may need to mount the limit switch closer to the center point of the part, where the machine motion will be more uniform. A variety of cams, levers, and rods are commercially available that can extend the range of motion that a limit switch can detect. Other applications may require that you fabricate special trip mechanisms that will activate the limit switch.

If the amount of machine travel that must be detected is minimal, you may select a precision limit switch. A precision limit switch is specially manufactured to activate on a very small amount of travel. Essentially, the part of the limit switch that is different is the amount of travel that is required to activate the switch. Most manufacturers will list the degree of travel or the minimum amount of movement that is required to activate the switch. These distances must be taken into consideration when selecting a switch for sensitive applications. Keep in mind that a proximity switch or photoelectric switch will also provide the function of detecting machine motion. In some applications the amount of machine travel is so small that a limit switch cannot accurately sense the motion repeatably and a proximity switch may be a suitable substitute. The operation of the proximity switch is explained in more detail in Chapter 11.

A limit switch can be mounted as a surface-mounted device, a flush-mounted device, or a duplex-mounted device. The location that you select should be based on the amount of wear the limit switch will encounter, the accessibility of the switch for troubleshooting and replacement, and the amount of adjustment area the switch can utilize during calibration or adjustment. Other considerations must be given to the amount of pretravel, total travel, and differential travel that the limit switch lever must make to activate the switch contacts. In some cases the amount of torque or force required to move the limit switch lever must also be calculated. If the amount of torque the machine movement can exert is limited, a precision limit switch should be considered.

When you are installing the limit switch, the base of the switch should be mounted securely to the location where the machine travel will be sensed. The mounting base has adjustable holes so that the switch can be aligned accurately. The mounting base also provides a

Figure 10–14 Correct methods for mounting limit switches.
(Courtesy of Rockwell Automation)

LIMIT SWITCHES

Application and Installation Data

Limit switches are used in electrical control systems to sense position. They are actuated by the predetermined motion of a cam, machine component or piece part. This mechanical motion is then converted to an electrical signal through the actuation of a set of contacts. These signals can be used in the control circuits of solenoids, control relays and motor starters to control the operation of conveyors, hoists, elevators, machine tools, etc. (typically motor driven machines, processes and systems).

Experience has shown that the mechanical and electrical operating life of a limit switch is influenced to a large degree by proper installation and application procedures. This publication is intended to be used as a guide.

WIRING

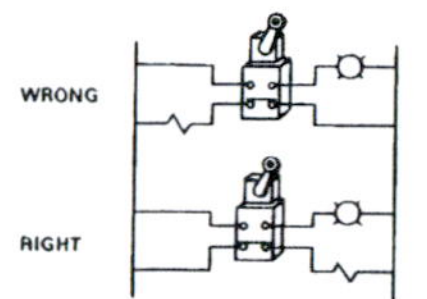

Opposite polarities should not be connected to the contacts of one limit switch unless the limit switch is specifically designed for such service.

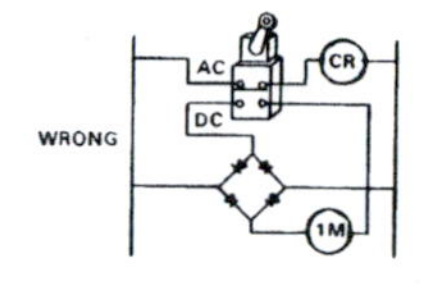

Power from different sources should not be connected to the contacts of one limit switch unless specifically designed for such service.

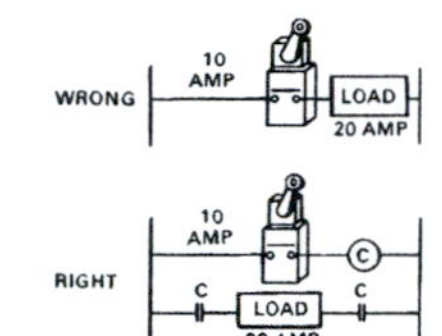

Limit switches should be used within their contact ratings.

ACTUATOR CONSIDERATION

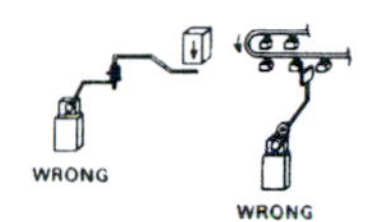

Limit switches are designed for proper performance with the actuators with which they are supplied. Supplementary actuators should not be used unless the limit switches are specifically designed for them.

Operating mechanisms for limit switches should be so designed that, under any operating or emergency conditions, the limit switch is not operated beyond its over-travel limit position. A limit switch should not be used as a mechanical stop.

For limit switches with lever actuators, the actuating force should be applied as nearly perpendicular to the lever as practical and perpendicular to the shaft axis about which the lever rotates.

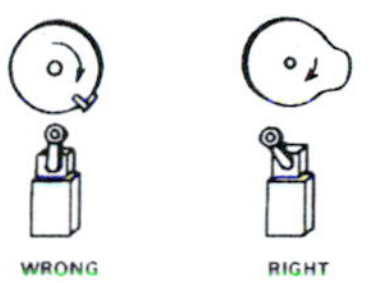

Where relatively fast motions are involved, cam arrangements should be such that the actuator does not receive a severe impact. Cams should be designed such that the limit switch will be held operated long enough to operate relays, valves, etc.

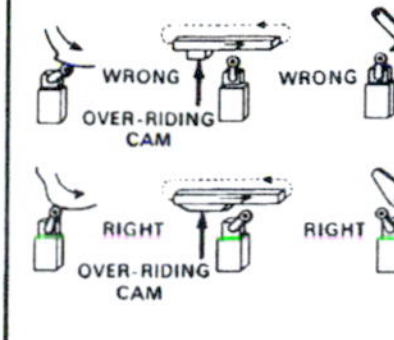

Cam or dog arrangements should be such that the actuator is not suddenly released to snap back freely.

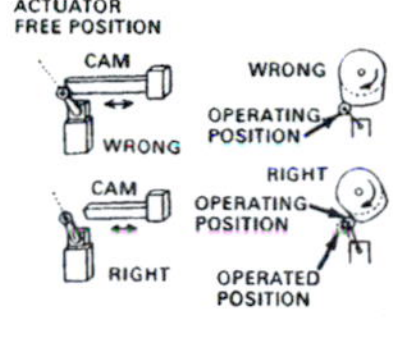

A limit switch actuator must be allowed to move far enough for positive operation of the contacts.

For limit switches with pushrod actuators, the actuating force should be applied as nearly as possible in line with the pushrod axis.

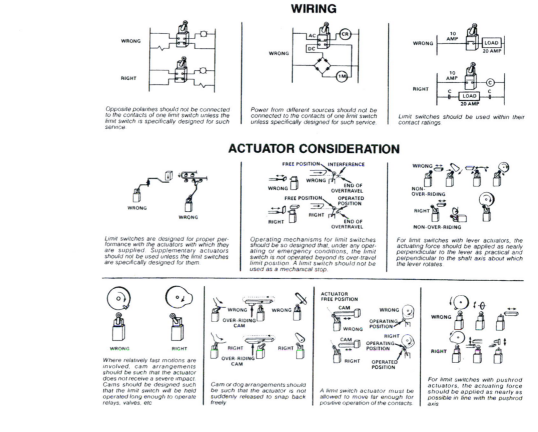

threaded hole for conduit fittings where the field wiring will enter the switch. The field wiring terminals will be connected to the terminal screws in the base. The second part of the assembly contains the normally open and normally closed switches and a set of pins that will make contact with the field wiring terminals when the switch assembly is screwed to the base assembly. Notice that this part of the assembly can easily be removed for quick change out of worn or faulty switch contacts. The switch assembly is available as spring-return or maintained contacts. After the field wiring is attached to the base unit, the switch assembly is screwed onto the front of the base assembly. The third part of the assembly is the actuator head. The actuator head can be any one of the types of actuators shown in Figure 10–12. After the head has been installed, the actuator arm or lever can be adjusted slightly for proper operation. Remember that this switch assembly is very similar to all other pilot devices. It is designed to provide simple installation and it is easy to troubleshoot and replace if problems develop later.

10.3.2 Troubleshooting the Limit Switch

The contacts of the limit switch can be tested in a manner similar to the contacts of other switches. If the switch is out of the circuit, an ohmmeter may be used to test for normally open or normally closed contacts. If the switch is in the circuit and under power, a voltmeter should be used to test the complete circuit prior to testing the limit switch at random. Keep in mind that the reason you will be called to troubleshoot the circuit is because the total machine operation is malfunctioning.

By checking the machine operation systematically, you will be able to narrow the troubleshooting tests to one circuit and later to an individual component. When you are ready to troubleshoot the faulty circuit, be sure to leave one of the meter probes at a reference point and test both the input and output terminals of the contacts of each pilot device in the circuit. If you suspect that the limit switch is faulty, the most important part of testing it is to duplicate the type of travel that causes the switch to activate. If the switch is mounted on a machine, be sure to jog the machine back and forth across the limit switch so that the contacts are activated during the time you are checking for the change of voltage. If the contacts are not activating, you should suspect the machine travel as well as the electrical part of the switch. Make adjustments to the cams or levers if necessary to provide adequate machine travel to trip the switch.

If the switch is causing intermittent problems, check the speed of the machine travel across the limit switch. If the travel is exceedingly rapid, the contact closure may not be long enough in duration to allow proper operation. In this situation you may be able to prolong the contact closure by lengthening the cam on which the limit switch is activating. This problem is critical when the limit switch is connected to a programmable controller or robot. In these applications the programmable controller may scan its input modules at a rate of 50 milliseconds. If the switch contacts activate faster than 50 milliseconds, the programmable controller will not sense this switch activation as a legitimate input signal.

When you have located the faulty component, it should be removed and replaced as quickly as possible. In some cases this means that only the contact portion or the actuator portion of the switch needs to be changed. Always be aware of electrical shock hazards when you are making the tests, and be sure to disconnect the power supply when the components are removed and replaced.

10.4 Selector Switches

Selector switches are similar to pushbutton switches in that they require manual movement to cause the switch contacts to change position. Instead of pressing the switch to cause operation, the selector switch is turned with its handle to cause the contacts to change. Figure 10–15 shows several pictures of selector switches. From this figure you can see that the selector switch can have more than the on and off positions of a pushbutton switch. Typical selector switches have two, three, or four positions.

Figure 10–15 Typical selector switches. (Courtesy of Eaton/Cutler Hammer)

2 Position Selector Switch

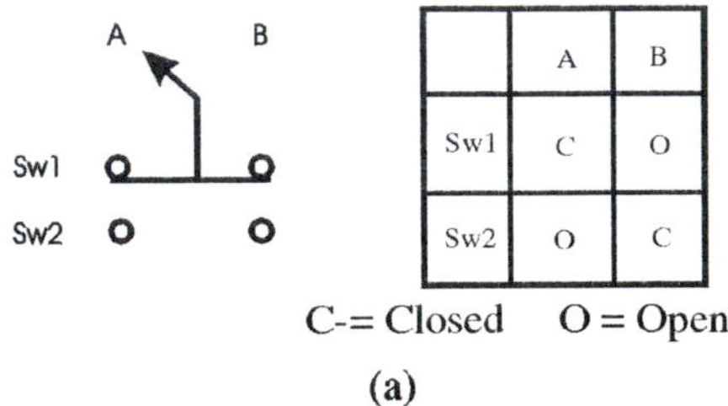

	A	B
Sw1	C	O
Sw2	O	C

C-= Closed O = Open

(a)

3 Position Selector Switch

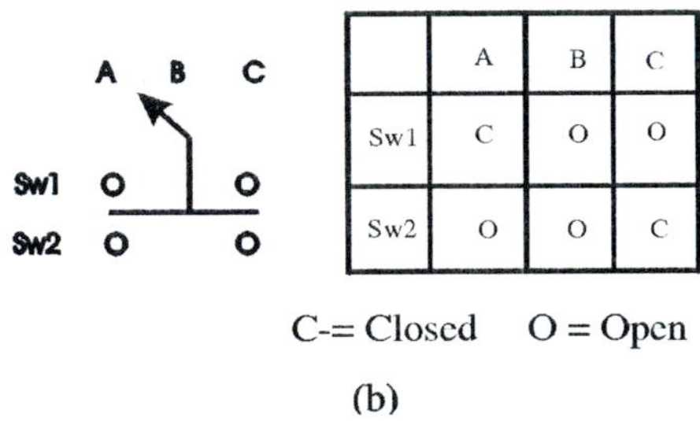

	A	B	C
Sw1	C	O	O
Sw2	O	O	C

C-= Closed O = Open

(b)

Figure 10–16 Electrical symbol and logic table for a two-position and a three-position selector switch.

The two-position selector switch is normally used for such applications as on/off, up/down, and other dual conditions. Figure 10–16 shows the electrical diagram and a logic table for the two-position and three-position selector switches. These diagrams are used as the switch symbols in all electrical diagrams. The selector switch abbreviation is SS. If multiple selector switches are used in a circuit, they will be identified as SS1, SS2, SS3, and so on. The input side of these switches is considered to be the left side of the switch or the point where the single terminal is located. If the switch is a two-position selector switch, either terminal may be used as the input of the switch. The three-position selector switch has one input terminal and two or three output terminals. If the middle position is used as the off position, only two output terminals are required; if the switch has three separate functions, three output terminals are required. In this type of application the top position of the switch is connected to a relay coil marked "hand." The middle position is set for the off position, with nothing connected to the middle output terminal. The lower terminal of the switch is connected to a relay marked "auto." In this application the switch must be in one of the three positions.

The selector switch's operation is similar to that of the maintained pushbutton switch, in that it will stay in the position to which it is changed. This function is quite useful in motor control circuits because it allows an operator to select an application and the switch will remain in that position. The selector switch requires manual motion from an operator to cause the switch to change position. The selector switch is installed in the operator's panel in a manner similar to the pushbutton switch. This means that all electrical connections are made on the contact block at the rear of the panel. Various legend plates may be installed to explain or define the function of the selector switch.

Several types of levers or knobs are also available. One of these is the keyed knob, which allows the switch to be locked into position. If a change is required, the switch must first be unlocked. This type of switch is useful in applications where security is needed to guard against unauthorized changes of machine operation.

10.5 Joystick Controls

Joystick control is used in many industrial applications. It is used primarily for multiaxis motion control, such as overhead hoist operation, moving molten iron in ladles for pouring operations, and semiautomatic robot operation. The joystick control actually consists of a multiple set of switch contacts located so that the stick's movement will activate one or more contacts at the same time. When the two-position control is used, the operator can move the stick forward to activate the upper set of contacts. When the stick is pulled back, the lower set of contacts are activated. When the stick is returned to the neutral position, all contacts are deactivated and will return to their normal position. Most joysticks use a spring to return the stick to the neutral position.

One application of a joystick control is for an overhead hoist operation. When the upper set of contacts is activated, the up motor starter is energized and the motor will turn in a clockwise direction, moving the hoist hook upward. When the stick is returned to the neutral position, the up motor starter is deenergized and the motor is turned off. This type of motor includes an electrical brake that would be energized any time the motor is deenergized. The brake would prevent the motor from creeping and heavy loads from dropping when the motor is not running. When the joystick is pulled back, the lower set of contacts is energized and the motor starter to allow the load to be lowered is activated. When the motor starter is energized, the motor will reverse and turn in the counterclockwise direction. In this sense the motor starter is wired for forward/reverse operation, which will cause the crane's cable to move up or down.

Another variation in this application is the operation of an overhead crane that can raise and lower heavy loads such as rolls of steel. The crane in this application can also move forward and reverse along a set of overhead rails. Another motor allows the crane to move back and forth across the width of the rails. These four motions allow the crane to be positioned anywhere on the tracks. They also allow the crane to move heavy loads to any machine that is located under the rails. This includes all of the area from one end of the rail to the other, and all of the area side to side between the rails.

The best way to explain the operation of this joystick is to use the directions of the compass. If the joystick is moved north or south (pushed ahead or pulled back), the crane will move forward or reverse along the length of the rail. If the joystick is moved east or west (left or right), the crane will move back and forth across the rail, left and right. This motion allows the crane to reach all the positions from side to side between the rails. These four motions are controlled by two reversing-type motor starters that control the motor for forward and reverse travel, and the motor for left-and-right travel.

If the joystick were pushed in the northeast direction, the forward motor starter and the right motor starter would be energized. This would cause the crane to move down the track and to the right in one smooth simultaneous motion. If the joystick were pushed in the northwest direction, the forward motor starter and the left motor starter would be activated, causing the crane to move forward and to the left. When the joystick is pulled southeast the reverse and right motor starters are energized, causing the crane to move in reverse and to the right. Each motor can be activated individually by moving the joystick to the north, south, east, or west location. An up/down pushbutton could also be added to raise or lower the load. Other safety features, such as key lock and indicator lamps, could be added to ensure safe operation. This type of switch allows total operation of the crane with minimum operator motion.

10.5.1 Troubleshooting the Joystick Control

If you think the joystick control is not operating correctly, it can be tested with power applied or with power removed. The first step in this procedure is to determine the number of contacts the switch has. After you have located each set of contacts, move the switch in each direction and test for proper switch operation.

Sometimes the switches will have to be adjusted slightly for the joystick to activate them fully. After each individual motion has been tested, move the joystick so that the compound moves are active. If the stick is moved northwest, the upper and left-side circuits should be activated. Remember: The contact sets may be located in the opposite position on the switch from the function they provide. This means that when the stick is moved forward, the lower set of contacts is closed, and when the switch is moved to the right, the left set of contacts is activated. Your tests of the contacts will help you determine the location of each set. If any one of the sets of contacts is inoperative, adjust the contact actuator and retest the contacts. If the contacts remain inoperative, they must be replaced. After the single contact motions have been checked, continue the test of the multiple-contact moves. When the joystick is moved in the northwest direction, two sets of contacts will be activated. If only one of the two is activated, adjust the contacts to remedy this fault. If each contact set passes the individual test, the fault must be in the adjustment of the contact actuator for the multiple-contact move. When you are certain that the contacts in the joystick are operating correctly, continue testing for faults in the motor starter circuits.

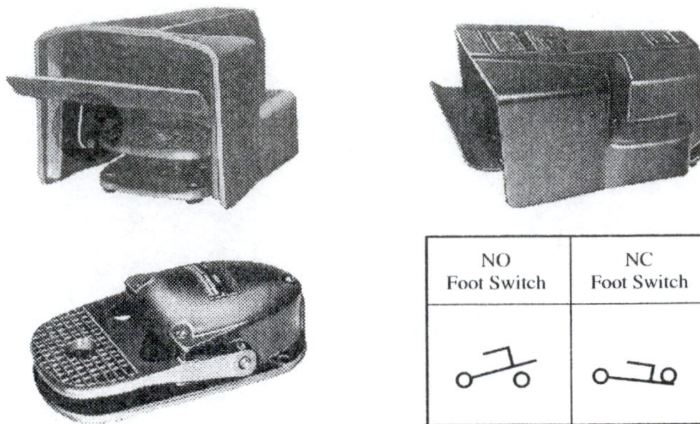

Figure 10–17 Pictures and electrical symbols for a foot switch.
(Courtesy of Square D/Schneider Electric)

10.6 Foot Switches

Foot switches are pilot devices that are similar to a pushbutton switch, except that they are activated by pressing on them with your foot. Figure 10–17 shows several pictures of the foot switch and symbols for the foot switch. The activator portion of the switch is manufactured with heavy-duty material to withstand repeated activation from someone stepping on it. The switch portion of the foot switch is very similar to that of other pilot contacts. The contacts are available as single-pole, single-throw or double-pole, double-throw combinations. If the switch contacts are shown above the terminal point in the diagram, the switch is a normally open switch and the contacts will be closed when the foot switch is activated. If the contacts are shown below the terminals, the switch is a normally closed type and the contacts will be opened when the switch is activated. The abbreviation for the foot pedal switch is FTS. If several foot pedal switches are used in a circuit, they are identified as FTS1, FTS2, and FTS3.

One typical application is for a foot switch to be used by a person unloading a conveyor. Parts are placed on the conveyor by robots that are unloading a plastics press. The conveyor is activated by the worker to bring parts to the end of the line, where they will be loaded into baskets. Since the switch can be activated by the operator's foot, the person's hands will be free to remove parts from the conveyor. Each time the operator is ready to remove a part from the conveyor, the pedal is depressed. The foot pedal activates the switch contacts, which energizes the conveyor motor starter. The motor starter energizes the conveyor motor, which moves the parts toward the person handling them. When the conveyor moves the parts close enough to be picked up, the operator releases the foot switch and the conveyor stops. After the operator places the part into the basket, the process is repeated. If the conveyor does not advance the parts to the end of the line, a jam-up will occur at the point where the robot is placing the parts on the line. If a jam occurs, the robot senses that a part is in its way and will indicate a fault condition, and a technician must be called to test the circuit.

Other applications that utilize the foot switch are manual spot-welding applications and manual stitching or stapling machines. In both of these applications it is important that operators have their hands free to move the parts under the machine head. In spot-welding applications the metal parts are moved into position by hand. When the parts are in the proper location, the operator can depress the foot switch and activate the spot welder. A similar operation is utilized with manual stitching machines. A cardboard box can be stapled manually by moving the box from point to point under the stapler head. Each time the operator moves the box into position, the foot switch is depressed and the stapler head injects a staple into the box at that point.

The operation of a foot switch is similar to that of other pilot devices. The switch operator must be depressed for the contacts to change position. The distance the activation portion of the switch must travel can be adjusted. Since the pedal is returned to its original position with spring tension, the amount of clearance between the pedal and the switch is critical. If the clearance is too close, the switch will not return correctly to its normally open position. It will also have problems with false activation. When the switch is adjusted correctly, the pedal can be activated with the proper amount of weight. If the switch requires too much weight to activate the pedal, the operator will become overly tired from the repetitive operation.

10.7 Float Switches

Float switches are specialized pilot devices used to measure the level of liquids and granular materials used in industry. For this reason, the float switch is also called the *level switch*. Figure 10–18 shows photographs and symbols for the float switch. The symbol for the float switch looks like a ball connected directly under the contact symbol. The ball in the symbol represents the float device that is connected to the end of the actuator rod. The contacts for this symbol can show the switch opening under the terminal point or above the terminal point. The abbreviation of the float switch is FS. If multiple float switches are used in a circuit, they are identified as FS1, FS2, FS3, and so on.

If the switch contact is shown above the terminal point in the symbol, it indicates that the switch contacts will open as the level increases and will close when the level decreases. This type of contact arrangement, called *normally closed*, would be used to keep a water tank from overflowing. When the water level increased to the proper level, the contacts would open and turn power to the motor off. When the tank level is lowered to a predetermined set point, the contacts would close again and the pump would turn on. Figure 10–19 shows an example of this application with the electrical diagram of the pump motor and level control.

If the symbol shows the contacts below the terminal point, it indicates that they will close when the level that is being measured increases and open when the level decreases. These contacts, called *normally open*, are used to activate a pump that is used to empty a sump. This application is also shown in Figure 10–19. When the level in the sump increases, the float will rise and cause the contacts to close and turn on the pump motor. When the pump has lowered the sump level to the predetermined level, the float will drop and cause the switch to open and turn off the pump motor.

10.7.1 Float Switch Operation

Several types of float switches are used in industrial control: the open switch, the closed switch, and rod and chain control. Figure 10–19 shows diagrams of the operation of these controls. The open float control (Figure 10–19a) has a set of contacts activated by a lever actuated by the float and rod. Several types of floats are used with these controls. The simplest type is made from plastic. It is shaped like a ball and provides a threaded connector where the actuator rod can be connected. The float is completely sealed so that liquid cannot enter it and cause it to sink. The top end of the rod has an adjustable collar that can be located to cause the

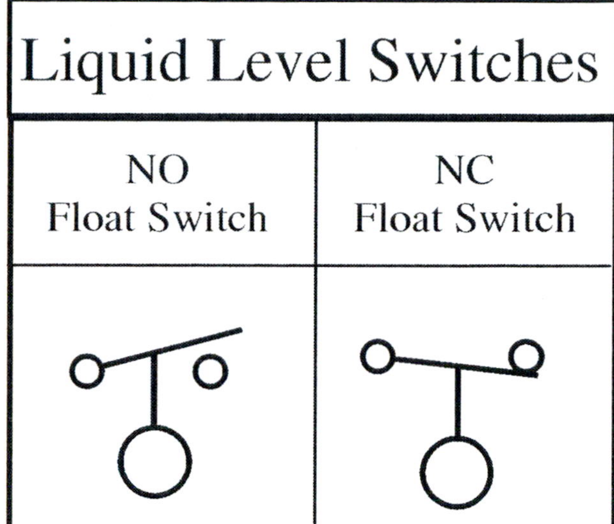

Liquid Level Switches
NO Float Switch
NC Float Switch

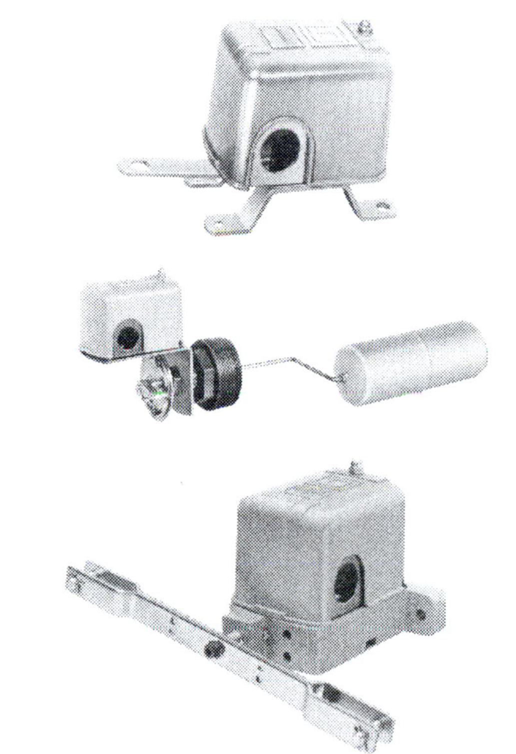

Figure 10–18 Picture and electrical symbols for float switches.
(Courtesy of Square D/Schneider Electric)

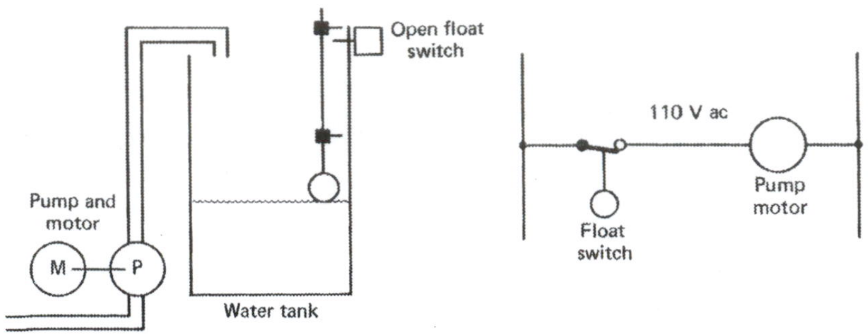

(a) Open float switch application and electrical diagram

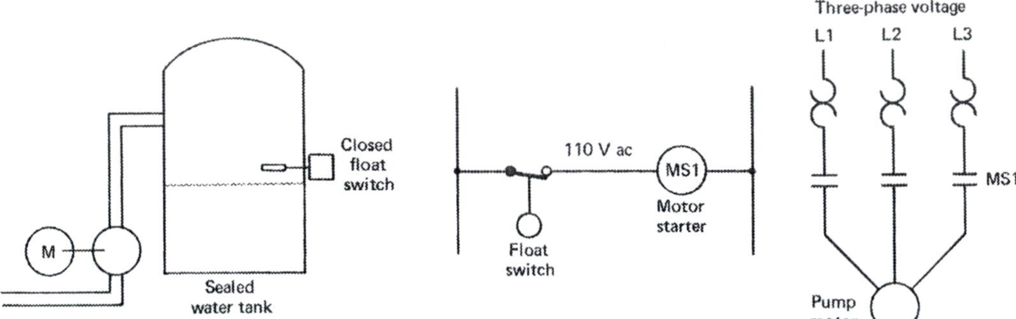

(b) Closed float switch application and electrical diagram

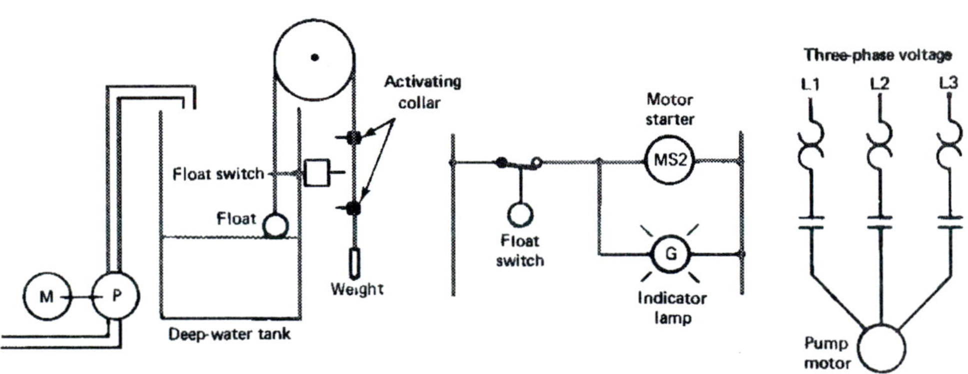

(c) Rod and chain float switch application and electrical diagram

Figure 10–19 Diagram of (a) an open-type float switch; (b) a closed-type float switch; (c) a rod-and-chain-type float control. (Courtesy of Square D/Schneider Electric)

switch to activate when the liquid level reaches a predetermined point. The length of the rod can be varied so that the switch can be located well above the liquid level. This allows the switch to be mounted in a dry location where it is easily accessible for installation, troubleshooting, and adjustment. After this type of control is installed, the float should be adjusted on a trial-and-error basis to ensure that the switch is activated when the liquid reaches the proper level. If the switch is not activating at the correct level, the collar on the activating rod can be adjusted up or down. Since most switches include both the normally open or normally closed contacts, this control can be used for both filling and emptying applications.

Open level control is named for the type of tank application whereby the tank is not sealed or pressurized.

The control is available with the switch portion open for dry locations or sealed for watertight, corrosion-resistant, and explosive-proof applications.

Another type of level control is called *closed tank control*. The operation of this control is shown in Figure 10–19b. The switch contacts are located in the same fixture as the level control. The switch contacts are isolated from the float with a seal. This allows the float control to be mounted directly onto a tank and still maintain the tank's seal. This is especially useful for applications where the level of a pressurized vessel is controlled. It is vitally important that the seal be maintained when the control is installed.

The *rod and chain float control* is shown in Figure 10–19c. This type of actuator allows the switch to be located remotely from the float mechanism. The switch is activated by collars attached to the chain. The chain has a float attached to one end and a weight attached to the other. This assembly is passed over a pulley so that the weight will counterbalance the float. When the tank level decreases, the float will move lower in the tank. This action will cause the chain to pull the weighted end higher. When the chain has moved far enough, the collar will activate the switch mechanism and engage the contacts. When the level of the tank increases, the float will rise with the liquid level and provide slack in the chain. The weight on the end of the chain will cause it to press the collar down against the switch mechanism and deactivate it. Since the chain can be varied in length, the switch can be mounted any distance from the tank to provide easy access for maintenance and troubleshooting. This type of assembly is also useful for very deep tanks where an extended rod assembly may be too bulky or expensive.

10.7.2 Resistive-Sensitive Level Indicators

Another type of float switch or level indicator uses the difference of resistance between two probes to activate a relay when the level of a liquid in a tank changes. A solid-state circuit is used to sense the change of current that is caused by a change in resistance. In this type of control, current will flow between two sensors through a liquid as long as the liquid is covering both probes. If the liquid level falls below one of the probes, the current will not flow because the air gap between the lower liquid level and the probe causes a higher resistance. When the liquid level rises and covers the probe again, the resistance of the liquid is small enough to provide adequate current to allow the control to activate the relay again. This type of control can also be designed to operate with only one probe. In this application, the metal sides of the tank will carry current like the second probe would. The single probe would be placed at a point where it can sense the liquid level. A very small amount of current will flow through the liquid between the two sensors or between a single sensor and the side of the tank any time the liquid level is high enough. Even though the current flow is very small, the solid-state part of the control is able to distinguish between the small current flow and no current flow when liquid level is below the sensor. The electronic part of the control requires 12 VDC to operate correctly. The contacts of the relay are isolated and are rated for 5 A at 125 VAC, 1 A at 120 VDC, and 2.5 A at 230 VAC or VDC for non-inductive loads.

10.7.3 Installing and Troubleshooting the Float Switch

The contacts for float switches are very similar to the contacts in other pilot devices. Figure 10–20 shows a float switch with its cover removed. Since normally open and normally closed contacts are available, this type of switch is easily adapted to a variety of control applications. The installation process involves two parts. The first part is the installation of the float apparatus. The float must be installed so that it can move with the liquid level of the tank or sump. You must be sure that the activator rod is long enough to reach the switch mechanism at both the high and low levels of the liquid. It is also important to determine the dead band for the application. The dead band is the distance between the point where the switch is activated at the high level and the point where it is activated at the low level. The dead band can also be measured as *dwell time*, which is the amount of time the switch contacts are deactivated between the high-level point and the low-level point. If the dwell time is not sufficient, the pump motor may be required to start too many times per hour, which will cause it to overheat and wear out prematurely.

After the float mechanism has been installed and adjusted, the switch contacts can be wired. Use the wiring diagram to determine to which switch terminal the field wiring should be connected. Be sure to observe wire color codes and utilize wire markers. Field test the complete float control to ensure that it is operating correctly.

If the float control is not operating correctly, you must execute troubleshooting procedures to find the problem as quickly as possible. Test the complete circuit to determine the exact location of the fault. If you have determined that the float switch is at fault, the first step in troubleshooting it is to determine that the float mechanism is operating correctly. Remember that the switch contacts cannot be activated if the float mechanism does not move far enough. If the switch is covered, you may need to remove the cover so that the assembly

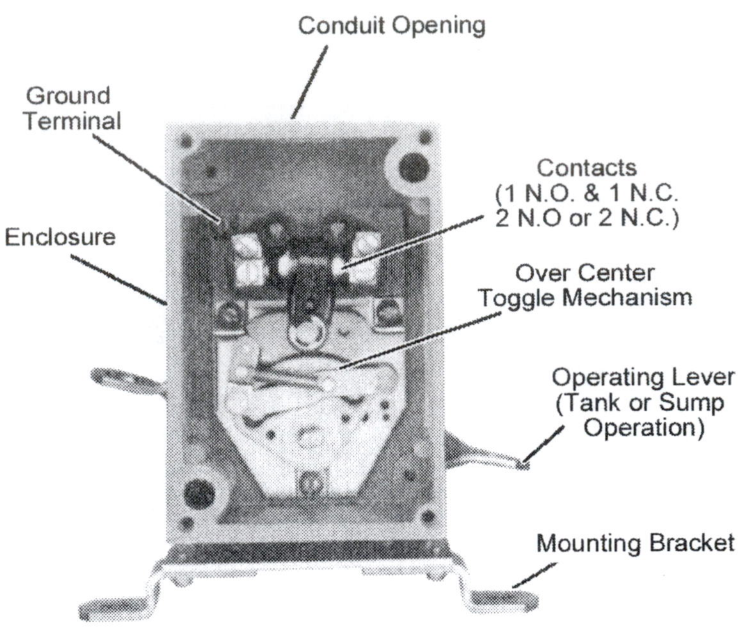

Figure 10–20 Internal parts of a level (float) switch.
(Courtesy of Rockwell Automation)

can be clearly observed. If the float mechanism is not traveling correctly or if it has not moved far enough to activate the switch, correct this problem prior to continuing the troubleshooting procedure. After you have determined that the float mechanism is operating correctly, proceed to the electrical part of the test. If the switch is powered, use a voltmeter to test the contacts for proper operation. The procedure for locating a problem in a control circuit will be the same regardless of the actuating device.

If the troubleshooting procedure determines that the problem is in the electrical contacts, be sure to test them again while you activate the switch manually. In this case you can manually activate the switch without harm to the pump motor. Remember that some pumps may be damaged if they are operated dry. If this is the case, be sure that the pump has an adequate supply of liquid during the testing procedures. If the motor control application involves a closed tank application, be sure to test the float control for leakage after all repairs have been completed.

10.8 Pressure Switches

Pressure switches are used in a variety of motor control applications. These applications include detecting high pressures, low pressures, differential pressures, and vacuums. These devices can be used as safety controls to protect systems from high pressures or the lack of pressure, or as operational devices to cycle an air compressor to maintain constant pressure in a system.

Pictures and electrical symbols for a pressure switch are shown in Figure 10–21. Figure 10–22 shows a diagram of a pressure switch. The symbol for a vacuum switch, differential switch, and pressure switch is the same. The symbol is very similar to those for other pilot devices, in that it shows a contact and an operator. The operator in this case is shown as a sealed chamber. The symbol that shows the contacts above the terminal point indicates that this pressure switch opens when pressure increases and closes when pressure decreases. This indicates that the device is used as a high-pressure control. It could also be used to maintain pressure in a tank at a predetermined point. The symbol that shows the contact below the terminal point indicates that the switch will open when pressure decreases and close when pressure increases. This indicates that the device is used as a low-pressure control. A low-pressure switch may be used to indicate when a sealed pressurized system has lost pressure. This type of control is used extensively in air-conditioning and refrigeration systems to deenergize the compressor motor if the system loses its refrigerant. PS is the abbreviation for a pressure switch. If additional pressure switches are used in a circuit, they will be identified as PS1, PS2, PS3, and so on.

10.8.1 Diaphragm Pressure Switches

The diaphragm pressure control is used to sense relatively low pressures. It is manufactured with two halves separated by a rubber diaphragm. The bottom chamber is connected to the source of pressure. The top chamber is vented to the atmosphere (atmospheric pressure is 14.7 psi). The chamber that is connected to the pressure source will be manufactured with a threaded fitting to provide a means of connecting the operational part of the switch to the pressure part of the system. The pressure can be supplied from an air system or a fluid system, such as water or hydraulic fluid. This fitting must be made to withstand the pressure the switch is sensing without leaking. When air pressure increases, the diaphragm moves upward against the switch mechanism and causes it to activate the switch contacts. The amount of pressure required to activate the switch contacts can be adjusted by spring tension. The contacts can be single-pole, double-throw or double-pole, double-throw. They can be set against the switch mechanism to open when the pressure decreases or when the pressure increases beyond the switch's set point. You must be very careful when you adjust the pressure setting for this type of switch so as not to damage the spring mechanism.

After the switch contacts are activated by the movement of the diaphragm, they will remain in that state until the pressure returns to its preset value. The switch and diaphragm mechanism is manufactured so that it will not activate both off and on at the same pressure.

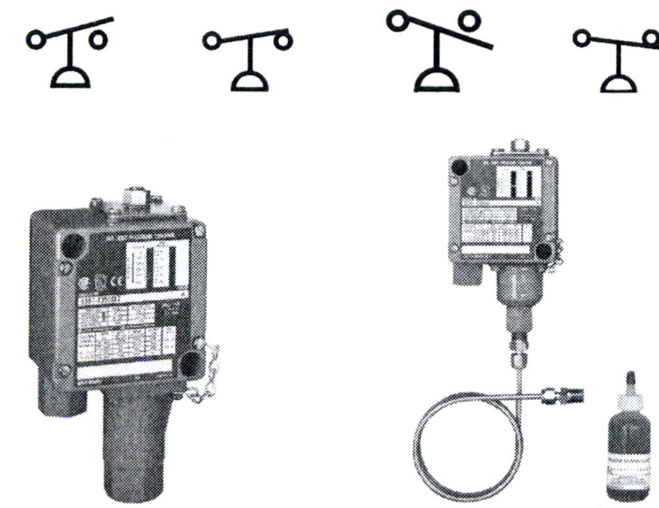

Figure 10–21 Pictures and electrical symbols of a pressure switch.
(Courtesy of Rockwell Automation)

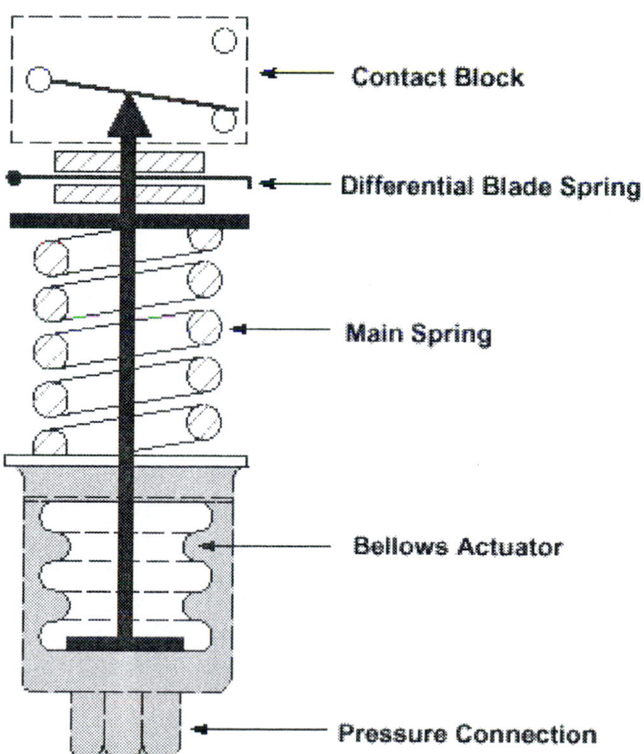

Figure 10–22 Internal parts of a pressure switch.
(Courtesy of Rockwell Automation)

For instance, if a high-pressure switch is used as an operational control to regulate the amount of pressure in an air compressor tank at 50 psi, the switch may have a range of 0 to 150 psi and a differential of 1 to 15 psi. The range is the highest value at which the switch will cut out and deactivate the switch contacts. In this switch, the pressure can be set at any value between 0 and 150 psi. To maintain the pressure at 50 psi, the range part of the switch should be set at 55.

The differential is the amount of pressure drop the system should have before the compressor is switched on again. If the differential is set at 0 or 1 psi, the pressure switch would deenergize the compressor motor when the tank pressure reached 55 psi, and reenergize the compressor motor when the pressure dropped to 54 psi. This would cause the compressor motor to cycle off and on continually any time air was used in the system. If the range was set to 55 psi and the differential was set at 5 psi, the switch would turn the compressor motor off at 55 psi and back on at 50 psi. This would allow a time delay between the turn-off and turn-on pressure, which would extend the life of the compressor motor. The differential pressure setting in the pressure switch will also extend the life of the switch's electrical contacts.

10.8.2 Vacuum and Extra-Low-Pressure Switches

The diaphragm switch can also be manufactured to sense very small changes in air pressure or vacuum. In this type of switch, the amount of movement the diaphragm provides is amplified through mechanical means. This allows a very small change of pressure of less than 2 psi to be detected. In a vacuum switch, the top side of the diaphragm is vented to atmospheric pressure and the lower side is connected to the air line that has the vacuum pressure. The springs and cams are used to set the amount of vacuum that will activate the switch contacts. Vacuum is measured in feet of water column or inches of mercury, which means that the switch may be labeled in inches rather than in psi.

10.8.3 Bellows Pressure Switch

Diaphragm control is useful only for applications where the maximum pressure does not exceed 150 psi. Since the pressure presses against the rubber diaphragm, a pressure that is too high may rupture it. For applications where pressures exceeding 150 psi must be sensed and controlled, a metal bellows is used. The pressure sensor may also be operated by a bellows. In this type of pressure switch, the pressure line is connected to the inside of the bellows. Atmospheric pressure will exert pressure on the outside of the bellows. When the pressure increases, the bellows expands upward until it activates the contact mechanism. The amount of pressure required to activate the contact mechanism can be adjusted by spring tension. An external adjustment knob or screw is provided to set the pressure that is required for activation. Since the bellows is made of metal, it can withstand pressures of up to several thousand pounds. This type of pressure switch is generally used for high-pressure air systems or high-pressure hydraulic systems.

The Bourdon tube actuator is very similar to the bellows operator except that it will operate more accurately with high repeatability. The Bourdon tube is sealed at one end and the air or liquid that is being sensed is allowed to enter the other end. Since the tube is wrapped in a spiral, it will try to straighten out as pressure increases. The amount of movement is directly related to the amount of pressure inside the tube. A set of springs, gears, and cams are used in conjunction with the tube to cause a set of contacts to activate. There will also be a setscrew and an offset spring that allows the set point and differential pressures to be adjusted. The contacts in these types of switches can be set by the manufacturer to activate on an increase or a decrease of pressure. The contacts can be single-pole, double-throw or double-pole, double-throw, which provides both normally open and normally closed contacts. The contact rating will be similar to the ratings listed in the table in Figure 10–1.

10.8.4 Differential Pressure Switch

The differential pressure switch is activated by a diaphragm mechanism similar to a low-pressure switch, for the operation of this switch checks the difference between two pressures. The lower pressure of the two is connected to the top of the diaphragm chamber, where it will press down on the diaphragm. The higher pressure is then connected to the lower chamber, where it will press upward on the diaphragm. A set of springs and cams are used to set the exact amount of pressure differential that will activate the switch contacts.

This type of pressure switch is widely used to detect loss of oil pressure in an air-conditioning compressor. Since the compressor oil sump is part of the sealed system, it will have 70 to 80 psi in it when the compressor is not running. When the compressor begins to run, the oil pump in its sump will begin to pressurize the lubrication system and the compressor's pistons will increase pressure on the refrigerant. The oil pressure is typically 15 to 30 psi and will be in excess of the refrigeration pressure. A differential pressure switch is used to compare the refrigeration system pressure to the oil pressure. For application the differential pressure would be set to activate at a pressure difference of 15 psi. This means that the low-pressure side of the switch is activated by the refrigerant pressure. The high-pressure side of the switch is connected to the compressor oil sump, where it will see the same amount of refrigerant pressure plus the 15 to 30 psi of oil pressure. This causes enough pressure differential to activate the switch. If the oil pressure drops below 15 psi, the differential will become too low and the switch contacts will drop out and turn the compressor off. Since refrigerant oil may foam slightly or the oil pump may cavitate for a few seconds, a 30-second time delay is generally added to this switch. The time delay ensures that the pressure differential must exist for over 30 seconds before the switch activates and turns the compressor motor off.

The oil pressure seems to float with the refrigerant pressure. This means that if the refrigerant pressure increases by 40 or 50 psi, the other side of the switch will also see the same increase in refrigerant pressure. The differential is still due to the additional pressure caused by the oil pump.

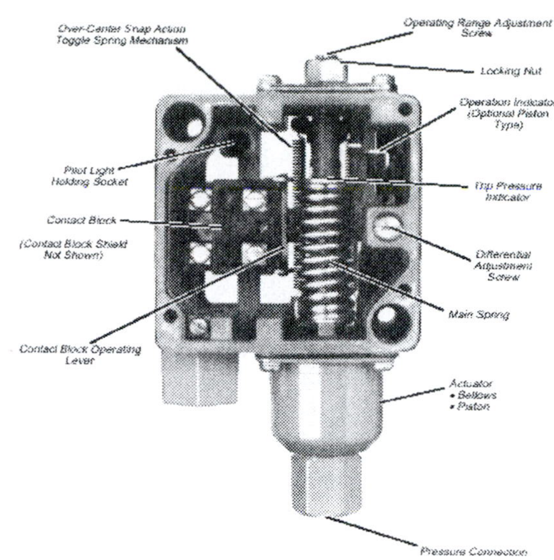

Figure 10–23 Picture of internal parts of a pressure switch.
(Courtesy of Rockwell Automation)

10.8.5 Installing the Pressure Switch

The installation procedure involves two distinct parts: the field wiring of the switch contacts and the connection and adjustment of the pressure side of the control. The field wiring should be completed after the switch has been connected to the pressure source. Figure 10–23 shows a pressure switch with its cover removed so you can see the electrical terminals on the switch. During the installation process, you will need to adjust the pressure side of the control with a gauge. For this type of adjustment, connect an ohmmeter across any set of contacts. If the control is a high-pressure switch, increase the pressure until the switch contacts are activated. If the contacts do not activate at the correct pressure, release all pressure and adjust the range on the control. Retest the control by increasing the pressure. After the contacts activate at the correct pressure, the differential should be adjusted. The differential is adjusted by lowering the pressure slightly after the contacts have activated. Watch the ohmmeter to determine when the switch contacts close again. If the differential is too small or too large, adjust it by turning the differential screw. Some pressure switches have a fixed differential; only the range can be adjusted.

If the control is a low-pressure control, test the contact activation as you release pressure from the control. Be sure that the pressure is higher than the activation pressure plus the differential. After you have determined the point where the switch contacts are activated, allow the pressure to rise slightly, to the point where the differential reactivates the contacts.

After you have adjusted the range and differential on the pressure switch, connect the field wiring to the switch contacts in accordance with the field wiring diagram. After the field wiring has been completed, apply power and test the switch for proper operation. Be sure to allow the pressure to rise and fall naturally with system operation while you observe the switch cycling. If the control is used as a safety device, such as for a low-pressure or high-pressure switch, manipulate the system so that these events will occur while you are observing the system to be sure that the safety switches operated properly.

10.8.6 Troubleshooting the Pressure Switch

The pressure switch can be tested while it is in or out of the circuit. During the troubleshooting procedure, begin troubleshooting the total machine operation and try to narrow the problem down to one circuit. Then concentrate your troubleshooting tests on that circuit. When you have determined the origin of the fault, you will need to perform diagnostic tests on that component. If the component is a pressure or vacuum switch, the most important point to remember when testing is to use the electrical symbol to determine whether the switch contacts open with an increase of pressure or whether the contacts will close when the pressure increases. After you have determined this, be sure that the pressure is changing enough to activate the contacts and return past the point where the differential will return the contacts to their normal state. You will need to install a pressure gauge to make this determination. In some cases a pressure gauge remains in the circuit at all times for this reason. If the pressure is not changing enough to activate the contacts, the problem may exist in the pressure part of the system rather than in the electrical part of the system.

If you have determined that the pressure part of the system is operating correctly, use a voltmeter to test the switch while it is connected to power. Leave one probe of the meter on the line 2 terminal of the circuit and move the other probe from point to point along the contacts in the circuit. Remember that the point where the voltage reading is zero indicates the point where the open circuit exists.

After you locate the faulty part, be sure to turn the power off before you remove the field wires. Also remember to check the pressure side of the control so that you do not remove the pressure line with pressure on it. When you are ready to reinstall the pressure switch, use the information on installation procedures provided earlier in this section.

10.9 Flow Switches

Flow switches are used in a variety of applications to indicate the presence or absence of flow. This type of pilot device can be used to detect the flow of water, air, and other fluids. Figure 10–24 shows the electrical symbol of a flow switch. The symbol for the flow switch is a small flag or sail attached to the switch operator. The sail represents the device used to sense the flow. In some cases this device will be a small metal paddle that will move as fluid flows past it. The movement of the paddle activates the switch contacts. The abbreviation for the flow switch is FLS. In some diagrams a number

NO Flow Switch	NC Flow Switch

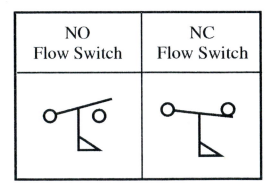

Figure 10–24 Symbols of flow switches.

is used in addition to the abbreviation to identify each flow switch (for example, FLS1, FLS2, and FSL3).

From the electrical symbol you can see that the contacts can be shown as closing when flow is detected, or as opening when flow is detected. If the switch contact is shown above the terminal points, it indicates that the flow switch is normally closed and will open when fluid flow is established. If the switch contact is shown below the terminal point, the flow switch is normally open and will close when fluid flow is established.

10.9.1 Applications and Flow Switch Operation

In a simple application a flow switch is mounted in a cooling-water line to detect sufficient cooling water flow. The cooling water is used to keep the temperature of a plastics press at a predetermined level. The system is set so that the press cannot be placed in the automatic mode if cooling-water flow is not established. If the flow is interrupted at any point during press operation, a time-delay circuit will activate and deactivate the press after 30 seconds. The time-delay circuit allows the flow to cavitate for short periods without tripping the switch. Notice that the 30-second time delay is also used to allow the system to start. The flow switch will seal in this circuit after flow is established.

The switch will have an adjustment screw to set the amount of flow required to activate the switch. Some switches may not have this type of adjustment; instead, the paddle must be cut or bent slightly to adjust the amount of flow required to activate the switch. If material is removed from the paddle, the switch becomes less sensitive. This means that more flow is required to cause the switch to activate.

In another application a flow switch is mounted where it can detect the presence of airflow in a chimney. For this application, airflow is required through the chimney for 5 minutes prior to the ignition process. It is very important to be sure that the purge blower is operating correctly prior to ignition, since most gas valves in industrial boilers and other gas-fired heaters have a possibility of leaking gas or fuel while they are turned off. The fuel tends to accumulate near the lower portion of the chimney close to the ignition source. If the ignition is turned on before the chimney is purged, a tremendous explosion could result. The flow switch in this case ensures that a proper amount of fresh air has moved through the chimney to remove any residual fuel vapors prior to ignition. A flow switch is used instead of simply monitoring the operation of the motor, because it is possible that the motor may be running and not enough air is moving. This could happen if the system had loose belts, a broken motor shaft or fan shaft, a blocked air passage, or a dirty fan. If the ignition system does not detect sufficient airflow for 5 minutes, the circuit will time out and lock out the ignition system. An indicator lamp will be energized to identify the system fault. The system must be reset manually to restart the fan motor. If the flow switch does not detect sufficient airflow for a full 5 minutes, the circuit will continue to lock out.

The flow switch operates a set of contacts when the paddle has moved a predetermined distance. The contacts can be single-pole, double-throw or double-pole, double-throw. These contacts can be any assortment of normally closed or normally open sets. The voltage and current ratings for these contacts are the same as for other pilot devices; in fact, some manufacturers use the same set of contacts in all their pilot devices. This allows users to stock one or two types of contacts that can be used for replacement in all their pilot devices.

Some flow switches detect very small amounts of airflow. A microswitch is used since it can be activated with a very small amount of travel from the switch operator. One application of this type of switch involves a small continuous flow of air direct from a nozzle at the flow switch operator. This switch assembly is used to detect the presence of moving plastic film. The airflow nozzle is placed on one side of moving film, and the paddle of the flow switch is located on the other side of the film. As long as the film is traveling correctly, the airflow cannot get to the nozzle. If a film breaks or runs out, the air from the nozzle will cause the flow switch to activate and signal that the plastic has broken.

It is very important with flow switches that they can activate without binding. It is also important that the flow not be reversed against the flow switch, as this will alter the activation point during normal operation.

The pilot device is basically an on/off switch. Flow switches are also available as analog devices, which can indicate the amount of flow from zero to maximum.

10.9.2 Installing the Flow Switch

Flow switches are available for waterproof, explosion-proof, and other NEMA applications. This means that you can locate a flow switch in virtually any environment. The actual installation process should be completed in two distinct steps, similar to the pressure switch. The first step involves actual placement of the flow-detecting part of the switch. Be careful that the paddle is free to move and return as the flow is established and stopped. You should inspect the flow detection mechanism for proper direction of flow. The paddle is meant to detect flow in only one direction. If the flow is allowed to move against the paddle in the opposite direction, the switch will be damaged.

After the flow detection part of the switch has been installed, place an ohmmeter across the contact set you are planning to use and activate the flow. Be sure to check the amount of flow with some other type of indicator or flow meter to ensure that the proper amount of flow is obtained before you try to adjust the flow switch. The flow switch may be adjusted by a spring and screw mechanism or by adjusting the amount of the paddle that is in the flow. This can be altered by raising or lowering the paddle into the flow or by removing part of the paddle. The farther the paddle is inserted into the flow, the more it will disturb the flow. Be aware that this turbulence may affect other flow-measuring devices in the system.

After you are assured that the flow switch is operating correctly, continue the installation by installing the field wiring. Use the field wiring diagram to complete the field wiring process. After the switch has been wired correctly, apply power and test the switch under actual conditions. Be sure to allow the flow to reach the low or high level of flow that you are trying to detect. Execute these tests several times to ensure that the switch installation is operating correctly.

10.9.3 Troubleshooting the Flow Switch

The flow switch can be tested like other pilot devices. It can be tested in or out of the circuit. In the case of this and other pilot devices, it is important to do some preliminary testing to determine if the pilot device is actually the faulty part of the system. To execute this test, leave power connected to the circuit and use a voltmeter to determine if the load has power across its terminals. If power is present, continue testing the load. If power is not present, leave one meter probe on the line 2 (L2) terminal of the circuit and continue moving the other probe from point to point, starting at the line 1 (L1) side of the circuit. The point where you lose the voltage reading indicates the fault. If this test indicates that the flow switch is at fault, continue with this procedure to test it.

The most important part of the test procedure is to ensure that the switch is actually detecting flow. This means that the problem may be in the system, not in the switch itself. Remember: the flow switch is used to detect the presence or absence of flow. If the flow stops, the switch will actuate its contacts and indicate that flow has been lost. Electricians often assume that since the switch contacts are open, the switch is faulty, when in reality the switch is operating correctly and indicating that flow has been lost. There is also a strong possibility that the paddle that activates the flow switch contacts has become inoperative or is stuck in one position. This occurs because the fluid flow that the paddle is detecting can also corrode its mechanism. Sometimes the paddle will accumulate dirt or debris that is in the fluid and become immovable. When this occurs the system will actually have the proper amount of flow, but the paddle cannot move to indicate this. In these cases it is suggested that the paddle be removed from the fluid and inspected carefully for proper travel. Also inspect the location where the paddle is installed. Sometimes a buildup of corrosion exists around the installation site, which will prevent the paddle from proper movement.

After you are satisfied that the operational part of the switch is functioning properly, continue the test to the electrical side of the switch. Test the contacts for proper operation. If they activate but do not pass power, they are faulty and must be replaced. Be sure to turn the power off prior to removing the switch contacts. It is important to replace the contacts with a set that have the same voltage and current rating. Also make sure that they will activate at the same point where the previous set activated. To confirm this, execute the adjustment procedures listed in the earlier section covering installation. In some cases you may want to replace the entire flow switch. But sometimes this involves draining down the system or dealing with leakage of the fluid during the change. In these cases, the flow meter paddle is normally left in the system and only the switch contacts are changed. Prior to making your decision, you may need to check with the hydraulic or pneumatic specialist in your shop, who should be able to tell you which method can be accomplished most easily.

10.10 Temperature Switches

Temperature controls are widely used in industrial applications. Typical temperature pilot devices and their symbols are shown in Figure 10–25. The abbreviation for the temperature switch is TS. If more than one temperature switch is used in a system, each switch is identified with a number in addition to the letters (for example, TS1, TS2, TS3). From the symbols in this figure you can see that the switch contact can be shown above the terminal point or below the terminal point. If the switch contact is shown above the terminal contact, it is identifying a normally closed temperature switch that will open when the temperature increases. This type of control is used as an overtemperature switch or as a heating thermostat. If the switch contact is shown below the terminal, it is considered to be a normally open temperature switch, which will close on temperature rise. This type of temperature switch is used to turn on a fan when the temperature gets too hot, or to turn on an air-conditioning system when the temperature increases.

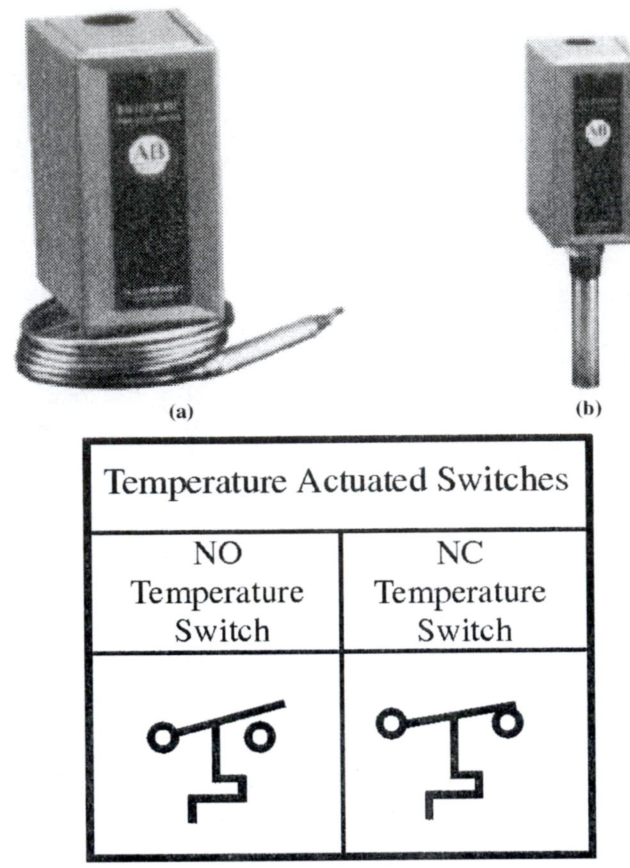

Figure 10–25 (a) Temperature switch with capillary tube; (b) temperature switch with well adapter; (c) electrical symbol of normally open and normally closed temperature switches.
(Courtesy of Rockwell Automation)

10.10.1 Types of Temperature Operators

A variety of operators are used to cause the temperature switch contacts to activate. The simplest of these to understand is the *bimetal strip*. A bimetal operator is manufactured by bonding two strips of dissimilar metal together so that their surfaces touch. One end of the bimetal switch is mounted to a fixture, and the other end will move upward when it is heated. After the heat is removed, the free end will move back to its original position.

The simplest temperature switch is made by mounting a set of contacts where the movement of the bimetal strip can activate them. This type of switch can use springs to cause the contacts to snap over center when they open or close so that an arc cannot be drawn across them. Since the amount of movement can be predicted, the switch can be set for a single set-point temperature, or an adjustment screw and spring can be added to make the set point variable.

The motion of the bimetal can be amplified by bending the bimetal strip into a spiral. The spiral will open as the temperature applied to it increases. When the temperature is returned to its original value or lowered, the spiral will tend to close. Since this movement is also predictable for any given change of temperature, hardware can be added to make an adjustable set point. A set of cams and gears can also be added to activate a set of snap-over-center contacts, or a mercury switch can be used as contacts. Since this spiral operator is very sensitive to temperature and has to be mounted to take advantage of the maximum amount of travel as the temperature changes, you should not touch the spiral or attempt to adjust it. It is also important to keep the spiral clean and free of grease or oil, which might collect dirt and cause the operator to become out of adjustment.

Another type of temperature operator is the *sensing bulb control*. This type of control is also called *capillary tube control*. It uses a bulb filled with a liquid, such as alcohol or refrigerant. The bulb is connected by a capillary tube to a pressure-sensitive switch activator. The switch activator can use a rubber diaphragm, a bellows operator, or a Bourdon tube, which are all similar to traditional pressure switches. When the temperature on the bulb increases, the liquid will experience a corresponding increase of pressure, which will cause the bellows or diaphragm to activate the switch contacts.

The bulb operator is considered a temperature switch even though the switch is activated by the pressure change that is caused by the temperature change. If a liquid is confined in a bulb, the pressure that it exerts on the bulb and switch operator is directly proportional to the temperature. This means that for any given temperature, the amount of pressure can be determined and predicted. From this relationship it is easy to design a temperature switch that will be activated by the change of pressure on the fluid in the bulb.

The main advantage of bulb control is that the switch contacts can be mounted some distance from the bulb and away from the source of heat. The bulb can be mounted directly on the source of heat, or it can be mounted in a well where heat transfer can easily occur. This type of temperature switch is also useful where very cold temperatures must be measured. If the switch is mounted in very cold temperatures, any lubricants used on the switch mechanism may solidify and hamper the operation of the switch mechanism, which causes the switch to become inaccurate and unreliable. The thermal bulb allows the switch to be mounted in a cabinet at normal room temperature while the bulb is mounted remotely in a low-temperature environment.

Another type of temperature control uses a thermocouple to sense the temperature and activate a set of contacts. The thermocouple is made of two dissimilar wires that will produce a small amount of voltage when they are heated. The two metals are connected together at a point called the *hot junction*. Thermocouples can use several different types of metals to make sensors for a variety of ranges. For example, a type J thermocouple is made from iron and constantan. Its temperature range is 32 to 1400°F. A type K thermocouple is made of chromel and alumel and can sense temperatures in the range of 1000 to 2000°F.

Thermocouples are generally used as an analog device. This means that the output signal that they produce is proportional to the temperature that is being sensed. The thermocouple produces a millivolt signal. This signal must be amplified to be useful for control purposes.

The thermocouple can also be used as the sensor for an on/off controller even though the original signal is analog. In this type of controller, a thermocouple is connected to the control as a sensor. The set-point temperature is entered into the control by adjusting the set-point indicator on the face of the control. When the temperature reaches the set-point value, the signal from the thermocouple will be strong enough to activate the contacts. These terminals provide a set of contacts, and terminals 5 and 6 are normally open and normally closed, respectively. A heating contactor coil will usually be connected across these terminals, and the contacts of this contactor will be used to energize a heating element. In the normal operational mode, the normally open contacts will be closed any time the temperature is below the set-point value. This means that if the set point is adjusted for 350°F and the temperature that the thermocouple is sensing is 340°F, the normally closed contacts will be closed and the heater coil will be energized, which continues to cause the temperature to increase.

10.10.2 Solid-State Temperature Controllers Using RTDs and Thermistors

Another way to control a solid-state sensor is to use a resistive temperature detector (RTD) or thermistor as a temperature-sensing input. The RTD uses a wire whose resistance will change when a change of temperature occurs. For support this wire is wound around a nonconductive core. The amount of resistance change is sensed by a Wheatstone bridge circuit. This means that the amount of milliampere current flow will change with a change in temperature. The RTD provides a fairly linear response to the temperature change. RTDs are made from three types of wire, to respond to a variety of temperature ranges. The platinum type has a sensing range of −300 to 1300°F, the nickel type has a sensing range from −100 to 300°F, and the copper type has a sensing range of −60 to 300°F. The solid-state control for the RTD is essentially the same as thermocouple control. A set point can be entered into the control, and any time the temperature is below the set point, the control will close a set of normally open contacts. Since the contacts also have a set of normally closed contacts, the control can be used as an overtemperature control and deenergize a circuit when the temperature exceeds the set point.

The thermistor sensor uses a temperature-sensitive resistor as the temperature detector. The thermistor usually has a negative temperature coefficient, which means that as the temperature of the thermistor increases, the electrical resistance of the material decreases. This relationship is linear enough to provide accurate temperature indication.

10.10.3 Installing the Temperature Pilot Device

The temperature pilot device is installed in a manner similar to that for other pilot devices. Figure 10–26 shows a temperature switch with its cover off. The main concern in the installation process is to position the temperature sensing element correctly. It is very important to position the sensor close enough to the heat source so that it can monitor the temperature accurately. It is also important to keep the sensor located where it will not be damaged by excessive heat. Most sensors will provide instructions indicating the best location to mount the sensor. Remember that bulb (capillary tube) sensors must be mounted so that the liquid will not fall directly on the sensing diaphragm. This means that the sensing bulb should be mounted horizontally or in a vertical position so that the liquid will be in the bottom of the bulb. In this location, only the vapor of the liquid will provide the pressure to activate the diaphragm. If the liquid reaches the diaphragm, it will cause the switch to become inaccurate.

After the temperature portion of the switch is installed, the temperature should be changed to cause the contacts to activate. Allow the temperature to move above and below the set point several times while you have a meter on the contacts to observe their operation. This exercise will also allow you to identify the set of normally open or normally closed contacts that you intend to use.

Once you have identified the contacts and are sure that they are operating correctly, continue the installation procedure by installing the field wiring. Be sure to provide the proper source voltage to the control. If the device is a mechanical operator, the contacts can be wired in series as a pilot device. If the device is a solid-state control, follow the field wiring diagram that is provided on the back of the control and attach the thermocouple or RTD at the terminals indicated. Also

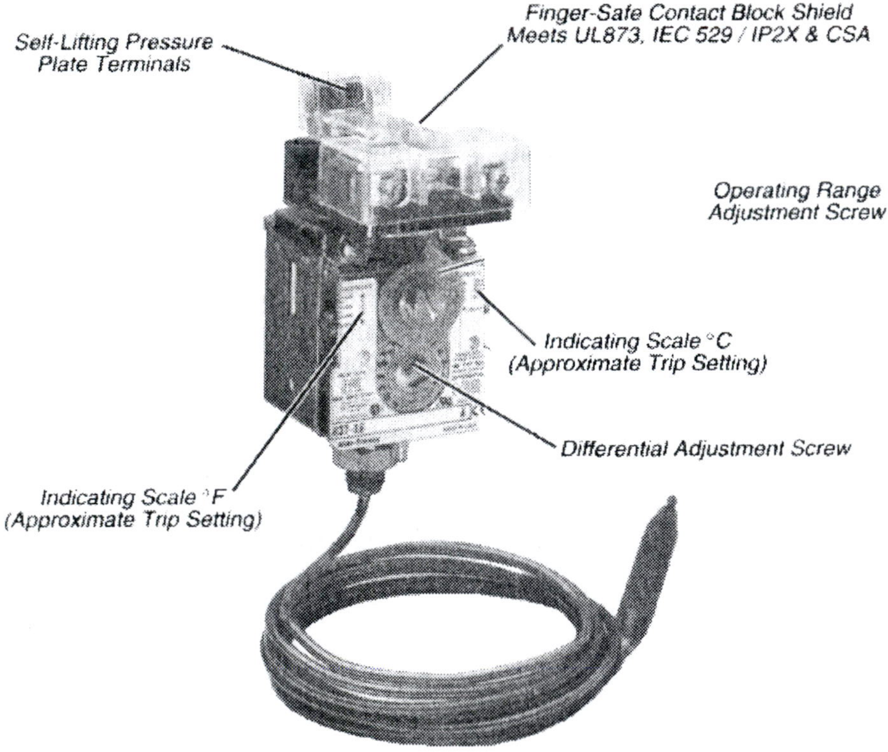

Figure 10–26 Internal parts of a temperature switch.
(Courtesy of Rockwell Automation)

be sure to provide line 1 and line 2 potential at the correct terminals. After the field wiring is completed, allow the temperature to change above and below the set point so that the control can be thoroughly tested.

10.10.4 Troubleshooting the Temperature Pilot Device

A temperature pilot device should be tested in the circuit when possible. It is important to determine if the temperature has changed enough to activate the switch contacts. In some cases it will be useful to identify where the problem exists. This means that you should use a voltmeter to test whether voltage is present across the load terminals. If voltage is not present across the load terminals, continue to test the circuit to look for the open circuit. Use the procedure listed earlier in this chapter where one of the meter's probes is left on the line 2 terminal of the load as a reference. The other meter probe is moved from point to point in the circuit until the open is located.

If this procedure indicates that the temperature switch is causing the open, be sure to check the temperature where the sensor portion of the switch is located. Allow the temperature to increase and decrease to determine that the switch contacts are being activated. If the switch contacts are not activated, the switch operator is inoperative and must be replaced.

If the control is a solid-state control, the thermocouple and RTD should be tested separately to determine that they are producing the proper amount of voltage or current for the temperature. The control can be tested by supplying the proper amount of voltage or current with a separate power supply. A temperature–voltage curve should be used to determine the proper amount of voltage for any given temperature. If the voltage from the external supply cannot cause the solid-state control to activate its contacts, the control should be replaced. Be sure to turn the power off prior to replacing the control.

10.11 Indicator Lamps

Some indicator lights with pushbuttons have another feature that is called press-to-test. The press-to-test feature allows a technician to press the light. Indicator lamps are also considered pilot devices, since they are located in the control portion of the circuit and because they draw a minimal amount of current. The abbreviation for the indicator lamp is LT. Multiple lamps are identified as LT1, LT2, or LT3. Some of these lamps are used as stand-alone devices, while others are incorporated into other pilot devices such as pushbuttons and selector switches. Indicator lamps are available as flush-mounted units, press-to-test units, flasher units, extended mount units, and master test units. Several types of lamps are used in the indicator assemblies, including the incandescent lamp, the neon bulb, and the LED indicator.

When the indicator lamp is shown in a electrical diagram, the electrical symbol is used in combination with a letter indicating the color of the lamp lens. The code for these colors are shown in Figure 10–27. The meaning of each of the colors has been standardized for motor control panels. A typical panel layout of indicator lamps is shown in Figure 10–28. The lens for the indicator lamp can be made of plastic or glass. Special lenses can allow the lamp assembly to be used in hazardous

Pushbutton Color Code

Color	Typical Function	Examples
Red	Stop. Emergency Stop	Stop one or more motors, master stop.
Yellow	Return. Emergency Return	Return machine elements to start position.
Black	Start Motors, Cycle, etc. Any operation for which no other color is specified	Start of one or more motors, start cycle or partial sequence.

Pilot Light Lens Color Code

Color	Typical Function	Examples
Red	Danger. Abnormal Condition. Fault Condition.	Voltage applied, cycle in automatic, faults in air, water, lubricating or filtering systems, ground detector circuits.
Amber (Yellow)	Attention	Motors running, machine in cycle, unit or head in forward position.
Green	Safe Condition (Security)	End of cycle, unit or head returned, motors stopped, motion stopped, contactors open.
White or Clear	Normal Condition	Normal pressure of air, water lubrication.

Multiple Station Pilot Light Requirements

Color	Typical Function	Examples
Red	Power On, Emergency On, Automatic Cycle	Ground detectors, lubrication failure, master relay on, pressure failure (water, air, gas).
Amber (Yellow)	Motors Running, Machine in Cycle, Full Depth,	Machine elements in advanced position, manual cycle.
Green	End of Cycle, Heads in Returned Position	
White or Clear	Parts in Place, Lubrication Normal, Pressure Normal (Water, Air Gas)	

Figure 10–27 Color codes for indicator lamps.

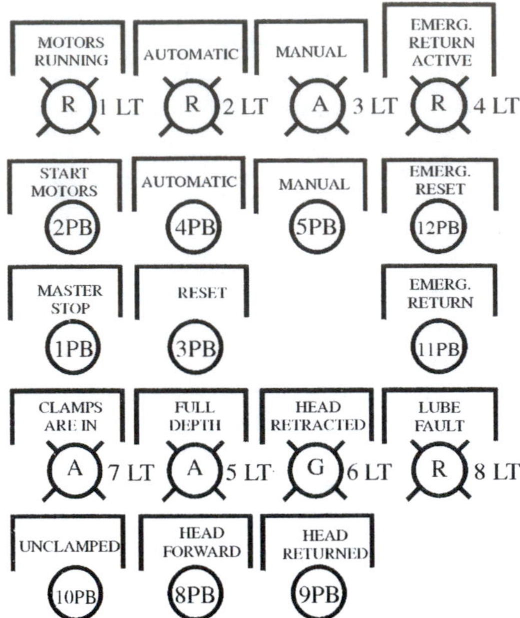

Figure 10–28 Typical panel layout of indicator lamps for control station.

environments where moisture or corrosive materials may be present.

The indicator lamp assembly used in most applications utilizes an incandescent lamp. The indicator lamp is actually an assembly. Figure 10–29 is an exploded view of a typical indicator lamp assembly. This assembly includes the terminal block where the wires are connected; the light unit, which is the receptacle where the bulb is secured; the mounting ring and mounting nut, which secure the light unit under the electrical panel; the locating ring, which prevents the lamp from twisting after it is mounted into the panel; the legend plate, which is used to identify the lamp; the sealing washer; and the extended lens. This assembly allows the indicator lamp to be easily mounted into most electrical panels and allows the device to be tested and serviced with a minimum of effort.

Indicator lamps are also classified by the voltage that the assembly requires. Typical voltages include 6, 12, 24, 48, 120, 208, and 220 V AC and DC. Some assemblies require a transformer to be used to provide the lower voltages. The bulb must be matched to the voltage at the lamp terminals.

10.11.1 Installing the Indicator Lamp

Indicator lamps are usually installed into electrical control panels or operator panels. In many applications the panel layout is designed to fit the machine operation. For some machines, switches and indicator lamps are mounted into the front or top of the electrical control panel, which serves as the operator's control unit. The first step for installing a pilot device into this type of panel is to lay out and cut the holes into the panel. After the holes have been cut into the panel, the indicator assembly can be mounted through the metal surface. All field wiring is attached to the terminal block. The terminal block is then connected to the rest of the lamp assembly. Be sure to follow the electrical diagram for the field wiring. This diagram will be slightly different for press-to-test lamps and master test lamps, since they require three wires instead of the normal two.

10.11.2 Troubleshooting the Indicator Lamp

The indicator lamp assembly can be tested either with power applied or with power removed. The preferred testing method is with power applied to the assembly. If the lamp is a press-to-test or is part of a master test system, simply press the test lamp and see if the lamp illuminates. If the lamp illuminates during the test operation but does not light up during normal machine operation, the control circuit should be suspected and tested for an open circuit or malfunctioning part. Check the circuit diagram for a list of components in the lamp's circuit and use it for the test procedure.

If the lamp assembly does not illuminate during the press-to-test operation, change the light bulb or test for proper voltage. If you have an extra bulb, or if one is available in another indicator lamp on the panel, changing the bulb is the preferred test. If an extra lamp is not available and you have a voltmeter, test the circuit for proper voltage. Be sure to start at the lamp terminals and work your way back to the voltage supply. Remember that the lamp is in the system as an indicator. Voltage may be interrupted by a pilot switch indicating that a specific condition exists in the system. In this case, the lamp circuit is not faulty; rather, it is operating correctly as it is designed.

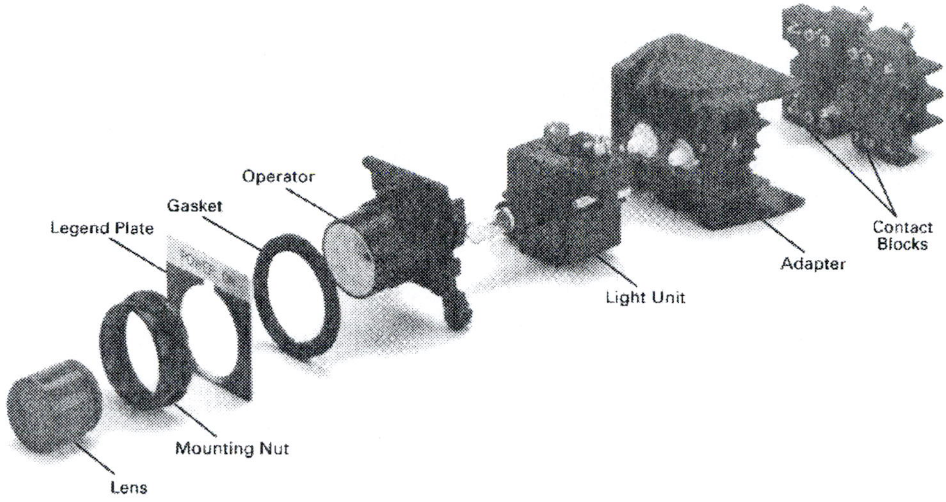

Figure 10–29 Exploded view picture and diagram of an indicator lamp. (Courtesy of Eaton/Cutler Hammer)

Questions

1. Explain the operation of normally open and normally closed contacts.
2. Explain how the contacts of a reed switch are different from those of normal contacts.
3. Explain how reed switch contacts are moved from the open to the closed position.
4. Explain the difference between a momentary pushbutton and a maintained pushbutton.
5. Explain why a key-lock type of pushbutton might be used.
6. Explain the operation of a limit switch. What is used to make the switch turn on or off?

True and False

1. A legend plate is provided on a selector switch to indicate which contacts are opened or closed.
2. The contacts used in pilot devices can be either normally open or normally closed.
3. The shape of the pushbutton head is also used for classification of pushbutton switches.
4. The fastest and most accurate test for locating an open wire, loose terminal, or faulty pilot switch is to test for a voltage drop (loss).
5. Pressure-type switches are used to detect high pressures, low pressures, differential pressures, and vacuums.

Multiple Choice

1. A limit switch can be tested by __________
   a. an ohmmeter if the switch is out of circuit.
   b. a voltmeter if the switch is in circuit and has power applied.
   c. manually activating the switch.
   d. All of the above
2. The differential pressure switch is _________
   a. activated by a diaphragm mechanism similar to a low-pressure switch and checks the difference between two pressures.
   b. activated by a rubber bellows.
   c. activated by a spring and balance mechanism.
   d. activated by a metal bellows.
3. The joystick control ____________
   a. consists of a multiple set of switch contacts located so that the stick's movement will activate one or more contacts at the same time.
   b. is mainly a video game controller.
   c. is a type of limit switch.
   d. is a type of pushbutton.
4. The foot switch is _______
   a. a type of limit switch.
   b. similar to a pushbutton switch, except that it is activated by pressing on it with your foot.

c. similar to a pressure switch, since the pressure of your foot activates the switch.

d. similar to a joystick control, since it is operated with a variety of motions.

5. A float switch is __________

a. a specialized pilot device used to measure the level of liquids or granular materials used in industry.

b. a specialized switch that measures the presence of liquid flow.

c. a specialized switch that measures the presence of granular flow.

d. Both b and c

Problems

1. Show the electrical symbol for normally open and normally closed contacts.
2. Show the symbol for normally open and normally closed pushbutton switches. Include the abbreviation for the switch.
3. Draw a diagram that shows how you would add additional start and stop pushbuttons to a circuit controlling a motor starter.
4. List the types of heads you may find on pushbuttons.
5. List several conditions you must be aware of when you select a place to mount a limit switch.
6. Explain how you would test the circuit you provided for a bad pushbutton switch. Remember that power is applied to the circuit.
7. Show the electrical symbol for a normally open and closed limit switch. Include the abbreviation for the limit switch.
8. List six types of actuators that you would find on limit switches.
9. Draw a sketch of each of the six limit switch actuators listed in the previous question.
10. Draw the electrical symbol for a normally open and a normally closed selector switch. Be sure to include the abbreviations for this switch.

◀ Chapter 11 ▶

Photoelectric and Proximity Controls

Objectives

After reading this chapter you will be able to:

1. Explain how each type of light has a different waveform frequency.
2. Explain the term modulated light source.
3. Explain the operation of through-beam, retroreflective, and diffuse scan photoelectric devices.
4. Explain the operation of the ECKO proximity switch.
5. Explain the operation of the capacitive proximity switch.

Photoelectric switches are used extensively as motor control devices. They provide on/off (discrete) control that can be amplified to control larger AC and DC voltages at pilot-level currents. Some photoelectric devices control relay and motor starter coils so that even larger load currents, such as motors, can be controlled through its contacts. Photoelectric switches are very versatile in that several different models are available to operate in a variety of industrial environments that have differing amounts of background lighting. These switches are made to be used as stand-alone controls, or they can be connected directly to robots or programmable controllers.

Two basic classes of photoelectric controls are used today: the incandescent type and the modulated light source (MLS). *Incandescent controls* have been in use for many years, while the MLS type has been developed around solid-state electronic technology. The incandescent type uses a beam of incandescent light focused on a receiver that is light sensitive, such as a photovoltaic cell or photoresistive cell. The earliest types of these controls used photoemissive vacuum tubes.

When solid-state technology provided devices such as light-emitting diodes and phototransistors, new controls using the MLS technology were designed. There are three basic types of photoelectric controls: the retroflective scan, the through-beam scan, and the diffuse scan. You may need to get additional information on solid-state electronic components in Chapter 19 to learn more about the components discussed in this chapter.

The basic theory of operation for a solid-state control includes a light-emitting diode that creates a light source and a phototransistor that acts as a light receiver. You will learn more about these types of photoelectric systems later in this chapter.

Figure 11–1 shows diagrams of the three basic MLS types of photoelectric controls. The *retroflective scan* device sends out a beam of light that is reflected off a target back to the receiver. When something breaks the beam of light, the contacts in the photoelectric switch change position. The contacts of the switch may be connected to devices such as a counter to count parts being manufactured, or a robot to indicate that a part is in place, or a programmable controller to detect the presence of a part. The *through-beam scan* sends a beam of light directly to a receiver unit instead of using a reflected beam. When the beam of light reaches the receiver it will trigger the contacts there, which can be used to provide control to pilot devices. As with retroflective control, when something interrupts the beam of light, the contacts will change position. The *diffuse scan* switch works on the principle of reflecting light in a diffuse pattern off a specific type of material

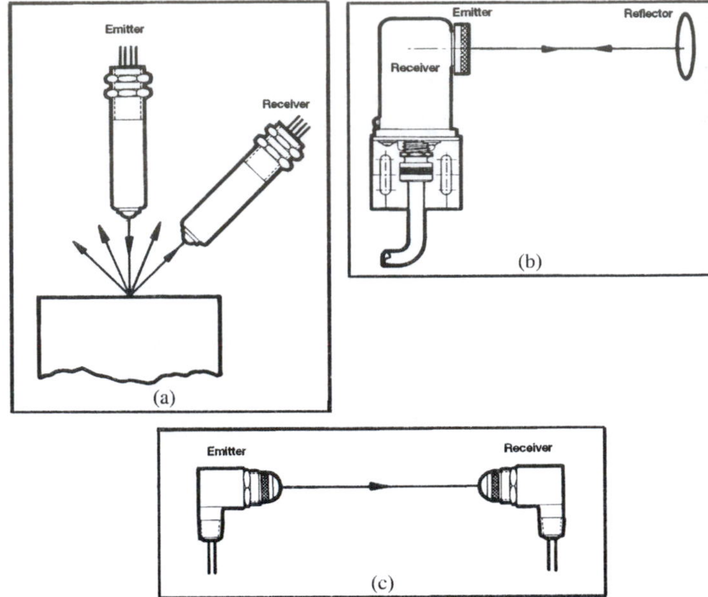

Figure 11–1 Modulated light source types of photoelectric controls.
(Courtesy of Pepperl + Fuchs, Inc.)

used as a background. Since different material will reflect light in a variety of ways, the receiver can be adjusted to the proper angle to collect light that is reflecting off the material at a predetermined angle.

Each of these types of switches can be adjusted to energize the controlling contacts to the on position in either the presence or absence of the light beam. A detailed discussion of the incandescent and MLS types of photoelectric controls is presented in this chapter. We include the theory of operation, wiring diagrams for installation and interfacing, adjustment and calibration methods, and troubleshooting procedures for all types of photoelectric controls.

Various manufacturers have developed slightly different technology to create the three basic types of photoelectric switches. Figure 11–2a shows a polarized reflective switch where light is reflected off the surface of a target. Figure 11–2b shows a diagram of a reflective scan switch. Figure 11–2c shows a diagram of a wide-angle diffuse sensor, and Figure 11–2d shows a diagram of a fiber optic photoelectric switch. Figure 11–2e shows a diagram of a transmitted beam photoelectric type switch. Figure 11–3 shows pictures of these types of switches.

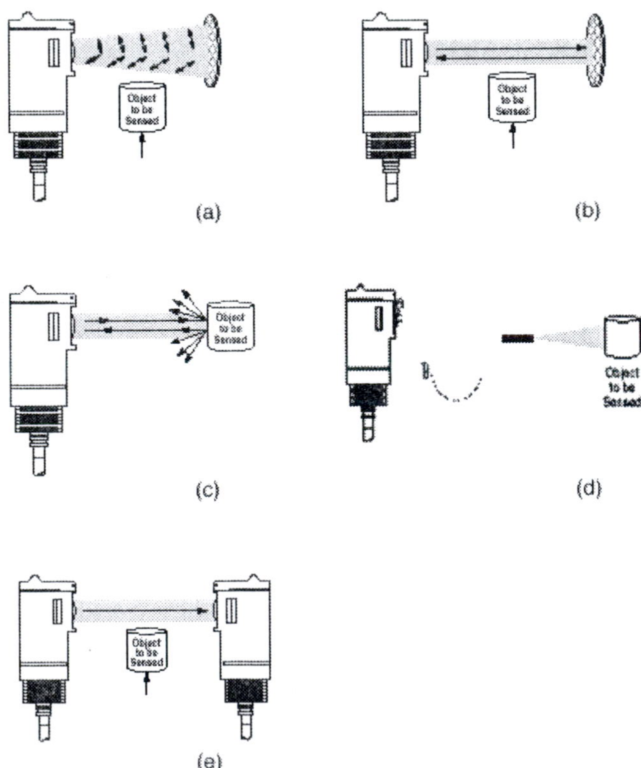

Figure 11–2 Types of photoelectric switches: (a) polarized reflective photoelectric switch where light is reflected off the surface of the target; (b) retroflective photoelectric switch; (c) wide-angle diffuse sensor; (d) fiber optic photoelectric switch; (e) transmitted beam photoelectric switches.
(Courtesy of Rockwell Automation)

11.1 Incandescent Photoelectric Controls

Early photoelectric switches used a variety of methods of sensing the presence or absence of light. One of these devices was the photoemissive tube. The tube works on

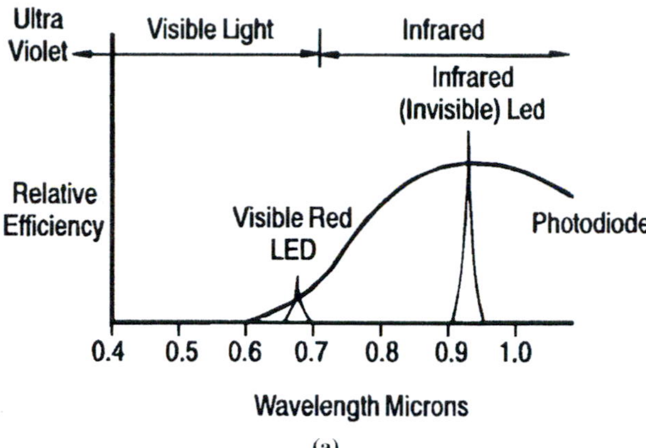

(a)

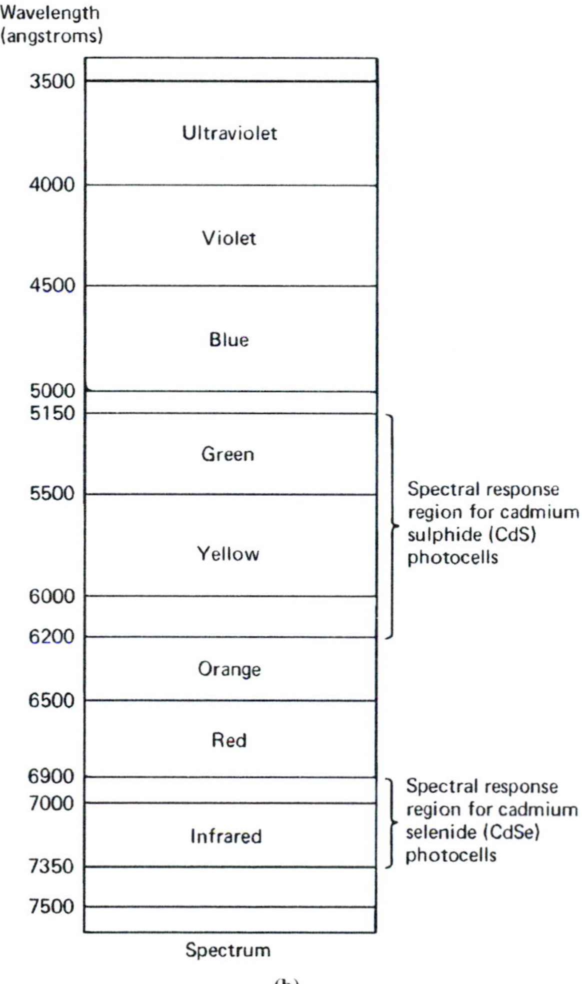

(b)

Figure 11–3 (a) The invisible infrared LED is a spectral match for this silicon phototransistor, and has much greater efficiency than the visible (red) LED (courtesy of Rockwell Automation); (b) Spectrum that shows the wavelength of various types of light sources used in photoelectric sensors.
(courtesy of Pepperl & Fuchs, Inc.)

the principle of photoemission. When light strikes the cathode element of the tube, it emits electrons. The stronger the light source, the more electrons are emitted from the cathode. Another element is placed very near the cathode to collect the emitted electrons. This element is the anode. The anode will begin conducting current as it absorbs more electrons. This means that the anode will conduct current in proportion to the strength of the light source until it is saturated.

In most applications of this type of switch a beam of light is focused on the cathode, which causes current to flow in the anode circuit. The anode current was used as the control current to power a relay coil. If something interrupted the beam of light, the relay contacts would change from closed to open to perform the actual control function.

This type of control was very popular before the advent of solid-state devices, but it presented some problems when background light caused false triggering. Early solid-state devices, such as cadmium sulfide cells, photovoltaic cells, and transistors, were easily adapted for use as photoelectric devices and began to replace photoemissive tubes because their sensitivity could be adjusted to account for background lighting.

11.1.1 Photovoltaic Cells

The photovoltaic cell is used as a light detector because of its ability to produce a small voltage when light is shined on it. The photovoltaic cell is also called the *solar cell*. The small voltage it produces is in the microvolt range, so like a transistor, it must be connected to an amplifier to provide control of pilot-level voltages. Since this type of switch includes a transistor for control, some amount of adjustment can be made to compensate for background light or other problems, such as suspended dirt and dust in the air. The adjustment involves applying a small amount of bias voltage to the transistor base. The bias can be positive or negative to change the sensitivity. If the bias voltage is positive, less voltage is required from the photovoltaic cell, thus making the switch more sensitive. If the bias voltage is negative, the photovoltaic cell needs to produce more voltage to trigger the transistor into conduction.

11.1.2 Cadmium Sulfide Cells

A similar type of control is the photoresistive cell. This device is made from cadmium sulfide (cad) cells. When light strikes a cad cell, it changes its internal electrical resistance. In total darkness the cad cell has very high resistance, in the range of several hundred thousand ohms. In direct light the resistance in the cad cell drops below 100 Ω. In most applications the cad cell is connected to the base of a transistor. When the cad cell is sensing direct light, the resistance in the transistor base circuit drops and base current flow causes the transistor emitter–collector circuit to go into conduction. The collector current is large enough to power most pilot devices. This type of sensor can also be adjusted for sensitivity by applying a small bias voltage to the base.

The cad cell has difficulty distinguishing control light and ambient light or background light. This is a problem in most industrial areas, since background light is ever-present from a variety of sources: overhead lights used to light work areas, reflection of light off shiny surfaces, and light from processes, such as pouring molten steel, arc welding, and opening and closing outside doors. A new generation of photoelectric switches are required that are more sensitive and more accurate than previous types. They are used in controlling robotic work cells and totally automated processes and must have response times fast enough to interface with programmable controllers and other microprocessor controls.

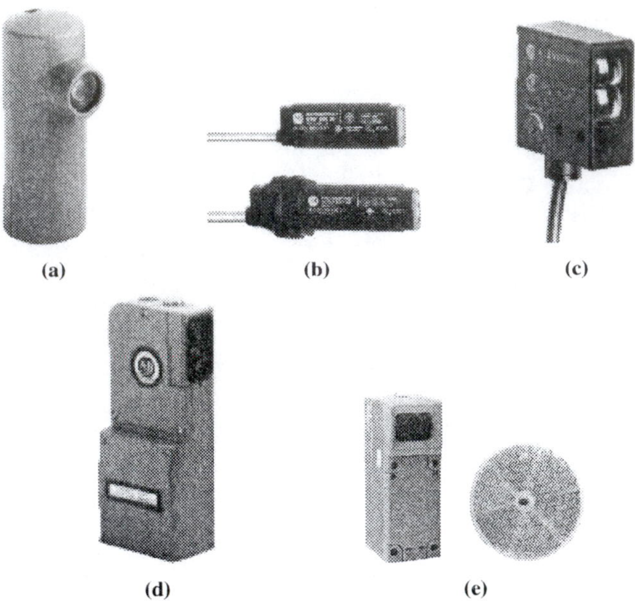

Figure 11–4 Types of photoelectric switches: (a) long range (courtesy of Rockwell Automation); (b) color registration (courtesy of Rockwell Automation); (c) Series 7000 miniature rectangular (courtesy of Rockwell Automation); (d) Series 5000 (courtesy of Rockwell Automation); (e) reflective with reflector (courtesy of Banner Engineering Corp.).

11.1.3 Analysis of Light

To fully understand the design differences of photoelectric switches, you must first understand some basic concepts about light. Light travels around us in the form of waves, and it is identified by its wavelength. Figure 11–4 shows a table indicating the various types of light and their wavelengths. The wavelength spectrum shows that ultraviolet light is at the low end and infrared light is at the high end of the spectrum. You can determine the response range for sensors from this graph. The cad cell is sensitive to green and yellow wavelengths, which means that it can distinguish light with this wavelength from all other light. It is very important for the cad cell to sense only the source of light (green or yellow) that is emitted from its source and not the light that is in the background, such as overhead lighting. Another type of cad cell, the cadmium selenide (CdSe) cell, is sensitive only to light in the infrared spectrum range. This allows the CdSe cell to be used where the cadmium sulfide cell cannot be used.

11.1.4 Modulated Light Source Devices

The majority of photoelectric devices used today operate on a principle known as modulated light source or polarized light source. Figure 11–5 shows an example of the light-emitting diode and the modulated light source. This involves the use of a light-emitting diode (LED) as

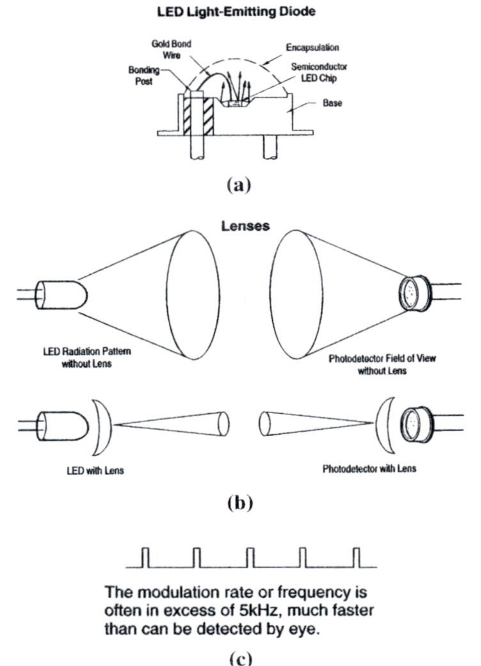

Figure 11–5 (a) A light-emitting diode used in photoelectric switches. (b) The effects of using a lens with photoelectric switches. (c) Example of modulation signal used in photoelectric switches. Modulated light sources inhibit false triggers.
(Courtesy of Rockwell Automation)

a light source, a phototransistor as a receiver, and a control circuit driven by the phototransistor. The LED is modulated at a specified frequency by a solid-state circuit called an oscillator. The phototransistor utilizes a circuit that tunes in the oscillated frequency that the LED is sending and filters out all other frequencies of light waves received. Since each type of light, such as ultraviolet, infrared, and regular colors, from the color spectrum oscillates at a different frequency range, it is rather simple to filter out unwanted light signals. This is known as *light rejection* or *light polarization*. The phototransistor and the LED have such a good frequency match that MLS sensors can be used in the presence of strong ambient light sources in an industrial environment. Pulsing the LED also allows the light source to be more powerful, since less power is consumed during pulsing than if the source was powered continuously. Heat from the devices is also dissipated more easily when the LED is pulsed. Actual on time for the LED may be only several milliseconds, while the off time may be over 100 milliseconds. This provides an off-to-on duty cycle of better than 10:1. This also means that the signal pulse may have the strength of several hundred milliamperes, yet consume an average current of only 30 mA. Figure 11–6 shows pictures of all of the types of photoelectric switches. You will learn more about them in the next sections.

11.1.4.1 Through-Beam Scan Photoelectric Switches

The through-beam scan photoelectric switch may also be called direct scan or separate control. As you saw in Figure 11–1c and 11–2e, this type of photoelectric switch is made of two separate parts. One part is the source for a beam of light, which is developed from an LED-modulated light source. This part is called the emitter or source. The source focuses the light beam on a phototransistor in the second part, called the receiver or detector. When the beam is focused on the receiver, it triggers the phototransistor into conduction. The phototransistor may control a solid-state or relay output. The output will be able to operate solenoids, motors, or other types of loads. When the beam of light is interrupted, the phototransistor drops out of conduction and the output that it is driving changes state.

The through-beam photoelectric switch also works very well for applications where clear objects, such as glass bottles or clear plastic containers, must be sensed. Light from a through-beam scan source may be strong enough to penetrate the clear material and trigger the switch. By comparison, a reflected light will be scattered and its signal strength will be diminished when it passes through clear objects, which will allow them to be detected.

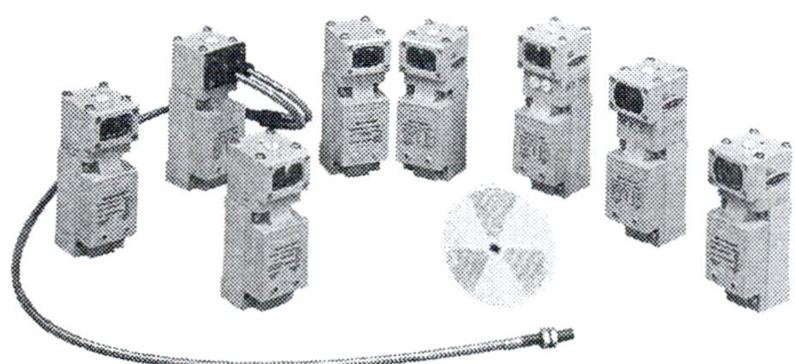

Figure 11–6 Examples of programmable modular self-contained sensors that have the power supply and logic module mounted inside the sensor.
(Courtesy of Banner Engineering Corp.)

11.1.4.2 Retroreflective Photoelectric Switches

The retroreflective photoelectric switch has the modulated light source and the receiver in the same housing. Examples of this type of switch were shown in Figures 11–1b, 11–2a, and 11–2b. An LED is used to develop the modulated light source, which is focused on a target that has light-reflecting properties. The target can be a reflector or the shiny surface of the part being detected and it will reflect and refocus the light back to a lens, where it is directed to the detecting circuit to be filtered and demodulated. Since the source and receiver are in the same housing, installation and wiring is easier. Figure 11–4e showed a typical reflector for this type of switch and is rather inexpensive and is easily mounted to most surfaces. The main property of the target is that it will reflect and refocus the beam of light that has been sent from the source.

Once a retroreflective switch is installed, it can easily be adjusted to focus the reflected light beam back into the receiver. If the reflector is large enough, the alignment may be up to 15 degrees out of perpendicular to the receiver. Since the alignment is not crucial for this type of switch, it makes it a very good choice for applications where vibration of the sensor is a concern. After the control is mounted you should check the exact location where the object will break the beam and trigger the output, and make minor adjustments if needed. In this type of switch only one power source is required, since the source circuit and receiver circuit are in the same housing. The source (emitter) circuit receives power from the regulated power supply. The main function of this circuit is to establish a frequency to modulate the LED. The pulsing will allow the LED to be pulsed at full power to provide the strongest beam of light, yet not overheat the device. This also allows power consumption of the source circuit to be as small as possible. The modulating beam of light is directed through the lens toward the reflector. The lens for this type of device actually has two separate sections. One part of the lens is ground to make it a perfect emitter of light. The other part is ground to make it a perfect receiver of light. Once the source beam is adjusted and focused, the reflector will direct the reflected light back into the receiver part of the lens. This reflection process does not disturb the frequency on which the beam is modulated.

Once the beam is focused back into the receiver, it is amplified and filtered. The filter does the same job that it did in the through-beam device: It eliminates any signals from light that is outside the frequency bandwidth. From this section the signal is passed through a second filter. This filter is digital and improves the selection process, so that the device will have greater sensitivity. If the light causing this signal has originated from the modulating LED source, it will be allowed to pass through the filter and trigger the output. If the light is modulating at a different frequency, it will be rejected by one of the two filters.

After the signal is passed through the digital filter it is demodulated and sent to the output amplifier. The output section is very similar to the output section in the through-beam device and requires that the load be connected in series with it. Light operation or dark operation is selectable through an externally mounted switch as in the through-beam type. The retroreflective switch also has built-in protection against short circuits in the output stage, reverse polarity of DC power supply voltage, and a circuit that provides the voltage-limiting characteristic of inductive spikes in the control circuit.

11.1.4.3 Diffuse Scan Photoelectric Switches

The diffuse scan control and specular scan control are very similar in operation. Some switch manufacturers consider them the same device for applications involving different types of reflective surfaces and mounting positions. These types of photoelectric switches operate on the principle of emitting a beam of light on an object rather than on a reflector. The receiver is mounted at a specified angle to the surface to collect the reflecting light.

Figure 11–1a showed the emitter and receiver for the specular scan application are both mounted at a specified angle. The angle must be identical so that the receiver will receive only light that is reflected from the surface of the conveyor. When the object being detected moves into position, the emitted beam of light will stroke the top of the object and reflect at a different angle and miss the receiver. Since this type of control operates from light reflected from the surface of a part, it is also called *proximity* photoelectric control.

Operation of the diffuse scan switch shown in Figure 11–2c is very similar to that of the specular scan in that the source light beam is focused where it will be reflected by either the background or the material being sensed. The receiver portion of the switch is mounted at an angle where it will receive the reflected light. For diffuse scan the source part of the switch is mounted so that the beam of light strikes the reflective surface at exactly 90 degrees. This will provide the best reflective action for surfaces that do not reflect well. The diffuse scan works well in applications where the object being sensed is made of paper, mat, cardboard, or similar materials. Since these materials do not reflect light as well as does shiny material, such as foil, the sensitivity on the switch will need to be adjusted for proper switch triggering.

The biggest problem with the diffuse scan and specular scan photoelectric switches is that the beam of light must travel twice as far as with through-beam devices. Also, the reflected light may lose some of its intensity when it is reflected from material that may absorb part of the light. Generally, the sensitivity of the switches can be adjusted close enough to provide precision control action. A lens can be included with this type control to increase the sensing distance.

The circuitry required for operation of the diffuse scan and specular scan devices is very similar and in some cases identical to the circuitry for retroreflective devices. These controls come in several styles for various applications, just like the other types of controls. These styles include self-contained types and discrete types using a remote base for control.

11.1.4 Selecting a Photoelectric Switch for Your Application

Figure 11–7 shows a table that compares all of the different types of photoelectric switches. The installation of any photoelectric switch starts with selecting the correct type of switch for your application. The selection of a switch type involves determining the object to be sensed, the distance the source must be mounted from the receiver or reflector, the response time, the type of ambient conditions (dust, dirt, or other particulate in the air), and the voltage type and level of the load devices. Each of these conditions can be addressed by carefully checking the specification sheets and data sheets listing switch characteristics for each of the applications being considered. The specification and data sheets are provided in manufacturers' catalogs and product literature.

Part of the selection process includes determining the ambient conditions where the switch must operate. Tables are available to show the amount of contaminant

Photoelectric Sensing Modes Advantages and Cautions

Sensing Mode	Applications	Advantages	Cautions
Transmitted Beam	General purpose sensing Parts counting	• High margin for contaminated environments • Longest sensing distances • Not affected by second surface reflections • Probably most reliable when you have highly reflective objects	• More expensive because of separate light source and receiver required, more costly wiring • Alignment important • Avoid detecting objects of clear material
Retroreflective	General purpose sensing	• Moderate sensing distances • Less expensive than transmitted beam because simpler wiring • Ease of alignment	• Shorter sensing distance than transmitted beam • Less margin than transmitted beam • May detect reflections from shiny objects (use polarized instead)
Polarized Retroreflective	General purpose sensing of shiny objects	• Ignores first surface reflections • Uses visible red beam for ease of alignment	• Shorter sensing distance than standard retroreflective • May see second surface reflections
Standard Diffuse	Applications where both sides of the object cannot be accessed	• Access to both sides of the object not required • No reflector needed • Ease of alignment	• Can be difficult to apply if the background behind the object is sufficiently reflective and close to the object
Sharp Cutoff Diffuse	Short-range detection of objects with the need to ignore backgrounds that are close to the object.	• Access to both sides of the object not required • Provides some protection against sensing of close backgrounds • Detects objects regardless of color within specified distance	• Only useful for very short distance sensing • Not used with backgrounds close to object
Background Suppression Diffuse	General purpose sensing Areas where you need to ignore backgrounds that are close to the object	• Access to both sides of the target not required • Ignores backgrounds beyond rated sensing distance regardless of reflectivity • Detect objects regardless of color at specified distance	• More expensive than other types of diffuse sensors • Limited maximum sensing distance
Fixed Focus Diffuse	Detection of small targets Detects objects at a specific distance from sensor Detection of color marks	• Accurate detection of small objects in a specific location	• Very short distance sensing • Not suitable for general purpose sensing • Object must be accurately positioned
Wide Angle Diffuse	Detection of objects not accurately positioned Detection of very fine threads over a broad area	• Good at ignoring background reflections • Detecting objects that are not accurately positioned • No reflector needed	• Short distance sensing
Fiber Optics	Allows photoelectric sensing in areas where a sensor cannot be mounted because of size or environment considerations	• Glass fiber optic cables available for high ambient temperature applications • Shock and vibration resistant • Plastic fiber optic cables can be used in areas where continuous movement is required • Insert in limited space • Noise immunity • Corrosive areas placement	• More expensive than lensed sensors • Short distance sensing

Figure 11–7 Comparison table for photoelectric switches, including advantages of each type and cautions for each type. (Courtesy of Rockwell Automation)

in a particular environment and an example of where this type of environment may be found. The suggested amount of gain required for each application is also provided. This type of table is used to determine the estimated amount of particulate and what type of photoelectric switch will operate best in these conditions. You may find that several different types of photoelectric switches will work well for your application. If this is the case, other conditions, such as pricing, stocking, and parts replacement, should enter into your decision for selection.

11.1.5 Installing the Photoelectric Switch

After you have determined which type of switch to use, the mounting part of the installation can begin. For this example a three-piece photoelectric switch is being used (Figure 11–8). The three parts are the upper cover, sensor block with the photoelectric sensor head, and the logic module that controls the switch. Figure 11–9 shows an example of mounting the photoelectric switch. The sensor block is the part that is actually mounted solidly to the bracket or support. All wiring to the switch is connected to screw terminals in the receptacle. The terminals are numbered or identified to match the wiring diagram of the switch. The sensor body is the part of the switch that contains the electrical controls, power supply, and output devices. It is connected to the receptacle part of the switch once the control wiring has been connected to the receptacle's terminals. The sensor body has electrical pins that mate with pins in the receptacle to complete the control circuitry. The sensor body is connected to the receptacle in this manner so that the electrical part of the switch can be changed quickly, without rewiring any part of the switch. The photoelectric head is the part of the switch that contains the photoelectric sensors. This part of the switch is also connected in a way that it can easily be changed if problems arise. This also allows the sensor body and receptacle to be used as a base for other types of switches, such as inductive or capacitive proximity switches. This allows the user to stock fewer parts and still be able to provide a quick change of any inoperative components.

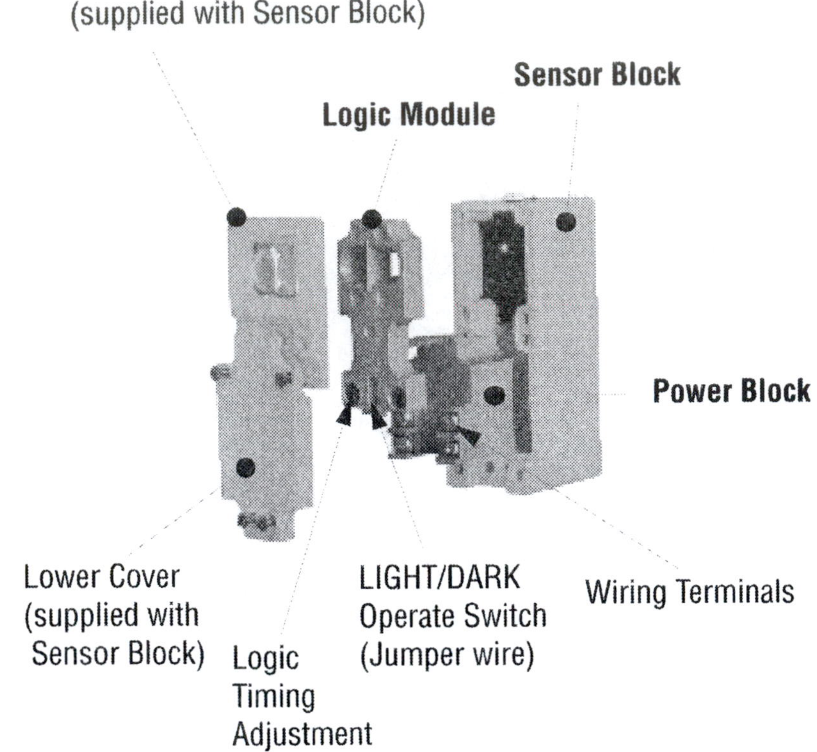

Figure 11–8 Exploded view of a photoelectric switch.
(Courtesy of Banner Engineering Corp.)

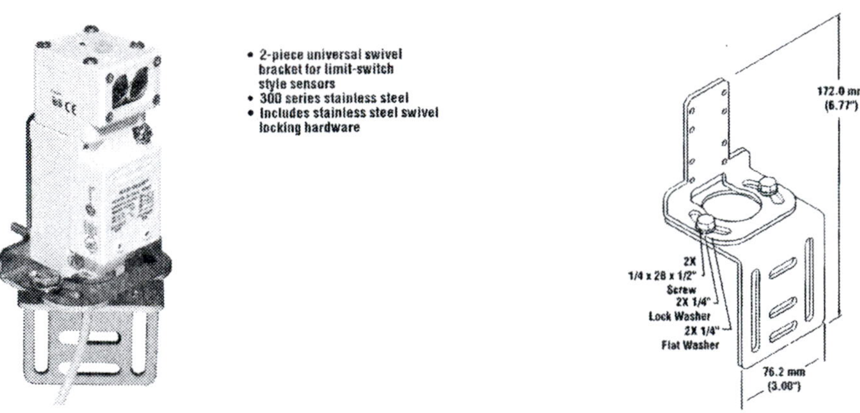

Figure 11–9 (a) A photoelectric switch mounted on a bracket, which allows it to be adjusted precisely toward its target. (b) The mounting bracket for a photoelectric switch.
(Courtesy of Banner Engineering Corp.)

11.1.5.1 Fiber Optic Photoelectric Controls

Another solution to the problem that requires both the source and the receiver to have a power supply is the use of fiber optic technology. The fiber optic through-beam scan photoelectric switch operates very similarly to the regular type with a source and a receiver. Instead of having two separate housings for the source and receiver, the circuitry for both are contained in one housing. Figure 11–10 shows a variety of these types of switches. The light from the source is generated in the module housing and is transmitted to a lens through a fiber optic cable that will focus it. The fiber optic cable is very durable, yet is pliable enough to be routed through conduit. It also keeps the signal from receiving electrical interference or noise. The fiber optic switch has circuits that are similar to those of the regular through-beam switch. The main difference between fiber optic photoelectric switches and regular modulated-light-source photoelectric switches is in the fiber optic cables. The cables come in precut lengths that must be specified when ordering. Since the cable is made of a glass fiber, you must be careful not to step on the cable or bend it sharply during installation, as the glass will shatter.

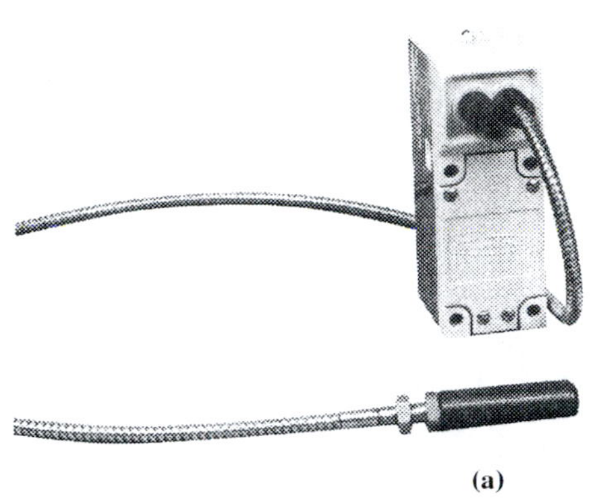

MULTI-BEAM Ambient Light Receiver, model SBARIGHF (shown) is equipped with an upper cover assembly (model UC-RF) which allows an individual glass fiber optic assembly to be attached to the receiver optoelement

(a)

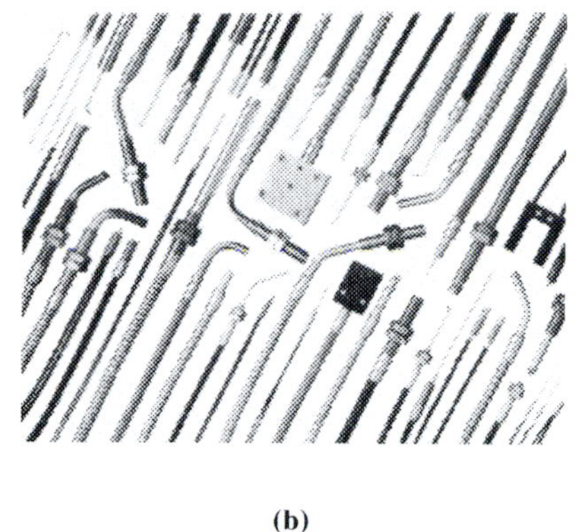

(b)

Figure 11–10 (a) Example of a fiber optic photoelectric switch (courtesy of Banner Engineering Corp.); (b) A variety of fiber optic cables for photoelectric switches.
(Courtesy of Rockwell Automation)

11.1.5.2 Wiring the Switch

Once the switch is mounted, the control wiring must be completed according to the control wiring diagram. The wiring diagram uses symbols to show the location of terminals where field wiring is connected and shows whether the switch contacts are normally open or normally closed. The symbol for the photoelectric switch can be shown in several ways. One of the symbols in Figure 11–11 shows a box sitting on its corner with switch contacts inside. This is the symbol for all types of proximity switches. Since photoelectric switches are used to detect the location of objects, they are considered proximity switches.

Another type of diagram used by the electrician is the ladder diagram of the photoelectric switch. Remember:

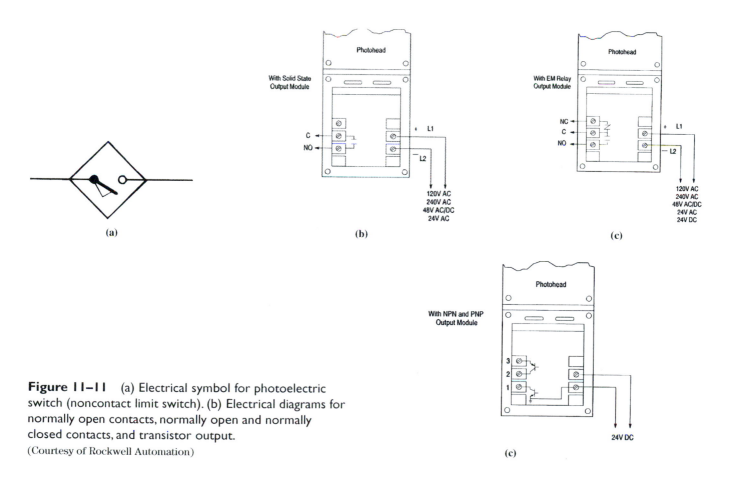

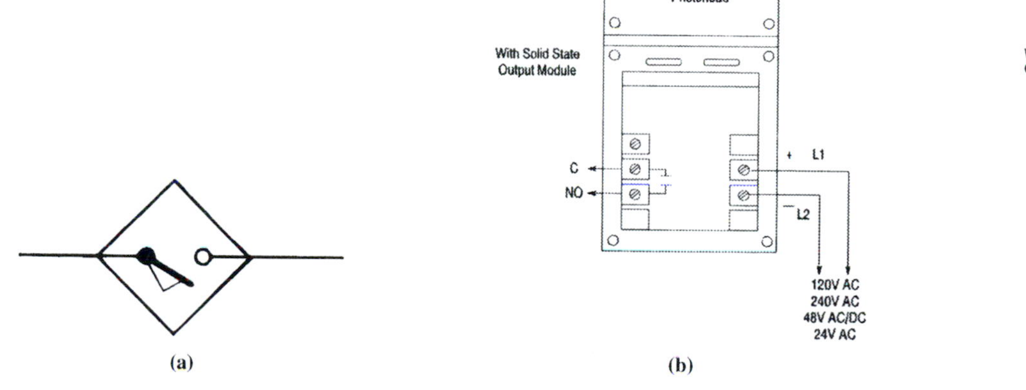

(a) (b) (c)

(c)

Figure 11–11 (a) Electrical symbol for photoelectric switch (noncontact limit switch). (b) Electrical diagrams for normally open contacts, normally open and normally closed contacts, and transistor output.
(Courtesy of Rockwell Automation)

The purpose of the wiring diagram is to show the location of the switch and its terminals, while the purpose of the ladder diagram is to show the sequence of operation of the circuit. This means that the contacts of the photoelectric switch are located on the line of the load (relay coil, solenoid, or motor starter) that they are controlling.

11.1.5.3 Output Stage for Photoelectric Switches

The output stage of the photoelectric switch is the most important part of the device when it comes to interfacing the switch to relays, motor starters, programmable controllers, or robots. The reason the output is so important is that it is the active part of the control that must switch voltage on or off to meet specific logic conditions. It is also the part of the device that must match the voltage, polarity, and signal type of the rest of the control circuit. The voltage may be AC or DC and its level may vary from TTL (transitor-transitor logic) levels of DC through 220 VAC. The device may also need to interface with any number of logic level signals for robot or programmable controller applications. These may range from current sourcing or current sinking to a complementary metal oxide semiconductor interface. In this part of the chapter we provide examples of all the common types of outputs provided and some examples of common circuits.

11.1.5.3.1 AC Voltage. Figure 11–11 shows examples of normally open and normally closed contacts for the output. One type of output is made for controlling 100 to 240 VAC at currents up to 200 mA. On/off control of AC current can be obtained several different ways. An indepth explanation of the solid state controls and circuits used these photoelectric controls is provided in Chapter 19. The first way uses a bridge rectifier, SCR, zener diode, and varistor. The load is connected in series with the AC part of the bridge rectifier and the AC supply voltage (100 to 250 V), while the SCR is connected in series with the DC side of the bridge rectifier. The SCR has control over the bridge rectifier. If current is flowing through the SCR, AC current can flow through the AC side of the rectifier and through the load device. The load device will be some type of AC relay, motor starter, or solenoid. The gate to the SCR will receive a signal any time the photo portion of the switch is triggered. The signal to the gate comes directly from the amplified circuits in the photoelectric switch. The zener diode will provide protection against overvoltage on the DC side of the bridge, and the varistor will protect against overvoltage on the AC side.

The second type of AC control is similar to the first, except that the output from the photoelectric switch's amplifier goes directly to an optocoupler, which will isolate the amplifier and switch from the AC circuit. The output of the optocoupler sends a drive signal to the gate of the SCR. The SCR has control over the DC side of a bridge rectifier, just as in the previous circuit. The AC part of the bridge rectifier will be wired in series with the AC supply voltage (100 to 250 V) and AC load device, which again could be an AC relay, motor starter, or solenoid.

Another type of output circuit for controlling AC loads uses a triac controlled by an optocoupler. The AC load and power is connected in series with the triac main terminals. When the triac's gate receives a signal from the optocoupler it causes the triac to go into conduction and pass current to the AC load. The optocoupler receives the signal from the amplifier of the photoelectric device. The optocoupler provides isolation between the AC load circuit and the amplifier and optical portion of the photoelectric switch. It also allows the amplifier signal, which is DC, to interface with the triac, which is an AC device.

11.1.5.3.2 DC Voltage. Another type of output control is for DC voltages. Figure 11–12 shows examples of these types of outputs. One variation of the DC-controlled circuit uses a PNP transistor for current-sourcing applications, while another uses an NPN transistor for current-sinking applications. The current-sourcing circuit has a DC load connected in series with a PNP transistor and diode. The amplifier stage of the photoelectric switch provides a small DC voltage signal to the base of the transistor any time the switch has been triggered. The signal on the base of the transistor causes current to flow in the emitter–collector circuit. The diode provides reverse polarity protection.

Circuits for interfacing the photoelectric output to transistor–transistor logic (TTL) or CMOS (complementary metal oxide semiconductor) circuits are also available. The voltage for this circuit must be supplied outside the photoelectric switch to isolate the switch from the TTL or CMOS devices. This type of circuit is very useful to interface the rugged photoelectric switch with small microprocessor-controlled circuits. Since programmable controllers and industrial robots are specifically designed to interface with traditional motor controls, they will have input/output modules available for direct interface to all ranges of AC and DC voltages, and this type of circuit would be unnecessary.

Another type of output uses a built-in relay for control. This type of photoelectric switch is different in that the relay's coil is wired directly to the transistor in the output stage and the relay is actually mounted and housed in the switch. The contacts of the relay are isolated from the coil and are available for control of load currents up to 10 A. This type of control allows the photoelectric switch to be used without an additional relay. Remember that if an extra relay is required, as in the AC and DC solid-state-controlled outputs, it will require cabinet space for mounting.

OPBT2 Power Blocks with Attached Cable
Current Sinking (NPN) Configuration

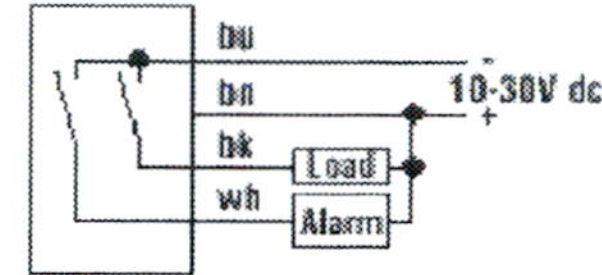

OPBT2 Power Blocks with Quick-Disconnect
Current Sinking (NPN) Configuration
(4-Pin Mini- or Euro-Style)

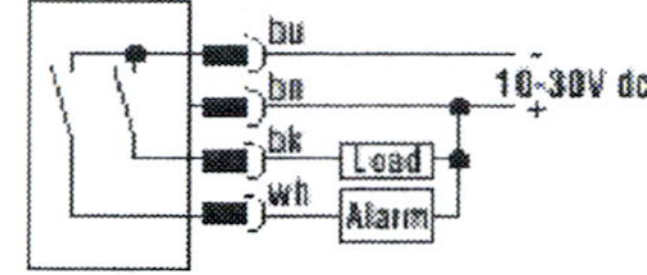

OPBT2 Power Blocks with Attached Cable
Current Sourcing (PNP) Configuration

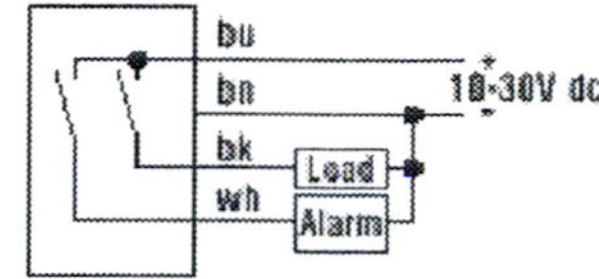

OPBT2 Power Blocks with Quick-Disconnect
Current Sourcing (PNP) Configuration
(4-Pin Mini- or Euro-Style)

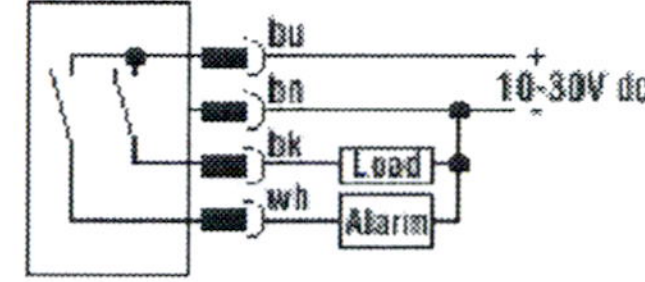

Figure 11–12 (a) Current sinking circuits (with NPN transistor) for photoelectric switches. (b) Current sourcing (with PNP transistor) for photoelectric switches.
(Courtesy of Rockwell Automation)

Some companies, such as Micro Switch, use plug-in relays for output control. The types of relays available for plug-in mounting are the reed relay, solid-state relay, and electromechanical relay. Operation of these relays was covered in previous chapters. These relays provide control for a variety of different voltages and applications with easy mounting for installation.

The plug-in relay requires a control base for proper mounting. The control base will also house a power supply, which is a step-down transformer that will convert 110 or 230 VAC to low-voltage AC or regulated 12 VDC. This is the same control base that is required for photoelectric switches that do not have self-contained power supply and control circuitry.

The control base can also support a logic card that provides a wide variety of logic functions. This means that the photoelectric switch can execute complex control functions as a stand-alone control or in conjunction with robots and programmable controllers. These functions include time-delay-on, time-delay-off, latched output, nonrepeat electronic latch one-shot pulsed output, zero speed detector, and division counter (similar to a four-input AND gate or OR gate, where all four inputs are considered to get an output). Each of the types of cards except the zero speed detector also includes a switch to select light or dark operation and a sensitivity adjustment that must be set at the full clockwise position if the card is used with modulated LED controls. The cards use an edge connector for easy push installation.

Figure 11–13 shows a table that compares all of the different types of outputs used with photoelectric devices.

11.1.5.4 Setting Light-Activated or Dark-Activated Switches

You can set the photoelectric switch to sense the switch condition under light–activated or dark-activated conditions. Figure 11–14 shows an example of setting the light-activated or dark-activated switch in the photoelectric switch while it is installed in an application. This switch can be set after the switch is installed.

11.1.5.5 Two-Wire and Three-Wire Controls

Another way to classify outputs for photoelectric controls is by the number of wires that are required to make them operate correctly (Figure 11–15). The two-wire control (Figure 11–15a) is self-contained. The retroreflective and diffuse scan controls are available as small, stand-alone, two-wire controls. This control is connected in series with the load and will receive operating power from this voltage. When the proper amount of light is received, the switch portion of the control will become energized and the current needed to power the load will be allowed to flow.

A diagram of three-wire control is shown in Figure 11–15b. The control receives power to operate from a separate wire than that from which the load receives its power. When the photoelectric switch energizes, the load current will flow through the control to power the load. Each of these types of controls provides a smaller control that is usable in stand-alone applications. In each control the amount of load current is limited to 300

Output Type	Strengths	Weaknesses
Electromechanical Relay *AC or DC switching*	• Output is electrically isolated from supply power • Easy series and/or parallel connection of sensor outputs • High switching current	• No short circuit protection possible • Finite relay life
FET *AC or DC switching*	• Very low leakage current • Fast switching speed	• Low output current
Power MOSFET *AC or DC switching*	• Very low leakage current • Fast switching speed	• Moderately high output current
TRIAC *AC switching only*	• High output current	• Relatively high leakage current • Slow output switching
NPN or PNP Transistor *DC switching only*	• Very low leakage current • Fast switching speed	• No AC switching

Figure 11–13 Table comparing the strengths and weaknesses of different types of outputs.
(Courtesy of Rockwell Automation)

NOTE: The power indicator will turn off when the output indicator is on.

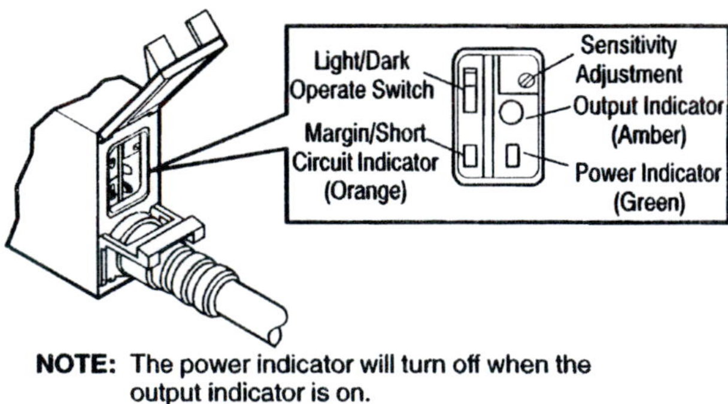

Figure 11–14 Switches to select light/dark operation, sensitivity adjustment, and indicators for margin/short circuit, output indicator, and power indicator. These switches and indicators are available on most photoelectric switches.
(Courtesy of Rockwell Automation)

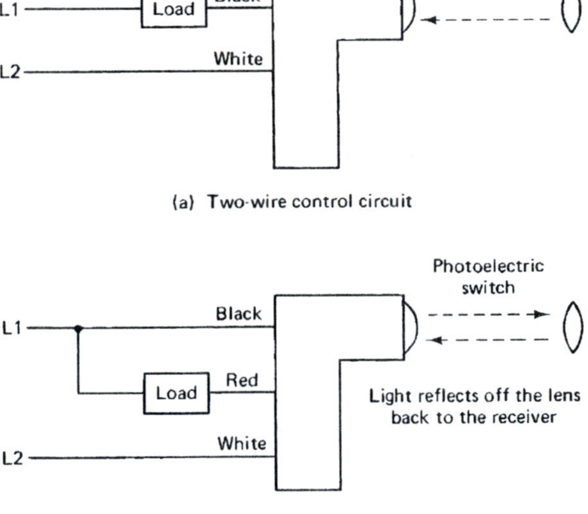

Figure 11–15 (a) A diagram of a two wire photoelectric control. (b) a diagram of a three wire photoelectric control.

to 500 mA. This type of device is ideal for interface to programmable controllers and robots.

`11.1.6 Analog Photoelectric Switches

Another type of photoelectric switch is called an analog switch. The analog device can sense the exact distance in values of percent or the number of feet an object is to a sensor. The digital type senses the object and turns on or off. The analog device can measure all the distances between the first point where an object is. Figure 11–16 shows a diagram of an analog sensor.

11.1.7 Applications for Photoelectric Switches

Figures 11–17 and 11–18 show several applications for photoelectric switches. You will see that photoelectric switches can be used to detect all types of products. Figure 11–17a shows an application where a photoelectric switch is used to detect labels on boxes. Figure 11–17b shows an application where photoelectric switches are used to detect large and small boxes on a conveyor. Figure 11–17c shows an application where photoelectric switches are used to detect the separation line on clear plastic bags. Figure 11–18a shows an application where photoelectric switches are used to detect the caps on spray paint cans. Figure 11–18b shows an application where photoelectric switches are used to detect baked goods on a conveyor. Figure 11–18c shows an application for sensing a gear tooth for quality control.

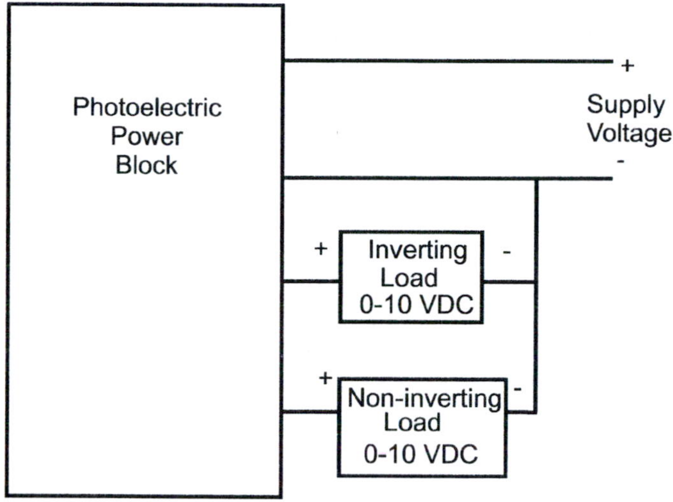

Figure 11–16 Wiring connection diagram for an analog photoelectric switch.

11.1.8 Troubleshooting Photoelectric Switches

The photoelectric switch is easy to troubleshoot. You must consider problems concerning both the electrical and the optical parts of the switch. You can test the part of the photoelectric switch that emits light by checking for the light beam. You should remember that if the light beam is not in the visible range, you may need to use a special sensor to determine if a beam is emitted. If the beam is not emitted, you can check the power supply or the photoelectric switch itself. If the beam is emitted, but the switch is still not activating, you should suspect the

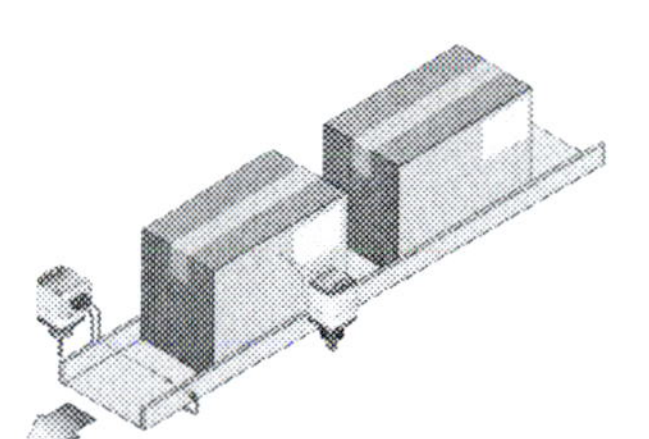

Label on Box Detection

Application: Detect the presence or absence of a white label on a corrugated kraft cardboard box

Sensor Models: OMNI-BEAM model OSBCVB sensor head with OPBT power block; OMNI-BEAM model OSBFP with OPBT power block and a pair of PIL46U opposed mode plastic fiber optic assemblies

Application Notes: OMNI-BEAM model OSBCVB converges the light from a blue LED source at 38 mm (1.5") to reliably sense a white label on a kraft cardboard box. A model OSBFP uses individual plastic fiber optics in the opposed sensing mode to gate the label sensor to check for the label when the leading edge of a box is sensed.

(a)

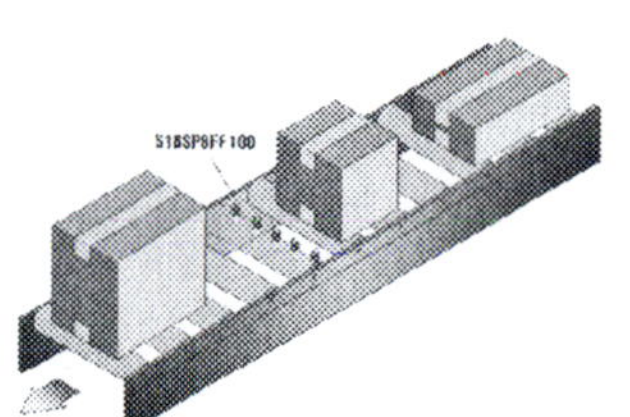

Box Sensing

Application: Sense boxes anywhere across a roller conveyor, where sensing is possible only from under the conveyor

Sensor Model: Five EZ-BEAM model S18SP6FF100 fixed-field sensors

Application Notes: The optics of this fixed-field sensor are ideal whenever sensing must be accomplished from underneath a conveyor. Excess gain is very high at the optimum sensing distance of about 25 mm (1"). Also, excess gain is low right at the lens of the sensor (at 0 mm or 0"). The result is that sensor performance is not affected by moderate amounts of dirt and dust falling on the lens.

(b)

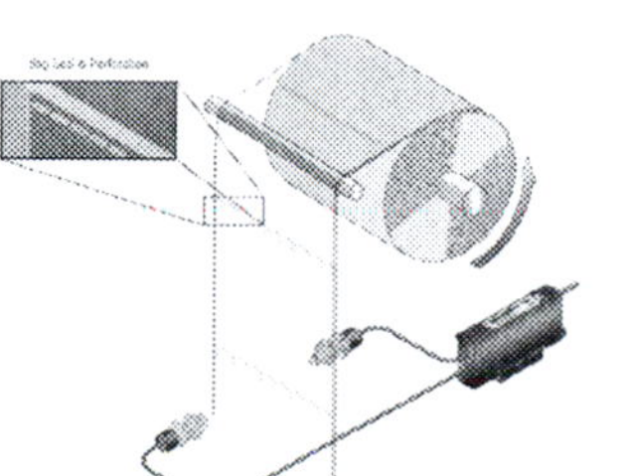

Clear Bag Separation

Application: Sense the perforations in a continuous clear web to trigger a separation mechanism

Sensor Models: D12EN6FP *Expert*™ with a pair of PIL46U plastic fiber optics

Application Notes: The low contrast capability of the Expert sensor, combined with the lensed plastic fiber optics, can detect perforations in clear materials.

(c)

Figure 11–17 (a) Photoelectric application for label detection on cardboard box. (b) Box-sensing application. (c) Clear bag separation application for photoelectric.
(Courtesy of Banner Engineering Corp.)

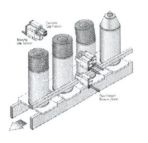

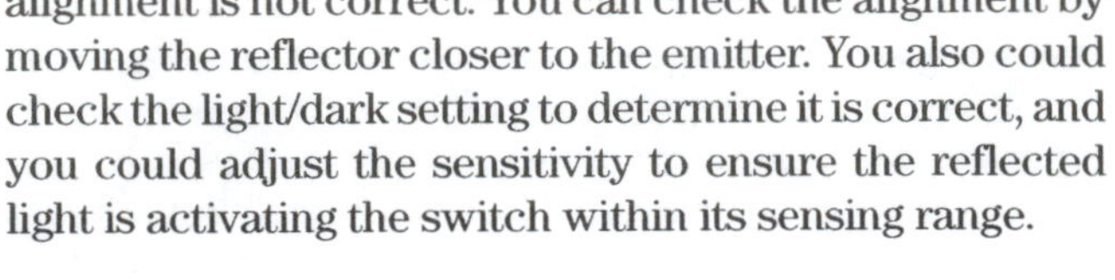
Spray Can Inspection
Application: Inspect spray cans with crooked or missing caps
Sensor Models: Two pairs of MINI-BEAM SM31E and SM31R opposed mode sensors, and one MINI-BEAM SM312CV convergent mode sensor
Application Notes: The convergent sensor provides a gate signal for one of the two opposed mode receivers which checks for a missing cap. The other opposed mode pair is fitted with rectangular apertures to sense any excess can height caused by a crooked cap.

(a)

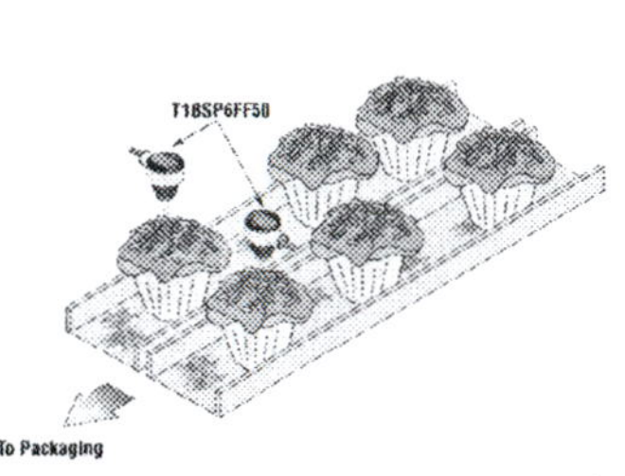
Sensing Baked Goods on a Conveyor
Application: Sense baked goods on adjacent, parallel flat conveyors
Sensor Models: Two, EZ-BEAM T18SP6FF50 fixed-field sensors
Application Notes: Fixed-field technology permits reliable reflective-mode sensing of irregular-shaped objects, while ignoring background surfaces such as the conveyor.
T18SP6FF50
To Packaging

(b)

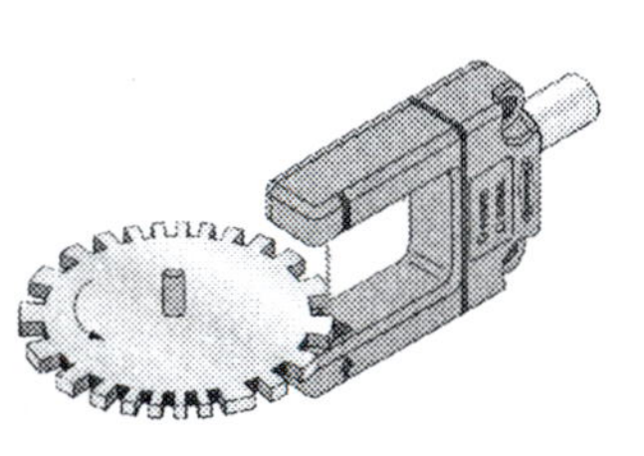
Gear Tooth Sensing
Application: Sense teeth of timing gear
Sensor Model: SL30VB6V slot sensor
Application Notes: SL Series slot sensors provide a clean, economical way to reliably produce pulses from timing gears used in automated production machinery.

(c)

Figure 11–18 Photoelectric applications for detecting (a) caps on spray paint cans; (b) baked goods on a conveyor; (c) a gear tooth for quality control.
(Courtesy of Banner Engineering Corp.)

alignment is not correct. You can check the alignment by moving the reflector closer to the emitter. You also could check the light/dark setting to determine it is correct, and you could adjust the sensitivity to ensure the reflected light is activating the switch within its sensing range.

11.2 Proximity Switches

Proximity switches are used in a variety of applications for sensing the presence of parts and locating fixtures. These applications include sensing distance or location, and sensing the presence or absence of parts at high speeds. One application is the sensing of missing parts. The proximity switch in this application is set up to sense missing aluminum caps on a bottling line. This particular switch senses the presence of the caps. If a cap is missing, the sensor trips the output and the bottle is moved off the main line to a recapping line. Proximity switches are particularly well suited for these applications because they can sense a variety of ferrous and nonferrous metal products as well as a variety of nonmetallic materials and liquids. The proximity switch provides functions that are similar to a limit switch in that it can detect the presence or absence of an object without making contact with the part. For this reason it is called a noncontact limit switch. These switches are based on several different theories of operation used for a variety of sensing techniques and include the eddy current killed oscillator, high-frequency capacitive oscillator, Hall effect transducer, and reed relay. Figure 11–19 shows several examples of proximity switches and a typical wiring diagram.

Proximity switches are readily adaptable to today's control systems because they are compact and totally solid-state devices. This allows them to be mounted very close to the material they are sensing, in

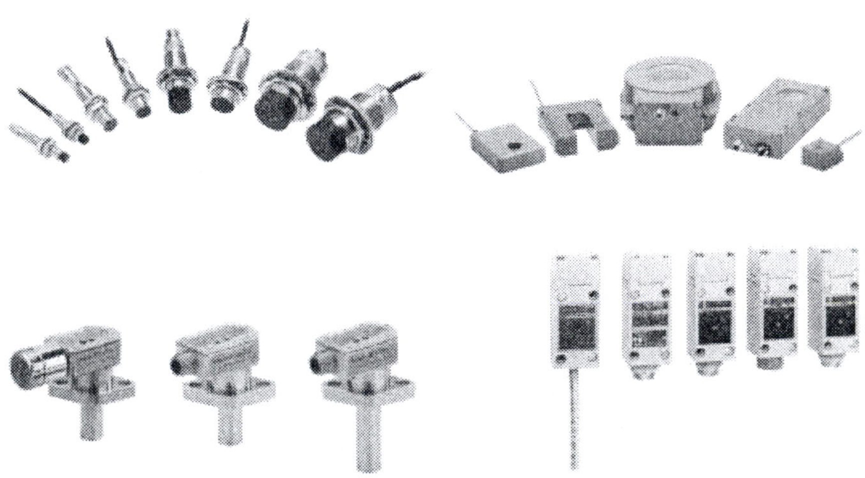

Figure 11–19 Examples of proximity switches and noncontact position sensors.
(Courtesy of Rockwell Automation)

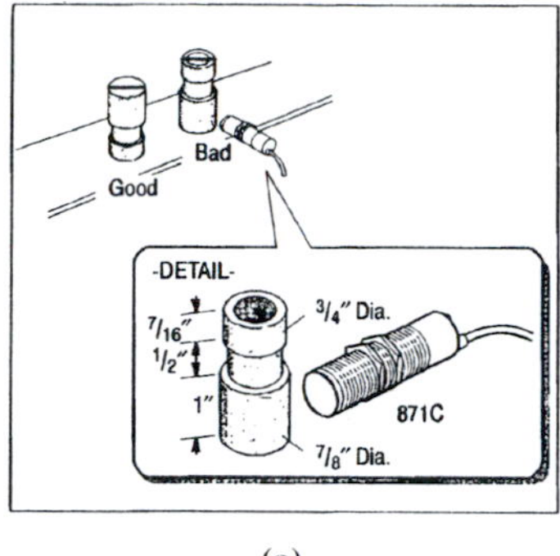

(a)

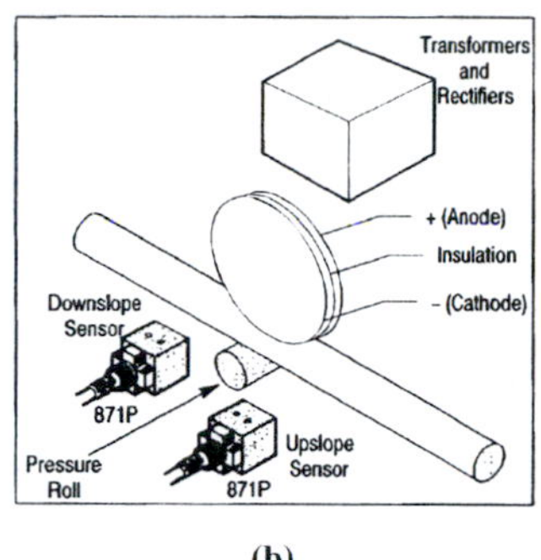

(b)

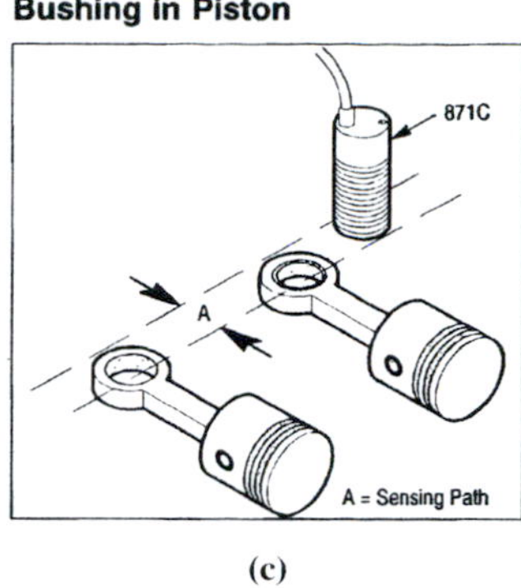

(c)

Figure 11–20 (a) Proximity switch application for online parts sorting. (b) Application of up and downslope control of continuous pipe welding. (c) Application for detecting the presence of a bushing in a piston.
(Courtesy of Rockwell Automation)

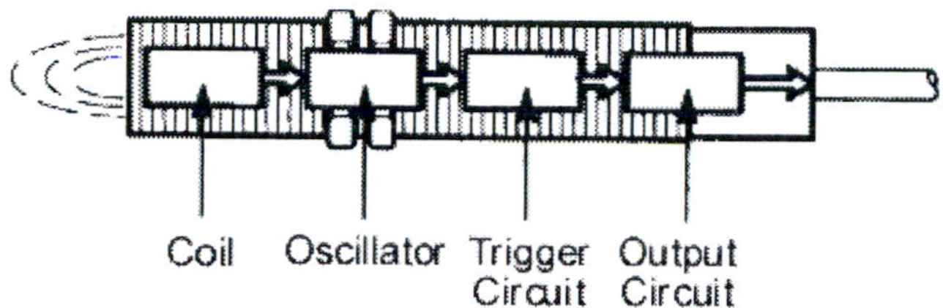

Figure 11–21 Cutaway diagram of a proximity switch.
(Courtesy of Rockwell Automation)

highly automated locations. They come in a variety of self-contained sensor bodies as well as separate sensors with remote bases for providing power supplies, amplifiers, and outputs. Figure 11–20 shows examples of using proximity switches in sensing applications.

11.2.1 Proximity Switch Operation

Proximity switches are actually a broad family of switches that are used as sensors in industrial control systems. They can sense a variety of various materials' inductive and capacitive targets. The different classes of material require different sensing circuits.

One class of material that is commonly sensed by proximity switches is ferrous and nonferrous metals, including iron and steel products, aluminum, brass, and copper. The main type of circuit used to sense these materials is the *eddy current killed oscillator* (ECKO). An ECKO proximity switch is made of three basic parts: the sensor (which is part of a tuned inductive-capacitive tank and oscillator circuit), a solid-state amplifier, and a switching device.

Figure 11–21 shows a cutaway picture and a diagram of an ECKO proximity switch. Figure 11–22 shows several diagrams explaining the operation of the proximity switch. The operation of the switch can be described in this way. DC voltage enters the switch through the regulator circuit. The regulator will maintain a constant supply voltage. The oscillator generates a radio-frequency (RF) field out the end of the sensor. This is accomplished by mounting the coil part of the tank circuit near the end or head inside the proximity sensor. The tuned tank circuit oscillates at a predetermined frequency to produce the RF field. When metal is brought near the RF field (at the sensor end of the switch), eddy currents are formed in the metal that kill the oscillations in the tank circuit. The integrator part

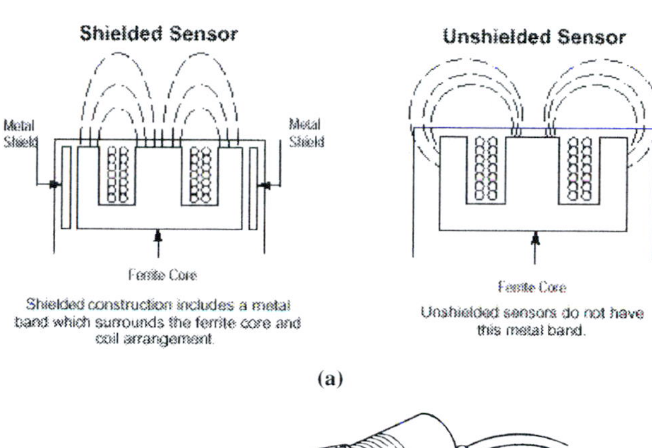

(a)

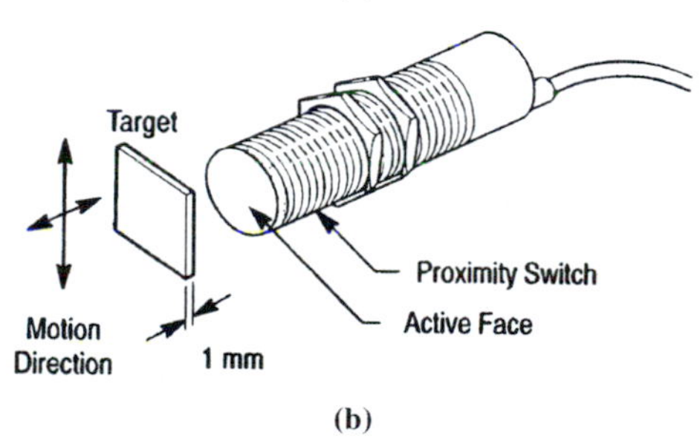

(b)

Figure 11–22 Diagrams that show (a) the location of the sensing field for a shielded and unshielded proximity switch; (b) how close a target must be to the proximity switch for it to be sensed.
(Courtesy of Rockwell Automation)

of the circuit converts the sine-wave signal generated by the oscillator into a DC signal. The DC signal, which varies in amplitude with the amplitude of the oscillation, is sensed by the Schmitt trigger and converted to a digital (off/on) signal. The digital signal from the Schmitt trigger is used to power the output transistor.

A similar type of proximity switch is powered by AC voltage and is connected in series with the AC load it is controlling. The hot line of the AC voltage is wired to the load. From the load the voltage is sent to a noise-suppression circuit in the proximity switch. The noise-suppression part of the switch consists of a resistive-capacitive circuit for filtering and a metal-oxide varistor that is used to shunt unwanted spikes on the AC line.

The voltage passes from the suppression circuit to a full-wave bridge rectifier. In this case the rectifier is acting as an AC switch that is being controlled by the DC side of the circuit. An SCR is used to switch the current in the DC circuit on or off. If DC voltage is allowed to flow through the bridge rectifier, AC current will also flow through the AC side of the bridge. If current is interrupted on the DC side of the bridge by the SCR, it will also interrupt current on the AC side of the bridge and thus switch the current off and on to the AC load.

The oscillator and sensing coil at the end of the switch operate like a DC ECKO proximity switch. Whenever a metal object enters the rf field produced at the end of the sensor, the oscillations are killed and the Schmitt trigger activates the amplifier circuit. The amplifier in the AC switch is used to trigger or gate the SCR. An LED is connected with the SCR circuit to indicate when the switch is on. The LED in this case will cause a small voltage drop to the load, but it is very useful as an external indicator of the switch's condition and can be seen at some distance by the electrician or technician during troubleshooting. The output circuit for the proximity switch can be DC as shown in Figure 11–23, or current sourcing, current sinking, or AC as shown in Figures 11–24 and 11–25.

You may also need to take into account that a small amount of leakage current will pass to the load when the switch is in the off state. This leakage current is passing through the suppression circuit. If it is important to have absolutely no leakage current at the load, you can use the coil of a relay as the load to the proximity switch and, through the relay's contacts, control the load that cannot tolerate leakage current. Another type of switch is the *capacitive proximity switch.* This switch works on a principle similar to that of the inductive switch. In the capacitive proximity switch, the tank circuit, which includes the sensor, is tuned to the dielectric field immediately surrounding the outside of the sensor head. Any changes in the dielectric field (ca-

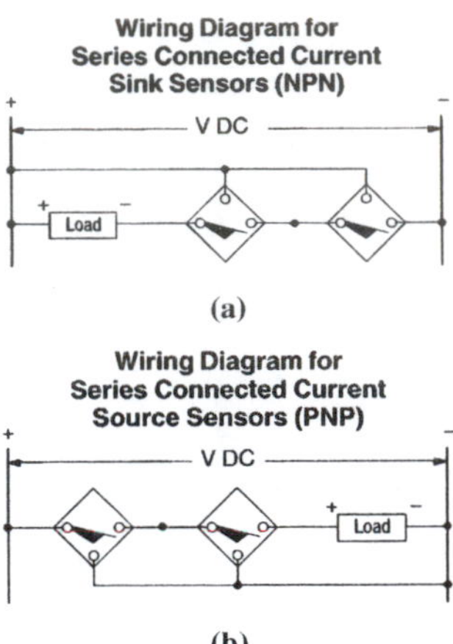

Figure 11–23 Wiring diagram for connecting two proximity switches with (a) current sinking outputs, in series; (b) current sourcing outputs, in series.
(Courtesy of Rockwell Automation)

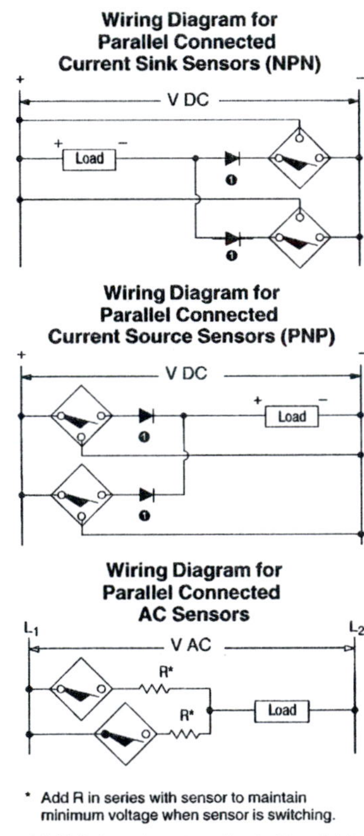

Figure 11–24 Wiring diagram for connecting two proximity switches with (a) current sinking outputs, in parallel; (b) current sourcing outputs, in parallel; (c) AC outputs, in parallel.
(Courtesy of Rockwell Automation)

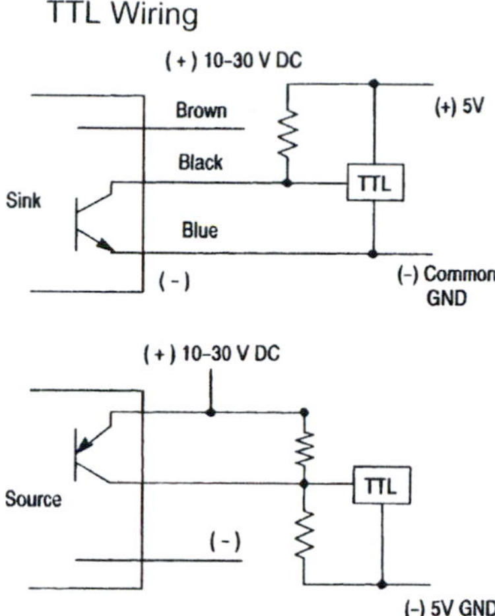

Figure 11–25 Wiring diagrams for TTL sinking and sourcing outputs for proximity switches.
(Courtesy of Rockwell Automation)

pacitance of the area around the sensor) will cause the switch to trigger. During installation, the sensitivity of this switch can be set to sense the presence or absence of the material as it moves through the sensor's field. As material (the target) moves through the field, the field's capacitance is changed.

The sensitivity of the switch is adjusted so that the switch will not trigger in ambient air. This means that when no part or material is in the sensor field, the switch will not trigger. As soon as material enters the field and changes the dielectric characteristics, the switch can be adjusted to trigger on that material. This type of proximity sensor is well suited to measuring liquids, since the sensor can be fully immersed in the liquid. Since liquids have different dielectric characteristics than those of air, the switch can be sensed to trigger on the liquid when it covers the sensor. If the level lowers, and the sensor is now exposed only to the air, the dielectric field will change and the switch will turn off.

The sensor does not have to be immersed to work well with liquids. A glass tube can be used to make the measurement. When the liquid level in the tank rises, it will also rise in the glass tube. When the liquid level in the tube rises to a point where it moves past the sensor, the capacitance in the field in front of the sensor changes and triggers the switch. The sensitivity of the switch must be set so that it will not trigger when the liquid level in the glass tube is below the sensor.

The sensing distance of the capacitive proximity switch may be affected by elements in the air, such as humidity and dust concentration. If the material being sensed is moving very rapidly, the pulse will be of short duration and may not be long enough to trigger the output circuit. By adjusting the sensitivity, the capacitive proximity switch can be set to operate correctly under these conditions. If the pulse is still too short, a latching output can be used to overcome the problem.

Since the capacitive proximity switch can be adjusted to trigger on any change in the dielectric field around the sensor, it is capable of detecting ferrous and nonferrous metals as well as nonmetallic materials. This makes the capacitive proximity switch a universal type of sensor.

The main problem with capacitive proximity switches is that they may be very sensitive to background or ambient conditions. This means that changes in the air due to contamination or electrical fields or even strong RF signals may cause the switch to trigger falsely. This problem also prohibits positioning sensors too close to each other. The main correction for this condition is to use sensors that have built-in shielding, which helps to focus the capacitive field where the switch is sensing. In this way the sensor can be focused directly on the area where the target (parts or material that is being sensed) will pass through.

11.2.2 Applications for Proximity Switches

Proximity switches presently used as industrial controls for detecting motion on machinery and robotic work cells are also being used to detect parts in place for automated applications. Motion detectors are also available in bistable styles. The bistable proximity switch is capable of sensing motion in two directions. This is useful in automated applications where cylinders must move out and back or where robots, mills, and lathes use *X*, *Y*-axis motion. Other applications include label detection, parts counting, missing parts detection, broken tool detection, and machine on/off control. A full list of applications would involve almost every parameter of automation control.

Proximity switches are available in a variety of models and are usable under nearly all conditions in industry. Among these are watertight, oiltight, and totally submersible models. Other switches are available for high-pressure applications (0 to 5075 psi), high-temperature applications (32 to 392°F), and for use where extremely large magnetic fields are present, such as in welding applications (30,000 A). Proximity switches have also been very successful in applications in hazardous locations, such as providing explosion-proof environments.

Questions

1. Explain the operation of a modulated light source photoelectric control.
2. List the three types of photoelectric controls.
3. Explain how a cadmium sulfide cell works.
4. Explain how fiber optics are used in photoelectric controls.
5. Explain the operation of a proximity control.

True and False

1. You can quickly test an inductive proximity switch to determine if it is operating correctly by moving a piece of ferrous iron near its head, which will cause it to activate.
2. You can adjust a proximity switch to ensure that it is set correctly by loosening its locking nuts and turning the switch clockwise to move it closer to its target or turn it counterclockwise to move it away from its target.
3. You should use a relay as the output stage of a photoelectric control when you want to provide electrical isolation.
4. A photoresistive cell is a device that is made from cadmium sulfide (cad) cells.
5. The cadmium selenide (CdSe) cell is sensitive only to light in the infrared spectrum range.

Multiple Choice

1. The term ECKO means _______
   a. electrical circuit kinetic oscillator.
   b. eddy current killed oscillator.
   c. electrical current kinetic oscillator.
   d. a proximity switch that bounces signals off its target and receives the echo.
2. The difference between an inductive proximity switch and a capacitive proximity switch is_____
   a. the inductive proximity switch is activated by nonmetal parts and the capacitive proximity switch is activated by metal parts.
   b. the inductive proximity switch is activated by metal parts and the capacitive proximity switch is activated by nonmetal parts.
   c. the inductive switch is more accurate.
   d. the capacitive switch is more accurate.
3. A modulated light source for a photoelectric device consists of ______
   a. a light-emitting diode (LED).
   b. a phototransistor.
   c. an oscillator circuit.
   d. All of the above
4. The retroreflective photoelectric switch ________
   a. has the modulated light source and the receiver in the same housing.
   b. has the modulated light source and the receiver in separate housings.
   c. uses a reflector to send light back to the receiver.
   d. All of the above
   e. Both a and c
5. A diffuse scan photoelectric switch _________
   a. operates on the principle of emitting a beam of light on an object rather than on a reflector.
   b. operates on the principle of emitting a beam of light on a reflector rather than on an object.
   c. must use fiber optic cables.
   d. All the above

Problems

1. Sketch a diagram of the components used in a throughbeam photoelectric control.
2. Sketch a diagram of the component parts of a retroreflective photoelectric control.
3. Draw an electrical diagram of a two-wire and a three-wire control used in photoelectric switches.
4. Draw the electrical symbol for a photoelectric and proximity control.
5. Draw an electrical diagram of a two-wire and three-wire proximity control connected to a motor starter as a load.

◀ Chapter 12 ▶

Timers, Counters, and Sequencers

Objectives:

After reading this chapter you will be able to:

1. Explain the operation of a pneumatic time-delay relay.
2. Explain how to mount an add-on timer and adjust its time delay.
3. Explain how a motor-driven timer operates.
4. Explain how a solid-state timer operates.
5. Explain the operation of a counter.

Timers, counters, and sequencers provide complex control applications for today's automated systems. These devices can be used as add-on controls or stand-alone controls, or can be directly interfaced with programmable controllers and robots. In this chapter we cover the complete line of available timing devices. First we explain the operation and use of timers. Then, we discuss the use of counters. In the final part of the chapter, we discuss the use of timer/counter devices and sequencers. The information provided in this chapter will help you identify the different types of controls and will explain typical applications, theory of operation, diagrams and installation, and troubleshooting techniques.

12.1 Timers

Timers have been used to control industrial applications for many years. Timers provide time delay by using motors, gears, and cams to control sets of contacts. Other types of timers and time-delay devices use air or liquid in a chamber that empties slowly to cause a delay in the action of contacts. As technology has advanced, new timing devices have incorporated modern synchronous motors, solid-state technology, and microprocessor technology. Some devices fill applications where the requirement is mainly low cost and simple operation. These devices are typically added on to standard relays, contactors, and motor starters to provide some degree of time delay.

Four basic principles of operation are used in modern timing devices: pneumatic, electromechanical motor-driven, solid-state electronic, and microprocessor, also called digital. These devices come in a variety of models that are available with the following types of control for the outputs they activate: delay-on, delay-off, cycle times, and logic timers. These timers can be used as reset timers where the timer is activated and then times out and actuates a set of contacts, or they can be used as a repeat cycle timer that continually repeats its cycle and actuates a set of contacts at the end of each cycle. The final way to use a timer is as a totalizing timer, where the amount of time that a device operates is recorded. A comparison of the synchronous motor, solid-state, and the digital timers is presented in Figure 12–1.

12.1.1 Symbols and Electrical Diagrams for Time-Delay Circuits

The symbols for on-delay and off-delay timing circuits are shown in Figure 12–2. The symbol for the time-delay contacts is half an arrow. The time-delay-on symbol uses the lower half or tail of the arrow, and the time-delay-off uses the upper half or arrowhead portion of the arrow. When the arrowhead is used for the time-

• **CLASSIFICATION BY OPERATING PRINCIPLE**

Classification	Operating Principle	Merits	Demerits
Synchronous motor timer	The motor timer employs a synchronous motor, actuated by an electrical input signal, produces a mechanical motion, by which the contacts are caused to make and break after the lapse of a pre-determined time. In general, a desired time is obtained, through a clutch and reduction gear mechanism, by actuating a small synchronous motor by applying commercial 50 or 60Hz power as the input signal. The speed (r.p.m.) of the synchronous motor in principle is proportional to the input power frequency and is not affected by a change in the supply voltage or ambient temperature. The DC type motor timer is available in various types including the one employing a DC motor and the type with a similar mechanism to an AC timer, which actuates the synchronous motor by the built-in DC-AC converter.	1. Wide time setting range from short to long time. 2. Elapsed time indication by moving pointer. 3. Less affected by temperature and voltage fluctuations.	1. Precise, minute time setting impossible. 2. Short mechanical service life because of its construction with mechanical moving parts. 3. Manufacturing of DC operation type is difficult.
Solid-state timer	In the RC oscillation counting system, the RC oscillation circuit consisting of a resistor and a capacitor starts oscillating upon application of power. When the counter circuit counts up the set value by taking the oscillated signal as the reference signal, an output signal is generated. This output signal is then amplified to operate the output relay. The basic circuitry of the timer consists of a power circuit, a constant voltage circuit, an RC oscillation circuit, a counter circuit, an output amplifier circuit and a relay circuit. In the RC charging system, a desired time delay is obtained by utilizing the charge and discharge characteristics of the RC (resistor and capacitor) network. When the terminal voltage of the capacitor reaches the specified value, an output signal is generated from the relay or semi-conductor. Solid-state timers with less mechanical moving parts are highly resistant to vibration and shock, and are capable of highly frequent operation, thus boasting long service life.	1. Highly frequent operation with long mechanical service life. 2. Both DC and AC operations possible. 3. Solid-state output system and separate time setting system are possible.	1. Manufacturing of long-time versions is difficult. 2. Susceptible to temperature and voltage fluctuations. 3. Elapsed time indication impossible.
Digital timer	A constant frequency signal (usually, commercial power frequency) is employed as a time-limit element. This signal is counted by the counting circuit to obtain a time delay. The shorter and the more accurate the duration of the reference signal, the higher the timer accuracy. However, the duration of the reference signal must be such that permits its discrimination from externally occurring noises. The timer employs a thumbwheel switch in its time setting section to permit highly accurate time control, and also indicates the elapsed time digitally. The version employing an electromagnetic counter is also available. The counter advances mechanically at each input signal by the electromagnetic force of the electromagnetic coil. However, this version is not suitable for practical application because of its low counting speed and short mechanical service life.	1. High precision. 2. Error-free digital time setting. 3. Elapsed time indication also possible digitally.	1. Larger in size and more costly than other types. 2. Susceptible to electrical noise.

Figure 12–1 Comparison of types of timers.
(Courtesy of Omron Electronics)

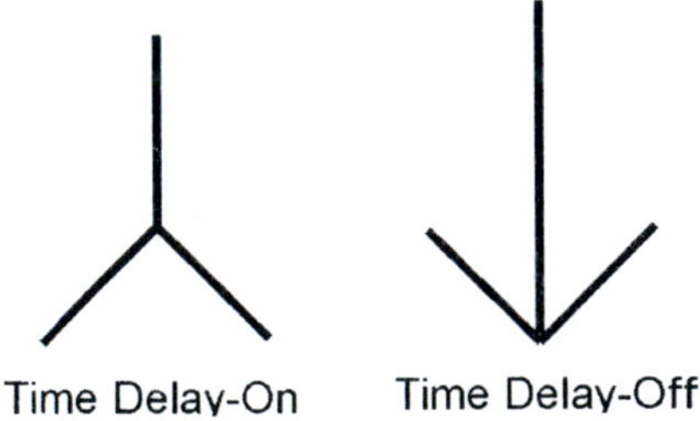

Figure 12–2 Example of a time-delay-on and time-delay-off electrical symbol.

delay-off symbol, it is usually pointing down; in the time-delay-on symbol, the tail is usually pointing up. Time-delay-on and time-delay-off controls can both have normally open and normally closed contacts.

Figure 12–3 shows the electrical diagram of a time-delay control that will delay turn-off time for a motor. A limit switch in the line will open and deenergize the control relay's coil, and the contacts for that relay are controlled by a pneumatic timing device. This is indicated by the time-delay-off symbol showing control of the relay contacts, and the contacts are identified by the same label as the relay coil. When the electrical control to the relay coil is deenergized, the contacts push on the pneumatic actuator, causing air to begin to leave its chamber. When all the air has been pushed out of the chamber, the contacts are allowed to open. The net result is that the motor is shut off some time after the limit switch has deenergized the relay coil. The exact amount of time delay will be determined by the adjustment of the needle valve on the pneumatic add-on time-delay control.

12.1.2 Simple Time-Delay-On and Time-Delay-Off Examples

The time-delay devices can show a variety of applications through their wiring diagrams. The symbols for the timing functions are divided into two categories,

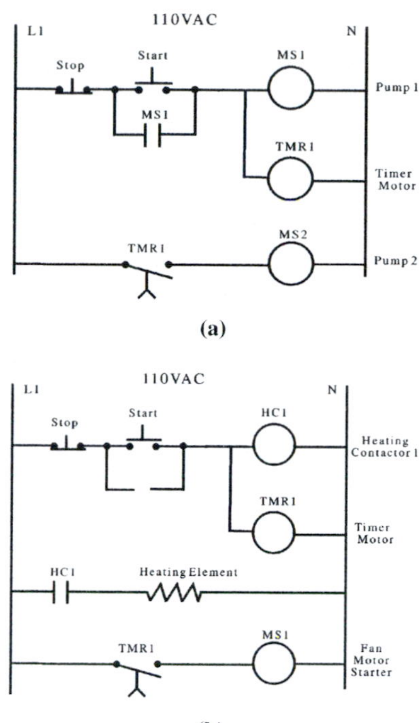

Figure 12–3 (a) Time-delay-on timer used to delay the time two pump motors start to limit the effects of inrush current. (b) Time-delay-off timer used to delay the time a furnace fan stays on after the heating element is turned off. The additional time the fan runs allows extra heat to be removed from the heating chamber.

which refer to delay on and delay off. These may also be called on-delay and off-delay, delay-on make and delay-on break, or TD-on and TD-off. A switch turning on a motor is an example of a delay-on application. Instead of the switch turning on the motor directly, it would energize the time-delay element, which would allow its contacts to close some time later and energize the motor. This type of application could be used where three motors are started from the same circuit and a slight delay of 15 seconds is needed between the time each motor is started, to avoid the large power surge (inrush) that occurs when multiple motors are started. For this application, the first motor starter would energize directly from the control circuit. The second motor starter would be controlled by an on-delay timer that would have a preset time of 15 seconds. The third motor would be controlled by an on-delay timer that would have a preset time of 30 seconds. When the control circuit is powered, it would energize the first motor starter immediately and begin the timing cycle for each of the on-delay timers. Fifteen seconds after the first motor started, the second motor's on-delay timer circuit would energize the second motor starter. Fifteen seconds later the third motor's timer circuit would time out and start the last motor. Each motor in this application would have time to start and come up to full speed, and the inrush current would return to normal limits, before the next motor would draw its large starting current.

In an off-delay circuit, the timing circuit is actuated when the control switch is turned off. The timing device begins the timing cycle when power has been removed from it by the control switch. An example of an off-delay circuit is an overhead lighting application controlled by a switch with a time-delay device. When the switch is turned off, it does not turn the light off directly; instead, it activates the off-delay element, which allows the light to stay on for a short delay after the switch has been turned off. This type of application allows personnel a few extra minutes to leave an area after turning off the switch before the lights go out. This is especially useful in large industrial areas. Many other motor control circuits that use time-delay-off applications are discussed in this chapter.

12.1.3 Pneumatic Timers

Pneumatic time-delay devices are generally used as add-on or stand-alone controls. Examples of add-on pneumatic timers are shown in Figure 12–4. The basic operation of the pneumatic device is shown in the diagram in Figure 12–5. From this diagram you can see that the pneumatic control has a flexible diaphragm that seals

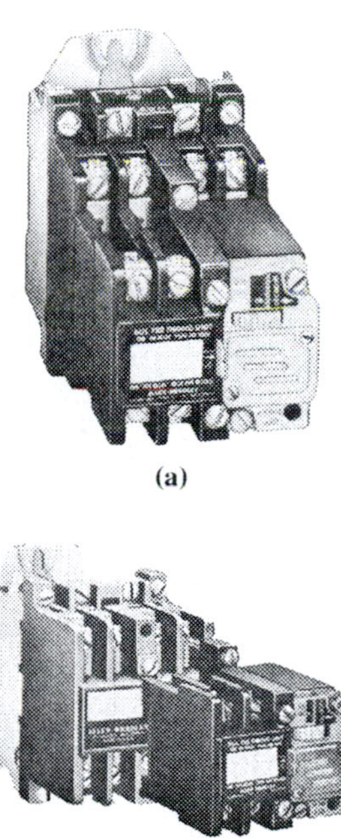

Figure 12–4 (a) Pneumatic time-delay element. (b) Pneumatic element is added to a traditional relay to create a time-delay relay.
(Courtesy of Rockwell Automation)

an air chamber. This diaphragm has a metal plate attached to it that activates the electrical switch contacts when the diaphragm has moved far enough. A spring keeps constant pressure against the diaphragm during the timing cycle. Air is released from the chamber through an orifice controlled by a needle valve. The needle valve controls the rate at which air can leave the chamber. When a sufficient amount of air has been released, the spring will cause the diaphragm to move far enough to activate the electrical switch element. When the pneumatic switch is reset, the chamber is refilled with air. The amount of time delay is set by adjusting the needle valve and testing the response time. Several trials may need to be completed to get the time adjustment exactly right. The amount of time delay can be adjusted between 0 and 180 seconds. The accuracy can be controlled to about 0.5 second.

This type of device can be used to stand alone or as an add-on timing control. This type of control is added to a standard relay or contactor. It has an adjustment knob on the front to set the amount of time delay. The actual adjustment should be made using trial-and-error methods. This means that you should set the adjustment and then check the actual time with a watch. If the time delay is critical, minor corrections to the time delay can be made by adjusting the knob slightly.

The pneumatic add-on control can also be used as a time-delay-on device. Figure 12–6 shows how to change the device from time-delay-off control to time-delay-on control. This can be done in the field while the timer-delay relay is wired in a circuit by turning the module upside-down when it is installed onto the relay. The pneumatic add-on control has a marker with an arrow showing which way to mount the device for time-delay-on and time-delay-off. When the device is installed for time-delay-on, the limit switch energizes the relay coil immediately after it closes. The magnetic pull from the relay coil causes the relay contacts to push on the bellows in the air chamber. Since the chamber is full of air, the contacts cannot close yet. As the air is removed from the chamber, the contacts are allowed to close and turn on the motor. The time it takes the air to leave the chamber provides the amount of time delay.

12.1.4 Electromechanical Timers

Another type of timer control uses a synchronous motor to turn a shaft on which cams are mounted. Figure 12–7 shows an example of a synchronous motor-driven timer, the electrical diagram for the timer, and the electrical diagram of the base the timer is plugged into. The timer motor causes a shaft to rotate and the shaft has a number of cams that rotate to a point where they cause sets of contacts to open or close. Spring tension will cause the contacts to return to their original state (open or closed) when the timer is reset.

The synchronous motor turns the shaft at a predetermined speed. The amount of time delay will be

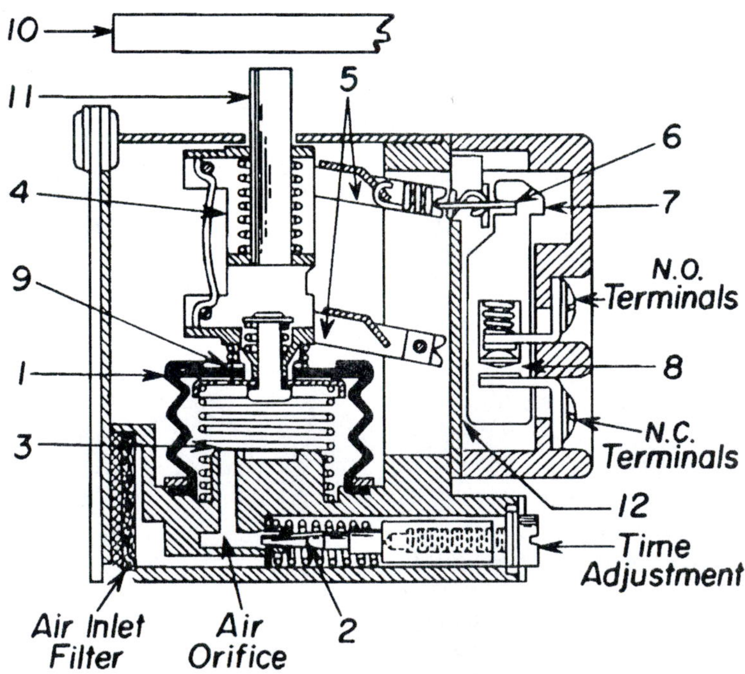

Figure 12–5 Cutaway diagram of a pneumatic timer showing internal operation.
(Courtesy of Rockwell Automation)

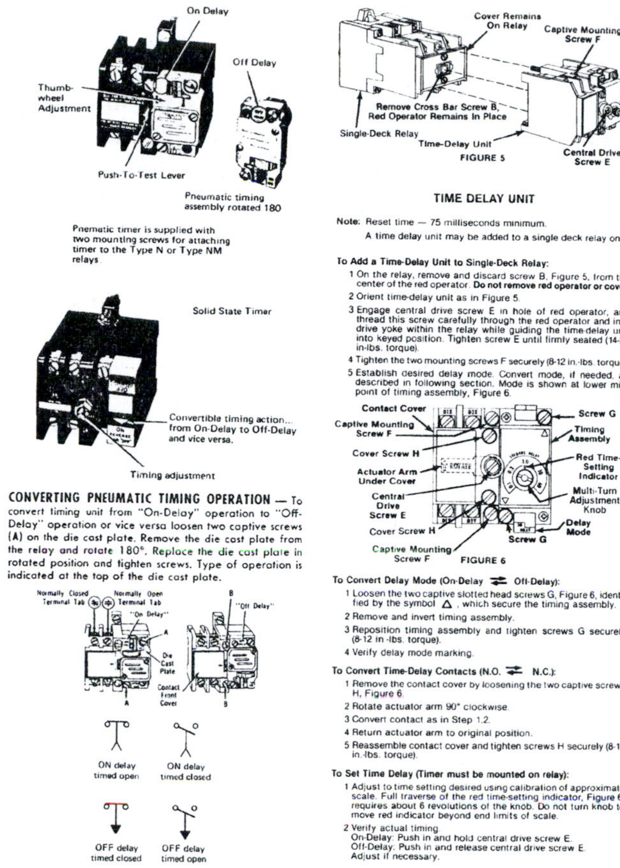

Figure 12–6 Changing time-delay elements for on-delay and off-delay applications.
(Courtesy of Rockwell Automation)

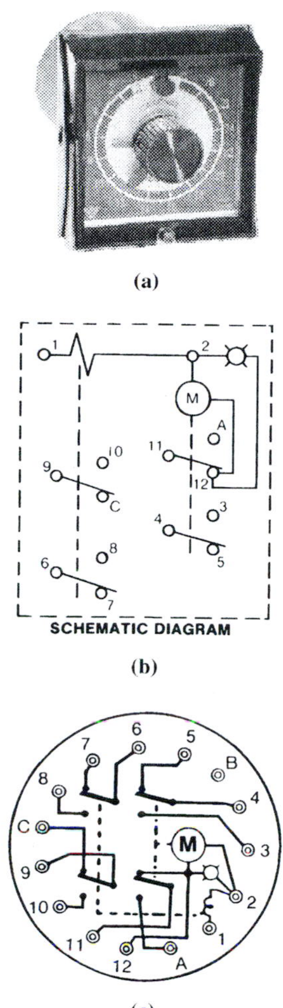

Figure 12–7 (a) Synchronous motor-driven timer; (b) electrical diagram of motor-driven timer; (c) diagram of the base that the synchronous timer plugs into.
(Courtesy of Danaher Controls)

determined by the speed of the motor and the distance the cams are from the switch contacts when the control is set. The amount of time delay can be extended by adding reduction gears between the motor shaft and the cam shaft.

A synchronous motor-driven timer has two separate controlling devices. The first is shown as a solenoid, called a *clutch solenoid,* which controls an electromechanical clutch and two sets of isolated contacts. These contacts are called *instantaneous contacts.* The second part of the switch is controlled by the timing motor. The motor is identified by an M in the motor symbol. The timing motor provides the drive for the camshaft that actuates the timing contacts. Two sets of isolated contacts are controlled by the timer motor. Each has a normally open circuit and a normally closed circuit. An indicator is wired in parallel with the timer motor and is illuminated whenever the timer motor is energized.

A dashed line extends from the solenoid symbol down through the two sets of contacts, which indicates that these contacts are under the control of this solenoid. A similar dashed line extends from the timer motor down through the remaining two sets of contacts, indicating that these contacts are under the control of the timer motor and will be affected by the time delay.

12.1.5 Motor-Driven On-Delay Timer Operation

The operation of the motor-driven timer will be explained in terms of the electrical diagram shown in Figure 12–8. This diagram shows the internal wiring of the timer and the field-installed wiring. For this example the timer is set up as an on-delay timer.

At the bottom of the diagram the field wiring shows a limit switch connected in series with line 1 (L1) and terminal 1 on the timer. A field-installed jumper is connected between terminals 1 and 11 on the timer. This will provide L1 power to the timer motor through normally closed contacts 11 to 12. Line 2 (L2) is connected directly to terminal 2. Terminal 2 inside the timer is

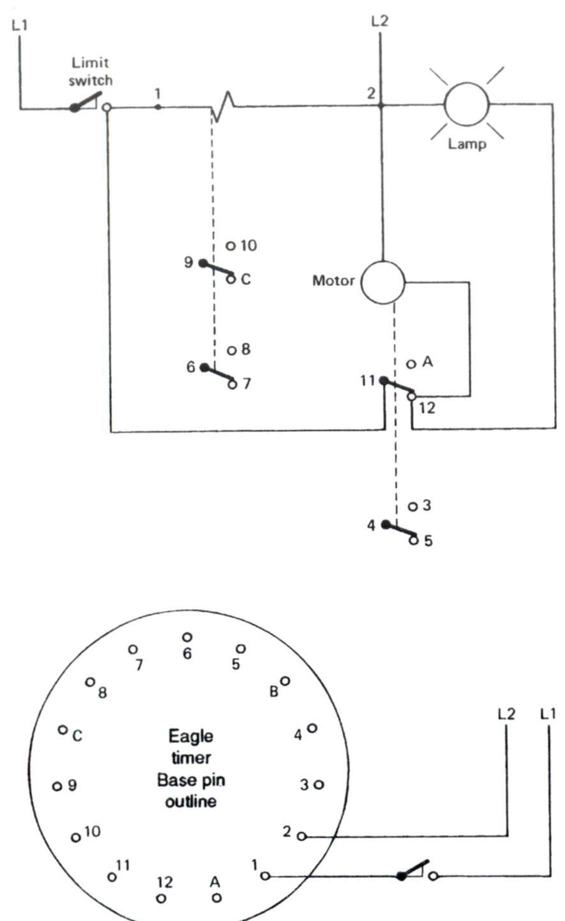

Figure 12–8 Field wiring diagram of motor-driven timer.

connected to the clutch solenoid, timer motor, and the indicator lamp. This provides the L2 potential side of the circuit to all of these loads inside the timer. When the limit switch is closed, line 1 power is sent to the left side of the clutch solenoid and through the normally closed contacts at 11–12 to provide power to the timer motor and indicator. Since L2 power is connected directly to the other side of each of these loads from terminal 2, each load becomes energized.

When the clutch solenoid is energized, it pulls the gears into position to be driven by the timer motor shaft. The clutch solenoid also pulls in the two sets of instantaneous contacts. This causes normally closed contacts 9 to C to open and normally open contacts 9 to 12 to close. Similarly, normally closed contacts 6 and 7 are opened and normally open contacts 6 to 8 will close. These contacts can be used to control any load that you want to operate instantaneous with the limit switch.

At the same time, the timer motor is powered through normally closed contacts 11 and 12 and begins to turn its shaft at the synchronous speed. This makes the timing gears (that have been engaged by the clutch) begin to turn. The timing gears drive the camshaft, which will cause the 11–12–A contacts and the 4–3–5 sets of contacts. Normally closed contacts 11 and 12 will open and break the power circuit to the timer motor, which will stop its shaft from turning. This keeps the camshaft in position where it is activating the contacts.

This action would also turn on any electrical load such as a motor starter and three-phase motor through contacts 4 and 3. Through this action, the limit switch would close, and the three-phase motor would turn on after the timer motor has completed its time delay cycle and closed contacts 4 and 3.

Whenever the limit switch is opened, the clutch solenoid and timer motor deenergize. When the clutch solenoid is deenergized, it disengages the gears from the timer motor and returns the two sets of contacts under its control to their normal position. As the drive gears disengage, a rewind spring returns the timer cam to reset the original preset time. This prepares the switch for the next timing cycle.

12.1.6 Timing Diagrams

Two types of timing diagrams are used to show the operation of timing devices. Figure 12–9 shows both. The first diagram uses a timing diagram and the second type uses a matrix-type diagram to indicate when each part of the timer control is activated or deactivated. An X is used to indicate that a part of the device is on or activated, and an O is used to indicate that a part of the device is deactivated or off. The parts are listed down the left side of the matrix, and include the clutch, timer motor, and contacts 9–10, 9–C, 4–3, 4–5, 11–12, and 11–A.

The sections of the timing cycle (reset, timing, and timed out) are indicated across the top of the matrix. The first block at the top, labeled "on delay," indicates that the matrix diagram is for an on-delay timer. The term "reset" indicates that the timer clutch is disengaged and the rewind spring has turned the camshaft back to its starting point. At this point the timer control is reset, and the full amount of time delay is ready to run down when the timer begins timing.

The term "timing" indicates that power has been applied directly to the timer motor and the clutch has engaged the motor shaft to the camshaft. The length of the timing period will be determined by the preset value that was set at the face of the control by adjusting the hands on the dial.

The term "timed out" indicates that the timer motor has moved the camshaft to a point where the cam activates the switch contacts. When this occurs, the timer motor has been deenergized by contacts 11–12 opening. Any other load that is connected to normally open contacts 4–3 will be energized and any load connected to

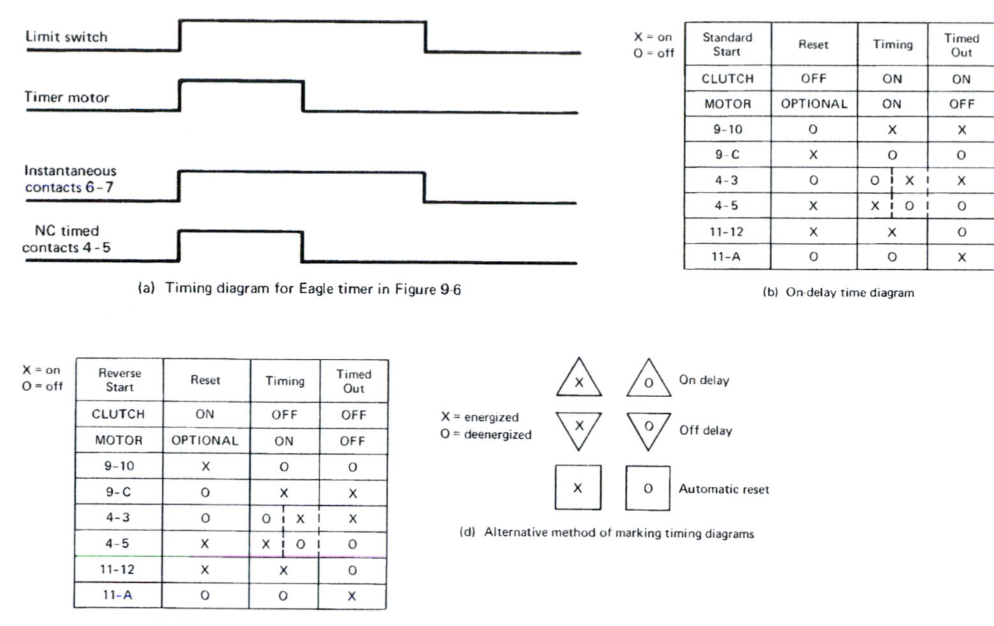

Standard Start	Reset	Timing	Timed Out
CLUTCH	OFF	ON	ON
MOTOR	OPTIONAL	ON	OFF
9-10	O	X	X
9-C	X	O	O
4-3	O	O ¦ X	X
4-5	X	X ¦ O	O
11-12	X	X	O
11-A	O	O	X

Reverse Start	Reset	Timing	Timed Out
CLUTCH	ON	OFF	OFF
MOTOR	OPTIONAL	ON	OFF
9-10	X	O	O
9-C	O	X	X
4-3	O	O ¦ X	X
4-5	X	X ¦ O	O
11-12	X	X	O
11-A	O	O	X

Figure 12–9 (a) A timing diagram for a motor-driven timer. (b) Matrix diagram indicating when contacts are on or off for an on-delay timer. (c) Matrix diagram indicating when contacts are on or off for an off-delay timer. (d) Alternative methods of marking timing diagrams.

contacts 4–5 will be deenergized. The timing device is ready for reset after this step.

If the device is an off-delay timer, the three categories at the top of the matrix will again be reset, timing, and timed out. In the off-delay device, power must be removed from the clutch solenoid to cause the clutch to engage the timer motor shaft to the camshaft. This means that during the time before timing, the timer motor is not energized and the clutch is not engaged. Once the input switch is opened, power is removed from the clutch and the timer motor is energized. This time is considered timer timing.

Figure 12–9a shows the timing sequence in a timing diagram. The on and off cycle of each part of the timer control are shown in relationship to each other. This type of diagram is also called a logic diagram, because it shows the outputs relationship to the input signals. Both of these types of diagrams are used extensively in timing diagrams and in specification sheets for timing controls.

The X and O symbols used in the matrix diagram (Figure 12–9b) are also used to indicate the timing sequence of the load in a timing diagram. Another type of diagram (Figure 12–9c) uses triangles located above the load to indicate that the load is controlled by a timing device. An X is placed inside the triangle to indicate that the load is energized, and an O is used to indicate that the load is deenergized. If the triangle is pointing up, it means that the load is controlled by an on-delay device. If the triangle is pointing down, it means that the load is controlled by an off-delay device. The triangles will be displayed in groups of three, which indicate the three phases of the timing control sequence from left to right: reset, timing, and timed out.

12.1.7 Timing Terms

You have seen the diagram of the normally open and normally closed timer contacts in the previous section. Some manufactures will refer to the movement of their contacts in slightly different terms and use some additional symbols on electrical prints to indicate the types of operation the timing contacts will provide. For example, Figure 12–10 shows four timing symbols that represent the four types of timers. Figure 12–10a shows the symbol for a normally open, timed closed (NOTC). This symbol uses the tail of the arrow, so it is a time-delay-on timer and the contacts are normally open and they will time close. Figure 12–10b shows a normally closed, timed open (NCTO). This symbol uses the tail of the arrow, so it is a time-delay-on timer and the contacts are normally closed and they will time open.

Figure 12–10c shows the normally open, timed open (NOTO) timer. The head of the arrow is used to indicate that this is an off-delay timer, and the contacts are normally open, timed open. At first this seems like a contradiction in that you can see that the contacts are drawn as though they are open, yet the timer is called a timed-open timer. The way to think about this type of timer is that it is normally open when no power is applied to the circuit. When power is applied to the circuit, the contacts will "pick up" and go closed, and when the power is turned off to the timer, it will start timing and the contacts will go back to their open state when the timer times out.

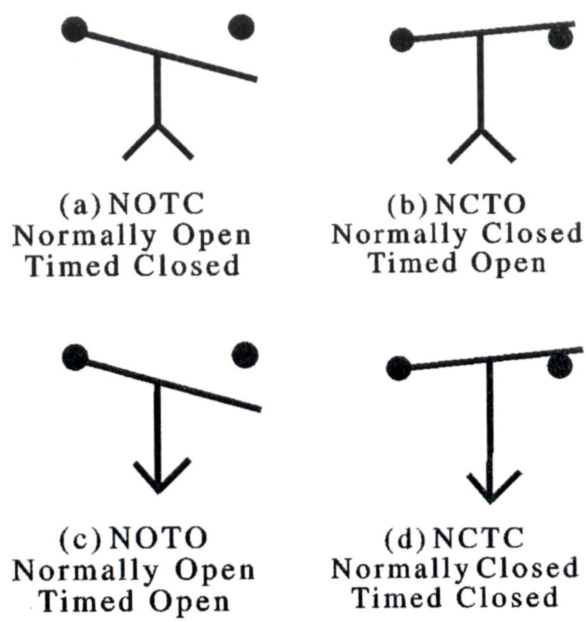

Figure 12–10 Symbols for timers.

Figure 12–10d shows a normally closed, timed closed (NCTC) timer. This timer is an off-delay timer since the symbol uses the head of the arrow. Again you may think the symbol is a contradiction, but the timer contacts start out as normally closed, and when power is applied, the contacts immediately go to the open condition; when power is turned off, the time delay starts and the contacts will move to the closed position.

Figure 12–11 shows examples of these fours types of timers in circuits. This time the triangles are added to the timing diagram. The timing conditions for the on-delay timer are reset, during timing, after timing, and reset again. The timer function is always described as starting in the reset condition when no power is applied to the circuit. The second set of timing conditions are for the off-delay timer and the terms are reset, before timing, during timing, and after timing. Again the term reset is used to indicate the period before power is applied to the timing cycle. Figure 12–12 shows a typical diagram for a motor-driven timer. The X and O symbols are used to indicate when a load is on or off, respectively. The three symbols are used to indicate the condition of the load before timing starts, during timing, and after the timer times out.

As you can see, there are many ways for the timing functions to be shown in diagrams and electrical prints. You may need to refer to this chapter to interpret the symbols in these diagrams when you encounter timers in diagrams on the job.

Figure 12–11 Symbols for NTOC, NCTO, NOTO, and NCTC timers as they would be found in typical timing circuits. Timing diagrams are provided for each type of timer.

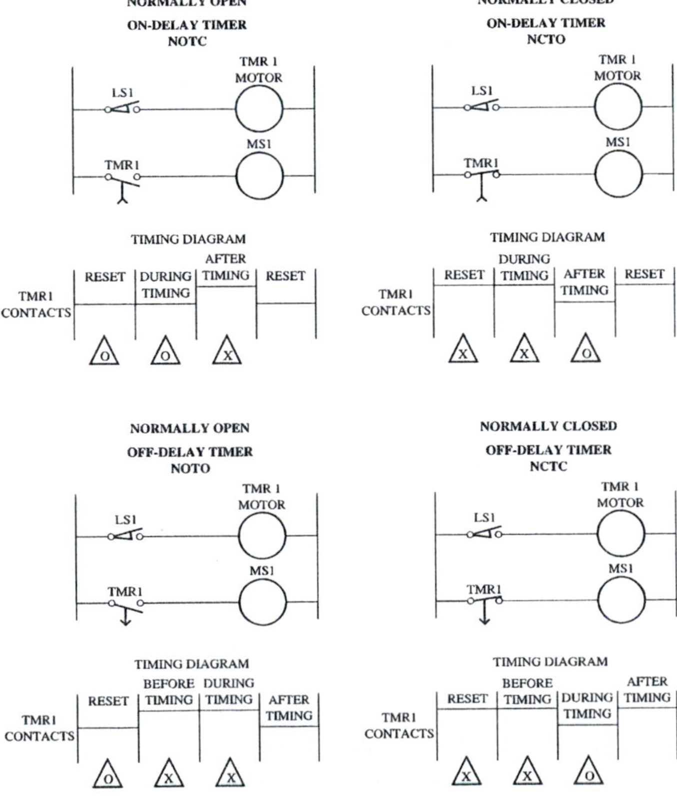

WIRING DIAGRAMS — Bold Lines are Internal Wiring

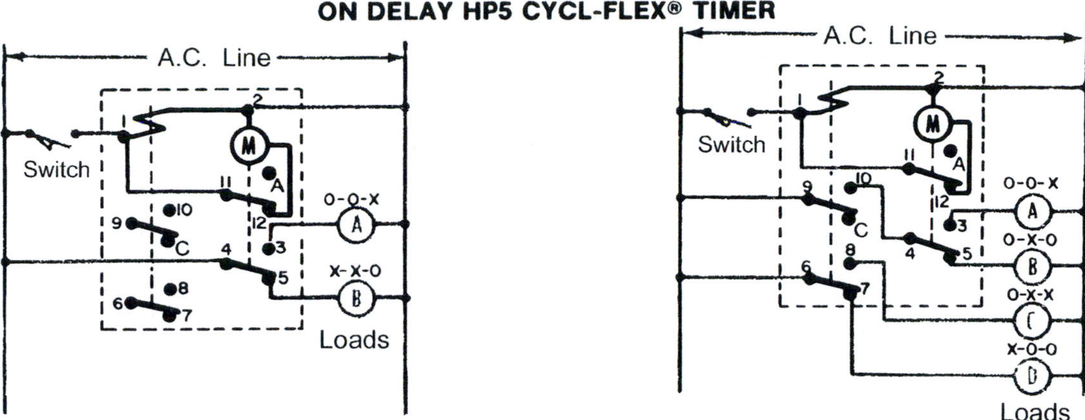

Sustained Control Switch. Close to Start, Open to Reset - Simple delayed closing and opening of load circuits.

Sustained Control Switch. Close to Start, Open to Reset - Additional load circuit operations obtained by connecting contacts in series.

Figure 12–12 Timing application diagrams for a synchronous motor-driven timer: (a) on-delay type; (b) off-delay type. (Courtesy of Danaher Controls)

12.1.8 Thermal Time-Delay Devices

A simple and inexpensive time-delay device is the thermal time-delay element. This element is made of two dissimilar metals that have a predetermined movement when heated. The two strips are bonded together during manufacturing and are mounted in the switch mechanism directly over a heating element. One end of the bimetal is secured so that the other end will move a maximum distance when current is applied to the heating element. When the free end of the bimetal strip has traveled its maximum distance, it will activate a set of contacts. These contacts can be normally open or normally closed. When current is removed from the heating element, the bimetal strip begins to cool down. As the bimetal is cooling, the end that is free to travel will begin to move back to its original location directly above the heating element.

The bimetal element will move when its element is heated and again after it has cooled. The bimetal is mounted in the switch in such a way that its movement will cause it to snap open or closed as the contacts are opening and closing. This will allow the switch to take advantage of the slow travel that the bimetal provides, yet have quick enough action at the contacts to keep them from pitting and arcing.

The NO or NC contacts are insulated from the bimetal element, which allows the contacts to be connected to a control circuit using pilot voltages. The time it takes the heating element to move the contacts far enough to open or close the contacts is determined by the amount of wattage the heating element uses. In some thermal switches the wattage is fixed, which causes the time delay to be fixed (usually 5 to 10 seconds). In other switches the heating element is a variable resistor that can be adjusted externally. If the resistor is variable, the amount of heat produced can be increased by adjusting the resistance to a lower value. From Ohm's law we know that if you keep voltage constant and reduce the resistance, the current will increase. When the current increases with the voltage staying constant, the wattage will increase. In this type of device the amount of time delay will be shortened as wattage is increased.

12.1.9 Solid-State and Digital Timing Devices

Since the advent of solid-state devices, solid-state time-delay controls have provided a variety of applications as add-on, stand-alone, and programmable devices. Figure 12–13 shows an add-on-type solid-state timer. These devices can utilize circuitry as simple as a variable resistor and a capacitor to provide the time delay or something as complex as a 16–bit microprocessor chip. The timing controls used today provide high degree of accuracy and repeatability, in the range of 0.1 second, and can be reset as quickly as 20 milliseconds for the next cycle.

The solid-state timing device can provide inexpensive control as add-on devices that work in conjunction

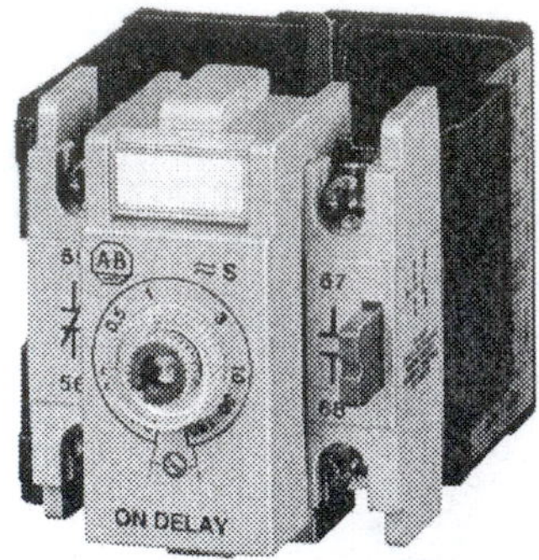

Figure 12–13 Solid-state timer element used as add-on timer.
(Courtesy of Rockwell Automation)

with standard relays, or as stand-alone devices that provide pilot-type control. More advanced models provide digital readout and digital preset instead of a dial-type indicator. These devices also provide a variety of preset times, up to 120 minutes, and timing intervals (time base) of 0.1 second. Time bases and preset time ranges can be selected by moving jumper wires within the device. With the use of microprocessors, digital timers can now be fully programmable from a touch pad. The programmable features include time-delay-on, time-delay-off, time bases, preset times, and other conditions, such as autoresetting and number of cycles.

Solid-state time control devices use several types of circuits to produce exact time delays. The simplest type of circuit uses a resistor and capacitor with other solid-state devices to trigger the base of a triac, the gate of a silicon-controlled rectifier, or a relay coil. The resistor and capacitor provide a time-constant circuit. When voltage is applied to the circuit, the resistor controls how fast the capacitor charges up. When the capacitor's voltage is large enough, it can be used to fire a variety of solid-state devices, such as a unijunction transistor. If a unijunction transistor is used, it will not fire until the voltage has hit the predetermined level that has been set by the size base 1 and base 2 resistors. Other electronic devices could be used in a similar manner. Generally, all of these devices are referred to as blackbox circuits, and all that is important is the output signal, which will turn on a triac or SCR.

A potentiometer (pot) is usually used as the resistor in this type of circuit. The size of the pot will determine the maximum amount of time delay. As the pot is adjusted, a smaller amount of resistance is put in the circuit with the capacitor, which causes the time delay to become shorter. The adjustment knob for the pot is usually accessible on the time-delay device so that the time delay can be adjusted while the device is still in the circuit.

12.1.10 Cube Timers

The time-delay control shown in Figure 12–14 is called a *cube timer*. It is a very accurate way of providing time delay to a load such as a small motor or to a motor starter coil for larger motor control. This type of device can be used in both an AC and a DC circuit without modification. This is possible because a triac is being used as the solid-state control device and a triac can control either AC loads or DC loads. If the timer control was rated for DC circuits only, it would be using an SCR as the solid-state device. Some of these types of devices allow the resistor for the time constant to be mounted remotely from the control in an operator's panel. In this way the operator can adjust the exact amount of time delay that is required for the operation. This is especially useful for batch mix timers or for plastic press operations. The load can be either 110 or 220 V. Again this is possible because the electronic solid-state circuit uses a voltage regulator circuit to set the incoming voltage to the time-delay part of the circuit at a constant level regardless of the incoming value.

Figure 12–15 also shows several other diagrams for the cube timer. It can provide either two-wire or three-wire control. If two-wire control is used, it will get the supply voltage for its electronic circuit through the pilot device and the return part of the circuit will be completed through the load. The time delay will begin when the pilot switch is closed. This also means that if the load fails to open or the pilot switch does not close, the timer will look like it is inoperative.

If the three-wire device is used, its timing circuit is connected across the line at terminals 2 and 3. The pilot devices that will initiate the timing circuit will be connected across terminals 6 and 7. The load is connected between terminal 1 and line 2. The operation of the timing begins when the circuit between terminal 6 and 7 has continuity. Notice that no external power is applied to these two terminals. The terminal where the load is connected will be capable of controlling loads up to 1 A. Remember that since a solid-state device is used as the switch, leak current of up to 2 mA is normal.

12.1.11 Adjusting the Time Delay of a Cube Timer

The time delay for some applications must be adjusted during installation and is not routinely readjusted during machine operation. In these applications the manufacturer will provide a formula to calculate the amount of resistance required for the specified time delay. The resistance can then be set and measured with an ohmmeter to give the preliminary setting. Trial-and-error tests will provide the fine adjustment.

Q1 Series

Operating Logic:
Upon application of input voltage, the time delay starts. At the end of the time delay the load is energized. Reset is accomplished by removing input power.

Note: 1) The load may be located on either side of the line. 2) R_t and terminals 4 and 5 are used for external time adjustment.

Logic Function Diagram

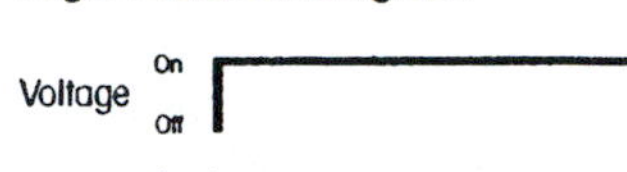

(b)

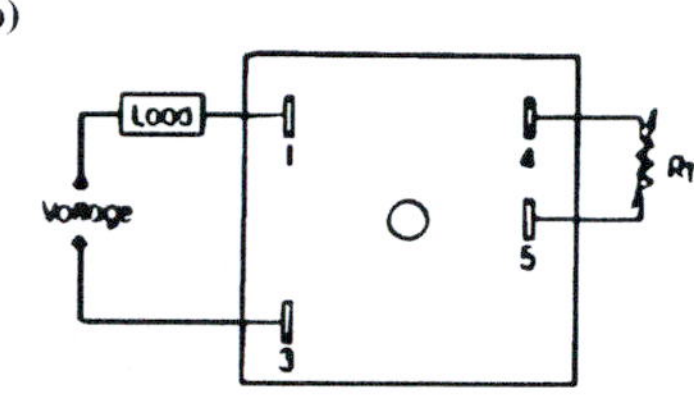

External Resistance/Time Delay Relationship
1 megohm external resistance is required to obtain the maximum time for all ranges. To determine the actual resistance needed to obtain the required time delay, use the following formula.

$$R_t = \frac{T\ req - T\ min.}{T\ max - T\ min.} \times 1{,}000{,}000\ \text{ohms}$$

Note Due to component tolerances, the actual time obtained will normally be within 5% of desired time

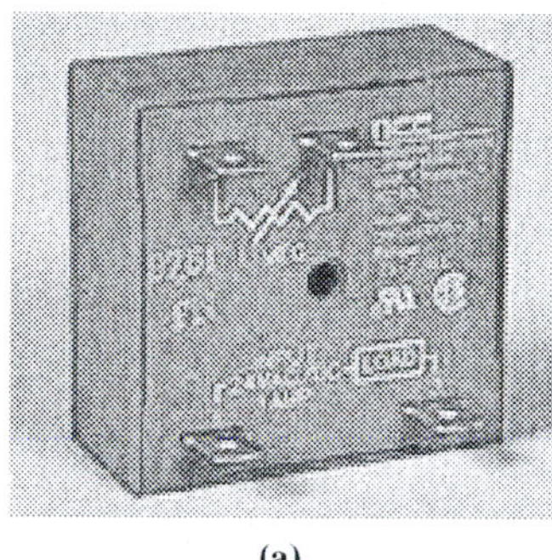

(a)

(c)

Figure 12–14 Cube-type timer: (a) picture; (b) electrical diagram; (c) timing diagram. (Courtesy of AMETEK National Controls Corporation)

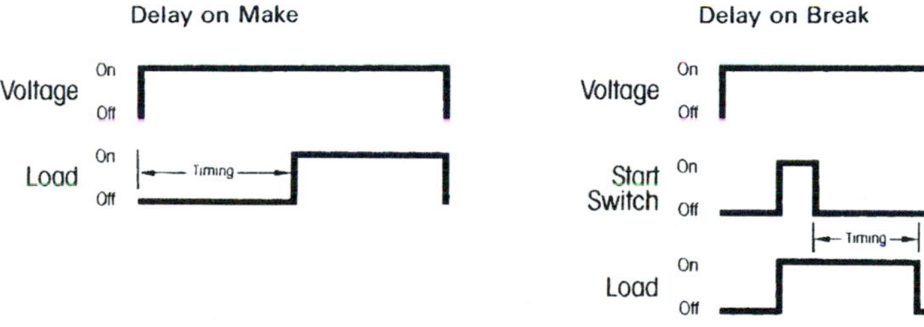

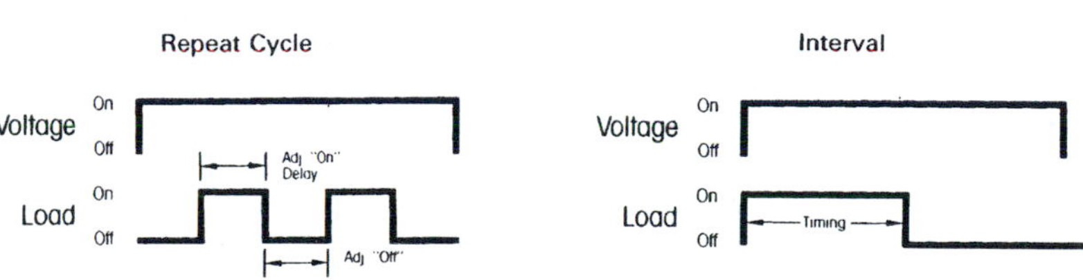

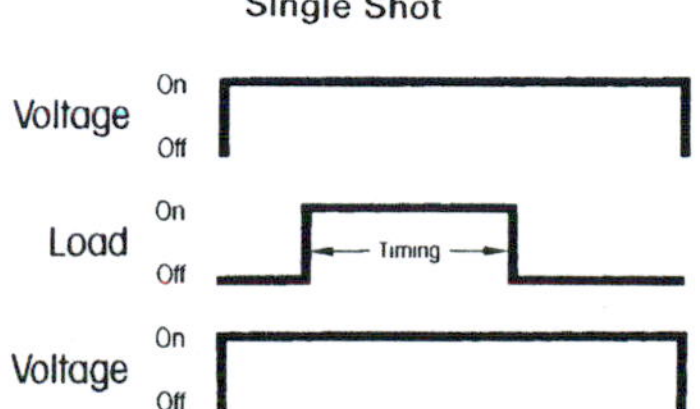

Figure 12–15 Electrical diagram of delay-on make, delay-on break, repeat cycle, interval, and single-shot timers.

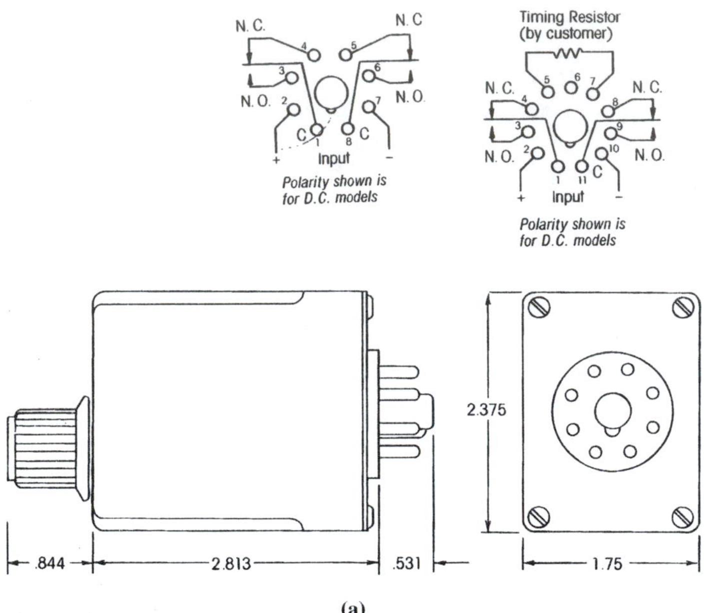

(a)

(b)

Figure 12–16 Octal base-type plug-in timer: (a) diagram; (b) picture. (Courtesy of AMETEK National Controls Corporation)

12.1.12 Eight-Pin (Octal)-Base Time-Delay Relays

Another type of solid-state timer is enclosed in a plastic case with eight pins protruding from the bottom. These eight pins allow the timer to be plugged directly into an eight-pin base. All field wiring is connected to the base, and if any problems are encountered with the timer circuit, the complete timer can be exchanged in several seconds. This is a very big advantage in industrial applications, where downtime must be kept to a minimum. Figure 12–16 shows a picture and a wiring diagram of this type of solid-state timer.

The operation of this type of timer control is very similar to that of the cube timer. A resistor and capacitor circuit controls solid-state devices that switch on and off for the output control. The output can also use a single-pole, double-throw relay. The time base can be set for divisions of 0.1 and 1.0 second, which provides time delay up to 10 hours. These controls can have an internal potentiometer for adjusting the amount of time delay. The knob and dial on the top of the control show the approximate amount of time delay. Final adjustment of the time delay can be made by trial and error.

The input to start the timer can be a pushbutton switch. Any type of momentary pilot device would work as well. Several switches can be mounted in parallel or series for additional logic if necessary. These terminals should be connected to voltage or ground circuitry. Since they are looking for a reduction in resistance, which will be supplied when the momentary switch contacts close, the switch can be mounted remotely.

The output of the timer can control up to 10-amp resistive loads or could control the coil of large motor starters, which could start any size motor. A similar model of time-delay control has 11 pins that plug directly into a base. Again, as in an eight-pin base, all field wiring is connected to the base so that the time-delay device can quickly be unplugged and replaced. The three extra pins in the 11-pin device provide connections to a second single-pole, double-throw set of contacts. Another model of this type of timer uses slip-on clip connectors as terminals. These also provide easy removal and replacement of the timing control when it is suspected of operating incorrectly. Figure 12–15 shows examples of available logic outputs for their timers.

12.1.13 CMOS Solid-State Controls

Newer models of time-delay controls use complementary metal oxid semiconductor (CMOS) circuitry to provide accurate and repeatable time delays. The CMOS control uses a crystal oscillator that is sent to sets of divider networks. The divider networks are used to divide the crystal frequency into increments that will be used to provide a variety of time bases for the timing control. Newer designs have incorporated this circuitry into existing timer control housing and packages so that they can easily be interchanged and mounted on existing bases for use with standard electromechanical timer controls.

The crystal and CMOS circuits are very reliable because of their low power consumption and their repeatability of millions of operations. This type of timing control is also reliable because of its construction on an integrated circuit (IC), and has been used in computers to provide clock and timing pulses for many years. Integrated-circuit, or chip, construction has allowed high-quality controls to become available at a much lower price. This type of construction allows the timing device to be more compact at a time when space in the electrical control panel is a major concern. These devices have also proven to be easily interfaced to other electrical mechanical control circuit devices, such as relays, solenoids, and motor starters.

One problem of early CMOS circuits was their inability to exist in the industrial electricity environment where uncontrolled voltage spikes and strong magnetic fields caused undesirable operation or circuit failure. The CMOS circuits that are used in the newer industrial controls have been refined to a point where they are isolated from the voltage transients and magnetic fields. This allows them to be mounted in the same electrical cabinets as their electromechanical counterparts.

12.1.14 Examples of CMOS Solid-State Controls

Solid-state controls have been incorporated for use in some traditional applications, such as add-on devices. The original add-on time-delay control used a pneumatic actuator. The solid-state add-on relay uses solid-state circuitry to control the time-delay portion of the control. The outputs for this control are sealed contacts that will allow the control of either AC or DC loads.

Since the timing circuitry is encapsulated, it is immune to vibrations, humidity, and harsh environments with atmospheric contaminants. These devices use separate models to provide on-delay and off-delay control. A wide range of delay times is also available by selecting different models.

Solid-state time-delay controls are also available to fit into the same housing as that of the synchronous motor-driven timer. This allows the newer solid-state technology to be incorporated in existing installations with a minimum of changes. This type of control has physical dimensions that exactly fit the synchronous motor-driven time-delay wiring base. The solid-state device uses CMOS circuitry to provide accuracy and repeatability to 0.1 second. Reset time is also decreased to several milliseconds for repeat operations. The preset time is set through the use of three thumbwheels, representing units of ones, tens, and hundreds, which allow preset times up to 999. By moving internal jumper wires, these three values can represent several timing ranges. The timing ranges are 0.1 to 99.9 seconds, 1 to 999 seconds, and 0.1 to 99.9 minutes.

Three seven-segment light-emitting diodes (LEDs) are used to display the time remaining. This display will be synchronized with the time base that was selected with the jumpers. The digital readout on the display allows for the control to be observed from a greater distance than with a dial readout. This type of timer control is also available without the digital readout for applications where the timer control may be mounted inside an electrical cabinet.

Time-delay-on and time-delay-off functions are set with internal jumper wires. This versatility allows the user to stock only one timer control and get the functions of time-delay-on or time-delay-off with multiple time bases.

12.1.15 Microprocessor-Controlled Timers

Microprocessor chips have become very usable in industrial control because of their size, versatility, and price. They have been incorporated into timer controls to provide a variety of timer ranges and operating logic that is selectable in one device. This control allows the user to select the timer range with a selector switch on the front of the control. The range selector allows the timer device to be set for ranges of 0.05 to 9.99 seconds, 0.1 to 99.9 seconds, 1 to 999 seconds, 0.1 to 99.9 minutes, and 1 to 999 minutes. The timer has five different logic modes that are also selectable from the front of the control: repeat cycle with 50-percent fixed duty cycle, single shot, delay-on break, interval, and delay-on make.

The preset time is dialed into the timer control through digital switches (thumbwheels) on the front of the control. An LED is also located on the front of the control to indicate when the timer is timing. This control has an eight-pin connector on the bottom that will plug directly into any octal base control. The timer operation will be initiated by the pilot device connected between terminals 5 and 6. The timer's operation will be identical to that of similar solid-state timer controls.

The microprocessor-controlled timer allows for a maximum number of functions in one control, which will allow industries to stock one control for all their timer functions. Since the timer range and logic control is user selected, this control can be used for all existing applications. It is also easily changed out during downtime because of its pin and socket mounting. All of these features are making the microprocessor timer the control of the future.

12.1.16 Programmable Timers

New advances in technology have allowed microprocessors to be incorporated into programmable timer controls. The programmable timer provides timer control for continuous periods of time up to 365 days. This control is very similar to the 24-hour time clock controls used in business lighting and security systems. The programmable timer can control up to 16 circuits and use 60 separate programs. Each program can control any or all of the 16 outputs. Typical applications for this type of control are for water treatment facilities to add water chemicals, small business control of lighting, security and fire protection, and energy management systems. Each program can be set to repeat at differing intervals, such as hourly, daily, weekly, monthly, or annually. The programmable timer can also account for differences in the program cycle caused by holidays and changes for daylight savings time and even leap year.

The timer is programmed from a keypad and the program can be displayed on a printer for editing. Changes can be made from the keypad or from a modem (modulator/demodulator) over ordinary telephone lines. This allows for the timer to be installed at one location and serviced from another location. The program is backed up by batteries in case of power failure. The outputs that are controlled by the timer can control resistive loads up to 15 A and inductive loads up to 2 hp at 250 V AC. Standard pilot-duty loads can also be controlled, which allows the programmable timer to be used in conjunction with other control devices for a wide variety of applications.

12.1.17 Applications for Timers

Many applications for timers are in use in industry today. Some of these circuits allow the timing device to be interfaced with robots and other automated systems. One example includes a robot in a loading and unloading application. The robot will transfer parts from the end of the first conveyor and place the parts in a mill for machining. When the mill is finished with the parts, the robot is signaled to pick up the part and place it on the second conveyor. Each conveyor is controlled by a stand-alone cycle timer activated by the robot.

When the robot has picked up a part from the first conveyor, it will signal the timer to jog that conveyor for 2 seconds to bring the next part into position. After the robot places the part on the second conveyor it will send a signal to the timer that controls that conveyor to jog it for 2 seconds to move the parts ahead, which will provide adequate space for the next part to be placed.

Stand-alone timers are used so that maintenance personnel can adjust the timers slightly without getting into the robot program. If the timers in the robot were used, the robot program would need to be edited each time an adjustment had to be made. This jogging technique causes the parts to be placed on the conveyor with the same space between each part. This accurately locates the parts for the next robot that must handle the parts to place them on a pallet.

12.1.18 Testing the Timer Control

After the timer control has been installed, it must be field tested and adjusted. The first step in this process is to check the load portion of the circuit to make sure that any unwanted signal will not harm personnel or damage equipment. This may be accomplished by turning off the main disconnect to the motor or hydraulics that the coils and solenoids may activate. If the timer is controlling a motor starter, you will still be able to see the motor starter coil pull in, even though the motor will not turn when the disconnect is open, and if the timer turns on the motor starter at the wrong time, it will not damage the motor or other equipment.

After the equipment safety check is made, power may be applied to the control circuitry. At this point, close the enabling switch to the timer and check to see that it is operating in the on-delay or off-delay mode as you have designed it. Also check the operation of the instantaneous contacts to make sure that the loads are connected to the proper set of normally open or normally closed contacts as indicated in the circuit diagram. Do not be overly concerned about the amount of time delay during the first test.

After the first test has been completed and you are satisfied that the control is operating as the control circuit diagram indicates, trials can be made to adjust the time delay. You can use a watch that displays seconds to time the device. If a remote potentiometer is used to set the time delay, it would be helpful to calculate the amount of resistance required for the amount of time delay that you desire. Adjust the resistance in the potentiometer with an ohmmeter prior to the first trial. Additional trials can be used to make final adjustments.

After you are satisfied that the timer control is energizing all its load to the proper time, you can apply power to the motor disconnects and allow the complete application to run through a full cycle. During this test, be ready to disconnect power or hit the emergency stop buttons if any unwanted machine motion occurs. After you have completed these tests, be sure to add any notes to the electrical diagram, indicating any changes or adjustments that will be useful during troubleshooting. If you encounter any problems during these tests or at any time after the system has been operating for some time, use the troubleshooting instructions to aid in your diagnostics.

12.1.19 Troubleshooting the Timer Control

The most important tool in diagnosing problems with any automated controls or circuitry is the knowledge of how it is supposed to operate when it is working correctly. At times you may need to review the theory of operation for the particular control that is malfunctioning. This will help you remember how it was designed to operate, and then through observation and electrical testing you can determine any differences. You should also review the electrical diagrams for the circuit prior to beginning the diagnostic procedure to see if the device is wired according to the field wiring diagram, and determine what the sequence of operation should be.

The diagnostic tests for timer controls will be grouped according to the type of control. Since there are some similarities, there will be some duplication in the tests. To use these tests effectively, you will need to determine which type of device is controlling the circuit. You must also determine if the control is wired for a delay-on or delay-off application. Use only the tests that apply to your device and application.

Some problems will be unique to a newly installed timer control, such as being wired improperly. If the device has failed after it has been operating correctly for some time, you may not need to worry about the device being wired incorrectly, and you can concentrate on a failed or broken component in the timer control. The timer control can operate correctly only when it receives the proper amount of voltage. If pilot switches are connected in series with the timer, they must be energized and supply the control with the proper amount of voltage for it to work correctly. The final point to remember is that the timer can only switch voltage to the load at the proper time. If the timer switches the proper amount of voltage to a load at the proper time it is considered to be operating correctly. If the electrical load (relay, solenoid, or motor starter) does not operate after the timer has switched power to it, then the electrical load should be checked instead of the timer.

12.1.20 Troubleshooting the Solid-State Timer Control

Solid-state timer controls are very easy to troubleshoot. They essentially operate as a switch that is in line with the load being controlled. They require voltage to the operational part of their circuit and will act as a switch from L1 or the positive DC line to the load. These devices have a variety of field wiring assemblies, such as the octal base or push-on terminals. In some cases it will be faster to remove and replace the device from its base than to make elaborate field tests. At one time removing and replacing components without field testing was considered a poor practice. In today's industrial applications, the time that equipment is malfunctioning and not operating may cost several thousands of dollars per hour. In these cases it is prudent to keep a variety of spare parts and simply try removal and replacement as the first order of operation.

Many solid-state controls are encapsulated and cannot be repaired. They are relatively inexpensive and are easily replaced. This means that most of the tests in troubleshooting these controls involves testing for the presence of the proper amount of voltage at the supply terminals of the device and checking to see if the voltage is switched to the proper output terminals after the amount of time delay has occurred. Since most of the devices have a variable time adjustment, minor changes in the amount of time delay can be compensated for by small adjustments during trial-and-error timing tests.

12.1.21 Troubleshooting the Synchronous Motor-Driven Timer

The electromechanical motor-driven timer is also very easy to troubleshoot. You may need to review the theory of operation prior to making these tests. You will also need to check the wiring diagram to see if the control has been wired as a time-delay-on or a time-delay-off control. Remember that the off-delay and on-delay timer applications are achieved by using two different models of timer controls.

The timer motor and clutch for this type of control require power from lines 1 and 2 for AC control and from the positive and negative lines for DC control. Any number of pilot devices can be used to enable timer control. These switches may include photoelectric switches, proximity switches, and limit switches. To diagnose problems correctly with this type of control, you

COUNTER OPERATION

INSTANTANEOUS CONTACTS 6-7 A-C are open and 6-8 A-B closed when the counter is in reset. Contacts 6-8 A-B are open and 6-7 A-C closed when the counter is in the counting or counted out position.

DELAYED CONTACTS 4-3 and 9-10 close, 4-5 and 9-11 open when the red progress pointer reaches zero. Contacts 4-5 and 9-11 close, 4-3 and 9-10 open when the counter is reset.

NEON PILOT LIGHT is built into dial to indicate counter clutch coil is energized.

SWITCH	RESET	COUNTING	COUNT OUT
6 - 7	O	X	X
6 - 8	X	O	O
A - C	O	X	X
A - B	X	O	O
4 - 5	X	X	O
4 - 3	O	O	X
9 - 11*	X	X	O
9 - 10*	O	O	X
HZ170A6 Clutch Coil	Deener.	Ener.	Ener
HZ170A601 Clutch Coil	Ener.	Deener.	Deener.
Switches trip to count-out position on last deenergized stroke of count solenoid.			

X — Switch Closed O — Switch Open

SCHEMATIC DIAGRAM

Standard Clutch

CLUTCH COUNT

1 2 12 6 7 8 A C B 4 3 5 9 10 11

INSTANTANEOUS CONTACTS DELAYED CONTACTS

*THIS SWITCH IS NOT FURNISHED ON HZ171 (12 COUNT) COUNTER

Reverse Clutch (01)

CLUTCH COUNT

1 2 12 6 8 7 A B C 4 3 5 9 10 11

INSTANTANEOUS CONTACTS DELAYED CONTACTS

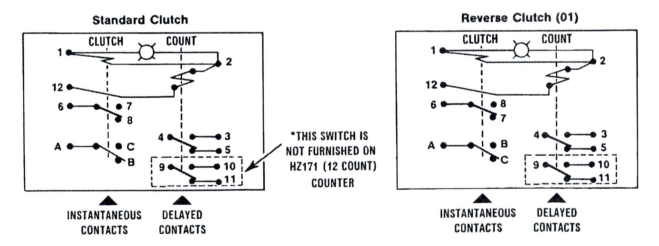

Figure 12–17 A typical counter with its wiring diagram.
(Courtesy of Danaher Controls)

may also need to observe and test the pilot devices that enable the timer and the electrical loads that the timer output controls. Another troubleshooting technique involves starting at the electrical load end of the circuit and work backward to the time control. If you would rather use this method of troubleshooting, the tests in each procedure should be executed in reverse order for each timer control.

12.2 Counters

Counters are used in industrial control systems to count (totalize) parts being made and to keep track of sequential steps in machine operations, such as in plastic presses and punch presses. Electromechanical counters have been in use for many years. Recently, solid-state components and microprocessor technology have been integrated into counters. These advances have made counters easy to install, operate, troubleshoot, and replace. In this section we explain the operation of various types of counters, such as cycle counters, used to control sets of contacts, and totalizing counters, used to keep count of parts being made. Typical applications, installation, and troubleshooting are also covered.

12.2.1 Mechanical Counters

The simplest type of counter, which has been in use for many years, is the mechanical counter used to keep track of or totalize the number of parts being made. The internal operation of this type of counter is still used on the electromechanical types of counter that you will find in industry today. A mechanical counter consists of a set of wheels with the numbers 0 through 9 listed on them. As parts move past the counter arm, the arm is moved to trip the counter and provide an input to the counter clutch. When the arm moves to register an input, each wheel rotates and causes the number to show up in a window. As many as four of these wheels are placed side by side to display counts up to 9999.

Each wheel is connected to the one next to it by gears. The first wheel will display the ones column. The wheel that is next to the ones column will display the tens column. When 10 input signals are counted, the gears from the wheel displaying the ones column will cause the wheel displaying the tens column to move ahead one count. If you were looking at the counter display it would show a 10, indicating that a total of 10 counts had been registered. As each new count is received the counter will display 11, 12, 13, and so on. After the counter receives 10 additional inputs, the gears on the ones column wheel will move the tens column wheel ahead another count so that it is displaying a 2. The count will now be displaying 20. Each additional input will now display 21, 22, 23, and so on.

The wheel that is displaying the tens column will continue this operation until it reaches the value 9. After 99 counts are received by the counter, the gear on the wheel displaying the tens column will move the wheel next to it ahead one count. This will cause the counter to display the value 100. After 999 counts have been received, the wheel that displays the hundreds value will engage the gears of the wheel next to it and the counter will display a value of 1000. This will continue until a value of 9999 is counted. At this point, the next count that is received will cause the counter to reset to the value of 0000. There will also be a reset button on the counter that will allow it to be reset at any time. When the counter is reset, a spring returns all the numbers in the window to zero. This type of counter is used to provide the total number of parts that move past the arm. The total count will be displayed in the window, where personnel view the count and write it down.

12.2.2 Electromechanical Counters

Another type of counter, called a *reset* or *cycle counter*, operates on electromechanical principles. The signal that causes the counter to advance one count comes from an electrical signal instead of through mechanical movement of an arm. This allows devices such as proximity switches, limit switches, and photoelectric switches to be used to provide the input signal. This type of control is generally used to count a specified number of input signals (counts) and to activate its outputs. This type of counter control has both instantaneous and controlled contacts.

Figure 12–17 shows a picture and diagram of this type of counter. From the picture you can see that the counter is mounted in a housing that is identical to the timer controls. This allows the timer and counters to share a common field wiring base, which cuts the cost of stocking spare parts. The counter face is also similar to the timer control, which makes the control easy to read.

From the electrical diagram you can see that the counter has a clutch and motor. There are also several sets of single-pole, double-throw contacts. Two sets of the contacts are instantaneous and are controlled by the clutch. The other sets of contacts are single-pole, double-throw, which are controlled by the counter control. These contacts are actuated when the counter reaches the specified number of counts.

The clutch is energized by voltage applied to terminal 1. The count signal is input to terminal 12. When the clutch is energized it engages the gears so that the count signal will be acknowledged. This type of control uses a dial with numbers printed on its face instead of wheels with numbers 0 to 9. Two hands are used to indicate the preset count and the actual count.

A preset value is set into the counter by adjusting the preset hand on the face of the control. This value can be any value between 0 and 100. Each time a count is registered the counter subtracts one count from the preset value by energizing the counter motor and causing it to advance one step on the hand that is indicating the actual count. This causes the counter to decrement the count (subtract one count from the preset value) each time an input signal is received. When the counter counts down to zero, the output contacts are activated. This type of counter is called a *down-counter*. While the counter motor is decrementing the count, it is also winding the reset spring. This spring will cause the hand that is indicating the actual count to return to the preset value when the clutch is deenergized. The counter can be reset to the preset value at any time by deenergizing the counter clutch.

This type of counter is also available as an *up-counter*. In this type of counter the actual count starts at zero and increments (adds 1) when an input signal is received. When the actual count equals the preset count, the contacts will be activated.

This type of counter control is generally used to keep track of input signals and control outputs. The counter is used in conjunction with a palletizing machine. A photoelectric switch detects boxes as they pass by the end of the conveyor to be stacked on each pallet. This switch inputs a signal to the counter, which will energize the wrapper solenoid after four boxes have been stacked on the pallet. The wrapper solenoid energizes a plastic wrapping machine that wraps in plastic the pallet containing the four boxes. When the palletizing machine is changed to place six boxes on each pallet, the counter can be adjusted to accommodate this change by setting the preset count to 6. Each change to the palletizing machine can be made by changing the counter preset.

12.2.3 Solid-State Counters

Several types of solid-state counters are available for use in modern industrial control systems. These include up-counters and down-counters that will energize outputs when the preset count is reached, and totalizing counters that keep track of the total number of parts that have been made. These types of counters utilize CMOS solid-state circuitry to execute the counter function. The counter has a thumbwheel on the face of the counter that is used to input the preset count value. Machine operators or production engineers can dial in the preset count by using the thumbwheels. This type of counter will be used as an up-counter. When the counter has recorded the number of input signals (counts) that is equal to the preset value, it will energize its outputs.

The counter control uses four lines that operate the counter. One of these lines is connected to terminal B, which is the electrical common for the control. The other lines are called *counter start*, *counter input*, and *inhibit input*. The counter start line is used to enable the counter control, similar to the action of the clutch solenoid in the electromechanical counter. When this line receives voltage, it will enable the counter control to record any count signals that it receives at the counter input line. Whenever this line is deenergized, the counter is reset to zero.

The counter input line is sensitive to the on-to-off transition of the input signal. This means that the input signal will be registered as a count when it transitions from full voltage to turning off. When the actual count becomes equal to the preset count, the counter's outputs will be activated. This means that any set of open contacts will be closed and any set of closed contacts will be opened. This type of counter has two sets of single-pole, double-throw contacts.

The inhibit input line can be used to cause the counter to ignore signals that are received on the count input line. This line is especially useful to deenergize the counter from outside controllers, such as robots, programmable controllers, or press controls. The internal operation of this counter utilizes existing CMOS chips that can increment or decrement a count. When the preset value is reached, the circuit activates its outputs. The outputs can be triacs or relay outputs.

Another type of solid-state counter utilizes a large-scale integrated (LSI) circuit. The LSI chip provides a counter with a thumbwheel input to set the preset count and LED seven-segment displays to show the current count. This type of counter is also available without the LED display, for applications where the control may be mounted inside a cabinet where the face cannot be seen. This type of control has preset ranges up to 9999 counts. It can read input signals of 40-millisecond on duration and 60-millisecond off duration as counts. This allows the counter to be interfaced with photoelectric and proximity controls for high-speed counting applications such as bottling lines and packaging lines. The input line can receive up to 1000 counts per minute. It is designed to fit the standard timer/counter mounting case, which makes it easy to install. This type of installation also makes the control easy for panel mounting.

From these diagrams you can see that control has two sets of single-pole, double-throw relay contacts that are activated when the actual count decrements to zero. These contacts are called *delay contacts.* There is also one set of single-pole, double-throw instantaneous contacts that are energized when power is applied to terminals 1 and 2. This counter is reset immediately after the actual count reaches zero, or any time that power is removed from the counter control relay (CR1).

12.2.4 Programmable Timer/Counters

Programmable timer/counters allow the user to purchase one control device that can be programmed to be a counter or timer. There are also several programmable options, such as up-counter or down-counter, counter ranges, and types of outputs. If the control is programmed as a timer, options are provided for timing ranges, time bases, time up or time down, and types of outputs.

As a programmable counter, the control provides a wide variety of options, which makes it very versatile. This allows industries to purchase only one model of control for all their timer and counter applications, which saves money and time when the control must be kept in stock as a replacement part.

The programmable counter uses a microprocessor chip similar to those used in computers. This chip provides the wide range of programmable options that makes the control so versatile. The programmable functions also provide the same options as those found in most programmable controllers, but is much cheaper, thus providing the greatest function at the lowest price.

The control has a seven-segment LED display that can show up to four digits. The front of the control also has a membrane keypad that allows set points to be programmed into the control right on the factory floor. This is accomplished by pressing the set key and then the up or down arrow to increase or decrease the value in the display. When the desired value is attained, the enter key is used to store this value in the control's memory. These controls can use a battery to provide memory failure protection during a power outage.

The control options are programmed or selected by setting dip switches on the main circuit board. The control must be removed from its housing to obtain access to these switches. These switches can also be changed on the factory floor to provide optimum use of the control.

Operation of the programmable counter is very similar to that of other counter controls discussed in this chapter. The control CR1 coil acts like a clutch coil, in that it controls the instantaneous contacts. The control also has an input line between terminals B and C that accepts the count signals. One of the terminals is connected to the common and the other is connected to the device, such as a proximity or photoelectric switch that is providing the count pulse. The third part of the control is the CR2 coil, which controls the delay outputs. CR2 energizes these output contacts after the counter has recorded the number of counts entered in the preset value. The controlled outputs are also called the *delay outputs.* These contacts are numbered 11, 12, and A for the first set and 4, 3, and 5 for the second set. The contacts can be used as outputs to control loads or can be wired to provide several variations of logic for the counter itself, such as automatic resetting. The contacts are rated to carry up to 10 A at 120 VAC. This means that the control can be used to drive small loads directly, or as a pilot for relay coils or motor starter coils for larger loads.

The counter will accept a preset value programmed into its keypad. This value can be retained in the counter's memory through battery backup. The battery is generally good for approximately two years of service even though it is rechargeable. It will ensure that the control will maintain its program even in the event of a power outage. The counter control will be enabled once power is applied to the CR1 relay. This enable condition also activates all sets of the instantaneous contacts. After the control is enabled it will register any signal on the counter input line that exhibits an on-to-off transition of 40 milliseconds or greater as a legitimate input signal (count). Every time the signal transitions from on to off, another count is registered in the counter control's memory.

After the number of counts in the control's memory equals the preset value, the delay contacts are energized. At this point the control will reset automatically. The time required for resetting is approximately 30 milliseconds. The reset can also be executed any time by removing power from CR1. Output diagrams can be shown with X's and O's to indicate when they are receiving power and when they are deactivated.

12.3 Sequencers

Sequencers are used in a variety of automated control systems where events must occur in a predetermined sequence. This type of control is usually an electromechanical switch that has multiple cams and contacts that are opened and closed as the switch's shaft is rotated with an electric solenoid. Figure 12–18 shows an example of a sequencer switch. In this section we explain the operation of solid-state sequencers and how they are used in control systems. The sequencer shown in Figure 12–18 is called a *step switch.* It has 19 sets of contacts that can be open or closed during the 16 steps that the cam actuates during one rotation. The number of actual steps the cam uses can be adjusted by chang-

SYSTEM OPERATION	CONTROL DEVICE TO ADVANCE TO NEXT STEP	STEP	LOAD SWITCHES ON STEP SWITCH CLOSED = X OPEN = O						
			1	2	3	4	5	6	7
Off	Pushbutton	1	O	O	O	O	O	O	O
Solenoid valve no. 1 energized. Hydraulic Ram hits LS1 at end of movement	Limit Switch No. 1	2	X	O	O	O	O	O	O
Solenoid valve no. 1 energized. Timer holds ram in position for adjustable time.	Timer No. 1	3	X	X	O	O	O	O	O
Solenoid valve no. 1 energized. pump and timer no. 2 starts.	Timer No. 2	4	X	O	X	O	O	O	O
Solenoid valve no. 1 energized. Solenoid valve no. 2 energized. 2nd ram hits limit switch 2.	Limit Switch No. 2	5	X	O	X	X	O	O	O
Design the control circuit by making a chart as above. List the system operation step-by-step, indicating which load circuits are closed in each step. Next, list the control device which is to terminate each step and to advance the switch to the next position. The chart above is easily translated into the electrical circuit diagram shown to the right. The input signal devices are connected to their respective tap switch terminals. The load circuits are connected to the switch terminals indicated by the chart.		LOADS	Solenoid Valve No. 1	Timer No. 1	Pump and Timer No. 2	Valve No. 2			

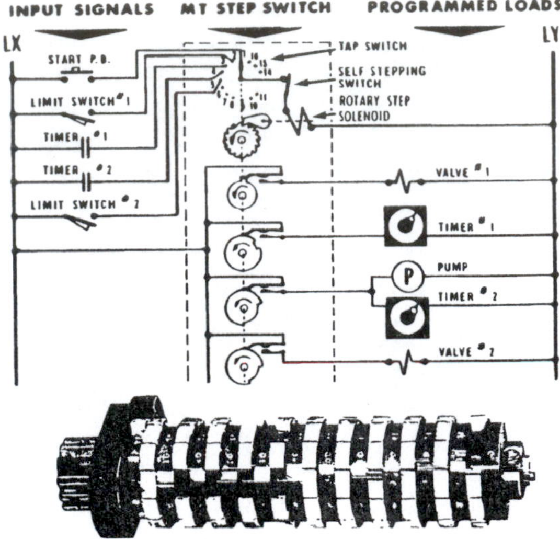

The cam shaft can be removed by loosening a lockscrew and removing a stud shaft in the right end plate. Another cam shaft with a different program can be inserted or the cams may be reprogrammed on the existing shaft. The cams are keyed so they must be correctly assembled on the hex cam shaft.

Figure 12–18 (a) Sequencer switch that controls electrical loads in a sequential operation. (b) A diagram of the matrix showing the sequence of operation provided by the sequencer switch.

ing the programming pins in the cam. This switch can control 120- or 240-V devices at up to 10 A for the lower voltage and 5 A for the higher voltage.

The switch requires a one tenth-second pulse as an input to advance the cam to the next step. The switch is capable of switching at a rate of 300 steps per minute from an external input provided by a pilot device or other control. The switch uses a solenoid stepping motor to rotate the camshaft and cams through 360 degrees. Each of the 19 sets of contacts can be activated to close on any particular step by inserting a cam pin at that location on the camshaft for that step. The activator for each set of contacts rides on the camshaft as it is rotated through its steps. If a pin is encountered on a particular step, the activator will move and close that set of contacts for that step. If a pin is not used on a particular step, the contacts will remain open for that step. In this manner, each set of contacts can be closed for one or any number of steps during the rotation. If 16 steps are not required, the switch can be adjusted to use fewer steps.

A timer control can be used to adjust the amount of time between steps on the sequencer. This type of device can be advanced by events in the logic circuits (contacts closing) or it can be advanced by the output from a timer. Some circuits use a combination of events and timer outputs to control the step increment of the sequencer device.

The step switch (sequencer) can be returned to the first step, called the *home position*, at any time by energizing the reset terminal.

12.3.1 Programmable Sequencers

The sequencer switches we have discussed thus far are all operated as electromechanical devices. Since the device has switch contacts and cams, it will tend to break down from extended operation. If the switch is making one complete revolution every 2 minutes, the contacts may be opened and closed up to 15 times a minute. When you calculate the number of operations per year (over 7,000,000), it is easy to understand why the contacts tend to wear out. For this reason, microprocessors have been used for extended operations. Some programmable logic controllers (PLCs) also provide sequencer operations through their program and input and output devices. Since the microprocessor controls the sequencer operation, and the solid-state components in the output modules control the on/off switching of the load, the PLC is better suited for sequencer applications. More information concerning the sequencer is provided in Chapter 18.

12.3.2 Installing and Troubleshooting the Sequencer

The installation of the sequencer switch should be accomplished with a detailed electrical diagram. This diagram will indicate what circuits (contacts) in the switch should be connected to each output, and it will also provide a matrix that indicates during which steps the contacts are closed. You will need to translate this information into locations on the cam so that you can insert the activator pins in the correct locations.

After the cam switch is wired and the pins are installed, the input signal should be activated, so that the switch can be stepped through a complete cycle. Pay particular attention to the sequence in which each load is energized. If any load is energized out of sequence, the activator pins should be checked for the correct locations.

If any other conditions occur that indicate that the sequencer is malfunctioning, the switch will need to be troubleshooted. The troubleshooting procedure can be broken into two distinct steps. First, you can check to see if the input signal is causing the sequencer to advance correctly. If the switch is advancing, you can begin to focus on the individual switch circuits activated by the cam. If the sequencer does not advance through its steps correctly, you should test the input circuit for the correct signal voltage and determine that the signal's duration is sufficient to cause the solenoid to advance the switch.

When you are testing the individual circuits of the switch, it is important to determine if the contacts close at any time and if the load operates during any of the steps. If the load operates correctly at some steps but is not energizing during other steps, the problem will be located in the area of the activator pins and cams. If the loads do not energize on any step, you should advance the switch to a step where you are certain the contacts are closed and use the voltage-drop method of troubleshooting to look for the problem. If the circuit can provide the correct voltage to the load, be sure to test the load for proper operation. If a set of contacts is faulty, it is possible to move the load to another set of contacts on the switch if they are not all used. Since all the contact sets are the same, the load can operate from any set as long as the activator pins have been adjusted to the same sequence. If all the sets of contacts are being used on the sequencer, you will need to change the faulty set of contacts when they are found. Whenever you make repairs or adjustments to the sequencer, be sure to test the complete control circuit before placing it back into operation.

Questions

1. Explain how the change of time delay is accomplished with a pneumatic timer.
2. Show the electrical symbol for normally open and normally closed time-delay-on and time-delay-off contacts.
3. Explain the operation of a counter.
4. Explain what occurs inside the solid-state timer to delay its operational time.
5. Explain how a one-shot timer operates and how it would be used in industry.

True and False

1. An add-on timer can be mounted directly onto a control relay to provide time delay.
2. In a timing diagram, an X is used to indicate that part of the device is on or activated, and an O is used to indicate that a part of the device is deactivated or off.
3. The length of the timing period will be determined by the preset value that was set at the face of the control by adjusting the hands on the dial.
4. Newer models of solid-state time-delay controls use CMOS circuitry to provide accurate and repeatable time delays.
5. The simplest type of counter, which has been in use for many years, is the mechanical counter used to keep track or totalize the number of parts being made.

Multiple Choice

1. The main difference between a reset timer and a cycle timer is that ______
   a. the reset timer is activated and then it times out and actuates a set of contacts and the repeat cycle timer continually repeats its cycle and actuates a set of contacts at the end of each cycle.
   b. the repeat cycle timer is activated and then it times out and actuates a set of contacts and the reset timer continually repeats its cycle and actuates a set of contacts at the end of each cycle.
   c. the repeat cycle timer uses a pneumatic controller while the reset timer uses a mechanical-type timer.
   d. There is no real difference between the two types of timers.
2. Timers provided time delay by using __________
   a. motors, gears, and cams to control sets of contacts.
   b. air or liquid in a chamber that empties slowly to cause a delay in the action of contacts.
   c. solid-state components.
   d. All of the above
3. When discussing the electrical symbol for the time, which of the following is true?
   a. The time-delay-on symbol uses the lower half or tail of the arrow.
   b. The time-delay-off symbol uses the upper half or arrowhead portion of the arrow.

c. The time-delay-on timer can have both normally open and normally closed contacts.

d. The time-delay-off timer can have both normally open and normally closed contacts.

e. All of the above

f. Only a and b

4. When discussing the pneumatic timer, which of the following is true?

   a. Air is released from the chamber through an orifice controlled by a needle valve to control the amount of time delay.

   b. The needle valve controls the rate at which air can leave the chamber.

   c. When the pneumatic switch is reset, the chamber is refilled with air.

   d. When a sufficient amount of air has been released, the spring will cause the diaphragm to move far enough to activate the electrical switch element.

   e. All of the above

   f. Only a and b

5. What type of counter could you encounter when you are on the job?

   a. A mechanical counter that consists of a set of gears and wheels with the numbers 0 through 9 listed on them.

   b. A *reset* or cycle counter that operates on electromechanical principles where the signal that causes the counter to advance one count comes from an electrical signal instead of through mechanical movement of an arm.

   c. A solid-state counter that utilizes CMOS solid-state circuitry to execute the counter function.

   d. All of the above

   e. Only a and b

Problems

1. Draw a diagram of a pneumatic diagram and explain its operation.

2. Show the electrical diagram for a synchronous motor timer and explain the operation of the instantaneous contacts and time-delay contacts.

3. Draw a timing diagram that shows a motor starter for conveyor 1 starting immediately when the circuit's start button is energized and conveyor 3 starting 20 seconds later.

4. Draw a diagram of a cube timer controlling a motor in a time-delay-on application.

5. Draw a diagram of an eight-pin (octal base) timer and identify each terminal.

Chapter 13

DC Motors

Objectives

After reading this chapter you will be able to:

1. Identify the basic parts of a DC motor and explain its operation.
2. Explain how to change the speed of a DC motor.
3. Explain how to change the rotation of a DC motor.
4. Explain the difference between series, shunt, and compound DC motors.

DC motors are commonly used to operate machinery in a variety of applications on the factory floor. DC motors were the first type of energy converter used in industry. The earliest machines required speed control and the speed could be set on DC motors by varying the voltage sent to them. The earliest speed controls for DC motors were nothing more than large resistors.

DC motors required large amounts of DC voltage for operation, so a source for the DC voltage is needed at the factory. Because DC voltage cannot be generated and distributed over a long distance, AC voltage is the industry standard. One way to provide the DC voltage is to use generators that are set up at the factory site where large AC motors are used to turn them to produce the amount of DC voltage that the DC motors require. This system uses a large AC motor to drive a DC generator directly at a constant speed. The field current in the generator is regulated to adjust the level of DC voltage from the generator, which in turn is used to vary the speed of any DC motor that the generator powers. This system, called a Ward–Lennard system, was popular until solid-state diodes became available for rectifying large amounts of AC voltage to DC for use in motor-driven circuits. Once solid-state diodes and SCRs became available, DC motors became more usable in industry.

During the 1950s and 1960s the use of DC motors became more prevalent in machinery control because their speed and torque were easy to control with simple SCR controllers. The SCR could rectify AC voltage to DC, provide current and voltage control at the same time, and were capable of being paralleled for larger loads up to 1000 A. As solid-state controls became more reliable in the late 1960s and the 1970s a wide variety of low-cost AC motor speed controls became available.

During these years transistors could handle larger loads, and microprocessors became relatively inexpensive so that they could be used to make variable-frequency AC motor controls. At this time, you had a choice of using good-quality AC or DC motors for all types of special speed and torque applications.

As you read this chapter you will learn about the concepts of controlling a DC motor's speed and torque, and being able to reverse the direction of its rotation. It is also important to be able to recognize the features that make the series, shunt, and compound DC motors different from each other. You should have a good understanding of the basic parts of the DC motor so that when you must troubleshoot a DC motor circuit, you will be able to recognize a malfunctioning component and make repairs or replace parts as quickly as possible.

It is also important to understand the methods of controlling the speed of AC motors, their direction of rotation, and the amount of torque they can develop, since these are the principles that motor drives use to control motors. Operation of special drive controllers is very easy to understand if you know what electrical principle they are trying to alter to provide control for the motor. If you do not understand the motor principle, it is doubtful that you will fully understand the motor control device, and this will make the system nearly impossible to troubleshoot. If you understand concepts, you will easily be able to understand the next generation of controls, those that will be produced during the next 10 years. If you do not understand principles, each new type of control device that you encounter will seem impossible to understand and repair.

Magnetic theories are provided with a discussion of basic DC motor components. Additional information will explain the difference between series, shunt, and compound motors. Diagrams are provided to explain the methods of controlling speed, rotation, and torque. These diagrams are useful for making field wiring connections and for testing motors during troubleshooting procedures.

13.1 Magnetic Theory

DC motors operate on the principles of basic magnetism. A coil of wire can be magnetized when current is passed through it. When this principle was used in relay coils, the polarity of the current was not important. When the current is passed through a coil of wire to make a field coil for a motor, the polarity of the current will determine the direction of rotation for the motor.

The polarity of the current flowing through the coil of wire will determine the location of the north and south magnetic poles in the coil of wire. Another important principle involves the amount of current that is flowing through the coil. The amount of current was not important as long as enough current was present to move the armature of the relay or solenoid. In a DC motor, the amount of current in the windings will determine the speed (rpm) of the motor shaft and the amount of torque that it can produce.

You should remember from basic magnetic theory that the left-hand rule of current flow through a coil of wire helps you understand that the direction of current flow will determine the magnetic polarity of the coil. The left-hand rule is used to show you a principle from which several facts can be determined. The first is that the direction of current flow will determine which end of a coil of wire is negative or positive. This will determine which end of the coil will be the north pole of the magnet and which end will be the south pole. It is also easy to see from this diagram that by changing the direction of the current flow in the coil of wire, the magnetic poles will be reversed in the coil. This is important because the direction of the motor's rotation is determined by the changing magnetic field.

Another basic concept about magnets is the relationship between two like poles and two unlike poles. When the north pole of two different magnets are placed close to each other, they will repel each other. When the north pole of one magnet is placed near the south pole of another magnet, the two poles will attract each other very strongly.

The third principle is that the strength of the magnetic field can be varied by changing the amount of current flowing through the wire in the coil. If a small amount of current is flowing, a small number of flux lines will be created and the magnetic field will be relatively weak. If the amount of current is increased, the magnetic field will become stronger. The strength of the magnetic field can be increased to the point of saturation. A magnetic coil is said to be saturated when its magnetic strength cannot be increased by adding more current.

Saturation is similar to filling a drinking glass with water. You cannot get the level of the glass any higher than full. Any additional water that is put into the glass when it is full will not increase the amount of water in the glass. The additional water will run over the side of the glass and be wasted. The same principle can be applied to a magnetic coil. When the strength of the magnetic field is at its strongest point, additional electric current will not cause the field to become any stronger.

13.2 DC Motor Theory

The DC motor has two basic parts: (1) the rotating part, which is called the *armature*, and (2) the stationary part, which includes coils of wire called the *field coils*. The stationary part is also called the *stator*. Figure 13–1 shows a picture of a typical DC motor, Figure 13–2 shows a picture of a DC armature, and Figure 13–3 shows a picture of a typical stator. From Figure 13–2 you can see that the armature is made of coils of wire wrapped around the core, and the core has an extended shaft that rotates on bearings. Notice that the ends of each coil of wire on the armature are terminated at one end of the armature. The termination points are called the commutator, and this is where the brushes make electrical contact to bring electrical current from the stationary part to the rotating part of the machine.

Figure 13–3 shows the location of the coils that are mounted inside the stator. These coils will be referred to as field coils in future discussions and they may be connected in series or parallel with each other to create changes of torque in the motor. The size of wire in these coils and the number of turns of wire in the coil will depend on the effect that is trying to be achieved.

Figure 13–1 Typical DC motor.

Figure 13–2 Armature (rotor) of a DC motor has wire coils wrapped around its core. The ends of each coil are terminated at commutator segments located on the left end of the shaft. The brushes make contact on the commutator to provide current for the armature.

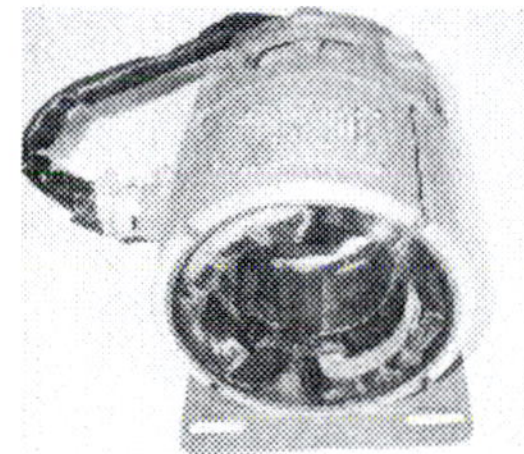

Figure 13–3 The stationary part of a DC motor has the field coils mounted in it.

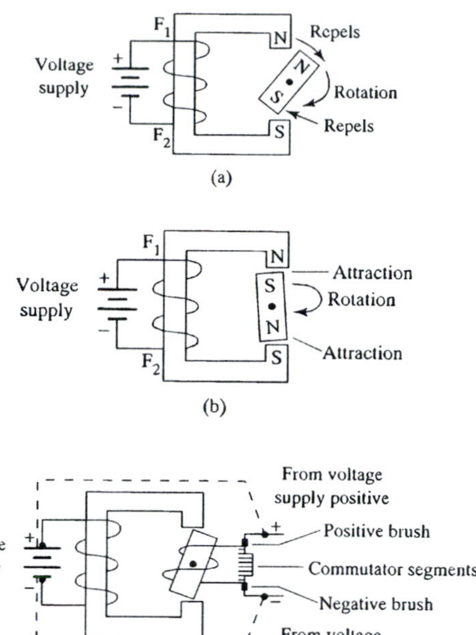

Figure 13–4 Magnetic diagram that explains the operation of a DC motor. (a) The rotating magnet moves clockwise because like poles repel. (b) The rotating magnet is being attracted because the poles are unlike. (c) The rotating magnet is now shown as the armature coil, and its polarity is determined by the brushes and commutator segments.

It will be easier to understand the operation of the DC motor from a basic diagram that shows the magnetic interaction between the rotating armature and the stationary field coils. Figure 13–4 shows three diagrams that explain the DC motor's operation in terms of the magnetic interaction. In Figure 13–4a a bar magnet has been mounted on a shaft so that it can spin. The field winding is one long coil of wire that has been separated into two sections. The top section is connected to the positive pole of the battery and the bottom section is connected to the negative pole of the battery. It is very important to understand that the battery represents a source of voltage for this winding. In the actual industrial-type motor this voltage will come from the DC voltage source for the motor. The current flow in this direction makes the top coil the north pole of the magnet and the bottom coil the south pole of the magnet.

The bar magnet represents the *armature* and the coil of wire represents the *field*. The arrow shows the direction of the armature's rotation. Notice that the arrow shows the armature starting to rotate in the clockwise direction. The north pole of the field is repelling the north pole of the armature, and the south pole of the field coil is repelling the south pole of the armature.

As the armature begins to move, the north pole of the armature comes closer to the south pole of the field, and the south pole of the armature is coming closer to the north pole of the field. As the two unlike poles near each other, they begin to attract. This attraction becomes stronger until the armature's north pole moves directly in line with the field's south pole, and its south pole moves directly in line with the field's north pole (Figure 13–4b).

When the opposite poles are at their strongest attraction, the armature will be "locked up" and will resist further attempts to continue spinning. For the armature to continue its rotation, the armature's polarity must be switched. Since the armature in this diagram is a permanent magnet, it would lock up during the first rotation and not work. If the armature is an electromagnet, its polarity can be changed by changing the direction of current flow through it. For this reason, the armature must be changed to a coil (electromagnet) and a set of commutator segments must be added to provide a means of making contact between the rotating member and the stationary member. One commutator segment is provided for each terminal of the magnetic coil. Since this armature has only one coil, it will have only two terminals, so the commutator has two segments.

Since the armature is now a coil of wire, it will need DC current flowing through it to become magnetized. This presents another problem; since the armature will be rotating, the DC voltage wires cannot be connected directly to the armature coil. A stationary set of carbon brushes is used to make contact to the rotating armature. The brushes ride on the commutator segments to make contact so that current will flow through the armature coil.

In Figure 13–4c the DC voltage is applied to the field and to the brushes. Since negative DC voltage is connected to one of the brushes, the commutator segment that the negative brush rides on will also be negative. The armature's magnetic field causes the armature to begin to rotate. This time when the armature gets to the point where it becomes locked up with the magnetic field, the negative brush begins to touch the end of the armature coil that was previously positive and the positive brush begins to touch the end of the armature coil that was negative. This action switches the direction of current flow through the armature, which also switches the polarity of the armature coil's magnetic field at just the right time so that the repelling and attracting continues. The armature continues to switch its magnetic polarity twice during each rotation, which causes it to continually be attracted and repelled with the field poles.

This is a simple two-pole motor that is used primarily for instructional purposes. Since the motor has only two poles, the motor will operate rather roughly and not provide too much torque. Additional field poles and armature poles must be added to the motor for it to become useful for industry.

13.3 DC Motor Components

The armature and field in a DC motor can be wired three different ways to provide different amounts of torque or different types of speed control. The armature and field windings are designed a little differently for different types of DC motors. The three basic types of DC motors are the *series motor*, the *shunt motor*, and the *compound motor*. The series motor is designed to move large loads with high starting torque, in applications such as a crane motor or lift hoist. The shunt motor is designed slightly differently since it is made for applications such as pumping fluids, where constant-speed characteristics are important. The compound motor is designed with some of the series motor's characteristics and some of the shunt motor's characteristics. This allows the compound motor to be used in applications where high starting torque and controlled operating speed are both required.

It is important that you understand the function and operation of the basic components of the DC motor since motor controls will take advantage of these design characteristics to provide speed, torque, and direction of rotation control. Figure 13–5 shows a cutaway picture of a DC motor and Figure 13–6 shows an exploded view diagram of a DC motor. The basic components include the armature assembly, which includes all rotating parts; the frame assembly, which houses the stationary field coils; and the end plates, which provide bearings for the motor shaft and a mounting point for the brush rigging. Each of these assemblies is explained in depth so that you will understand the design concepts used for motor control.

Figure 13–5 A cutaway picture of a DC motor.

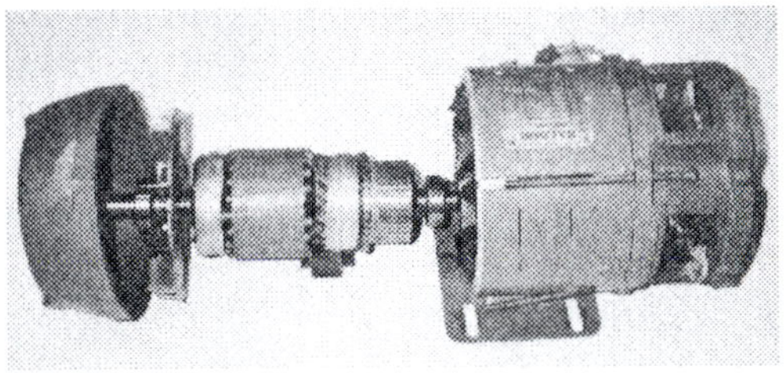

Figure 13–6 An exploded view of a DC motor, showing the relationship of all of the components.

13.3.1 Armature

The armature is the part of a DC motor that rotates and provides energy at the end of the shaft. It is basically an electromagnet, since it is a coil of wire that has to be specially designed to fit around core material on the shaft. The core of the armature is made of laminated steel and provides slots for the coils of wire to be pressed onto.

Figure 13–7a shows a sketch of a typical DC motor armature, and Figure 13–7b shows the laminated steel core of the armature without any coils of wire on it. This gives you a better look at the core. The armature core is made of laminated steel to prevent the circulation of eddy currents. If the core were solid, magnetic currents would be produced that would circulate in the core material near the surface and cause the core metal to heat up. These magnetic currents are called *eddy currents.* When laminated steel sections are pressed together to make the core, the eddy currents cannot flow from one laminated segment to another, so they are effectively canceled out. The laminated core also prevents other magnetic losses called *flux losses.* These losses tend to make the magnetic field weaker so that more core material is required to obtain the same magnetic field strengths. The flux losses and eddy current losses are grouped together by designers and called *core losses.* The laminated core is designed to allow the armature's magnetic field to be as strong as possible since the laminations prevent core losses. Notice that one end of the core has commutator segments. There is one commutator segment for each end of each coil. This means that an armature with four coils will have eight commutator segments. The commutator segments are used as a contact point between the stationary brushes and the rotating armature. When each coil of wire is pressed onto the armature, the end of the coil is soldered to a specific commutator segment. This makes an electrical terminal point for the current that will flow from the brushes onto the commutator segment and finally through the coil of wire. Figure 13–7c shows the coil of wire before it is mounted in the armature slot, and Figure 13–7d shows the coil mounted in the armature slot and soldered to the commutator segment.

The shaft is designed so that the laminated armature segments can be pressed onto it easily. It is also machined to provide a surface for a main bearing to be pressed on at each end. The bearing will ride in the end plates and support the armature when it begins to rotate. One end of the shaft is also longer than the other since it will provide the mounting shaft for the motor's load to be attached. Some shafts have a key way or flat spot machined into them so that the load that is mounted on it can be secured. You must be careful when handling a motor that you do not damage the shaft since it must be smooth to accept the coupling mechanism. It is also possible to bend the shaft or cause damage to the bearings so that the motor will vibrate when it is operating at high speed. The commutator is made of copper. A thin section of insulation is placed between each commutator segment. This effectively isolates each commutator segment from all others.

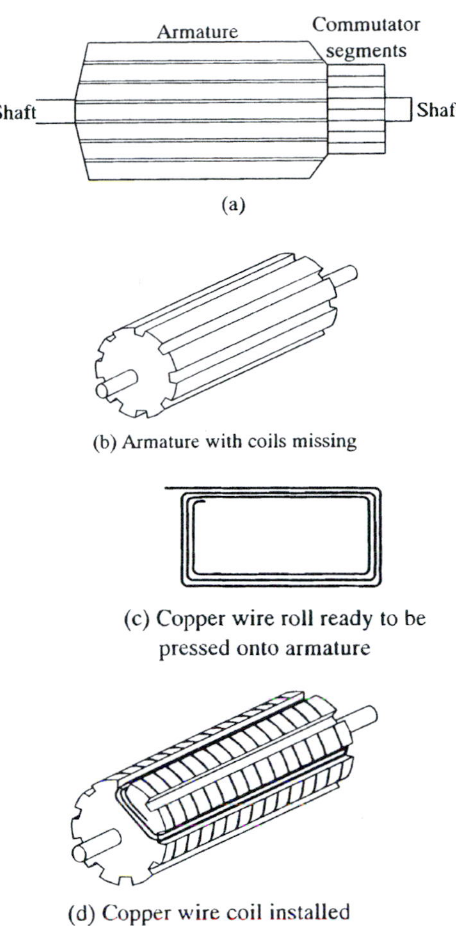

Figure 13–7 (a) Armature and commutator segments. (b) Armature prior to the coils of wire being installed. (c) Coil of wire prior to being pressed into the armature. (d) A coil pressed into the armature. The end of each coil is attached to a commutator segment.

13.3.2 Motor Frame

The armature is placed inside the frame of the motor where the field coils are mounted. When the field coils and the armature coils become magnetized the armature will begin to rotate. The field winding is made by coiling up a long piece of wire. The wire is mounted on laminated pole pieces called field poles. Similar to an

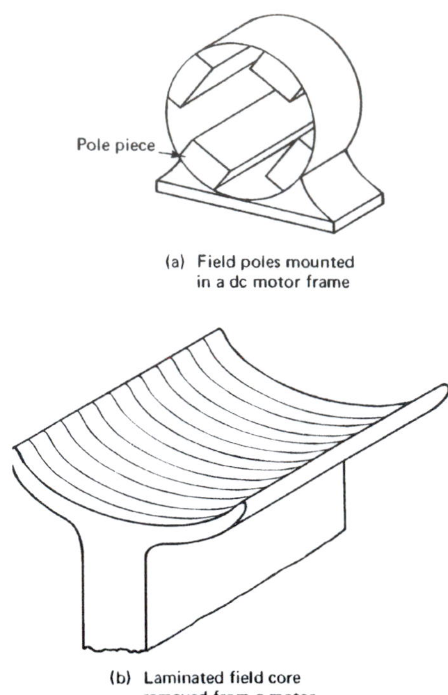

Figure 13–8 (a) The location of the pole pieces in the frame of a DC motor. (b) An individual pole piece. You can see that it is made of laminated sections. The field coils are wound around the pole pieces.

armature, these poles are made of laminated steel or cast iron to prevent eddy current and other flux losses. Figure 13–8 shows the location of the pole pieces inside a DC motor frame.

The amount of wire that is used to make the field winding will depend on the type of motor that is being manufactured. A series motor uses heavy-gauge wire for its field winding so that it can handle the very large field currents. Since the wire is a large gauge, the number of turns of wire in the coil will be limited. If the field winding is designed for a shunt motor, it will be made of very small gauge wire and many turns can be used.

After the coils are wound, they are coated for protection against moisture and other environmental elements. After they have been pressed onto the field poles, they must be secured with shims or bolts so that they are held rigidly in place. Remember that when current is passed through the coil, it will become strongly magnetized and attract and repel the armature's magnetic poles. If the field poles are not rigidly secured, they will be pulled loose when they are attracted to the armature's magnetic field and then pressed back into place when they become repelled. This action will cause the field to vibrate and damage the outer protective insulation and cause a short circuit or a ground condition between the winding and the frame of the motor.

The ends of the frame are machined so that the end plates will mount firmly into place. An access hole is also provided in the side of the frame or in the end plates so that the field wires can be brought to the outside of the motor, where DC voltage can be connected.

The bottom of the frame has the mounting bracket attached. The bracket has a set of holes or slots provided so that the motor can be bolted down and securely mounted on the machine it is driving. The mounting holes will be designed to specifications by frame size. The dimensions for the frame sizes are provided in tables printed by motor manufacturers.

Since these holes and slots are designed to a standard, you can predrill the mounting holes in the machinery before the motor is put in place. The slots are used to provide minor adjustments to the mounting alignment when the motor is used in belt-driven or chain-driven applications. It is also important to have a small amount of mounting adjustment when the motor is used in direct-drive applications. It is very important that the motor be mounted so that the armature shaft can turn freely and not bind with the load.

13.3.3 End Plates

The end plates of the motor are mounted on the ends of the motor frame. Figures 13–5 and 13–6 show the location of the end plates in relation with the motor frame. The end plates are held in place by four bolts that pass through the motor frame. The bolts can be removed from the frame completely so that the end plates can easily be removed for maintenance. The end plates also house the bearings for the armature shaft. These bearings can be either sleeve or ball type. If the bearing is a ball bearing type, it is normally permanently lubricated, and if it is a sleeve type, it will require a light film of oil to operate properly. The end plates that house a sleeve-type bearing will have a lubrication tube and wicking material. Several drops of lubricating oil are poured down the lubrication tube, where they will saturate the wicking material. The wicking is located in the bearing sleeve so that it can make contact with the armature shaft and transfer a light film of oil to it. Other types of sleeve bearings are made of porous metal that can absorb oil to be used to create a film between the bearing and the shaft. It is important that the end plate for a sleeve bearing be mounted on the motor frame so that the lubricating tube is pointing up. This position will ensure that gravity will pull the oil to the wicking material. If the end plates are mounted so that the lubricating tube is pointing down, the oil will flow away from the wicking and it will become dry. When the wicking dries out, the armature shaft will rub directly on the metal in the sleeve bearing which will cause it to quickly heat up, and the shaft will seize to the bearing. For this reason it is also important to follow lubrication instructions and oil the motor on a regular basis.

13.3.4 Brushes and Brush Rigging

The brush rigging is an assembly that securely holds the brushes in place so that they will be able to ride on the commutator. It is mounted on the rear end plate so that the brushes will be accessible by removing the end plate. An access hole is also provided in the motor frame so that the brushes can be adjusted slightly when the motor is initially set up. The brush rigging uses a spring to provide the proper amount of tension on the brushes so that they make proper contact with the commutator. If the tension is too light, the brushes will bounce and arc, and if the tension is too heavy, the brushes will wear down prematurely.

The brush rigging is shown in Figures 13–5 and 13–6. Notice that it is mounted on the rear end plate. Since the rigging is made of metal, it must be insulated electrically when it is mounted on the end plate. The DC voltage that is used to energize the armature will pass through the brushes to the commutator segments and into the armature coils. Each brush has a wire connected to it. The wires will be connected to either the positive or negative terminal of the DC power supply. The motor will always have an even number of brushes. Half of the brushes will be connected to positive voltage and half will be connected to negative voltage. In most motors the number of brush sets will be equal to the number of field poles. The voltage polarity will remain constant on each brush. This means that for each pair, one of the brushes will be connected to the positive power terminal, and the other will be connected permanently to the negative terminal.

The brushes will cause the polarity of each armature segment to alternate from positive to negative. When the armature is spinning, each commutator segment will come in contact with a positive brush for an instant and will be positive during that time. As the armature rotates slightly, that commutator segment will come in contact with a brush that is connected to the negative voltage supply and it will become negative during that time. As the armature continues to spin, each commutator segment will be alternately powered by positive and then negative voltage.

The brushes are made of carbon-composite material. Usually, the brushes have copper added to aid in conduction. Other material is also added to make them wear longer. The end of the brush that rides on the commutator is contoured to fit the commutator exactly so that current will transfer easily. The process of contouring the brush to the commutator is called seating. Whenever a set of new brushes are installed, they should be seated to fit the commutator. The brushes are the main part of the DC motor that will wear out. It is important that their wear be monitored closely so that they do not damage the commutator segments when they begin to wear out. Most brushes have a small mark on them called a wear mark or wear bar. When a brush wears down to the mark, it should be replaced. If the brushes begin to wear excessively or do not fit properly on the commutator, they will heat up and damage the brush rigging and spring mechanism. If the brushes have been overheated, they can cause burn marks or pitting on the commutator segments and also warp the spring mechanism so that it will no longer hold the brushes with the proper amount of tension. Figures 13–5 and 13–6 show the location of the brushes riding on the commutator.

If the spring mechanism has been overheated, it should be replaced and the brushes should be checked for proper operation. If the commutator is pitted, it can be turned down on a lathe. After the commutator has been turned down, the brushes will need to be reseated.

After you have an understanding of the function of each of the parts or assemblies of the motor, you will better be able to understand the operation of a basic DC motor. Operation of the motor involves the interaction of all the motor parts. Some of the parts will be altered slightly for specific motor applications. These changes will become evident when the motor's basic operation is explained.

13.4 DC Motor Operation

The DC motor you will find in modern industrial applications operates very similarly to the simple DC motor described earlier in this chapter. Figure 13–9 shows an electrical diagram of a simple DC motor. The DC voltage is applied directly to the field winding and the brushes. The armature and the field are both shown as a coil of wire. In later diagrams, a field resistor will be added in series with the field to control the motor speed.

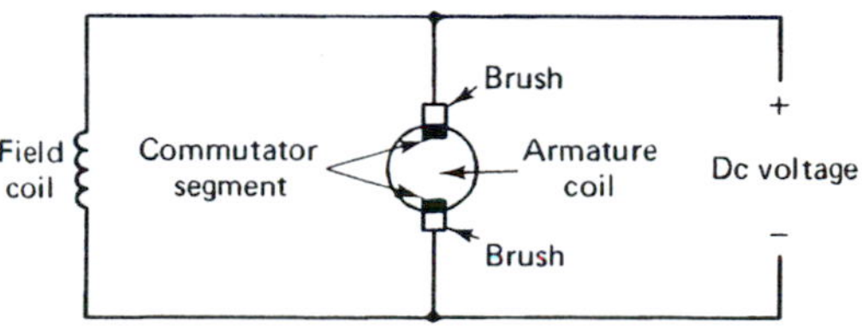

Figure 13–9 Simple electrical diagram of a DC shunt motor, showing the electrical relationship between the field coil and armature.

When voltage is applied to the motor, current begins to flow through the field coil from the negative terminal to the positive terminal. This sets up a strong magnetic field in the field winding. Current also begins to flow through the brushes into a commutator segment and then through an armature coil. The current continues to flow through the coil back to the brush that is attached to other end of the coil and returns to the DC power source. The current flowing in the armature coil sets up a strong magnetic field in the armature.

The magnetic field in the armature and field coil causes the armature to begin to rotate. This occurs by the unlike magnetic poles attracting each other and the like magnetic poles repelling each other. As the armature begins to rotate, the commutator segments will also begin to move under the brushes. As an individual commutator segment moves under the brush connected to positive voltage, it will become positive, and when it moves under a brush connected to negative voltage it will become negative. In this way, the commutator segments continually change polarity from positive to negative. Since the commutator segments are connected to the ends of the wires that make up the field winding in the armature, the magnetic field in the armature will change polarity continually from north pole to south pole. The commutator segments and brushes are aligned in such a way that the switch in polarity of the armature coincides with the location of the armature's magnetic field and the field winding's magnetic field. The switching action is timed so that the armature will not lock up magnetically with the field. Instead, the magnetic fields tend to build on each other and provide additional torque to keep the motor shaft rotating.

When the voltage is deenergized to the motor, the magnetic fields in the armature and the field winding will quickly diminish and the armature shaft's speed will begin to drop to zero. If voltage is applied to the motor again, the magnetic fields will strengthen and the armature will begin to rotate again.

13.5 Types of DC Motors

Three basic types of DC motors are used in industry today: the series motor, the shunt motor, and the compound motor. The series motor is capable of starting with a very large load attached, as in lifting applications. The shunt motor is able to operate with rpm control while it is at high speed. The compound motor, a combination of the series motor and the shunt motor, is able to start with fairly large loads and have some rpm control at higher speeds. In the remaining sections of this chapter we show a diagram for each of these motors and discuss their operational characteristics. As a technician you should understand methods of controlling their speed and ways to change the direction of rotation, because these are the two parameters of a DC motor you will be asked to change as applications change on the factory floor. It is also important to understand the basic theory of operation of these motors because you will be controlling them with solid-state electronic circuits and you will need to know if problems that arise are the fault of the motor or the solid-state circuit.

13.5.1 DC Series Motors

The series motor provides high starting torque and is able to move very large shaft loads when it is first energized. Figure 13–10 shows the wiring diagram of a series motor. From the diagram you can see that the field winding in this motor is wired in series with the armature winding. This is the attribute that gives the series motor its name.

Since the series field winding is connected in series with the armature, it will carry the same amount of current that passes through the armature. For this reason the field is made from heavy-gauge wire that is large enough to carry the load. Since the wire gauge is so large, the winding will have only a few turns of wire. In some larger DC motors, the field winding is made from copper bar stock rather than the conventional round wire that you use for power distribution. The square or rectangular shape of the copper bar stock makes it fit more easily around the field pole pieces. It can also radiate more easily the heat that has built up in the winding due to the large amount of current being carried.

The amount of current that passes through the winding determines the amount of torque the motor shaft can produce. Since the series field is made of very large conductors, it can carry very large amounts of current and produce very large torques. For example, the starter motor that is used to start an automobile's engine is a series motor that may draw up to 500 A when it is turning the engine's crankshaft on a cold morning. Series motors used to power hoists or cranes may draw currents of thousands of amperes during operation.

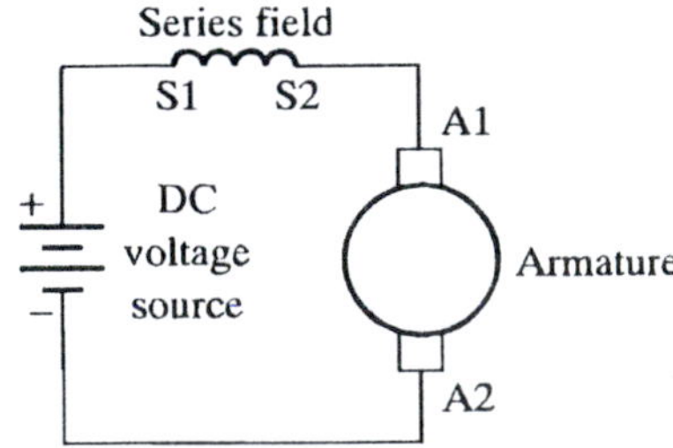

Figure 13–10 Electrical diagram of series motor. The series field is identified as S1 and S2.

The series motor can safely handle large currents since the motor does not operate for an extended period. In most applications the motor will operate for only a few seconds while this large current is present. Think about how long the starter motor on the automobile must operate to get the engine to start. This period is similar to that of industrial series motors.

13.5.1.1 Series Motor Operation

Operation of the series motor is rather easy to understand. In Figure 13–10 you can see that the field winding is connected in series with the armature winding. This means that power will be applied to one end of the series field winding and to one end of the armature winding (connected at the brush).

When voltage is applied, current begins to flow from negative power supply terminals through the series winding and armature winding and back to the negative terminal. The armature is not rotating when voltage is first applied, so the only resistance in this circuit is provided by the large conductors used in the armature and field windings. Since these conductors are so large, they will have a very small amount of resistance. This causes the motor to draw a very large amount of current from the power supply. When the large current begins to flow through the field and armature windings, it causes a very strong magnetic field to be built. Since the current is so large, it will cause the coils to reach saturation, which will produce the strongest magnetic field possible.

13.5.1.2 Producing Back EMF

The strength of these magnetic fields provides the armature shafts with the greatest amount of torque possible. The large torque causes the armature to begin to spin with the maximum amount of power. When the armature begins to rotate, it begins to produce voltage like a generator. This concept is difficult for some students to understand since the armature is part of the motor at this time.

You should remember from the basic theories of magnetism that any time a magnetic field passes a coil of wire, a current will be produced. The stronger the magnetic field is or the faster the coil passes the flux lines, the more current will be generated. When the armature begins to rotate, it will produce a voltage that is of opposite polarity to that of the power supply. This voltage is called *back voltage* or *back EMF* (electromotive force) or *counter EMF.* The overall effect of this voltage is that it will be subtracted from the supply voltage so that the motor windings will see a smaller voltage potential.

When Ohm's law is applied to this circuit, you will see that when the voltage is slightly reduced, the current will also be reduced slightly. This means that the series motor will see less current as its speed is increased. The reduced current will mean that the motor will continue to lose torque as the motor speed increases. Since the load is moving when the armature begins to pick up speed, the application will require less torque to keep the load moving. This works to the motor's advantage by automatically reducing the motor current as soon as the load begins to move. It also allows the motor to operate with less heat buildup.

This condition can cause problems if the series motor ever loses its load. The load could be lost when a shaft breaks or if a drive pin is sheared. When this occurs, the load current is allowed to fall to a minimum, which reduces the amount of back EMF that the armature is producing. Since the armature is not producing a sufficient amount of back EMF and the load is no longer causing a drag on the shaft, the armature will begin to rotate faster and faster and will continue to increase rotational speed until it is operating at a very high speed.

When the armature is operating at high speed the heavy armature windings will be pulled out of their slots by centrifugal force. When the windings are pulled loose, they will catch on a field winding pole piece and the motor will be severely damaged. This condition is called *runaway* and you can see why a DC series motor must have some type of runaway protection. A centrifugal switch can be connected to the motor to deenergize the motor starter coil if the rpm exceeds the set amount. Other sensors can be used to deenergize the circuit if the motor's current drops while full voltage is applied to the motor. The most important part to remember about a series motor is that it is difficult to control its speed by external means because its rpm is determined by the size of its load. (In some smaller series motors, the speed can be controlled by placing a rheostat in series with the supply voltage to provide some amount of change in resistance to control the voltage to the motor.)

Figure 13-11 shows the relationship between series motor speed and armature current. When current is low (at the top left) the motor speed is maximum, and when current increases, the motor speed slows down (bottom

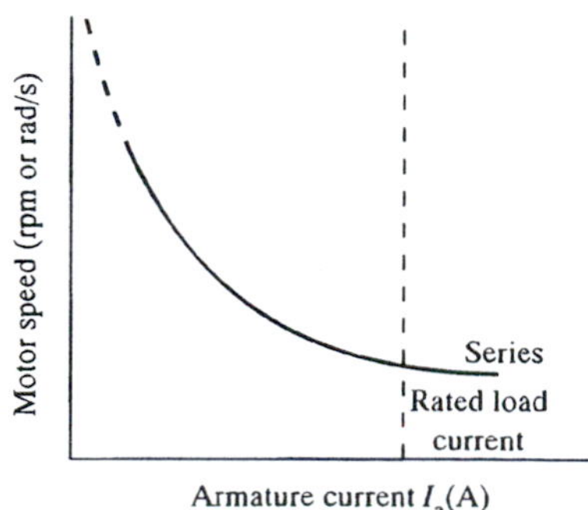

Figure 13–11 The relationship between series motor speed and armature current.

right). You can see from this curve that a DC motor will run away if the load current is reduced to zero. (In larger series machines used in industry, the amount of friction losses will limit the highest speed somewhat.)

13.5.1.3
Reversing the Rotation of the Motor

The direction of rotation of a series motor can be changed by changing the polarity of either the armature or the field winding. If you simply changed the polarity of the applied voltage, you would be changing the polarity of both field and armature windings and the motor's rotation would remain the same.

Since only one of the windings needs to be reversed, the armature winding is typically used because its terminals are readily accessible at the brush rigging. Remember that the armature receives its current through the brushes, so that if their polarity is changed, the armature's polarity will also be changed. A reversing motor starter is used to change wiring to cause the direction of the motor's rotation to change by changing the polarity of the armature windings. Figure 13–12 shows a DC series motor that is connected to a reversing motor starter. In this diagram the armature's terminals are marked A1 and A2 and the field terminals are marked S1 and S2.

When the forward motor starter is energized, the top contact identified as F closes so the A1 terminal is connected to the positive terminal of the power supply and the bottom F contact closes and connects terminal A2 to S1. Terminal S2 is connected to the negative terminal of the power supply. When the reverse motor starter is energized, terminals A1 and A2 are reversed. A2 is now connected to the positive terminal. Notice that S2 remains connected to the negative terminal of the power supply terminal. This ensures that only the armature's polarity has been changed and the motor will begin to rotate in the opposite direction.

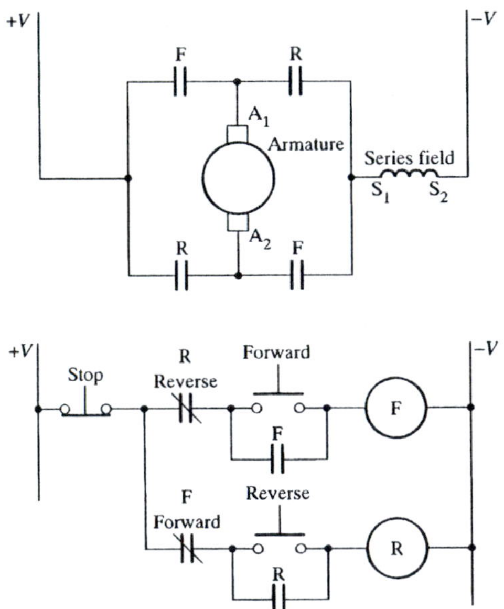

Figure 13–12 DC series motor connected to forward and reverse motor starter.

Also notice in the control circuit that the normally closed set of R contacts that are connected in series with the forward pushbutton, and the normally closed set of F contacts connected in series with the reverse pushbutton. These contacts provide an *interlock* that prevents the motor from being changed from forward to the reverse direction without stopping the motor. The circuit can be explained as follows: When the forward pushbutton is depressed, current will flow from the stop pushbutton, through the normally closed R interlock contacts, and through the forward pushbutton to the forward motor starter coil. When the forward motor starter coil is energized, it will open its normally closed contacts that are connected in series with the reverse pushbutton. This means that if someone depresses the reverse pushbutton, current could not flow to the reverse motor starter coil. If the person depressing the pushbuttons wants to reverse the direction of rotation of the motor, he or she will need to depress the stop pushbutton first to deenergize the forward motor starter coil, which will allow the normally closed F contacts to return to their normally closed position. When the reverse motor starter coil is energized, its normally closed R contacts that are connected in series with the forward pushbutton will open and prevent the current flow to the forward motor starter coil if the forward pushbutton is depressed. You will see a number of other ways to control the forward and reverse motor starter in later discussions and in the chapter on motor controls.

13.5.1.4
Installing and Troubleshooting the Series Motor

Since the series motor has only two leads brought out of the motor for installation wiring, this wiring can be accomplished rather easily. If the motor is wired to operate in only one direction, the motor terminals can be connected to a manual or magnetic starter. If the motor's rotation is required to be reversed periodically, it should be connected to a reversing starter.

Most DC series motors are used in direct-drive applications. This means that the load is connected directly to the armature's shaft. This type of load is generally used to get the most torque converted. Belt-drive applications are not recommended, since a broken belt would allow the motor to run away. After the motor has been installed, a test run should be used to check it out. If any problems occur, the troubleshooting procedures should be used.

The most likely problem that will occur with the series motor is that it will develop an open in one of its windings or between the brushes and the commutator. Since the coils in a series motor are connected in series, each coil must be functioning properly or the motor will not draw any current. When this occurs, the motor cannot build a magnetic field and the armature will not turn. Another problem that is likely to occur with the motor circuit is that circuit voltage will be lost due to a blown fuse or circuit breaker. The motor will respond similarly in both of these conditions.

The best way to test a series motor is with a voltmeter. The first test should be for applied voltage at the motor terminals. Since the motor terminals are usually connected to a motor starter, the test leads can be placed on these terminals. If the meter shows that full voltage is applied, the problem will be in the motor, and if it shows that no voltage is present, you should test the supply voltage and the control circuit to ensure that the motor starter is closed. If the motor starter has a visual indicator, check to see that the starter's contacts are closed. If the overloads have tripped, you can assume that they have sensed a problem with the motor or its load. When you reset the overloads the motor will probably start again but remember to test the motor thoroughly for problems that would cause an overcurrent situation.

If the voltage test indicates that the motor has full applied voltage to its terminals but the motor is not operating, you can assume that you have an open in one of the windings or between the brushes and the armature, and continue testing. Each of these sections should be disconnected from each other and voltage should be removed so that they can be tested with an ohmmeter for an open. The series field coils can be tested by putting the ohmmeter leads on terminals S1 and S2. If the meter indicates that an open exists, the motor will need to be removed and sent to be rewound or replaced. If the meter indicates that the field coil has continuity, you should continue the procedure by testing the armature.

The armature can also be tested with an ohmmeter by placing the leads on the terminals marked A1 and A2. If the meter shows continuity, rotate the armature shaft slightly to look for bad spots where the commutator may have an open or the brushes may not be seated properly. If the armature test indicates that an open exists, you should continue the test by visually inspecting the brushes and commutator. You may also have an open in the armature coils. The armature must be removed from the motor frame to be tested further. When you have located the problem you should remember that the commutator can be removed from the motor while it remains in place and it can be turned down on a lathe. When the commutator is replaced in the motor, new brushes can be installed and the motor will be ready for use.

It is possible that the motor will develop a problem but still run. This type of problem usually involves the motor overheating or not being able to pull its rated load. This type of problem is different from an open circuit because the motor is drawing current and trying to run. Since the motor is drawing current, you must assume that there is not an open circuit. It is still possible to have brush problems that would require the brushes to be reseated or replaced. Other conditions that will cause the motor to overheat include loose or damaged field and armature coils. The motor will also overheat if the armature shaft bearing is in need of lubrication or is damaged. The bearing will seize on the shaft and cause the motor to build up friction and overheat.

If either of these conditions occur, the motor may be fixed on site or be removed for extensive repairs. When the motor is restarted after repairs have been made, it is important to monitor the current usage and heat buildup. Remember that the motor will draw DC current, so that an AC clamp-on ammeter will not be useful for measuring the DC current. You will need to use an ammeter that is specially designed for very large DC currents. It is also important to remember that the motor can draw very high locked-rotor current when it is starting, so the ammeter should be capable of measuring currents up to 1000 A. After the motor has completed its test run successfully, it can be put back into operation for normal duty. Any time the motor is suspected of faulty operation, the troubleshooting procedure should be rechecked.

13.5.1.5 DC Series Motor Used as a Universal Motor

The series motor is used in a wide variety of power tools, such as electric hand drills, saws, and power screwdrivers. In most of these cases, the power source for the motor is AC voltage. The DC series motor will operate on AC voltage. If the motor is used as in a hand drill that needs variable speed control, a field rheostat or other type of current control is used to control the speed of the motor. In some newer tools, the current control uses solid-state components to control the speed of the motor. The motors used for these types of power tools have brushes and a commutator, and these are the main parts of the motor to wear out. You can use the same theory of operation provided for the DC motor to troubleshoot these types of motors.

13.5.2 DC Shunt Motors

The shunt motor is different from the series motor in that the field winding is connected in parallel with the armature instead of in series. You should remember from basic electrical theory that a parallel circuit is often referred to as a shunt. Since the field winding is placed in parallel with the armature, it is called a shunt winding and the motor is called a shunt motor. Figure 13–13 shows a diagram of a shunt motor. The field terminals are marked F1 and F2, and the armature terminals are marked A1 and A2. The shunt field is represented in the diagram with multiple turns using a thin line.

The shunt winding is made of very small-gauge wire with many turns on the coil. Since the wire is so small, the coil can have thousands of turns and still fit in the slots. The small-gauge wire cannot handle as much current as the heavy-gauge wire in the series field, but since this coil has many more turns of wire, it can still produce a very strong magnetic field. Figure 13–14 shows a picture of a DC shunt motor.

13.5.2.1 Shunt Motor Operation

The shunt motor has slightly different operating characteristics than the series motor. Since the shunt field coil is made of fine wire, it cannot produce the large current for starting like the series field. This means that the shunt motor has very low starting torque, which requires that the shaft load be rather small. When voltage is applied to the motor, the high resistance of the shunt coil keeps the overall current flow low. The armature for the shunt motor is very similar to the series motor and it will draw enough current to produce a magnetic field strong enough to cause the armature shaft and load to start turning. Like the series motor, when the armature begins to turn, it will produce back EMF. The back EMF will cause the current in the armature to begin to diminish to a very small level. The amount of current the armature will draw is directly related to the size of the load when the motor reaches full speed. Since the load is generally small, the armature current will be small.

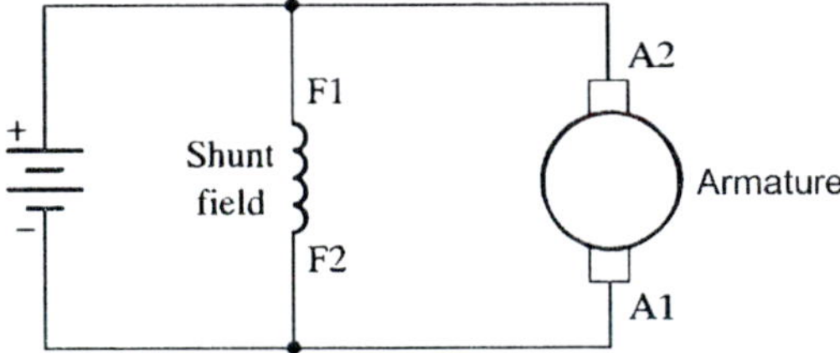

Figure 13–13 Diagram of DC shunt motor. Notice that the shunt coil is identified as a coil of fine wire with many turns that is connected in parallel (shunt) with the armature.

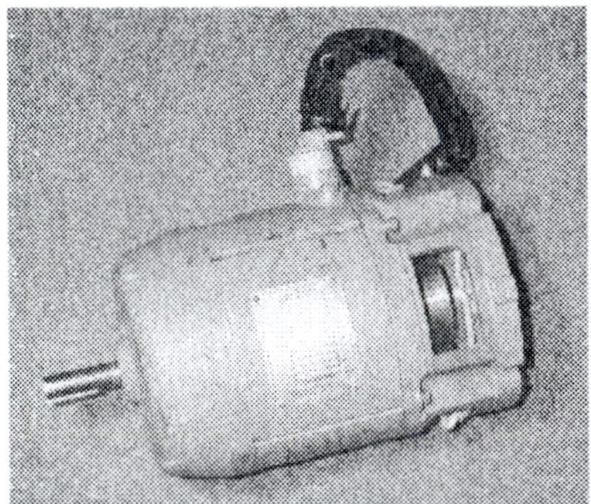

Figure 13–14 Typical DC shunt motor. These motors are available in a variety of sizes. This motor is 1 hp (approximately 8 inches tall).

13.5.2.2 Controlling the Speed of the Motor

When the shunt motor reaches full rpm, its speed will remain fairly constant. The reason the speed remains constant is due to the load characteristics of the armature and shunt coil. Remember that the speed of a series motor cannot be controlled since it is totally dependent on the size of the load in comparison to the size of the motor. If the load is very large for the motor size, then the speed of the armature will be very slow. If the load is light compared to the motor, the armature shaft speed will be much faster, and if no load is present on the shaft, the motor can run away.

The shunt motor's speed can be controlled. The ability of the motor to maintain a set rpm at high speed when the load changes is due to the characteristic of the shunt field and armature. Since the armature begins to produce back EMF as soon as it starts to rotate, it will use the back EMF to maintain its rpm at high speed. If the load increases slightly and causes the armature shaft to slow down, less back EMF will be produced. This will allow the difference between the back EMF and applied voltage to become larger, which will cause more current to flow. The extra current provides the motor with the extra torque required to regain its rpm when this load increased slightly.

The shunt motor's speed can be varied in two different ways: by varying the amount of current supplied to the shunt field and by controlling the amount of current supplied to the armature. Controlling the current to the shunt field allows the rpm to be changed 10 to 20 percent when the motor is at full rpm. This type of speed control regulation is accomplished by slightly increasing or decreasing the voltage applied to the field. The armature continues to have full voltage applied to it while the current to the shunt field is regulated by a rheostat that is connected in series with the shunt field. When the shunt field's current is decreased, the motor's rpm will increase slightly. When the shunt field's current is reduced the armature must rotate faster to produce the same amount of back EMF to keep the load turning, and if the shunt field current is increased slightly, the armature can rotate at a slower rpm and maintain the amount of back EMF to produce the armature current to drive the load. The field current can be adjusted with a field rheostat or an SCR current control.

The shunt motor's rpm can also be controlled by regulating the voltage that is applied to the motor armature. This means that if the motor is operated on less voltage than is shown on its data plate rating, it will run at less than full rpm. The shunt motor's efficiency will drop off drastically when it is operated below its rated voltage. The motor will tend to overheat when it is operated below full voltage, so motor ventilation must be provided. You should also be aware that the motor's torque is reduced when it is operated below the full voltage level.

Since the armature draws more current than the shunt field, the control resistors are much larger than those used for the field rheostat. SCRs are used for this type of current control. The SCR is able to control the armature current since it is capable of controlling several hundred amperes. In Chapter 17 we provide an in-depth explanation of the DC motor drive.

13.5.2.3
Torque Characteristics

The armature's torque increases as the motor gains speed due to the fact that the shunt motor's torque is directly proportional to the armature current. When the motor is starting and speed is very low, the motor has very little torque. After the motor reaches full rpm its torque is at its fullest potential. In fact, if the shunt field current is reduced slightly when the motor is at full rpm, the rpm will increase slightly and the motor's torque will also increase slightly. This type of automatic control makes the shunt motor a good choice for applications where constant speed is required, even though the torque will vary slightly due to changes in the load. Figure 13–15 shows the torque/speed curve for the shunt motor. From this diagram you can see that the speed of the shunt motor stays rather constant throughout its load range and drops slightly when it is drawing the largest current.

13.5.2.4
Reversing the Rotation of the Motor

The direction of rotation of a DC shunt motor can be reversed by changing the polarity of either the armature coil or the field coil. In this application the armature coil is usually changed, as was the case with the series motor. Figure 13–16 shows the electrical diagram of a DC shunt motor connected to a forward and reversing motor starter. The F1 and F2 terminals of the shunt field are connected directly to the power supply, and the A1 and A2 terminals of the armature winding are connected to the reversing starter.

When the forward starter is energized its contacts connect the A1 lead to the positive power supply terminal and the A2 lead to the negative power supply terminal. The F1 motor lead is connected directly to the positive terminal of the power supply and the F2 lead is connected to the negative terminal. When the motor is wired in this configuration it will begin to run in the forward direction.

When the reversing starter is energized, its contacts reverse the armature wires so that the A1 lead is connected to the negative power supply terminal and the A2 lead is connected to the positive power supply

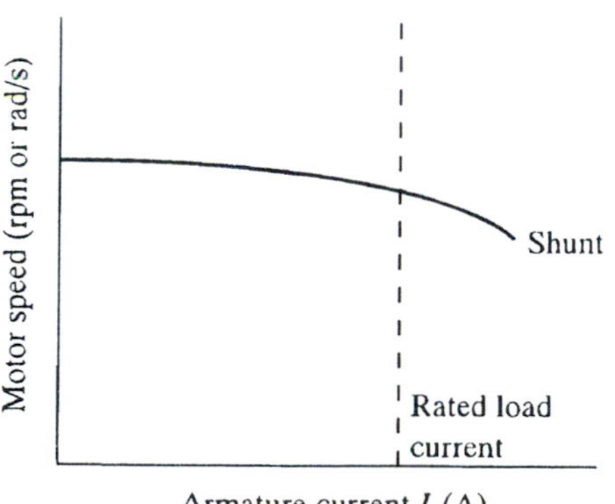

Figure 13–15 A curve that shows the armature current versus the armature speed for a shunt motor. Notice that the speed of a shunt motor is nearly constant.

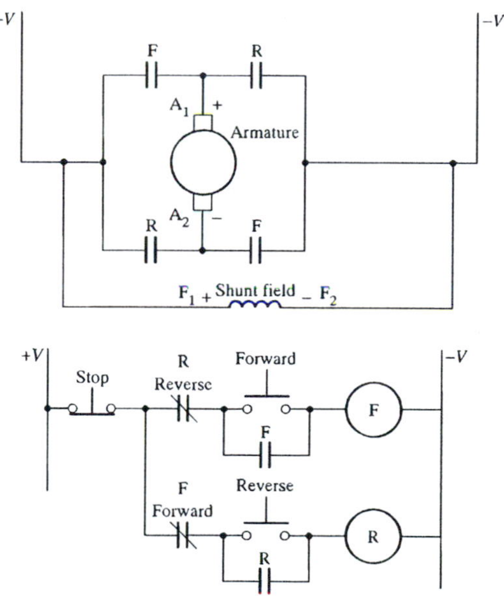

Figure 13–16 Diagram of a shunt motor connected to a reversing motor starter. Notice that the shunt field is connected across the armature and it is not reversed when the armature is reversed.

terminal. The field leads are connected directly to the power supply, so their polarity is not changed. Since the field's polarity has remained the same and the armature's polarity has reversed, the motor will begin to rotate in the reverse direction. The control part of the diagram shows that when the forward motor starter coil is energized, the reverse motor starter coil is locked out.

13.5.2.5
Installing the Shunt Motor

The shunt motor can be installed rather easily. The motor is generally used in belt-drive applications, which means that the installation procedure should be broken into two sections: the mechanical installation of the motor and its load, and the installation of electrical wiring and controls.

When the mechanical part of the installation is completed, the alignment of the motor shaft and the load shaft should be checked. If the alignment is not true, the load will cause an undue stress on the armature bearing and there is the possibility of the load vibrating and causing damage to it and the motor. After the alignment is checked, the tension on the belt should also be tested. As a rule of thumb, you should have about 1/2 to 1/4 inch of play in the belt when it is properly tensioned. Several tension measurement devices are available to determine when a belt is tensioned properly. The belt tension can also be compared to the amount of current the motor draws. The motor must have its electrical installation completed to use this method.

The motor should be started, and if it is drawing too much current, the belt should be loosened slightly but not enough to allow the load to slip. If the belt is slipping, it can be tightened to the point where the motor is able to start successfully and not draw current over its rating.

The electrical installation can be completed before, after, or during the mechanical installation. The first step in this procedure is to locate the field and armature leads in the motor and prepare them for field connections. If the motor is connected to magnetic or manual across the line starter, the F1 field coil wire can be connected to the A1 armature lead and an interconnecting wire, which will be used to connect these leads to the L1 terminal on the motor starter. The F2 lead can be connected to the A2 lead and a second wire, which will connect these leads to the L2 motor starter terminal.

When these connections are completed, field and armature leads should be put back into the motor and the field wiring cover or motor access plate should be replaced. Next, the DC power supply's positive and negative leads should be connected to the motor starter's T1 and T2 terminals, respectively.

After all of the load wires are connected, any pilot devices or control circuitry should be installed and connected. The control circuit should be tested with the load voltage disconnected from the motor. If the control circuit uses the same power source as the motor, the load circuit can be isolated so the motor will not try to start by disconnecting the wire at terminal L2 on the motor starter. Operate the control circuit several times to ensure that it is wired correctly and operating properly. After you have tested the control circuit, the lead can be replaced to the L2 terminal of the motor starter and the motor can be started and tested for proper operation. Be sure to check the motor's voltage and current while it is under load to ensure that it is operating correctly. It is also important to check the motor's temperature periodically until you are satisfied the motor is operating correctly.

If the motor is connected to a reversing starter or reduced voltage starting circuit, their operation should also be tested. You may need to read Chapter 16 to fully understand the operation of these methods of starting the motor using reduced-voltage methods. If the motor is not operating correctly or develops a fault, the troubleshooting procedure should be used to test the motor and locate the problem.

13.5.2.6
Troubleshooting the Shunt Motor

When the DC shunt motor develops a fault, you must be able to locate the problem quickly and return the motor to service or have it replaced. The most likely problems to occur with the shunt motor include loss of supply voltage, or an open in either the shunt winding or the armature winding. Other problems may arise that will cause the motor to run abnormally hot even though it continues to drive the load. The motor will show different symptoms for each of these problems, which will make the troubleshooting procedure easier.

When you are called to troubleshoot the shunt motor it is important to try and determine if the problem occurs while the motor is running or when it is trying to start. If the motor will not start, you should listen to see if the motor is humming and trying to start. When the supply voltage has been interrupted due to a blown fuse or a deenergized control circuit, the motor will not be able to draw any current and it will be silent when you try to start it. You can also determine that the supply voltage has been lost by measuring it with a voltmeter at the starter's L1 and L2 terminals. If no voltage is present at the load terminals, you should check for voltage at the starter's T1 and T2 terminals. If voltage is present here but not at the load terminals, it indicates that the motor starter is deenergized or defective. If no voltage is present at the T1 and T2 terminals, it indicates that supply voltage has been lost prior to the motor starter.

You will need to check the supply fuses and the rest of the supply circuit to locate the fault.

If the motor tries to start and hums loudly, it indicates that the supply voltage is present. The problem in this case is probably due to an open field winding or armature winding. It could also be caused by the supply voltage being too low.

The most likely problem will be an open in the field winding, since it is made from very small-gauge wire. The open can occur if the field winding draws too much current or develops a short circuit between the insulation in the coils. The best way to test the field is to remove supply voltage to the motor by opening the disconnect or deenergizing the motor starter. Be sure to use a *lockout* when you are working on the motor after the disconnect has been opened. The lockout is a device that is placed on the handle of the disconnect after the handle is placed in the off position, and it allows a padlock to be placed around it that cannot be removed until the technician has completed the work on the circuit. A lockout may have extra holes so that additional padlocks can be placed on it by other technicians who are also working on this system. This ensures that the power cannot be returned to the system until the last technician has removed his or her padlock. The lockout was shown in Chapter 2 in this text.

After power has been removed, the field terminals should be isolated from the armature coil. This can be accomplished by disconnecting one set of leads where the field and armature are connected together. Remember that the field and armature are connected in parallel and if they are not isolated, your continuity test will show a completed circuit even if one of the two windings have an open.

When you have the field coil isolated from the armature coil, you can proceed with the continuity test. Be sure to use the $R \times 1k$ or $R \times 10k$ setting on the ohmmeter, because the resistance in the field coil will be very high since the field coil may be wound from several thousand feet of wire. If the field winding test indicates the field winding is good, you should continue the procedure and test the armature winding for continuity.

The armature winding test may show that an open has developed from the coil burning open or from a problem with the brushes. Since the brushes may be part of the fault, they should be visually inspected and replaced if they are worn or not seating properly. If the commutator is also damaged, the armature should be removed, so the commutator can be turned down on a lathe.

If either the field winding or the armature winding have developed an open circuit, the motor will have to be removed and replaced. In some larger motors it will be possible to change the armature by itself instead of removing and replacing the entire motor. If the motor operates but draws excessive current or heats up, the motor should be tested for loose or shorting coils. Field coils may tend to come loose and cause the motor to vibrate and overheat, or the armature coils may come loose from their slots and cause problems. If the motor continues to overheat or operate roughly, the motor should be removed and sent to a motor rebuilding shop so that a more in-depth test may be performed to find the problem before the motor is permanently damaged by the heat.

13.5.3 DC Compound Motors

The DC compound motor is a combination of the series motor and the shunt motor. It has a series field winding that is connected in series with the armature and a shunt field that is in parallel with the armature. The combination of series and shunt winding allows the motor to have torque characteristics of the series motor and regulated speed characteristics of the shunt motor. Figure 13–17 shows diagrams of the compound motor. Several versions of the compound motor are shown.

13.5.3.1 Cumulative Compound Motors

Figure 13–17a shows a diagram of the cumulative compound motor. It is called cumulative because the shunt field is connected so that its coils are aiding the magnetic fields of the series field and armature. The shunt

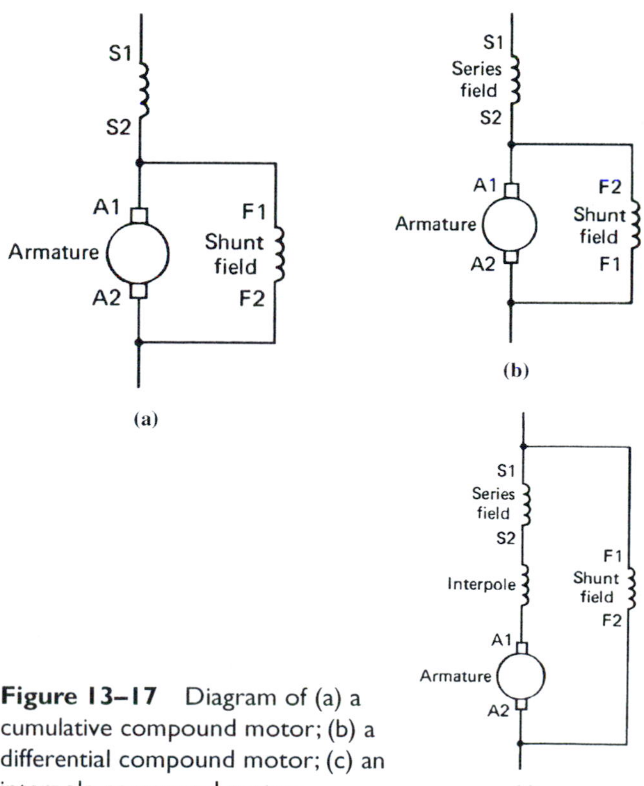

Figure 13–17 Diagram of (a) a cumulative compound motor; (b) a differential compound motor; (c) an interpole compound motor.

winding can be wired as a *long shunt* or as a *short shunt*. Figures 13–17a and 13–17b show the motor connected as a short shunt where the shunt field is connected in parallel with only the armature. Figure 13–17c shows the motor connected as a long shunt where the shunt field is connected in parallel with both the series field, interpoles, and the armature.

Figure 13–17a also shows the short shunt motor as a cummulative compound motor, which means that the polarity of the shunt field matches the polarity of the armature. The top of the shunt field is positive polarity and is connected to the positive terminal of the armature. In Figure 13–17b the shunt field has been reversed so that the negative terminal of the shunt field is now connected to the positive terminal of the armature. This type of motor is called a differential compound, because the polarities of the shunt field and the armature are opposite.

The cumulative compound motor is one of the most common DC motors because it provides high starting torque and good speed regulation at high speeds. Since the shunt field is wired with similar polarity in parallel with the magnetic field aiding the series field and armature field, it is called cumulative. When the motor is connected this way, it can start even with a large load, and then operate smoothly when the load varies slightly.

You should recall that the shunt motor can provide smooth operation at full speed but it cannot start with a large load attached, and the series motor can start with a heavy load, but its speed cannot be controlled. The cumulative compound motor takes the best characteristics of both the series motor and shunt motor, which makes it acceptable to most applications.

13.5.3.2 Differential Compound Motors

Differential compound motors use the same motor and windings as the cumulative compound motor but they are connected in a slightly different manner to provide slightly different operating speed and torque characteristics. Figure 13–17b shows the diagram for a differential compound motor with the shunt field connected so its polarity is reversed to the polarity of the armature. Since the shunt field is still connected in parallel with only the armature, it is considered a short shunt.

In this diagram F1 and F2 are connected in reverse polarity to the armature. In the differential compound motor the shunt field is connected so that its magnetic field opposes the magnetic fields in the armature and series field. When the polarity of the shunt field is reversed like this, its field will oppose the other fields and the characteristics of the shunt motor are not as pronounced in this motor. This means that the motor will tend to overspeed when the load is reduced just like a series motor. Its speed will also drop more than the cumulative compound motor when the load increases at full rpm. These two characteristics make the differential motor less desirable than the cumulative motor for most applications.

13.5.3.3 Compound Interpole Motors

The compound interpole motor is built slightly differently from the cumulative and differential compound motor. This motor has interpoles added to the series field (Figure 13–17c). The interpoles are connected in series between the armature and series winding. They are physically located behind the series coil in the stator. They are made of wire that is the same gauge as the series winding and it is connected so that its polarity is the same as the series winding pole it is mounted behind. Remember that these motors may have any number of poles to make the field stronger.

The interpole prevents the armature and brushes from arcing due to the buildup of magnetic forces. These forces are created from counter EMF called *armature reaction*. They are so effective that normally all DC compound motors that are larger than 1/2 hp will utilize them. Since the brushes do not arc, they will last longer and the armature will not need to be cut down as often. The interpoles also allow the armature to draw heavier currents and carry larger shaft loads.

When the interpoles are connected, they must be tested carefully to determine their polarity so that it can be matched with the series winding. If the polarity of the interpoles does not match that of the series winding it is mounted behind, it will cause the motor to overheat and may damage the series winding.

13.5.3.4 Reversing the Rotation of the DC Compound Motor

Each of the compound motors shown in Figure 13–17 can be reversed by changing the polarity of the armature winding. If the motor has interpoles, the polarity of the interpole must be changed when the armature's polarity is changed. Since the interpole is connected in series with the armature, it will be reversed when the armature is reversed. The interpoles are not shown in diagrams to keep them simple. The armature winding is always marked as A1 and A2 and these terminals should be connected to the contacts of the reversing motor starter.

13.5.3.5 Controlling the Speed of the Motor

The speed of a compound motor can be changed very easily by adjusting the amount of voltage applied to it. In fact, it can be generalized that at one time any industrial application that required a motor to have a constant speed would be handled by an AC motor, and any application that required the load to be driven at variable speeds would automatically be handled by DC motors.

This statement was true because the speed of a DC motor was easier to change than that of an AC motor. Since the advent of solid-state components and microprocessor controls, this condition is no longer true. In fact, today, a solid-state AC variable frequency motor drive can vary the speed of an AC motor as easily as that of DC motors. This brings about a condition where you must understand methods of controlling the speed of both AC and DC motors. Information about AC motor speed control is provided in Chapter 17 DC and AC Drives.

Figure 13–18 shows the characteristic curves of the speed vs. armature current for compound motors. You can see that the speed of a differential compound motor increases slightly when the motor is drawing the armature's highest current. The increase in speed occurs because the extra current in the differential winding causes the magnetic field in the motor to weaken slightly because the magnetic field in the differential winding opposes the magnetic in series field. As you learned earlier in the speed control of shunt motors, the speed of the motor will increase if the magnetic field is weakened.

Figure 13–18 also shows the characteristic curve for the cumulative compound motor. This curve shows that the speed of the cumulative compound motor decreases slightly because the field is increased, which slows the motor because the magnetic field in the shunt winding aids the magnetic field of the series field.

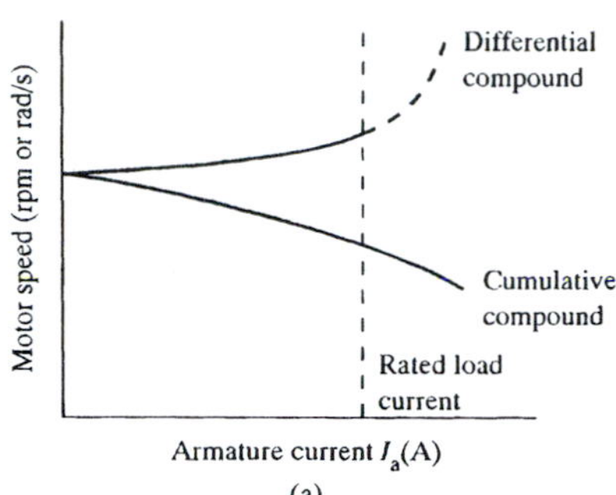

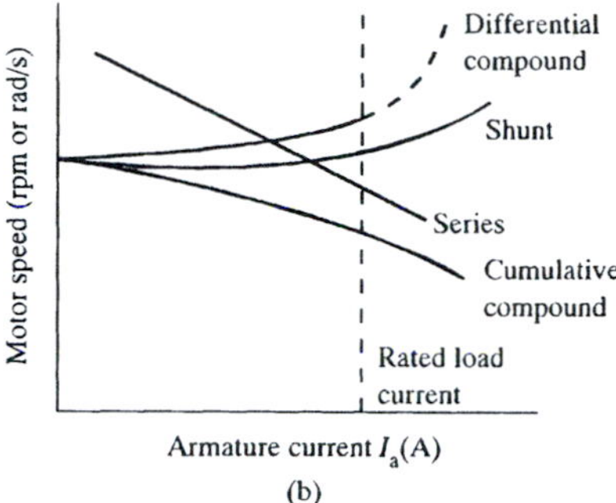

Figure 13–18 (a) Characteristic curve of armature current versus speed for the differential compound motor and cumulative compound motor. (b) Composite of the characteristic curves for all of the DC motors.

Questions

1. What is the function of the armature in a DC motor?
2. What is the function of the shunt field in a DC shunt motor?
3. What is the function of the series field in a DC series motor?
4. Explain how the forward and reverse motor starter circuit shown in Figure 13–12 provides lockout protection so the motor cannot be switched directly from the forward direction to the reverse direction.
5. Explain how you can change the speed of a DC shunt motor.

True and False

1. The series field of a DC series motor is made of very fine wire so that its magnetic field will be as strong as possible.
2. The armature is the rotating part of a DC motor.
3. The brushes in a DC motor make contact between the stationary part of the motor and the armature by touching the commutator segments.
4. The speed of a DC shunt motor is adjusted by changing the frequency of the supply voltage for the motor.
5. The direction of rotation of a series motor can be changed by changing the polarity of either the armature or field winding.

Multiple Choice

1. The armature in a DC motor is _____
   a. the rotating part of the motor.
   b. the part of the motor that makes contact with the brushes.
   c. made of heavy wire and is mounted in the stationary part of the motor.
   d. made of light-gauge wire and is mounted in the stationary part of the motor.
   e. All of the above
2. The series field in a DC motor is ______
   a. the rotating part of the motor.
   b. the part of the motor that makes contact with the brushes.
   c. made of heavy wire and is mounted in the stationary part of the motor.
   d. made of light-gauge wire and is mounted in the stationary part of the motor.
   e. All of the above
   f. Only a and b

3. The commutator in a DC motor is ______
   a. the rotating part of the motor.
   b. the part of the motor that makes contact with the brushes.
   c. made of heavy wire and is mounted in the stationary part of the motor.
   d. made of light-gauge wire and is mounted in the stationary part of the motor.
   e. All of the above
   f. Only a and b
4. The function of the end plates on a DC motor is to _______
   a. provide a mounting for the armature bearings.
   b. provide support for the armature shaft.
   c. provide a mounting point for the brush rigging.
   d. All of the above
   e. Only a and b
5. You can reverse the direction of a DC motor by _________
   a. reversing the field winding if the motor is a series motor.
   b. reversing the field winding if the motor is a shunt motor.
   c. reversing the armature winding if the motor is a series motor.
   d. All of the above
   e. Only a and b

Problems

1. Identify the three magnetic principles used to make a DC motor operate.
2. Draw an electrical diagram of a DC series motor.
3. What will happen to a DC series motor if the shaft breaks between the motor and the load?
4. Draw an electrical diagram of a cumulative, differential, and interpole compound motor.
5. Draw an electrical diagram to a DC series motor.

◀ Chapter 14 ▶

AC Motors

Objectives

After reading this chapter you will be able to:

1. Explain the term induction.
2. Explain the operation of an AC single-phase and three-phase motor.
3. Explain how to wire a wye and delta motor.
4. Explain why a single-phase AC motor needs a start winding and run winding.
5. Explain the operation of a capacitor-start motor and a permanent split-capacitor motor.

Today, AC motors are more widely used in industrial applications than DC motors. They are available to operate on single-phase or three-phase supply voltage systems. This allows the motor control designer to choose the type of motor to fit the application. Most single-phase motors are less than 3 hp; some larger ones are available, but they are not as common. Three-phase motors are available up to several thousand horsepower, although most of the motors that you will be working with will be less than 50 hp.

The AC motor provides several advantages over DC motors. First, its design eliminates the need for brushes and commutators, and second, its rotating member is made of laminated steel rather than wire that is pressed on a core, which reduces maintenance. The AC motor does not need brushes and commutators, since it creates the flux lines in its rotating member by *induction*. The induction process that is used to get the current into the rotating member is similar to the induction that occurs between the primary and secondary windings of a transformer. This is possible in an AC motor because supply voltage is sinusoidal.

The rotating field in the AC motor is called the *rotor*, and the stationary field is called the *stator*. The design of the rotor is different from the rotating armature in the DC motor because it is made completely of laminated steel rather than having copper coils pressed on a laminated steel core. This allows the AC motor to operate longer than the DC motor with less periodic maintenance, which means that more AC motors are used in industry than DC motors. The main reason DC motors were used in industry is that their speed could be controlled more easily than the speed of AC motors. With the advent of variable frequency drives, the speed of all AC motors can be adjusted more easily than that of DC motors, and the AC motor requires less maintenance, since it does not have brushes. All of these things have led to conditions where the AC motor is much more prevalent than the DC motor in modern industrial applications.

In this section we introduce each of the different types of AC motors and explain their basic parts, theory of operation, methods of controlling their speed and torque, changing the direction of rotation, and procedures for installation and troubleshooting. This basic information will also allow you to better understand methods that motor controls use to take advantage of the motor's design to provide control. If you understand how the motor operates, you will understand what the motor control is trying to control.

We will also introduce the basic parts that are found in all AC motors and explain their operation and function. After the operation of a basic three-phase motor is explained, each type of AC motor is introduced and its special design features and applications for which it is best suited are discussed. You will be able to use this information to recognize the type of AC motor you are working with and to understand the theory of its operation, which will allow you to install and interface it to motor controls and to troubleshoot the motor and quickly determine what faults it has. The three-phase motor is presented in the first part of the chapter, since some of the parts of a single-phase motor are designed specifically to compensate for the differences between three-phase and single-phase voltage. If you fully understand the characteristics of three-phase voltage and how three-phase motors take advantage of them, you will easily understand single-phase motors.

14.1 Characteristics of Three-Phase Voltage

The three-phase AC motor is an induction motor. It is designed specifically to take advantage of the characteristics of the three-phase voltage that it uses for power. For this reason it is important to review the characteristics of three-phase voltage. Figure 14–1 shows a diagram of three-phase voltage. Each of the three phases represents a separately generated voltage. The three-phase generator has three separate windings that produce the three voltages slightly out of phase with each other. The units of measure for this voltage are electrical degrees. One sine wave has 360 degrees. The sine wave is produced once during each rotation of the generator's shaft or twice during each generator's shaft rotation. If the sine wave is produced by one rotation of the generator's shaft, 360 electrical degrees are equal to 360 mechanical degrees. If the sine wave is produced twice during each shaft's rotation, 360 electrical degrees are equal to 180 mechanical degrees. Since this can be confusing, all electrical diagrams are presented in terms of 360 electrical degrees being equal to 360 mechanical degrees. In this way the degrees will be the same and you will not have to try and figure out if electrical or mechanical degrees are being used in the example.

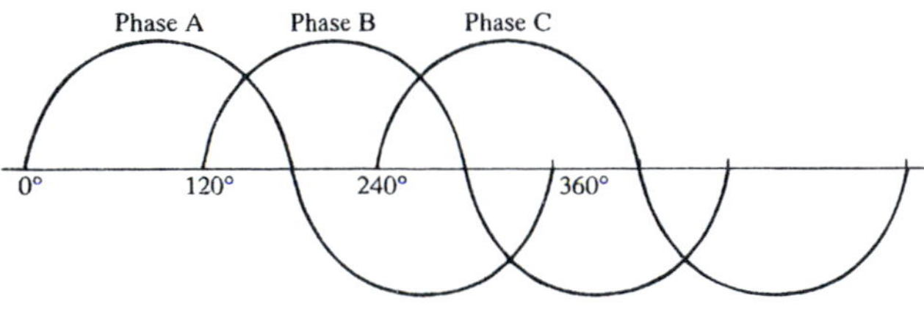

Figure 14–1 Diagram of three-phase voltage. Notice the phase shift between each voltage sine wave is 120 degrees.

The first voltage shown in the diagram is called A phase and it is shown starting at 0 and peaking positively at the 90-degree mark. It passes through 0 V again at the 180-degree mark and peaks negatively at the 270-degree mark. After it peaks negatively, it returns to 0 V at the 360-degree mark, which is also the 0-degree point. The second voltage is called B phase and starts its zero voltage point 120 degrees later than A phase. B phase peaks positive, passes through zero voltage, and passes through negative peak voltage as A phase does, except that it is always 120 degrees later than A phase. This means that B phase is increasing in the positive direction when A phase is passing through its zero voltage at the 180-degree mark.

The third voltage shown on this diagram is called C phase. It starts at its zero voltage point 240 degrees after A phase starts at its zero voltage point. This puts B phase 120 degrees out of phase with A phase and C phase 120 degrees out of phase with B phase.

The AC motor takes advantage of this characteristic to provide a rotating magnetic field in its stator and rotor that is very strong because three separate fields rotate 120 degrees out of phase with each other. Since the magnetic fields are induced from the applied voltage, they will always be 120 degrees out of phase with each other. Do not worry about the induced magnetic field being 180 degrees out of phase with the voltage that induced it. At this time this phase difference is not as important as the 120-degree phase difference between the rotating magnetic fields.

Since the magnetic fields are 120 degrees out of phase with each other and are rotating, one will always be increasing its strength when one of the other phases is losing its strength by passing through the zero voltage point on its sine wave. This means that the magnetic field produced by all three phases never fully collapses and its average is much stronger than that of a field produced by single-phase voltage.

14.2 Three-Phase Motor Components

The AC induction motor has three basic parts: the stator, which is the stationary part of the motor; the rotor, which is the rotating part of the motor; and the end plates, which house the bearings that allow the rotor to rotate freely. This section provides information about each of these parts of an AC motor. Figure 14–2 shows a cutaway diagram of a three-phase motor, and Figure 14–3 shows an exploded view picture of a three-phase motor. These figures will provide you with information about the location of the basic parts of a motor and how they work together.

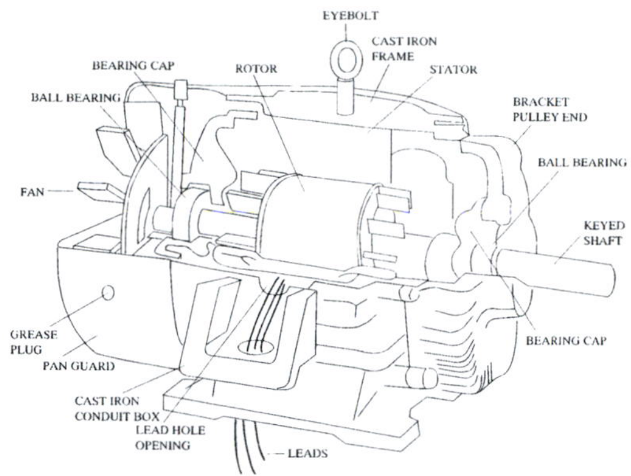

Figure 14–2 Cutaway diagram of the three-phase motor. (Courtesy of A.O. Smith Electrical Products Company)

Figure 14–3 Exploded view of a three-phase motor, showing the relative location of all the parts of the motor.

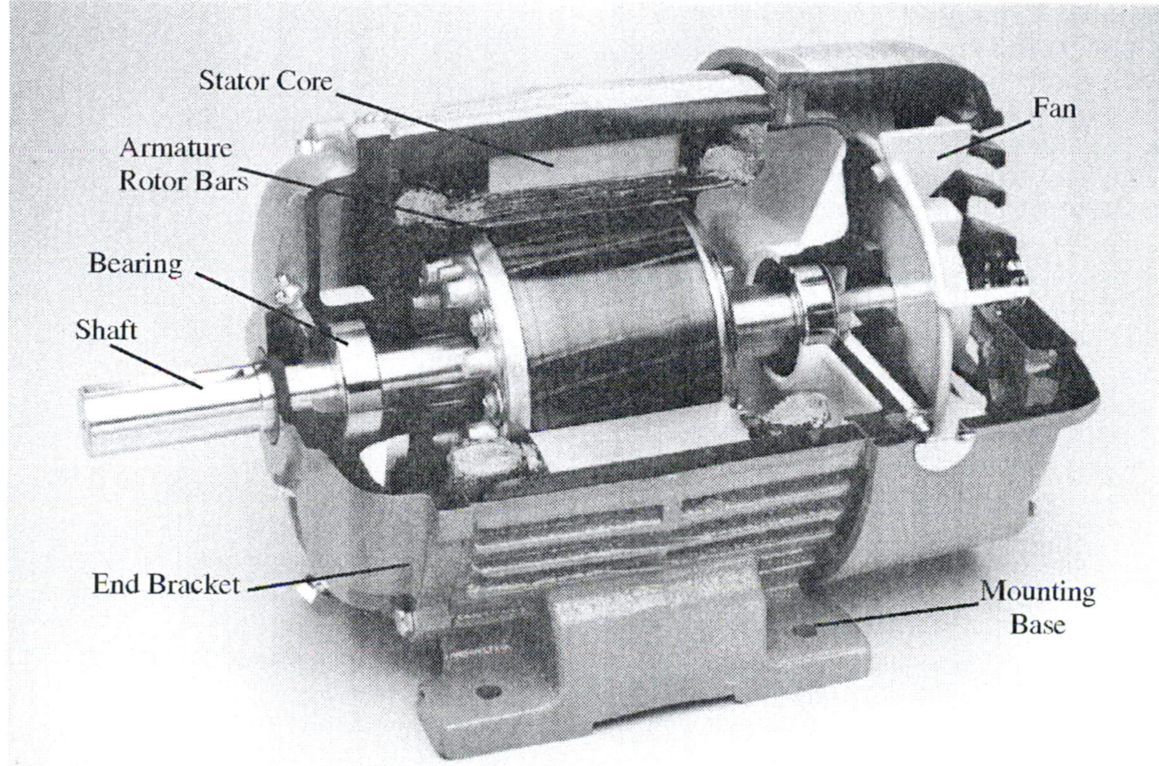

Figure 14–4 Example of a typical stator for an AC motor. (Courtesy of A.O. Smith Electrical Products Company)

14.2.1 Stator

The stator is the stationary part of the motor and is made of several parts. Figure 14–4 shows the stator and other major parts of a typical induction motor. The stator is the frame for the motor and it houses the stationary winding and provides mounting holes for installation. The mounting holes for the motor are sized according to NEMA standards for the motor's frame type. Some motors will also have a lifting ring in the stator to provide a means for handling larger motors. The lifting ring and mounting holes are actually built into the frame or housing part of the stator.

An insert is set inside the stator that provides slots for the stator coils to be inserted into. This insert is made of laminated steel to prevent eddy current and flux losses in the coils. The stator windings are made by wrapping a predetermined length of wire on preformed brackets in the shape of the coil. These windings are then wrapped with insulation and installed in the stator slots. A typical four-pole, three-phase motor will have three coils mounted consecutively in the slots to form a group. The three coils will be wired so that they each receive power from a separate phase of the three-phase power supply. Three groups are connected together to form one of the four poles of the motor. This grouping is repeated for each the other three poles, so that the motor has a total of 36 coils to form the complete four-pole stator. It not essential that you understand how to wind the coils or put them into the stator slots; rather you should understand that these coils are connected in the stator, and 3, 6, 9, 12, or 15 wires from the coil connections will be brought out of the frame as external connections. The external connections wires can be connected on the factory floor to allow the motor to be powered by 208/230 or 480 V, or they allow the motor to be connected to provide the correct torque response for the load. Other changes can also be made to these connections to allow the motor to start so it uses less locked-rotor current (LRA).

After the coils are placed in the stator, their ends (leads) will be identified by a number that will be used to make connections during the installation procedure. The coils are locked into the stator with wedges that keep the coils securely mounted in the slots and allow them to be removed and replaced easily if the coils are damaged or become defective due to overheating.

14.2.2 Rotor

The rotor in an AC motor can be constructed from coils of wire wound on laminated steel or it can be made entirely from laminated steel without any wire coils. A rotor with wire coils is called a *wound rotor* and it is used in a wound rotor motor. The wound rotor motor was a very popular type of AC motor in the 1950s since it could produce more torque than a similar size induction motor. The main drawback of the wound rotor motor is that it requires the use of brushes and slip rings to transfer current to it. Since a modern AC induction motor can now produce adequate torque, the wound rotor motor is not used very often because the brushes and slip rings require too much maintenance.

Motors that use a laminated steel rotor are called *induction motors* or *squirrel-cage induction motors*. The core of the rotor is made of die-cast aluminum in the shape of a squirrel cage. Laminated sections are pressed onto this core or the core is molded into laminated sections when the squirrel-cage rotor is manufactured. Figure 14–5 shows a diagram and picture of a squirrel-cage rotor and you can see the skeleton of the squirrel-cage core. The fins or blades are built into the rotor for cooling the motor. It is important that these fan

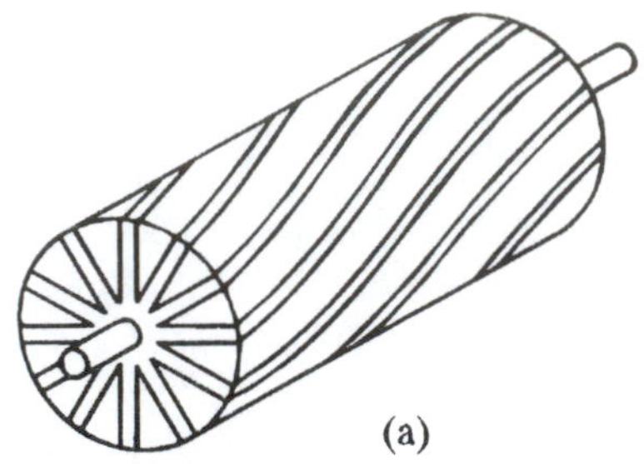

(a)

(b)

Figure 14–5 Squirrel-cage rotor for an AC motor: (a) diagram; (b) picture.

blades are not damaged or broken, since they provide all of the cooling air for the motor and they are balanced so that the rotor will spin evenly without vibrations.

14.2.3 Motor End Plates

The end plates house the bearings for the motor. The end plate and bearing can be seen in the picture of the rotor that is shown in Figure 14–5b. If the motor is a fractional-horsepower motor, it will generally use sleeve-type bearings, and if the motor is one of the larger types, it will use ball bearings. Some ball bearings on smaller motors will be permanently lubricated, while the larger motor bearings will require periodic lubrication. All sleeve bearings will require a few drops of lubricating oil periodically.

The end plates are mounted on the ends of the motor and held in place by long bolts inserted through the stator frame. When nuts are placed on the bolts and tightened, the end plates will be secured in place. If the motor is an open type, the end plates will have louvers to allow cooling air to circulate through the motor. An access plate may also be provided in the rear end plate to allow field wiring if one is not provided in the stator frame.

If the motor is not permanently lubricated the end plate will provide an oiler tube or grease fitting for lubrication. It is very important that the end plates are mounted on the motor so that the oiler tube or grease fitting is above the shaft so that gravity will cause lubrication to reach the shaft. If the end plate is rotated so that the lubrication point is mounted below the shaft, gravity will pull all of the lubrication away from the shaft and the bearings will wear out prematurely. If you need to remove the end plates for any reason, they should be marked so that they will be replaced in the exact position from where they were removed. This also helps to align the bolt holes in the end plate with the holes in the stator so that the end plates can be reassembled easily.

14.3 Operation of an AC Induction Motor

The basic principle of operation of an inductive motor is based on the fact that the rotor receives its current by induction rather than with brushes and slip rings or brushes and commutators. Current can be induced into the rotor by being in close proximity to the stator. This action is very similar to the action of the primary and secondary coils of a transformer. You should remember from your study of transformers and generators that when a coil of wire is allowed to pass across magnetic flux lines, a current will be generated in the coil. This current will be 180 degrees out of phase with the current that produced it.

When the induced current begins to flow in the squirrel-cage conductors imbedded in the laminated segments of the squirrel-cage rotor, a magnetic field will be built in the rotor. The magnetic field produced by the induced current will be very similar to the magnetic field produced by current that is provided to the rotor with brushes. Since the current is induced into the rotor, no brushes are required. The strength of the field in the rotor is not as strong as a field that could be produced by passing current through brushes to a coil of wire on the rotor. This means that the magnetic field in the stator must be much stronger to compensate for the weaker field in the rotor to produce sufficient horsepower and torque.

When three-phase voltage is applied to the stator windings, a magnetic field is formed. The natural characteristic of three-phase voltage will cause the magnetic field to move from coil to coil in the stator, which appears as though the field is rotating. Figure 14–6 shows an experiment your instructor can perform in your laboratory to prove that the magnetic field in the stator will actually rotate. From this figure you can see that the end plates and rotor have been removed from the stator. The three stator windings are connected to three-phase voltage through a switch. A large ball bearing, approximately 1 inch in diameter, is placed in the stator, and the switch is closed to energize the stator winding. When the magnetic field begins to form in the stator, it will begin to pull on the ball bearing. You may need to give the ball bearing a slight push with a plastic rod to get it to begin to rotate. Do not use a metal object such as a screwdriver to get the ball bearing to move because the tip of the screwdriver will be attracted to the stator by its magnetic field.

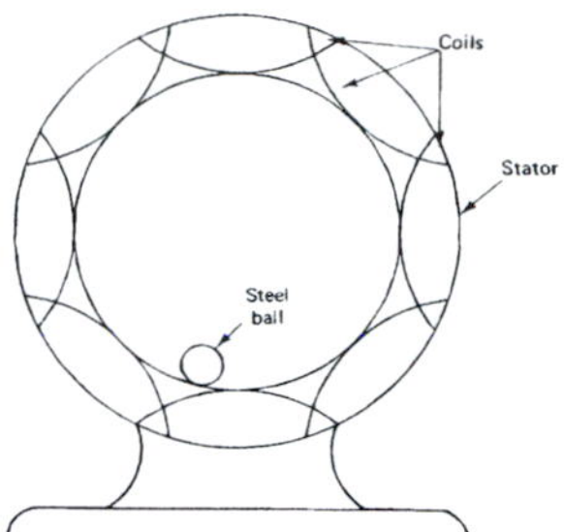

Figure 14–6 Experiment that shows a ball bearing placed in the stator of a three-phase AC motor. When the three-phase voltage is applied to the rotor, a rotating magnetic field will be established, and the ball bearing will close the rotating field around the stator. When the rotor is placed in the stator it will rotate in step with the rotating magnetic field.

Once you get the ball bearing to rotate, it will continue to rotate at the speed of the rotating magnetic field in the stator until the switch is opened and the stator's magnetic field has collapsed. If you reverse two of the three-phase supply voltage wires, the magnetic field will begin to rotate in the opposite direction, and when you give the ball bearing a push to start it, it will also rotate in the opposite direction.

Important Safety Notice . . .
This experiment is intended as a demonstration that your instructor will set up for you. It is recommended that a motor stator that is rated for 460 or 550 VAC be used for this experiment and that 208 or 230 VAC is applied to the windings so they do not overheat. It is also important that the voltage is not provided to the windings for more than 1–2 minutes at a time. A variable frequency drive could also be used to provide the voltage to the stator for this experiment. The frequency of the drive could be adjusted during the experiment to show the change in speed of the rotating magnetic field in the stator.

14.3.1 Induced Current in the Rotor

When the magnetic field in the stator cuts across the poles of the squirrel-cage rotor, a current is induced in the rotor. This current is out of phase with the applied current, but it is strong enough to cause the rotor to start to turn. The speed of the rotor is determined by the number of poles in the stator and the frequency of the incoming AC voltage. A formula is provided to determine the operating speed of the motor:

$$\text{Operating speed of motor} = \frac{120F}{P}$$

where F is the frequency of the applied voltage, 120 is a magnetic constant, and P is the number of poles. This formula calculates the speed of the rotating field, but the actual speed of the rotor will be slightly less due to slip. The concept of slip will be explained in the next sections.

The full rpm is called synchronous speed. From this formula we calculate that a two-pole motor will operate at 3600 rpm, a four-pole motor will operate at 1800 rpm, a six-pole motor will operate at 1200 rpm, and an eight-pole motor will operate at 900 rpm. These speeds do not include any slip or losses due to loads. From this example you can see that the only way that the speed of an AC induction motor can be changed is to change the number of poles it has or change the frequency of the voltage supplied to it.

When power is first applied, the stator field will draw very high current since the rotor is not turning. This current is called *locked-rotor amperage* (LRA) and is sometimes referred to as *inrush current*. When LRA moves through the stator, its magnetic field is strong enough to cause the rotor to begin to rotate. As the rotor starts moving, it will begin to induce current into its laminated coils and build up torque. This causes the rotor to spin faster, until it begins to catch up with the rotating magnetic field.

As the rotor turns faster, it will begin to produce voltage of its own. This voltage is called *back EMF* or *counter EMF*. The counter EMF opposes the applied voltage, which has the effect of lowering the difference of potential across the stator coils. The lower potential causes current to become lower when the motor is a full load. The full-load amperage is referred to as FLA and will be as much as 6 to 10 times smaller than the inrush current (LRA). The stator will draw just enough current to keep the rotor spinning.

When the load on the rotor increases, it will begin to slow down slightly. This causes the counter EMF to drop slightly, which makes the difference in potential greater and allows more current to flow. The extra current provides the necessary torque to move the increased load and the rotor's speed catches up to its rated level. In this way, the squirrel-cage induction motor is allowed automatically to regulate the amount of current it requires to pull a load under varying conditions. The rotor will develop maximum torque when the rotor has reached 70 to 80 percent of synchronous speed. The motor can make adjustments anywhere along its torque range. If the load becomes too large, the motor shaft will slow to the point of stalling and the motor will overheat from excess current draw. In this case the motor must be wired for increased torque, or a larger-horsepower motor should be used.

14.3.2 Connecting Motors for Torque Speed and Horsepower Conditions

The squirrel-cage induction motor can be connected in several different ways to produce constant torque, a change of speed, or a change of horsepower ratings. It can also be connected for variable torque. The load conditions or applications will dictate the method you select to connect the motor. These connections can be made after the motor is installed on the factory floor or when the motor is being installed in the machinery it is driving. The next sections provide examples of each way the motor can be connected to make these changes.

14.3.2.1 Wye-Connected Motors

The squirrel-cage induction motor may have 6, 9, or 12 terminal leads from the ends of its coils brought out of the motor frame for field wiring connections. Field wiring connections are the connections that you will change as a technician right on the factory floor with the motor mounted in place. Figure 14–7 shows the diagram of a nine-lead motor. Notice that the coils are positioned in the shape of a the letter Y. This motor is called a wye-connected motor or star-connected motor. The wye motor terminals are numbered in the clockwise direction. The two ends of the first coil are num-

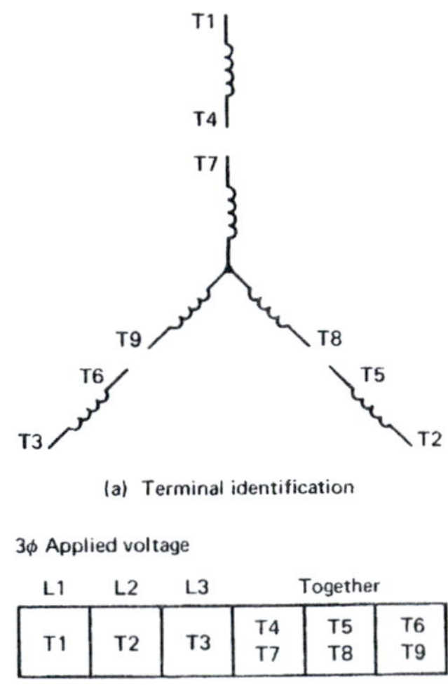

Figure 14–7 Nine-lead, wye-connected motor.

bered 1 and 4, the second coil is numbered 2 and 5, and the third coil is numbered 3 and 6. The outside coils are isolated from each other, while the inside coils are connected together at the wye point. The second diagram shows the motor connected to three-phase line voltage at T1, T2, and T3. A table is also provided to show which terminals are connected together and which ones are connected to line voltage for the motor to operate correctly. The information provided in this table is similar to data provided on the data plate of the motor.

You will need to make these connections in the field when you are installing the motor. Wire nuts or lugs should be used to make the connections for the coil terminals and supply voltage wires. The diagram is shown for the motor so it can be connected for high voltage. The diagram indicates that the high voltage is 480 VAC. In your area of the country the exact amount of high voltage may vary from 440 to 480 VAC. The diagrams will be marked so that 480 VAC represents any high voltage (440, 460, or 480 VAC) and 240 VAC represents any low voltage (220, 230, or 240 VAC) for a delta system and 208 VAC represents the low-voltage wye system.

14.3.2.2 Delta-Connected Motors

Another method of connecting the nine leads of the squirrel-cage induction motor is in a series circuit called a *delta configuration*. Figure 14–8 shows the nine-lead motor connected in a delta configuration. The term delta is used because the formation of the coils in this diagram resembles the Greek capital letter delta (Δ). These connections are also made when the motor is being installed on the factory floor.

A table is also shown in this figure that provides you with the proper terminal connections as they would be shown on the motor's data plate. The motor is also shown connected to three-phase line voltage at terminals T1, T2, and T3. The terminals are numbered in a clockwise direction starting at the top of the delta. Each tip of the delta is numbered 1–3. The other ends of each of the first coils are numbered 4 and 9, and the ends of the second coils are marked 5 and 7, while the ends of the third coil are marked 6 and 8. Terminals 1, 2, and 3 are the midpoints in these coils.

When the motor terminals are wired for operation, terminals 1, 2, and 3 are connected to the power supply at terminals T1, T2, and T3. Terminal 4 is connected to 7, 8 to 5, and 6 to 9 to complete the series circuit. Since these coils are in a series circuit, when three-phase power is supplied, it will come in T1 through the winding and go out terminal T2. Another of the phases will come in T3 and go out T2, and the final phase will come in T2 and go out T3. When the sine wave reverses itself, the currents will reverse and come in from the opposite terminals. This means that at any one instant in time, two of the three wires from the power supply will be used to make a complete circuit. Since the AC three-phase voltage is 120 degrees out of phase and the windings in the motor are also 120 degrees out of phase, the three-phase current will energize each coil in such a way as to cause the magnetic field to rotate.

14.3.3 Reversing the Rotation of a Three-Phase Induction Motor

The rotation of a wye- or delta-connected motor can be changed by exchanging any two of the three phases of the incoming voltage. Figure 14–9 shows diagrams for a wye-connected motor and Figure 14–10 shows the diagrams for a delta-connected motor for clockwise (forward) and counterclockwise (reverse) rotation. You can see that T1 and T2 supply voltage terminals have

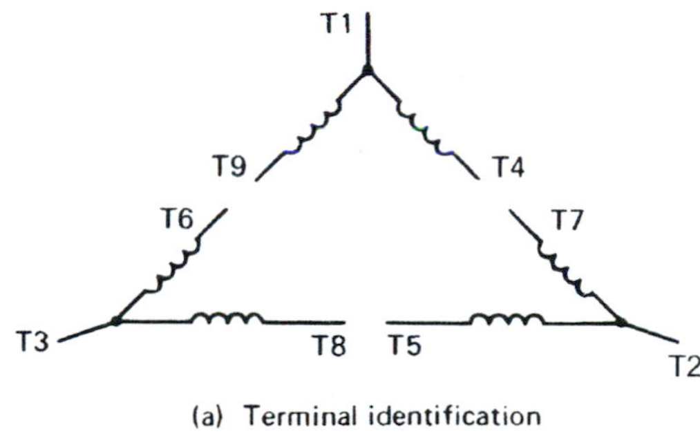

Three-phase voltage

L1	L2	L3	Together		
T1	T2	T3	T4 T7	T5 T8	T6 T9

Figure 14–8 Nine-lead, delta-connected motor.

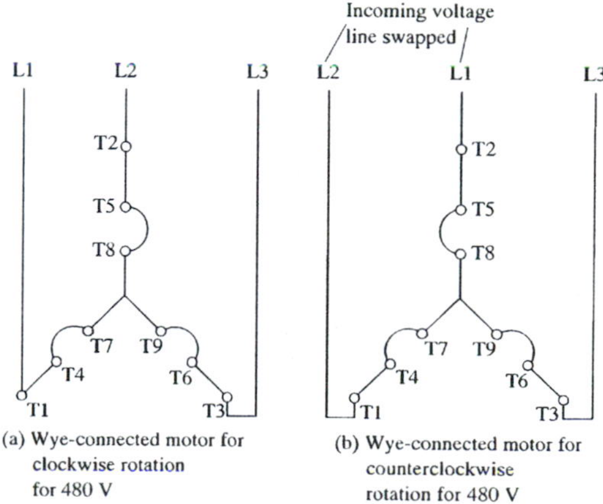

Figure 14–9 Diagram above the wye-connected motor for clockwise and counterclockwise operation.

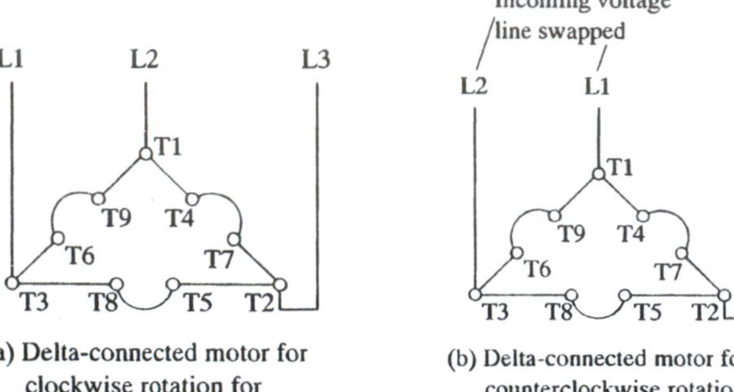

(a) Delta-connected motor for clockwise rotation for 480 V

(b) Delta-connected motor for counterclockwise rotation for 480 V

Figure 14–10 Diagram of a delta-connected motor for (a) clockwise rotation; (b) counterclockwise rotation.

been exchanged in the diagram for motor reversal. In industrial applications, terminals T1 and T2 are generally switched by the contacts of the reversing motor starter. These diagrams will be useful for installation connections and troubleshooting.

This wiring configuration is also used when the motor is connected for *plugging* use. When the motor is used for plugging, it is reversed while running at full rpm. When the motor's stator is reconnected for reverse rotation by switching two of the three input voltage lines, it will quickly build up a reverse magnetic field that will begin to rotate in the opposite direction. The rotor will follow this quick change in rotation and begin to rotate in the opposite direction. This will cause the load to decrease its rpm from full in the clockwise direction to zero, and begin to rotate counterclockwise. The moment the rotor begins to rotate in the opposite direction, the power is deenergized and the rotor shaft is stopped from rapid reverse torque.

14.3.4 Connecting Motors for a Change of Voltage

Delta- and wye-connected motors may have numerous sets of coils so that they can be wired to operate at two separate voltages. The extra windings are connected in series for the motor to operate at the higher voltage, and they are connected in parallel for the lower voltage. The higher voltage in these applications is usually 440 or 480 V, while the low voltage is usually 208 or 220 V. The actual voltage will be specified on the motor's nameplate.

The dual-voltage option provides a larger variety of choices when the motor must be connected to power distribution systems. You should remember from Chapter 6 that it is very important to balance the loads on the power distribution system and not overload any of the transformers. When the motors are connected to operate at higher voltages, their current draw is reduced by

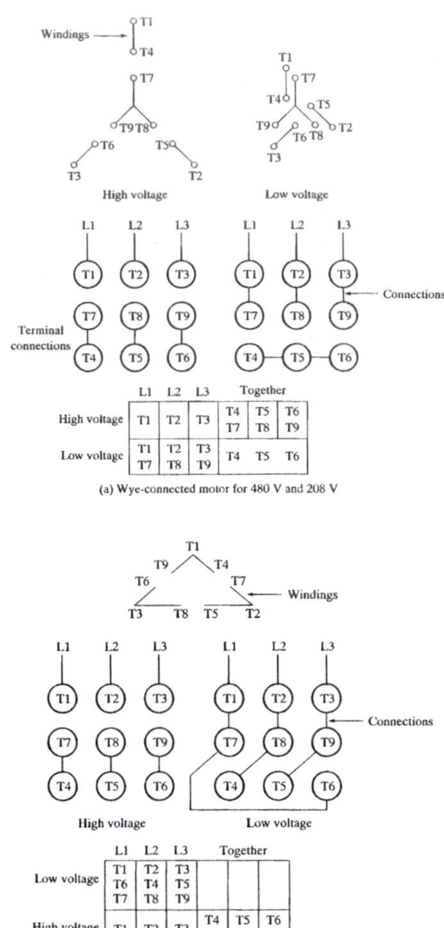

Figure 14–11 (a) Diagram of a wye-connected motor for high and low voltages. (b) Diagram of a delta-connected motor for high and low voltages.

half. The motor will use the same amount of wattage in both configurations, but smaller-gauge wire and smaller contacts and switchgear can be used throughout the circuit when the motor is connected to the higher voltage.

The lower voltage is also useful when no high voltage is available in the area of the factory where the motor must be installed. The dual-voltage motor can be connected for the lower voltage and a savings can be realized by not having to install extra-long power cables to reach the remote source.

Figure 14–11 shows a set of diagrams and tables that indicate the proper connections for the motor to operate at both high and low voltage. The diagrams are presented as wiring diagrams with the numbers on the terminal leads. The motor can be connected as either a delta motor or a wye motor, which will affect its starting torque and LRA characteristics, so high- and low-voltage diagrams are presented for each of these types of motors.

Fig. 14–11a shows a wye-connected motor wired for high voltage and for low voltage. The six coils are configured as three sets of parallel coils when the motor is connected for low voltage and as a series circuit when they are connected for high voltage. Since all the coils are used in the configurations for high and low voltage, the motor will have the same amount of torque and horsepower in both cases.

The diagram that shows the coils is presented to give you a picture of how the coils look in series and parallel, but this diagram is sometimes confusing when you must make changes in the field. For this reason a second diagram is presented that shows only terminals. Each of the terminals are numbered as you would find them on a motor, and the heavy line shows the connections that must be made to complete the connections.

A table is provided in this figure that indicates the terminals that are connected to each other and to the power supply, and those that are left open. When a terminal is left open, it means that a wire nut should be placed over the terminal end and wrapped with electrical tape to secure it. Do not cut this wire short, as it may be needed when the motor is reconnected later. Remember, machinery is moved around the factory rather frequently. The motor may need to be connected for a different voltage when it is moved in a few years, or a new service entrance and power distribution system may be installed and several motors may be changed to operate on the new voltage.

Figure 14–11b shows the coils of a delta-connected motor wired for high and low voltage. This diagram shows the coils for the motor connected for low voltage wired in parallel and the coils for a high-voltage connection wired in series. A terminal diagram and table are presented for the delta motor. These diagrams are presented to show you the variety of methods that manufacturers use to indicate the terminal connections for wiring their motor for high or low voltage.

All squirrel-cage motors that have nine leads brought out of their frame can be connected for high or low voltage. The diagrams presented in this figure are usable on all name brands of motors. This is very important, because the motor data plate is usually painted over, damaged, or removed when you need it to make field wiring connections. You will be able to use the diagrams found in this figure to make the connections. Figure 14–12 also shows several diagrams that indicate various ways to wire motors for a change of voltage.

14.3.5 Connecting Three-Phase Motors for a Change of Speed

Some delta-connected and wye-connected motors can be wired to operate at two different speeds. Unlike the nine-lead motor, which allows all motors to be wired for high or low voltage, not all motors are manufactured to be reconnected for a change of speeds. The motor must be specially manufactured with enough leads brought out of the frame to make the changes required to allow the motor to operate on the different speeds.

Some motors have enough leads brought out to operate at two different speeds, while other motors can be reconnected to operate at up to four speeds. A motor's speed is changed by changing the number of poles that are used. When the motor uses eight poles it will operate at 900 rpm, with six poles it will operate at 1200 rpm, with four poles it will operate at 1800 rpm, and the two-pole motor will operate at 3600 rpm. This means that the motor would provide less horsepower when poles are removed from the circuit completely to allow the motor to operate at a higher speed.

Some motors provide a means to reconnect the extra poles back into the circuit to keep the overall horsepower rating of the motor constant. Changes in the connections can be made to allow the motor to provide constant torque regardless of the speed at which it is operating.

14.3.6 Interpreting the Wiring Diagrams and Tables

The diagrams in Figure 14–12 show two ways of representing the connection that must be made to the motor terminals. One way shows the connections with an electrical diagram, and the second way is to show a table that indicates which terminal numbers should be connected. The terminals in the motor are marked with numbers that are stamped into the wire material or identified with a metal tag that has a number on it. The metal tag is crimped on each wire near the terminal end. After you have located all the terminals and have identified them by their numbers, you are ready to make the connections shown in the diagram for your application.

The table lists the terminals that should be connected to the three input voltage wires. These are identified as L1, L2, and L3, and any terminal number listed in the category under the line number should be connected to that line. Be sure to check the row indicating the speed for which the motor will be wired. The second column lists the wires that are left open. The wires listed in this column should not be connected to anything. They are supposed to remain unconnected and they should have an insulated wire nut or cap placed over the end of the wire securely so that it does not come in contact with any metal parts of the motor or other energized wires. In some tables any lead that is not listed in another column should be left open.

The fourth column lists the leads or wires that should be connected together. The wires listed in this column should be connected together and no power should be connected to these leads. The leads must be secured together with a wire nut and wrapped with insulating tape because they will be energized. In some

TERMINAL MARKINGS AND CONNECTIONS
PART WINDING START

NEMA NOMENCLATURE—6 LEADS

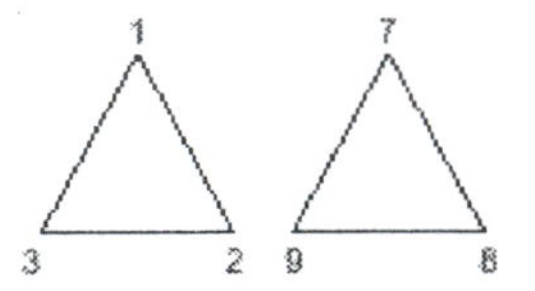

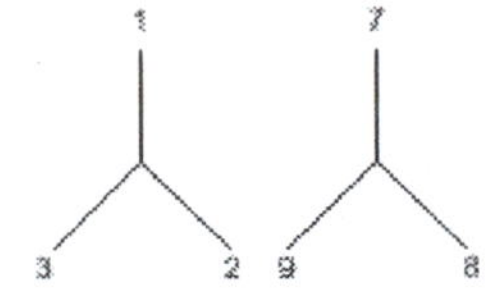

OPER. MODE	L1	L2	L3	OPEN
START	1	2	3	7,8,9
RUN	1,7	2,8	3,9	—

NEMA NOMENCLATURE—9 LEADS
WYE CONNECTED (LOW VOLTAGE ONLY)

	T1	T2	T3	T7	T8	T9	Together
MOTOR LEADS	1	2	3	7	8	9	4&5&6

NEMA AND IEC NOMENCLATURE—12 LEADS
SINGLE VOLTAGE OR LOW VOLTAGE OF DUAL-VOLTAGE MOTORS

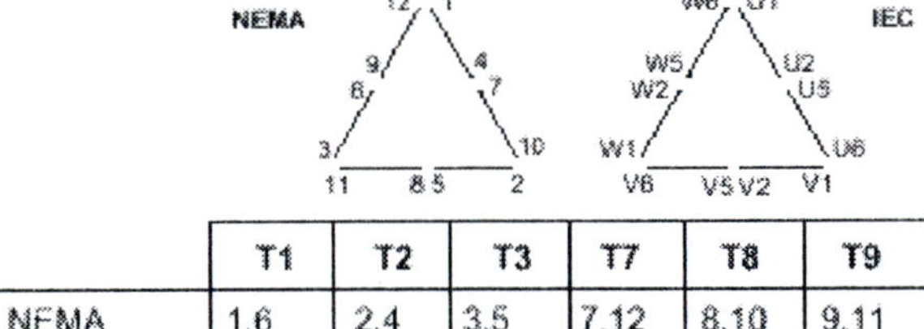

	T1	T2	T3	T7	T8	T9
NEMA	1,6	2,4	3,5	7,12	8,10	9,11
IEC	U1,W2	V1,U2	W1,V2	U5,W6	V5,U6	W5,V6

(a)

TERMINAL MARKINGS AND CONNECTIONS
THREE-PHASE MOTORS–SINGLE SPEED

NEMA NOMENCLATURE—9 LEADS

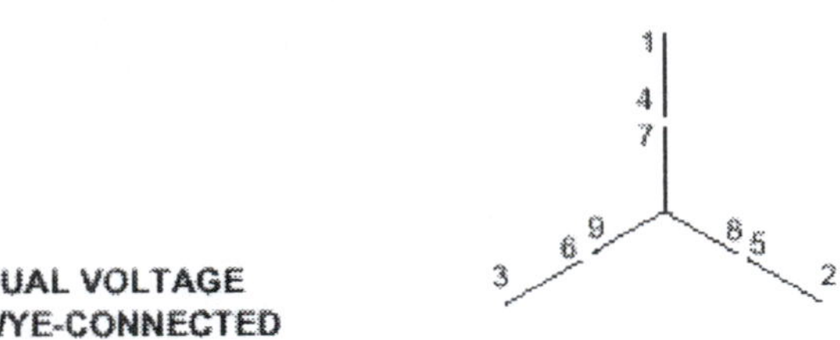

DUAL VOLTAGE
WYE-CONNECTED

VOLTAGE	L1	L2	L3	JOIN
HIGH	1	2	3	4&7,5&8,6&9
LOW	1,7	2,8	3,9	4&5&6

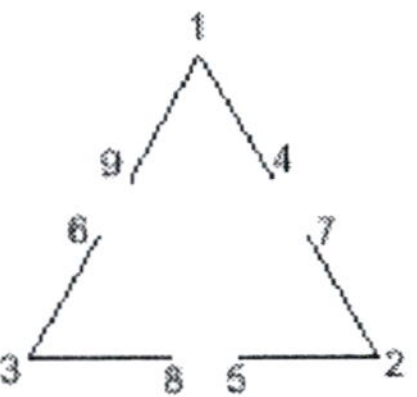

DUAL VOLTAGE
DELTA-CONNECTED

VOLTAGE	L1	L2	L3	JOIN
HIGH	1	2	3	4&7,5&8,6&9
LOW	1,6,7	2,4,8	3,5,9	—

(b)

Figure 14–12 (a) Part winding start diagram for a 6-lead motor and dual voltage diagram for a 12-lead motor. (b) Diagrams of a single speed 9-lead motor connected as a wye-connected motor and a delta-connected motor.
(Reprinted with permission of Electrical Apparatus Service Association)

diagrams no terminals will be listed in this column, which means that all the leads are used in one of the other columns. Be sure that you account for every lead before you put all the leads back into the motor and replace the field wiring access cover.

These diagrams are extremely useful when the application you are working with requires the motor to be reconnected on the factory floor. Many times these diagrams are not readily available when you need them, so this provides that much-needed reference. These motors can also be reversed by exchanging two of the three supply voltage lines. This allows the motors to be used in the widest possible number of applications.

You will be required to make these changes yourself or direct someone to make them for you, so you must understand the concept of changing the connections of motor leads to make the motor fit the application.

14.3.7 Motor Data Plates

The motor data plates list all the pertinent data concerning the motor's operational characteristics. It is sometimes called the name plate. Figure 14–13 shows an example of a data plate for a typical AC motor. The data plate contains information about the ID (identification number), FR (frame and motor design), motor type, phase, horsepower rating, rpm, volts, amps, frequency, service factor (SF), duty cycle (time), insulation class, ambient temperature rise, and NEMA design

TERMINAL MARKINGS AND CONNECTIONS
THREE-PHASE MOTORS–SINGLE SPEED

NEMA NOMENCLATURE–12 LEADS

DUAL VOLTAGE
EXTERNAL WYE CONNECTION

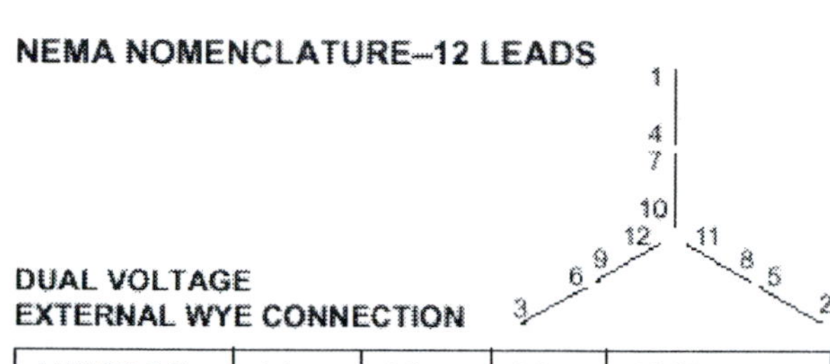

VOLTAGE	L1	L2	L3	JOIN
HIGH	1	2	3	4&7,5&8,6&9, 10&11&12
LOW	1,7	2,8	3,9	4&5&6, 10&11&12

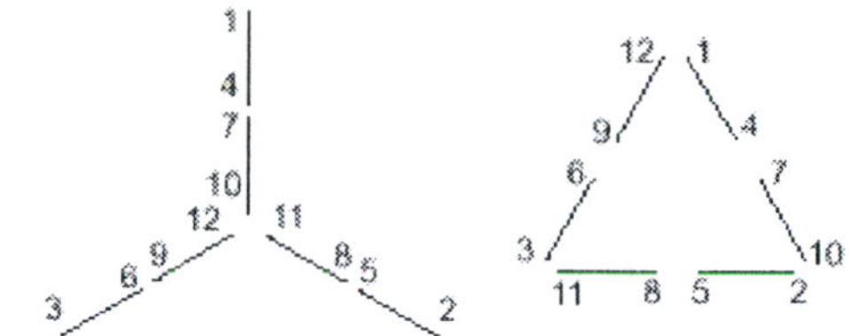

DUAL VOLTAGE
WYE-CONNECTED START
DELTA-CONNECTED RUN

VOLTAGE	CONN.	L1	L2	L3	JOIN
HIGH	WYE	1	2	3	4&7,5&8,6&9, 10&11&12
	DELTA	1,12	2,10	3,11	4&7,5&8,6&9
LOW	WYE	1,7	2,8	3,9	4&5&6, 10&11&12
	DELTA	1,6,7, 12	2,4,8, 10	3,5,9, 11	—

(c)

Figure 14–12 (continued) (c) Diagrams of a 12-lead motor connected for dual voltage as a wye-connected motor and a delta-connected motor.

ACE MOTOR CORP.

ID	FR	TYPE	PH.

HP	RPM	VOLTS	AMPS	HZ	SF

TIME RATING	INS. CL.	INS. TYPE	AMB. C.	NEMA DESIGN	CODE

Figure 14–13 Data plate for a typical AC motor.

and code. Each of these features are discussed in detail in the next sections.

14.3.7.1 Identification Number (ID)

The identification number for a motor will basically be a model and serial number. The model number will include information about the type of motor, and the serial number will be a unique number that indicates where and when the motor was manufactured. These numbers will be important when a motor is returned for warranty repairs, or if an exact replacement is specified as a model for model exchange.

14.3.7.2 Frame Type

Every motor has been manufactured to specifications that are identified as a *frame size*. Figure 14–14 shows a table with frame size data. These data include the distance between mounting holes in the base of the motor, the height of the shaft, and other critical data about physical dimensions. The data are then given a number such as 56. The frame number indicates that any motor that has the same frame number will have the same dimensions even though it may be made by a different company. This allows users to stock motors from more than one manufacturer or replace a motor with any other with the same frame information, and be sure that it will be an exact replacement.

14.3.7.3 Motor Type

The motor type category on the data plate refers to the type of ventilation the motor uses. These types include the open type, which provides flow-through ventilation from the fan mounted on the end of the rotor. In some motors that are rated for variable speed duty (used with variable frequency drives) the fan will be a separate motor that is built into the end of the rotor. The fan motor will be connected directly across the supply voltage, so it will maintain a constant speed to provide constant cooling regardless of the motor speed.

Another type of motor is the *enclosed type*. The enclosed motor is not air cooled with a fan; instead it is manufactured to allow heat to dissipate quickly to and from the inside of the motor outward to the frame. In most cases, the frame has fins built into it on the outside to provide more area for cooling air to reach.

14.3.7.4 Phase

The phase of the motor will be indicated as single phase or three phase. The number of phases may be indicated as 1 phase or 3 phase, or the number may be listed by itself.

IEC MOUNTING DIMENSIONS*
FOOT-MOUNTED AC AND DC MACHINES

* Dimensions in inches

FRAME NUMBER	H	A	B	C	K	BOLT OR SCREW
56 M	2.20	3.55	2.80	1.40	0.23	M5
63 M	2.48	3.95	3.15	1.55	0.28	M6
71 M	2.79	4.40	3.55	1.75	0.28	M6
80 M	3.14	4.90	3.95	1.95	0.40	M8
90 S	3.54	5.50	3.95	2.20	0.40	M8
90 L	3.54	5.50	4.90	2.20	0.40	M8
100 S	3.93	6.30	4.40	2.50	0.48	M10
100 L	3.93	6.30	5.50	2.50	0.48	M10
112 S	4.40	7.50	4.50	2.75	0.48	M10
112 M	4.40	7.50	5.50	2.75	0.48	M10
132 S	5.19	8.50	5.50	3.50	0.48	M10
132 M	5.19	8.50	7.00	3.50	0.48	M10
160 S	6.29	10.00	7.00	4.25	0.58	M12
160 M	6.29	10.00	8.25	4.25	0.58	M12
160 L	6.29	10.00	10.00	4.25	0.58	M12
180 S	7.08	11.00	8.00	4.75	0.58	M12
180 M	7.08	11.00	9.50	4.75	0.58	M12
180 L	7.08	11.00	11.00	4.75	0.58	M12
200 S	7.87	12.50	9.00	5.25	0.73	M16
200 M	7.87	12.50	10.50	5.25	0.73	M16
200 L	7.87	12.50	12.00	5.25	0.73	M16
225 S	8.85	14.00	11.25	5.85	0.73	M16
225 M	8.85	14.00	12.25	5.85	0.73	M16
250 S	9.84	16.00	12.25	6.60	0.95	M20
250 M	9.84	16.00	13.75	6.60	0.95	M20
280 S	11.02	18.00	14.50	7.50	0.95	M20
280 M	11.02	18.00	16.50	7.50	0.95	M20
315 S	12.40	20.00	16.00	8.50	1.11	M24
315 M	12.40	20.00	18.00	8.50	1.11	M24
355 S	13.97	24.00	19.70	10.00	1.11	M24
355 M	13.97	24.00	22.05	10.00	1.11	M24
355 L	13.97	24.00	24.80	10.00	1.11	M24
400 S	15.74	27.00	22.05	11.00	1.38	M30
400 M	15.74	27.00	24.80	11.00	1.38	M30
400 L	15.74	27.00	27.95	11.00	1.38	M30

IEC 72-1 Standards.
Dimensions, except for bolt and screw sizes, are shown in inches (rounded off). Bolt and screw sizes are shown in millimeters.
For tolerances on dimensions, see IEC 72-1, 6.1, Foot-Mounted Machines, Table 1. (Note: Data in IEC tables is shown in millimeters.)

Figure 14–14 NEMA frame dimensions table for physical size of motors.
(Reprinted with permission of Electrical Apparatus Service Association)

14.3.7.5
Horsepower Rating (hp)

The horsepower rating will be indicated as a fractional horsepower (number less than 1.0) or a larger horsepower. Fractional horsepower numbers may be listed as fractions (1/2) or decimals (.5).

14.3.7.6
RPM (Speed)

The motor speed will be indicated as rpm. This will be the rated speed for the motor, and it will not account for slip. The actual speed of the motor will be less because of slip. As you know, the speed of the motor is determined by the number of poles and the frequency of the AC voltage. Typical speed for a 2-pole motor is 3600 rpm, a 4-pole motor is 1800, and an 8-pole motor is 1200. The actual speed of the motor rated for 3600 rpm will be approximately 3450, for the 1800-rpm motor it will be approximately 1750, and for the 1200-rpm motor it will be approximately 1150. The actual amount of slip will be indicated by the *motor design letter.* The motor design letter is explained in Section 14.3.8.

14.3.7.7
Volts

Voltage ratings for single-phase motors will be listed as 115, 208, or 230 V. Three-phase motors have typical voltage ratings of 208, 240, 440, 460, 480, and 550 V. Other voltages may be specified for some special motors. You should always ensure that the power supply voltage rating matches the voltage rating of the motor. If the rating and the power supply voltage do not match, the motor will overheat and be damaged.

14.3.7.8 Amps

The amps rating is the amount of full-load current (FLA) the motor should draw when it is under load. This rating will help the designer calculate the proper wire size, fuse size, and heater size in motor starters. The supply wiring for the motor circuit should always be larger than the amps rating of the motor. The NEC (National Electric Code) provides information to help you determine the exact fuse size and heat sizes for each motor application.

14.3.7.9 Frequency

The frequency of a motor will be listed in hertz (Hz). The typical frequency rating for motors in the United States is 60 Hz. Motors manufactured for use in some parts of Canada and all of Europe and Asia will be rated for 50 Hz. You must be sure that the frequency of the motor matches frequency of the power supplied to the motor.

14.3.7.10 Service Factor (SF)

The *service factor* is a rating that indicates how much a motor can be safely overloaded. For example, a motor that has an SF of 1.15 can be safely overloaded by 15%. This means that if the motor is rated for 1 hp, it can actually carry 1.15 hp safely. To determine the overload capability of a motor, you multiply the rated hp by the service factor. The motor is capable of being overloaded because it is designed with ways of dissipating large amounts of heat.

14.3.7.11 Duty Cycle

The *duty cycle* of a motor is the amount of time the motor can be operated out of every hour. If the motor's duty cycle is listed as continuous, the motor can be run 24 hours a day and does not need to be turned off to cool down. If the duty cycle is rated for 20 minutes, the motor can be safely operated for 20 minutes before it must be shut down to be allowed to cool. A motor with this rating should be shut down for 40 minutes of every hour of operation to be allowed to cool.

Another way to specify the duty cycle of a motor is called the *motor rating*. The motor rating on the data plate refers to the type of duty the motor is rated for. The types of duty include continuous duty, intermittent duty, and heavy duty, which includes jogging and plugging duty. Continuous duty includes applications where the motor is started and allowed to operate for hours at a time. The intermittent duty includes operations where the motor is started and stopped frequently. This type of application allows the motor to heat up because it will draw LRA more often than will a motor rated for continuous duty.

Motors that are rated for jogging and plugging are built to withstand very large amounts of heat that will build up when the motor will draw large LRA during starting and stopping. Since the motor can be reversed when it is running in the forward direction for plugging applications, it will build up excessive amounts of heat. Motors with this rating must be able to get rid of heat as much as possible to withstand the heavy-duty applications.

14.3.7.12 Insulation Class

The insulation class of a motor is a letter rating that indicates the amount of temperature rise the insulation of the motor wire can withstand. The numbers in the insulation class are listed in degrees Celsius (°C). The table in Figure 14–15 shows typical insulation classes for motors. The insulation class and other temperature-related features of a motor will help determine the temperature rise the motor can withstand.

Class	Temperature Rise °C
A	105
B	130
F	155
H	180

Figure 14–15 Insulation class for motors. This table indicates the amount of temperature for which the wire insulation of a motor is rated.

14.3.7.13 Ambient Temperature Rise

The ambient temperature rise is also called the Celsius rise. It is the amount of temperature rise the motor can withstand during normal operation. This value is listed in degrees Celsius. A typical open motor can withstand a rise of 40°C (104°F) and an enclosed motor can withstand 50°C (122°F) rise. This means the motor should not be exposed to environments where the temperature is 104°F above the ambient. If the ambient temperature is considered 72°F, then the motor is limited to temperatures of 176°F. The insulation class will also help determine the amount of temperature a motor will be able to withstand.

14.3.8 NEMA Design

The National Electrical Manufacturers Association (NEMA) provides design ratings that may also be listed as motor design. The motor design is listed on the data plate by a letter A, B, C, or D. This designation is determined by the type of wire, insulation, and rotor that is used in the motor and is not affected by the way the motor might be connected in the field.

Type A motors have low rotor circuit resistance and have approximate slip of 5 to 10 percent at full load. These motors have low starting torque with a very high LRA. This type of motor tends to reach full speed rather rapidly.

Type B motors have low to medium starting torque and usually have slip of less than 5 percent at full load. These motors are generally used in fans, blowers, and centrifugal pump applications.

Type C motors have a very high starting torque per ampere rating. This means that they are capable of starting when the full load is applied for applications such as conveyors, crushers, and reciprocating compressors, such as air conditioning and refrigeration compressors. These motors are rated to have slip of less than 5 percent.

Type D motors have a high starting torque with a low LRA rating. This type of motor has a rotor made of brass rather than copper segments. It is rated for slip of 10 percent at full load. Normally, this type of motor will require a larger frame to produce the same amount of horsepower as a type A, B, or C motor. These motors are generally used for applications with a rapid decrease of shaft acceleration, such as a punch press that has a large flywheel.

These standards are set by NEMA, and a motor must meet all the requirements of the standard to be marked as a type A, B, C, or D. This allows motors made by several manufacturers to be compared on an equal basis according to application.

14.3.9 NEMA Code Letters

The NEMA code uses letters of the alphabet to represent the amount of LRA (in kVA per horsepower) that a motor will draw when it is started. These letters are listed in a table in Figure 14–16. Letters in the front of the alphabet indicate low LRA ratings, and letters in the back of the alphabet indicate higher LRA ratings. The number in the table is not the amount of LRA the motor will draw, but rather it is the number that must be multiplied by the horsepower rating of the motor.

NEMA Code Letter	Locked-Rotor kVA per hp
A	0–3.15
B	3.15–3.55
C	3.55–4.00
D	4.00–4.50
E	4.50–5.00
F	5.00–5.60
G	5.60–6.30
H	6.30–7.10
J	7.10–8.00
K	8.00–9.00
L	9.00–10.0
M	10.0–11.2
N	11.2–12.5
P	12.5–14.0
R	14.0–16.0
S	16.0–18.0
T	18.0–20.0
U	20.0–22.4
V	22.4 and up

Figure 14–16 Table with locked rotor and breach (LRA) ratings. The ratings are listed as amount of kVA per horsepower.
(Reprinted with permission of Electrical Apparatus Service Association)

14.4 Three-Phase Synchronous Motors

A synchronous motor is an AC motor designed to run at synchronous speed without any slip. As you know, the induction motor must have slip of approximately 9 to 10 percent to operate a maximum torque. Slip is required in an induction motor to allow the rotor to draw enough current to carry its load.

The synchronous motor is designed to operate with no slip by exciting the rotor with DC current once the motor reaches operating rpm. The motor can have DC applied from a DC power source or it can be developed through the use of diodes from a separately generated AC current. This current is produced by a small generator located on the end of the synchronous motor shaft. Prior to the use of modern variable frequency drives, the synchronous motor was used in industrial applications where the loss of rpm due to slip in an induction motor could not be tolerated and the extra rpm was needed for efficiency. Today the variable frequency drive can be used to increase the speed of an induction motor to make up the lost rpm, so the synchronous motor is not needed as much as it once was.

Figure 14–17 shows a diagram of a synchronous motor with the DC voltage developed from an outside source and from an internal source. The diagram, which shows diodes being used to rectify the AC from an internal source, is the most common type of synchronous motor used today. This motor also provides an additional function in that it is capable of correcting the power factor. The amount of DC current used to excite the rotor also determines the amount of improvement in the power factor.

Figure 14–17 Electrical diagram of (a) an externally excited synchronous motor; (b) an internally excited synchronous motor.

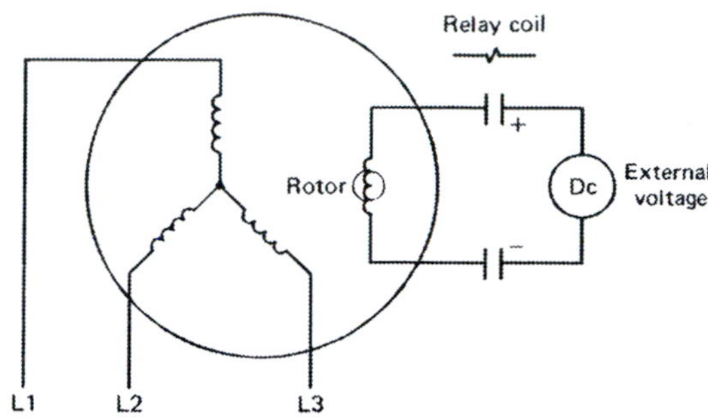

(a) Externally excited synchronous motor

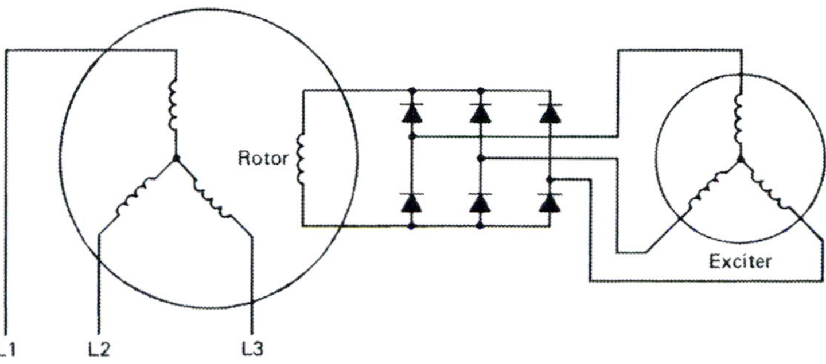

(b) Internally excited synchronous motor

When it is started, operation of the synchronous motor is similar to that of an induction motor. This means that the motor is started as an induction motor until the motor reaches full rpm. When the motor is at its highest inductive rpm, a switch is closed to provide exciter current to the rotor from the external or internal source. When the rotor is energized, it will cause its magnetic poles to lock, in step with the rotating magnetic field of the motor's stator. Since the speed of the magnetic field is determined by the number of poles in the motor and the frequency of the applied voltage, the speed of the rotor will be locked into this magnetic field's speed and they will rotate in unison. The amount of current in the rotor can be adjusted to provide more torque for the rotor. For this reason, it is important that the motor be started without a heavy load applied. The load can be increased as the motor is synchronized.

14.5 AC Single-Phase Motors

Single-phase motors are used quite frequently in large and small industries. They are especially useful in applications where motors of less than 1 hp are required and in locations where a motor is required and three-phase voltage is not available. In these applications the single-phase motor can be installed on the single-phase voltage source and produce the required horsepower.

The single-phase motor uses a theory of operation that is similar to that of the three-phase motor. There are several minor changes in the single-phase motor design to achieve the same function as a three-phase motor. These changes include the addition of several components and the modification of several others. Figure 14–18 shows a picture of a typical single-phase motor. A single-phase voltage source may consist of L1 and L2 for 208 and 230-V systems, or L1 and neutral for a 115-V system.

14.5.1 Single-Phase Motor Components

Like the three-phase motor, the single-phase motor has three basic components: the *stator*, *rotor*, and *end plates*. Figure 14–19 shows a cutaway picture of a sin-

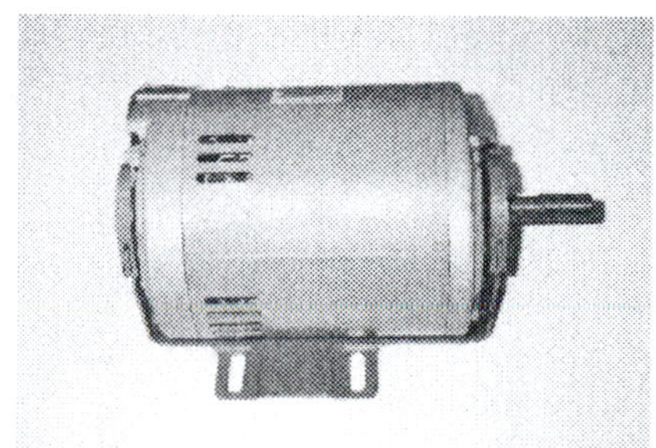

Figure 14–18 A typical single-phase motor.

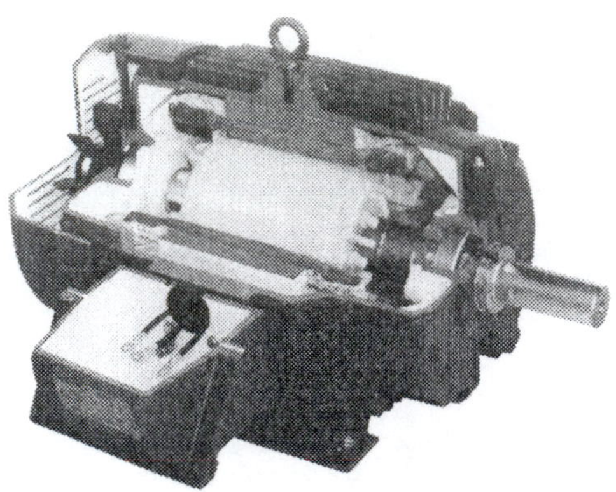

Figure 14–19 A cutaway picture of typical single-phase motor.
(Courtesy of A.O. Smith Electrical Products Company)

gle-phase motor and its components. Figure 14–20 shows a picture of a stator with its rotor removed, which gives you a better idea of the relationship between the stator and rotor.

14.5.1.1 Stator

The stator is the frame of the motor, which houses the windings. Since the single-phase motor uses single-phase voltage, it will need a way to produce the starting torque that three-phase voltage produces naturally in a three-phase motor. The single-phase motor has a special starting winding that is used to provide sufficient phase shift to provide starting torque. The motor also has a run winding that is very similar to the windings in a three-phase motor. The start winding is made of very fine-gauge wire, which has many more turns than the run winding. The run winding is made from wire that is sized to carry the current for the motor at full-load amperage. This means the run winding wire will be much larger than the start winding, usually in the range of 14- to 16-gauge wire.

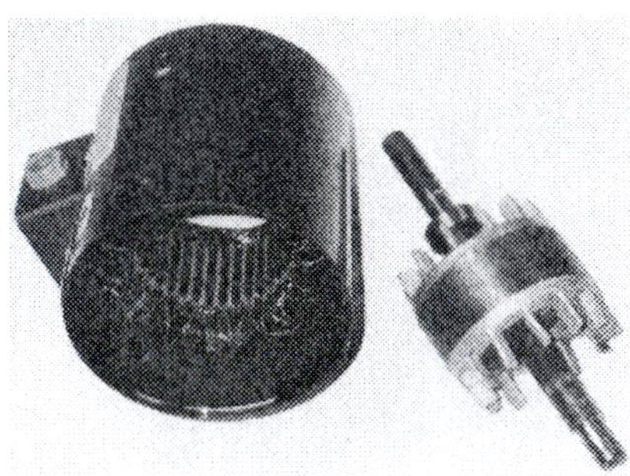

Figure 14–20 A single-phase AC motor stator with its rotor removed.

The start winding is also placed in the stator offset from the run winding to give a phase difference of 90 degrees. This physical phase shift will cause a shift in the magnetic field produced by two windings. Since the start winding is made of very fine wire that has many turns, it can produce a very strong magnetic field for a short period of time. Figure 14–21a shows the locations of the start and run windings in the stator. Four run and four start windings are located toward the inside of the stator, where they will be closer to the rotor, and the run winding is shown behind the start winding. The start and run windings are connected together in parallel in the motor to provide the magnetic phase shift.

Several electrical diagrams are also presented in this figure to show you methods of representing the single-phase motor. In Figure 14–21b the windings are shown placed at right angles to each other. This is done to remind you that the windings are physically offset in the stator to produce more of a magnetic phase shift. The run winding is always represented by the larger coil and its terminals are numbered 1 and 4. The start winding is shown as the smaller coils and its terminals are numbered 5 and 8. The rotor is shown in these diagrams as a circle in the middle of the windings.

In Figure 14–21c the windings are shown connected in parallel with each other, which is how you would indicate their electrical relationship. You should also notice that in this diagram the run winding is identified with the letter R and the start winding is identified with the letter S. The point where the two windings are connected at the bottom end of the parallel circuit is called the common point and is identified by the letter C. In some motor theory, it is referred to as terminal C, even though it is the point where the two windings are connected together. This type of diagram is used frequently to show the windings of single-phase compressor motors used in air conditioning systems.

Figure 14–21d shows the windings of a single-phase motor as you would normally see them in diagrams of open-type motors that are used to power small machines in a factory. The windings are shown as two sets of windings. The run winding is shown in two parts. The first part is a winding that is identified by terminal numbers T1 and T2, and the second part is identified as T3 and T4. The run winding consists of two parts so that the motor can be connected for high voltage (230 V) or low voltage (115 V). If the motor is connected for high voltage, the two parts of the run winding are connected in series, and if the motor is connected for low voltage, the two run windings are connected in parallel. A detailed discussion of high and low voltage connections is presented later in this chapter.

The start winding is also shown in two parts in Figure 14–21d. The first segment of the start winding is identified as terminals T5 and T6 and the second set is

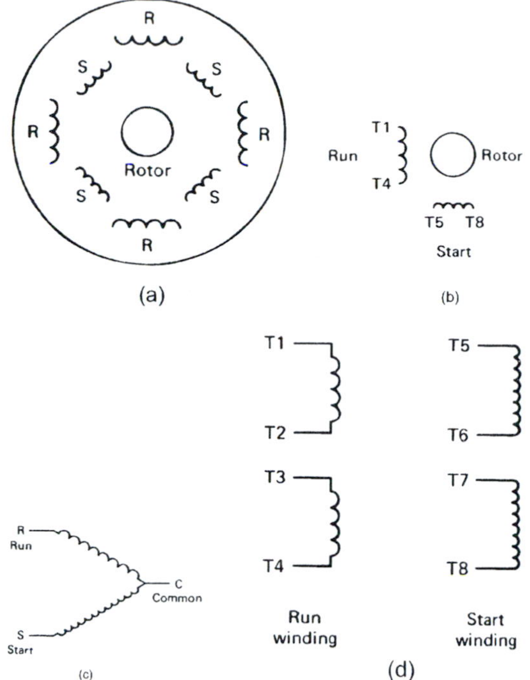

Figure 14–21 (a) Diagram of a start and run winding located inside the motor stator. (b) Diagram that shows the stator winding placed at 90 degrees to the run winding in the stator. (c) Diagram that shows one end of the start and the run winding connected together at a point called C. (d) The run winding shown in two parts and the start winding shown in two parts.

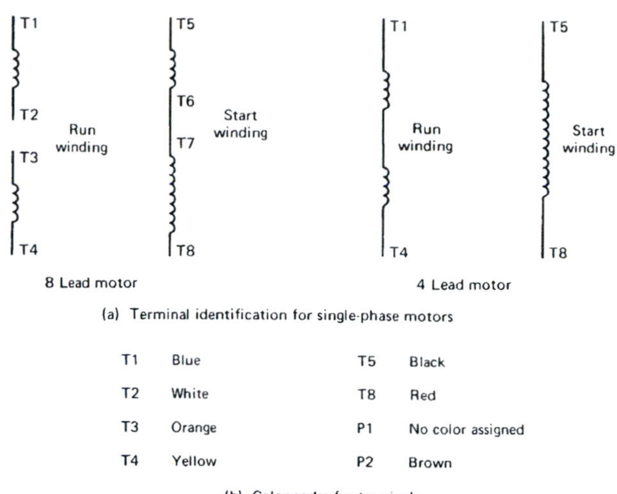

T1	Blue	T5	Black
T2	White	T8	Red
T3	Orange	P1	No color assigned
T4	Yellow	P2	Brown

(b) Color codes for terminals

Figure 14–22 Terminal identification and color codes for a single-phase motor.

identified as terminals T7 and T8. If T6 and T7 are permanently connected inside the motor, the two ends of the start winding will be identified as T5 and T8.

14.5.1.2 Terminal Identification

The terminals on a single-phase motor have a standard identification method. The ends of the start and run windings are numbered to help you identify and locate them when you must install or troubleshoot the motors. Figure 14–22 shows the standard numbering method for a split-phase motor. Two diagrams are presented to show an eight-lead and a four-lead motor. The two windings are shown connected in parallel as they operate electrically. The run winding is shown in two sections. The numbering starts at the top of the diagram, with the terminals of the first section being numbered 1 and 2. The second section's terminals are numbered 3 and 4. The start winding is also shown in two sections, with the terminals of the top section numbered 5 and 6, while the bottom terminals are marked 7 and 8.

The run and start winding terminals can also be identified by the amount of resistance each has. Since the start winding is made of many turns of very fine wire, its resistance will be much higher than that of the run winding. If the terminal identification is missing, you can use a continuity test to group the terminal leads into coils, and then by measuring their resistance you can compare the readings. The highest readings will belong to the start windings, while the lower readings indicate that the winding is the run winding. This figure also shows a table with the color code for each of the motor terminal conductors for single-phase motors. The color codes may be used if the terminal identification is not used. Some manufacturers identify their leads with terminal markers, so the color of wires that are used in their motors have no meaning. Some manufacturers do not adhere to the color codes, so you must always confirm the terminal markings with an ohm test.

14.5.1.3 Rotor and End Plates

The rotor in the single-phase motor is very similar to the rotor in the three-phase induction motor. The single-phase rotor also has the basic shape of a squirrel cage, so it is also called a squirrel-cage rotor. It has fan blades cast into the aluminum frame to provide cooling air for the motor.

The ends of the rotor provide the shaft for the load and the bearings. The rear part of the shaft is machined to mount inside the shaft bearing, and the front part of the shaft is extended 3 to 4 inches beyond the front bearing to provide a means of mounting pulleys or gears to drive the load.

The end plates are located on each end of the stator. They are secured in place by four bolts that are inserted completely through the stator. They house the bearings for the rotor shaft to ride on. The bearings can be the sleeve type or ball type. The ball bearing is usually lubricated for life and sealed, while the sleeve bearing must be lubricated frequently with several drops of high-grade electric motor oil. The sleeve bearing uses felt wicking to hold the excess lubricating oil in contact to the shaft. The end plates must be mounted with the lubricating port pointing upward so that the oil will be pulled to the wicking by gravity.

14.5.1.4
Centrifugal Switch for the Start Winding

The start winding can stay in the circuit for only a short time because its wire is too small and it will heat up rapidly, so a switch is provided to disconnect it from the circuit as soon as the motor is started. This switch is a centrifugal switch that mechanically senses the speed of the shaft and opens when the shaft reaches approximately 90 to 95 percent of full rpm. Figure 14–23a shows a picture of the end switch and Figure 14–23 shows the end switch the end plate, actuator that moves the centrifugal switch open when the proper speed is reached, and allows the switch to close when the motor stops.

The centrifugal switch is mounted in the rear end plate, and is commonly referred to as the end switch. The switch has two distinct parts, the switch and actuator. The switch is mounted in the end plate, and the actuator is mounted on the rotor shaft so that it will come in contact with the end switch when the rotor reaches full rpm.

The end switch is made of spring steel, which provides tension to keep the switch contacts closed. Whenever the centrifugal actuator is not pressing on the switch, the contacts will remain in the closed position. When the actuator moves along the shaft slightly, it will provide enough force to cause the switch contacts to snap open.

The actuator has a weight built into its outer edges. These weights, called fly weights, are hinged on the inside near the rotor and allowed to move or swing at the outer edge. Since the outer edge is heavier, the centrifugal force caused by the shaft rotation will cause them to move away from the shaft. Since the actuator is hinged to the inside, this action will cause the actuator to move along the length of the shaft slightly in the direction of the switch. The movement is only 1/2 to 3/4 inch, but it is sufficient to actuate the end switch to the open position.

Figure 14–23 (a) Centrifugal switch; (b) end plate; (c) fly weight mechanism; (d) motor shaft.

Since the fly weights snap over center to overcome the return spring's tension, you will hear a distinct snap when the motor reaches approximately 95 percent full speed, which indicates that the end switch has opened, and again after the motor is deenergized and the rotor shaft is coasting to a stop. When you hear the snap as the motor is coasting to a stop, it indicates that the end switch has returned to its closed position.

When the motor is deenergized, the rotor will deaccelerate to a stop and the centrifigual actuator will return to its original position with the aid of return springs. When the fly weights return to their normal position, the actuator moves back away from the switch and allows it to return to its closed position so that it is ready for the next time the motor is started.

14.5.1.5
Thermal Overload Protector

Single-phase motors provide a bimetal switch for use as a built-in overload device. This overload is mounted in the rear end plate near the centrifugal switch assembly and terminal board. The thermal overload consists of a heater and contacts. When the motor is starting and running, all its current is pulled through the heating element. If the motor draws excessive current, the heating element will become warm enough to cause the bimetal contacts to snap open and deenergize the motor windings. When the bimetal cools down, the bimetal will cool again and snap closed, which will reenergize the motor. If the same fault still exists, the motor will overheat again and continue to cycle off through its overload.

Some overloads do not reset automatically. Instead, they have a reset button that must be depressed manually to close the overload contacts. This necessitates someone going to the motor when it trips the overload and resetting it manually. At that time this person should inspect the motor and its loads to ensure that it is operating correctly.

14.5.2
Changing Voltage and Speeds of Single-Phase Motors

The single-phase motor is available for connection on 110 or 230 V. If the motor is connected for 115 V, it can be reconnected on the factory floor for 230 V. It is also possible to reconnect a motor from 230 to 115 V. This allows the motor to be used in any voltage application.

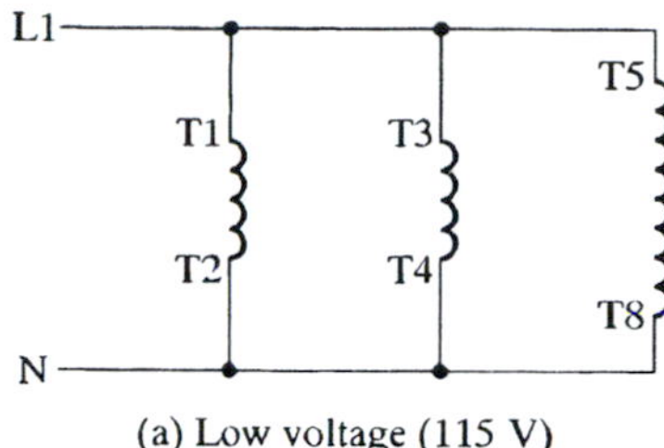

(a) Low voltage (115 V)

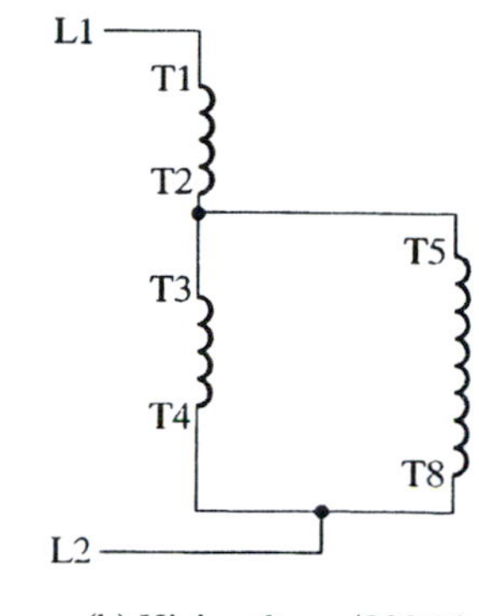

(b) High voltage (230 V)

Figure 14–24 Diagram for a single-phase motor wired for (a) 115 V (low voltage); (b) 230 V (high voltage).

Figure 14–24a shows a single-phase motor connected for low voltage and Figure 14–24b shows a single-phase motor connected for high voltage. Figure 11–24a shows the two coils of the run winding connected in parallel for low voltage. The start winding is then connected in parallel with these coils. The diagram in this figure shows that leads T1, T3, and T5 are connected to L1 and that T2, T4, and T8 are connected to L2. Figure 14–24b shows the connections for high voltage. From this diagram you can see that the two coils for the run winding are connected in series. The start winding is connected in parallel across the lower coil (T3–T4) of the run winding. When 230 V is applied to the two coils of the run winding, the voltage divides equally, 115 V across each coil.

The start winding will also receive 115 V from the terminals to which it is connected. Each coil section is rated for 115 V. If the motor is connected for low voltage, all the coil sections are connected in parallel so they will each have 115 V applied to them. When the motor is connected to 230 V, the run winding acts as a voltage divider so that each coil still receives 115 V. The single-phase motor is also available for dual speeds. This type of motor must have the number of poles reduced to increase the speed of the motor. In most cases the motor will lose half of its horsepower rating when it is operated on the faster speed, since fewer poles are used.

14.5.3 Increasing the Starting Torque of a Single-Phase Motor

The torque of a single-phase motor can be changed by adding capacitors to the start or run winding of the motor. When the single-phase motor is used without capacitors, as has been shown in the diagrams presented so far in this section, the motor is called a split-phase motor.

When a start capacitor is connected in series with the start winding and centrifugal switch, the motor is called a *capacitor-start, induction-run (CSIR) motor*. When the motor has a start capacitor in series with the start winding and a run capacitor is connected permanently across the run and start terminal, the motor is called a *two-capacitor motor* or a *capacitor-start, capacitor-run (CSCR) motor*. If the motor has only a run capacitor connected permanently across the start and run winding, it is called a *permanent split capacitor (PSC) motor*. If the rotor of the single-phase motor is made of copper wire rather than a squirrel-cage rotor, it is called a *wound rotor motor* or *repulsion-start motor*.

The following sections will explain the operation of each of these motors. Methods of reversing these motors are also presented with their diagrams. At the end of this section, methods of troubleshooting each of these types of motor are provided. You should gain an understanding of how these motors operate and how their rotation is reversed. It will also help to understand the methods of reconnecting the motor to operate on dual voltage or dual speeds. This information is important when you must connect motor control devices such as reversing starters or dual-voltage starters to them.

14.6 Split-Phase Motors

The *split-phase motor* is a single-phase motor that does not have any capacitors or other devices in its circuit to alter its torque characteristics. Diagrams of this motor are presented in Figure 14–25. This motor is also called the split-phase motor or the ISIR (induction-start, induction-run) motor, since it uses only induction to start and run.

This type of motor has the lowest starting torque of all single-phase motors. It uses the physical displacement of the run and start windings in the stator to provide the phase shift required to start the rotor moving. Remember that the three-phase motor uses the 120-degree phase shift that naturally occurs in the three-phase voltage to cause starting torque. Since the single-phase motor does not have a natural phase shift,

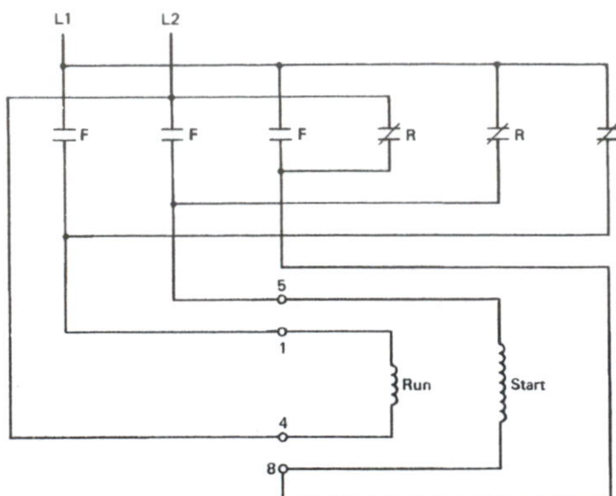

Figure 14–25 Diagram of a split-phase motor connected to a forward and reversing motor starter. The run winding (T1–T4) remains connected to the same way in both forward and reverse operation. The start winding (T5–T8) gets reversed to make the motor run in the opposite direction.

the split-phase motor uses the difference of the coil size to create a phase difference along with physically locating the start winding out of phase with the run winding to cause a magnetic phase shift that is large enough to cause the rotor to start spinning.

When voltage is first applied to the motor's stator, the rotor is not turning and the windings will draw maximum current. This current is called inrush current or locked-rotor amperage. After the rotor starts to turn, it will induce current from the stator and produce its own magnetic field. This field will cause the rotor to increase speed until it reaches its rated speed. The rated speed is determined by the number of poles the motor uses and the frequency of the applied voltage. This means that a two-pole motor will operate at 3600 rpm, a four-pole motor will operate at 1800 rpm, a six-pole motor will operate at 1200 rpm, and an eight-pole motor will operate at 900 rpm.

14.6.1 Counter EMF

When the rotor increases its rpm it will begin to generate a back voltage or counter voltage. This voltage is also called counter EMF (electromotive force). The difference between the counter EMF and the applied voltage at full speed may be only 1 to 2 V. This voltage is called the potential difference between the two voltages and it is responsible for causing the current flow required to keep the rotor spinning at its rated speed. The current required to keep the rotor spinning at its rated speed when it is driving a load is called full-load current or FLA.

If the load is increased, the rotor's rpm is slowed slightly, which will cause the counter EMF to be reduced slightly. Since the counter EMF has decreased and the applied voltage has remained the same, the difference of potential will increase and cause additional current to flow. The increased current will cause the motor to produce more torque, which will cause the rotor to come back up to speed. This feature makes the motor self-regulate its load and helps single-phase motors operate at fairly constant speed throughout their load range. If the motor is overloaded too heavily, the rotor will slow down to the point where it will stall. During this time it will continue to draw excessive current and overheat severely.

Methods of reversing the split-phase motor are presented in Figure 14–25. Terminals 5 and 8 of the start winding are reversed to get the rotor spinning in the opposite direction. Once the rotor begins spinning in one direction, the magnetic phase shift that is created will cause the motor to continue to rotate in that direction. If you find that the application requires more starting torque, a start capacitor should be added to the split-phase motor to make it a capacitor-start, induction-run motor.

14.6.2 Applications for Split-Phase Motors

The split-phase motor is used for general purpose loads. The loads are generally belt driven or small direct-drive loads like small drill presses, shop grinders, air conditioning and heating belt driven blowers, and small belt-driven conveyors. The main feature of the split-phase motor is that it can be used in areas of the factory where three phase has not been distributed, or on small loads on the factory floor where a fractional horsepower motor can handle the load. The motor does not provide a lot of starting torque, so the load must be rather small or belt driven, where mechanical advantage can be utilized to help the motor start.

The motor is very inexpensive and it can be replaced when it wears out rather than trying to rewind it. It is also available in a variety of frame sizes, which allows it to be mounted easily in most machinery. If the application requires too much torque for the split-phase motor, one of the other motors, such as the capacitor-start motor, can be used.

14.7 Capacitor-Start, Induction-Run Motors

The capacitor-start, induction-run motor is a split-phase motor with a starting capacitor connected in series with its start winding. This modification to the split-phase motor's design is accomplished when the motor is manufactured, so the capacitor-start motor is a separate choice in the manufacturer's catalog when the motor is being specified and selected for the application. Since

Figure 14–26 Picture of capacitor-start, induction-run (CSIR) motor. Notice that the start capacitor is mounted on the top of the motor.
(Courtesy of GE Industrial Systems)

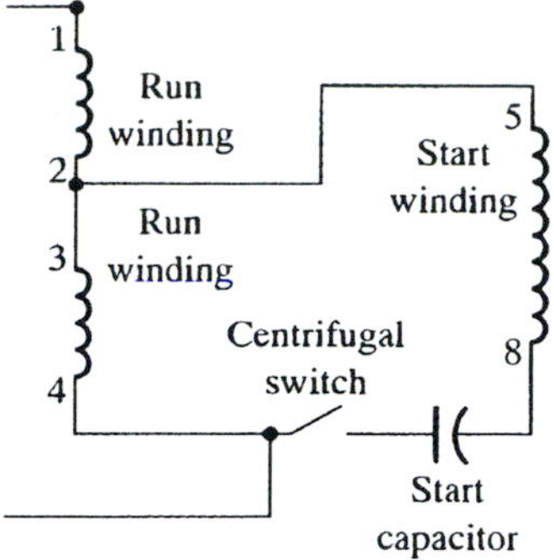

(a) Capacitor start, induction run (CSIR) motor wired for forward direction

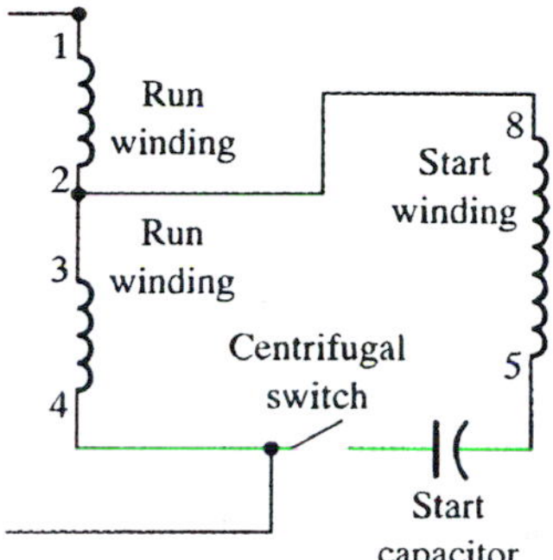

(b) CSIR motor wired for reverse rotation

Figure 14–27 Electrical diagram of a capacitor-start, induction-run (CSIR) motor connected for (a) forward rotation; (b) reverse rotation.

the CSIR motor can provide more starting torque than the split-phase motor, it would be selected for direct drive and other applications that require more power during start-up. Figure 14–26 shows a picture of a typical CSIR motor with the start capacitor mounted on top of the motor. The start capacitor is about the size of a small orange juice can and the case for the start capacitor is made of plastic.

14.7.1 Electrical Diagram for a CSIR Motor

A start capacitor is connected in series with the start winding of the CSIR motor to provide a larger phase shift when voltage is first applied to the motor. The increased phase shift causes a stronger magnetic field to pull on the rotor to cause it to begin to spin. The electrical diagrams for the CSIR motor connected for forward rotation and for reverse rotation are provided in Figure 14–27. The start capacitor is usually rated for over 100 microfarads (μF) (up to 1000 μF).

The start capacitor can provide large amounts of capacitance for a very short period of time, such as during starting. After the motor is started, the capacitor must be removed from the circuit so that it will not overheat. This is accomplished with the end switch. The electrical diagram in this figure shows the capacitor connected in series with the start winding. When voltage is first applied to the motor, both the start winding and the run winding will be energized. When the rotor reaches nearly full rpm, the end switch will open and disconnect the start winding and capacitor from the circuit. This means that no current will flow through either the start winding or the capacitor and they can cool down and be ready for the next time the motor is started. When the motor is deenergized and the rotor slows to a stop, the end switch is closed again and the start winding and capacitor are reconnected to the circuit for the next start.

Since the start capacitor is physically mounted on top of the motor, a metal cover is placed over it to protect it from damage. The capacitor's case is made of plastic and can easily be damaged if the motor is used in a harsh industrial environment.

14.7.2 Connecting the CSIR Motor for Dual-Voltage or Dual-Speed Applications

The capacitor-start motor is very similar to the split-phase motor since its terminals are identified in the same way and it can be connected to operate on either 115- or 230-V single-phase voltage. Like the split-phase motor, the CSIR motor will generally lose horsepower when it is operated at its higher speed, since two poles are deleted from the motor to gain the speed. Some factories prefer to stock dual-voltage, two-speed motors, which will cover nearly every application with their equipment. This means that they do not have to stock four separate types of motors for different applications, such as high speed, low voltage; high speed, high voltage; low speed, low voltage; and low speed, high voltage. By stocking the variable-speed, variable-voltage motor, they only need to stock different horsepower sizes, such as 1/2 and 1 hp; the other changes can be made by reconnecting the motor for the correct voltage or speed at the time the motor is installed.

14.7.3 Connecting the CSIR Motor for a Change of Rotation

The rotation of all of these motors can easily be reversed by changing the start winding leads. Diagrams that show the CSIR motor connected for both the forward and reverse operation are provided in Figure 14–27. The end switch (centrifugal switch) and start capacitor are connected near the end where terminal 8 is located. When the motor is reversed, terminals 5 and 8 are switched. When the motor is operating in the clockwise direction, T5 is connected to T3 and T8 is connected to T4. When the motor is connected for counterclockwise rotation, T8 is connected to T3 and T5 is connected with T4. This reverses the current flow in the start winding with respect to the current flow in the run winding, and the rotor will begin to spin in the opposite direction. Remember: Motor rotation is determined by looking at the motor shaft from the end opposite where the load is connected on the shaft. Also remember that the start winding must be connected so that it receives only 115 V. For this reason the start winding is always connected across the bottom part of the run winding.

14.7.4 Applications for CSIR Motors

The CSIR motor can be used for a wider variety of applications where more motor starting torque is required than a split-phase motor can supply. These applications include direct-drive water pumps, air compressors, and larger conveyors. The capacitor-start motor is also used to drive hermetic compressors for single-phase air conditioning systems on rooftops and in window units. The hermetic compressor is the sealed motor that is used to drive the refrigeration pump. Since this is a direct-drive application, the CSIR motor is generally used. The split-phase motor may be used on the smaller hermetic motors and systems that use a capillary tube for its refrigerant metering valve.

Since the compressor is hermetically sealed, it is not practical to place the end switch inside the compressor housing with the motor because it is likely to wear out and require replacement. It is also dangerous to allow the end switch to be mounted inside the compressor housing because it may cause a spark that could ignite fumes from the oil that is used in the compressor for lubrication and cause an explosion. Since an end switch cannot be used, a current relay is used to disconnect the start winding after the motor starts.

A sketch and electrical diagram of the current relay is provided in Figure 14–28. The current relay is generally connected directly on the motor terminals. You can see from the electrical diagram that the current relay consists of a coil and a set of contacts. The coil is connected between terminals L1 and R and the normally open contacts are connected between terminals L1 and S. The motor is shown as a run winding and a start winding. One end of the run winding and start winding are connected together at the right side and this point is identified as the common (C).

The diagram shows the current relay connected to the motor. Notice that the coil of the current relay is connected in series with the run winding and it will remain in this part of the circuit even when the motor is running. The contacts of the current relay are connected in series with the start winding. Since a start capacitor is used with this motor, the contacts (terminal S) of the relay are connected to the start capacitor and the other terminal of the start capacitor is connected to the start winding of the motor.

14.7.5 Current Relay Operation

From Figure 14–28a you can see that when voltage is applied to the compressor, the run winding will pull locked-rotor amperage of up to 40 A. Since the coil of the current relay is in series with the run winding, it will

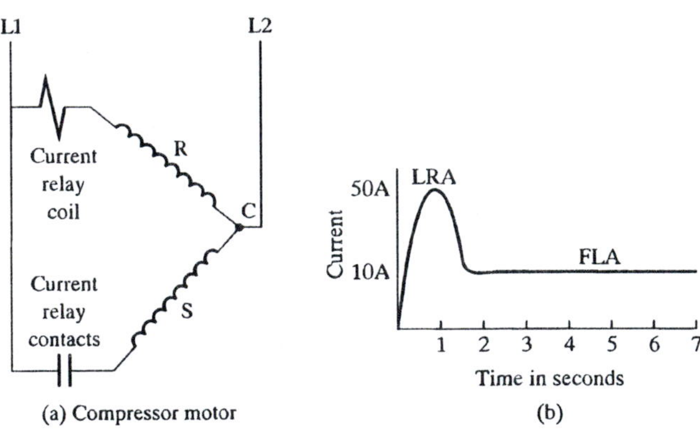

Figure 14–28 (a) Diagram of a current relay connected to a hermetic-type compressor motor. The current relay is used in place of a centrifugal switch since the compressor is a sealed motor. The current relay provides the same function as the centrifugal switch. (b) Diagram of a locked-rotor current (LRA) and full-load current (FLA) for the compressor motor.

also see this large current, which will be strong enough to pull the current relay's contacts closed. When the contacts close, it will provide a path for voltage to reach the start winding. This voltage will also pass through the start capacitor, which causes a phase shift in the magnetic field in the start winding and the torque will be strong enough to start the compressor motor even though it is under 70 to 80 psi of pressure.

The current relay has a return spring connected to its contacts that tries to pull them open. After the motor starts and picks up speed, the rotor will produce enough counter EMF to allow the full-load current to drop substantially. When the motor's current drops to the FLA level of 3 to 4 A, the spring will pull the contacts open. This drop in current from LRA to FLA occurs in 1 to 2 seconds as the motor reaches full speed. When the contacts open, the start capacitor and start winding are removed from the circuit and their current drops to zero. Since the run winding continues to be energized, it will draw FLA and the motor will continue to run. A graph of the locked rotor current and full-load current is also shown in Figure 14–28b. LRA exists for only several seconds while the rotor is coming up to speed. This current is strong enough to pull the current relay's contacts closed for the few seconds the motor requires to start.

Since the current relay is mounted on the outside of the compressor, it can easily be replaced if it becomes faulty. It can also be tested easily by checking voltage from L2 to the S terminal of the motor during the time the motor is starting. If no voltage is present at terminal S during start, the start capacitor should be removed from the circuit and an attempt made to restart the motor. Allow the motor to try and start for only a few seconds during this test because continued LRA current will damage the motor. If the terminal S on the motor receives voltage when the capacitor is removed, you can assume that the capacitor is open. If the terminal still does not receive voltage, the current relay is faulty and should be replaced.

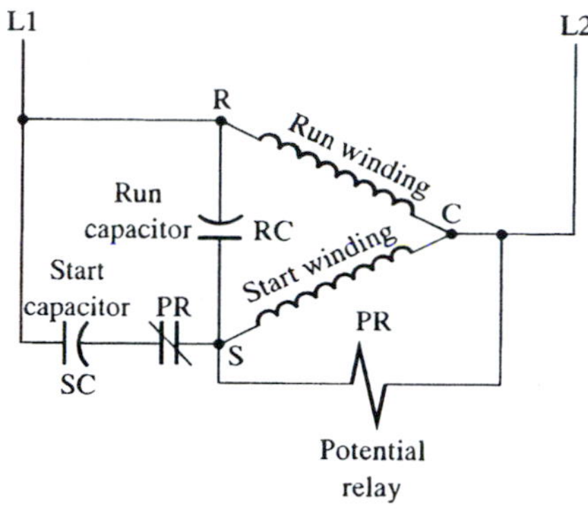

(a) Diagram of potential relay connected to a capacitor start, capacitor run motor

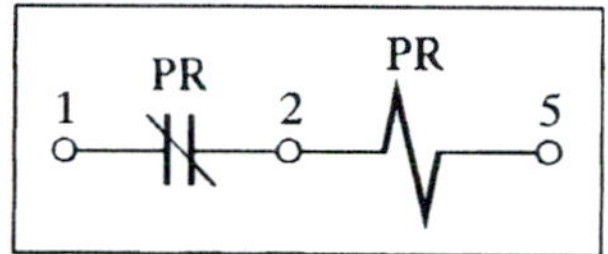

(b) Diagram of potential relay

Figure 14–29 (a) Capacitor-start, capacitor-run motor connected to a potential relay. The potential relay is used to energize and deenergize the start winding. (b) Diagram of a potential relay.

14.8 Capacitor-Start, Capacitor-Run Motors

The capacitor-start, capacitor-run (CSCR) motor adds a run capacitor to the start capacitor, which provides the motor with better torque characteristics when the motor is operating at full speed. The run capacitor is usually oval or square and has a metal housing rather than plastic. The metal housing allows the run capacitor to radiate any heat that is built up inside it, since it is connected to remain in the run winding circuit at all times.

From Figure 14–29a you can see that the run capacitor is essentially connected to the capacitor-start, induction-run motor. This type of motor is used almost exclusively for hermetic compressor motors in air conditioning systems. You will find air conditioning systems connected to most modern electronic panels today to provide the additional cooling that is required when computers, motor drives, and other amplifiers are placed in an enclosed cabinet.

For this motor, the run capacitor is connected in parallel with the run winding; it will be in the circuit during starting and remain in the circuit while the motor is running. During the time the motor is starting, the additional capacitor provides a little more phase shift than the start capacitor alone, which gives the motor more starting torque.

After the motor is running, the start capacitor is disconnected from the start winding circuit when the potential relay contacts open, but the run capacitor remains in the circuit because it is wired directly across the R and S terminals. If the load increases slightly because the compressor is trying to pump more refrigerant, the run capacitor will provide a small phase shift to give the rotor more torque and regain the loss of rpm quickly. The larger the load increase is, the more speed the rotor will lose, which will cause additional current to be drawn. The increase in current will pass through the run capacitor and cause it to provide a larger phase shift, which in turn provides the rotor with more torque. The run capacitor allows the speed of the motor to remain rather constant when the load is constantly varying.

14.8.1 Using a Potential Relay to Start a CSCR Motor

The capacitor-start, capacitor-run motor is used primarily in starting large, single-phase, hermetic compressors. Since the hermetic motor cannot use an end switch, a starting relay similar to the current relay must be used. The current relay requires that its coil be connected in series with the run winding of the compressor motor. In the CSCR motor, the run current (LRA) can be exceedingly large when the motor is starting. This current is too large for the current relay coil and would tend to burn it out. For this reason a different type of starting relay, which senses the amount of counter EMF for actuation, is used to start the CSCR motor. Since this relay uses the difference of potential between the applied voltage and the counter EMF produced by the rotor, it is called a potential relay.

Figure 14–29b shows a diagram of a potential relay whose terminals are identified as 1, 2, and 5. It has a normally closed set of contacts between terminals 1 and 2 and a high-resistance coil between terminals 2 and 5. The second diagram in this figure shows the potential relay connected to a motor with a start and run capacitor. The potential relay's contacts are connected in series with the start capacitor and the motor's start winding. The run capacitor is connected between the S and R terminals of the motor, where it will remain even when the motor is running. The dot on the curved side of the capacitor symbol represents the side of the run capacitor that should be connected to the line side of the power supply. This dot corresponds to a red mark on one of the run capacitor terminals. This terminal is marked because it is the terminal that is closest to the outside of the metal can that the capacitor is mounted in. If the capacitor is shorting out, it will short to this side of the circuit and cause a fuse to blow and the motor would not be damaged. If this capacitor is connected in the circuit so that the identified terminal is connected to the start winding, short-circuit current would be drawn through the motor windings and the motor would be damaged.

The operation of the potential relay is controlled by the amount of counter EMF that the rotor produces. When voltage is applied to the motor through L1 and L2, the run winding is energized directly and the start winding is powered through the start capacitor and normally closed set of contacts of the potential relay. The applied voltage is not strong enough to activate the potential relay's coil.

When the motor starts and the rotor begins to spin, a counter EMF is produced between motor terminals S and C. The counter EMF will become large enough to energize the potential relay coil when the motor reaches approximately 75 percent rpm. When the coil is energized, it pulls its normally closed contact open, which deenergizes the start capacitor and start winding. This effectively removes the start winding from the source of applied voltage.

The run capacitor is connected to the R and S terminals on the motor so that it is in circuit between the run winding and the start winding and will allow a very small amount of current to flow through from L1 to the start winding. When the load increases, such as when the compressor must pump more refrigerant, the rotor will begin to slow down and the counter EMF will be reduced slightly. This allows the run capacitor to provide a small phase shift, as it adds more current to the start winding. This increased current with the phase shift provides additional torque and the rotor's speed will again be increased to full rpm. This allows the CSCR motor to operate at a rather constant speed. The CSCR motor is not used in open motor applications, but the run capacitor concept is used frequently in a motor called the permanent split-capacitor motor.

14.9 Permanent Split-Capacitor Motors

The permanent split-capacitor (PSC) motor uses only a run capacitor to provide the phase shift required to start the motor. Figure 14–30 shows the PSC motor, and Figure 14–31a shows a diagram of the PSC motor. The run capacitor is connected between the run and start winding and no disconnecting switch or relay is required to deenergize the start winding from the applied voltage when the motor has started. The run capacitor is oval. It has a metal case, which allows it to dissipate extra heat that is built up in the capacitor, since it remains in the circuit at all times.

When voltage is applied to the motor, current will flow through the run winding to the common terminal. At the same time, current will flow through the run capacitor to the start winding. When the current flows through the run capacitor, it will provide a phase shift that is large enough to start the motor. As the rotor's speed increases, a counter EMF will be produced in the start winding that will limit current through it to less than 1 A when the motor reaches full speed. The small amount of current in the start winding when the motor is operating at full speed is small enough so that it will not cause the start winding to overheat.

When the motor shaft sees an increase in its load, it will slow down slightly. The decrease in the rotor's rpm causes a decrease in the counter EMF, which makes a larger potential difference between it and the applied

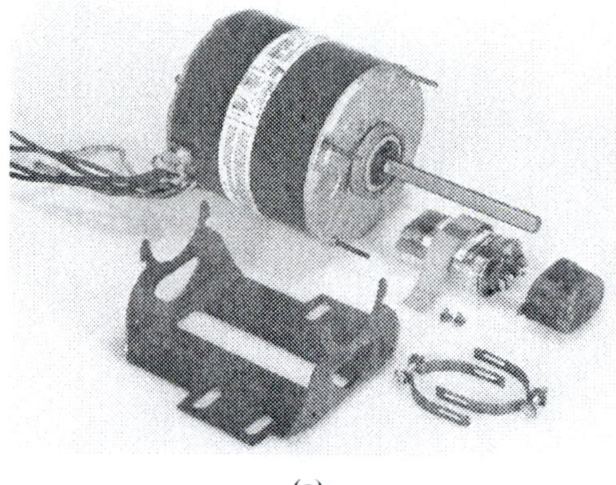

(a)

(b)

Figure 14–30 (a) A permanent split capacitor motor removed from its mounting bracket. (b) a permanent split capacitor motor shown in its mounting bracket and with its capacitor mounted on top.
(Courtesy of GE Industrial Systems)

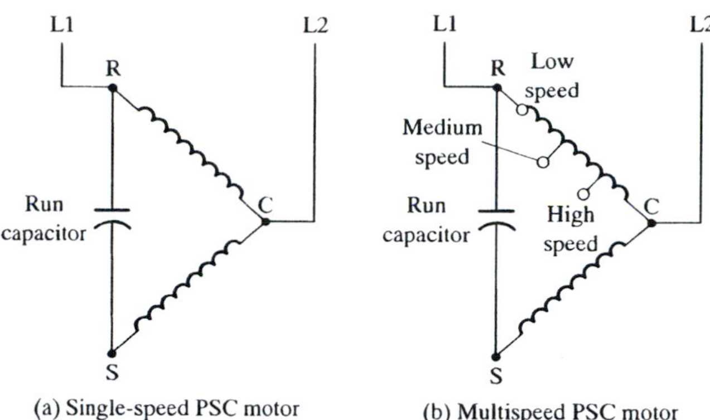

Figure 14–31 Electrical diagram of (a) a PSC motor; (b) a multispeed PSC motor.

voltage. The larger potential difference causes an increase in the current in the start winding, which will cause an increase in rotor torque that increases the rotor's rpm.

This characteristic allows the PSC motor to operate with a constant speed under varying load conditions without using any mechanical devices. The PSC motor is generally used for applications such as small hermetic compressors, blade fan loads, and other loads that require constant speed.

14.9.1 Connecting a PSC Motor for a Change of Speeds

The PSC motor is available as a variable-speed motor. Since the PSC motor does not require a centrifugal switch, it does not have an access plate in the end of the motor. Instead, all the leads are brought out of the motor together near the end plate. These leads are generally color coded to identify their speed. Figure 14–31b shows a diagram for a multispeed PSC motor. The color leads are connected at various points along the run winding coil. If power is applied at the very end of the run winding, all the poles of the run winding are used, and the motor will operate at its lowest speed. If power is connected at the terminal marked high speed, only one of the poles is used, and the motor will operate at its highest speed. The motor will lose torque as its speed is increased because fewer windings are used to gain the extra speed.

14.9.2 Changing Voltages or Direction of Rotation

Since the PSC motor is rather specialized and inexpensive, it generally cannot have its direction of rotation changed or be reconnected to operate on a different voltage. Instead, it is common practice to stock a clockwise and counterclockwise motor for each voltage application that you have in the facility. Since these motors are generally used for blade fan applications, they are commonly used for small air conditioning units used to cool small offices erected on the factory floor or the main offices of the facility. They are also used in the air conditioning units used to cool electrical control panels on larger equipment that have numerous electronic boards or motor drives mounted directly in the cabinet.

One problem with the PSC motor is that it may run in the wrong direction when it is used to drive a condenser fan on an outdoor air conditioning unit. This condition is caused when it is exceedingly windy and the wind blows across the fan blade when the motor is deenergized and causes the motor rotor to spin in the wrong direction. If the wind is blowing the fan blade in the wrong direction, it will continue to spin in that direction when voltage is applied. If the fan is running in the wrong direction it will cause insufficient air movement across the coils, which will cause the air conditioner to overheat and cause high pressure in the refrigerant coils. A ratchet mechanism is mounted on the motor shaft of condenser fans when this is a problem. The ratchet allows the motor to spin in the correct direction and prevents the motor shaft from spinning in the wrong direction when the wind is blowing.

14.10 Shaded-Pole Motors

Shaded-pole motors are commonly found in applications that require light-duty fans, such as small window air conditioners and exhaust fans used in rest rooms. If you are a maintenance electrician or technician, you may be requested to service all of the electrical equipment in the factory, including the office areas. If this is the case, you will run into shaded-pole motors. Figure 14–32 shows a diagram of the shaded-pole motor. You can see that the motor has only one winding. It does not have a start winding and a run winding like other single-phase motors. Instead, it has a shading pole that provides the magnetic field phase shift that is required to start the motor. The shaded pole motor has a copper bar that is inserted around the front of the run winding. The bar is connected at the ends to make a complete circuit called a *pole*. The shading pole may also be called a shading coil.

When voltage is applied to the motor to start it, current will flow through the run winding and build up a magnetic field. A current will be induced in the single winding of the shading pole, and it will cause a phase shift to occur that is large enough to make the rotor start to spin. Once the rotor starts to spin, it will begin to build its own magnetic field and come up to full rpm. The shading pole also helps the motor when its load changes at full rpm. If the motor shaft begins to slow down, the phase shift in the shading coil becomes stronger and provides enough torque to bring the rotor back up to full speed. Another unique feature of the shaded-pole motor is that it can withstand LRA for an extended period. Since the motor does not have a start winding, the run winding is large enough to carry locked-rotor current if the rotor becomes stuck. This is important, since it provides burnout protection without any additional devices or equipment being added to the motor. In industrial applications the shaded-pole motor is used as a level indicator. A paddle is attached to the shaft of the shaded-pole motor and turned very slowly. The paddle is mounted in a bin where granular material is stored. When the height of the material increases to the point where it covers the paddle, the paddle will stop turning and stall the motor. Since the motor is a shaded-pole motor, its current will increase, but the extra current will not damage the motor. A sensor is used to detect the change in current, which indicates the level of material is covering the paddle.

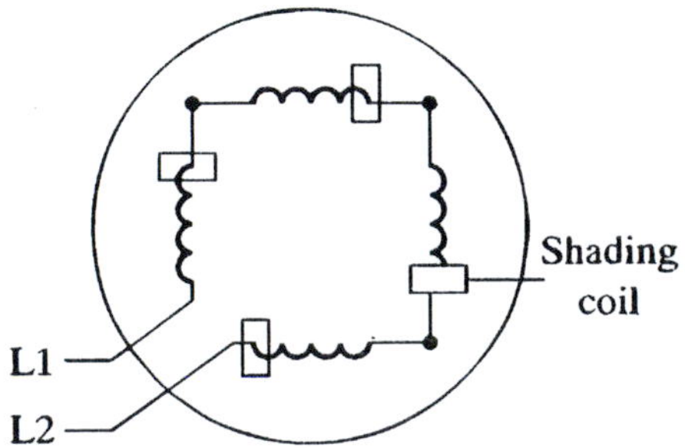

Figure 14–32 Diagram of a shaded-pole motor. Notice the shading pole (coil) located near each winding.

Since the shaded-pole motor has these characteristics, it is commonly used for small fan applications. If the fan becomes immovable for any reason, such as dirt or lack of lubrication, the motor will become warm, but it will not overheat and destroy itself like the split-phase or capacitor-start motors.

14.11 Repulsion-Start Motors

The repulsion-start motor was the most common single-phase motor in use prior to the squirrel-cage motor. After the 1960s very few repulsion-start motors were installed because they require brushes and a commutator to operate. The rotor for this type of motor is slightly different from the rotor of the squirrel-cage rotor since it uses copper wire to make its magnetic field. Another feature that makes the rotor different is that it has a wire that connects the commutator segments with a shorting mechanism, which is used in conjunction with the brushes. Since this motor was designed before squirrel-cage motor theory and technology became prevalent, the rotor was patterned after the wound rotor that is used in DC motors. The rotor was made of laminated sections with coils of wire pressed into place and their terminal ends brought out to commutator segments.

When the motor was being started, current was directed to the rotor coils through the brushes. After the rotor was spinning fast enough, the brushes were disconnected from the applied voltage and shorted so that the rotor would act like an inductive rotor. In some motors, the brushes remained connected to the applied voltage, but they were lifted slightly so that they would not make contact with the commutator. At the same time, a shorting mechanism would short the commutator segments to complete the circuit on each coil so that it could conduct the induced current like a squirrel-cage rotor in an induction motor.

In both of these types of motors the rotor would start the motor as a repulsion-start motor, and after the rotor came up to speed the motor would operate like an induction motor. This would give the motor the maximum amount of starting toque.

Since the rotor required brushes and some kind of lifting or shorting mechanism, it would require an excessive periodic maintenance. This made these motors too expensive to maintain and they were soon replaced with squirrel-cage motors.

14.12 Troubleshooting Three-Phase and Single-Phase Motors

Three-phase and single-phase motors are similar in operation and should be troubleshooted in a similar manner. Since each of the motors has some parts that are different, there are a few differences in the tests that should be made for each motor. The troubleshooting procedure should be broken into several sections, based on the types of symptoms the motor presents when it is not operating correctly. The procedures listed in this section are presented in the sequence of the problems that are most likely to occur to problems that are least likely to occur. You should adopt the same philosophy when you begin to troubleshoot the motor. Always begin by looking for the faults that are most likely to occur, then move on to faults that are least likely to occur. You should also perform simple tests first and complex tests later. This procedure will allow you to find most of the problems or faults with the simple tests.

14.12.1 Motor Will Not Turn When Power Is Applied

The most common problem that you will encounter with a motor is when voltage has been applied and the shaft does not turn. This can occur with the three-phase or single-phase motor. This symptom is caused by the loss of voltage to the motor or an open circuit somewhere in the circuit, including the voltage supply and the motor windings. You can test for voltage at all terminal points and determine where the loss has occurred.

14.12.2 Motor Hums But Will Not Turn When Power Is Applied

If the motor hums but will not turn when power is applied, it indicates that one phase of a three-phase motor is open or that the start winding of a single-phase motor is open. You can use a clamp-on ammeter to locate the line that is not drawing current in the three-phase motor. After you have determined which line is not drawing current you can look for a problem such as an open wire or open fuse. Note that if the motor continues to try and start, additional fuses may also blow and the motor will no longer hum.

If the motor is a single-phase motor and it hums but will not start, you can remove the voltage source and test the start winding for resistance. If the winding has high current or is open, you will need to change the centrifugal switch or change the motor. Be sure to test the capacitor also if one is used to start the single-phase motor. This problem may be caused by having too large a load for the motor or a bearing that needs lubrication.

Questions

1. What does the term induction mean in regards to an induction-type AC motor?
2. Explain how you can use a resistance test to detect the start and run winding of a split-phase AC motor.
3. What is counter EMF in a motor and what function does it provide?
4. Explain how the centrifugal switch operates in a split-phase motor to help it start.
5. Explain why a PSC motor does not need a centrifugal switch.

True and False

1. The three-phase, wye-connected motor has more starting torque than the three-phase delta-connected motor.
2. The synchronous AC motor uses DC voltage to provide extra field current so the motor does not have slip.
3. An induction-type AC motor needs slip to provide torque for the rotor.
4. Inrush current (LRA) is always larger than full-load current (FLA).
5. The speed of a DC shunt motor is adjusted by changing the frequency of the supply voltage for the motor.
6. The direction of rotation for a split-phase AC motor can be changed by changing the polarity of the supply voltage.
7. The direction of rotation for a three-phase AC motor can be changed by exchanging L1 and L2 of the supply voltage.

Multiple Choice

1. The _____ is connected in series with the start winding for a permanent split-capacitor (PSC) motor.
   a. start capacitor
   b. run capacitor
   c. potential relay
2. The start winding is connected in ____ with the run winding for a split-phase AC single-phase motor.
   a. series
   b. parallel
   c. series parallel
3. The _____ motor has the most torque of all of the single-phase AC motors.
   a. split-phase
   b. capacitor-start, induction-run
   c. capacitor-start, capacitor-run
4. The speed of an AC motor can be increased by _____
   a. decreasing the frequency of the supply voltage to the motor.
   b. increasing the frequency of the supply voltage to the motor.
   c. increasing the amount of voltage to the motor.
5. The single-phase induction motor needs a start winding _____
   a. to create a phase shift that will provide starting torque.
   b. to increase the speed of the motor.
   c. to create additional running torque.
6. The shaded-pole, single-phase motor _____
   a. uses a relay to remove its start winding from the circuit after the motor starts.
   b. uses a capacitor to remove the start winding from the circuit after the motor starts.
   c. does not have start winding, since it has a shading pole to provide the phase shift for starting.
7. The rated speed of a four-pole AC motor is _____
   a. 3600 rpm.
   b. 1800 rpm.
   c. 1200 rpm.

Problems

1. Calculate the rpm of a two-pole AC motor.
2. Draw the diagram of a delta-connected, three-phase motor that is connected for high voltage and low voltage and identify their terminals. Explain how you would reverse the direction of rotation for these motors.
3. Draw the wiring diagram of a wye-connected, three-phase motor that is connected for high voltage and low voltage and identify their terminals. Explain how you would reverse the direction of rotation for these motors.
4. Use the table in Figure 14–16 to determine the locked-rotor current for a 2 hp motor that has a NEMA code letter C.
5. Draw the wiring diagram for a split-phase, capacitor-start, induction-run, single-phase AC motor and identify their terminals.
6. Draw a sketch of a basic single-phase induction motor and identify the rotor, stator, and centrifugal switch.
7. Draw the wiring diagram for a PSC single-phase AC motor.
8. Draw the wiring diagram of a DC series, DC shunt, and any one of the DC compound motors and identify their terminals.
9. Determine the distance between the bolt holes for a 56 frame motor. (Use the table in Figure 14–14.)
10. Explain how you would troubleshoot a three-phase AC motor that fails to start, and whose motor windings do not make any noise or draw any current.
11. Explain how you would troubleshoot a single-phase AC motor that has a centrifugal switch and fails to start.

◀ Chapter 15 ▶

Motor Control Circuits

Objectives

After reading this chapter you will be able to:

1. Explain the operation of a two-wire circuit.
2. Explain the operation of a three-wire circuit.
3. Explain the operation of a circuit with multiple stop and start switches.
4. Explain the operation of a sequencing circuit.

In this chapter we explain basic motor control circuits commonly in use on the factory floor, along with their theory of operation and the flow of current through each circuit. We also provide laboratory exercises that allow you to wire components commonly found on the factory floor so that you can see the operation of these circuits. Later in the chapter you will be able to have faults entered into the circuits as they would occur on the machinery, which will provide you with realistic circuits to troubleshoot and repair.

The circuits in this chapter are identified so that you will be able to come back and study them if you need to review. This will be extremely useful when you find machinery that incorporates four or five of these basic circuits with additional complex circuits. In these circuits you will be able to gain knowledge that you can transfer to more difficult circuits for troubleshooting and repair. Another important feature of this chapter is that it is the first time that all the components of the circuit are used, including pilot devices, motor starters, as well as different types of motors. You should be able to deal with the operation of any of these devices as they are used in these circuits. If you do not understand their function or operation, you will need to review the chapters where they were introduced.

15.1 Two-Wire Control Circuits

The two-wire control circuit is commonly used in applications where the operation of a system is automatic. This may include such applications as sump pumps, tank pumps, electric heating, and air compressors. In these systems you normally close a disconnect switch or circuit breaker to energize the circuit, and the actual energizing of the motor in the system is controlled by the operation of the pilot device. The circuit is called a two-wire control because only two wires are needed to energize the motor starter coil. The circuit controls prior to the coil will provide the operational and safety features, while the circuit on after the coil will contain the motor starter overloads.

Figures 15–1 and 15–2 show a typical circuit of the pumping station and a circuit that controls an air compressor. In the pumping station diagram (Figure 15–1) you can see that the circuit is energized by a fused disconnect. After the disconnect is closed, the float switch in the diagram is in complete control of the motor starter. Notice that the float control is identifying a high-level control. When the level of the sump rises past the set point on the control, its contacts will energize and the motor starter coil will become energized. The coil of the motor starter will energize and pull in the contacts, which will energize the motor. As the motor operates, it will pump down the level of the sump until it is below the low set point of the float control. When the float control reaches this level, it will open and deenergize the motor starter coil, and its contacts will open and deenergize the pump motor.

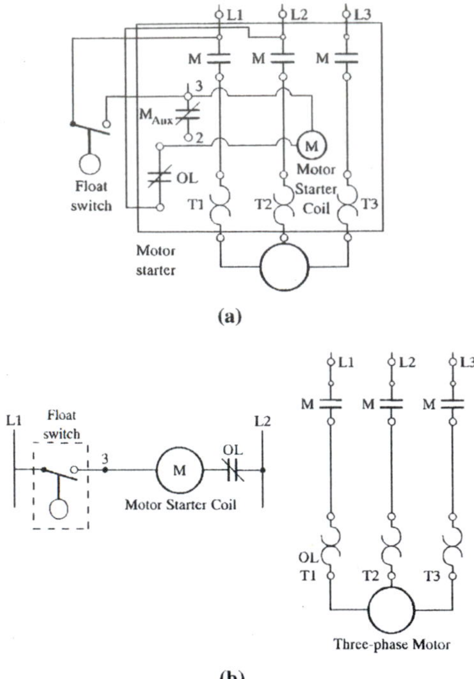

Figure 15–1 Two-wire control circuit for a pumping station: (a) wiring diagram; the level control switch will energize the pump motor when the level in the sump rises to the predetermined set point; (b) electrical ladder diagram.
(Courtesy of Rockwell Automation)

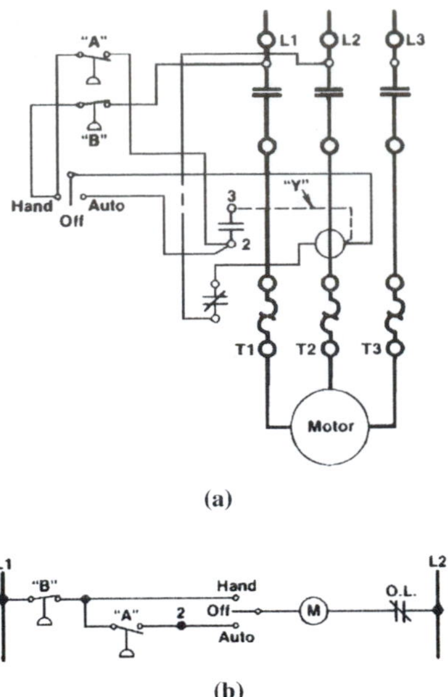

Figure 15–2 Two-wire control circuit for two compressors: (a) wiring diagram; (b) electrical ladder diagram.
(Courtesy of Rockwell Automation)

When the level of the sump rises again, the float control will close again and turn the motor on. This on/off sequence will continue automatically until the disconnect is switched off or unless the motor starter overloads are tripped. If the motor draws too much current when it is pumping, the heaters will trip the overloads and open the circuit. Since the overloads must be reset manually, the circuit will not become energized automatically when the overload is cleared. In this case, someone must physically come to the motor starter and press the reset button. At this time, the system should be thoroughly inspected for the cause of the overload. It would also be important to use a clamp-on ammeter to determine the full-load amperage (FLA) of the motor and cross reference this to the size of the heaters in the motor starter to see that they match. The FLA should be checked against the motor's data plate. You should also take into account the service factor listed on the data plate when you are trying to determine if the pump motor is operating correctly.

Figure 15–2 shows a control circuit for an air compressor. This two-wire control circuit is used to turn the air compressor on automatically when the pressure drops below 30 psi and to turn off the compressor when the pressure reaches 90 psi. Pressure switch A in this circuit controls the operation of the control at these pressures. Its high and low pressures are adjustable so that the system could be energized and deenergized at other pressures if the need arises. A hand switch is provided in this circuit that allows the system to be pumped to a predetermined pressure. The hand switch is intended to be used when the auto switch is not functioning properly or if you need to test the system.

Pressure switch B in this diagram acts as a safety for this circuit. This pressure switch is set at 120 psi and is not adjustable. It is in the circuit to prevent the tank pressure from rising too high. This could occur if the operational pressure switch became faulty and would not open when the pressure reached 90 psi. It could also be used to protect the system against overpressure when the switch is in the hand position. Generally, the safety switch is meant to protect the system against component failure or control failure. If a pressure control would fail, the compressor would continue pumping air into the tank, which would allow its pressure to rise to an unsafe level. Since the operational switch is still closed, the air pressure in the tank could be increased to a level where the pump would cause the motor to stall if the safety switch were not in the circuit. This could build up pressures to several hundred pounds, which could cause lines and fittings in the air system to explode.

The nonadjustable pressure switch would act as a backup to the operational switch and trip off any time the pressure reaches 120 psi. This switch is also different in that it is interlocked when it trips, so that it must be reset manually by having someone pressing the reset button. If you find the rest button activated on the safety switch, you must test the system thoroughly to determine why the operational switch did not control the circuit.

In normal operation, the motor starter would cycle the air compressor on and off to keep the pressure in the tank between the high and low set points that are set on the operational control. The motor starter overloads could also trip this circuit and require manual reset. This could occur if the motor was incurring overcurrent problems. The overcurrent could occur due to bad bearings or the pressure being set too high on the systems. It is very important to identify the cause of an overload condition and repair it before the circuit is put back into operation.

15.2 Three-Wire Control Circuits

Three-wire controls are so named because they have three wires connected to the motor starter coil. The extra wire in this case is the wire that connects the motor starter auxiliary contacts in parallel across the start pushbutton. A diagram of this circuit is shown in Figure 15–3. The circuit is shown as a wiring diagram and a ladder diagram. The operation of the three-wire circuit is easier to understand with the ladder diagram. From this diagram you can see that the start pushbutton is a normally open momentary pushbutton switch. This means that its contacts will stay closed only as long as someone physically depresses the switch. When the pushbutton is released, the normally open contacts return to their normally open state.

The normally open auxiliary contacts from the motor starter will close when the motor starter coil is energized. When they close they will provide an alternative path around the pushbutton contacts for current to get to the coil. These contacts are called *seal* or *seal-in contacts* when they are used in this manner. Sometimes the seal contacts are said to have memory, since they will maintain the last state of the pushbuttons in the circuit.

The three-wire control circuit is the most widely used motor control circuit. You should fully understand this circuit and learn to recognize it when it is shown as a ladder diagram and as a wiring diagram. In this way you will be able to understand the operation of the circuit wherever it appears. One variation of the three-wire circuit is shown in Figure 15–4. In this figure, multiple start and stop buttons are used, which allow additional control stations to be installed for the system. This type of application is usable where the operational system is installed over a large area, such as in a long conveyor system. If the conveyor is several hundred feet long, it would be inconvenient and unsafe to have one start and stop button at one end of the system. Pushbutton

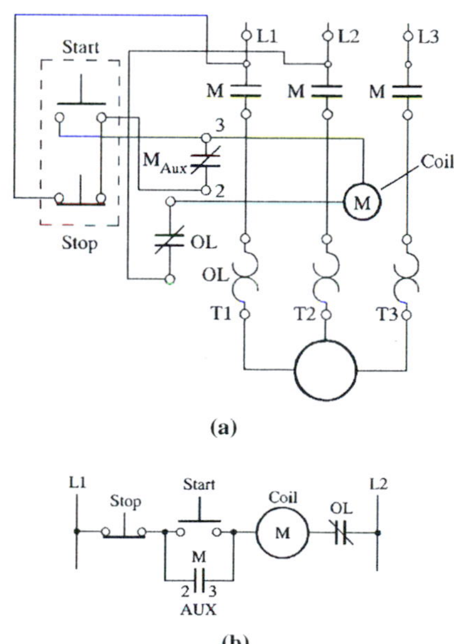

Figure 15–3 Three-wire control circuit of a start/stop circuit: (a) wiring diagram; (b) electrical ladder diagram.
(Courtesy of Rockwell Automation)

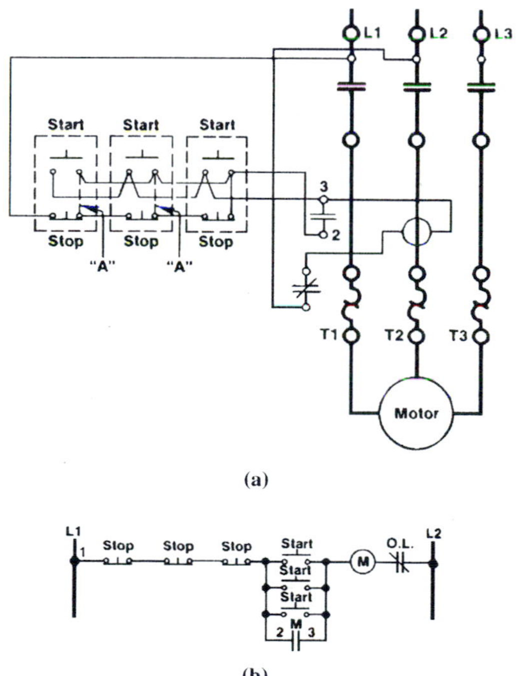

Figure 15–4 Three-wire control circuit with multiple start and stop switches: (a) wiring diagram; (b) electrical ladder diagram.
(Courtesy of Rockwell Automation)

stations can be installed every 50 ft using the diagram shown in Figure 15–4. Each additional start button should be connected in parallel with the first start button, and all additional stop buttons should be connected in series with the original stop button.

15.2.1 Step-Down Transformer in the Control Circuit

In some systems it is important that the control circuit be powered by 120 or 24 VAC, so that all the pilot switches will be connected to low voltage. This can be a problem in circuits where large motors are powered by 440, 550, or larger voltages. It would be impractical and unsafe to connect the control circuit directly to the high voltage.

In these applications, a control transformer is used to provide a lower voltage for the control circuit. The primary winding of the control transformer is specified to operate on line voltage and provide to the 120 or 24 V at its secondary. Figure 15–5 shows an electrical wiring diagram and a ladder diagram using a control transformer. The primary winding of the control transformers (H1 and H2) are connected to AC line voltage at terminals L1 and L2. Terminals H3 and H4 allow the control transformer to be connected to the high or low voltage specified on its data plate. Refer to Chapter 6 if you need help with this conversion.

Operation of the control circuit provided by the step-down transformer is identical to the circuit shown in Figure 15–4. This allows the technician to troubleshoot the circuit with the same procedures developed for the normal three-wire control circuit.

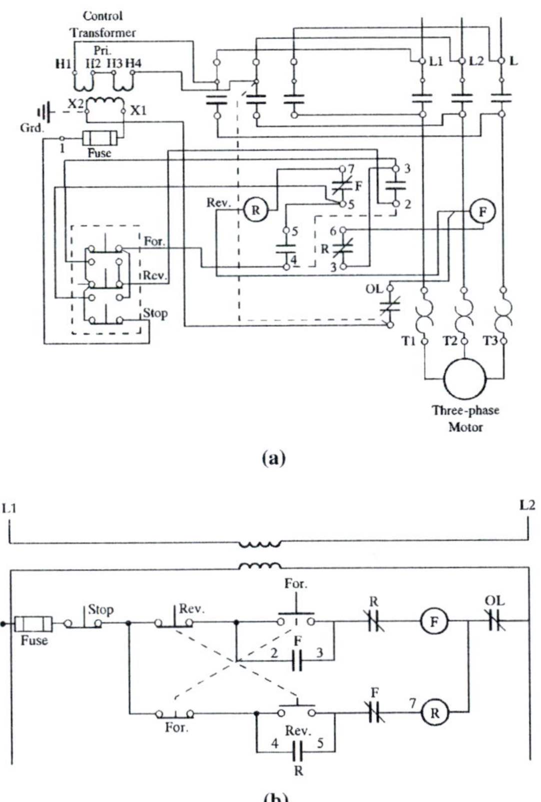

Figure 15–5 Forward/reverse motor starter circuit: (a) wiring diagram; (b) electrical ladder diagram. (Courtesy of Rockwell Automation)

15.2.2 Three-Wire Circuit with an Indicator Lamp

An indicator lamp can be added to a three-wire circuit to show when the circuit is energized or deenergized. The indicator lamp can be green or red to show when the system is energized and red to show when the circuit is deenergized. The lamp is usually mounted where personnel can easily see it at a distance. Sometimes the indicators are used where one operator must watch four or five large machines. After the machine has been set up, the operator will move on to the next machine. Since the installation is very large and spread out over a distance, the operator can watch for the indicators to see if the machine is still in operation.

The indicator lamps can also be used by maintenance personnel. The indicator lamps are wired into the circuit in such a way as to indicate that the motor starter is either energized or deenergized. This will help the maintenance person to begin testing for faults in the correct part of the circuit when the system has stopped or is not operating correctly. This is especially useful if the motor starter is mounted in a NEMA enclosure where its status is not easily verified.

Figure 15–6 shows a ladder diagram and a wiring diagram of a circuit with an indicator lamp connected in the control circuit. The lamps in this circuit are connected in parallel with the seal in contacts on the motor starter. When the motor starter closes, the lamp will be energized to indicate that the motor starter contacts are closed. Any time the lamp is deenergized, the operator and maintenance personnel know that the motor starter is not energized. A press-to-test lamp could also be used in this circuit to allow the operator and maintenance personnel to press the lamp at any time to see if it is operational. If the indicator is energized most of the time, such as in a continuous operation, the press-to-test lamp may not be necessary since the indicator will be energized most of the time.

Indicator lamps are available for 120, 240, 480, and 600 V, which provides them for any control circuit voltage. A wide variety of colored lenses are also available to indicate other conditions with the machine. These in-

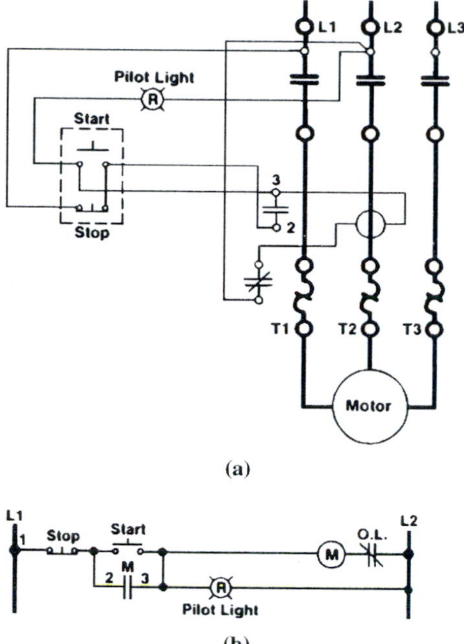

Figure 15–6 Three-wire control circuit with an indicator lamp: (a) wiring diagram; (b) electrical ladder diagram.

(Courtesy of Rockwell Automation)

dicators can be connected across different individual motor starters in the machine to provide information such as hydraulic pump running, heaters energized, conveyor in operation, and other conditions that are vital to the machine.

15.2.3 Three-Wire Circuit with the Indicator Lamp Showing That Circuit Is Deenergized

Another variation of this circuit would be to connect a red indicator lamp connected in series with a normally closed set of contacts. Since the lamp is connected in series with the normally closed set of auxiliary contacts, it will be energized any time the coil is not energized. When the coil becomes energized and pulls the contact carrier to the closed position, the normally closed set of auxiliary contacts will be opened and the lamp will become deenergized.

This circuit is useful in applications where it is important to know when the circuit is off, such as in a pumping station. The circuit shown in Figure 15–6 could be used, but a problem could arise if the indicator lamp burned out. In this case it may be useful to use a lamp to show when the circuit is energized and deenergized. If both lamps are out, it indicates that one of the lamps is burned out. Generally, this circuit is used when the system spends more time in the deenergized condition than in the energized condition.

The indicator lamp in this circuit can be mounted remotely or on the cover of the enclosure, where it will be close to the motor starter. This allows the lamp to be used in a variety of applications. In fact, a second or third indicator could be connected in the circuit in parallel with the existing lamp, and these could be mounted at a remote terminal or remote panel board to provide other personnel with information regarding the status of the circuit.

The circuits shown in these figures show the components as a ladder diagram and as a wiring diagram. The ladder diagram shows the operation of this circuit, while the wiring diagram shows the location of the terminals where these connections should be made. The wiring diagram will also be useful when technicians troubleshoot the circuit because it will show the location of terminal connections that are necessary for making voltage tests.

15.3 Reversing Motor Starters

From previous chapters you have found that DC and AC motors can be reversed. These chapters provided the terminal connections for each type of DC and AC single-phase and AC three-phase motors. Chapter 9 showed the control circuit required to operate the forward and reverse motor starter coils. When you must install or troubleshoot this circuit, you must treat it as an entity rather than as separated control and load circuits. The control circuit has a forward and a reverse pushbutton, each of which has a normally open and a normally closed set of contacts. In the ladder diagram, each button switch shows a dashed line, which indicates that both sets of contacts are activated by the same button. The forward and reverse pushbuttons are better defined in the wiring diagram, which shows that each switch has an open set and a closed set of contacts. The stop button is wired in series with the open contacts of both of these switches, so that the motor can be stopped when it is operating in either the forward or the reverse direction.

The operation of the pushbuttons is best understood through the use of a ladder diagram. The ladder diagram is used to show the sequence of the control circuit, while the wiring diagram shows the operation of the load circuit, which includes the motor and heaters for the overloads. The overload contacts are connected in the control circuit, where they will deenergize both the forward and reverse circuit if the motor is pulling too much current.

From the wiring diagram you can also see that two separate motor starters are used in the circuit. The forward motor starter is shown on the right side of the reverse motor starter. Each starter has its own coil and auxiliary contacts that are used as interlocks. The location of the auxiliary contacts is shown in the wiring diagram, but their operation is difficult to determine there. When you see the auxiliary contacts in the ladder diagram, their function can be more clearly understood.

This comparison of the ladder diagram and the wiring diagram should help you understand that you need both diagrams to work on the equipment. The ladder diagram will be useful in determining what should be tested, and the wiring diagram is useful in showing where the contacts you want to test are located. Notice that the control circuit is powered from a control transformer that is connected across L1 and L2. The secondary side of the transformer is fused to protect the transformer from a short circuit that may occur in either coil.

15.3.1 Other Methods of Reversing Motors

Several other motor control circuits allow you to reverse the direction in which a motor is turning. You studied the operation of the drum switch in Chapter 7 and studied methods of reversing three-phase and single-phase AC motors and DC motors in Chapters 13 and 14. As you know, the switch contacts are opened and closed manually by moving the drum switch from the off to the forward or reverse position. After you understand the operation of the drum switch in its three positions and the methods of reversing each type of motor, these concepts can be combined to develop manual reversing circuits for any motor in the factory as long as its full-load and locked-rotor amperage (FLA and LRA) do not exceed the rating of the drum switch.

These diagrams are especially useful for installation and troubleshooting of these circuits. The drum switch can be tested by itself or as part of the reversing circuit. The motors can also be disconnected from the drum switch and operated in the forward and reverse directions for testing or troubleshooting if you suspect the switch or motor of malfunctioning.

15.3.2 Adding Pilot Devices to the Control Circuit of a Motor Starter

Pilot devices can be added to the control circuit of any motor starter. They are generally added in series with the coil of the starter they are trying to control. When the contacts of the pilot device are opened, the coil is deenergized and the motor will also be deenergized. Since the pilot devices are located in the control circuit and have complete control over the motor, they are sometimes called *control devices*. A limit switch can be connected to the reversing motor starter control circuit. A limit switch is placed in series with the forward and reverse motor starter coils. The operation of this circuit is difficult to determine from the wiring diagram for this circuit, since it shows the limit switches and pushbuttons where they are located.

Two limit switches can be used and they will open and deenergize the motor when the load travels too far in either the forward or reverse direction. This circuit is very useful for applications where ball screw mechanisms are used for critical placement of a load, such as a gantry robot arm. When the arm is moving in the forward direction, the motor will continue to move it until it strikes the forward limit switch. When the limit switch is opened, the forward motor starter is deenergized and the motor is turned off, which stops the load at the precise location. As long as the arm stays in this location, it will continue to keep the limit switch open, which disables the forward coil. When the reverse pushbutton is depressed, the reverse coil can be enabled since its limit switch is closed when the arm is against the forward limit switch.

As soon as the reverse coil is energized, the reverse starter will close and the motor will begin to move the ball screw in the reverse direction. This action moves the arm in the reverse direction until it moves against the reverse limit switch and opens it. When the reverse limit switch is opened, the reverse coil circuit is deenergized, which turns off the motor. This action stops the motor in the exact location that is required.

If the arm is slightly out of position when the motor is deenergized and the ball screw stops, the limit switch can be adjusted to stop the motor in the correct position. The position of the limit switch can be used to aid an operator who may not be able to see the exact arm location from the operator's station where the pushbuttons are located. Other types of pilot devices could also be used in this type of application. A proximity switch or photoelectric switch could also be used to open when the mold is in the proper location. The operation, installation, and adjustment of these types of pilot devices are described in Chapter 10.

15.3.3 Using Indicator Lamps to Indicate Direction of Operation of the Reversing Starter

Concepts from several diagrams presented in this section can be combined to provide unique control applications. If you understand the operation and function of each circuit as it operates on its own, it is easy to see how you could combine several of the controls to design a new circuit. The main points to remember are whether the components should be connected in series or parallel with the coil in the circuit, and if the normally open or normally closed contacts should be utilized.

Figure 15–6 shows an example of this. In this circuit a lamp is used to indicate when the motor is energized. When the normally open contacts are energize the lamp, is energize to indicate the motor is energized.

You should start to get an idea of how other features can be combined to design the circuit required for your applications. When you must troubleshoot a circuit that combines concepts from several separate circuits, you should consider each part of the circuit's operation individually, and then as they come together to function as one circuit. You should try to isolate each individual section to make troubleshooting easier, but you may have to look at the entire circuit at times.

15.4 Jogging Motors

In some applications, such as motion control, machine tooling, and material handling, you must be able to turn the motor on for a few seconds to move the load slightly in the forward or reverse direction. This type of motor control is called *jogging*. The jogging circuit utilizes a reversing motor starter to allow the motor to be moved slightly when the forward or reverse pushbutton is depressed. Another requirement of the jogging circuit is that the motor starters do not seal in when the pushbuttons are depressed to energize the motor when it is in the jog mode, yet operate as normal motor starters when the motor controls are switched to the run mode. A diagram of a jogging circuit is provided in Figure 15–7. The motor is shown in the electrical wiring diagram, which shows the location of each component, and the ladder diagram, shows the sequence of operation.

The wiring diagram gives you a very good idea of the way the jog/run switch operates. This switch is shown to the left of the motor starter in the diagram. It is part of the start/stop station. The jog/run button is a selector switch that is mounted above the forward/reverse/stop buttons. When the switch is in the jog mode, the selector switch is in the open position. From the ladder diagram, you can see that the jog switch is in series with both of the seal-in circuits, which prevents them from sealing in the forward or reverse pushbuttons when they are depressed. This means that the motor will operate in the forward direction for as long as the forward pushbutton is depressed. As soon as the pushbutton is released, the motor starter will become deenergized. This jog switch also allows the motor to be jogged from one direction directly to the other direction without having to use the stop button. The motor is protected by the overloads that are connected in series with the forward and reverse motor starter coils. If the overload trips, the overload contacts in the control circuit will open and neither coil can be energized until it is reset.

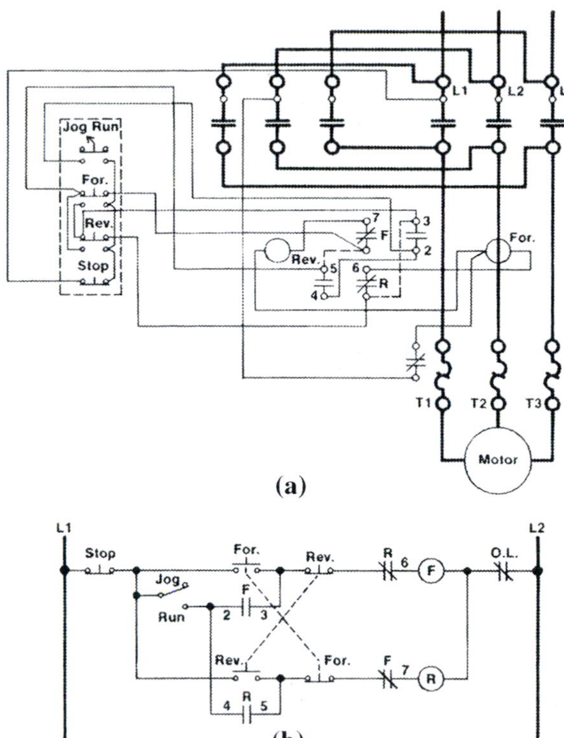

Figure 15–7 (a) Wiring diagram for a typical jogging circuit. (b) ladder diagram for a jogging circuit.
(Courtesy of Rockwell Automation)

These two diagrams will allow you to understand the operation of the jog circuit. You can make a forward and reversing motor starter circuit into a jogging circuit by adding the jog switch, but you must be sure that the motor and the motor starters are rated for jogging duty. Some motor starters and motors cannot take the heat that will be built up when the motor is started and stopped continually during the jogging operation. The motor and the motor starters will be rated for jogging or plugging if they can withstand the extra current and heat.

15.5 Sequence Controls for Motor Starters

Sequence control allows a motor starter to be utilized as part of a complex motor control circuit that uses one set of conditions to determine the operation of another circuit. Figure 15–8 shows an example of this type of

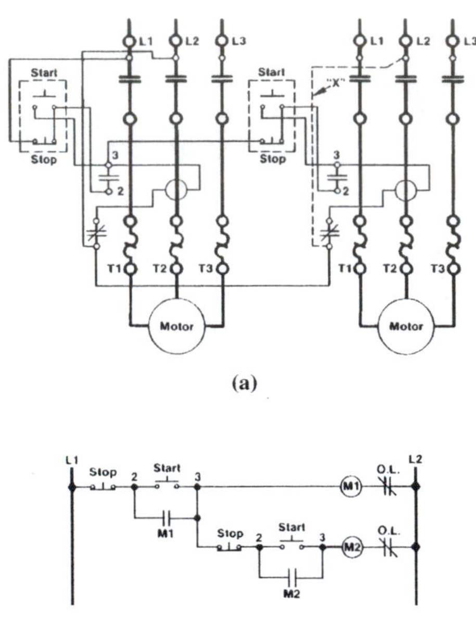

Figure 15–8 Sequence control circuit: (a) wiring diagram; (b) electrical ladder diagram.
(Courtesy of Rockwell Automation)

circuit. The circuits in this figure are presented in wiring diagram and ladder diagram form. You will really begin to see the importance of the ladder diagram as it shows the sequence of operation, which would be very difficult to determine from the wiring diagram. The wiring diagram is still very important, since it shows the field wiring connections and the locations of all terminals that will need to be used during troubleshooting tests.

The operation of the circuit in Figure 15–8 shows two conveyors that are controlled by two separate motor starters. Conveyor 1 must be operating prior to conveyor 2 being started, because conveyor 2 feeds material onto conveyor 1 and material would back up on conveyor 2 if conveyor 1 was not operating and carrying it away. The ladder diagram shows a typical start/stop circuit with an auxiliary contact being used as a seal-in around the start button. When the first start button is depressed, M1 will be energized, which will start the first conveyor in operation. The M1 auxiliary contacts will seal the start button and provide circuit power to the second start/stop circuit.

Since the second circuit has power at all times after M1 is energized, its start and stop buttons can be operated at any time to turn the second conveyor on and off as often as required without bothering the first conveyor motor starter. Remember that this circuit requires conveyor 1 to be operating prior to conveyor 2, since conveyor 2 feeds material onto conveyor 1. The circuit also protects the sequence if conveyor 1 is stopped for any reason. When it is stopped, the M1 motor starter becomes deenergized, and the M1 auxiliary contacts return to their open condition, which also deenergizes power to the second conveyor's start/stop circuit.

The power for this control circuit comes from the L1 and L2 of the first motor starter. This means that if supply voltage for the first conveyor motor is lost for any reason, such as a blown fuse or opened disconnect, the power to the control circuit is also lost and both motor starters will be deenergized, which will stop both conveyors. If the first motor draws too much current and trips its overloads, it will cause an open in the motor starter's coil circuit, which will cause the auxiliary contacts of the first motor starter to open and deenergize both motor starters.

If you need additional confirmation that the belt on the first conveyor is actually moving, a motion switch can be installed on the conveyor, and its contacts would be connected in series between the first start button and M1 coil. This would cause the first motor starter coil to become deenergized any time the conveyor belt was broken or slipping too much.

Another application of sequence control has several motors that require a pump or fan to be in operation any time one of them is running. An example of this application would be an exhaust fan that is required to remove fumes from an area any time manufacturing equipment is in operation. Another example would be multiple air-conditioning compressors that utilize the same water-cooled condenser. In this application, the water pump on the condenser must be operating prior to allowing any of the compressors to run.

This application would also be useful where water is required for a food process. In this system the water under pressure must be provided to each processing machine. The water is pumped into a tank under pressure. The operation of the water pump is controlled by motor starter M1 and the pressure switch. A conveyor is also used in this system and it must be operating any time either of the process machines is operating. The conveyor is operated by motor starter M4, and the process machinery is controlled by motor starters M2 and M3.

The ladder diagram in this figure shows the sequence of this operation. You can see that the master start circuit must be energized to allow any of the motors to operate. A control relay is used to create the master start circuit. When it is energized, its seal-in contacts keep it energized as long as the stop button is not depressed. A second set of normally open CR contacts are also used to apply power to the remainder of the control circuit.

After the control relay is energized, the water pump will be energized any time the pressure in the tank drops below the set point. Motor starters M2 and M3 control two separate process machines and they can be started or stopped by their individual start/stop controls. When either M2 or M3 is energized, their auxiliary contacts will energize motor starter M4 and cause the conveyor to operate. If the master stop button is depressed, the control relay will deenergize and all the other circuits will become disabled.

Another variation of sequence control uses a time-delay relay to prevent two motor starters from energizing their motors at the same time. If the two motors are powered from the same distribution system, they may draw too much LRA if they both try to start at the same time. This application uses one start/stop station to start both motors, and a small time delay is used to allow the first motor to start and come up to speed before the second motor is started.

From the ladder diagram in Figure 15–8 you can see that both motor starters are controlled by one set of start/stop switches. A timer motor (TR) is connected in parallel with the coil of the first motor starter and the coil of the second motor starter is connected in series with the contacts of the timer. When the start button is depressed, the first motor starter coil (M1) is energized, which starts the first motor and closes the auxiliary contacts that act as a hold-in for the start button. Since the timer motor is connected in parallel with the coil M1, it will become energized and begin to run its cycle. The time delay of this cycle will be adjusted so that it provides enough time for the first motor to reach full rpm.

After the time delay has expired, the timer contacts will close and energize the second motor starter. This allows the second motor to draw locked-rotor current after the first motor is fully up to speed. This causes less current demand on the power distribution system than if both motors were to draw locked-rotor current at the same time.

Questions

1. Explain the operation of a two-wire circuit.
2. Explain the operation of a three-wire circuit.
3. Explain the operation of a circuit with multiple stop and start switches.
4. Explain the operation of a sequencing circuit.
5. Explain the operation of a jog circuit.

True and False

1. The two-wire control circuit is commonly used in applications where the operation of a system is automatic.
2. Three-wire control circuits have a set of auxiliary connects that are wired in parallel across the start pushbutton.
3. The coil in a two-wire circuit will stay energized after the start button is released.
4. The coil in a three-wire circuit will stay energized after the start button is released.
5. In the sequencing circuit shown in Figure 15–8, the coil of M2 can be energized only if the coil of M1 is energized.

Multiple Choice

1. If you add multiple stop pushbuttons to a three-wire control circuit, the extra stop buttons should be _____
   - **a.** wired in parallel with the start pushbutton.
   - **b.** wired in series with the start pushbutton.
   - **c.** wired in parallel with the stop pushbutton.
   - **d.** wired in series with the stop pushbutton.
2. If you add multiple start pushbuttons to a three-wire control circuit, the extra start buttons should be _____
   - **a.** wired in parallel with the start pushbutton.
   - **b.** wired in series with the start pushbutton.
   - **c.** wired in parallel with the stop pushbutton.
   - **d.** wired in series with the stop pushbutton.
3. In a jog circuit the jog switch should be _____
   - **a.** wired in series with the start pushbutton.
   - **b.** wired in parallel with the coil of the motor starter.
   - **c.** wired in series with the auxiliary contacts that are used to seal in the start pushbutton.
   - **d.** wired in parallel with the auxiliary contacts that are used to seal in the start pushbutton.
   - **e.** All of the above
4. A motor reversing circuit should have _____
   - **a.** mechanical interlocks built into the reversing motor starter.
   - **b.** logic interlocks designed into the circuit by wiring NC contacts from the forward motor starter in series with the reversing coil.
   - **c.** wired interlocks designed into the circuit by wiring NC contacts on the start pushbutton in series with the reversing coil.
   - **d.** All of the above
5. If you add an indicator lamp to a circuit, the lamp should be wired _____
   - **a.** in series with the motor starter coil.
   - **b.** in parallel with the motor starter coil.
   - **c.** in series with the auxiliary contacts.
   - **d.** in parallel with the auxiliary contacts.

Problems

1. Draw an electrical diagram of a two-wire control system and explain its operation.
2. Draw an electrical diagram of a three-wire control system and explain its operation.
3. Draw a three-wire control circuit that uses multiple start and stop buttons.
4. Draw an electrical diagram of a three-wire control circuit that uses an indicator lamp to show when a motor is energized.
5. Draw an electrical diagram of a forward and a reverse motor starter.

◀ Chapter 16 ▶

Advanced Motor Control Circuits: Accelerating and Decelerating Circuits

Objectives

After reading this chapter you will be able to:

1. Explain the operation of a primary resistor reduced voltage starter.
2. Explain the operation of an acceleration circuit.
3. Explain the operation of a deceleration circuit.
4. Explain the operation of an autotransformer starter.
5. Explain the operation of a wye–delta starter.
6. Explain the operation of a solid-state reduced-voltage starter.

At times you will be asked to install or troubleshoot more advanced motor control circuits that are used to control the way a motor is started or stopped. These circuits are called acceleration and deceleration circuits. *Acceleration circuits* are used to start a motor without allowing the locked-rotor current to become too large. The locked-rotor current for a motor during starting can be up to six times the normal operational current. This can cause problems if the distribution system is loaded to near capacity, because the excessive current draw can cause interruption to the whole system. The excessive starting current can also cause the demand factor on the electric meter to become too large, which doubles or triples the electric bill. Another problem created by the large locked-rotor currents is the wear and tear on switchgear. When motors are allowed to draw maximum current, they cause arcing and heat buildup that stress contacts and switchgear. This stress causes the equipment in bus ducts, disconnects, and motor starters to wear out prematurely. If locked-rotor current is limited during starting, the life of the switchgear can be extended more than enough to pay for the more complex circuits required to control the motors. A problem may also arise when loads are started with full torque. The starting torque of a squirrel-cage motor can be as large as 140 percent of the normal operating torque. This may become a problem with loads such as a conveyor. When the large torque is applied during starting, the material on the conveyor belt may be spilled.

It is also important to control the time it takes a motor to stop. In some applications it is important that the load stop at exactly the time and location when the motor is deenergized. In normal motor operation, when a motor is deenergized, the load is allowed to coast to a stop, which means that the larger the load is, the longer the coasting time. This causes the load to be located at random, which may be unacceptable in motion control applications. In other applications it is unsafe to allow the load to coast to a stop. This is true where large cutting blades are turned at high speeds in machine tools or wood-cutting applications and are allowed to continue to rotate after power to the motor has been deenergized. In all these cases, motor control circuits or hardware must be provided to bring the motor to a stop quickly when power is deenergized. These circuits, called *deceleration circuits*, may involve shorting terminals of the motor to cause regeneration of voltage back into the rotor, or may involve hardware such as some type of mechanical or electrical brakes.

In this chapter we provide information with diagrams to explain operation of these types of circuits. Other information regarding wiring of basic components will aid in installation and troubleshooting of each circuit.

16.1 Acceleration Circuits

Acceleration circuits allow motors to come up to speed without drawing excessive locked-rotor current (LRA). These circuits are designed for DC and AC three-phase motors. They can utilize several theories of operation. One way to reduce the LRA is to reduce the amount of voltage applied to the motor while the motor is starting. This can be accomplished by using resistors in series with the supply voltage to the motor, or by reducing the amount of voltage supplied by the transformer. Another way to reduce the amount of LRA is to change the winding configuration of the motor during the starting. When three-phase windings are connected in a wye configuration, they will draw less current than when they are connected in delta. After the motor is up to speed, the motor windings can be converted to delta to provide better running torque and operation.

Another method of reducing current involves using only part of the motor winding when the motor is started. After the motor is running, the remainder of the winding is reenergized so that the motor is operating at full horsepower. This type of acceleration requires the motor load to be reduced during starting, and it can be increased after the motor is up to speed.

16.2 Reduced-Voltage Starters

Five types of reduced-voltage starting circuits are presented and explained in this chapter. These circuits are compared to each other and to an across-the-line starter with regards to the amount of voltage applied to the motor at starting, the line current, the starting torque, the type of transition, the cost, advantages and disadvantages, and applications.

Figure 16–1 shows a table with these comparisons. The solid-state reduced-voltage starter is not included in this table, but we discuss it at the end of this section, following our discussion of the primary resistor, auto-

TYPE OF STARTER	STARTING CHARACTERISTICS IN PERCENT OF FULL VOLTAGE VALUES			STANDARD MOTOR	TRANSITION	EXTRA ACCELER. STEPS AVAILABLE	COST OF INSTALLATION	ADVANTAGES	DISADVANTAGES	REMARKS	APPLICATIONS
	VOLTAGE AT MOTOR	LINE CURRENT	STARTING TORQUE								
ACROSS-THE-LINE A10	100%	100%	100%	Yes	None	None	Lowest	• Inexpensive • Readily available • Simple to maintain • Maximum starting torque	• High inrush • High starting torque		Many and various
AUTO-TRANSFORMER A400	80% 65% 50%	64% 42% 25%	64% 42% 25%	Yes	Closed	No	High	• Provides highest torque per ampere of line current • 3 different starting torques available through auto-transformer taps • Suitable for relatively long starting periods • Motor current is greater than line current during starting	• In lower hp ratings is most expensive design • Low power factor • Large physical Size	• Most flexible • Very efficient	Blowers Pumps Compressors Conveyors
PRIMARY RESISTOR A430	65%	65%	42%	Yes	Closed	Yes	High	• Smooth acceleration — motor voltage increases with speed • High power factor during start • Less expensive than autotransformer starter in lower HP's • Available with as many as 5 accelerating points	• Low torque efficiency • Resistors give off heat • Starting time in excess of 5 seconds requires expensive resistors • Difficult to change starting torques under varying conditions	• Can be designed so starting characteristics closely match requirements of load	Belt and gear drives Conveyors Textile machines
PART WINDING A460	100%	65%	48%	❶	Closed	Yes (but very uncommon)	Low	• Least expensive reduced voltage starter • Most dual voltage motors can be started part winding on lower voltage • Small physical size	• Unsuited for high inertia, long starting loads • Requires special motor design for voltage higher than 230 • Motor will not start if the torque demanded by the load exceeds that developed by the motor when thet first half of the motor is energized • First step of acceleration must not exceed 5 seconds or else motor will overheat	• Not really a reduced voltage starter. Is considered an increment starter because it achieves objective by reconnecting motor winding.	Reciprocating compressors Pumps Blowers Fans
WYE DELTA A490	100%	33%	33%	No	Open ❷	No	Medium	• Suitable for high inertia, long acceleration, loads • High torque efficiency • Ideal for especially stringent inrush restrictions. • Ideal for frequent starts.	• Requires special motor • Low starting torque • During open transition there is a high momentary inrush when the delta contactor is closed	• Same as part winding (above) • Very efficient	Centrifugal compressors Centrifuges

❶ Standard dual voltage 230/460 volt motor can be used on 230 volt systems.
❷ Closed transition available for about 30% more in price.

Figure 16–1 Table comparing reduced-voltage starting circuits. (Courtesy of Eaton/Cutler Hammer)

transformer starter, wye–delta starter, part winding starter, and secondary resistor. Two graphs are also presented in this figure to compare line current versus speed characteristics and torque versus speed characteristics. These ratings are based on a typical NEMA design B motor. The information in these graphs is presented in this section to provide you with a comparison of these types of starting circuits. Operation of these circuits is also explained.

16.2.1 Open and Closed Transition Starters

The classification of open and closed transition starters are used to explain the operation of the switching contacts in these types of starters. Each of these types of starters provides a means of starting the motor on the reduced-voltage circuit, and then switching to the normal across-the-line configuration to allow the motor to go into its run mode. If the operation of the contacts is such that the reduced-voltage circuit is completely disconnected before the run circuit is connected, the starter is called an *open transition starter.* During the time when the first set of contacts have been opened and the run set have not yet closed, the motor is completely disconnected from power, hence the name "open transition."

If the operation of the contacts allows the run set of contacts to close before the first set have been fully disconnected, the starter is called a *closed transition starter.* During the time when the run set of contacts are closed, the starting circuit is still connected. This means that the motor is never fully disconnected from the power source except when the control circuit is completely disconnected.

16.2.2 Primary Resistor Starters

The primary resistor reduced-voltage starter allows a large resistor to be connected in series with the supply voltage for a specified amount of time until the motor is running approximately 65 percent of full rpm. It provides approximately 42 percent of normal starting torque and limits starting current to 65 percent. This type of starting configuration provides smooth acceleration since voltage will increase with the motor's speed. It also provides a high power factor during starting and is less expensive than other types of reduced-voltage starters. This type of starting can utilize up to five separate resistors to provide stepped acceleration. The primary resistor starter is useful for belt and gear drives, conveyors, and textile machine application.

16.2.2.1 Primary Resistor Starter Operation

Figure 16–2 presents a wiring diagram and a ladder diagram that shows the basic components and the sequence of operation for this type of circuit. From the wiring diagram you can see that one resistor is used in each phase of the supply voltage. The resistors are called primary resistors because they are in the supply side of the motor circuit. The resistor in the first phase is identified by the

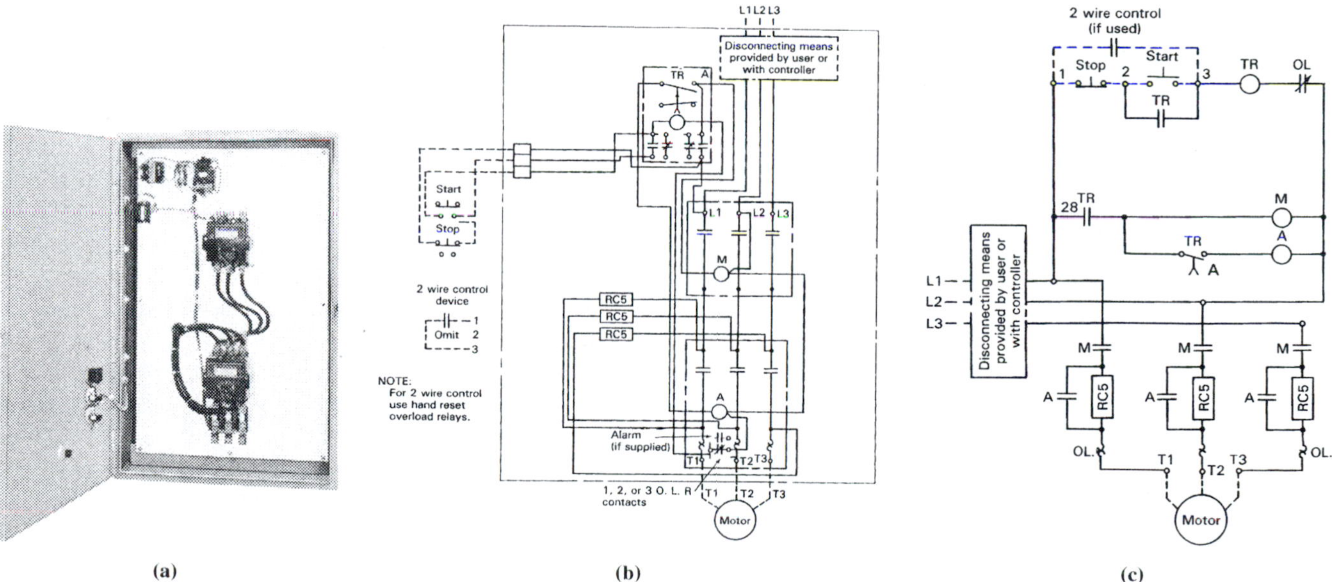

Figure 16–2 (a) Primary resistor reduced-voltage starter circuit. (b) Wiring diagram. (c) Ladder diagram.
(Courtesy of Square D/Schneider Electric)

numbers R1 and R2, while the second resistor is numbered R11 and R12. The last resistor in the third phase is numbered R21 and R22. It is difficult to see the sequence of operation in this diagram since it is only supposed to show the location of each component.

The ladder diagram shows the load circuit and the control circuit. The load circuit shows the connection of the resistors in reference to the motor. You can see that one resistor is in series with each phase of the voltage that is supplied to the motor. A set of shorting contacts are connected in parallel with each resistor. These contacts are identified by the letter A, which indicates that they are controlled by contactor A in the control circuit.

The control circuit has a normal start/stop circuit with a set of hold-in contacts around the start button. The contacts for the hold-in circuit are the instantaneous contacts (1B) of the timer motor. These contacts will close immediately when the timer is energized by momentary depression of the start button. A second set of instantaneous timer contacts (2B) are used to energize the main motor starter. When the coil of this motor starter is energized, it closes the M contacts and provides voltage to the motor through the primary resistors. Since this voltage flows through the fixed resistors, a voltage drop of about 20 percent is produced.

The amount of locked-rotor current is determined by the amount of voltage applied to the fixed resistance of the motor stator. Since the voltage has been reduced by the fixed resistors, the amount of starting current will be reduced proportionately. Because the amount of starting current determines the amount of starting torque the motor will be able to produce, the starting torque of this type of starting circuit will be reduced up to 56 percent.

After the reduced voltage has been applied to the motor, the rotor will begin to accelerate. As the rotor accelerates, it will produce an increased amount of counter EMF, which will cause the amount of current the motor is drawing to be reduced. The motor will accelerate to approximately 75 to 80 percent full rpm while the reduced voltage is applied. When the motor reaches this speed, the time-delay contacts from the timer are closed and the A contacts close to short out the primary resistors.

When the A contacts close, they provide an alternative path for current to reach the motor windings without going through the resistors. This connects the motor windings directly to the supply voltage, which allows them to receive full voltage. The increased amount of voltage provides an increased amount of current and torque while the motor is at full speed. The amount of time delay can be adjusted for more or less time to ensure that the motor is at the proper speed before the A contacts are closed and the motor is allowed to run on full voltage.

When LRA and the reduced-voltage starter are compared, you can see that the amount of current the motor uses to get up to full speed is much less with the reduced-voltage starter circuit than when the motor is started with full voltage.

16.2.2.2
Sizes of Primary Resistor Starters

The primary resistor reduced-voltage starter is available in a variety of voltages, including 220, 230, 380, and 575 V. The starters are rated for 5- to 600-hp loads. These starters provide control through motor starters and contactors rated in sizes ranging from a NEMA 1 through a NEMA 7. This provides a complete selection of controls to fit motors for a large variety of applications.

A photograph of a typical primary resistor-type starter is included in Figure 16–2. The resistors are mounted right in the enclosure with the contactors. Since they are mounted in an enclosure, a disconnect and fusing are also available as part of the control.

16.2.3
Autotransformer Starters

One problem that occurs with a primary resistor reduced-voltage starter is that all the voltage that is dropped through the resistors is turned into heat. The amount of heat may become very large and cause problems when it cannot be removed quickly enough. An autotransformer reduced-voltage starter is able to provide the same type of voltage reduction without building up large quantities of heat. This type of starting provides the largest amount of torque per ampere of line current. These controls have multiple taps that allow modification of the current or torque characteristics. They also have the ability to provide relatively long start-up times. Another advantage of this type of starter is that the autotransformer can produce currents on its secondary side to the motor that are larger than the current in the supply voltage side. Since the current in the supply side is measured by the electric demand meter, this means that the autotransformer starter can produce a larger current without loading up the demand meter.

Figure 16–3 shows an electrical diagram of the autotransformer starter. A multitapped autotransformer is provided for each phase of the supply voltage. One transformer is placed in series with each phase of the supply voltage. The taps on the transformer are set at 50, 67, and 84 percent of the supply voltage. The tap that is connected will drop the voltage to that percentage of supply voltage. The taps allow the voltage to be raised or lowered in reaction to the load and locked-rotor current. If the load needs more torque, a higher tap can be used to provide the motor with more voltage and more current. If the motor is drawing too much LRA but is still

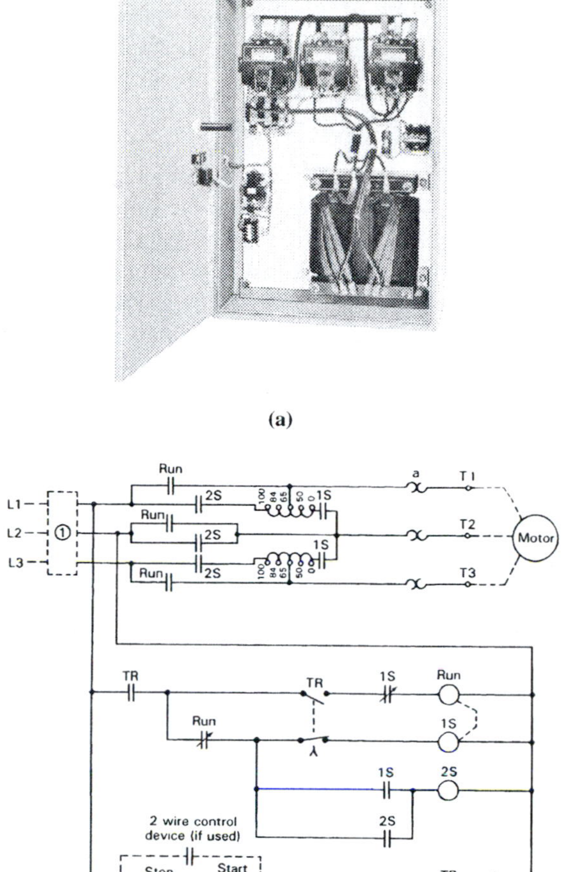

Figure 16–3 (a) Autotransformer-type reduced-voltage starter. (b) Wiring diagram for autotransformer type reduced-voltage starter.
(Courtesy of Square D/Schneider Electric)

starting satisfactorily, the next-lower tap can be tried. The control circuit for this application is very similar to the control circuit of the primary resistor, since it utilizes a motor timer to provide the time delay for the circuit. The timer controls the amount of time delay for the circuit. When the timer delay has elapsed, the timer contacts will close and energize the 2S contacts that are in parallel with the autotransformer. When these contacts close, they will short out the transformer circuit and apply full voltage to the motor. This is a shunt circuit, so this short circuit around the transformer is not dangerous or harmful to the autotransformer or the motor.

16.2.3.1 Autotransformer Starter Operation

You can follow the sequence of operation of this circuit by referring to the ladder diagram (elementary diagram). From this circuit you can see that when the start button is depressed, the timer motor is energized and the timer's instantaneous contacts at TR 1B and TR 2B are both closed immediately. The timer's delay contacts will close after the preset time delay has elapsed. The TR 2B contacts act as a hold-in circuit for the start button, while the TR 1B enables the rest of the control circuit and supplies power to the delay contacts. The time-delay contacts of the timer have direct control over the main motor starter and two auxiliary contactors. Their coils are identified as Run, 1S, and 2S.

The Run motor starter controls the contacts that supply full voltage to the motor after it has been started by the autotransformer. The 1S and 2S contacts are located at each end of the autotransformer, so that when they are open, the transformer is completely disconnected from the circuit and isolated. The contacts in this type of control operate in close transition. This means that the second set of contacts will close while those in the first set are still energized. This does not cause problems in this type of starter, since the Run contacts merely apply line voltage in parallel with the autotransformer.

After the timer TR 1B instantaneous contacts have closed, the normally closed Run contacts allow the 1S coil to be energized. The 1S coil closes all three sets of contacts that are connected to one end of each autotransformer. An auxiliary set of contacts are used in the control circuit to energize the 2S coil. A mechanical interlock is also provided between the Run motor starter and the 1S contactor so that they both cannot be energized at the same time.

When the 2S coil is energized, it closes its three sets of load contacts, which are connected to the opposite end of the autotransformer. These contacts complete the circuit and connect the autotransformer in series with the three main motor windings. A set of 2S auxiliary contacts are used to hold in the 2S coil in the control circuit.

After the motor has started running and reaches the percentage of predetermined speed, the time-delay contacts close and energize the Run motor starter coil. This coil closes its load contacts, which connects the motor directly across the line, which provides the windings with full voltage. The auxiliary Run contacts deenergize the 1S and 2S coils in the control circuit at the same time that the load contacts are connecting the motor to full voltage. This deenergizes the autotransformer and completely isolates it from the circuit. After the motor is connected to full voltage, it can operate for as long as necessary, just like a normal across-the-line starter.

When the motor is connected to the autotransformer, it will draw all its starting current through the transformer windings, which tends to heat up the autotransformer. If the motor takes too long to come up to speed or the time delay does not deactivate the starting circuit, the transformer could overheat and become damaged. To prevent this from occurring, a thermal overload device may be installed in the transformer winding when it is manufactured. This overload device is connected in series with the timer motor, which would deactivate the complete starting circuit and require someone to come to the control and depress the start button to restart the circuit after the transformer cooled down.

16.2.3.2 Sizes and Applications for Autotransformer Starters

The autotransformer starter is available in a variety of sizes for different applications. Typical supply voltages for the control include 200, 208, 440, 550, and 575 V. Typical horsepower ratings include sizes from 15 through 400 hp. The contacts are available in NEMA sizes from 1 through 6, which can carry up to 540 A. The coils for these controls are available in sizes from 208/240 through 600 V. These controls are available in a variety of enclosure types.

This type of reduced-voltage starter is used for applications where compressors, blowers, pumps, and conveyors are started infrequently, but allowed to run for long periods. One reason it is generally used for these types of applications is that the control has a low power factor rating, which would cause problems to the power distribution system's correction factor if the motor required frequent starting.

16.2.4 Wye–Delta Starters

Another type of motor starting circuit allows a three-phase AC motor to start without drawing excessive locked-rotor current by starting the motor when it is connected in wye configuration and then switching to delta configuration during run. You should recall that a three-phase motor will use less LRA when it is started as a wye-connected motor. It will have less starting torque in this configuration, but it is normally enough torque to move the load. This type of starting arrangement requires one motor starter and two contactors. This type of control is also called a *star–delta starter.*

An advantage of using this type of starting is that it is capable of supplying frequent starts to the motor. It is also well suited for high-inertia loads and long acceleration times. The ability to keep inrush or LRA to a minimum makes it ideal for many applications where the power distribution system is nearly overloaded. Some disadvantages may require close comparison to other types of starters, since this type of control requires that the motor be six-terminal delta-connected so that the changes can be made. Another problem involves the reduction of starting torque when the motor is connected in the wye configuration. Recall from Chapter 14 that the reduction in torque is caused by a change in the type of connection. When the motor is connected in a wye configuration, one end of each of the three motor windings is connected with the others at the wye point. When a supply voltage of 208 V is connected to any two of the three leads, the two windings that are powered make a voltage divider, so that each of the two coils splits the full 208 V. Instead of splitting the voltage directly in half, it is split at a rate of 3, since it is supplied by three-phase voltage. This means that each winding receives 58 percent of 208 V, which is equal to 120 V, and the windings will automatically draw less current since less voltage is applied. Since less current is flowing, the windings will have a reduction in the amount of torque that can be supplied. You can figure the torque will be 33 percent of what it normally would be.

If the motor is connected in delta, voltage is supplied at each of the three corners of the delta so that each winding of the delta receives the full 208 V. Since each winding of the motor is now receiving 208 V instead of 120 V, it will draw much more current and be able to supply much more torque. Remember that the load will determine the exact amount of torque and current the motor will draw. If the load is large, the torque will also be large, and if the load is reduced, the motor will automatically draw less current and provide less torque.

Figure 16–4 shows a wiring diagram and a ladder diagram for this circuit, and a photograph of the hardware that is used. From the wiring diagram you can see that the motor starter's contacts connect T1, T2, and T3 directly to L1, L2, and L3 of the power supply. It is difficult to determine the connection of the motor winding with this diagram, so two separate diagrams are provided to show only the connections that are used to connect the motor in wye and then in delta.

From the simplified wye diagram you can see that terminals T4, T5, and T6 all need be connected together, and terminal T1 should be connected to L1, T2 to L2, and T3 to L3 for the motor to operate as a wye-connected motor. This would involve the 1M motor starter and the S contactor being closed. The 1M motor starter takes care of the connection between T1, T2, and T3 with the supply voltage of L1, L2, and L3, and contactor S provides a set of jumpers to connect T4, T5, and T6 together. The ladder diagram also has been modified to show only the part of the circuit that is needed to make the starter and contactor energize at the same time during starting. This involves only the top two rungs of the circuit. When the start button is depressed, the S coil is energized immediately. The S auxiliary contacts provide voltage to energize the 1M coil, which will also become energized as soon as the S contacts have closed.

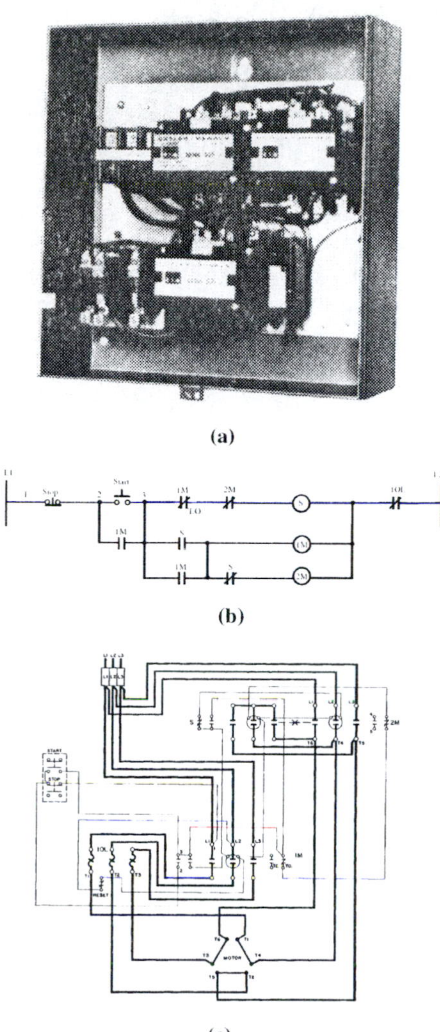

Figure 16–4 Wiring diagram of a wye–delta starter. (Courtesy of Eaton/Cutler Hammer)

The 1M contacts provide a hold-in circuit around the start button. This means that the 1M coil will remain energized even after the start button is released. Since the S coil is not part of this hold-in circuit, the S coil will stay energized only as long as the start button is depressed. When the start button is released, the S coil becomes deenergized, and the normally closed S contacts will return to their closed condition and energize the 2S coil.

The operation of the circuit requires that the operator remain at the start/stop station while the motor is brought fully up to speed since the motor will stay on the wye starting circuit only as long as the start button is depressed. As soon as the start button is released, the S contactor will deenergize and the 2S contactor will energize, causing the motor to switch to delta connection for running.

16.2.4.1 Simplified Delta Connections

You can see from the simplified delta connections that terminals T1 and T6 should be connected together with L1, and T2 and T4 should be connected together with L2, while T3 and T5 should be connected together with L3. The 2S contactor provides a set of contacts that connect T6 back to the terminal block of the supply voltage at terminal L1. Since 1M connects terminal T1 to the same supply voltage terminal, this effectively connects T1 to T6 and to the supply voltage at L1. The same is true of the contact connection that 2S provides between T4 and L2 and between T5 and L3. Since 1M connects T2 directly to L2, and T3 to L3, all the proper delta connections are made when the 1M motor starter is energized at the same time as the 2S contactor. This will occur after the start button has been released once the motor is brought up to speed with the wye winding. When the start button is released, the hold-in circuit provided by the 1M contacts also provides voltage to the 2S coil through the normally closed S contacts. Since the operator presses the start button to start the motor as a wye-connected motor and releases it to allow the motor to run in delta connection, the start/stop station does not have to be attended to allow the motor to continue to operate as a delta-connected motor.

The reduction in the amount of locked-rotor current during starting is significant enough to provide a savings in the maximum load a power distribution system will have to supply. This also prevents a large amount of heat from building up in the switchgear and contacts, which allows these components to last much longer. The reduction allows the fuses and circuit breaker to be set closer to the actual running current of the motor, which will provide adequate overcurrent protection.

16.2.4.2 Sizes of Wye–Delta Starters

The wye–delta starter is available for use on motors with voltages from 200 to 600 VAC. This provides control for motors that are rated from 10 through 1500 hp. The starters and contactors are rated from NEMA size 1 through NEMA size 8 to handle the currents for this control. The control is available with coil voltages rated at 120, 208, 240, 480, and 600 V. This allows the control circuit to operate on the voltage that is available at the installation location. If low voltage is required, control transformers can be used to step down the higher line voltages.

The start/stop controls for this type of circuit can be mounted directly in the door of the cabinet or at a remote station. Since there is such a large difference between the amount of current the motor will draw during starting and during run, this type of motor-starting control requires specially sized heaters for overload protection.

16.2.5 Part-Winding Starters

The part-winding reduced-voltage starter utilizes the fact that a motor can be connected so that the coils within each winding can be split into two groups and be connected in parallel with each other. Since these coils are in parallel, the motor can be started using one group from each winding, bringing in the remaining coil after the motor is up to speed. This can be accomplished when the motor is wired for wye or for delta. When only one group of motor windings are used, less current is drawn because the resistance of a single winding is larger than when the two equal groups are connected in parallel as one winding. (This is the same theory that causes total resistance to drop as resistors are connected in parallel with each other.) Since the amount of LRA is caused by the Ohm's law relationship of the supply voltage and the amount of resistance in the winding when the rotor is not turning, we know that the lower the resistance of the winding, the higher the LRA current will be.

This type of control provides the least expensive method of starting a motor with reduced voltage, since no hardware is required. It can be used with most dual-voltage motors as long as the lower voltage is used. Another advantage of not using any extra hardware is that this control will be smaller than other types of controls, which is important where panel space is restricted.

This type of starting does present several disadvantages. It is not suited for high-inertia or long-starting loads. It cannot start loads where the starting torque is very large; instead, the load must be energized with a clutch or by a valve if it is too large. Another problem associated with this type of control is that the motor will overheat if the motor starting time for acceleration exceeds 5 seconds.

The most typical applications for this type of starting include reciprocating compressors that can be unloaded by valves during start-up. After the motor is up to speed and under power from the full motor winding, the valves can be closed to bring the compressor under load. Other applications include pumps, blowers, and fans that can be started in an unloaded condition. It is possible to accomplish this with clutches or by arranging dampers and valves to unload the motor during start.

Figure 16–5 presents a photograph and a set of wiring diagrams, ladder diagrams, and terminal-connection diagrams for use with part-winding start circuits. The elementary (ladder) diagram shows the control circuit and the load terminals for this circuit. To prevent the diagram from becoming too complex, not all terminals are shown for each motor. Instead of showing the motors, a table is presented to show the motor terminal connections that you would make to connect the motor for half-wye or half-delta operations. The connections in this table list the motor terminals that should be connected together, and the connections that should be made to the terminals of the motor starter and contactors. The terminals for the S contactor are listed as A, B, and C; the terminals for the Run motor starter are listed as D, E, and F. For field wiring connections, you should look at the table and make the motor connections to the contact terminals as shown for terminals A–F.

The operation of energizing the coil of the S contactor and the Run motor starter is controlled by the top part of the ladder diagram, called the *control circuit*. The control circuit utilizes a motor timer to provide a time delay to allow the motor to start on the part-winding circuit and

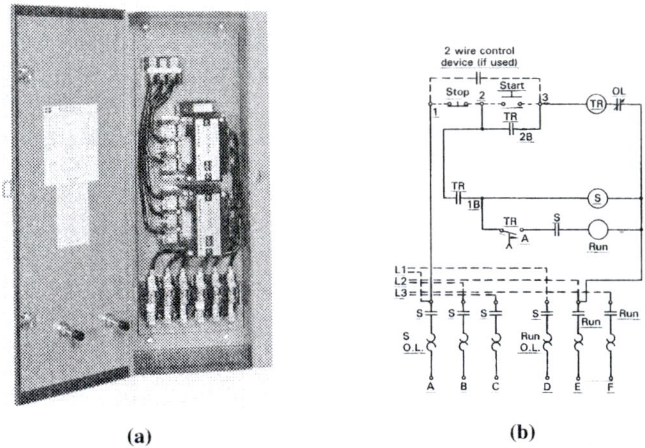

(a) (b)

MOTOR LEAD CONNECTIONS TABLE						
PART WINDING SCHEMES	LETTERED TERMINALS IN PANEL					
	A	B	C	D	E	F
1/2 Wye or Delta 6 Leads	T1	T2	T3	T7	T8	T9
1/2 Wye 9 Leads (1)	T1	T2	T3	T7	T8	T9
1/2 Delta 9 Leads (2)	T1	T8	T3	T6	T2	T9
2/3 Wye or Delta 6 Leads	T1	T2	T9	T7	T8	T3
2/3 Wye 9 Leads (1)	T1	T2	T9	T7	T8	T3
2/3 Delta 9 Leads (2)	T1	T4	T9	T6	T2	T3

(1) Connect terminals T4, T5 & T6 together at terminal box
(2) Connect terminals T4 & T8, T5 & T9, T6 & T7 together in three separate pairs at terminal box.

(c)

Figure 16–5 (a) A part-winding start motor control. (b) Diagram with terminal connection tables for a part-winding start motor circuit. (c) A table of connections for a part-winding start motor control.
(Courtesy of Eaton/Cutler Hammer)

then shift to the full winding after the motor is running approximately 50 to 60 percent of full rpm. The timer motor will be energized when the start button is depressed. The timer's instantaneous contacts, TR 1B, will provide a hold-in circuit for the start button. This means that the start button only needs to be depressed momentarily and then released for the motor to begin its starting operation on the part winding.

The second set of instantaneous timer contacts, at TR 2B, provide voltage to the S coil as soon as the timer motor is energized. After the motor has started running and the timer delay has expired, the timer's delay contacts will close and energize the Run motor starter coil. When the Run motor starter contacts are closed, the remainder of the motor windings are connected back into the circuit and the motor will operate as a normal wye or delta motor.

This type of starting configuration allows the motor to start with a minimum of locked-rotor current, but this also means that the motor will be able to provide only minimal torque during starting. For this reason this application is used only where the motor load can be switched on and off with a transmission, clutch, or by similar means. If the load is some type of pump or large compressor, solenoids can be used to bypass all the pump's flow or unload compressor cylinders until the motor is up to full speed.

16.2.5.1 Sizes of Part-Winding Starters

Part-winding motor starters are available in voltages of 200, 230, 460, and 575 V. These starters can control motors from 10 up to 1400 hp. They can contain NEMA-rated starters from size 1 through size 8, controlled by coils powered with voltages ranging from 120 through 600 V, which allows this type of control to be used in a large variety of applications. The control is available with a fusible or circuit breaker disconnect.

16.2.6 Secondary-Resistor Starters

Secondary-resistor starters are used exclusively for starting wound rotor motors. Although there are not a large number of wound rotor motors in use, this information will help you install, troubleshoot, and repair these controls when you come across them.

A diagram of a secondary-resistor motor starter is also provided in Figure 16–5. The stationary winding of the motor is connected directly across the line with S contacts and it will receive full voltage when the S contacts are closed. The rotor is a three-phase, wye-connected rotor that has two sets of resistors connected to it. When the rotor is shorted out, it will operate like a squirrel-cage rotor, but when resistors are connected between the ends of the windings, it lowers the amount of current that is allowed to flow in the circuit and the motor's speed is reduced. Since all current in the rotor is induced, the resistors will directly control the amount of induced current that is allowed to flow. The resistors in this type of motor are called *secondary resistors* because they are located in the rotor circuit rather than in the line circuit.

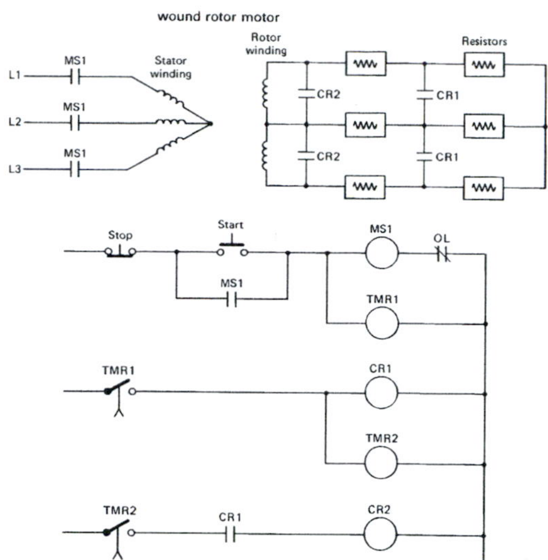

Figure 16–6 Diagram of wound rotor-type reduced-voltage starter.

When the motor is started, all the resistance is connected in the circuit to keep the current flow to its lowest value. Since the rotor's current is very low, the amount of torque the motor can produce will be very low, and the amount of current the motor will draw will be very small. This is controlled by closing the S motor starter, which applies line voltage to the main motor winding, keeping the R1 and R2 contactors deenergized, which allows all the resistance to be in series with the rotor winding.

The timer motor 1 is energized at the same time as the S motor starter coil. This timer provides approximately 5 to 10 seconds of time delay before the first set of resistors is shorted out, which bypasses the amount of resistance they provide to the rotor. When the TMR1 times out, its contacts will energize contactor R1, whose contacts will close to short out the set of resistors farthest from the rotor. Since the rotor now has less resistance in its circuit, more current is allowed to flow and the motor will pick up speed. When TMR1 contacts close, TMR2 is energized. This causes TMR2 to run for approximately 10 more seconds before it closes its contacts, which shorts out the remainder of the resistors. When this occurs, the three rotor windings are connected directly to each other, which allows maximum current to flow in the rotor. This allows the motor to reach maximum rated speed.

A selector switch could be used to allow the motor to be started manually rather than with the timers. The selector switch would also allow the motor to be operated continually at any of the speeds provided by the secondary resistors. This provides a multispeed control for this type of motor.

Since the wound rotor uses slip rings and technology has improved the amount of current that can be induced in a squirrel-cage rotor, the round rotor has largely been replaced with other types of three-phase motors. Replacement has been hastened by the introduction of solid-state controls.

16.2.7 Troubleshooting Reduced-Voltage Starters

All reduced-voltage starters can be troubleshooted similarly since they all operate on similar principles. The first step in the troubleshooting process is to determine if the motor is having problems when it is running or when it is starting. It is also possible that the motor is having trouble during starting and when it is running. Each type of starter has two separate circuits that are used to make the starter operate the way it was designed. The control circuit will provide current to the motor starter and contactor coils at the appropriate time, while the load circuit will contain the multiple sets of contacts to direct current to the correct starting component or motor winding.

When you begin the troubleshooting procedure, you should break down the type of problem into one of the following categories, which will help you eliminate some of the possible problems. The first determination that you should make is whether the motor is trying to start when the start pushbutton is depressed. If the motor does not try to start, you should use the diagram to determine which type of reduced-voltage starting circuit the system is using. Next you should make several tests of the control circuit to see if the first coil is pulled in. You may need to refer to the sequence of operation for the type of control that is being used. This test would include the loss of three-phase voltage to the system.

If the control circuit has voltage and the coils will not pull in, you should use the test method to determine where the voltage to the controls is being lost. Other problems to test for include a bad motor starter coil and open overloads. If you have voltage at the coil of the motor starter but no voltage through its contacts, you may have a faulty contact or starting device, such as primary resistors or the autotransformer. Also test for the presence of voltage at the motor terminals. Be sure to measure the exact amount, since this is a reduced-voltage circuit. If you have any voltage, you can assume that the starting control is operating and that the problem may be caused by an open winding in the motor.

If the motor will start at the reduced voltage but will not come up to full speed, or trips out prior to reaching run speed, you should check the control circuit to determine what controls are used to place the motor across the line. After you have determined what contactors must be energized to switch the motor to the run mode, try to restart the motor and watch these devices closely to see if they are being energized at the correct time. Use a voltmeter and ammeter to watch the voltage and current to determine if the timers are set for the proper amount of time delay. Be sure that you are not trying too many starts too quickly, as the motor and control are rated for the maximum number of starts that they tried per hour.

It is also possible that the load or switching contacts are not allowing the motor to come up to full speed. One way to test these possibilities is to connect the motor temporarily for across-the-line operation. Check to make sure that other loads are off the power distribution system and that safeguards have been provided where high starting torque can cause damage to the load. *Do not attempt to run the motor in this configuration for more than one or two starting tests, since the contacts in the starter are rated for the reduced-voltage application and cannot accept the high inrush current on a continual basis.*

Once you have made these precautions, start the motor with full applied voltage and watch its response. Be sure to monitor voltage and current on each phase. If the motor operates correctly during this test, you have narrowed the problem down to the switching circuit that changes the motor connections between the starting phase of the control and the running phase.

If the motor does not come up to speed properly during the across-the-line test, you have narrowed the problem down to the motor being faulty or its load has become too large. Remember that the torque for these applications is severely limited during starting and it is possible to have bad bearings or a load that has become too large from overloading. If the application uses valves, solenoids, or clutches to apply full load to the motor once it is started, be sure to test them for proper operation. If the load is a belt-driven load, you may loosen the belts temporarily to reduce torque during the troubleshooting to pinpoint the problem. If the load is a conveyor or machine that should start with no load, be sure that the load has been removed during these tests. After you have determined the problem, make corrections and test the circuit several times for proper operation before you sign off the job.

16.2.8 Solid-State Reduced-Voltage Starters

Reduced-voltage starting can also be provided by solid-state controls. Early models of these controls used three pairs of SCRs (silicon-controlled rectifiers) to control the amount of voltage that is provided to start the motor. The reduced voltage from the SCR control will have the same effect on the motor as primary resistor control. Newer models of solid-state control use insulated gate, bipolar transistors (IGBT) or AC solid-state controls. If you want to know how these devices operate or how to troubleshoot them, you can see them in Chapter 19. These newer devices are better suited for this application, since they can control the voltage through an entire range of voltages rather than two- or three-stepped voltages like the primary resistor control.

The solid-state starter can provide voltage to the motor in small increments to allow the motor to accelerate smoothly. These controls can provide ramping through a potentiometer, which increases the speed of the motor at a set rate. Other circuits allow the control to provide constant or varying current throughout the starting cycle. The solid-state starters are easily interfaced with microprocessor controls, which allow their ramp time and deceleration time to be adjusted in the field. The microprocessor controls also allow these devices to be interfaced to networks such as Device Net and Control Net so that their status can be checked from remote locations. These networks also allow technicians to determine what has happened to the solid-state control prior to it becoming faulty.

Figure 16–7 shows a picture and Figure 16–8 shows a diagram of a solid-state reduced-voltage starter. From this diagram you can see that the solid-state part of this control is identified as a box, so you do not know the exact type of solid-state control that is in the circuit. Solid-state controls are usually encapsulated, so you cannot repair them; rather you will need to change the entire control when it is faulty. The simplest way to troubleshoot this control is to determine if voltage is present to the input side of the control. The second test is to determine if the coil has voltage. Finally, you need to test to see if the output terminals have voltage. If they do not, or if voltage is low, you will need to check the settings on the acceleration or deceleration controls. If the controls are set properly, and you do not have voltage at the output terminals, you can replace the control.

If the control uses SCR technology, you can use this information to understand its operation. The SCR is used to control voltage in one direction, such as a DC signal. If the SCR is used to control AC current, the circuit will have two SCRs mounted back-to-back in a circuit called inverse parallel. The current that is allowed to flow through the SCR is controlled by adjusting the time the SCR is in conduction. Since one SCR of the pair controls the positive half-cycle of the AC and the other one controls the negative half-cycle, the maximum amount of voltage will be conducted when the SCR is in conduction for the full 180 degrees. If the SCR is turned on at the 90-degree point, it will conduct half the voltage. The turn-on point is referred to as the *conduction angle*.

Since two SCRs are paired off to control each AC phase, their control is matched so that each SCR will fire at the same conduction angle. This ensures that each of the paired SCRs is conducting the same amount of voltage. The control circuit can adjust the conduction angle to control the amount of voltage the reduced-voltage starter will provide to the motor.

Control of the SCRs is accomplished through the control module box. This box contains several solid-state circuits that use feedback from current transformers and preset values from potentiometers on the control panels. This control also provides several sets of contacts that can be connected into the hardwired control circuit. These contacts will energize when current is too high or when the motor hits the end of limit. The end-of-limit circuitry senses the current the motor is drawing and indicates when the current is returning to FLA after the motor is up to full speed at the end of the starting cycle. This is useful if a series of multiple motors must be started in succession. As soon as the first motor is up to speed and its current is near normal

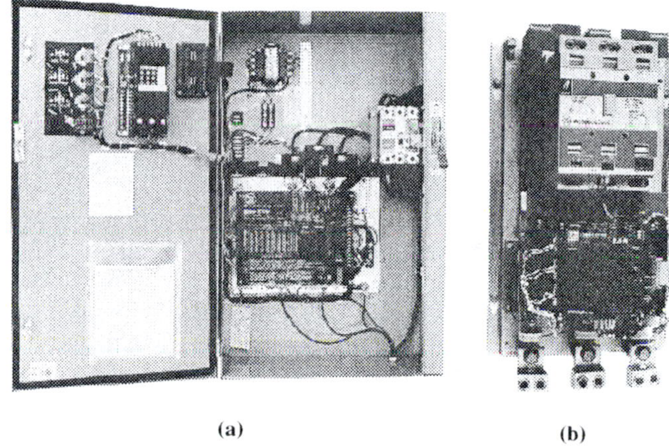

Figure 16–7 A solid-state-type reduced-voltage starter (a) mounted in a cabinet; (b) removed from a cabinet for easy viewing.

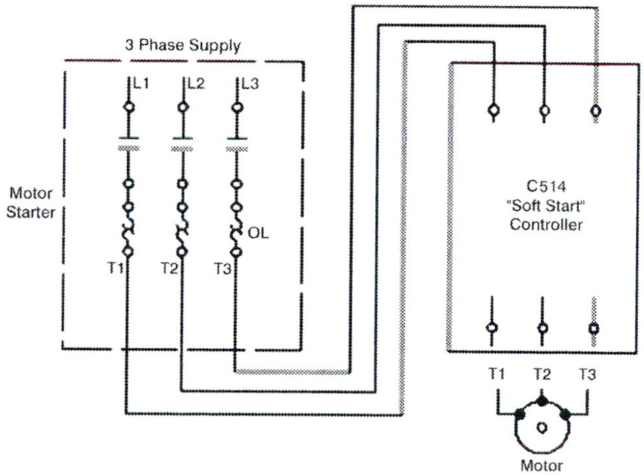

Figure 16–8 Electrical diagram of a solid-state reduced-voltage starter connected between a motor starter and a motor.
(Courtesy of Rockwell Automation)

FLA, these contacts are closed, which can energize the control of the next solid-state starter.

A shunt trip is also provided that will allow the motor to be disconnected when an SCR has shorted. If the SCR is shorted, the amount of current it is conducting cannot be controlled, and it acts as a shunt or parallel conductor (wire). When an SCR shorts, the phase of voltage it is controlling will act like an across-the-line contact, and it will allow more current to flow.

The amount of voltage that is supplied to the motor can be preset into the controller in one of several ways. One way is to determine the rate of change of current flow, which is called *ramping*, and another way is to use a fixed turn-on time. You can also set the maximum or minimum current that is allowed to flow. The theory of operation of each of these controls is explained below so that you will understand the total operation of the solid-state reduced-voltage starter.

16.2.8.1 Solid-State Starter Operation

Figure 16–7 shows a typical control diagram used to start the solid-state control. The control module provides a set of terminals marked "close to run," which have a set of contacts connected across them. Since these terminals do not need an input voltage, the contacts will simply complete the control circuit inside the control module. The contacts are identified as PR, which means they are controlled by the PR relay in the control circuit. When the start button is depressed, the PR relay is energized and its auxiliary contacts will hold in or seal the start button. Any other pilot devices can be connected in series with this circuit for safety or operation. The overload contacts in this circuit can be controlled by the current trip from the control.

Another method of starting a motor with solid-state control requires a contactor to be connected in series. The contactor should always be connected between the solid-state control and the power supply. In this case, the "close to run" terminals should be jumpered so that control will begin to operate as soon as the contactor is closed. The contactor's coil can be connected to a two-wire or three-wire control circuit. Since the solid-state reduced-voltage starter provides motor protection against overloads, the only other protection that should be supplied to the circuit is fuses for short-circuit protection for the supply voltage wires. If the motor has internal thermal protection, its contacts can be connected in the control circuit for the coil.

When the solid-state starter is energized, it will begin to allow only the preset amount of current to flow to the motor. This current will cause the motor to come up to full speed, in accordance with the preset conditions. The current transformers act as current sensors for the control, which will react automatically to the amount of current the motor is drawing at each step of the acceleration process. If the load presents any changes to the motor during the acceleration process, the motor will react with a change of current, which will be detected by the current transformers. The control can respond to these changes by adjusting the amount of current that is allowed to flow through the SCRs.

16.2.8.2 Specialized Control Circuits

One of the specialized control circuits provided by the control module is the fixed turn-on ramp. In older solid-state controls this ramp was not adjustable, since it is designed to provide an increasing amount of current at a fixed rate that will increase torque slowly to remove slack or backlash from the load. Other control functions are adjustable and can override the fixed ramp. On newer controls the ramp current is fully adjustable by changing the amount of time for acceleration.

The starting current can be adjusted to provide enough current to allow the load to break away. This is called *breakaway torque*. Since this torque will vary from load to load, breakaway torque is adjustable through a potentiometer that allows starting current to be adjusted 100 to 400 percent.

If currents in excess of the fixed turn-on ramp are required, another circuit, called the *adjustable ramp*, may be used. The potentiometer for this circuit allows the current to be steadily increased over a longer period. The adjustable ramp time can be varied from 2 to 30 seconds to allow larger loads more time to come up to speed. The amount of time it takes the motor to come up to speed will determine the slope of the ramp. This adjustment should allow the motor to start the load to remove slack or backlash with a minimum of disruption.

The current limit adjustment controls the maximum amount of current that the control will allow the motor to draw. The maximum current can be set at any amount from 100 to 400 percent. The limit affects only the amount of current flow during starting. This is a useful feature where the maximum amount of current must be limited because of an overloaded distribution system, or because several motors must be started at the same time.

Pulse start allows an extra pulse of full current to be sent to the motor for 2 seconds when current is first applied. The full current is rated at 400 percent, which allows the motor to provide full torque to the load to help it break away. At the end of 2 seconds, the current is returned to the ramping level and the motor is brought up to speed under the ramp control.

Other features provided in the solid-state control include a current trip circuit, transient protection, and a power-saver circuit. The current trip circuit monitors the amount of current the control is sending to the motor. During the starting operation, the current trip circuit is set to 50 percent above the maximum allowable current that the control will allow to flow. This makes the circuit inoperative during start-up. Once the motor is running on full-load current, the circuit is again connected into the circuit. It is adjustable between 50 and 400 percent of the full-load current. This allows the control to sense either overcurrent or undercurrent conditions. An overcurrent condition can occur when the load is excessive or through a malfunction that causes the load to jam. An undercurrent may occur when a belt breaks or a clutch drive malfunctions. In either case, the current trip contacts can be used to disconnect power to the motor. The transient protection circuit uses metal-oxide varistors (MOVs) to provide protection against line voltage surges. This provides protection to the SCRs and to sensitive solid-state components in the control unit. The power-saver circuit reduces the amount of voltage to the motor after it reaches full speed. This option may be used only with motors rated for increased temperatures, since the reduced voltage will cause the motor to operate at a slightly higher temperature. The voltage reduction will not be large enough to affect the operation of the load. If the motor is slightly loaded, the power-saver circuit will provide a small power factor correction.

16.2.8.3 Troubleshooting the Solid-State Starter

After the starter is installed and operating, it may have problems. The source of the problem must be identified so that corrective action can be taken. As a troubleshooting technician, you will naturally suspect the solid-state portion of the system because its operation cannot be observed without using test equipment. This is why it is important to use the troubleshooting procedure to identify the problem.

Figure 16–9 is a troubleshooting table for the solid-state reduced-voltage starter. This table will lead you

Figure 16–9 Troubleshooting chart for a solid-state reduced-voltage starter.
(Courtesy of Eaton/Cutler Hammer)

Symptom	Most Likely Causes	Recommended Action
HMCP trips (electrically operated disconnecting means opens) as it is closed.	Motor is not connected Three phase power not connected or terminal loose Shorted SCR SHUNT TRIP contact between terminals closes due to shorted SCR Shunt trip latching relay not reset (if used)	Connect motor to starter. Switch two incoming power lines, corresponding motor leads must also be switched to maintain the same motor rotation. Perform shorted SCR check. Perform shorted SCR check. Reset STR relay by pushing reset button on relay.
HMCP trips while motor is running HMCP trips when STOP is pressed Starter does not pick up and maintain	Microprocessor detected non-standard operation. Voltage regeneration of motor causing zero voltage drop across SCR (looks like shorted SCR). Line voltage is not applied Overload is not reset 120 V control voltage is not present Heatsink overtemperature switches are open	Install time delay relay — consult factory. Check incoming lines for proper voltage. Reset overload. Check control transformer fuse. Check control circuits. Check for continuity through overtemperature switches. Check that all fans are rotating freely. Check for excessive motor current draw.
Starter maintains but motor does not accelerate or does not attain full speed (stalls)	Current Limit is too low	Increase Current Limit setting clockwise. Set Ramp Time to minimum (counterclockwise).
Insufficient Torque	Improper current feedback High breakaway torque required	Check Current Feedback Resistor for proper calibration. See page 6. See Pulse Start Section on page 12. Load is not suitable for reduced voltage starting.
Motor accelerates too slowly	Current Limit is too low Ramp time is too long Broken current feedback resistor	Increase Current Limit setting clockwise. Decrease Ramp Time setting counterclockwise. Check by recalibrating CFR. See page 6.
Motor accelerates too quickly	Current Limit is too high Improper current feedback Pulse start setting is too long Broken current feedback transformer wire	Decrease Current Limit setting counterclockwise. Check current calibrator for proper resistance value. See page 6. Decrease Pulse Start time counterclockwise. Check for a broken Current Feedback Transformer wire. Frame Size A and B the Current Feedback Transformer is contained within the Logic Module.
Current Limit adjustment has no effect during acceleration	Improper current feedback	Check current calibrator for proper resistance value. See page 6.
Starter or motor is noisy or vibrates when starting	Single phasing due to open phase Single phasing or unbalanced current due to non-firing SCR Defective motor	Check wiring and overload heater coils. Check gate lead wiring to SCR's. Check motor for shorts, opens, and grounds.
Mechanical shock to machine	Current increases too quickly	Decrease Starting Current counterclockwise. If necessary increase Ramp Time clockwise.
End of Limit contact does not close	Starter is in current limit	Perform SCR Full Voltage Test. See page 15.
Current Trip contact does not close when current goes above the trip setting	Improper current feedback	Check current calibrator for proper resistance value. See page 6.
Current Trip contact closes when current is below trip setting	Improper current feedback	Check current calibrator for proper resistance value. See page 6.
Motor current, voltage and speed oscillate	Power Saver is misadjusted	Turn Power Saver adjustment CCW until oscillations cease.
Overload relay trips when starting	Incorrect heater coils (melting alloy) Loose heater coil Long starting time (high inertia applications may require slow trip overload and oversize starter) Mechanical problems Single phasing Excessive starting time (current limit may be set too low)	Check heater coil rating. Tighten heater coil. Motor and starter thermal capabilities must be evaluated before extending overload trip times. Check machinery for binding or excessive loading. See "starter or motor is noisy" symptom in troubleshooting chart. Increase Current Limit setting clockwise.
Overload relay trips when running	Incorrect heater coils (melting alloy) Mechanical problems Single phasing	Check heater coil rating. Check machinery for binding or excessive loading. See "starter or motor is noisy" symptom in troubleshooting chart.
Heatsink overtemperature switch opens	Excessive current Defective heatsink fan	Check motor current draw. Check that all fans are rotating freely.
Erratic operation	Loose connections	Check all connections.

through tests from the most likely to the least likely cause. It will also provide you with a list of suggested actions to take to make repairs. You should always have a strategy in mind when you are troubleshooting so that each test you make will verify the operation of the portion of the circuit that you are testing. Most of these tests lead you through the control circuit in the same way that current is passed. This will help you verify if each section of the control is operating correctly.

16.3 Acceleration Control for DC Motors

DC motors can be accelerated slowly in much the same way as AC motors by changing the amount of voltage and current that they use during starting. The speed of a series motor is controlled by the amount of load, but the speed of shunt and compound motors can be controlled in this manner. The current can be controlled by using primary resistors, rheostats, or SCR solid-state controllers. Since these types of controls can also be used to control the speed of the motor, the circuits are generally used as double purpose for starting and controlling the speed of the DC motor.

In recent years, SCR control has been used extensively, since it serves two distinct purposes. First, SCRs are used to rectify the AC current to DC current for use in the motor circuit. Prior to the use of the SCR or diode, DC current was produced by a motor–generator set. The motor was powered by AC current and the generator produced DC current. Current in these systems was controlled by adjusting the field current of the generator, which determined the amount of voltage the generator produced. The voltage generated was sent directly to the DC motor. If the motor speed was controlled, a field rheostat was placed in the DC generator field to adjust the amount of generated current. The DC motor in these systems was accelerated with primary resistors. Usually, several banks of resistors were used to provide smooth acceleration. When SCRs were first introduced they were not rated large enough to control voltage and current to a large DC motor individually. In these early applications, several SCRs would be paralleled to carry the larger currents. Today, SCRs can be produced that will easily carry several hundred amperes of current.

The second duty of the SCR is to adjust the amount of DC voltage and current that is sent to the motor. This type of control is very similar to the solid-state reduced-voltage starter except that SCRs are connected as a bridge rectifier, which produces a full-wave pulsing DC current. The firing angle of the SCRs is controlled to adjust the amount of voltage they will allow to conduct. As the firing angle is increased, more voltage is conducted to the motor.

The ramp and slope of the motor acceleration can be fixed or controlled by external potentiometers in the control. This and similar features, such as current limit, provide the SCR controller with the flexibility to control the motor's acceleration and speed under a variety of different loads.

16.4 Deceleration and Braking Methods

It is just as important to bring a motor to a safe stop as it is to start with acceleration control. Some large motor loads develop high inertial forces when they are operating at full speed. If voltage is simply disconnected from the motor, the load may coast several minutes before the shaft comes to a full stop. This is true in applications such as those involving large saw blades and grinding wheels. If someone becomes injured or stuck in the machinery, the motor load must be stopped immediately. It is also important to bring these loads to a quick, smooth stop when the operator does not have time to stay with the equipment as it comes to a stop. In other load applications, such as elevators and cranes, the location where the load stops is as important as moving the load. This means that the motor shaft must stop moving at the precise time to place the load in its proper location.

In some of these applications, the motor can be decelerated quickly by reconnecting the motor to operate in the reverse direction at the time when the stop button is depressed. This connection is allowed for only a second, while the motor shaft comes to a stop. The circuit is disconnected before the motor actually begins to run in the opposite direction. This type of circuit is called a *plugging circuit.*

In another type of circuit, DC current is applied to the stationary field of an AC motor when the stop button is depressed. Since the field is fixed and it replaces the rotating AC field, the rotor is quickly stopped by the alignment of the unlike magnetic fields between the rotating and stationary winding. The attraction between the unlike fields is so strong that the rotor is stopped quickly. This type of deceleration is called *DC braking.* A similar type of braking, called *regenerative braking,* is used in DC motors and some AC motors. The motor is reconnected as a generator when the stop button is depressed. When the stop button is depressed, it disconnects the motor leads from the power supply and reconnects them as a generator. The armature of the motor (generator) is connected directly to a small resistor to load the generator as heavily as possible. This heavy load brings the rotor to a smooth stop within several rotations.

The third type of deceleration circuit used in industry today is the electromechanical *clutch and brake assem-*

bly. The clutch and break assembly uses combinations of disk brakes, drum brakes with shoes, and electromagnets to bring the motor to a quick stop. Since the mechanism can control the deceleration of the motor by allowing a specific amount of slip, the speed of the load can be adjusted through this type of clutch mechanism.

16.4.1 Plugging Circuits

One of the easiest ways to bring an AC or DC motor to a quick, safe stop is to disconnect power from the motor and reconnect it so that the motor will rotate in the opposite direction. This can be accomplished in applications where the motor is operating in only one direction or in motors that are operated in either direction. This type of circuit can also be used to apply current to the reconnected motor after the load has coasted for several seconds, so that the shock to the load is not so severe.

Figure 16–10 shows a diagram of a plugging circuit. The motor in this diagram is an AC motor, so only two of the three leads need to be reversed to cause the motor to rotate in the opposite direction. When the motor is started, the contacts of the motor starter connect the motor directly to the supply voltage so that the motor will run in the forward direction. The forward motor starter controls the set of contacts on the right side of the diagram.

When the stop button is depressed, the forward motor starter is deenergized and the reverse motor starter is energized. The reverse motor starter switches L1 and L3, which causes the magnetic fields in the stator to rotate in the opposite direction. The reversed magnetic field quickly brings the rotor to a stop and tries to start it in the reverse direction. If this circuit did not have any other controls, the motor would begin to run in the reverse direction.

Since we are interested in stopping the shaft rotation, a centrifugal switch is added to the circuit. This switch, which senses the rpm of the shaft and opens its contacts when the rotor approaches low rpm, is called a *plugging switch* or *speed switch*. Its symbol incorporates a double-pole switch with an arrow to indicate that the switch is rotary in nature. The centrifugal switch is adjustable so that its contacts deenergize the reverse motor starter coil at just the right time to make the shaft stop rather than coast or begin to turn in the

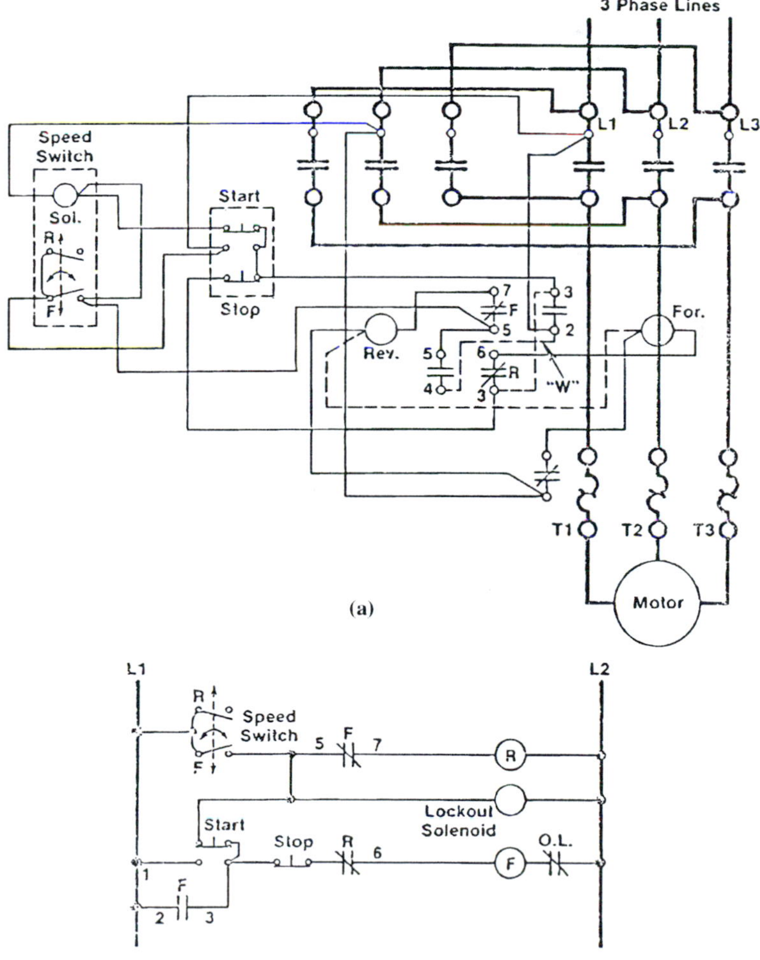

Figure 16–10 (a) Wiring diagram for a plugging circuit for motor that can run in forward or reverse direction. (b) Ladder diagram for a plugging circuit. (Courtesy of Rockwell Automation)

reverse direction. Since the size of the load will also determine how fast the shaft can be brought to a stop, the plugging switch can be adjusted to fit the application.

The circuit shown in Figure 16–10 allows the motor to be stopped any time the stop button is depressed. For this reason it is also called a *continuous plugging circuit.* The circuit could be modified so that the plugging circuit would be energized only when the emergency stop button was depressed, and would coast to a stop if the normal stop button was depressed.

Other modifications can be made to the circuit so that the motor can be operated in the reverse direction without the plugging circuit stopping the motor shaft. In this type of circuit, a lockout solenoid can disable the plugging switch so that when the reverse pushbutton is depressed, the motor is allowed to change direction and accelerate in the reverse direction. It is important to remember that split-phase and capacitor-start motors cannot be used with plugging circuits because they need to use their own centrifugal (or end) switch for starting, and it will not close until the motor is nearly stopped.

Figure 16–11 shows another plugging circuit. In this circuit, the motor can be plugged while it is operating in either direction. The first plugging circuit would operate correctly only when the motor was rotating in the forward direction. In some applications, the motor must be operated and stopped while it is running in either direction.

In this type of circuit, the plugging switch is set to sense the motor rpm in either direction. The motor can be started in either direction by pressing the forward or the reverse pushbutton. When the stop button is depressed, the interlocks set up a condition where the opposite motor starter is energized to bring the motor to a rapid stop. When the rotor decelerates to near zero, the plugging switch disconnects all power from the motor, and the control circuit is ready to start the motor in either direction.

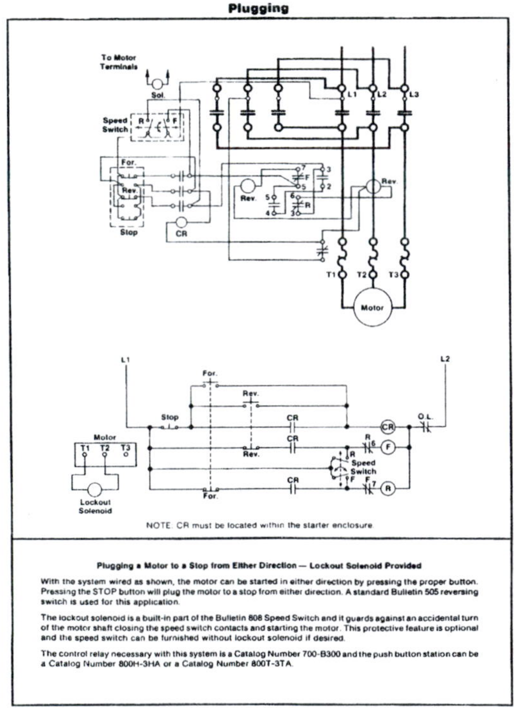

Figure 16–11 Electrical and ladder diagram that shows a method of plugging a motor in either the forward or reverse direction.
(Courtesy of Rockwell Automation)

16.4.2 Using a Time-Delay Relay for Plugging

The circuits shown in Figures 16-10 and 16-11 could use a time-delay relay instead of the plugging switch. The time-delay contacts would be located where the plugging switch contacts are in these diagrams. The timer motor would be connected to the reverse motor starter coil where the motor was designed to operate in only one direction, and it would be connected where it would be energized by either motor starter when the stop button was depressed in circuits where the motor is allowed to switch. The amount of time delay can be adjusted for different-size loads.

16.4.3 Antiplugging Circuits

Another method of bringing a motor to a smooth stop is called *antiplugging.* This type of circuit allows the motor to coast for a short period before reverse voltage is applied. The short time the load is allowed to coast prior to applying the reverse current provides a small amount of cushion to bring the load to a smooth stop.

The centrifugal switch in this circuit prevents the reversing circuit from energizing until the motor's speed has slowed to the predetermined rpm. Since the centrifugal switch is used to prevent the reversing circuit from energizing until the motor has reduced its rpm, the switch is called an *antiplugging switch.*

The motor is allowed to coast before the reversing circuit is energized. This is accomplished by the set of contacts that the switch opens, as long as the rpm is too fast. When the motor is deenergized, it is allowed to coast while its rpm slows. The centrifugal switch will close the open contacts when the speed is reduced to the preset level. At this time, the reversing circuit is energized and the motor is brought to an immediate halt.

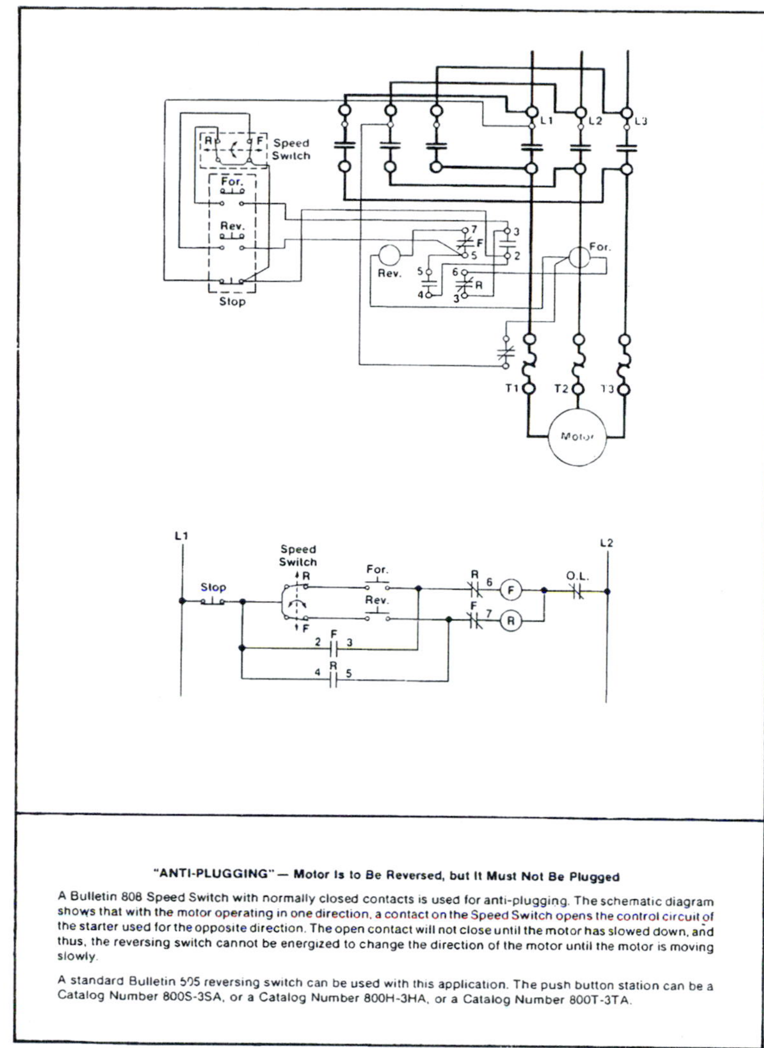
Speed Switch
For.
Rev.
Stop
L1
L2
L3
T1
T2
T3
Motor
O.L.
"ANTI-PLUGGING" — Motor Is to Be Reversed, but It Must Not Be Plugged
A Bulletin 808 Speed Switch with normally closed contacts is used for anti-plugging. The schematic diagram shows that with the motor operating in one direction, a contact on the Speed Switch opens the control circuit of the starter used for the opposite direction. The open contact will not close until the motor has slowed down, and thus, the reversing switch cannot be energized to change the direction of the motor until the motor is moving slowly.
A standard Bulletin 505 reversing switch can be used with this application. The push button station can be a Catalog Number 800S-3SA, or a Catalog Number 800H-3HA, or a Catalog Number 800T-3TA.

Figure 16–12 Electrical wiring and ladder diagram of an antiplugging circuit.
(Courtesy of Rockwell Automation)

Figure 16–12 shows a diagram of an antiplugging circuit. From this circuit you can see the operation of the reversing switch. When the motor is turning at full rpm, the centrifugal switch disables the reversing circuit. When the motor is deenergized, the load will cause the rotor to coast as the rpm is reduced significantly enough to allow the switch contacts to close. At this point the reversing circuit is energized and the motor is stopped. The centrifugal switch is adjustable to energize the reversing circuit at any rpm below full speed. A time-delay relay could also be used in this circuit to delay the reversing circuit until the motor has slowed significantly.

16.4.4 Electric Braking

Another popular method of decelerating a motor or providing instantaneous stopping is called *electric braking*. The principle of electric braking is based on the theory of disconnecting the applied voltage that produces the rotating magnetic field and replacing it with a fixed magnetic field produced by DC current. When the DC current is applied to the motor's stationary field, the fixed north and south polarity of the magnetic field will attract the opposite field of the rotor and quickly bring it to a stop.

Figure 16–13 shows an electrical diagram of a DC electric braking circuit. The DC voltage for the braking circuit is produced through a set of diodes or SCRs, which are connected as a single-phase bridge rectifier. If the circuit uses regular diodes, the amount of DC voltage provided for the braking circuit will be fixed. If the DC voltage is provided by SCRs, the gate firing circuit can be regulated to adjust the amount of braking action.

From the diagrams you can see that the DC current is applied to the stationary winding of a single-phase and a three-phase motor. This circuit requires a contactor or motor starter with one or two sets of normally closed auxiliary contacts. When the motor starter is deenergized, the applied voltage is removed from the stator as the main contacts are opened, and

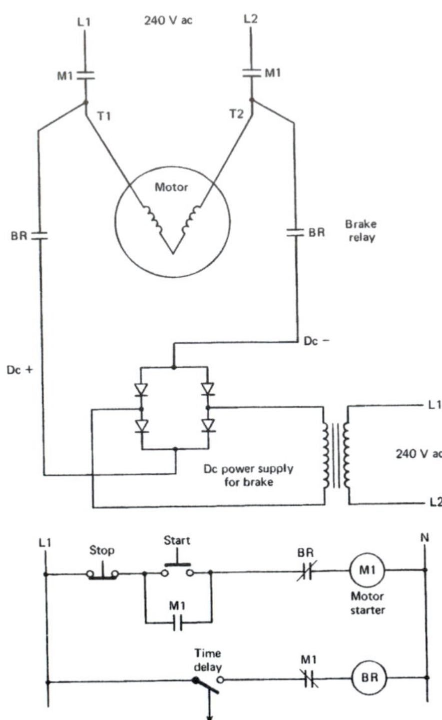

Figure 16–13 Circuit diagram of a DC electric braking circuit connected to a three-phase motor.

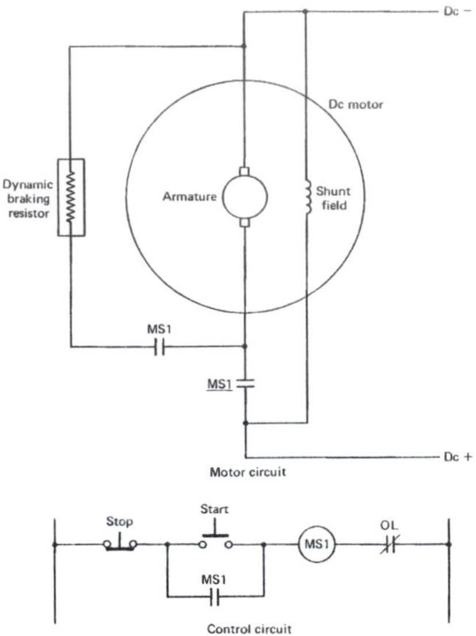

Figure 16–14 Electrical diagram of a dynamic braking circuit.

the DC current is connected through the auxiliary contacts. If the full DC voltage is applied, the rotor will be brought quickly to a stop. If the motor needs some time to coast so that the load is stopped more smoothly, the amount of DC voltage may be reduced by resistors or by adjusting the gate firing circuit of the SCR. A time-delay circuit could also be used to allow the motor to coast for a time before the braking circuit is applied, which will make it operate similar to the antiplugging circuit.

16.4.5 Dynamic Braking

Dynamic braking uses a principle that is similar to that of DC electric brakes. This circuit is used primarily on DC motors, but it could be used on wound-rotor and some synchronous motors. In this type of braking circuit, the applied voltage is removed from the motor when the motor starter is deenergized. Since the armature will continue to spin as it coasts to a stop, it will produce voltage in a manner similar to a generator.

The generator action can be used to bring the rotor to a quick stop by loading it with a small resistor. The small resistor will cause the rotor to generate very high levels of current, which produces reversed magnetic forces on the shaft and causes it to stop quickly. The inertia that is built into the rotating shaft is quickly dissipated by the strong generated current since the rotor is no longer under power. If the shaft needs to be stopped more quickly, a jumper can be used to short the output of the rotor, which will cause a much larger current and a stronger magnetic field to be produced. A diagram of this circuit is shown in Figure 16–14. The resistor is connected in the circuit by the normally closed set of auxiliary contacts in parallel with the armature. These contacts will disconnect the resistor when power is applied to the motor, and reconnect at the instant when the motor starter coil is deenergized. This circuit is called a *dynamic braking circuit* since it utilizes the energy that is stored in the rotor to cause the braking action.

16.4.6 Brakes and Clutches

Brakes and clutches allow the speed and torque of a load to be controlled while the motor is operating at full rpm. In these types of systems, the speed and torque of the motor is not controlled; instead, the motor is allowed to run at full speed while adjustments are made to the clutch or brakes to accelerate, control operating speed, or decelerate and stop the load. Mechanical and electromechanical systems are used between the shaft of the motor and the shaft of the load to provide this control.

One of the most simple methods of controlling the deceleration and stopping of a load is with an electrically operated mechanical friction brake. Figure 16–15 shows a photograph and electrical diagram and Figure 16–16 shows an operational diagram of this type of friction brake. The brake mechanism utilizes brake shoes

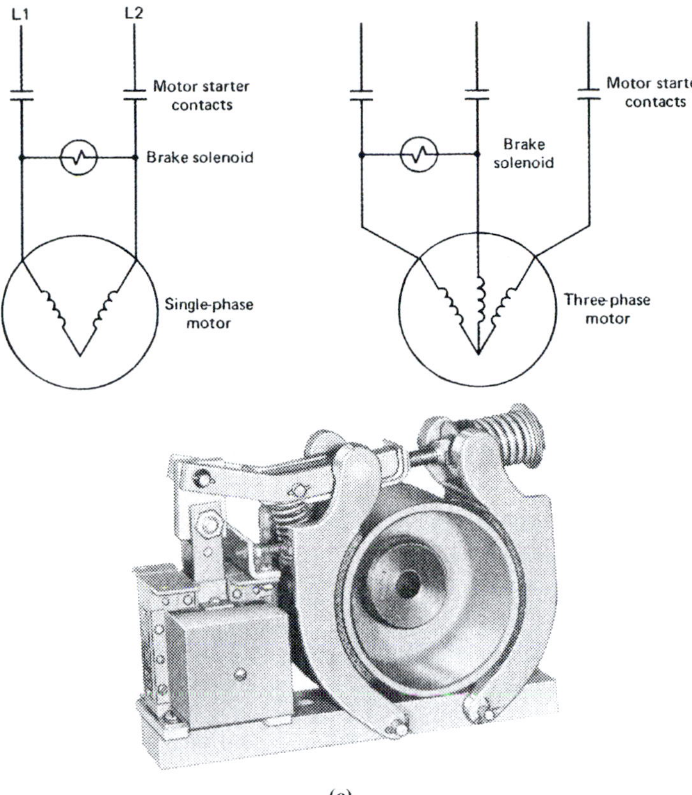

Figure 16–15 (a) Diagram of an electrical brake connected to a single-phase circuit. (b) Diagram of an electrical brake connected to a three-phase circuit. (c) Picture of friction brakes. Notice the adjustment screw to set the amount of friction and the linkage to the solenoid that opens the shoes during the run mode.
(Courtesy of Eaton/Cutler Hammer)

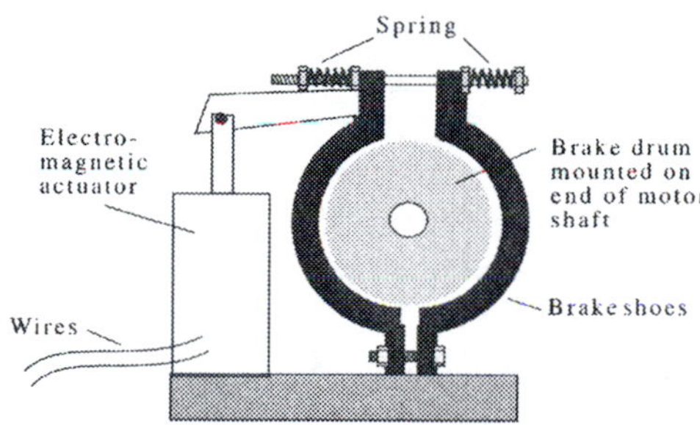

Figure 16–16 Mechanical diagram of drum and shoe friction brakes.

that are very similar to the type found on the rear brakes of an automobile. This system is slightly different in that these shoes apply force inward against a drum, whereas automotive brakes apply force outward against a drum. In this system, the brake shoes are fitted around a drum that is fitted onto the shaft of the load or motor. A large spring is provided for adjustment to set that amount of pressure that the shoes will apply against the drum. A linkage is designed to connect an actuator arm to a powerful solenoid coil that will open the brake shoes slightly when it is energized. When the coil is deenergized, the spring will again cause the shoes to apply force to the drum.

16.4.6.1
Friction Brake Operation

The operation of the friction brakes can best be understood by an electrical diagram, provided in Figure 16–15. The solenoid coil for the brakes is connected across the supply voltage lines for the motor, which means that the coil will be energized any time the motor is energized. When the solenoid is energized, it pulls the linkage to cause the shoes to open slightly and remove all friction to the drum. This allows the motor to operate freely and to accelerate to normal speed and run for as long as the application requires. When the motor starter is deenergized, the voltage to the motor is disconnected, which also deenergizes the solenoid to the brakes. When the solenoid is deenergized, the spring is allowed to apply full force to the shoes so that they create maximum friction with the drum. This friction causes the drum to slow down and stop quickly. Since the drum is rotating with the motor and load shaft, they are also stopped. A screw is provided to adjust the amount of pressure the spring can apply to the brake shoes. If the pressure is maximum, the drum will stop immediately when power is removed from the motor and coil, and if the pressure is reduced, the motor can be allowed to coast slightly as it comes to a smooth stop. This configuration of control also provides fail-safe operation, so that if power is lost for any reason, the brakes will be applied automatically by the spring tension.

The adjustment of brake shoes is very important, since they determine the amount of friction that is applied to the drum when the brakes are applied. If they are set too loose, they will not apply enough friction to stop the load correctly, and if they are set too tight, they will become overheated and wear prematurely.

16.4.6.2
Applications for Drum and Shoe Brakes

The drum and shoe mechanical braking system is generally used on cranes and elevators, where positioning is critical and slippage when power is disconnected cannot be tolerated. Many other applications also require that the motor hold a load in position after it has been deenergized. Some larger robots and transport automation may use DC electric braking to stop the motor and brakes to hold it in location when the emergency stop button has been depressed or when the motor has been deenergized.

16.4.6.3
Other Types of Friction Brakes

Other types of friction brakes utilize a slightly different theory of operation. Their design makes use of a rotating disk and pads to cause friction. This brake uses a rotating disk and friction pads. The pads apply friction to the disk with electromagnetic force. The magnetic force can be varied with current. In some styles of brakes a permanent magnet is used to create the magnetic field, and electric current is applied to nullify the strength of the permanent magnet. This means that the stronger the current, the weaker the pressure is to provide friction to the brakes.

In other styles, an electromagnet is used to cause the pressure against the pads. In this type of brake, the stronger the current, the more friction the brakes will apply. This system is very similar to the front disk brake system on an automobile. The rotating disk is connected directly to the motor or load shaft by a keyway. This means that it will rotate at the same speed as these shafts. The pads are connected to the stationary part of the brake shown at the top of this system.

In this system, the pad applies pressure to the rotating disk when the electromagnetic coil is energized. The more current that is applied to the electromagnet, the more pressure the pads will apply to the rotating disk, which will bring it to a stop faster. If a small amount of current is applied to the magnet, the pads will apply a small amount of pressure and the friction between the pad and disk will be small. This will allow the load to coast slightly as it comes to a stop. The electromagnet assembly is available with 1 to 16 magnets to apply the amount of force required to stop the load. In smaller applications the disk will have one friction face, and for larger loads it will have two.

The friction face of the disk in this type of brake is replaceable. The face is bolted to the disk with four small bolts. This allows them to be replaced easily when they begin to show excessive wear. The brake pads that rub against the friction face on the disk are also easily replaceable while the brake is installed on the factory floor.

The brakes shown in Figures 16–16c and 16–16d are called *fail-safe brakes* because they operate similarly to shoe brakes. This type of brake operates slightly differently, in that a very powerful ceramic magnet is used to apply the braking force. This magnet is a permanent magnet and current is applied to nullify its field when the brakes are dissengaged to allow the load to turn freely. When power is removed from the brake, the permanent-magnet field is allowed to activate the brake disks so that they rub against each other and cause sufficient friction to stop the load.

16.4.6.4
Other Styles of Brakes

Other styles of brakes are also available for a wide variety of applications. One type has a motor with an electric brake mounted directly to the motor shaft. One end of the shaft for the brake slips over the motor shaft when it is mounted, and the other end provides a mounting for the load to be mounted. This type of brake is compact and air cooled and will operate in a fail-safe mode, providing braking action as well as holding power when the brake is deenergized.

In another style the brake is connected to the end of the motor that is opposite the drive end. This allows the drive end of the motor to have pulleys connected to operate a belt-driven load, and have a brake mounted to the opposite end. This makes maintenance of the belts and the brakes much simpler, since they are not in each other's way. This type of application also prevents the two systems from adding heat to each other.

The brake module can also be mounted on the motor shaft prior to it entering a right-angle speed reducer. This provides braking action and holding action to loads that are driven by the motor through a speed reducer. Since the brake is self-adjusting, it does not require much maintenance, and it can easily be replaced by unbolting it from the motor and speed reducer.

The brake module can also be mounted on the through shaft. This allows it to hold the load at the shaft end where there may be more room. In some applications it is also more desirable to have the brake on the load shaft rather than on the motor shaft, since a belt may break or slip and allow the load to run free. Because the brake in this application is mounted on the load shaft, the load will always be held, regardless of the condition of the belts.

These types of brakes can handle rapid cycling applications. The friction pads are designed to self-adjust as they wear. This provides a system that does not require constant maintenance for adjustments. Since the brake incorporates fail-safe operation, it will be able to stop and hold loads when the power is removed from the brake. This makes them useful for most applications.

16.4.6.5
Electromechanical Clutch Used as a Motor Drive

Another method of controlling the speed of a load during acceleration or deceleration is through the use of a clutch mechanism. The clutch operates on the theory that the motor's speed is not going to be controlled, so it will be assumed to operate at full speed. The clutch will allow the load to slip from 100 to 0 percent, which will provide speed control. If the application is looked at from the load's point of view, the operation of the speed control will be the same as if the motor speed were controlled.

The clutch consists of rotating members that are connected to the motor shaft and to the load shaft. The motor shaft drives the clutch, and as electric current is applied, the clutch plates are engaged and the load begins to turn. As control current is increased to the electromagnet in the clutch, more pressure is applied to the friction plates. The stronger the pressure is at this point, the smaller the amount of slip will be. If the clutch does not have any slip, the load will rotate at the same speed as the motor shaft. If the clutch provides 50-percent slip, the load will turn at half the rpm of the motor. The main function of clutches in industrial applications today is to allow the motor to start and accelerate in the unloaded condition. When the motor is operating at full speed, the clutch can be engaged and the load can smoothly be brought on line. This allows the motor to start the load with a minimum amount of locked-rotor current. This arrangement also allows acceleration systems to be used that do not provide a lot of starting torque.

16.4.6.6
Using Clutch and Brake Functions in the Same Device

Since the operation of the clutch and brake are very similar, they can be combined into one device and mounted on most motors or loads. This allows a compact device to be used to provide two functions rather than using two separate devices. The pads of the brake assembly are mounted very similar to the stand-alone brake. The clutch mechanism is mounted on the shaft of the device. When the clutch is energized, the two shaft sections begin to operate as one, and when the motor is stopped, the brake pads apply friction to stop and hold the load.

16.4.6.7
Sizing and Selecting the Proper Clutch and Brake Assembley

Several charts and graphs are available to aid in the selection of a clutch and brake, based on the measured rpm of the shaft at the clutch or brake and the size of the load (hp). The formulas allow you to calculate the amount of clutch torque and the amount of brake torque required. These selection charts show typical models and their overall ratings and dimensions. If you were to select a different brand name of clutch and brake, the size of the unit that is required will be the same, but each manufacturer will provide its own product selection charts, which list their products by size.

Questions

1. Explain why acceleration circuits are required for larger loads.
2. Explain the operation of a primary resistor reduced-voltage starter.
3. Explain why motors may need deceleration or braking controls.
4. Explain the operation of an acceleration circuit.
5. Explain the operation of a deceleration circuit.
6. Explain the operation of an autotransformer starter.
7. Explain the operation of a wye–delta starter.
8. Explain the operation of a solid-state reduced-voltage starter.

True and False

1. Acceleration circuits allow motors to come up to speed without drawing excessive locked-rotor current.
2. The primary resistor reduced-voltage starter allows a large resistor to be connected in series with the supply voltage for a specified amount of time until the motor is running approximately 65 percent of full rpm.
3. The wye–delta starting circuit allows the motor to start when it is connected in delta configuration and then switch to wye configuration during run.
4. The solid-state starter can provide voltage to the motor in small increments to allow the motor to accelerate smoothly.
5. The principle of electric braking is based on the theory of disconnecting the applied voltage that produces the rotating magnetic field and replacing it with a fixed magnetic field produced by DC current.

Multiple Choice

1. Acceleration circuits allow motors to come up to speed without drawing excessive locked-rotor current. Locked-rotor current can be controlled by _____
   a. reducing the amount of voltage supplied to the motor through primary resistors.
   b. reducing the amount of voltage supplied to the motor through the transformer.
   c. changing the winding configuration of the motor during the starting.
   d. All of the above
   e. Only a and b
2. The wye–delta starter controls starting current by allowing a three-phase motor to be started _____
   a. as a wye-connected motor and switched to a delta-connected motor after the motor is up to speed.
   b. as a delta-connected motor and switched to a wye-connected motor after the motor is up to speed.

c. as a delta-connected motor and switched to a star-connected motor after the motor is up to speed.
d. All of the above

3. Electromechanical brakes are used to slow or stop a motor by _____
   a. having the electric coil energized when the motor is running and deenergized to allow the mechanical brakes to engage.
   b. having the electric coil deenergized when the motor is running and energized to allow the mechanical brakes to engage.
   c. having the electric coil energized when the motor is running and energized to allow the mechanical brakes to engage.

4. A plugging circuit _____ to cause the motor to decelerate.
   a. connects a motor in reverse
   b. applies a mechanical brake
   c. applies an electromechanical brake
   d. uses a solid-state circuit

5. A regenerative braking system _____ to cause the motor to decelerate.
   a. applies a mechanical brake
   b. disconnects voltage from the motor windings and reconnects the windings as a generator
   c. applies an electromechanical brake
   d. uses a solid-state circuit

Problems

1. Draw an electrical diagram of a primary resistor reduced-voltage starter and explain its operation.
2. Draw an electrical diagram of an autotransformer reduced-voltage starter and explain its operation.
3. Draw an electrical diagram of a wye–delta reduced-voltage starter and explain its operation.
4. Draw an electrical diagram of a part-winding reduced-voltage starter and explain its operation.
5. Draw an electrical diagram of a secondary-resistor starter and explain its operation.
6. Draw a diagram of a solid-state reduced-voltage starter and explain its operation.
7. Draw a diagram of a plugging circuit and explain its operation.
8. Draw a sketch of mechanical brakes (drum and shoe type) and explain how they operate.
9. Draw an electrical diagram of a motor using electrical brakes. When are the brakes energized? When the brakes are energized, do they allow the motor shaft to rotate, or do they prevent the motor shaft from rotating?

◀ Chapter 17 ▶

AC and DC Drives

Objectives

After reading this chapter you will be able to:

1. Explain the operation of a variable frequency drive.
2. Explain how AC voltage is converted to DC voltage.
3. Explain how DC voltage is converted back to AC voltage.
4. Explain how external control switches are used to control the drive.
5. Explain how parameters are used to set up the drive.

Variable frequency drives for AC motors have been the innovation that has brought the use of AC motors back into prominence. The speed of an AC induction motor can be changed by changing the frequency of the voltage used to power it. This means that if the voltage supplied to an AC motor is 60 Hz, the motor will run at its rated speed. If the frequency is increased above 60 Hz, the motor will run faster than its rated speed, and if the frequency of the supply voltage is less than 60 Hz the motor will run less than its rated speed. The variable frequency drive is the electronic controller that is specifically designed to change the frequency of voltage that is supplied to the motor.

Early frequency drives had rather small solid-state components that limited the amount of current the drive could supply to the motor. This tended to limit the size of motor that could be controlled by a frequency drive and they were not commonly used. When larger transistors became available the variable frequency drives allowed the largest motors to have their speed controlled. These earliest drives utilized linear amplifiers to control all aspects of the drive. Jumpers and dip switches were used to provide ramp-up (acceleration) and ramp-down (deceleration) features by switching larger or smaller resistors into circuits with capacitors to create different slopes.

The advent of the microprocessor has allowed the variable frequency drive to become a very versatile controller that not only controls the speed of the motor, but protects the motor against overcurrent during ramp-up and ramp-down conditions. Newer drives also provide methods of braking, power boost during ramp up, and a variety of control during ramp down. The biggest savings that the variable frequency drive provides is that it can ensure that the motor does not pull excessive current when it starts, so the overall demand factor for the entire factory can be controlled to keep the utility bill as low as possible. This feature alone can provide payback in excess of the price of a drive in less than one year. When motors are started with a traditional motor starter they will draw locked-rotor amperage (LRA) while they are starting. When LRA occurs across many motors in a factory, it pushes the electrical demand factor too high, which results in the factory paying a penalty for all of the electricity consumed during the billing period. Since the penalty can be as much as 15 to 25 percent, the savings on a $25,000 electric bill can be used to purchase drives for virtually every motor in the factory even though the application does not require variable speed.

Today the variable frequency drive is perhaps the most common type of output or load for a control system. As applications become more complex the frequency drive has the ability to control the speed of a motor, the direction the motor shaft is moving, the torque the motor provides to the load, and any other motor parameter that can be sensed. Newer drives have a variety of parameters that can be controlled by

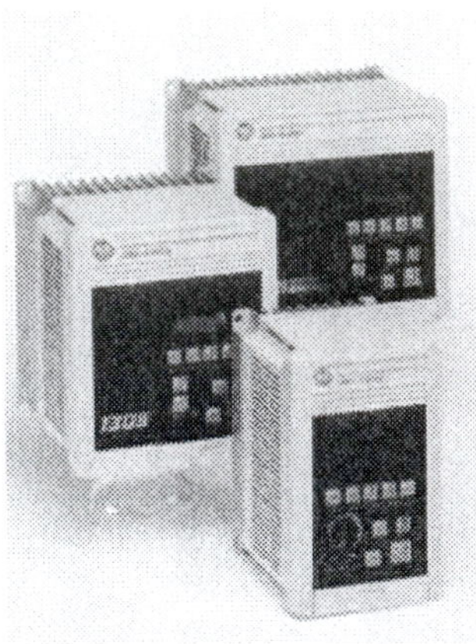

Figure 17–1 Pictures of variable frequency drives.
(Courtesy of Rockwell Automation)

numbers that are programmed into it or downloaded from another microprocessor-controlled system, such as a programmable controller. These drives are also available in smaller sizes that are cost efficient and take up less space in the electrical cabinet. Figure 17–1 shows a picture of a variable frequency drive.

17.1 Block Diagram of a Variable Frequency Drive

The block diagram of the variable frequency drive can be divided into three major sections: the power conversion section; the microprocessor control section (CPU) and the control section, which includes the external switches and signals that control the drive operations; and the power section, where AC voltage is converted to DC and then DC voltage is inverted back to three-phase AC voltage. Figure 17–2 shows a block diagram of the power section of the variable frequency drive as three separate sections to indicate the three main functions: the rectifier section, the filter section, and the switching section, which uses regular transistors, Darlington pair transistors, or insulated gate, bipolar transistor (IGBT) to *invert* the DC voltage back to AC voltage with the proper frequency. The individual electronic components in these sections are explained in detail in Chapter 19. Here we review the major components and explain how they work together to provide the major drive functions. You can refer to Chapter 19 to review any detailed information about the way the individual components operate.

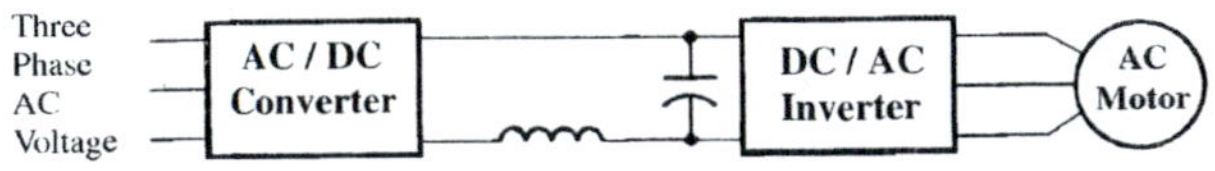

Figure 17–2 The variable frequency drive shown as three separate sections: the rectifier section, where AC is converted to DC; the DC intermediate circuit, which contains the capacitor and inductor for filtering; and the DC to AC inverter, where DC is turned back into three-phase AC.
(Courtesy of Rockwell Automation)

17.2 A Typical Drive Installation

Figure 17–3 shows a typical variable frequency drive installation, including the wires that supply power to the drive, the wires that provide voltage from the drive to the motor, and all of the necessary input and output signals that the drive needs for operation. The power source for the variable frequency drive is provided at terminals R, S, and T by three-phase AC voltage. The value of this voltage can be 208, 240, or 480 V. The three-phase voltage is converted to DC voltage in the rectifier section of the drive where six diodes are connected as a three-phase, full-wave bridge rectifier. On larger drives the diodes can be replaced with silicon-controlled rectifiers (SCRs).

The next components in this circuit are the choke (L1) and the capacitors that make up the filter section for the drive. The capacitors and choke provide a filter that removes all of the ripple and any trace of the original frequency. The voltage at this point in the drive is pure DC voltage and it will be approximately 670 V.

The output section of this drive contains three pairs of IGBTs. These transistors are turned on by a pulsewidth modulation (PWM) control circuit that times the conduction of each IGBT so that a PWM wave is produced that looks like a sine-wave output. The transistors are turned on and off approximately 12 times for each half-cycle. Each time the transistor is turned on, its amplitude will be adjusted so that the overall shape of the waveform looks like a sine wave. The time each transistor is turned on is adjusted as the frequency for the output signal is adjusted. The overall frequency for the drive output signal to the motor will be determined by the frequency of the PWM sine wave. The frequency can be adjusted from 0 to 400 Hz on some drives and typically it can be adjusted from 0 to 120 Hz. The amplitude of the signal will change to change the voltage of the signal. The voltage and current for the output signal will be adjusted to provide the correct amount of torque to the motor load. The drive will maintain a volts per hertz (V/Hz) ratio to ensure that the motor has sufficient power to provide torque to respond to changes in the load. The V/Hz ratio can be adjusted slightly to provide more voltage at lower frequencies if the motor is used

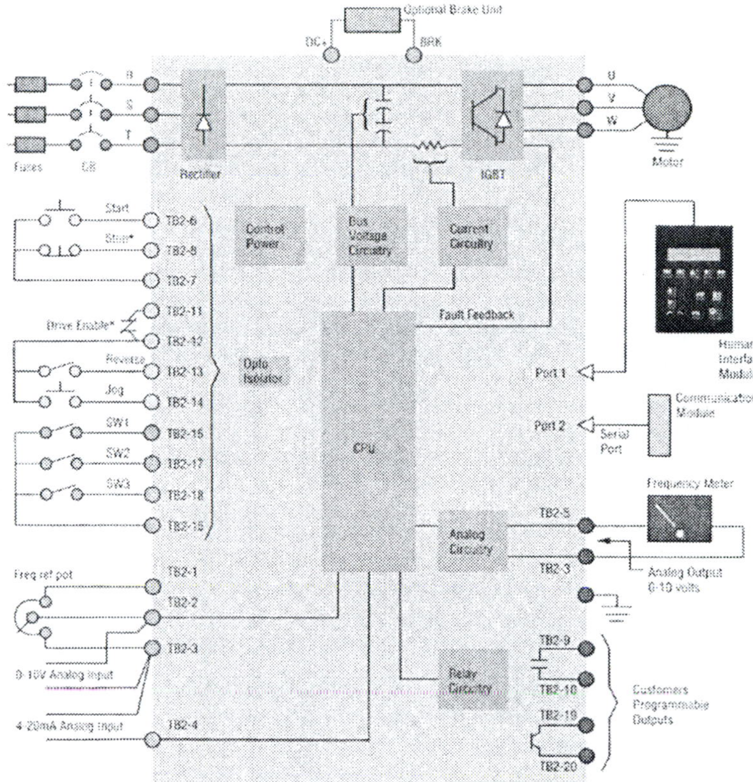

Figure 17–3 Block diagram of variable frequency drive that also shows the components that are connected to the drive to provide additional control.
(Courtesy of Rockwell Automation)

in applications where larger loads must be moved accurately at lower speeds. In the block diagram, a diode is connected in reverse bias across each IGBT to protect it from excess voltage spikes that may occur. The IGBTs are controlled as pairs so that one will provide the positive part of the PWM sine wave and the other will provides the negative part of the wave.

The output terminals of the drive provide a place to connect the three motor leads. These terminals are identified as U, V, and W. The markings R, S, and T for the input voltage and U, V, and W for the output terminals are prescribed by European and worldwide standards. Some drives manufactured in the United States prior to 1990 may still be identified as L1, L2, and L3 for input terminals, and T1, T2, and T3 for output terminals where the motor is connected.

17.3 External Control Switches and Contacts for the Drive

In the previous application the drive is enabled and controlled by external switches or contacts. The National Electric Code and local codes will specify the exact number of external controls required for each drive. The diagram in Figure 17–3 shows typical control switches that are used. You can see at the upper left side of the drive that a normally open start pushbutton is connected to terminals TB2-7 and TB2-6, and a normally closed stop pushbutton is connected between terminals TB2-7 and TB2-8. The voltage for these circuits is provided internally from TB2-7. Power must be received continually through the stop pushbutton, and when the start pushbutton is depressed, a pulse signal will be received and the drive will begin the start sequence. The start sequence will provide a ramp that is used to start the motor by slowly increasing the frequency and voltage. Any time the stop button is opened, the drive will be stopped and if a ramp-down sequence is programmed, the drive will come to a stop gradually. If a ramp-down sequence is not programmed, the motor can be stopped with braking, or it can be allowed to coast to a stop.

The second set of switches provide a set of terminals where an external set of contacts from a control relay can be connected to provide a *drive-enable function*. When the contacts are connected to terminals TB2-11 and TB2-12, power will flow from terminal TB2-12 through the closed contacts to terminal TB2-11 to enable the drive. When the drive is in the enable condition, power is allowed to flow through the drive to the motor. Any time these contacts are opened, the drive will become disabled, and the drive will not send voltage to the motor.

The drive can be switched from forward to reverse with a remote switch. The external switch is connected between terminals TB2-12 and TB2-13 to provide the reverse signal to the drive. Any time the switch is closed and this circuit is made, the drive will reverse the phase sequence to the motor, which provides the same effect as physically swapping leads U and W. Remember that any time that any two leads of a three-phase power supply to a motor are switched, the phase relationship between these two leads is changed and the motor will run in the opposite direction. This is an important feature of variable frequency drives in that they can reverse the direction of a motor without using expensive reversing motor starters.

A jog pushbutton can be connected across terminals TB2-14 and TB2-12 to provide the jog function. When this switch is closed, the drive will check the jog parameter and the drive will produce the frequency that is programmed for this parameter. For example, if you want the motor to turn at 20 percent rpm at jog speed, you would enter 20 percent into the program as the jog parameter and the drive will produce a frequency that causes the motor to rotate 20 percent any time the jog pushbutton is depressed.

The external switches identified as SW1, SW2, and SW3 are connected between terminal TB2-15 and terminals TB2-16, TB2-17, and TB2-18. If switch 1 is closed, the drive will provide the frequency to operate the drive at the frequency that is entered for program 1. This allows an operator to change fixed speeds for a drive if it is used as a stand-alone control. For example, you may have an application where you have a frequency drive controlling the speed of a conveyor. An operator can close switch 1, 2, or 3 to manually set the conveyor speed. If switch SW1 is closed, the drive will operate at 60 percent rpm, which is the value that is entered into the program for speed 1. If SW2 is closed, the drive will operate at 80 percent rpm, which is the value that is entered into the program for speed 2. If SW3 is closed, the drive will operate at 100 percent rpm, which is the speed entered into the program parameter for speed 3.

If you wanted to provide the operator the ability to set the speed at any value from 0 to 100 percent rpm, you could connect a remote potentiometer of 10 KΩ to terminals TB2-1, TB2-2, and TB2-3. When the drive is set for external signal, the voltage from the potentiometer will provide the reference signal for the drive. Terminals TB2-2 and TB2-3 are used for an external voltage signal (0–10 VDC), which could come from some other type of microprocessor controller such as a programmable logic controller. If the external control signal is going to be a milliampere signal, terminals TB2-3 and TB2-4 would be used.

Two ports are provided on the right side of the drive to accept serial inputs. A hand-held programmer is connected to port 1 and can be used for entering programming parameters or to check parameters for troubleshooting. The second serial port can be used to establish serial communications from a portable computer, which allows the parameters to be loaded from a disk or stored to a disk. This provides a means of reloading the parameters in case of programming problems or if a drive is changed.

17.4 External Outputs for the Drive

The drive has two types of on/off outputs available and they are shown on the right side of the diagram in Figure 17–3 for the drive. A set of normally open contacts are provided at terminals TB2-9 and TB2-10. These contacts are controlled by the drive and their function can be selected in the drive parameters. For example, these contacts can be selected to close when the drive is enabled, or they can be used to indicate the drive is in a fault condition. A solid-state-type output is available at terminals TB2-19 and TB2-20. This circuit is basically the emitter and collector of a transistor. When the drive controls this circuit, it sends a signal to the transistor base, which in turn causes the collector–emitter circuit to go to saturation.

Another output that is available from the drive is a frequency signal that can be sent to a frequency meter. The frequency meter can be located on the operator panel where it will be used to indicate the speed signal that the drive is sending to the motor.

17.5 Solid-State Circuits for Variable Frequency Drive

As you know, the solid-state circuitry of a variable frequency drive can be described as three sections. The first section of the drive is called the rectifier section or converter and consists of a three-phase bridge rectifier. The second section of the drive is called the DC intermediate section and contains the filter components. The third block of the drive is called the inverter section because this is where the DC voltage is turned back into three-phase AC voltage.

Figure 17–4 shows an electronic diagram of an Allen–Bradley 1336 drive. Each section of the drive is now shown with the actual components connected as you would find them when you opened the drive to troubleshoot it. The rectifier section of this drive utilizes a three-phase bridge rectifier, which is actually a module. This means it can be removed and replaced as a unit

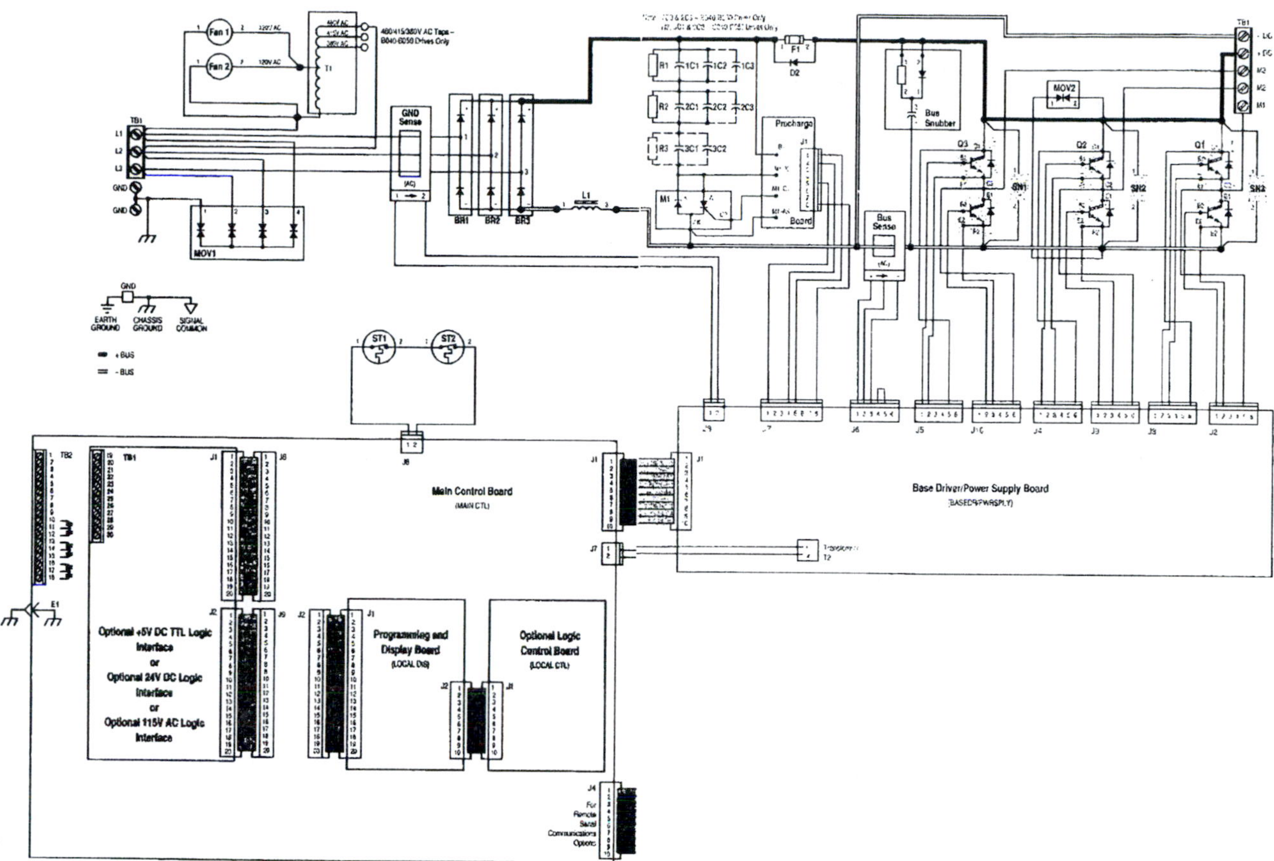

Figure 17–4 Electronic diagram of an AC variable frequency drive. The major parts of the drive include the rectifier, the filter, and the output transistors.
(Courtesy of Rockwell Automation)

rather quickly if it fails. The output of the rectifier section is six half-waves. A set of metal-oxide varistors are connected to the input of the rectifier section to protect against voltage surges.

When the input voltage to the drive is 480 VAC, the output DC voltage from the rectifier section will be approximately 670 VDC. The pulsing DC voltage is applied to the DC bus on this system. The DC bus is identified by the +DC and the −DC wires that run through the length of the drive circuit.

The filter section of the drive uses capacitors and an inductor to filter the voltage and current. The capacitors have a precharge circuit that allows the capacitors to reach full charge slowly so that they are not damaged. The capacitors are connected in parallel with the DC bus, and the inductor is connected in series with the negative DC bus wire. A set of resistors are provided to discharge the capacitors any time power is removed. You should always allow sufficient time for the capacitors to discharge before you try to work on the solid-state components in the drive. The filter allows the pulsing DC voltage to be changed to pure DC.

The output section of the drive converts the DC voltage back to three-phase voltage. This section uses pulse width modulation techniques to switch three pairs of transistors on and off up to 12 times during each half-wave to produce a three-phase output. The amplitude of the signal determines the amount of voltage for the AC voltage, and the frequency of the signal will determine the frequency of the output of the drive. In most cases the output frequency can be any value between 0 and 120 Hz. Some drives allow the upper frequency to reach 400 Hz.

The output transistors are connected across the DC bus. One transistor is connected to the positive DC bus wire, and when it is switched on and off, it will provide the positive half-cycle for one phase of the AC signal. The second transistor is connected to the negative DC bus wire, and when it is switched on and off, it will provide the negative half-cycle for one phase of the AC signal. The base drive board provides the PWM signals for all of the transistors. The drive has a microprocessor that accepts the command signal and determines the correct frequency and voltage for the output transistors.

17.6 Pulsewidth Modulation Waveforms for Variable Frequency Drives

Figure 17–5 shows a typical waveform for a pulsewidth modulation circuit in the AC variable frequency drive. The transistors in the PWM circuit are switched on and off approximately 12 times each half-cycle. The on and off cycles create the overall frequency waveform for the output signal. The output waveform of the PWM section looks like a traditional three-phase signal to the motor. If you place an isolated case-type oscilloscope across any two of the output leads of the drive, you will see a signal that looks very similar to the one in the diagram.

If you are using a digital voltmeter to measure the output voltage on a variable frequency drive, be aware that some digital voltmeters will not read the AC voltage from this section accurately because of the switching frequency of the transistors. Digital voltmeters tend to read the drive's output voltage higher than it actually is because the voltmeter may be fast enough to sample some of the individual waveforms created when the transistors are switched on and off rapidly. An analog meter may show the voltage more accurately, because the needle cannot change as fast as the transistor is switched on and off. For this reason, some drive manufacturers provide an LED display to show an accurate voltage reading right on the face of the drive.

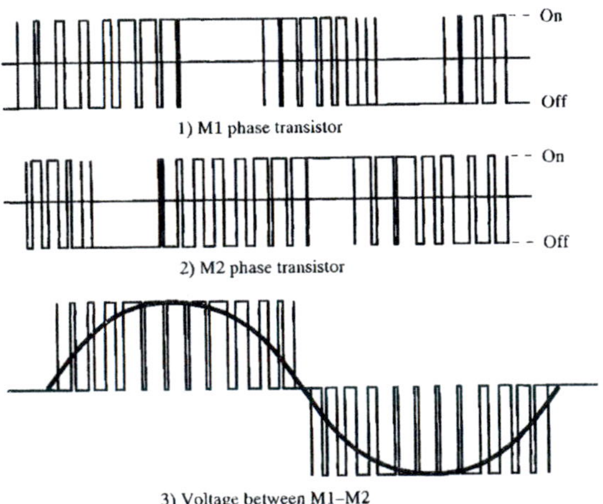

Figure 17–5 Output waveform of the PWM section of the variable frequency drive. Notice all of the points where the transistor is switched on and off inside each half-wave. (Courtesy of Rockwell Automation)

17.7 Insulated Gate, Bipolar Transistor Drives

In newer drives, the transistors may be replaced by insulated gate, bipolar transistors. The IGBTs are used because they can be switched on and off at much higher frequencies that do not conflict with other signals. The higher frequencies are also used because they are outside of the audible range for humans, so the drive will not emit a hum that humans can hear. The high frequency is divided so that the output signal for the drive is within the 0- to 120-Hz range.

Figure 17–6 shows three pairs of IGBTs used to produce three phases of a typical output section for a variable frequency drive that uses IGBTs instead of bipolar transistors. One IGBT of each pair is connected to the positive bus and a second is connected to the negative bus. The IGBTs operate similar to the transistors in that they are cycled on and off at high frequencies within the overall waveform of a sine wave.

17.8 Variable Frequency Drive Parameters

Variable frequency drives have had the ability to provide features such as ramp-up speed, ramp-down speed, boost voltage, and braking functions since they were first designed. Prior to having a microprocessor, these

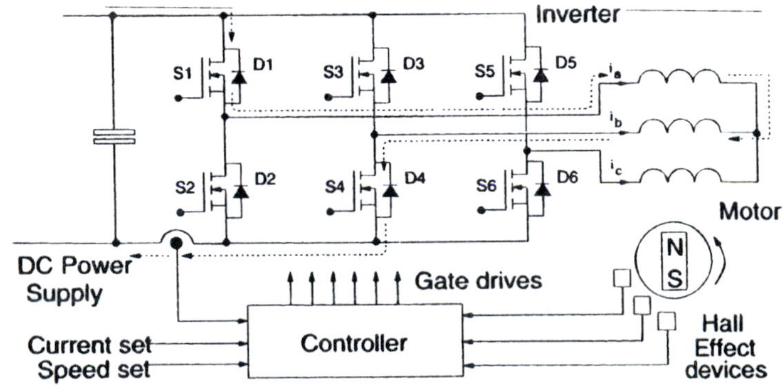

Figure 17–6 Insulated gate, bipolar transistors (IGBTs) connected to the output stage of the variable frequency drive. The IGBTs are used instead of traditional bipolar transistors in newer drives.
(Courtesy of Rockwell Automation)

Figure 17–7 List of parameters for a variable frequency drive.
(Courtesy of Rockwell Automation)

Parameter Number	Description	Units	Min/Max Values	Factory Setting
0	Parameter Mode	None	0/0	—
1	Output Volts	Volts	0/575	0
2	Output Current	% Rated	0/200	0
3	Output Power	% Rated	0/200	0
4	Last Fault	Code	0/37	0
5	Frequency Select 1	Code	0/5	0
6	Frequency Select 2	Code	0/5	0
7	Accel Time 1	Seconds	0.0/500	5.0
8	Decel Time 1	Seconds	0.0/600	6.0
9	DC Boost Select	Code	0/12	2
10	Stop Select	Code	0/2	0
11	Decel Frequency Hold	Off/On	0/1	0
12	DC Hold Time	Seconds	0/15	0
13	DC Hold Volts	Volts	0/115	0
14	Auto Restart	Off/On	0/1	1
15	Factory Set—Do Not Change	None	0/0	0
16	Minimum Frequency	Hertz	0/120	0
17	Base Frequency	Hertz	40/120	60
18	Base Volts	Volts	115/575	460/575 ○
19	Maximum Frequency	Hertz	40/250	60
20	Maximum Volts	Volts	115/575	460/575 ●
21	Local Run	Off/On	0/1	1
22	Local Reverse	Off/On	0/1	1
23	Local Jog	Off/On	0/1	1
24	Jog Frequency	Hertz	.0/120	0
25	Analog Output	Code	0.1	0
26	Preset/2nd Accel	Code	0/1	0
27	Preset Frequency 1	Hertz	0.0/250	0.0
28	Preset Frequency 2	Hertz	0.0/250	0.0
29	Preset Frequency 3	Hertz	0.0/250	0.0
30	Accel Time 2	Seconds	0.0/600	5.0
31	Decel Time 2	Seconds	0.0/600	5.0
32	Skip Frequency 1	Hertz	0/250	250
33	Skip Frequency 2	Hertz	0/250	250
34	Skip Frequency 3	Hertz	0.250	250
35	Skip Frequency Band	Hertz	0/15	0
36	MOPC	% Rated	50/150	150
37	Serial Baud Rate	Code	0/1	1
38	Overload Current	% Rated	50/115	100
39	Fault Clear	Off/On	0/1	1

features and functions were designed into the op-amp circuits and were enabled by the placement of jumpers or the setting of dip switches. After microprocessor chips were integrated into the drive, these features became programmable and, in many cases, proportional, so that, for example, you can ask for braking, but you can limit the braking to 60 percent for 3 seconds. If you select ramp up, you can select more than one ramp speed, and then you can integrate the selection with an external switch so that the ramp speed can be selected by external conditions. The numbers that are programmed into the drive to select these features are called *parameters*. The type and number of parameters will vary from drive to drive. The Allen–Bradley 1336 drive has 89 parameters, while the Allen–Bradley 1305 drive has 136 parameters. Figure 17–7 shows a typical list of parameters for a variable frequency drive.

The drive has a set of standard values that are used for the parameter. These parameters are called the *default settings* or *factory settings*. It is important to record the actual parameter settings if they are different from the factory settings, so that the proper settings can be put into the drive if it is ever removed and replaced with a new one. The parameters provide a means to customize the drive to any specific application. Some manufacturers provide software that can be used on a

Parameter Number	Description	Units	Min/Max Values	Factory Setting
40	Power Fault	On/Off	0/1	0
41	Motor Type	Code	0/2	0
42	Slip Compensation	Hertz	0.0/5.0	0.0
43	Dwell Frequency	Hertz	0/120	0
44	Dwell Time	Seconds	0/10	0
45	PWM Frequency	WM	0.4/2.0	0.4
46	Pulse Scale Factor	Ratio	1/255	64
47	Language	Code	0/5	0
48	Start Boost	Volts	0/115	0
49	Break Frequency	Hertz	0/120	0
50	Break Volts	Volts	0/230	0
	Parameters 51–69 Can Only Be Accessed Thru the Serial Port			
70	Base Driver Board Version	Code	—	—
71	Control Board Version	Code	—	—
72	Activate Parameters 73–76	Off/On	0/1	0
73	Preset Frequency 4	Hertz	0.0/250	0.0
74	Preset Frequency 5	Hertz	0.0/250	0.0
75	Preset Frequency 6	Hertz	0.0/250	0.0
76	Preset Frequency 7	Hertz	0.0/250	0.0
77	Above Frequency Contact	Hertz	0/250	0
78	Traverse Period	Seconds	0.0/30	0.0
79	Maximum Traverse	Hertz	0.0/100	0.0
80	Inertia Compensation	Hertz	0.0/20	0.0
81	Soft Start/Stop Enable	Off/On	0/1	0
82	Amp Limit Fault Enable	Off/On	0/1	0
83	Run Boost	Volts	0/115	0
84	Analog Inverse	Off/On	0/1	0
85	Restart Tries	Code	0/9	0
86	Fault Buffer 0	Code	0/37	0
87	Fault Buffer 1	Code	0.37	0

Figure 17–7 List of parameters (continued).
(Courtesy of Rockwell Automation)

portable computer to save and load the parameters from a drive. The parameters can also be saved and loaded from the PLC program, which means they can also be changed while the system is in operation to provide custom parameters for multiple recipes.

The parameters can be described in groups by their function. Some example of these groups include metering, setup, advanced setup, frequency settings, diagnostics, faults, and process displays. Examples of the setup group include minimum frequency and maximum frequency. This allows the minimum and maximum frequency to be fixed so that the motor does not run less than the minimum value and faster than the maximum value. For example, if the drive controls a pump and the pump should never be allowed to run below 10 percent rpm or it will loose its prime, the minimum frequency could be set to 6 Hz. If the maximum speed of a motor would cause the pump to create excess pressure, the maximum frequency could be set at 50 Hz. The maximum number could also be set in excess of 60 Hz if the motor needs to run above 100 percent rpm. Most drives allows speeds up to 200 percent or 120 Hz.

Another setup parameter is acceleration time. The acceleration time determines the acceleration ramp.

For example, if you select the ramp time as 10 seconds, the drive will increase the frequency proportionally from the programmed minimum frequency to the programmed maximum frequency in 10 seconds. If you wanted the motor to ramp up to speed faster, you would shorten the ramp time. If you wanted the motor to ramp up slower, you would increase the ramp time. Most drives have more than one acceleration ramp parameter and each ramp is enabled to the input switches that were previously discussed. This means that you could have up to three acceleration ramps, which would be selected by switch 1, 2, or 3. The deceleration time is also programmable like the acceleration time. Three deceleration ramps are also available for this drive.

Another setup parameter available for the drive is overload current limit. This parameter will act like a programmable fuse. You can select any value from 100 to 115 percent. The other variable that operates with the overload current is the amount of time this current can exist. Since the drive is controlled by a microprocessor, it monitors the current and voltage and turns off the drive if these values become excessive.

The drive also provides metering parameters. These can be integrated with protection, display, and fault functions. Most drives have displays that are built into the face of the drive, or in some cases the face for the drive can be mounted remote on the front of an electrical panel, so the drive can be monitored without opening the panel. Other drives have glass panels available so that the face of the drive can be viewed through the panel. The technician can use the display on the face of the drive to observe the amount of input voltage the drive has, the amount of current the drive is using, the frequency, the temperature, output voltage, and the last fault that was recorded.

Some of the advanced setup parameters include the type of braking the drive will use. The choices for this parameter are no brakes (coast to stop) or braking and the amount of braking voltage and the amount of time the braking voltage should be applied is determined. Another advanced setup parameter is called *DC boost voltage*. The DC boost voltage is DC voltage that can be applied with the AC frequency during starting or at times when the motor needs more torque. The DC boost voltage makes the magnetic field in the motor stronger, which will reduce the amount of slip the motor has. This will allow the motor to provide additional power to some applications where more starting torque is needed or when more torque is needed during specific loading conditions, such as when the motor is used to provide power for a conveyor whose load varies from light to extra heavy.

The drive has the ability to test hundreds of points in its circuit boards for changes in voltage, current, frequency, and temperature. The present value of each of these variables at both the input stage and the output stage of the drive can be compared against the value set into the parameter. If the value is exceeded, the drive can indicate each occurrence with a fault code. The drive also changes the state of contacts that can be used to enable a fault indicator lamp or horn. When this occurs the technician can come to the drive to check the problem, and if a fault has occurred, it will be stored in the drive where it can be brought to the display by pressing a series of keys on the front panel of the drive. The serial port connection provides a method to send the fault codes to an external controller, such as a PLC, where they can be logged with the date and time they occurred and they can also be printed as they occur. Some drives have the capacity to store multiple faults so that the technician can review the last 5 faults. This provides a means to detect and store multiple faults if more than one problem occurs during the fault condition. Typical fault conditions are low voltage, high voltage, high current, and overtemperature.

17.9 Typical Applications for Variable Frequency Drives

Variable frequency drives are widely used to control the speed of conveyor systems, blower speeds, pump speeds, machine tool speeds, and other applications that require variable speed with variable torque. In some applications, such as speed control for a conveyor, the drive is installed with a remote potentiometer that personnel can adjust manually to set the speed for the conveyor. In this type of application, the personnel that use the conveyor can manually set the motor speed with the minimum and maximum frequency that is programmed into the parameters.

In other applications, such as blower speed control or pump speed control, the variable frequency drive can be controlled by a 4- to 20-mA or 0- to 10-V input signal that comes from some type of microprocessor controller. When the input signal for the drive is provided from a controller, the system is considered to be closed loop. For example, in a system where the drive is controlling the speed of a blower, some type of temperature sensor can be used to determine the temperature of a room. If it is too cold and the blower is moving warm air, the speed of the blower can be increased. If the temperature becomes too hot, the speed of the blower can be slowed until the temperature returns to the correct set point. If the variable frequency drive is used to control the level of product in a tank by varying the speed of a pump, a level sensor can be connected to a controller, and if the level of the tank is becoming low, the drive can increase the speed of the motor and pump, and if the level is too high, the speed can be reduced.

Figure 17–8 shows an example of the field wiring connections at the terminal boards of a drive for a

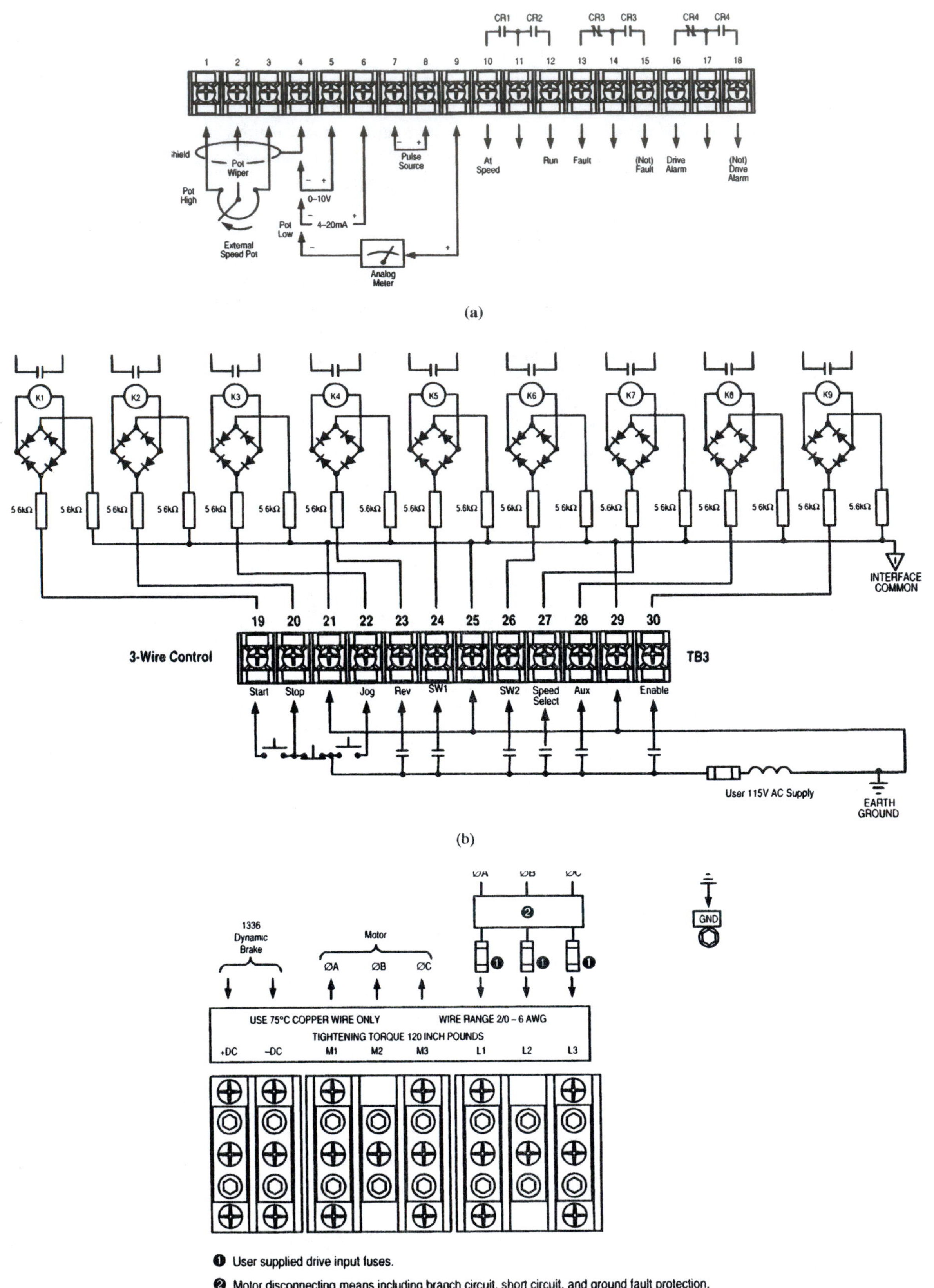

Figure 17–8 Diagram of (a) TB2 for a 1336 Allen–Bradley drive; (b) TB3 for a 1336 Allen–Bradley drive; (c) TB1, which shows the power connections to the drive and terminal connection for the motor.
(Courtesy of Rockwell Automation)

typical application. In this case the drive diagram is for an Allen–Bradley 1336 drive. The top diagram shows the external control signals that are connected to terminal board 2 (TB2) of the drive. You can see that the command signal can be an external potentiometer connected to terminals 1, 2, and 3, or it can be a 0- to 10-V signal connected to terminals 4 and 5, or it can be a 4- to 20-mA signal connected to terminals 4–6. This terminal board also provides normally open (NO) and normally closed (NC) contacts to indicate the drive is at speed, running, faulted, or has a drive alarm. These contacts are used as part of fault or safety circuits for the system.

Terminal board 3 (TB3) in this figure shows the start and stop switches that must be connected to the drive to provide the signals to cause it to start and become enabled. The signals voltage for this board can be selected as 5V TTL, 28 VDC, or 115 VAC supply. The diagram shown uses 115 VAC. Start, stop, and jog buttons are connected to terminals 19, 20, and 22. A 115-V signal must also be provided at terminal 30 to enable the drive. Terminals 24 and 26 provide the inputs for switch SW1 and SW2, which will be used to make a two-bit binary code that is used to indicate which acceleration and deceleration parameters to use.

Terminal 23 is used to set the motor in reverse from a remote switch. Terminals 21, 25, and 29 are the common for this board and since they are connected inside the board, the common from the transformer need only be connected to one of these terminals. After the 115-VAC signals that are used as inputs for this board are received, they must be isolated and converted before they are sent to the microprocessor section of the drive. The isolation and rectification portion of this board is shown at the top part of the diagram in Figure 17–8. From this diagram you can see that after each 115-VAC signal is received it is rectified and directed through a DC-powered relay to provide isolation before it is sent to the processor. The diagram also shows each signal connected to a set of resistors to drop the voltage and to a bridge rectifier to change the AC voltage to DC. The output of the bridge rectifier is connected to the coil of a DC relay. The contacts of the relay are connected to the processor inputs. Since this signal comes through a relay, there is total isolation between the 115 VAC that is used in the field switches connected to this board and the contacts that are connected directly to the processor section of the drive.

The high-voltage terminal board for the variable frequency drive is designated as TB1. This terminal board provides connections at the far right side for L1, L2, and L3 for three-phase 480 VAC. Terminal M1, M2, and M3 are connections for the output of the variable frequency drive to the three-phase motor. Terminals −DC and +DC are located at the far left side of the terminal block and they provide DC voltage directly from the DC bus. This voltage is provided for the dynamic brake that can be added to the three-phase motor. The dynamic brake provides a means to stop the motor shaft quickly. The operation of the dynamic brake is covered in Chapter 16.

17.10 Operating the Variable Frequency Drive

The variable frequency drive can be operated as a stand-alone unit or it can be controlled by a PLC or other electronic control system that sends a 0- to 10-V signal or a 4- to 20-mA signal. When the drive is operated as a stand-alone system, a potentiometer may be provided directly on the face of the drive. The operator of the machinery can set the speed for the drive by adjusting the potentiometer anywhere from 0 to 100%. In some applications, the potentiometer is mounted on the machine's front panel instead of on the face of the drive. The reason for this is that in some cases the drive must be mounted in an electrical cabinet, and the potentiometer on the face of the drive is not accessible when the door to the electrical cabinet is closed.

If the drive is controlled by a 0- to 10-VDC or 4- to 20-mA signal, the command signal will be changed by the logic of the controller it is connected to. The controller generally will have some type of sensor that indicates speed or position if the system is a closed-loop system, or the command signal can be generated from a predefined table in the controller that sets the motor to the proper speed for the application. In this type of application, the local speed potentiometer can be used to jog the motor or when the motor must be operated for maintenance.

17.11 Scalar Drives and Vector Drives

Some companies provide additional circuits inside the drive to bring in the closed-loop feedback signals from encoders or resolvers. These circuits will decode the signals from the sensors and scale them to the level required. This feature makes the drive operate more like a package system and makes it easier for design engineers to put a complete system together where all of the components match. Allen–Bradley calls their drive that has closed-loop capability a vector drive. They also have a drive called a scalar drive that can be operated as either an open-loop or closed-loop drive.

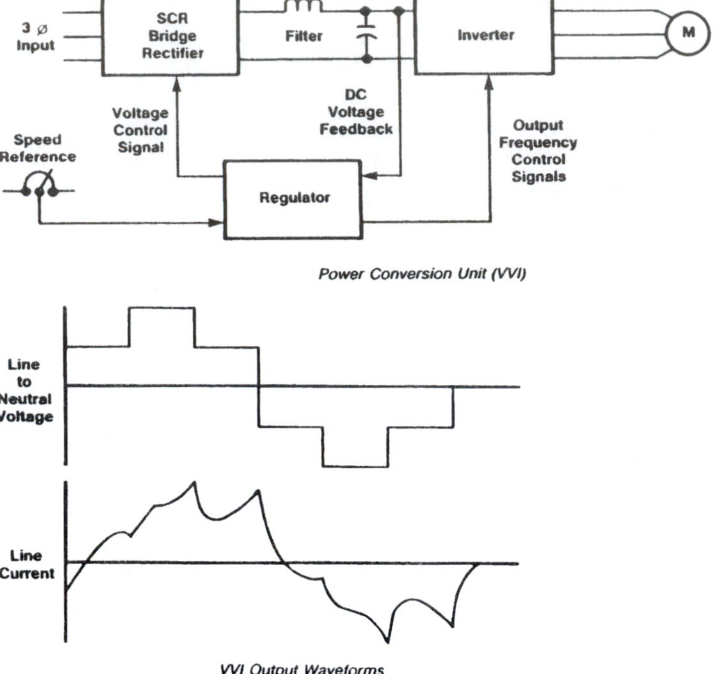

Figure 17–9 Block diagram and waveform for a variable voltage input (VVI) drive. This is one of the earliest types of variable frequency drives.
(Courtesy of Rockwell Automation)

Other manufactures provide similar drive packages. The vector and scalar drives are more expensive and more complex than a simple variable frequency drive. A simple fractional horsepower variable frequency drive costs less than $500 today.

17.12 Variable Voltage Input Drives

The variable voltage input (VVI) drive is the technology that was used in some of the earliest AC variable frequency drives. Since these earlier drives did not have microprocessor chips to establish the transistor driver signals, they used existing technology, such as oscillators. Figure 17–9 shows a block diagram of this type of drive and a typical waveform of the output.

From the block diagram you can see that the basic parts of the drive, such as the rectifier section, filter, and inverter sections, are much the same as modern drives. The major difference is that SCRs are used for the rectifier section instead of diodes. The reason for this is that diodes were not manufactured as large as SCRs at this time, so the SCRs were used because they were more durable. Also, it is possible to adjust the SCRs' timing with a signal from the regulator section of the drive to change the amount of voltage and current delivered to the drive. The amount of change provided sufficient change to meet most of the changing demands of torque.

From the waveform diagram you can see that the voltage waveform was a 6-step signal. The switching devices in the inverter section (usually SCRs) would be switched on and off at specific points to provide the 6-step signal. The current waveform looks more like a sine wave because the inductance of the motor helps to cause the phase shift needed to smooth out the square wave shape of the 6-step signal. You will encounter a few drives with this early technology because they have paid for themselves over and over. In some cases, the drives are changed out whenever there is a problem and replaced with a newer microprocessor programmable drive. It is also more practical to change to a modern drive that can provide a wider range of torque for all applications the motor may encounter.

17.13 Current Source Input-Type Drives

Another type of drive technology that was used in early variable frequency drives is called current source input (CSI). Figure 17–10 shows a block diagram of this type of drive and the voltage and current waveforms for the drive. From the block diagram you can see that this drive also did not have a microprocessor. This means that all of the drives for the waveforms were established from oscillators and other circuits that combined with op-amps to make the closed-loop portion of the drive.

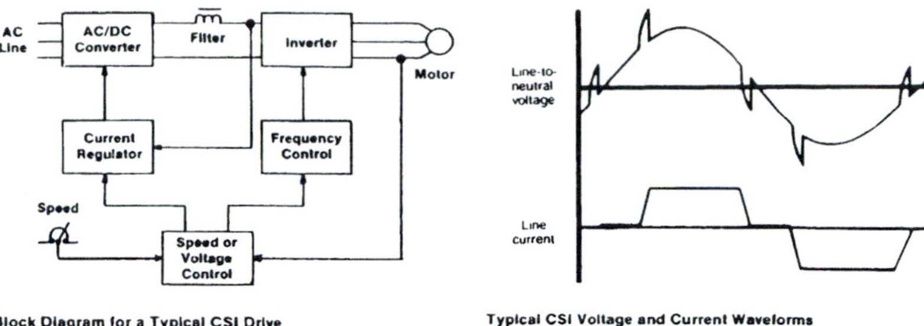

Figure 17–10 Block diagram with waveform of a current source input (CSI) drive.
(Courtesy of Rockwell Automation)

The input speed potentiometer set a reference voltage on the op-amps in the *speed* or *voltage control block*. The feedback signal from the output lines to the motor is compared to the set point in this block. The output of the speed control block is sent to the *current regulator block* that controls the firing angle on the SCRs in the rectifier section. The voltage and current are changed to meet the changing demands of the torque that is needed to move the load.

A companion signal is sent from the speed control block to the *frequency control block*. The frequency control adjusts the speed of the motor by changing the frequency that is sent to it. The speed potentiometer sets the percentage of speed that is needed and the frequency control adjusts the frequency of the inverter section. The inverter section of the drive uses large SCRs to switch the voltage and current.

The waveform for this drive is also shown in the figure. From this diagram you can see that the current is adjusted similar to the 6-step voltage waveform of the VVI-type drive. The voltage signal for the CSI-type drive is smoother and looks more like a traditional sine wave. As a technician you may find a few of these drives still operational, but as a rule, they are generally changed out as problems are encountered, because it is cheaper to purchase newer drives than to repair the older-type drive.

17.14 DC Drives

Variable DC drives have been used to control DC motors longer than variable frequency drives have been used to control AC motors. The first speed control of motors used DC motors because of the simplicity of controlling the voltage to the armature and field of a DC motor to change the speed of the DC motor. The main obstacle in using DC motors is the increased amount of maintenance that is caused because the DC motor has brushes and a commutator. Early speed control for DC motors consisted of large resistors that were switched in the motor circuit to reduce the amount of voltage supplied to it. The resistors created problems because of the heat buildup.

DC drives that were designed in the 1970s and 1980s combined op-amp circuits to provide ramping capability with SCR firing circuits to control large voltages. Today, modern DC drives utilize the latest solid-state power-switching technology that is combined with microprocessors to provide programmable features. When you analyze the diagrams for the DC drives you will notice that much of the circuitry looks very similar to that of the AC drive. The main difference is that the rectifier stage and output stage of the DC drive are combined because the DC drive simply adjusts the DC voltage and current rather than inverting it back to AC. Since the output voltage for the drive is DC, SCRs will be used in a rectifier circuit. The newest drives have programmable parameters that are similar to AC drives in that they set the maximum voltage, current, and speed, as well as provide protection against overcurrent, overtemperature, phase loss of incoming power, and field loss.

Figure 17–11 provides a picture of a typical DC drive as a stand-alone product, and as you would find it when it is mounted in a panel. You can see that it is difficult to distinguish a DC drive from an AC drive by its physical features. Figure 17–12 shows a diagram of the simple DC drive and you can see that the

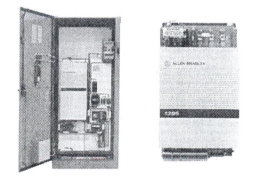

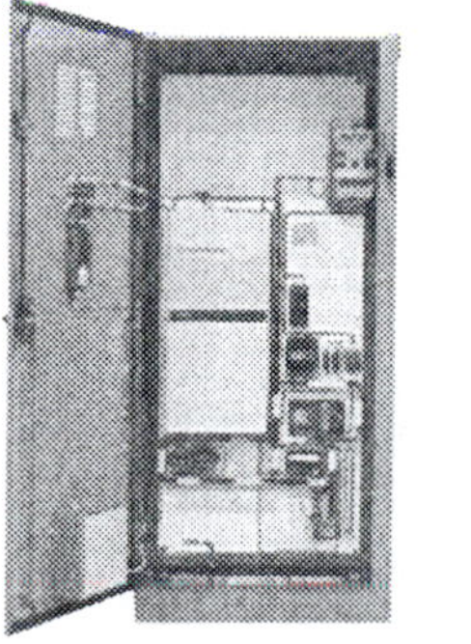

Figure 17–11 Typical DC drive shown mounted in a panel and as a stand-alone drive.
(Courtesy of Rockwell Automation)

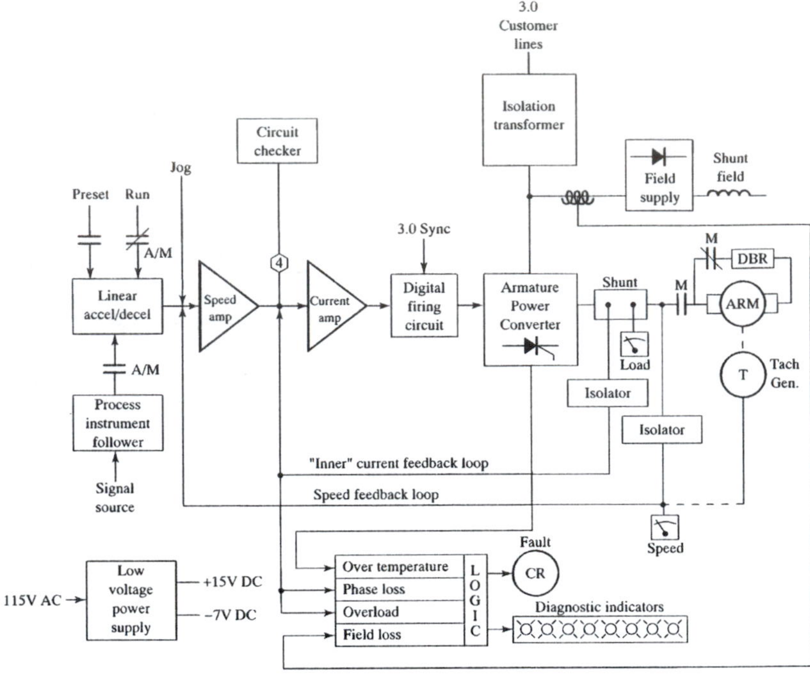

Figure 17–12 Block diagram of a DC drive. Three-phase voltage is supplied at the top of the diagram, and the DC motor is shown as a shunt field and armature.
(Courtesy of Rockwell Automation)

electronic circuits of the drive are slightly different from those of the AC drive. In the diagram of the DC drive, three-phase supply voltage is provided at the top of the diagram. This incoming voltage is sent directly to an isolation transformer and then to a three-phase bridge rectifier in the *armature power converter circuit*. The three-phase rectification in the armature power converter circuit is very similar to the rectifier section of the AC drive, except large SCRs are used instead of diodes.

The SCRs are used for the rectification section because they can provide voltage control as well as rectification. This simplifies the drive somewhat, since rectification and voltage control are combined in one circuit. Op-amps provide speed ramps and current ramps for the SCR firing (control) circuit. In the older drives, the op-amps were used as stand-alone ramping circuits. In modern drives that have microprocessors, the firing circuits are controlled by digital-to-analog (D/A) circuits that integrate linear circuits with the processor. The voltage from the armature power converter circuit is sent directly to the armature. The DC motor is shown in this diagram as a shunt field and armature. A tachometer is shown connected to the armature as a dotted line, which means it is physically connected to the motor shaft.

The rectifier section may use six SCRs as a bridge that is similar to the diode bridge rectifiers in AC drives, or larger drives may connect two SCRs in parallel for each of the six sections of the bridge to provide a 12-SCR full-wave rectifier circuit. When SCRs are connected in parallel, the current rating of the rectifier is nearly doubled.

The firing circuit for the SCRs is synchronized with the three-phase incoming voltage. The firing circuit also receives an input signal called a reference signal or command signal from the speed amplifier and the current amplifier. The speed amp receives a feedback signal from a tachometer, and the current amp receives a signal from a current transducer (shunt) that is connected in series with the armature. As the current in the wire to the armature increases or decreases, the voltage across the shunt will increase or decrease and provide a feedback signal to the current amplifier.

In the diagram you can also see that DC field voltage is provided by a smaller diode bridge. The AC voltage supply for this bridge rectifier is tapped off of the output of the isolation transformer prior to the main rectifier in the armature power converter. Since this voltage comes from a diode bridge rectifier, it will be constant. Speed control for the DC motor is provided by keeping the shunt field voltage constant, and varying the armature voltage and current.

Fault circuits are provided in the drive to test for overtemperature, phase loss, overload conditions, and the loss of field current in the motor. Indicator lamps are provided on the front of the drive to show when a fault has occurred. A speed indicator is also provided on the face of the drive to show the actual speed of the motor. The speed indicator receives its signal from the tachometer that is connected to the shaft of the DC motor.

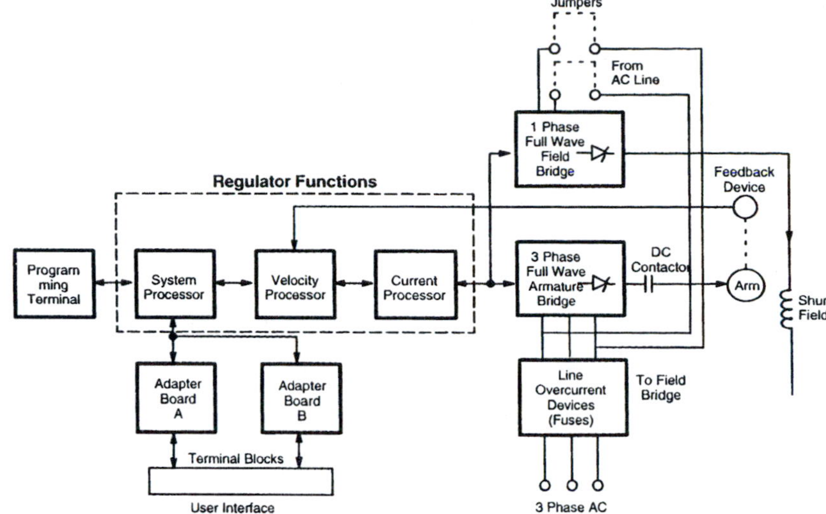

Figure 17–13 Block diagram of a microprocessor-controlled DC drive. This drive provides PID control of the DC motor speed.
(Courtesy of Rockwell Automation)

17.15 Modern Programmable DC Drives

Modern DC drives combine a microprocessor with the drive circuitry shown in the previous diagram. The major difference that the microprocessor provides is that it allows programmable parameters to be used with the drive to set maximum voltages and currents and to provide a variety of ramp-up and ramp-down signals. These parameters are very similar to the ones listed for the AC drive presented in the previous section. Another feature the microprocessor provides is closed-loop control features, such as PID (proportional, integral, and derivative) control. Figure 17–13 shows a block diagram of a modern DC drive that incorporates the closed-loop PID control that is used for speed (velocity) control. A separate control loop for current is also used. In this diagram, three-phase voltage is supplied at the bottom of the drive and is converted to variable DC voltage by the three-phase SCR rectifier section. An encoder is used as the feedback sensors to provide a velocity feedback signal to the velocity processor. Current sensors can also be used to provide a second loop for current control.

Another feature of the microprocessor-controlled drive is that it can be programmed from a programming terminal or programming software in a laptop computer. The programming parameters can also be loaded or *changed on the fly* from a PLC. This provides equal functions to the AC type of drive.

Questions

1. Explain the operation of a variable frequency drive.
2. Explain how AC voltage is converted to DC voltage.
3. Explain how DC voltage is converted back to AC voltage.
4. Explain how external control switches are used to control the drive.
5. Explain how parameters are used to set up the drive.
6. Explain the difference between PWM, CSI, and VVI drives.

True and False

1. The block diagram of the variable frequency drive can be divided into four major sections: the power conversion section, the microprocessor control section (CPU), the control section, and the output section.
2. Drive parameters allow maintenance personnel to set drive features, such as acceleration time, deceleration time, and type of braking.
3. The VFD provides external terminals that allow start, stop, and jog pushbuttons to be connected to allow external control.
4. In the VFD, a set of metal-oxide varistors (MOVs) are connected to the input of the rectifier section to protect against voltage surges.
5. Newer VFDs use insulated gate, bipolar transistors (IGBTs) instead of traditional bipolar transistors.

Multiple Choice

1. Which of the following is not controlled by parameters on a variable frequency drive?

a. Ramp-up speed
b. Ramp-down speed
c. Boost voltage
d. Braking functions
e. None of the above

2. Variable frequency drives are widely used to control the speed of _____
   a. conveyor systems.
   b. blower speeds.
   c. pump speeds.
   d. machine tool speeds.
   e. All of the above

3. The three main types of circuits used in a drive are called _____
   a. CSI, VVI, and PWM.
   b. CCS, VVI, and PWM.
   c. CSI, VVI, and PWI.
   d. CSI, VFD, and PWM.

4. Typical DC drives use _____ in their circuits for control.
   a. light-emitting diodes (LEDs)
   b. silicon-controlled rectifiers (SCRs)
   c. field-effect transistors (FETs)
   d. All of the above

5. The filter circuit of a VFD consists of _____
   a. an inductor (choke) and capacitor.
   b. a fuse and capacitor.
   c. a fuse and inductor.
   d. an MOV and fuse.

Problems

1. Draw a block diagram of the feedback loop used in drives.
2. Explain how a signal from a programmable controller can be used to control the motor drive system.
3. Draw a sketch of an eddy current drive and explain its operation.
4. Provide a block diagram of the PMW drive and show a typical output waveform. Explain how the motor speed is controlled by this type of drive.
5. Provide a block diagram of the CSI drive and show a typical output waveform. Explain how the motor speed is controlled by this type of drive.

◀ Chapter 18 ▶

Programmable Controllers

Objectives

After reading this chapter you will be able to:

1. Describe the four basic parts of any programmable controller.
2. Explain the four things that occur when the PLC processor scans its program.
3. Explain the function of an input module and describe the circuitry used to complete this function.
4. Explain the function of an output module and describe the circuitry used to complete this function.
5. Explain the function of an internal control relay.
6. Discuss the classifications of PLCs.
7. Describe what happens when a PLC is in run mode.
8. Identify the input instruction and output instruction for a PLC.
9. Explain the operation of a PLC on-delay timer.
10. Explain the operation of a PLC up-counter and down-counter.

The programmable controller is the most powerful change to occur in factory automation. The programmable controller (P/C) is also called the programmable logic controller (PLC). Since the personal computer is called a PC, the programmable controller is referred to as a PLC to prevent confusion. As a maintenance technician, you will run into the PLC in a number of places, such as on the factory floor. In this chapter you are going to study PLCs from two different perspectives. First, you will see how the PLC is programmed to perform logic functions that control machines much the same way that motor controls do. Second, you will see how simple it is to troubleshoot large machine controls and automation with a programmable controller. You will see how to use the status indicators on PLC modules to help you troubleshoot systems. Finally, you will learn to wire switches to PLC inputs and motor starter coils to PLC outputs.

Today in industry you may find PLCs as stand-alone controls or as part of a complex computer-integrated manufacturing (CIM) system. In these large integrated manufacturing systems the PLC will control individual machines or groups of machines. They may also provide the interface between machines and robots, or machines and color graphics systems, called man–machine interfaces (MMI).

In the 1970s and 1980s companies were able to hire both an electrician and an electronics technician to install, interface, program, and repair PLCs. Since the late 1980s the number of PLCs has grown so large that many companies have found that it is better to hire one individual as a maintenance technician to make minor programming changes and troubleshoot. If you have never heard of PLCs or have had only a minor introduction to them, this chapter will provide you with all of the information necessary to work successfully with them. Several typical types of PLCs are shown in Figure 18–1. This chapter will use generic PLC addresses wherever possible. When specific applications are provided, the Allen–Bradley Micrologix, PLC5, and SLC500 will be used as example systems. The early examples in this chapter will not be specific to any brand of PLC so that you may understand the functions that are generic to all controllers.

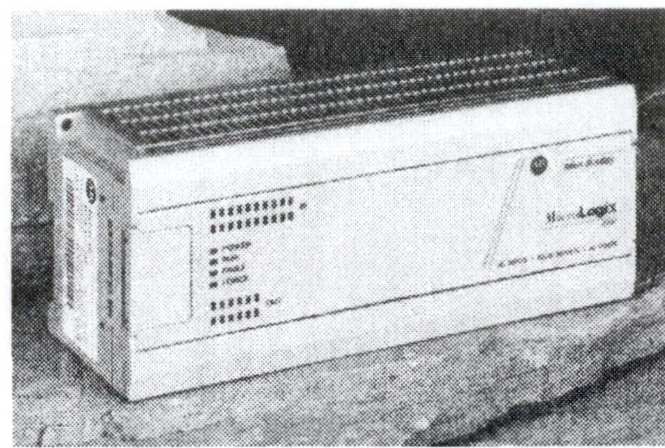

Figure 18–1 Typical programmable controller.
(Courtesy of Rockwell Automation)

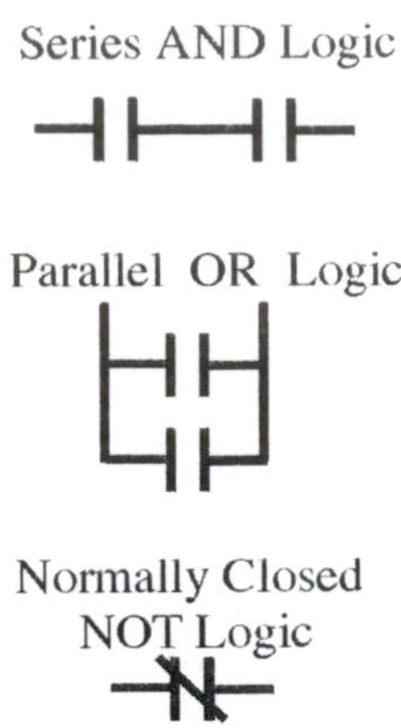

Figure 18–2 Logic for AND, OR, and NOT functions

18.1 The Generic Programmable Logic Controller

The programmable logic controller is a computer that is designed to solve logic (AND, OR, NOT) that specifically controls industrial devices, such as motors and switches, and allows other control devices of varied voltages to be easily interfaced to provide simple or complex machine control. Figure 18–2 shows examples of AND, OR, and NOT logic. Switches that are connected in series are called AND logic in a PLC, and switches that are connected in parallel are called OR logic. If closed contacts are used, they are called NOT logic. At this point do not let the term "logic" confuse you. Logic is the word that describes the action that switches and contacts will perform when they are wired in series or parallel.

The PLC is designed to have industrial-type switches, such as 110-V pushbutton and 110-V limit switches, connected to it through an optically isolated interface called an input module. The PLC can also control 480-V three-phase motors through 120-V motor starter coils with a similar interface called an output module. The computer part of the PLC, called a central processing unit (CPU), allows a program to be entered into its memory that will represent the logic functions. The program in the PLC is not a normal computer programming language like BASIC. Instead, the PLC program uses contact and coil symbols to indicate which switches should control which outputs. These symbols look very similar to the typical relay ladder diagram that was discussed in Chapter 4. You can view the program through a hand-held terminal or the program can be displayed on a computer screen where the action of switches and outputs are animated as they turn motor starter coils and other outputs on or off. The animation includes highlighting the input and output symbols in the program when they are energized.

In short, the PLC executes logic programs like the hardwired circuits you used in Chapter 4. The PLC was designed to provide the technology of logic control with the simplicity of an electrical diagram so that its inputs and outputs are easy to troubleshoot. Since the logic is programmable, it is easily changed. As you know, it may take several hours to make a change in a hardwired circuit, such as adding a limit switch. In a PLC, you will be able to make changes to the operation of a machine in a matter of minutes. The logic can also be easily stored on a disk so that it can be loaded into a PLC on the factory floor any time the program logic for a controller needs to be changed as different parts are made or when the machine function needs to be changed.

Today, in most factories, it is important for an expensive machine to be able to make more than one part or different sizes or styles of the same part. This requires the machine to have a different set of switches. The PLC is designed to make these changes as simple as possible.

When PLCs were first introduced, they allowed the factory electrician to use them to help troubleshoot large automated systems. PLCs have evolved and now provide many of the more complex logic functions, such as timing, up/down counting, shift registers, and first-in, first-out (FIFO) functions.

Figure 18–3 shows an example of a PLC program. The program looks exactly like an electrical relay diagram. As the program (diagram) gets larger, additional lines will be added. This gives the program the appearance of a wooden ladder that has multiple rungs. For this reason the program is called a ladder diagram or it may be referred to as ladder logic.

The original function of the PLC was to provide a substitute for the large number of electromechanical relays that were used in industrial control circuits. Early automation used large numbers of relays, which were difficult to troubleshoot. The operation of the relays

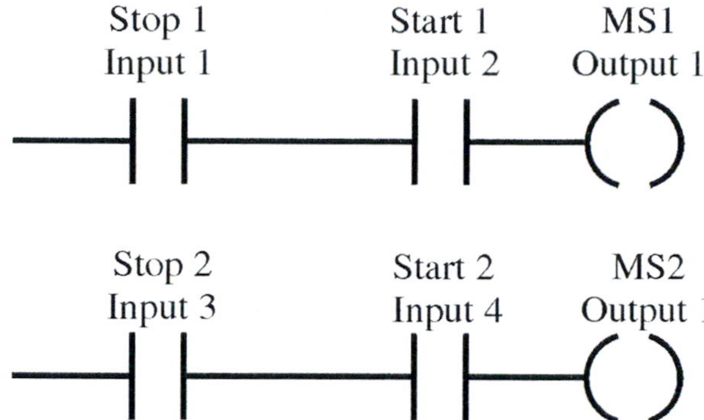

Figure 18–3 Example of typical programmable controller program that is called ladder logic. This program uses generic address for input 1, input 2, and output 1.

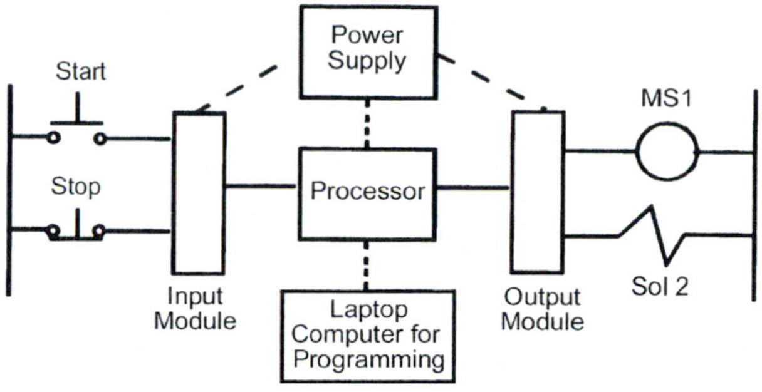

Figure 18–4 Block diagram showing the four major parts of a programmable controller. The programming panel is the fifth part of the system 1, but it is not considered a basic part of the PLC, because it can be disconnected when it is not needed.

was slow and they were expensive to rewire when changes were needed in the control circuit. In 1969 the automotive industry designed a specification for a reprogrammable electronic controller that could replace the relays. This original PLC was actually a sequencer device that executed each line of the ladder diagram in a precise sequence. Later in the 1970s the PLC evolved into the microprocessor controller used today. Several major companies, such as Allen–Bradley, Texas Instruments, and Modicon, produced the earliest versions. These early PLCs allowed a program to be written and stored in memory, and when changes were required to the control circuit, changes were made to the program rather than changing the electrical wiring. It also became feasible for the first time to design one machine to do multiple tasks by simply changing the program in its controller. Today, the major controllers are Allen–Bradley, Modicon, GE, Siemens (formerly Texas Instruments), and Square D.

18.2 Basic Parts of a Simple Programmable Controller

All PLCs, regardless of brand name, have four basic major parts: the power supply, the processor, input modules, and output modules. A fifth part, a programming device, is not considered a basic part of the PLC, since the program can be written on a laptop computer and downloaded to the PLC or the program can be written to an EPROM (erasable programmable read only memory) chip. Figure 18–4 shows a block diagram of the typical PLC. You can see that input devices such as pushbutton switches are wired directly to the input module, and the coil of the motor starter or solenoids are connected directly to the output module. Before the PLC was invented, each switch would be directly connected (hardwired) to the coil of the motor starter or solenoid it was controlling. In the PLC, the switches and output devices are connected to the modules and the program in the PLC will determine which switch will control which output. In this way, the physical electrical wiring needs to be connected only one time during the installation process, and the control circuitry can be changed unlimited times through simple changes to the ladder logic program.

18.3 The Programming Device

At the bottom of the diagram in Figure 18–4 you can also see a programming device. A programming device is necessary to program the PLC, but it is not considered one of the parts of a PLC because it can be disconnected after the program is loaded and the PLC will run by itself. The programming device is used by technicians to make changes to the program, to troubleshoot the inputs and outputs by viewing the status of contacts and coils to see if they are energized or deenergized, and to save programs to disk or load programs from disk.

The programming device in smaller PLCs is usually a hand-held programmer, and in larger systems it is a laptop computer with PLC programming software loaded on it. The ladder logic program is able to be displayed on the programming device where it can become animated. This feature is unique to the PLC and it helps the technician troubleshoot very large logic circuits that control complex equipment. When a switch that is connected to the PLC input module is turned on, the program display can show this on the screen by highlighting the switch symbol everywhere it shows up in the program. The display can also highlight each output when it becomes energized by the ladder logic program. When the on/off state of a switch on the display is compared with the electrical state of the switch in the real world, you can quickly decide where the problem exists.

Another feature that makes the PLC so desirable is that it has a wide variety of logic functions. This means that functions such as timing and counting can be executed by the PLC rather than using electromechanical or electronic timers and counters. The advantage to using timer and counter functions in the PLC is that they do not involve any electronic parts that could fail, and their preset times and counts are easily altered through changes in the program. Additional PLC instructions provide a complete set of mathematical functions as well as manipulation of variables in memory files. Modern PLCs such as the Allen–Bradley PLC-5 have the computing power of many small mainframe computers as well as the ability to control several hundred inputs and outputs.

18.4 An Example Programmable Controller Application

It may be easier to understand the operation of a PLC when you see it used in a simple application of sorting boxes according to height as they move along a conveyor. The address numbering schemes for inputs and outputs of some PLCs may be difficult to understand when you are first exposed to them. For this reason, the first examples of the PLC presented in this section will use generic numbering that does not represent any particular brand name. This will allow the basic functions of the PLC that are used by all brand names of PLCs to be examined in a simplified manner. Later sections of this chapter will go into specific numbering systems for several major brands of PLCs.

Figure 18–5 shows a top view of the conveyor sorting system. The location of the photoelectric switches that detect the boxes and the pneumatic cylinders used to push the boxes off of the conveyor are also shown. Each air cylinder is controlled by a separate solenoid. The conveyor is powered by a three-phase electrical motor that is connected to a motor starter. Start and stop pushbuttons are used with the photoelectric switches to energize the coil of the motor starter, which turns the conveyor motor on and off. Figure 18–6 shows all of the switches connected to an input module and solenoids and the motor starter connected to an output module. The inputs are generically numbered input 1, input 2, and so on. The outputs are numbered output 10, output 11, and so on. The ladder logic program that determines the operation of the system is shown in Figure 18–7. In

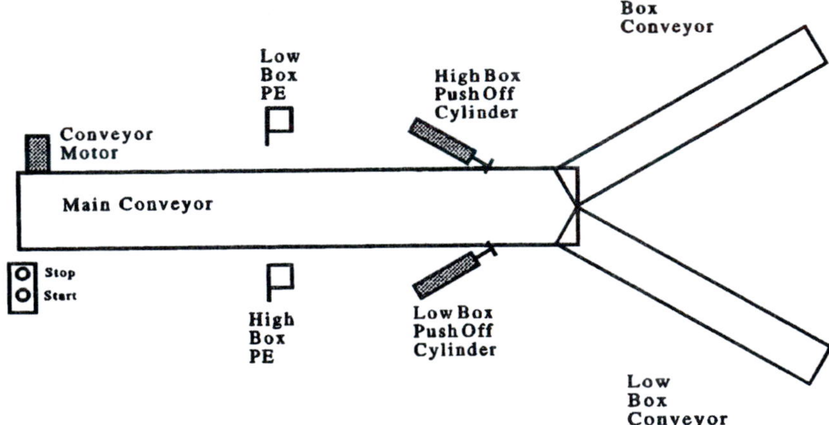

Figure 18–5 Diagram of the overhead view of a conveyor sorting system that is controlled by a generic programmable controller.

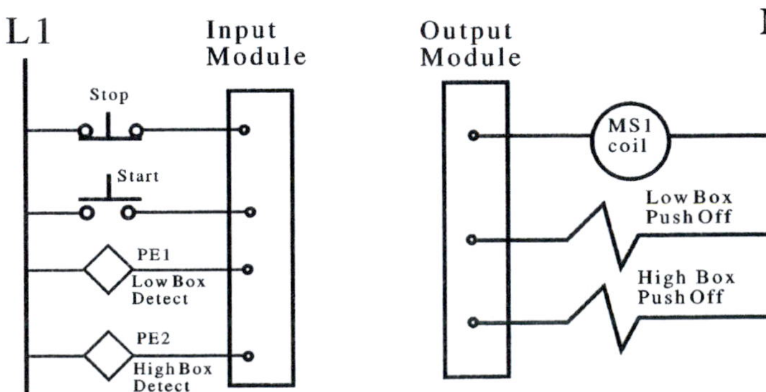

Figure 18–6 Input and output diagram that shows switches that are connected to the PLC input module, and the coil of a motor starter connected to the PLC output module.

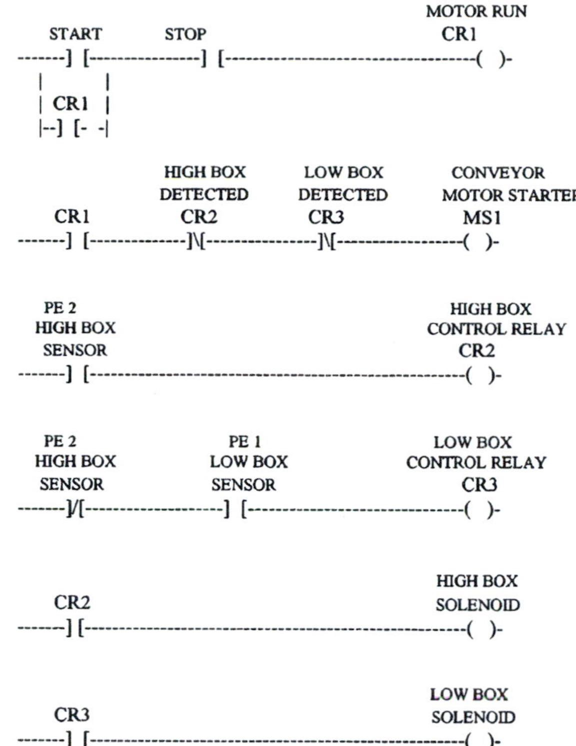

Figure 18–7 Ladder logic diagram showing the logic control for the conveyor sorting system. The diagram is generic and does not represent any particular PLC brand. Notice that all switches are represented by contact symbol and all outputs.

the diagram, the standard electrical symbol is used for each switch and output device.

In the ladder diagram you will see that each switch is represented as a set of open contacts or closed contacts (—] [—or—]\[—). Each motor starter or solenoid in the hardware diagram is represented by an output symbol —( )—. The same numbers that are assigned to the switches, motor starters, and solenoids in the hardware diagram are used in the logic diagram so that you can determine which switch is being used.

18.5 Operation of the Conveyor Sorting System

When the conveyor system is ready for operation, the start pushbutton must be depressed. Since the stop pushbutton is wired as a normally closed switch, power will pass through its contacts in the program to the start pushbutton contacts. When the start pushbutton is depressed, the "motor run" condition will be satisfied in the first rung of the program and the "motor run" control relay in the program will be energized. (The PLC can have hundreds of control relays in its program like the "motor run" relay. The advantage of having control relays in a PLC program is that it provides the logic functions of a control relay, which can have multiple sets of normally open and normally closed contacts, yet they exist only inside the PLC, so they can never fail or break.) A set of normally open contacts from CR1 are connected in parallel with the start pushbutton to seal it in because the start pushbutton is only momentarily closed. The photoelectric switches are used to energize and deenergize the conveyor motor starter in the second line of the logic. Line 2 of the diagram is designed so the start/stop circuit can turn the motor starter on or off without affecting the start/stop circuit. The photoelectric switches send out a beam of light that is focused on a reflector that is mounted on the opposite side of the conveyor. As long as a box does not interrupt the beam of light, the photoelectric switches will be energized.

When the motor run control relay coil in the first rung of the program is energized, its normally open contacts in the second rung of the program will become "logically" closed. Since no boxes are being sensed by the photoelectric switches, they will be energized and their contacts in the program, low box PE and high box PE, will pass power to the output in the second rung, called the motor starter, which directly controls the coil of motor starter 1. When motor starter 1 is energized, the conveyor motor is energized and the conveyor belt begins to move.

As boxes are placed on the conveyor, they will travel past the photoelectric switches. If the box is a low box, photoelectric switch PE1, which is mounted to detect the low box, will become energized. This will cause three things to happen. First, the normally open contacts of the PE1 low box sensor in rung 4 will close. Since the box is not a high box, the normally closed contacts of the PE2 high box sensor in rung 4 will remain closed to pass power to the PE1 low box sensor contacts (that are now closed) and on to the coil of CR3 (low box control relay). Second, the coil of CR3 is energized and all of the contacts in the program that are identified as CR3 will change state. This means the normally closed CR3 contacts in rung 2 will open and deenergize MS1, the conveyor motor starter. When MS1 is deenergized, the motor starter will become deenergized and the conveyor motor will stop. The third event to happen in this program when the coil of CR3 is energized by the PE1 low box sensor is that the output for the low box solenoid is energized, the rod of the low box push-off cylinder is extended, and the low box is pushed off of the main conveyor onto the low box conveyor.

After the low box is pushed off the main conveyor, it will no longer activate PE1, and the coil of CR3 will become deenergized in rung 3. When the coil of CR3 is deenergized, the normally closed CR3 contacts in rung 2 will return to their closed state, which will energize the output for MS1 and start the conveyor motor again.

When a high box moves into position on the conveyor, it will block both the high box photoelectric switch PE2 and the low box photoelectric switch PE1. This could create a problem of having both the low box control relay CR3 and the high box control relay CR2 energized at the same time. The logic in rung 4 is specifically designed to prevent this. Since the contacts of the high box sensor PE2 are programmed normally closed in rung 4, they will open when the high box is detected and not allow power to pass to the coil of CR3. This type of logic is called an exclusion, since it prevents both control relay coils from energizing at the same time even though both photoelectric switches are activated by the high box.

When the high box is detected by PE2, the normally closed PE2 contacts in rung 3 will close and energize the coil of the high box control relay CR2. When the coil of CR2 is energized, all of the contacts in the program that are identified as CR2 will change state. The normally closed CR2 contacts in rung 2 will open and deenergize the coil of MS1 in the conveyor motor starter. The normally open CR2 contacts in rung 5 will close at this time and energize the high box solenoid, air will be directed to the high box pneumatic cylinder, and the high box will be pushed off of the conveyor.

After the box has been pushed off of the conveyor, the photoelectric switch that detected the high box will return to its normal state, high box contacts in rung 2 will return to their normally closed logic state, and the motor starter will become energized again and start the conveyor motor. The high box contacts in rung 3 will return to their normally open logic state and the push-off cylinder will be retracted. This allows the conveyor to return to its normal operating condition.

18.6 How the PLC Actually Works

When the PLC is in the run mode its processor examines its program line by line, which is the way it solves its logic. The processor in the PLC actually performs several additional functions when it is in the run mode, such as reading the status of all inputs, solving logic, and writing the results of the logic to the outputs. When the processor is performing all of these functions it is said to be scanning its program. Figure 18–8 shows an example of the program scan.

When the PLC checks its inputs it only needs to determine if they are energized or deenergized. When the contacts of a switch are closed they allow voltage to be sent to the input module circuit. The electronic circuit in the input module uses a light-emitting diode and a phototransistor to take the 110-VAC signal and reduce it to the small voltage signal used by the processor.

The contacts in the program that represent the switch will change state in the program when the module receives power. If you use the programming panel to examine the contacts in the PLC program, they will highlight when the switch they represent is energized. The PLC will transition the state of the programmed contacts at this time. For example, if the contacts are programmed normally open, when the real switch closes and the module is energized, the PLC will cause the programmed contact to transition from open to closed. This means that the programmed contacts will appear to pass power through them to the next set of contacts. Since the contacts in line 1 of the program are in series, the output will become energized when both sets of contacts are closed and passing power.

Another important point to understand is that the term "set" is used to describe contacts or functions in the PLC that are energized, and the term "reset" is used to describe contacts or functions that are off.

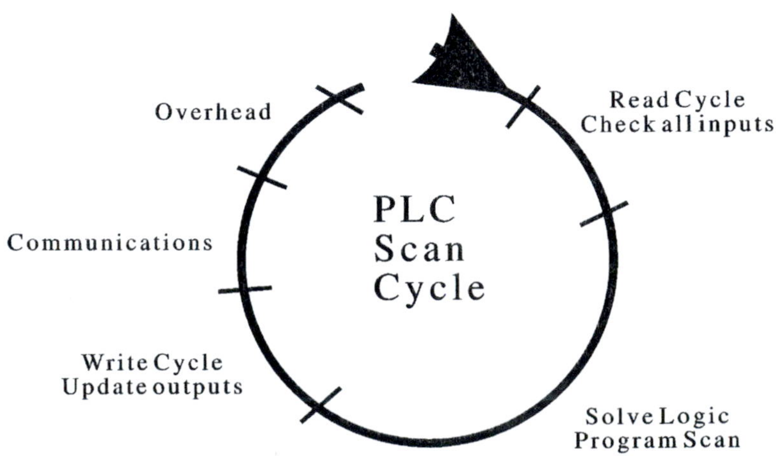

Figure 18–8 Scan cycle for a typical PLC.

18.7 Image Registers

One of the least understood parts of the PLC is the image register. The image register resides in the memory section that is specifically designed to keep track of the status (on or off) of each input. In some controllers this image register is also called an image table or an input data file. Figure 18–9 shows an example of the image register. All of the inputs and outputs for the PLC each have one memory location where the processor will place a 1 or 0 to indicate whether the input or output connected to that address is energized or deenergized. The processor determines the status of all inputs during each scan cycle by checking each one to determine if the switch connected to it is energized. If the switch is energized, and the input module sends a signal to the processor, the processor will place a binary 1 in the memory location in the image register that represents that input. If the switch is off, the processor will place a binary 0 in the memory location for the image register. The word "binary" is derived from the word meaning two, because each input switch can have only one of two states, on or off. The PLC will represent a number of things in its memory with 1's or 0's.

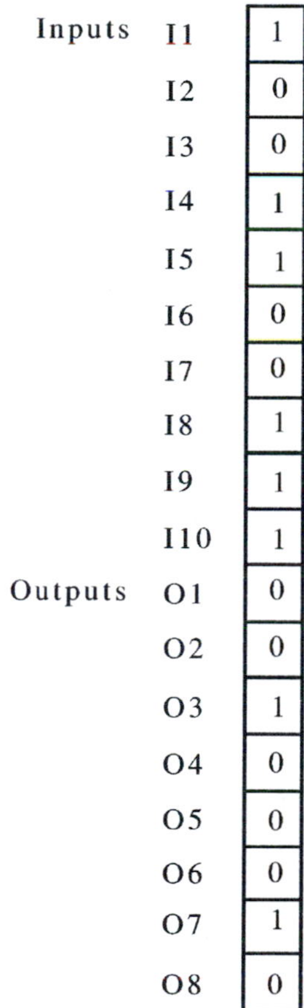

Figure 18–9 Example of an image register for a PLC. The image register is 1 bit wide and has a memory location to store a "1" if the bit is on, and a "0" if the bit is off.

Another important word is also derived from the word binary. This word is "bit," and it comes from the combination of the word "binary" and the word "digit." A bit is any value in the computer that is restricted to the value of 1 or 0.

Now that you have a better idea of what a bit is, you can understand how easy it is for the PLC to use the image register to keep an exact copy of the status of all its inputs during the read part of the scan. The image register is important to you as a troubleshooter since it will tell you if the PLC is receiving voltage from an input switch, or if the PLC is sending power to an output device like a motor starter coil. In a hardwired control system you would have to use a voltmeter to test all switches to tell which ones are on or off. In the PLC, you can view the image register and quickly see the status of all switches at one time: Those with a 1 in the image register are on, and those with a 0 are off.

In the second part of the PLC scan the processor uses the values in the image register to solve the logic for that particular rung of logic. For example, if input 1 is ANDed with input 2 to turn on output 12, the processor looks at the image register address for input 1 and input 2 and solves the AND logic. If both switches are high, and a 1 is stored in each of the image register addresses, the processor will solve the logic and determine that output 12 should be energized.

After the processor solves the logic and determines the output should be energized, the processor will move to the third part of the scan cycle and write a 1 in the image register address for output 12. This is called the write cycle of the processor scan. When the image register address for output 12 receives the 1, it immediately sends a signal to the output module address. This energizes the electronic circuit in the output module, which sends power to the device such as the motor starter that is connected to it.

The PLC processor updates all of the addresses in the image register during the read cycle even if only 1 or 2 switch addresses are used in the program. When the processor is in the solve logic part of the scan, it will solve the logic of each rung of the program from left to right. If a line of logic has several contacts that are ANDed with several more contacts that are connected in an OR condition, each of these logic functions are solved as the processor looks at the logic in the line from left to right. The processor also solves each line of logic in the program from top to bottom. This means that the results of logic in the first rung are determined before the processor moves down to solve the logic on the second rung. Since the processor continues the logic scan on a continual basis approximately every

10–20 milliseconds, each line of logic will appear to be solved at the same time. The only time the order of solving the logic becomes important is when the contacts from a control relay in line 1 are used to energize an output in line 2. In this case the processor will actually take two complete scan cycles to read the switch in line 1 and then write the output in line 2. In the first scan, the processor will read the image register and determine that the start pushbutton has closed. When the logic for rung 1 is solved, the control relay will be energized. The processor will not detect the change in the control relay contacts in the second line until the second scan when it reads the image register again. The output in the second line will finally be energized during the write cycle of the second scan and power will be sent to the motor starter coil that is connected to the output address. The processor will write the condition of every output in the processor during the write part of the scan even if only 1 or 2 outputs are used in the program.

The image register is also useful to the troubleshooting technician, since the contents of each address can be displayed on a hand-held programmer or CRT, and the technician can compare the image register with the status lights on the input and output modules and the actual switches and solenoids that are hardwired to the modules. If an input has voltage from a switch, its status light will be illuminated, and a 1 should show up in the image register location for the input. This gives the technician two places to tell if a switch is on or off without using a voltmeter.

A separate image register or data file is also maintained in the processor for the control relays. In some systems these relays are called memory coils or internal coils. A control relay has one coil and one or more sets of contacts that can be programmed normally open or normally closed. When the coil of a control relay is energized, all of the contacts for the relay will change state. Since control relays reside only in the program of the PLC and do not use any hardware, they cannot have problems such as dirty contacts or an open circuit in their coil. This means that if the contacts of a control relay will not change from open to closed, you would not suspect the coil of having a malfunction; instead, you would look at the contacts that control power to the control relay coil in the program, and one or more of them will be open, which causes the coil to remain deenergized.

When the processor is in the run mode, it updates its input, output, and control relay image registers during the write-I/O part of every scan cycle. Since the image registers are continually updated approximately every 20 to 40 milliseconds, it is possible to send copies of them to printers or color graphic terminals at any time to indicate the status (on or off) of critical switches for the machine process. This allows the machine operator to do some minor troubleshooting when a problem occurs, before a technician is called. For example, if an operator is running a press or machine that must have a door or gate closed before it will begin its cycle, a fault monitor on a color graphic display can show the status of the door switch. If the door is ajar and the cycle will not start, the operator can look at the fault monitor and see from the copy of the input image register from the PLC that the door switch is open. The operator could then take the appropriate action and close the door instead of calling a technician. In this manner, the PLC not only controls the machine operation, but also helps in troubleshooting problems.

18.8 The Run Mode and the Program Mode

When the PLC is in the run mode, it is continually executing its scan cycle. This means that it monitors its inputs, solves its logic, and updates its outputs. When the PLC is in the program mode, it does not execute its scan cycle. Perhaps a better name for the program mode is the not run mode or the no scan mode. The name "program mode" was originally given to this mode because older PLCs could not have their programming changed while they were executing their scan cycle. Modern PLCs have the ability to have their program edited and changed while the processor is in the run mode. At first glance, changing the PLC program while it is in the run mode looks dangerous, especially when the machine the PLC is controlling is in automatic operation. But you will find in larger control applications, such as pouring glass continually or continuous steel rolling mills, it is not practical to stop the machine process to make minor program changes such as the changes to a timer. In these types of applications the changes are made while the PLC is in the run mode. As a rule, you should switch the PLC to the program mode if possible when any program changes are being made.

18.9 On-Line and Off-Line Programming

The programming software for a PLC may allow you to write the ladder logic program in a personal computer, and later download the program from the personal computer to the PLC. If you are connected directly to a PLC and you are writing the ladder logic program in the PLC memory, it is called on-line programming. If you are writing the ladder logic program in a personal computer or programming panel that is not connected to a PLC, it is called off-line programming. Most modern PLC programming software allows the program to be written on a personal computer or laptop computer without being connected to the PLC. This allows new programs or program changes to be written at a location away from the PLC. For example, you may write a program in Detroit at an office, save the changes on a floppy disk, and mail the disk to a company in Chicago where the PLC is connected to automated machinery.

If you are using a hand-held programmer, it must be connected directly to a PLC so that you are writing the program in the PLC memory. Some small PLCs like the Allen–Bradley SLC100 do not have a microprocessor chip, so you cannot write the PLC program on-line to the processor with a laptop computer. Since the SLC100 uses a sequencer chip instead of a microprocessor chip, all programming must be completed off line in the laptop, and then you must download the program changes to the PLC memory. An alternative method would be to use a hand-held programmer to program the PLC on line.

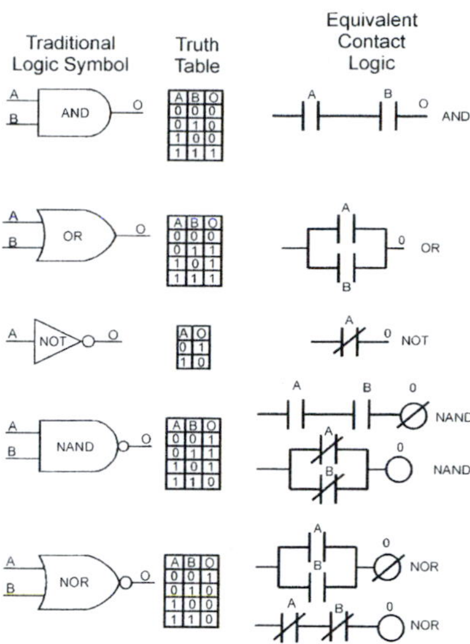

Figure 18–10 Traditional logic symbols for AND, OR, NOT, NAND, and NOR functions.

18.10 Features of the Programmable Controller

The examples in Figures 18-5 through 18-7 show several unique features of the PLC. First, the ladder diagram allows you to analyze the program in the sequence that it will occur. Second, the PLC allows you to program more than one set of contacts in the program for each switch. For example, the photoelectric switch may be in the program both normally open and normally closed and placed in the program as many times as the logic requires. The PLC gives you the capability of solving AND, OR, and NOT logic by connecting contacts in series or parallel, and using normally open and normally closed contacts as needed. Figure 18–10 shows examples of the five basic logic symbols: AND, OR, NOT, NAND, and NOR. This figure also shows the truth tables for these logic functions and the equivalent contact arrangements that provide the same function as you would see them in ladder logic. Notice that the NAND and NOR functions can be provided with either series contacts or parallel contacts. As complex as some ladder logic programs look, the logic program can consist only of combinations of these five basic functions.

Another feature of the PLC is the control relay or internal coils or memory relays. As stated in the discussion of image registers, the control relay acts like the control relays that were used in relay panels years ago. The relay has one coil and it can have any number of normally open or normally closed contacts. When the coil is activated, all the contacts in the program that have the same address number as the coil will change state. The control relay exists only inside the PLC, which means it can never break or malfunction physically. Also, since control relays exist only in the PLC memory, you can have as many of them as the memory size of the PLC can accommodate. This means you can add control relays to a PLC program to accommodate changes to the machine control logic as long as the PLC has space remaining in its memory.

It may help to see a case where a control relay is used in an actual circuit to better understand its function. For example in Figure 18–7, you can see that the low box detected control relay (CR3) is used to indicate that the logic has determined that a low box has been detected. The control relay in this rung of the program does nothing more than "represent" the condition that indicates that a low box has been detected. This condition exists when low box PE1 is on and the high box PE2 is off. Any time CR3 is energized, a low box has been detected, and it will change the state of all the contacts that have its number throughout the remainder of the program, to show that the low box has been detected.

In many cases, each control relay in a program represents a condition for the machine operation, such as "ready to go to auto," "all safety doors closed," or "ready to jog." If you think of the condition the control relay represents each time you see the control relay contacts, you will find it easier to read and interpret the program.

Another feature of the PLC that makes complex factory automation easier to troubleshoot is the way that each input switch or each output load is electrically connected to its interface module. In Figure 18–6 you can see that when a switch is connected to an input module, only two wires are involved. The outputs also use only two wires to connect the load device like the coil of a motor starter to the module. This means that when there is a malfunction in the electrical part of the circuit, the troubleshooter has only the switch and two wires to check for any individual input and the coil and two wires for any output. When you compare this type of wiring to a complex series parallel circuit in an older style relay panel, you can see it is much easier to troubleshoot two wires and a switch, or two wires and a solenoid. In the older style relay panels, a typical signal may pass through 10 to 15 relay contacts and switches and all the wiring that connects them together before it gets to a solenoid. This means that an open could occur in this complex circuit when any set of contacts open, when any switch opens, or if a wire has a fault where it becomes open.

To make troubleshooting the switch and two wires that connect it to an input module even easier, a small indicator light called a status indicator is connected to each input circuit and it is mounted on the face of the module so it is easy to see. The troubleshooter can look at the status indicator and quickly determine if power is passing through the contacts or not. If the switch contacts are closed, but the status indicator is not illuminated, the troubleshooter needs to check only the two wires and the switch for that circuit to see where the problem is. This feature is particularly valuable when the automated system the PLC is controlling has several hundred switches that could cause a problem. A status indicator is also connected to each output circuit and mounted on the face of the output module. Any time the PLC energizes an output, the status indicator for that circuit will be illuminated.

18.11 Classifications of Programmable Controllers

PLCs are generally classified by size. A small system costs around \$500 to \$1000 and will have a limited number of inputs and outputs. As a general rule, the small PLCs have less than 100 inputs and outputs with approximately 12 inputs and 8 outputs mounted locally with the processor. In some systems additional inputs and outputs can be added through remote I/O racks to accommodate the remaining inputs and outputs. In other systems, the number of inputs and outputs are limited, but several small PLCs can be networked together to provide a larger number of I/O that are controlled together. The smaller PLCs generally have 2K to 10K of memory that can be used to store the user's logic program.

Medium-size PLCs cost between \$1000 and \$3000 and have extended instruction sets that include math functions, file functions, and PID process control. These PLCs are capable of having up to 4000 to 8000 inputs and outputs. They also support a wide variety of specialty modules, such as ASCII communication modules, BASIC programming modules, 16-bit multiplexing modules, analog input and output modules that allow interfaces to both analog voltages and currents, and communication modules or ports that allow the PLC to be connected to a local area network (LAN).

Large PLCs were very popular before networks were perfected. The concept for the large PLC was to provide enough user memory and input and output modules to control a complete factory. Problems occurred when minor failures in the system brought the complete factory to a halt.

The advent of local area networks brought about the concept known as distributive control, where small and medium-size PLCs are connected together through the network. In this way, the entire factory is brought under the control of a number of PLCs, but a failure in one system will not disturb any other system. The majority of PLCs that you will encounter today will be small and medium-size systems that are used as stand-alone controls or as part of a network.

18.12 Operation of Programmable Controllers

Now that you have a better understanding how the basic PLC operates you are ready to understand some of the more complex operations. You may need to review several of the previous diagrams to help you get a better understanding of these functions. Back in Figure 18–5 you can see that the PLC has a processor, input modules, output modules, and a power supply. The processor has several sections, including resident memory that contains the instruction set, user memory reserved for programs, and addressable data memory that has room to store timer, counter, and other variables. The variables may be stored as binary data, as an integer (whole numbers), or as a floating point (numbers that have decimal points and can be expressed with exponents).

When the processor is in the run mode, it will continually scan the inputs to see if they are on or off and then update the input image register. Next it will sequentially solve the ladder logic that is stored in its user program area. It completes this part of its cycle by reading the ladder diagram. The processor goes through a fetch cycle to get each address found in the ladder diagram and puts it into the logic solving area. Each address is compared to the value found in its image register address and this value is used to solve the logic. During the next part of the cycle, the processor writes the results of the logic to the output part of the image register. If the output in the ladder logic has contacts assigned to it, like the run CR1 in the second rung of the diagram in Figure 18–7, the contacts will reflect the status of the output image register address when the processor solves that line of logic. If the output address is connected to the output module, the signal is sent to the module to ensure that the device connected to it is energized.

The final section of each processor scan cycle is called overhead. During the overhead part of the scan, the processor does all of its work outside of solving logic, such as updating the computer screen (monitor) if one is connected, sending messages to its printer, and interacting with the network. Since the main purpose of the PLC is to read inputs, solve logic, and write to its outputs, less than 10% of the scan cycle is devoted to the overhead. The amount of time for the typical scan cycle for a generic PLC is approximately 20–60 milliseconds. Newer enhancements to the PLCs try to lower the scan cycle to less than 10 milliseconds.

18.13 Addresses for Inputs and Outputs for the MicroLogix PLC

Up to this point in our discussion we have identified inputs and outputs in generic terms by name only. In reality the processor must keep track of each input and output by an address or number. Each brand name of PLC has devised its own numbering and addressing scheme. These addressing schemes generally fall into one of two categories: sequentially numbering each input and output, or numbering according to location of input and output modules in the rack. The sequential numbering system is used in most small PLCs and about half of the medium-size systems. Allen–Bradley, which is one of the most widely used brands, uses two different addressing systems. In its medium-size PLCs (PLC2, PLC5, and SLC500) the address for an input or output uses numbers that indicate the location of the input or output module in the rack. In the smaller PLCs like the MicroLogix a sequential numbering system for addresses is used.

From this point on, the applications will use the address and numbering scheme of the Allen–Bradley MicroLogix. If you are using a different brand name of PLC you will need to convert the addresses to your system if you wish to try the program examples. Figure 18–11 shows an example of the numbering system used in the Allen–Bradley MicroLogix small PLC. You can see that the input addresses start with address I0 (I stands for input and zero is the first address) and continue through address I11. Output addresses start with address O0 (O stands for output and zero is the first address) and continue through address 07. The first group of I/O are mounted locally on the processor. Each I/O address refers to a specific circuit in the hardware module, and to a specific address in the processor's memory. The addresses in memory are assigned to memory blocks. The diagram that displays all of the addresses in a memory block is called a memory map.

The Micrologix PLC is available in three sizes: a unit that has 10 inputs and 6 outputs, one that has 12 inputs and 8 outputs, and one that has 20 inputs and 12 outputs. At this time the Micrologix does not support expansion modules; instead, it provides a means to network multiple PLCs so that they can provide common control to a larger group of inputs and outputs.

The Micrologix also provides up to 512 internal relays. As you know, the internal relays act like a control relay in that they perform logic, but do not directly energize an output.

Inputs	*Outputs*
I0	O0
I1	O1
I2	O2
I3	O3
I4	O4
I5	O5
I6	O6
I7	O7
I8	O8
I9	09
I10	O10
I11	O11
I12	
I13	
I14	
I15	
I16	
I17	
I18	
I19	

Figure 18–11 Input and output addresses for the Allen–Bradley MicroLogix PLC.

You can view the entire User Manual for the Allen–Bradley MicroLogix PLC by visiting the Allen–Bradley site on the Internet at *http://www.ab.com/* and looking at Publications On Line and Manuals On Line.

18.14 Example Input and Output Instructions for the MicroLogix

Figure 18–12 shows an example of all of the input and output instructions the MicroLogix uses. From this figure you can see that normally open and normally closed contacts are provided for input instructions under a number of different names to indicate if they are connected in series or parallel in the circuit. The first column in this figure shows a symbol for each instruction. The symbol shows the way the instruction looks when it is displayed in a program as you would see it with a hand-held terminal display, or when the program is printed out. The second column shows the mnemonic (pronounced ni-mon-ic) for each instruction. A mnemonic is the code that you will enter in the hand-held programmer and it will be similar to an abbreviation for each instruction. The function code is the code you can use with the mnemonic. The fourth column is the name of the instruction. Notice that the name of the

Basic Instructions

HHP Display	Mnemonic	Code	Name	Purpose
	LD	20	Load	Examines a bit for an ON condition. It is the first normally open instruction in a rung or block
	LDI	21	Load Inverted	Examines a bit for an Off condition. It is the first normally closed instruction in a rung or block.
	AND	22	And	Examines a bit for an On condition. It is a normally open instruction placed in series with any previous input instruction in the current rung or block..
	ANI	23	And Inverted	Examines a bit for an Off condition. It is a normally closed instruction places in series with any previous input instruction in the current rung or block.
	OR	24	Or	Examines a bit for an On condition. It is normally open instruction placed in parallel with any previous input instruction in the current rung or block.
	ORI	25	Or Inverted	Examines a bit for an Off condition. It is a normally closed instruction placed in parallel with any previous input instruction in the current rung or block.
LDT	LDT	26	Load True	Represents a short as the first instruction in a rung or block.
ORT	ORT	27	Or True	Represents a short in parallel with the previous instruction in the current rung or block.
OSR	LD OSR	28	One-Shot Risisng	Triggers a one time event
OSR	AND OSR	29	One-Shot Rising	Triggers a one time event
( )	OUT	40	Output (Output Energize)	Represents an output driven by some combination of input logic. Energized (1) when conditions preceding it pemit power continuity in the rung, and de-energized after the rung is false.
(L)	SET	41	Set (Output Latch)	Turns a bit on when the rung is executed, and this bit retains its state when the rung is not executed or a power cycle occurs.
(U)	RST	42	Reset (Output Unlatch)	Turns a bit off when the rung is executed, and this bit retains its state when the rung is not executed or when power cycle occurs.
	TON	0	Timer-On Delay	Counts timebase intervals when the instruction is true.
	TOF	1	Timer-Off Delay	Counts timebase intervals when the instruction is false.
	RTO	2	Retentive Timer	Counts timebase intervals when the instruction is true and retains the accumulated value when the instruction goes false or when power cycles occur.
	CTU	5	Count Up	Increments the accumulated value at each false-to-true transition and retains the accumulated value when the instruction goes false or when power cycle occurs.
	CTD	6	Count Down	Decrements the accumulated value at each false-to-true transition and retains the accumulated value when the instruction goes false or when power cycle occurs.
	RES	7	Reset	Resets the accumulated value and status bits of a timer or counter. Do not use with the TOF timers.

Figure 18–12 Example input, output, timer, and counter instructions available in the Allen–Bradley MicroLogix PLC.

instruction is very similar to the mnemonic. The last column in this table is the purpose of the instruction, which explains the function of each instruction.

From the table in this figure you can see that there are several different instructions for normally open and normally closed contacts. If the normally open contact is the first contact in the line of logic, it is identified by the term LOAD and has the mnemonic LD. The LD mnemonic is used to indicate that the contact is the start of a new rung. The AND instruction is used for all other contacts in the line of logic that are connected in series. If a contact is connected in parallel in the line of logic, it will be identified by the OR instruction and mnemonic.

If the contacts are normally closed, the term "inverted" will be added to the AND instruction, so that a normally closed set of contacts that is connected in series will have the mnemonic ANI and a normally closed contact that is connected in parallel will have the mnemonic ORI.

Sometimes Allen–Bradley uses the terms "examine on" for the normally open contacts —] [—, and "examine off" for the normally closed contacts —]\[—. The term "examine" was chosen as part of this name to remind you that the processor looks at (examines) the electronic circuit in the input module to see if it is energized (on) or deenergized (off). The processor will keep track of the status of the circuit by placing a 1 in the image register address if it is energized and a 0 if it is deenergized. The word "examine" is also used to remind you to look at the status light on the front of the module so that you can also determine if the input circuit is energized or deenergized. The branch instructions are used to allow contacts to be connected in parallel to each other in the ladder diagram.

The output instruction in this figure includes the traditional coil, which uses the symbol -( )-. This instruction is named Output or Output Energize. One specialized output instruction is called the Output latch —(L)— and it has a mnemonic called Set. Its companion is called the Output unlatch —(U)—,which has the mnemonic Reset. The latch output will maintain its energized state after it is energized by an input even if the input condition becomes deenergized. The unlatch output must be energized to toggle the output back to its reset (off) state. The latch coil and unlatch coil must have the same address to operate as a pair.

Latch coils are used in industrial applications where a condition should be maintained even after the input that energized the coil returns to its deenergized state. The latch coil bit will remain HI even if power to the system goes off and comes back on. This feature is important in several applications such as fault detection. Many times a system has a fault such as an overtemperature or overcurrent condition. After the fault is detected, the machine is automatically shut down by the logic. By the time the technician or operator gets to the machine, the condition may return to near normal conditions and it would be difficult to determine why the system shut down. If a latch coil is used in the detection logic, the fault bit will remain HI until the fault reset switch energizes the unlatch instruction. Another application for the latch coil is for feed water pumps or fans that are located in remote locations in the building, such as in the ceiling. A latch coil is used in these circuits because the motors must return to their energized condition automatically after a power loss. In some parts of the country, it is common for power to be lost for several seconds during severe lightning storms. If a regular type of coil is used in the program, a technician would have to go to each fan or pump and depress the start pushbutton to start the motor again. If a latch coil is used, the coil will remain in the state it was in when the power went off. This means that if the latch coil was energized and the motor was running when power goes off, the latch coil will remain energized when power is returned and the motor will begin to run again automatically. The latch coil is very useful in these circumstances, especially when a factory has dozens of fans and pumps, and it would take several hours to reset them all.

Other output instructions have a rather specific function that is beyond traditional relay logic. These instructions include the timer, counter, sequencer, and zone control. Each of these instructions will be covered in detail.

18.15 More About Mnemonics: Abbreviations for Instructions

In the previous figure you can see that each instruction has a mnemonic, which is an abbreviation for the function of the instruction. The mnemonic for each instruction is selected because it sounds like the name of the function that the instruction provides. It also is limited to two or three letters so that it is easy to program. The mnemonic is the letters the PLC's microprocessor recognizes as instructions. You will need to learn the mnemonic for each instruction so that you can read and write PLC programs. The mnemonics for different PLC brand names may be slightly different but they will always describe the instruction they represent.

18.16 Example of a Program with Inputs Connected in Series

Figure 18–13 shows a typical PLC program with the inputs and output identified. In this program the first

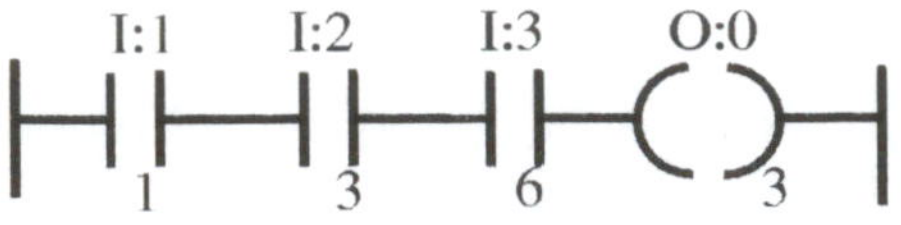

I:1 I:2 I:3 O:0

1 3 6 3

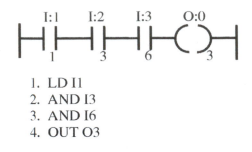

1. LD I1
2. AND I3
3. AND I6
4. OUT O3

Figure 18–13 Example of a PLC program with contacts in series using Allen–Bradley MicroLogix addressing.

input is number 1 and it is identified as I:1/1. The second input is I 3 and it is identified as I:1/3. The I indicates the contact is controlled by an input. The :1 indicates the processor keeps the status of this contact (on or off) in file number 1, and the /3 indicates this is input number 3. The output is identified in the program as O:0/3. The O:0 indicates this is an output whose status is stored in file 0, the /3 indicates this is output number 3. The format for numbering files is used because it is the format that Allen–Bradley uses in its other PLCs like the PLC5.

When you enter inputs with the hand-held programmer, you would simply enter I3 for the first input, I6 for the second input, and I6 for the third input. You would use O3 for the output. The PLC automatically does the remainder of the formatting, such as adding the file number.

Figure 18–14 shows the format Allen–Bradley uses for file types, identifiers, and numbers to store information about inputs, outputs, timers, and counters in the MicroLogix PLC.

File Type	*Identifier*	*Number*
Output	O	0
Input	I	1
Status	S	2
Bit	B	3
Timer	T	4
Counter	C	5
Control	R	6
Integer	N	7

Figure 18–14 File type indentifiers and numbers used in the Allen–Bradley MicroLogix PLC.

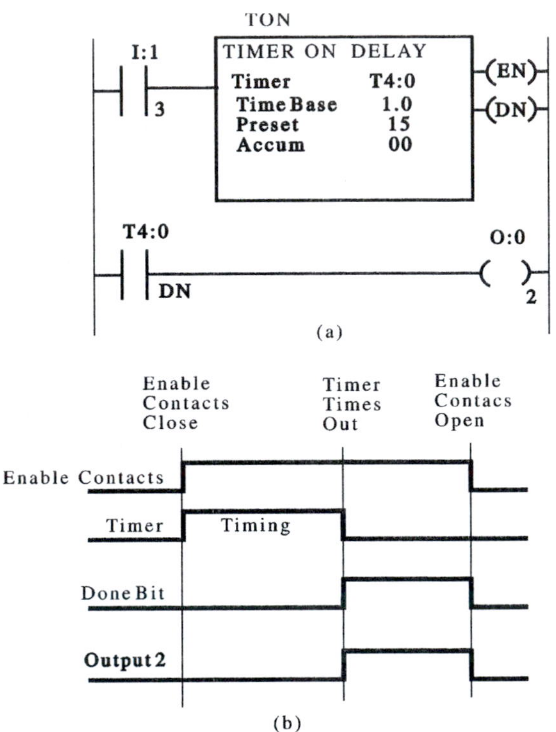

Figure 18–15 Example of a TON (delay-on) timer.

18.17 Using Timers in a PLC

Many industrial applications require time delay. The MicroLogix uses the mnemonic TON to represent the timer on-delay instruction. The timer will operate like the traditional hardware-type motor-driven timer that was discussed in Chapter 12. Figure 18–15 shows an example of a timer instruction used in a line of logic. The timers can use addresses T0–T39. Each timer has a preset value (PRE value) and accumulative value (ACC value). The preset value is the amount of time delay the timer will provide and the accumulative value is the actual time delay that has accumulated since the timer has been energized. The timer accumulative value increments (counts up) at a rate determined by the time base. The maximum preset value for timers in the MicroLogix is 32767.

The time base for the MicroLogix timers can be 1.0 or 0.01 second. When the timer accumulative value equals the preset value, the timer times out. When the timer times out it can cause a set of contacts to change state. These contacts need to have the same address as the timer and they should be identified with the mnemonic DN indicating they are controlled by the done bit for the timer.

The timer can control normally open or normally closed contacts with the done bit or two other status bits. Each of these bits is identified with a number as well as a mnemonic. The done bit is bit 13 and it is assigned the mnemonic DN. It will change state when the timer times out. The timer timing bit is bit 14 and it is assigned mnemonic TT. It will be set (ON) any time the timer is timing. The enable bit is bit 15 and it is assigned the mnemonic EN. It will be set (ON) any time the timer is enabled. You can use either the bit number or its mnemonic when you are entering a program. In Figure 18–15 the timer is timer T0, so it has the number T4:0 in the program. In the second line of the program you can see a set of contacts that are controlled by T4:0 DN, which is the done bit for this timer. When the timer times out, these contacts will close and energize output O:0/2.

Another important part of each timer is the reset function. When a timer is reset, its accumulative value is reset to 0. Each timer must be reset at some point to make it ready to be used for the next timing cycle. The TON timer is reset any time power provided to it is interrupted. In the example in Figure 18–15, contacts I:1/3 control power to the timer and any time these contacts are open, the timer will reset its accumulative value. Any time these contacts are closed, the timer is enabled and its accumulative value will increment (count up) at the rate determined by the time base, which in this example is a 1-second interval. This may also be described by saying that the timer's accumulative value will increase 1 second with every tick of the timer time base. The contacts I:1/3 in this example will be called enable/reset contacts since they control the enable and reset functions of the timer.

18.17.1 Retentive Timer

Another type of timer provided in the PLC is called the retentive timer and it has the mnemonic RTO. A retentive timer is different from the TON timer in that its accumulative value will not reset when the enable contacts are opened. In the retentive timer, the accumulative value is retained and when the enable contacts are closed again, the accumulative value in the timer will begin accumulating values again. This type of timer operates similar to an hour meter on a piece of equipment. For example, this type of timer can be enabled by a set of contacts that control an air compressor motor. If the contacts are closed and the motor running runs for 20 minutes, the timer will keep track of this time. When the contacts open and the motor stops, the timer also stops, but the accumulative value (20 minutes) does not reset. When the contacts close again and the motor runs an additional 30 minutes, the timer will show the total value of 50 minutes.

18.17.2 Resetting the RTO Timer

Since the enable contacts will not cause the RTO timer to reset, it will need a special reset instruction whose mnemonic is RES. Figure 18–16 shows an example of the RTO timer and the RES instruction that will cause its accumulative value to reset. In this example, the timer will add time to its accumulative value any time the enable contacts I:1/4 are closed. The timer will reset its accumulative value any time contacts I:1/5 are closed and pass power to the RES instruction.

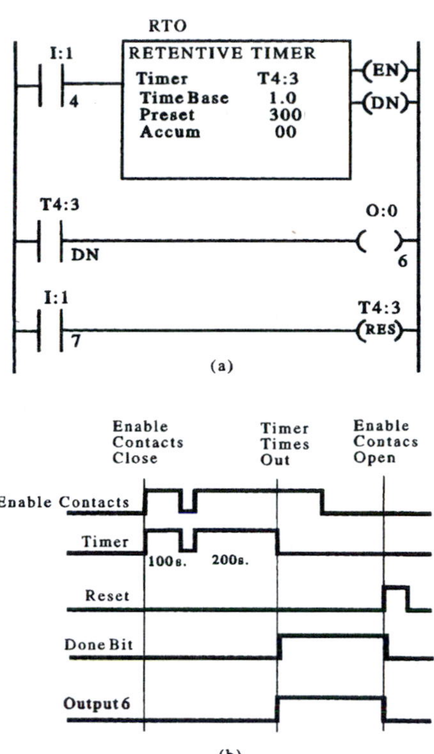

Figure 18–16 Example of an RTO (retentive) timer with a reset instruction.

18.17.3 The TON as an Automatic Resetting Timer

A TON timer can be made to automatically reset itself. The automatic reset part of this circuit is accomplished by using a TON timer instruction and placing a set of the normally closed timer contacts in series with it, and placing a set of normally open timer contacts in series with the timer reset instruction. Figure 18–17 shows an example of a timer programmed as an autoresetting timer with its contacts controlling a lubrication solenoid. In this example, the TON timer T4:3 instruction has a set of normally closed contacts controlled by the DN bit connected in series with it so that the timer begins to time any time the PLC is in the run mode. After 60 seconds it times out, and it will close the normally open T4:3/DN contacts in line 2 of the program to energize the lube solenoid. The normally closed T4:3/DN contacts that enable the timer will open and cause the timer to reset and start its timing cycle over again automatically. The lube system only needs to be solenoid pulsed to activate the lubrication system.

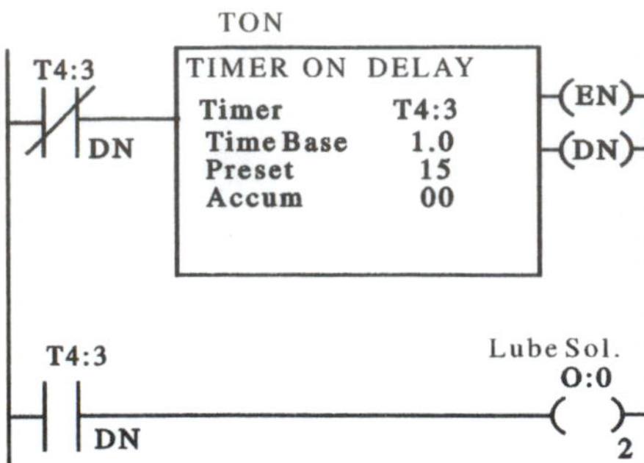

Figure 18–17 Example of a TON timer programmed as an automatic resetting timer that controls a lube solenoid.

18.17.4 Off-Delay Timers

The off-delay timer is used in applications where the amount of time delay is timed from the time a condition turns off. For example, in a heating system, at the end of a cycle, the fuel is turned off, and the fan remains running for several minutes to ensure that all of the residual heat in the heat exchanger is removed. The time-delay-off function of the heating system fan can be controlled by an off-delay timer. For the off-delay timer to operate, its enable contacts must be energized and then turned off. At the point where the enable contacts are turned off, the timer will begin its delay cycle, which will run until the timer's accumulative value becomes equal to the preset value.

The off-delay timer in the PLC is called timer-off-delay instruction and it has the mnemonic TOF. The TOF timer and the TON timer are similar in that they both can use any of the 40 timers (T0–T39), and they both have preset values, accumulative values, time base values, and status bits. An example of a TOF instruction is shown in the top diagram of Figure 18–18. The important difference between the TOF and TON timers is their timing cycle of when the done bit is on or off. The bottom diagram in Figure 18–18 shows an example of the timing cycle of the done bits (DN) for the on-delay timer and the off-delay timer. At the start of a cycle when the TON instruction is first energized by its enable contacts, the timer's done bit is off and remains off until the accumulative value becomes equal to the preset value and the timer times out. At that point, the done bit becomes energized and remains energized until the enable contacts to the timer are opened and the timer is reset. The sequence for the TOF timer shows that its done bit (DN) contacts are off until the enable contacts for the timer are closed. At that point, the done bit is energized and it remains energized until the timer cycle is complete. The TOF timer does not start its timing cycle until the enable

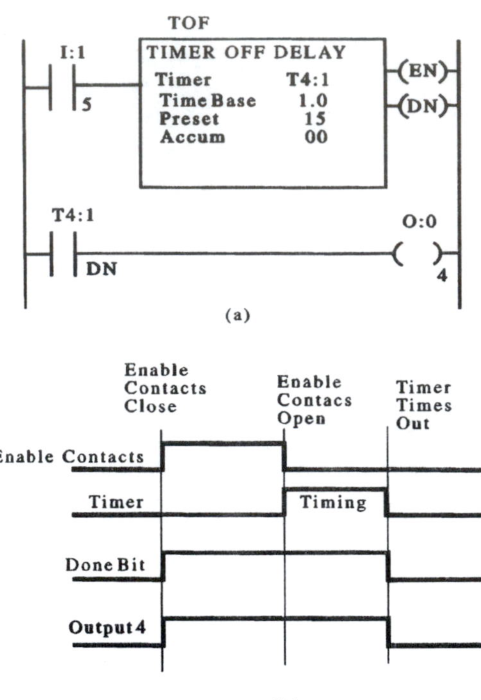

Figure 18–18 (a) A TOF (timer-off delay) instruction with its "DONE" bit controlling output O4. (b) The timing diagram for an off delay timer.

contacts are opened again. After the enable contacts are opened, the timer runs its cycle and times out, and the done bit will be turned off.

18.18 Counter Applications for Industry

Many industrial applications use counters. These applications can be separated into two basic operations: The first is called totalizing and refers to when a machine is making a large number of parts and the counter must keep track of the total. The second type of application is called sequencing. In this type of application the counter is used to keep track of a number of steps and then it resets itself to start the sequence all over again. An example of the sequence-type application is where a counter is used to keep track of the number of boxes being placed on each layer of a pallet and the number of layers the pallet has. For example, the pallet may require 4 boxes for each layer, which are stacked 3 layers high. One counter would count 1, 2, 3, 4 to keep track of the 4 boxes that are placed on each layer, and then it would reset to start the next layer. A second counter would count 1, 2, 3 and then reset to keep track of the layers on the pallet.

18.18.1 Counter Operation in a PLC

The PLC has two types of counter instructions available for use. They include the up-counter, whose mnemonic CTU stands for count up, and the down-counter whose mnemonic CTD stands for count down. The counters are similar to the timer instruction in that they have a preset and accumulative value and they each can control a number of status bits. The up-counter instruction increments (adds 1 to) its accumulative value when the enable contacts that allow power to flow to it transition from open to closed. This means that the enable contacts must return to their open state before they can transition for a second time to add another count to the counter accumulative value. The counter does not care how long the contact stays closed like the timer instruction does; rather the counter looks only for the transition of the enable contacts. The down-counter CTD decrements (subtracts 1) from its accumulative value each time power is transitioned to it.

Both types of counters require a reset instruction (RES) to set the accumulative value back to zero. Figure 18–19 shows the instruction for an up-counter that includes the reset operation. The enable contacts (I4) are not part of the counter, but they must be present for the counter to operate.

The preset value is the desired value for the counter and may represent how large a value the counter is expected to count. The maximum preset value for the up-counter is +32767 and for the down-counter is −32768. Each counter also has several status bits under its control. Both types of counters use bit 13 as a done bit. The done bit is energized (set) any time the accumulative value is equal to the preset value. The up-counter uses bit 12 as an overflow bit. The overflow bit is set any time the accumulative value is larger than +32767. The down-counter uses bit 11 as an underflow bit. The underflow bit is set any time the accumulative value goes below −32768. The up-counter uses bit 15 as an enable bit and the down-counter uses bit 14 as an enable bit. The bit number is used with the counter address (number) to create control logic. For example, if you want counter number 6 (C5:6) to control a solenoid that places a piece of cardboard between layers on a pallet, you would use contacts that have the number C5:6/13 on them. The /13 indicates that the done bit for the timer controls the contacts, and they will become energized each time the counter accumulative value reaches the preset value.

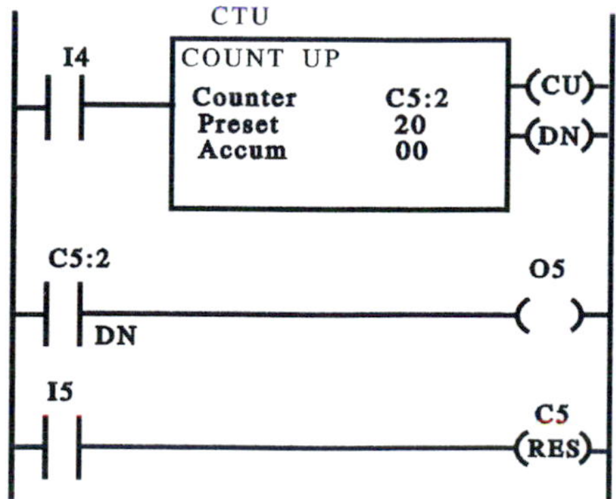

Figure 18–19 An example of a CTU (up-counter) with a reset instruction.

18.18.2 Up–Down Counters

The PLC can use an up-counter or a down-counter. Some applications are easier to understand when the up- and down-counters are used together in applications such as a good parts/bad parts counter. In this type of application the up-counter (CTU) and the down-counter (CTD) instructions are assigned to the same counter address. Each time the CTU instruction is transitioned, the counter will count up 1 value, and each time the CTD instruction is transitioned, the counter will count down 1 value. In this application both counters have the same accumulative value since they use the same counter. At the end of the day, the number displayed as the accumulative value will be the total difference between the good parts and the bad parts.

18.19 Entering Programs with the Hand-Held Programmer

After you understand how inputs, outputs, timers, and counters work, you are ready to enter a program into the PLC using the hand-held programmer. Figure 18–20 shows the hand-held programmer for the Allen–Bradley MicroLogix PLC. You can see that it has an LCD display and a set of number keys (0–9) and a set of instruction keys, such as contacts and coils. The programmer also has a set of function keys that are used for diagnostics and troubleshooting. These keys are named Menu, Mode, Force On, Force Off, Trace, Search, and Fault. Other keys are named Enter, Escape, Monitor, Overwrite, and Delete and they are used for general editing of the PLC program.

The Menu function has 8 choices, which are listed in the table in Figure 18–21. Figure 18–22 shows the six choices listed under the MODE function.

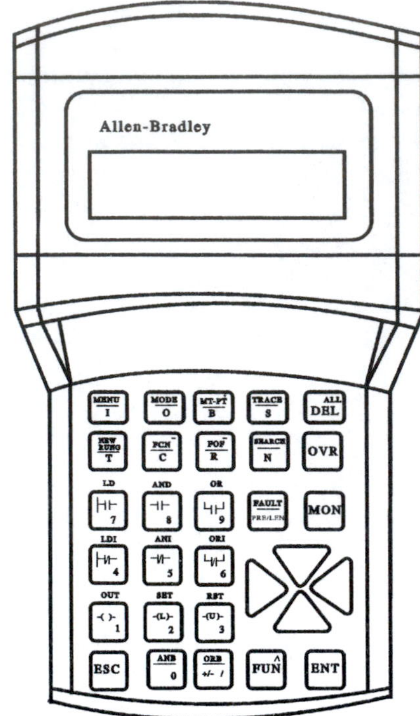

Figure 18–20 Hand-held programmer for the Allen–Bradley MicroLogix PLC.

1. LANGUAGE
2. ACCEPT EDITS
3. PROG. CONFIG
4. MEM MODULE
5. CLEAR FORCES
6. CLEAR PROG.
7. COMMS.
8. CONTRAST

Figure 18–21 Choices listed under the MENU function.

1. RPRG Remote Program
2. RRUN Remote Run
3. RCSN Remote Test-Continuous Scan
4. RSSN Remote Test-Single Scan
5. RSUS Remote Suspend
6. FLT Fault

Figure 18–22 Functions provided under the MODE function.

18.20 Programming Basic Circuits in the PLC

At times you will be requested to program simple programs into the PLC through the hand-held programmer. These simple programs include contacts that are connected in series or parallel, timers, and counters. Figures 18–23 through 18-29 show six simple programs

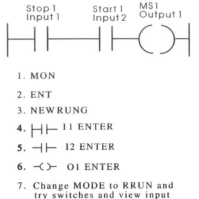

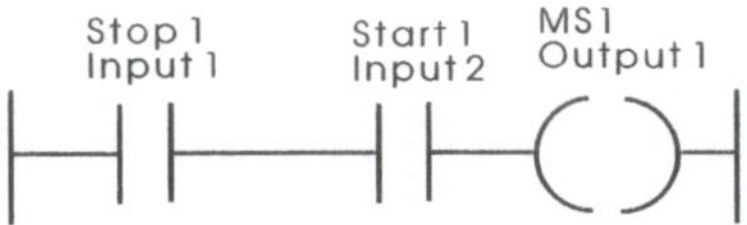

1. MON
2. ENT
3. NEWRUNG
4. |-| |- I1 ENTER
5. -| |- I2 ENTER
6. -( )- O1 ENTER
7. Change MODE to RRUN and try switches and view input and output status lights.

Figure 18–23 Simple circuit with two normally open contacts.

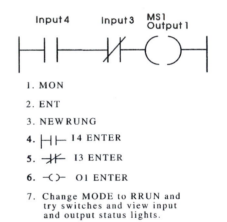

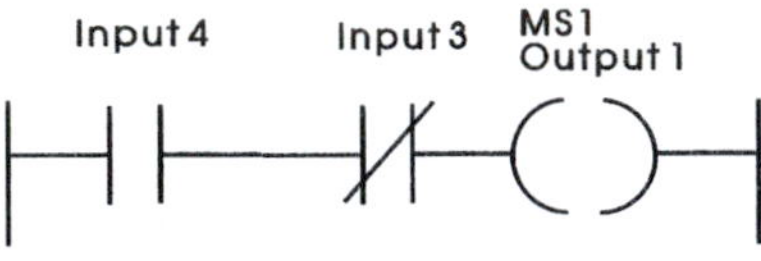

1. MON
2. ENT
3. NEWRUNG
4. |-| |- I4 ENTER
5. -|/|- I3 ENTER
6. -( )- O1 ENTER
7. Change MODE to RRUN and try switches and view input and output status lights.

Figure 18–24 Simple circuit with one normally open and one normally closed contact.

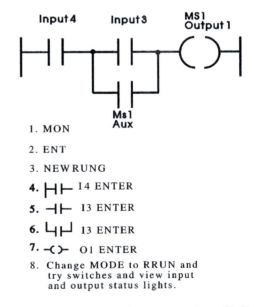

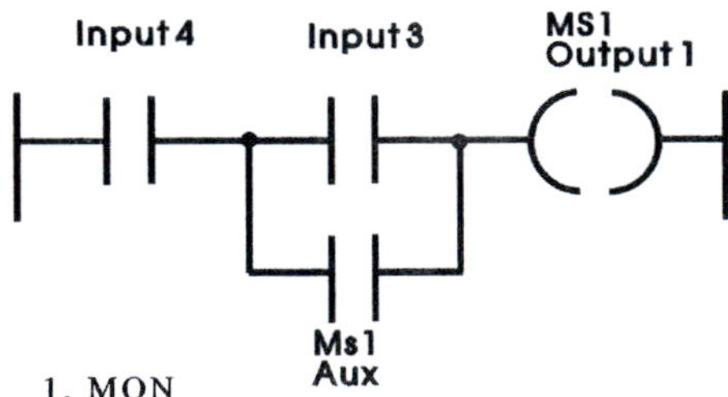

1. MON
2. ENT
3. NEWRUNG
4. |-| |- I4 ENTER
5. -| |- I3 ENTER
6. └| |┘ I3 ENTER
7. -( )- O1 ENTER
8. Change MODE to RRUN and try switches and view input and output status lights.

Figure 18–25 A basic start/stop circuit for a PLC.

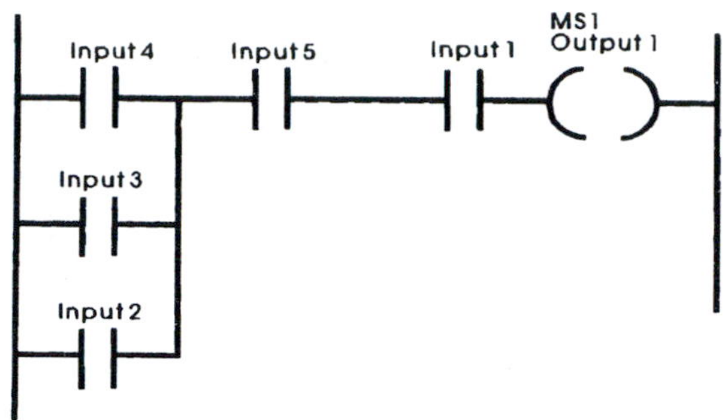

1. MON
2. ENT
3. NEWRUNG
4. ├| |├ I4 ENTER
5. └| |┘ I3 ENTER
6. └| |┘ I2 ENTER
7. ┤ ├ I5 ENTER
8. ┤ ├ I1 ENTER
9. ─( )─ O1 ENTER
10. Change MODE to RRUN and try switches and view input and output status lights.

Figure 18–26 Simple parallel circuit for a PLC.

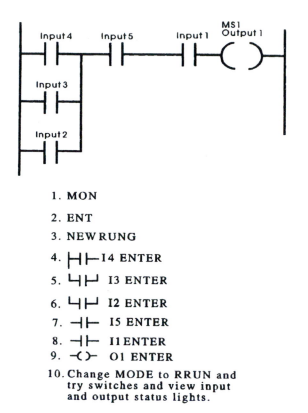

Input1 Input5 Input2 MS1 Output1

Output1 Input3 Input6 MS2 Output2

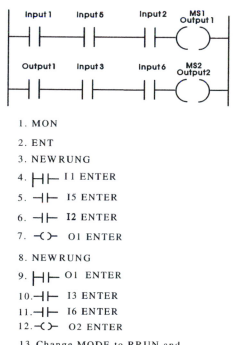

1. MON
2. ENT
3. NEWRUNG
4. ├| |├ I1 ENTER
5. ┤ ├ I5 ENTER
6. ┤ ├ I2 ENTER
7. ─( )─ O1 ENTER
8. NEWRUNG
9. ├| |├ O1 ENTER
10. ┤ ├ I3 ENTER
11. ┤ ├ I6 ENTER
12. ─( )─ O2 ENTER
13. Change MODE to RRUN and try switches and view input and output status lights.

Figure 18–27 Two series circuit creating a starting sequence where motor 1 must start before motor 2 can be started.

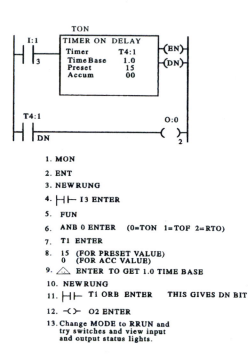

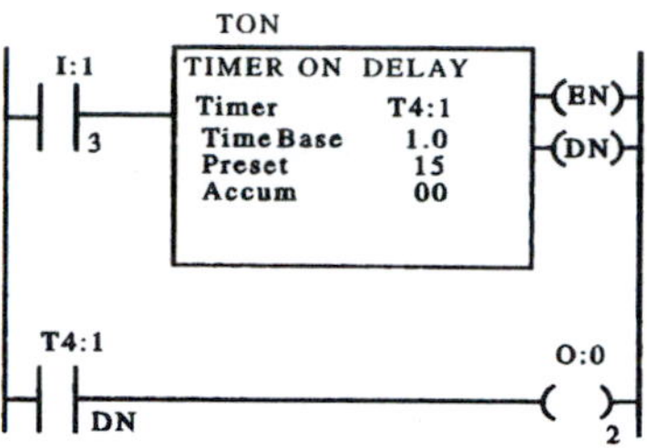

1. MON
2. ENT
3. NEWRUNG
4. ├| |├ I3 ENTER
5. FUN
6. ANB 0 ENTER (0=TON 1=TOF 2=RTO)
7. T1 ENTER
8. 15 (FOR PRESET VALUE)
   0 (FOR ACC VALUE)
9. △ ENTER TO GET 1.0 TIME BASE
10. NEWRUNG
11. ├| |├ T1 ORB ENTER THIS GIVES DN BIT
12. ─( )─ O2 ENTER
13. Change MODE to RRUN and try switches and view input and output status lights.

Figure 18–28 Simple timer program that delays the time the motor starts after the input switch is closed.

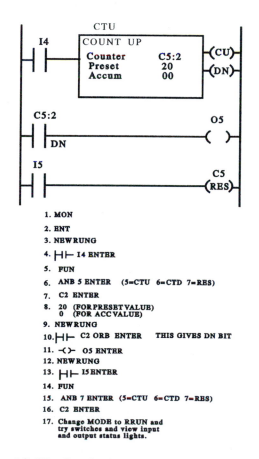

CTU

COUNT UP

I4 Counter C5:2 Preset 20 Accum 00 (CU) (DN)

C5:2 DN O5

I5 C5 (RES)

1. MON
2. ENT
3. NEWRUNG
4. ├| |├ I4 ENTER
5. FUN
6. ANB 5 ENTER (5=CTU 6=CTD 7=RES)
7. C2 ENTER
8. 20 (FOR PRESET VALUE)
   0 (FOR ACC VALUE)
9. NEWRUNG
10. ├| |├ C2 ORB ENTER THIS GIVES DN BIT
11. ─( )─ O5 ENTER
12. NEWRUNG
13. ├| |├ I5 ENTER
14. FUN
15. ANB 7 ENTER (5=CTU 6=CTD 7=RES)
16. C2 ENTER
17. Change MODE to RRUN and try switches and view input and output status lights.

Figure 18–29 Simple counter program.

and all of the keystrokes required to enter these programs. If you have a MicroLogix PLC available you can use these examples to try these programs.

18.21 Wiring a Start/Stop Circuit to Input and Output Terminals on the PLC

All of the inputs and outputs that are used in the PLC system must be connected to the input or output module hardware. The simplest circuit to understand is a start/stop circuit that controls a motor starter coil. Figure 18–30 shows an example of this circuit and you can see where the start pushbutton switch and the stop pushbutton switch are wired to two input addresses and to a motor starter coil connected to an output address. You can also see the location of the screws on the MicroLogix PLC where the wires from the switch and coil for the motor starter are connected.

The start pushbutton is wired as a normally open switch and the stop pushbutton is wired as a normally closed switch, just as in a hardwired control system. The major difference in the start/stop circuit in the PLC from a hardwired system is in the way the start and stop pushbuttons are programmed into the PLC. In the PLC program the stop pushbutton is programmed normally open. At first this seems to be incorrect, but if you look at the line of logic in the PLC you will notice that since the stop pushbutton is wired normally closed, it will pass power to the input module, which in turn sends a 1 to the image table. Since the stop contacts in the program are programmed normally open, they will change state and pass power to the start contacts whenever the input module sends a 1 to the image table. If the stop contacts were programmed normally closed, they would open when the input module received power from the hardwired stop switch and the circuit would not operate.

Imortant Notice:
The PLC program would work if you wired the stop pushbutton normally open and programmed its con-

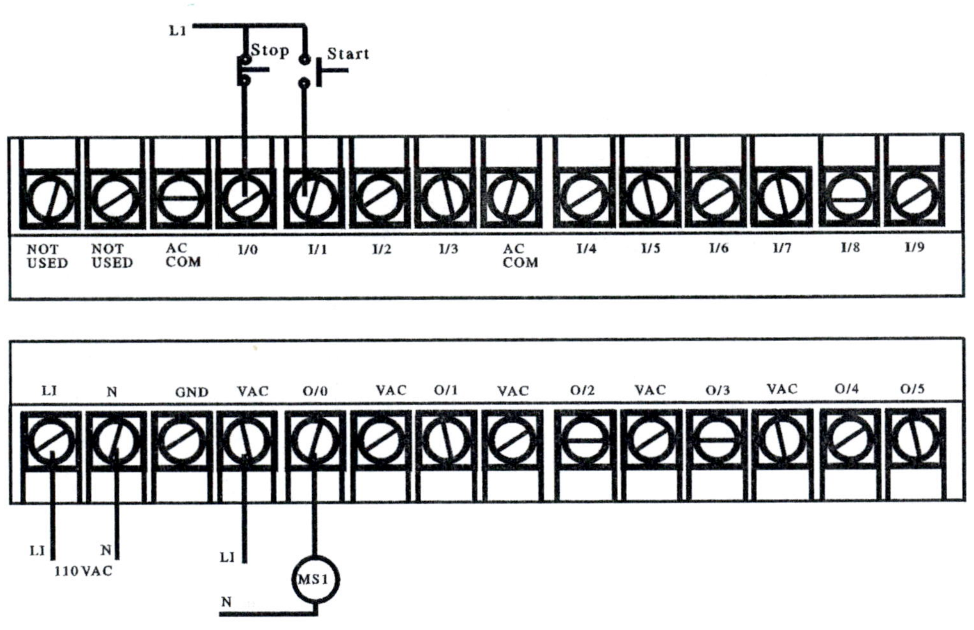

(a) Start/stop hardwired to PLC

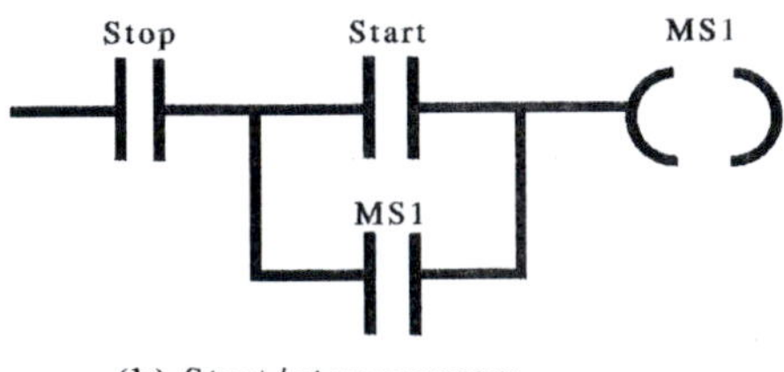

(b) Start/stop program

Figure 18–30 (a) Start and stop pushbuttons hardwired to a PLC. (b) Program for stop and stop circuit.

tacts normally closed. This creates a severe problem in that the stop pushbutton becomes a safety hazard if it is wired normally open and a wire comes loose from its terminal or if the stop button fails, because it cannot send a signal to the PLC input module and stop the circuit. It is important that you always wire the stop pushbutton as a normally closed switch so that if it fails, it will interrupt power to the input module and stop the circuit.

The hardwired circuit of a PLC must also be protected by a real relay called a master control relay. This relay will be identified as CRM (control relay master) and all of the voltage to the input and output module is controlled by the contacts of the CRM. The coil of the CRM is connected to a hardwired start–stop pushbutton station that is not part of the PLC program. This allows the power to be shut off to the inputs and output by depressing the stop button. Power is connected directly to the processor so that it stays energized regardless of the state of the CRM.

18.22 Typical Problems a Technician Will Encounter with PLCs

As a technician you may encounter problems with PLCs at either the software or hardware level. At the software level, you may be required to lay out and enter the original ladder logic program. You may also be involved at the start-up of a machine when a program has been written for the PLC and a technician is required to test the program against the machine operation. In other cases, you may be involved in using a program to troubleshoot the system when a fault occurs. Sometimes you will be requested to make changes in the documentation that describes the operation of the system and identifies all the inputs and outputs. In each of these cases, you may need additional information to satisfactorily accomplish these jobs. This additional information may be provided in manufacturers' manuals or in short 4- to 5-day training courses that are provided on specific PLCs.

You may also be involved with the hardware modules and I/O devices that are connected to PLCs. In some systems you will be requested to read the specification sheets for the signals that each module sends or receives so that you can select the proper types of electronic devices to interface with the PLC. Sometimes you will be requested to analyze a signal to see that it meets the proper criteria for data transmission. If system component failure occurs, you will be required to use all the knowledge you have about electronics to test the hardware and make decisions. You now have a better idea of how electronics have been blended with the programmable controller to provide industrial automation.

18.23 Troubleshooting the PLC

When you are called to troubleshoot the PLC, the first thing you need to understand is that very seldom is the PLC the problem; rather it is indicating a problem with hardware, such as switches and coils. Each input module has a status indicator (light) for each input switch, and an output status indicator for each output circuit. If the input switch is in the "on state" and passing 110 VAC to the module, the status light will be on. The status light is provided on the PLC module so that you do not need a voltmeter to determine if power is at the terminals. You can tell if the input switch is passing power to the PLC by observing the status light. On the output module, the status light will be on whenever the output circuit is receiving power. Again, the status light can be used instead of a voltmeter to quickly tell if the circuit has power. Figure 18–31 shows an example of input and output status lights.

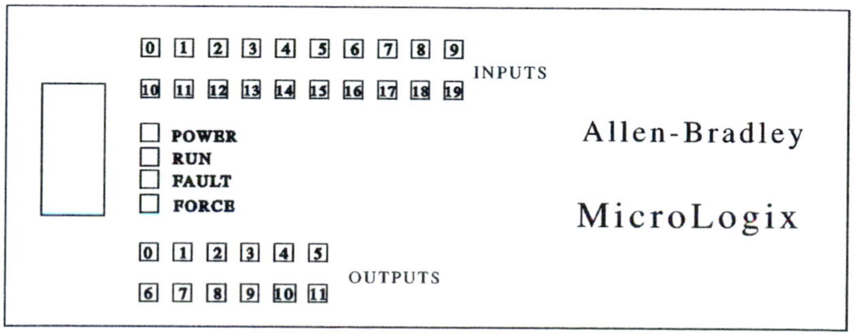

Figure 18–31 Example of the status indicators on input and output modules for the MicroLogix PLC.

Questions

1. Describe the four basic parts of any programmable controller.
2. Explain the operation of a PLC up-counter and down-counter.
3. Explain the four things that occur when the PLC processor scans its program.
4. Discuss the advantages a PLC-controlled system would have over a hardwired relay-type control system if you wanted to add one more photoelectric switch and output to detect a medium-size box in the box-sorting system described in Section 18.2.
5. The input module has a status indicator for each input circuit. Explain how the status indicator is used in troubleshooting.

True and False

1. One advantage of all PLCs is that you can use contacts with the same address more than once in the program.
2. When the PLC processor is in the run mode, it does not execute its scan cycle.
3. A mnemonic is a timer whose time delay is controlled by air pressure.
4. When a timer in a PLC program will not time, you should suspect that the timer is broken and that you will need to program a replacement timer.
5. A control relay (memory coil) in a PLC program is different from an output instruction in that its signal does not control a circuit in an output module and it is used mainly to determine logic conditions.

Multiple Choice

1. When the input contact that enables a TON-type timer is opened, the accumulated value in the timer will _____
   a. reset (go to zero).
   b. freeze (remain at its present value).
   c. go to an undetermined value and the timer must be reset manually.
2. An RTO-type timer has its accumulative value reset _____
   a. any time the enable contacts are opened.
   b. only when the reset for the timer is "HI."
   c. any time the accumulative value in the timer exceeds 99.
3. When the PLC processor is in the PROGRAM mode, _____
   a. it executes its scan cycle.
   b. it does not execute its scan cycle.
   c. it is impossible to tell whether the processor is executing its scan cycle because you are not on line.
4. When it is used as part of a start/stop circuit for a PLC system, the stop pushbutton should be wired _____ and programmed _____
   a. normally closed, normally open.
   b. normally open, normally closed.
   c. It doesn't matter.
5. You can use the _____ on the face of a PLC to determine whether an input switch is on or off.
   a. power light
   b. input status indicators
   c. output status indicators

Problems

1. Enter the conveyor box-sorting program in Figure 18–7 in a programmable controller and execute the inputs to watch the program operate. Describe the operation of each output.
2. Enter an on-delay timer and an off-delay timer in your PLC and describe the following: preset time, accumulative time, what must occur for the timer to run, what must occur for the timer to time out, and what contacts change when the timer times out.
3. Enter an up-counter and a down-counter in your PLC and describe the following: preset value, accumulative value, what must occur for each counter to count, what occurs when the counter's preset and accumulative values are equal, what happens when the counter's accumulative value exceeds the overflow value, what contacts change when the counter reaches its preset value, and what contacts change when the counter exceeds the overflow.
4. Program an autoresetting timer and describe its operation.
5. Program an autoresetting counter into your PLC and describe its operation.

◀ Chapter 19 ▶

Electronics for Maintenance Personnel

Objectives

After reading this chapter you will be able to:

1. Explain P-type and N-type material.
2. Identify the terminals of a diode and explain its operation.
3. Explain the operation of a PNP and NPN transistor.
4. Identify the terminals of a silicon-controlled rectifier (SCR) and explain its operation.
5. Explain the operation of a diac and a triac.

Electronic devices such as diodes, transistors, and SCRs have become commonplace in industrial circuits because they provide better control than electromechanical devices and are less expensive to manufacture. When you are troubleshooting you will find solid-state components in control boards and other controls. In this chapter you will gain an understanding of P and N material, which forms the building blocks of all electronic components. After you get a basic understanding of P and N material, you will be introduced to diodes, transistors, SCRs, and triacs and you will see application circuits of each of these types of components. The theory of operation and troubleshooting techniques for each type of device will also be presented.

19.1 Conductors, Insulators, and Semiconductors

In basic science you learned that atoms have protons and neutrons in their nucleus and that electrons move around the nucleus in orbits that are also called shells. The number of electrons in the atom is different for each element. For example, copper has 29 electrons and three of them are located in the outermost shell. The outermost shell is called the valence shell and the electrons in that shell are called valence electrons. The atoms of the most stable material have 8 valence electrons and these 8 valence electrons will be found as four pairs.

A *conductor* is a material that allows electrons (electrical current) to flow easily through it, and an *insulator* does not allow current to flow through it. An example of a conductor is copper, which is used for electrical wiring. An example of an insulator is rubber or plastic. The atomic structure of a conductor makes it easier for electrons to flow through it and the atomic structure of an insulator makes it nearly impossible for any electrons to flow through it.

Figure 19–1 shows the atomic structure of a conductor. Atoms of conductors can have 1, 2, or 3 valence electrons. The atom in this example has 1 valence electron. Since all atoms try to get 8 electrons (4 pairs) in their valence shell, it takes less energy for conductors to give up these electrons (1, 2, or 3) so that the valence shell will become empty. At this point the atom becomes stable because the previous shell becomes the new valence shell and it has 8 electrons. The electrons that are given up are free to move as current flow.

Figure 19–2 shows the atomic structure of an insulator. Insulators will have 5, 6, or 7 valence electrons. In this example the atom has 7 valence atoms. This structure makes it easy for insulators to take on extra electrons to get 8 valence electrons. The electrons that are

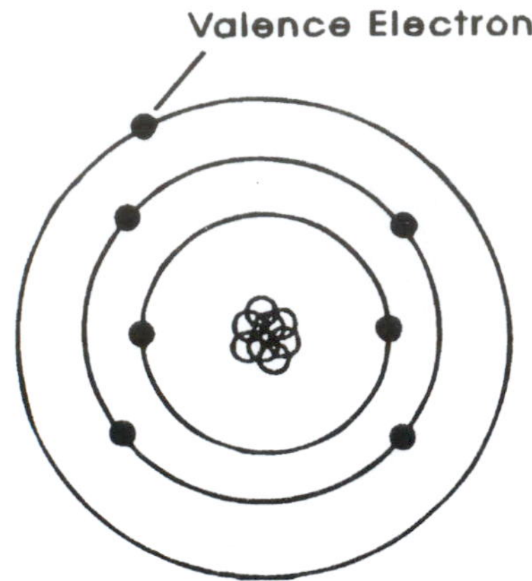

Figure 19–1 Atomic structure of a conductor.

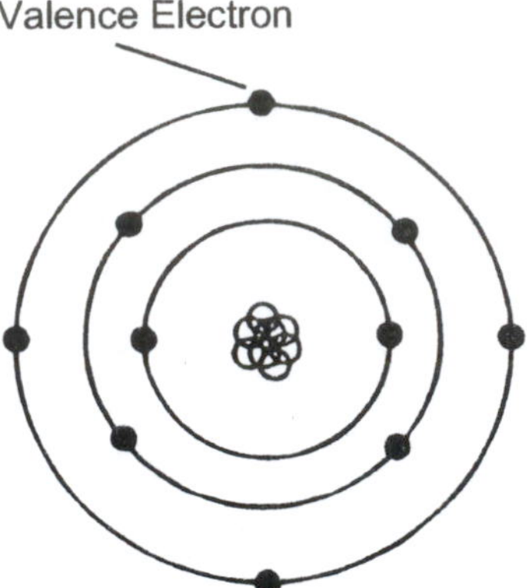

Figure 19–3 Atomic structure for a semiconductor material.

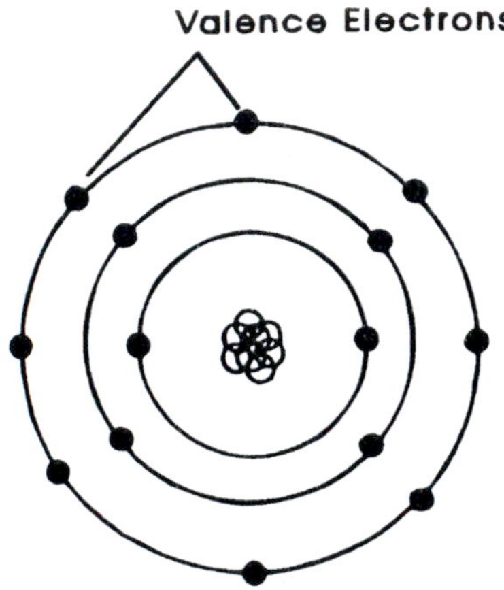

Figure 19–2 Atomic structure of an insulator.

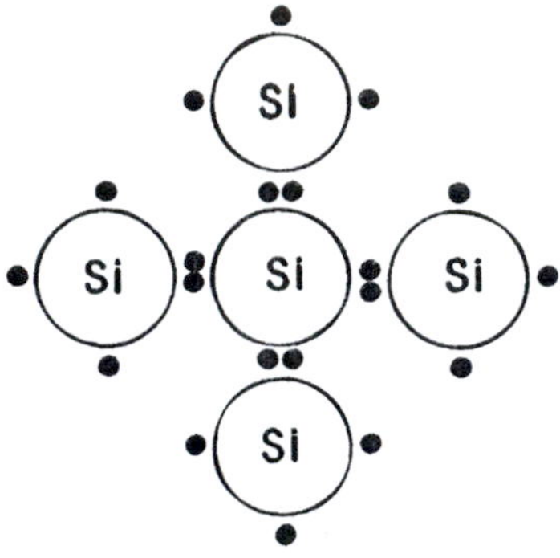

Figure 19–4 Atoms of silicon semiconductor material are combined to create a lattice structure.

captured to fill the valence shell are electrons that would normally be free to flow as current.

Semiconductors are materials whose atoms have exactly 4 valence electrons, which means they can take on 4 new valence electrons like an insulator to get a full valence shell, or they can give up 4 valence electrons like a conductor to get an empty outer shell, so the previous shell, which has 8 electrons, becomes the valence shell. Figure 19–3 shows an example of the atomic structure for semiconductor material.

19.2 Combining Atoms

When solid-state material or other material is manufactured, large numbers of atoms are placed together. The structure that becomes most stable at this point is called a lattice structure. Figure 19–4 shows an example of the lattice structure that occurs when atoms are combined. In this diagram atoms of silicon (Si), which is a semiconductor material with 4 valence electrons, are combined so that one valence electron from each of the neighbor atoms is shared so that all atoms look and act as if they each have 8 valence electrons.

19.2.1 Combining Arsenic and Silicon to Make N-Type Material

Other types of atoms can be combined with semiconductor atoms to create the special material that is used in solid-state transistors and diodes. Figure 19–5 shows a diagram of four silicon atoms that are combined with one atom of arsenic. Arsenic has 5 valence electrons, and when the silicon atoms are combined with it they create a very strong lattice structure. You can see that each silicon atom donates one of their valence electrons to pair up with each of the valence electrons of the arsenic atom. Since the

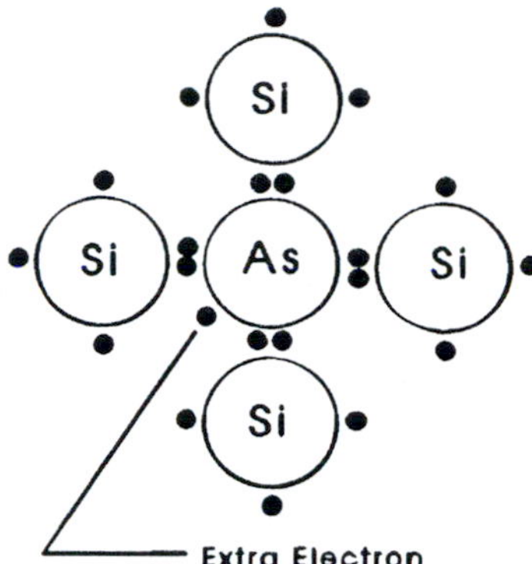

Figure 19–5 N-type material formed by combining four silicon atoms with a single arsenic atom.

arsenic atom has 5 valence electrons, one of the electrons will not be paired up and it will become displaced from the atom. This electron is called a *free electron* and it can go into conduction with very little energy. This new material with a free electron is called *N-type material.*

19.2.2
Combining Aluminum and Silicon to Make P-Type Material

An atom of aluminum can also be combined with semiconductor atoms to create the special material that is called P-type material. Figure 19–6 shows a diagram of four silicon atoms that are combined with one atom of aluminum. Aluminum has 3 valence electrons, and when the silicon atoms are combined with it they create a very strong lattice structure. Each silicon atom donates one of its valence electrons to pair up with each of the valence electrons of the aluminum atom. Since the aluminum atom has 3 valence electrons, one of the 4 aluminum electrons will not be paired up and it will have a space where any free electron can move into to combine with the single electron. This free space is called a *hole* and it is considered to have positive charge since it is not occupied by a negatively charged electron. Since this new material has an excess hole that has a positive charge, it is called *P-type material.*

19.3
The PN Junction

One piece of P-type material can be combined with one piece of N-type material to make a *PN junction.* Figure 19–7 shows a typical PN junction. The PN junction forms to make an electronic component called a *diode.*

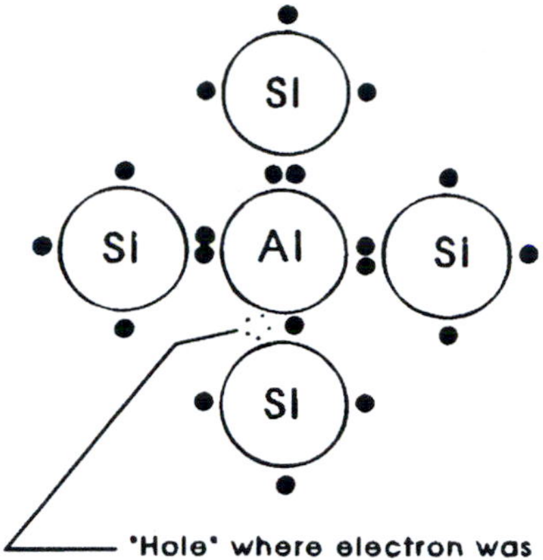

Figure 19–6 P-type material formed by combining four silicon atoms with a single aluminum atom.

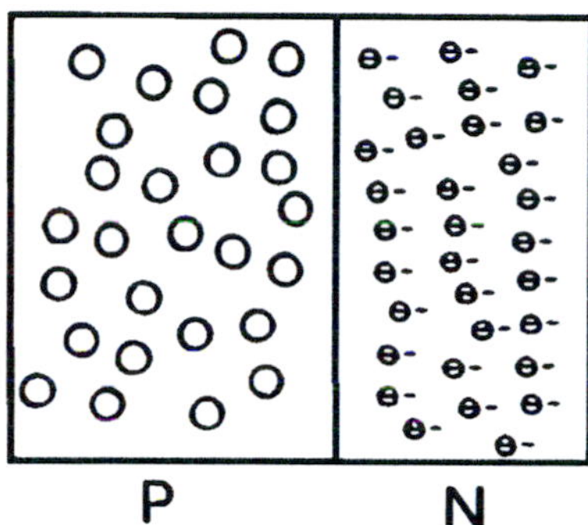

Figure 19–7 An example of a piece of P-type material connected to a piece of N-type material.

The diode is the simplest electronic device. When DC voltage is applied to the PN junction with the proper polarity it will cause the junction to become a very good conductor; if the polarity of the voltage is reversed, the PN junction will become a good insulator.

19.3.1
Forward Biasing the PN Junction

When DC battery voltage is applied to the PN junction so that positive voltage is connected to the P material and negative voltage is connected to the N material, the junction is forward biased. Figure 19–8a shows a battery connected to the PN junction so that it is forward biased. The positive battery voltage causes the majority of holes in the P-type material to be repelled so that the free holes move toward the junction, where they will come into contact with the N-type material. At the same time the negative battery voltage also repels the free electrons in the N-type material toward the junction. Since the holes and free electrons come into contact at the junction, the electrons recombine with the

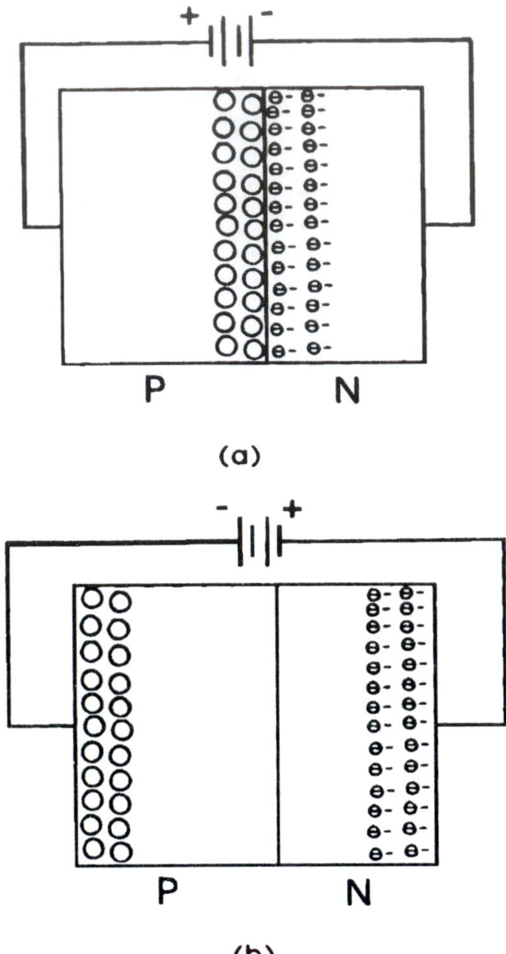

Figure 19–8 (a) A forward-biased PN junction. (b) A reverse-biased PN junction.

holes and cause a very low resistance junction, which allows current to flow freely through it. When a PN junction has low resistance it will allow current to pass, just as if the junction were a closed switch.

Up to this point in this text, all current flow has been described in terms of *conventional current flow*, which is based on a theory that electrical current flows from a positive source to a negative return terminal. You can see that this theory will not support current flow through electronic devices. For this reason *electron current flow theory* must be used when discussing electronic devices. In electron current flow theory, current is the flow of electrons and it flows from the negative terminal to the positive terminal in any electronic circuit.

19.3.2 Reverse Biasing the PN Junction

Figure 19–8b shows a battery connected to the PN junction so that it is reversed biased. The positive battery voltage is connected to the N-type material and the negative battery voltage is connected to the P-type material. The negative voltage on the P-type material attracts the majority of holes in the P-type material. They move away from the junction so they cannot come into contact with the N-type material. At the same time the positive battery voltage that is connected to the N-type material attracts the free electrons away from the junction. Since the holes and free electrons are both attracted away from the junction, no electrons can recombine with any holes, so a high-resistance junction is formed that will not allow any current flow. When the PN junction has high resistance it will not allow any current to pass, just as if the junction were an open switch.

Figure 19–9 (a) PN junction for a diode. (b) Electronic symbol of a diode. The anode is the arrowhead part of the symbol and the cathode is the other terminal.

19.4 Using a Diode for Rectification

The electronic diode is a simple PN junction. Figure 19–9 shows the symbol for a diode, which looks like an arrowhead that is pointing against a line. The part of the symbol that is the arrowhead is called the *anode*, and it is also the P-type material of the PN junction. The other terminal of the diode is called the *cathode* and it is the N-type material. Since the anode is made of positive P-type material it is identified with a plus sign (+). The cathode is identified with a minus sign (−) since it is made of N-type material.

Figure 19–10 shows a diode connected in a circuit that has an AC power source. The AC power source produces a sine wave that has a positive half-cycle and then a negative half-cycle. The diode converts the AC sine wave to half-wave DC voltage by allowing current to pass when the AC voltage provides a forward bias to the PN junction, and it blocks current when the AC voltage provides a reverse bias to the PN junction. The forward-bias condition occurs when the AC voltage sine wave provides a positive voltage to the anode and a negative voltage to the cathode. During this part of the AC cycle, the diode is forward biased and it has very low resistance so current can flow. When the other half of the AC sine wave occurs, the diode becomes reverse biased with negative voltage applied to the anode and positive voltage applied to the cathode. During the time the diode is reverse biased, a high-resistance junction is created and no current will flow through it.

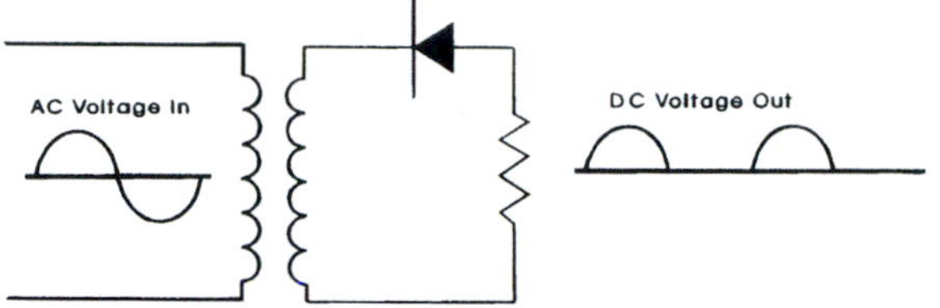

Figure 19–10 A single diode used in a circuit to convert AC voltage to half-wave DC voltage.

Rectification is the process of changing AC voltage to DC voltage. One of the main jobs of the diode is to convert AC voltage to DC voltage. Most electronic circuits used in industrial circuits need some DC voltage to operate. Since the equipment is connected to AC voltage, a power supply is required to provide regulated DC voltage for the solid-state circuits, and the diode is part of the power supply that rectifies the AC voltage to DC.

19.4.1 Half-Wave and Full-Wave Rectifiers

When one diode is used in a circuit to convert AC voltage to DC voltage, it is called a *half-wave rectifier*, since only the positive half of the AC voltage is allowed to pass through the diode, while the negative half is blocked. The rectifier shown in Figure 19–10 is a half-wave rectifier. The half-wave rectifier is not very efficient, since half of the AC sine wave is wasted.

If four diodes are used in the circuit, they can convert both the positive half-wave and the negative half-wave of the AC sine wave. Figure 19–11 shows a circuit with four diodes used to rectify AC voltage to DC voltage. Since the four diodes can convert both the positive half and the negative half of the AC sine wave, this type of rectifier is called a full-wave rectifier.

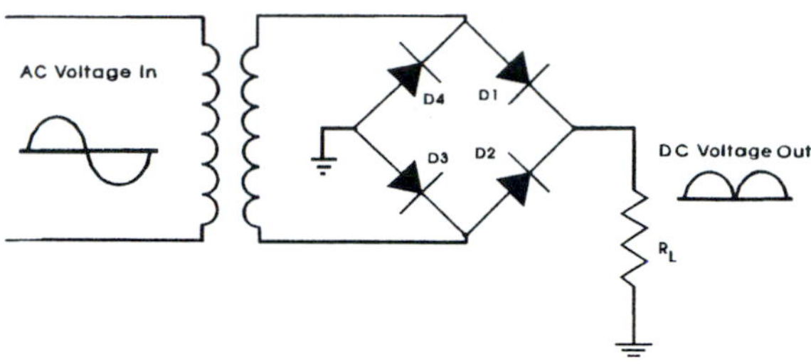

Figure 19–11 Four diode, full-wave rectifier. This type of rectifier is often called a full-wave bridge rectifier, because the diodes are connected in a bridge circuit.

19.4.2 Three-Phase Rectifiers

Larger three-phase AC compressor and fan motors used in air conditioning, heating, and refrigeration systems can have their speed changed to run more efficiently by changing the frequency of the voltage supplied to them. In these applications, six diodes are used to convert three-phase AC voltage to DC voltage, and then a microprocessor-controlled circuit converts the DC voltage back to three-phase AC voltage and the frequency of this voltage can be adjusted to change the speed of the motors. Figure 19–12 shows a six-diode, three-phase rectifier. Notice that the supply voltage to the diodes is three-phase AC voltage, and the output voltage from the rectifier consists of six positive half-waves.

19.4.3 Testing Diodes

One of the tasks that you must perform as a technician is to test diodes to see if they are operating correctly. One way to do this is to apply AC voltage to the input of

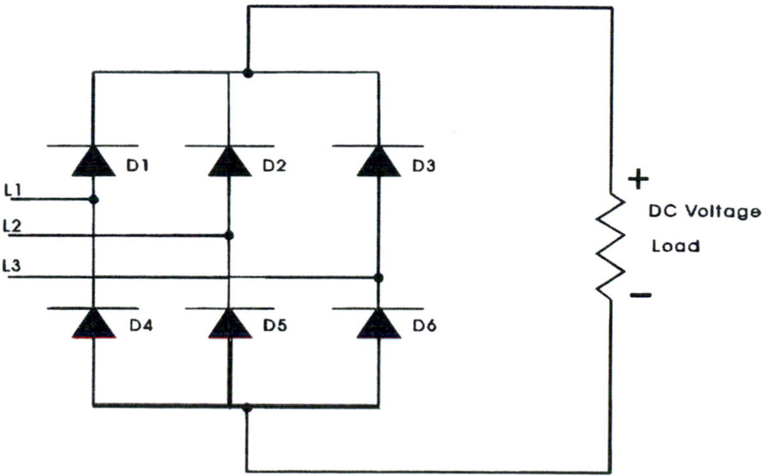

Figure 19–12 A six-diode bridge used to rectify three-phase AC voltage to DC voltage.

the diode circuit and test for DC voltage at the output of the diode circuit. If the amount of DC voltage at the power supply is half of what it is rated for in a four-diode bridge rectifier circuit, you can suspect one of the diode pairs has one or both diodes opened. If this occurs, you can use an ohmmeter to test each diode to determine which one is faulty.

When you are testing the diodes with an ohmmeter, it is important that all power to the diode circuit is turned off. You should remember from earlier chapters that the ohmmeter uses a battery as a DC voltage source. Since you know the diode can be tested for forward bias and reverse bias with a DC voltage source, you can use the ohmmeter as the voltage source and the meter to test for high resistance and low resistance through the diode junction. Figure 19–13a shows an example of putting the positive ohmmeter terminal on the anode of the diode, and the negative ohmmeter terminal on the cathode of the diode to cause the diode to go into forward bias. During this test the diode is forward biased and the ohmmeter should measure low resistance. When the ohmmeter leads are reversed as in Figure 19–13b so that the negative meter lead is connected to the anode of the diode, and the positive meter lead is connected to the anode of the diode, the diode is reverse biased. When the diode is reverse biased, the ohmmeter should measure infinite (4) resistance. If the diode indicates high and low resistance, it is good. If the diode indicates low resistance during the forward bias test and the reverse bias test, the diode is shorted. If the diode shows high resistance during both tests, it is open.

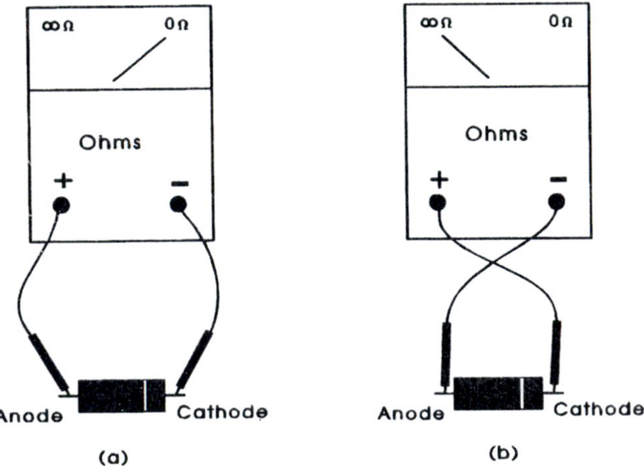

Figure 19–13 (a) Using the battery in an ohmmeter to forward bias a diode. The diode should have low resistance during this test. (b) Using the battery in the ohmmeter to reverse bias a diode. The diode should have high resistance during this test.

19.4.4 Identifying Diode Terminals with an Ohmmeter

Since the ohmmeter can be used to determine if a diode is good or faulty, the same test can be used to determine which lead of a diode is the anode and which lead is the cathode. When you use the ohmmeter to test the diode for forward bias and reverse bias you should notice that the ohmmeter will indicate high resistance when the diode is reverse biased and low resistance when the diode is forward biased. When the meter indicates low resistance, you know the diode is forward biased, so the positive lead is touching the anode and the negative lead is touching the cathode. This method will work when you are testing any diode. If the diode has markings, you can identify the cathode end of the diode because it has a strip around it. Figure 19–14 shows several types of diodes, and the anode is identified in each.

19.4.5 Light-Emitting Diodes

A light-emitting diode (LED) is a special diode that is used as an indicator because it gives off light when current flows through it. Figure 19–15 shows a typical LED and its symbol. From this figure you can see the LED looks like a small indicator lamp. You will likely encounter LEDs on various controls such as thermostats. The major difference between an LED and an incandescent lamp is that the LED does not have a filament, so it can provide thousands of hours of operation without failure.

The LED must be connected in a circuit in forward bias. Since the typical LED requires approximately 20 mA to illuminate, it will usually be connected in series with a 600Σ to 800Σ resistor, which will limit the current. Figure 19–16 shows a set of seven LEDs that are connected to provide a seven-segment display. The seven-segment display has the capability to display all numbers 0–9. Seven-segment displays are used to display numbers on thermostats and other electronic devices.

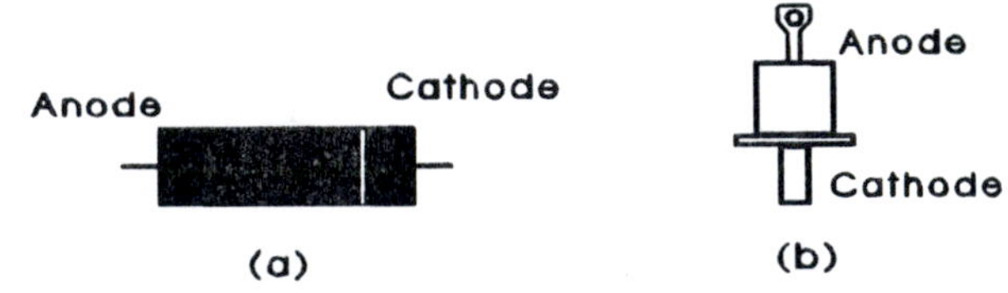

Figure 19–14 (a) Typical diode with anode and cathode identified. (b) Power diode with anode and cathode identified.

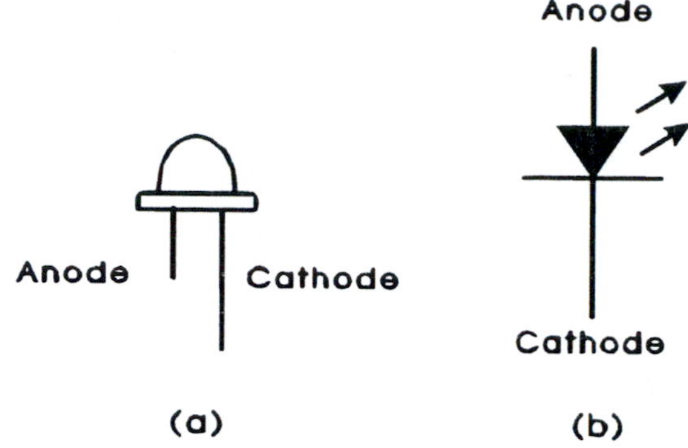

Figure 19–15 (a) A typical light-emitting diode (LED). Notice the LED looks similar to a small indicating lamp. (b) The symbol for an LED.

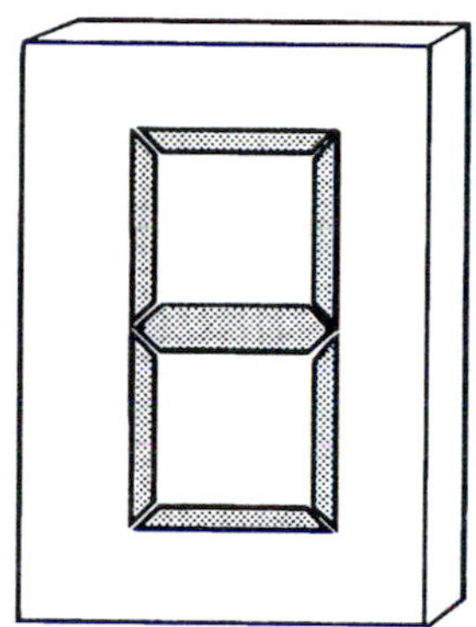

Figure 19–16 LEDs used in seven-segment displays. Each seven-segment display can display the numbers 0–9.

LEDs are also used in optoisolation circuits where larger field voltages are isolated from smaller computer signals. The LED is encapsulated with a photo transistor. When the input signal is generated, it will cause current to flow through the LED, and light from the LED will shine on the photo transistor, which will go into conduction and pass the signal on to the computer.

19.5 PNP and NPN Transistors

Two pieces of N-type material can be joined with a single piece of P-type material to form an NPN transistor. A PNP transistor can be formed by joining two pieces of P-type material with a single piece of N-type material. Figure 19–17 shows the electronic symbol and the material for both the PNP and the NPN transistors. The terminals of the transistors are identified

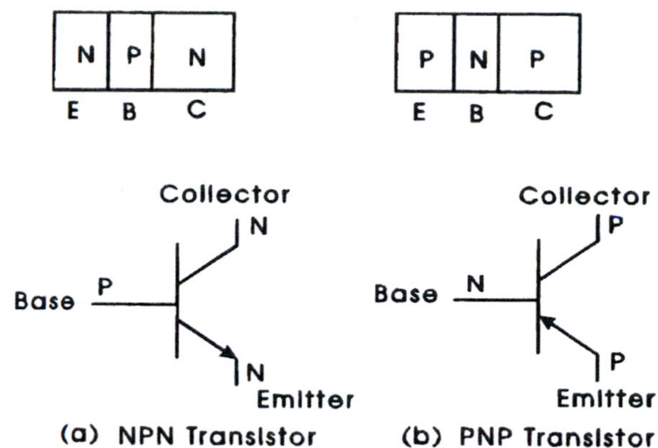

Figure 19–17 (a) Diagram of NPN material that is joined to form a transistor, and the electronic symbol for an NPN transistor. (b) Diagram of PNP material that is joined to form a transistor, and the electronic symbol for a PNP transistor.

as the emitter, collector, and base. The base is the middle terminal, and the emitter is the terminal identified by the arrowhead.

19.5.1 Operation of a Transistor

A transistor can be connected in a circuit to perform a wide variety of functions. The simplest is the function of an electronic switch (Figure 19–18). In this type of

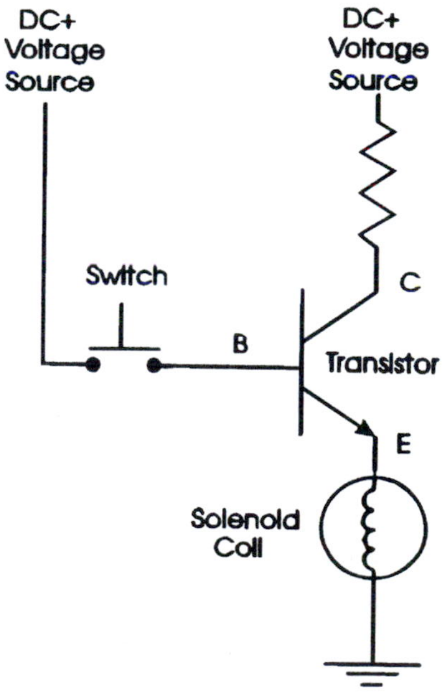

Figure 19–18 A transistor used as an electronic switch.

Figure 19–19 PNP and NPN transistors shown as their equivalent PN junctions. Each PN junction can be tested for forward and reverse bias like a diode.

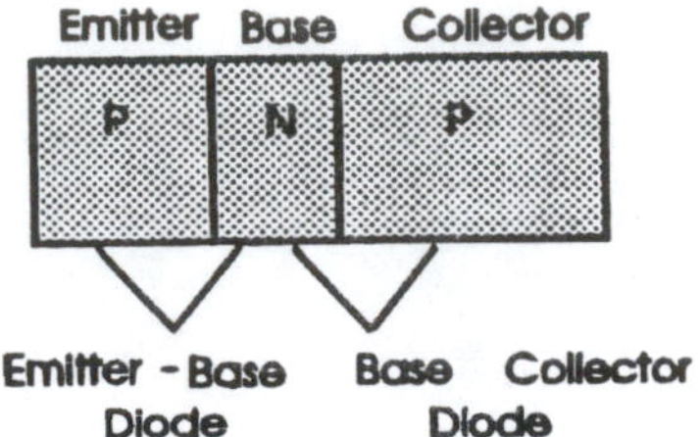

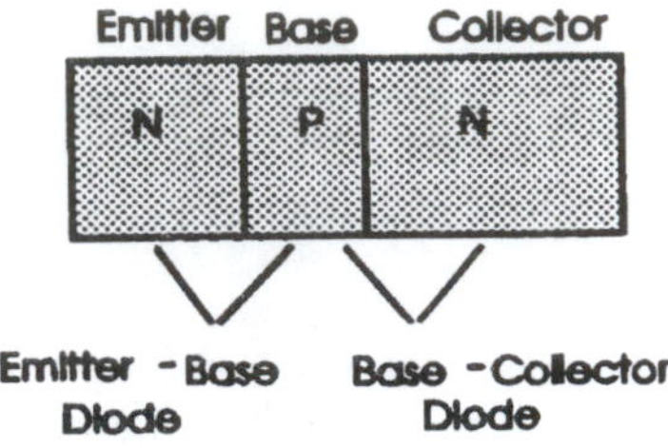

application, the base terminal of the transistor provides a function like the coil of a relay, and the emitter–collector circuit provides a function like the contacts of a relay. When the proper amount and polarity of DC voltage is applied to the base of the transistor, the resistance between the collector and emitter is relatively low, which allows the maximum amount of circuit current to flow through the emitter–collector circuit. The transistor at this time acts like a relay that has its coil energized.

When the polarity of the voltage on the base of the transistor is reversed, the emitter–collector circuit is changed to a high-resistance circuit, which acts like the relay when the coil is deenergized. The major advantage of the transistor is that a very small amount of voltage or current on the base can switch the transistor from high resistance to low resistance. Since the base current is very small and the current flowing through the collector is very large, the transistor is called an *amplifier*. Transistors are used in a variety of applications, including thermostats, compressors protection circuits, and variable frequency motor drives.

Figure 19–19 shows a PNP transistor and an NPN transistor as two PN diode circuits. The equivalent diode circuits are shown with each transistor to give you the idea how the two junctions work together inside each transistor. Since each transistor is made from two PN junctions, each junction can be tested just as in the single-junction diode for forward bias (low resistance) and reverse bias (high resistance). If you must work on a number of systems that have electronic circuits, you may purchase a commercial transistor tester that allows you to test the transistor while it is in the circuit or when it is out of the circuit.

19.5.2 Typical Transistors

You will be able to identify transistors by their shape. Small transistors are used for switching control circuits, and larger transistors will be mounted to heat sinks so that they can easily transfer heat. Figure 19–20 shows examples of several types of transistors.

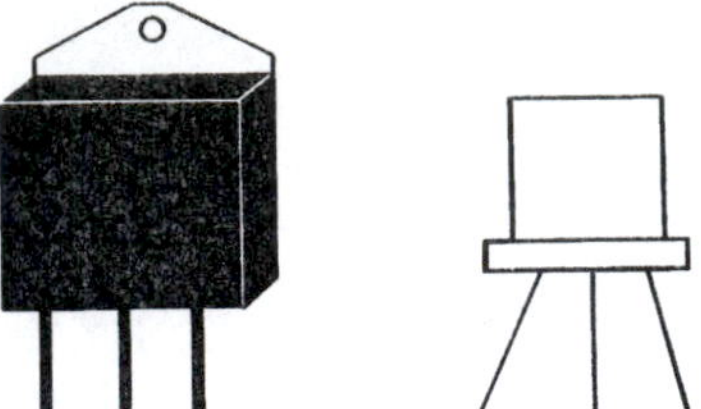

Figure 19–20 Typical transistors that are used for power control and switching.

19.5.3 Troubleshooting Transistors

Transistors can be tested by checking resistance at each P and N junction from front to back. Figure 19–21 shows these tests. You can see each time the battery in the ohmmeter forward biases a PN junction, the resistance is low, and when the battery reverse biases the junction, the meter indicates high resistance. You can test a transistor in this manner if it has been removed from the circuit. You can also test transistors while they are connected in circuit with a commercial transistor tester. The transistor tester performs similar front-to-back resistance tests across each junction.

19.6 Unijunction Transistors

Approximately two dozen different types of electronic devices are made by combining a number of different sections of P and N material. Some of these devices are designed specifically for switching larger voltages and currents, and other devices, like the unijunction transistor, are designed to produce a small pulse of voltage that can be used to turn on or bias other switching devices. The unijunction transistor (UJT) is used to produce a pulse of voltage that is used to turn on silicon-controlled rectifiers (SCRs). SCRs are used to control large DC voltages and currents and they will be introduced in Section 19.7. As a maintenance technician, you may be more familiar with the current switch-

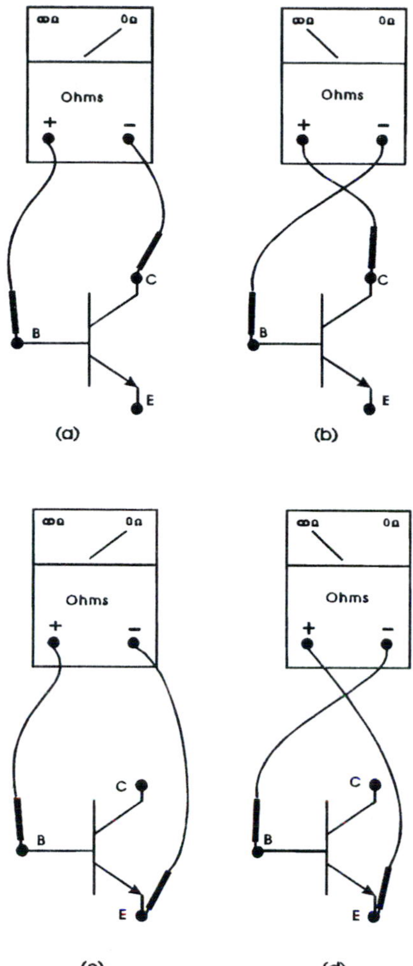

Figure 19–21 Testing the base-collector junction of (a) an NPN transistor for forward bias; (b) an NPN transistor for reverse bias; (c) a PNP transistor for forward bias; (d) a PNP transistor for reverse bias.

ing devices such as SCRs or triacs, but you must remember they will not operate without the devices such as UJTs that are used to turn them on and off.

The unijunction transistor is made of a single PN junction. Figure 19–22 shows a diagram of the PN material and the electronic symbol for the unijunction transistor. The terminals for the unijunction transistor are called base 1, base 2, and emitter.

From the diagram you can see that base 2 and base 1 leads are mounted in the large section of N-type material. This means that if you measured the resistance between base 2 and base 1 terminals, you would measure the same amount of resistance regardless of the polarity of the meter leads. When you measure the resistance between base 2 and the emitter, you would find it is like a diode PN junction, and the polarity of the ohmmeter battery would cause the PN junction to be

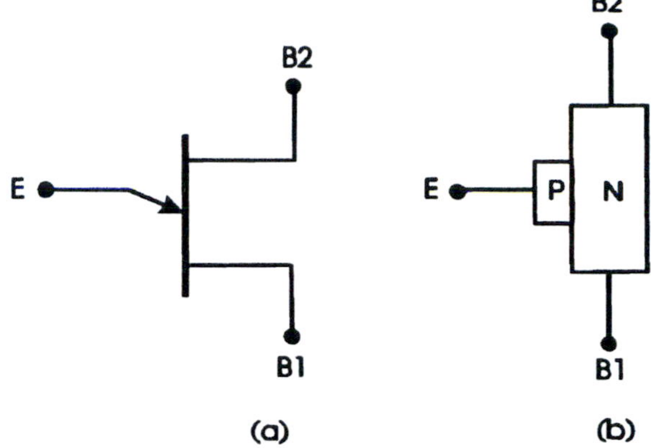

Figure 19–22 (a) Electronic symbol for the unijunction transistor. The terminals for the UJT are base 2, base 1, and emitter. (b) Diagram of P- and N-type material in the unijunction transistor.

forward biased one way, and reverse biased when you reversed the leads. The base 1 to emitter junction also acts like a diode junction.

19.6.1 Operation of the UJT

The simplest way to explain the operation of a UJT is to show it in a typical circuit. Figure 19–23 shows the UJT in a circuit with a resistor and capacitor connected to the emitter terminal, and a resistor connected to each of the base terminals. When voltage is applied to the resistor and capacitor that are connected to the emitter, the capacitor will begin to charge. The size of the resistor will control the time it takes for the capacitor to charge. This time is referred to as the *time constant* for the circuit.

The same circuit voltage is applied to the resistors that are connected to base 2 and base 1. The resistors connected to base 2 and base 1 create a voltage drop with the internal resistance of the UJT. When the charge in the capacitor grows to a value that is larger than the voltage drop across the UJT and the resistor connected to the base 1 lead, a current path will be developed through the emitter and through the resistor that is connected to base 1. When current flows through this path, it creates a voltage pulse that increases to its maximum voltage very quickly. The waveform of this pulse is shown in the diagram, and you can see that it turns on and increases to its maximum value immediately. The duration of the pulse will depend on the internal resistance of the UJT and the ratio of the size of the base 2 and base 1 resistors.

When the UJT is producing the pulse, the capacitor is discharging, and it will begin to charge up immediately

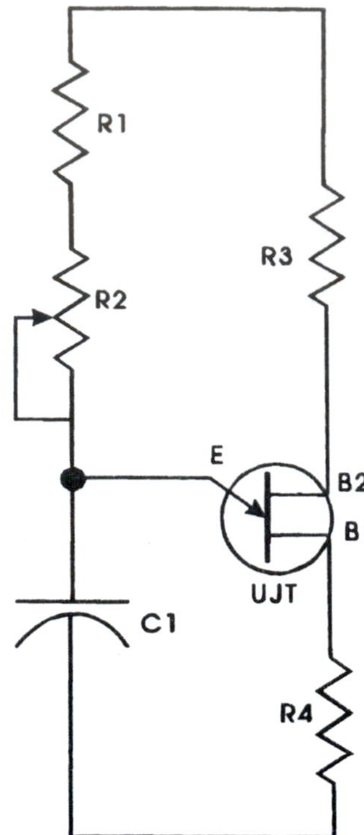

Figure 19–23 A UJT connected to an oscillator circuit to produce sharp output pulses. The size of the resistance and capacitance connected to the emitter controls the frequency of the pulses.

and repeat the process as its voltage charge increases to a point where it is larger than the voltage across the UJT. At this point the UJT will produce another pulse. The time between pulses will be determined by the size of the resistance and capacitance of the resistor and capacitor that are connected to the emitter. If the amount of resistance is increased, the time for the capacitor to charge will increase and the time between pulses will increase. If the amount of resistance at the emitter is reduced, the capacitor will charge more quickly and create pulses that are grouped closer together. This type of circuit is called an *oscillator* and the number of pulses the oscillator produces in 1 second is called the oscillator frequency.

19.6.2 Testing the UJT

The unijunction transistor can be tested like a PN junction. The UJT must be isolated from its circuit and you can test the PN junction between base 2 and emitter, and then test the PN junction between base 1 and emitter. Use an ohmmeter and switch the polarity of its probes so that you forward bias and reverse bias each junction. Remember that when you forward bias a PN junction it should have low resistance, and when you reverse bias a PN junction it should have high resistance.

19.7 The Silicon-Controlled Rectifier

The silicon-controlled rectifier is called an SCR and it is made by combining four PN sections of material. Figure 19–24a shows the PN material for the SCR and Figure 19–24b shows its electronic symbol. The terminals on the SCR are identified as anode, cathode, and gate. Since the SCR is basically a diode that is controlled by a gate, its symbol uses the arrow from the basic rectifier diode that you studied at the beginning of this chapter. When the SCR is turned on, it can conduct large amounts of DC voltage and current (over 1000 V and 1000 A) through its anode and cathode. The major difference between the SCR and the junction diode is that the junction diode is always able to pass current in one direction when the diode is forward biased. The SCR is forward biased by applying positive voltage to its anode and negative voltage to its cathode. At this point the SCR will still have high resistance between its anode–cathode junction. If a positive voltage pulse is applied to the SCR gate, the SCR's anode–cathode junction will have low resistance and the SCR will be in conduction. When the pulse is removed from the gate of the SCR, it will remain in conduction because positive current that comes through the anode will replace the voltage the gate provided. The only way to turn the SCR off is to provide reverse-bias voltage to the anode–cathode, or to reduce the current flowing through the anode–cathode to zero. You should remember that the AC sine wave has zero voltage right before it provides the negative half of its waveform. This means that if the SCR is powered with AC voltage, the SCR will be

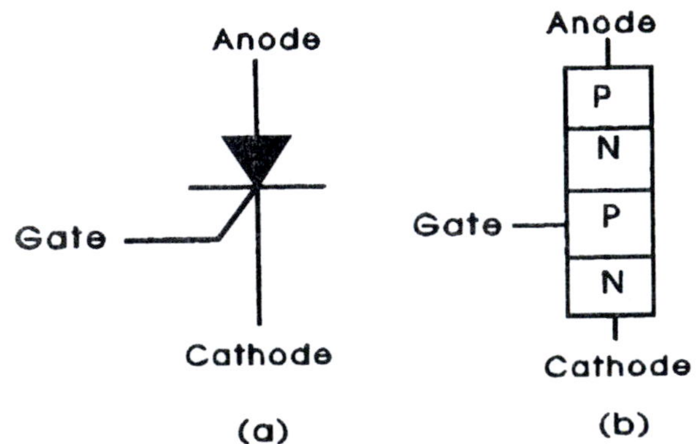

Figure 19–24 (a) Electronic symbol for the silicon-controlled rectifier. The terminals of the SCR are the anode, cathode and gate. (b) P- and N-type material in a silicon-controlled rectifier (SCR).

turned off when the AC waveform goes through 0 V and then to its negative half-cycle. When the AC voltage waveform goes positive again, a gate pulse can be provided and the SCR can go into conduction again. The gate is used to provide a pulse that is used to cause the SCR to go into conduction.

19.7.1 Operation of the Silicon-Controlled Rectifier

Figure 19–25 shows an SCR connected in a circuit to control voltage to a DC load. The source voltage for this circuit is AC voltage. The main advantage of the SCR is that it will not go into conduction until it receives a pulse of voltage to its gate. The timing of the pulse can be controlled so that it can be delivered any time during the half-cycle, which will control the amount of time the SCR will be in conduction. The amount of time the SCR is in conduction will control the amount of current that flows through the SCR to its load. If the SCR is turned on immediately during each half-cycle it will conduct all of the half-wave DC voltage just like a normal diode rectifier. If the gate delays the point where the SCR turns on and goes into conduction at the 45E point of the half-wave, the amount of voltage and current the SCR conducts will be 50 percent of the full applied voltage.

The other important feature of the SCR is that it will go into conduction only when its anode and cathode are forward biased. This means that if the supply voltage is AC, the SCR can go into conduction only during the positive half-cycle of the AC voltage. When the negative half-cycle occurs, the anode and cathode will be reverse

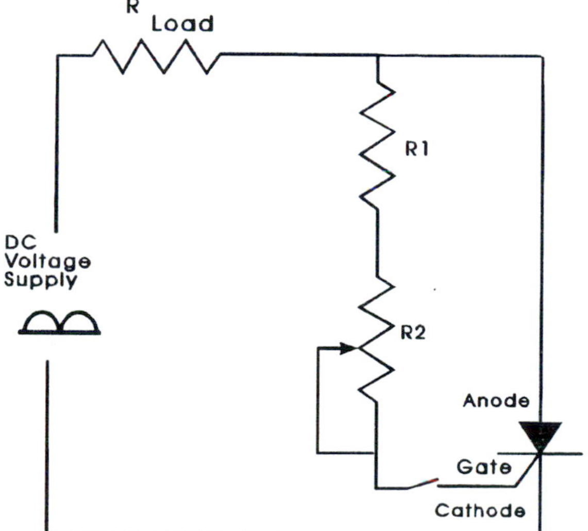

Figure 19–25 A silicon-controlled rectifier (SCR) shown in a circuit controlling DC voltage and current to a load.

biased and no current will flow. This means the SCR will automatically be turned off when the negative half of the AC sine wave occurs. Since the positive half of the AC sine wave occurs for 180E, the SCR can provide control of only 0–180 degrees of the total 360-degree AC sine wave.

Since the turn off point of the SCR is fixed to the point where the sine wave begins to go negative, the SCR can be controlled only by adjusting the point where it is turned on. The point where the SCR is turned on and goes into conduction is called the *firing angle.* If the SCR is turned on at the 10 degrees point in the AC sine wave, its firing angle is 10 degrees. If the SCR is turned on at the 45-degree point, its firing angle is 45 degrees. The number of degrees the SCR remains in conduction is called the *conduction angle.* If the SCR is turned on at the 10-degree point, its conduction angle will be 170 degrees, which is the remainder of the 180 degrees of the positive half of the AC sine wave.

19.7.2 Controlling the SCR

Figure 19–26 shows an SCR in a circuit with a UJT connected to its gate. The load in this circuit is a DC motor. This type of DC motor is often used as a damper control motor. The circuit is powered by AC voltage, and the variable resistor in the oscillator (capacitor–resistor) circuit sets the timing for the pulse that is used to energize the gate of the SCR. Notice that a diode rectifier provides pulsing DC voltage for the capacitor, which charges to set the timing for the pulse that comes from the UJT. Since this DC voltage comes from the original AC supply voltage, it will have the same timing relationship of the original sine wave. This makes adjusting the pulse of the UJT to turn on the SCR gate at just the right time to control the firing angle of the SCR from 0 to 180 degrees. In reality, the firing angle is usually controlled from 0 to 90 degrees, which gives sufficient range of control to adjust the output DC voltage that is sent to the DC motor. Remember that the speed of the DC motor can be controlled by adjusting the voltage sent to the armature and field. This diagram shows the waveform for the voltage at each point in this circuit. The load in this circuit could also be any other DC-powered load.

19.7.3 Testing the SCR

You will need to test the SCR to determine if it is faulty. Since the SCR is made of PN junctions, you can use forward-bias and reverse-bias tests to determine if it is faulty. In this test you should put the positive probe on the anode and the negative probe on the cathode. At this point the ohmmeter will still indicate the SCR has high resistance. If you use a jumper wire and connect

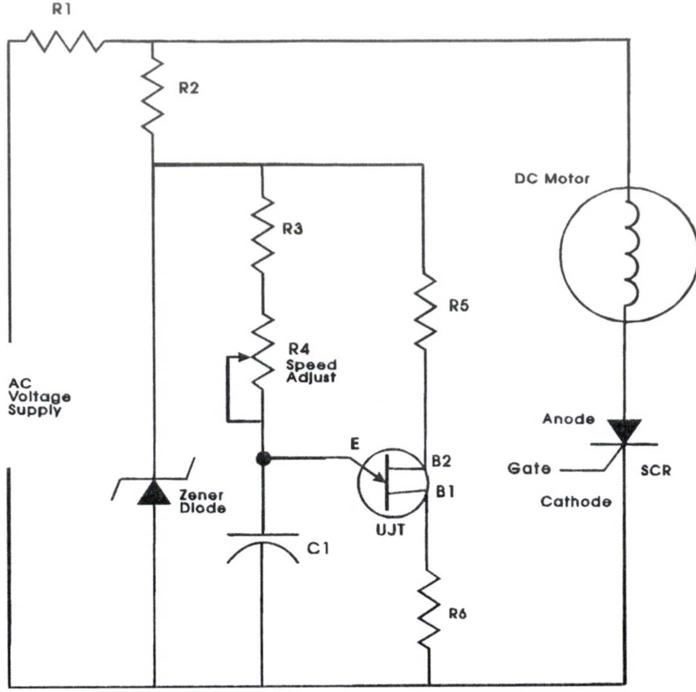

Figure 19–26 An SCR used to control a DC motor. A UJT is connected to the gate of the SCR to control its firing angle.

positive voltage from the anode to the gate, the SCR will go into conduction and have low resistance. The SCR will remain in conduction until the voltage applied to its anode and cathode is reverse biased, or if the voltage applied to the anode and cathode is reduced to zero. This means that you can turn the ohmmeter polarity switch to the opposite setting, or you can remove one of the probes and the SCR will stop conducting. The amount of current to keep the SCR in conduction is approximately 4 to 6 mA. This means that some high-impedance digital volt/ohmmeters will not have enough current when set to the ohms range to keep the SCR in conduction. If this is the case, you may need to test the SCR with an analog ohmmeter, which has a needle and scale. You should also test the SCR for reverse bias to ensure that it has high resistance. Sometimes an SCR will not go into conduction because it has developed an open in its anode–cathode circuit. Other SCRs may stay in conduction at all times, which means the SCR is shorted.

19.7.4 Typical SCRs

SCRs are available in a variety of packages and case styles that include three terminals and larger types that have threads so that they can be mounted directly into heat sink material. The heat sink material will be made from metal and in some cases it will include a fan that moves air over fins that are molded into the material to help remove the large amounts of heat that build up in the devices. Figure 19–27 shows examples of different types of SCRs.

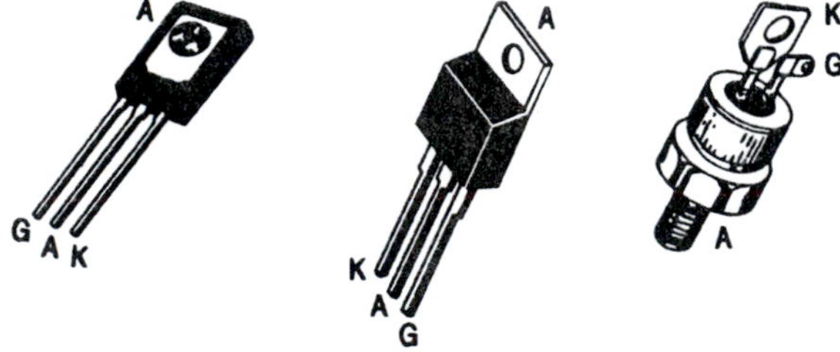

Figure 19–27 Typical SCRs shown in a variety of packages and case styles.
(Copyright of Motorola. Used by Permission)

19.8 The Diac

The unijunction transistor provides a positive pulse for the gate of the SCR that allows the SCR to control large DC voltages and currents. The *diac* is an electronic component that provides a positive and negative pulse that is used as trigger signal for a device called a triac. The triac is similar to the SCR except that it can conduct voltage and current in both the positive and negative directions. This means that the triac provides a control-

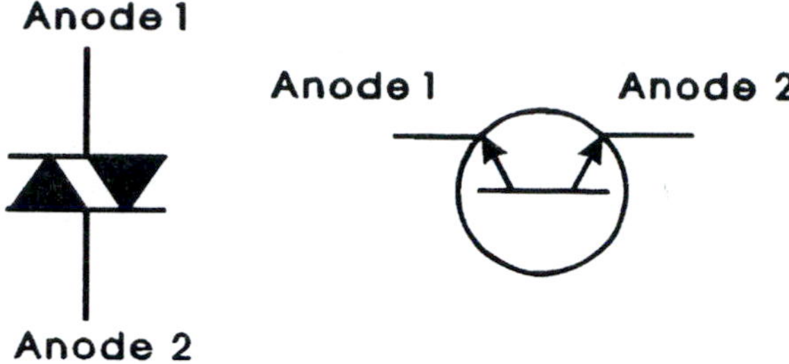

Figure 19–28 Electronic symbols for the diac. Notice the diac has two anodes and also has two symbols. It can be represented by either symbol.

ling function for AC voltage and current that is similar to that of the SCR. As a maintenance technician you may not notice the diac because it is used as a device to produce a firing pulse for the triac. You will learn in the next section how the triac is used in speed controls and other AC voltage control circuits. Figure 19–28 shows the electronic symbol for the diac. The symbol consists of two arrows, which show that voltage pulses can be positive or negative. Since the diac symbol has two arrows, it can be shown in two different styles.

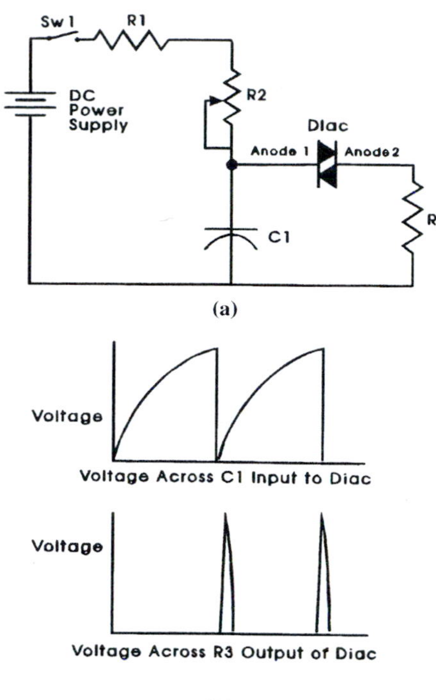

Figure 19–29 (a) A diac in a circuit and (b) the pulses that the diac produces.

19.8.1 Operation of the Diac

When voltage is applied to a diac in a circuit, its PN junction will remain in a high-resistance state and block the voltage until the voltage level reaches the breakover level. This means that when AC sine wave voltage is applied to the circuit, the sine wave voltage will increase to its peak level, then return to zero, and then increase to negative peak voltage. Each time the voltage increases toward its peak, and it exceeds the breakover level of the diac, the diac will provide a sharp output pulse of the same polarity as the voltage that is supplied to it. For example, if the breakover voltage level is 18 V, the diac will block voltage until the AC voltage exceeds the 18-V level. At this point, the diac will produce a pulse that is approximately 18 V. Figure 19–29 shows the diac in its circuit and the pulse that it creates.

19.9 The Triac

The *triac* is basically two SCRs that have been connected back to back in parallel so that one of the SCRs will conduct the positive part of an AC signal, and the other will conduct the negative part of an AC signal. As you know, the SCR can control voltage and current in only one direction, which means it is limited to DC circuits when it is used by itself. Since the triac acts like two SCRs that are connected inverse parallel, one section of the triac can control the positive half of the AC voltage and the other section of the triac can control the negative half of AC voltage. Figure 19–30a shows the electronic symbol of the triac and Figure 19–30b shows the arrangement of its P-type and N-type material. The terminals of the triac are called *main terminal 1 (MT1)*, *main terminal 2 (MT2)*, and *gate*. Since the triac is basically two SCRs that are connected inverse parallel, MT1 and MT2 do not have any particular polarity.

19.9.1 Using the Triac as a Switch

The triac can be used in an industrial circuit as a simple on–off type switch. In this type of application, the MT1 and MT2 terminals are connected in series with the AC load. When the gate gets a positive pulse signal, the triac will turn on for the positive half of the AC cycle. When the AC voltage waveform returns from its positive peak to 0 V, the triac will turn off. Next the negative half-cycle of the AC voltage waveform will reach the triac, and it will receive a negative pulse on its gate and go into conduction again.

This means that the triac will look as though it turns on and stays on when AC voltage is applied. The load

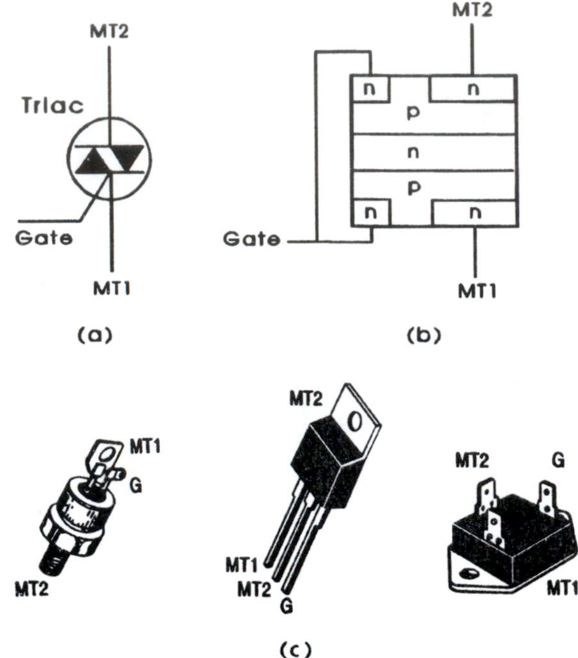

Figure 19–30 (a) Electronic symbol for the triac. (b) P-type and N-type material for the used in the triac. (c) Typical packages for triac semiconductor devices.
(Copyright of Motorola. Used by Permission)

that is connected to the triac will receive the full AC sine wave just as if it were connected to a simple single-pole switch. The major difference is that the triac switch can be used for millions of on–off switching cycles. The other advantage of the triac switch is that the gate pulse can be a very small amount of voltage and current. This allows the triac to be used in temperature control where the temperature-sensing part of a thermostat can be a small solid-state sensing element called a *thermistor.* The sensing element can also be a vary narrow strip of mercury in a glass bulb that is very accurate, but can carry only a small amount of voltage or current.

Another useful switching application for a triac is the solid-state relay. Section 19.10 will explain AC and DC solid-state relays. Figure 19–31 shows a triac in a thermostat connected to a 24-V AC gas valve in a furnace. The temperature-sensing element is connected to the gate terminal of the triac. When the temperature increases, it makes the mercury in the temperature-sensing element expand and make contact between the two metal terminals. The small amount of voltage flowing through the sensing element will be sufficient to cause the triac to go into conduction and provide voltage to the gas valve.

19.9.2 Using the Triac for Variable Voltage Control

The triac can also be used in variable voltage control circuits since it can be turned on any time during the positive or negative half-cycle, similar to the way the SCR is controlled for DC circuit applications. In this type of application, a diac is used to provide a positive and negative pulse that can be delayed from 0 to 180 degrees to control the amount of current flowing through the triac. This type of circuit can be used to control the amount of current and voltage supplied to electric heating elements. This allows the amount of current and voltage to be controlled from zero to maximum, which in turn allows the temperature to be controlled very accurately. Figure 19–32 shows a diagram of a triac used to control an electric heating element that is powered by an AC voltage source. Notice that a variable resistor is connected with a capacitor to provide an oscillator pulse to the diac. Since the resistor and capacitor are connected to an AC voltage source, the pulse will be both positive and negative as the AC sine wave changes polarity. The triac is connected to the same AC voltage source so that the timing of the pulse from the diac to its gate will always be synchronized with the polarity of the voltage arriving at the main terminals of the triac.

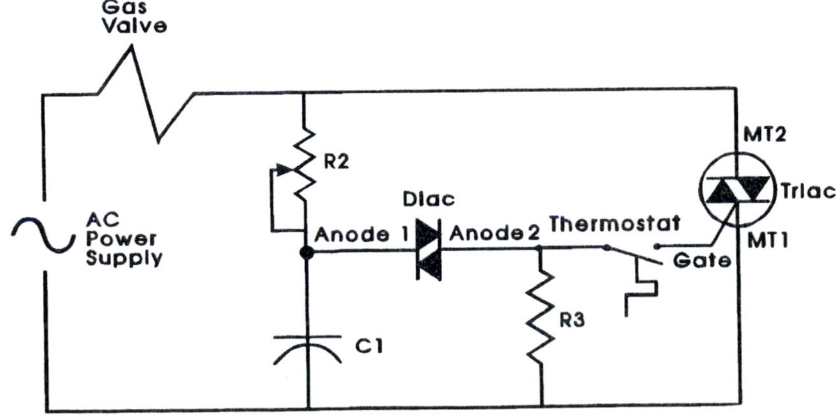

Figure 19–31 A triac used as a switch to turn on a solenoid. The triac receives its gate signal from a small amount of voltage that moves through the temperature-sensing element.

Figure 19–32 A triac used to control variable voltage to an AC heating element. Notice that a diac is used to provide a positive and negative pulse to the triac gate.

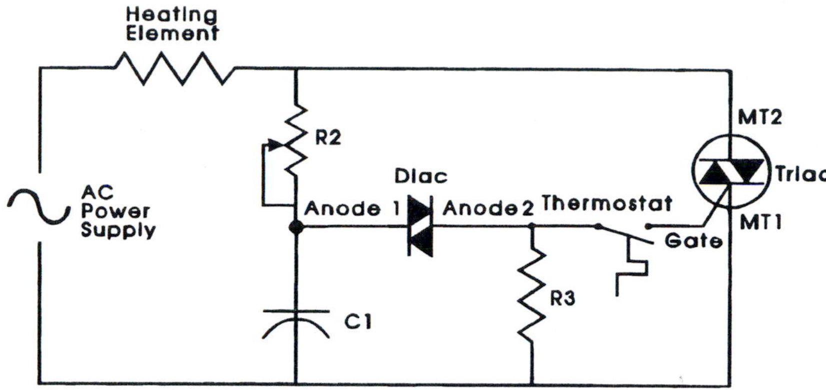

19.9.3
Testing the Triac

Since the triac is made from P-type material and N-type material, it can be tested like other junction devices. The only point to remember is that since the triac is essentially two SCRs mounted inverse parallel to each other, some of the ohmmeter tests will not be affected by the polarity of the ohmmeter leads. In the first test for a triac, an ohmmeter should be used to test the continuity between MT1 and MT2. When no gate pulse is present, the resistance between these terminals should be infinite, regardless of which ohmmeter probe is placed on each of the terminals. Figure 19–33 shows how the ohmmeter should be connected to the triac. In Figure 19–33a you can see that the positive ohmmeter probe is connected to MT1, and the negative probe is connected to MT2. Since no voltage is applied to the gate, the resistance should be infinite. When voltage from MT2 is jumped to the gate, the triac will go into conduction and the ohmmeter will indicate low resistance.

In Figure 19–33b you can see that the positive ohmmeter probe is connected to MT2, and the negative probe is connected to MT1. Since no voltage is applied to the gate, the triac is not in conduction and the ohmmeter will indicate that the resistance at this time is high. When voltage from MT2 is jumped to the gate, the triac will go into conduction and the ohmmeter will indicate that the resistance is low. The triac will stay in conduction only while the gate signal is applied from the same voltage source as MT2. As soon as the voltage source is removed, the triac will turn off.

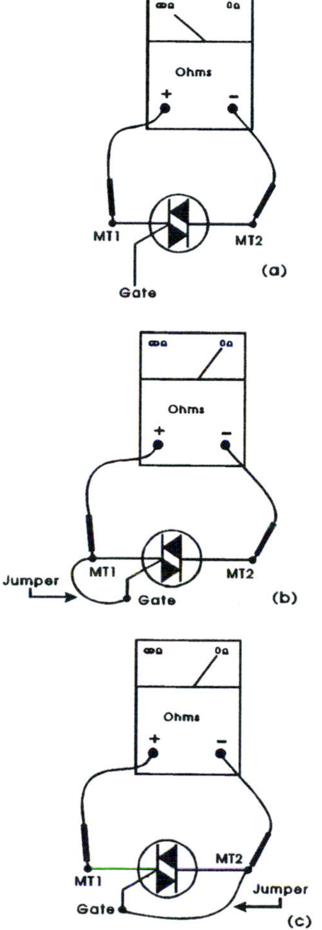

Figure 19–33 (a) Testing a triac by placing the ohmmeter's positive probe on MT2 and its negative probe on MTI. The triac is not in conduction at this time. (b) When positive gate voltage is applied from MT2 to the gate, the triac goes into conduction. (c) When negative voltage from MTI is applied to the gate, the triac goes into conduction.

19.10
Solid-State Relays

Another device that you will see frequently used in industrial control systems is called a *solid-state relay (SSR)*. The solid-state relay can use AC or DC voltage for the control part of the circuit that acts like the coil of a regular relay, and it can switch AC voltages through the part of the circuit that acts like the contacts. The solid-state relay uses SCRs, transistors, or triacs to control various loads. It operates very similarly to an

(a)

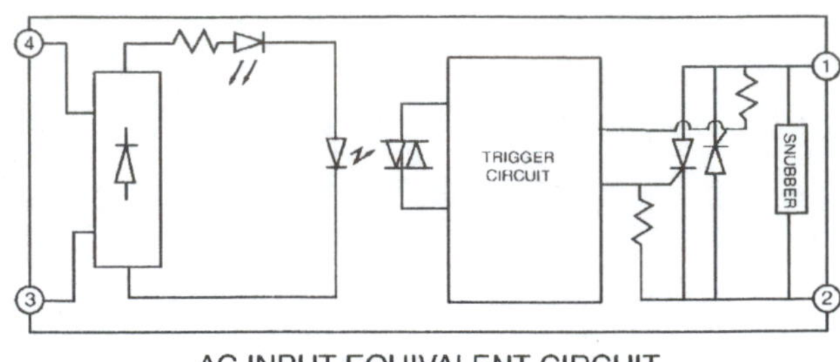

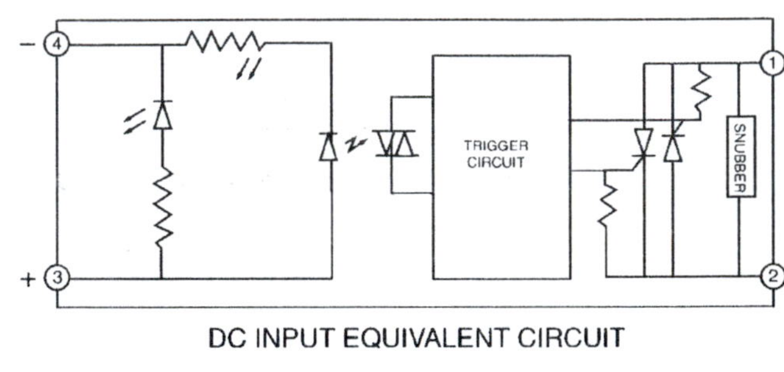

(b)

Figure 19–34 A solid-state relay. Diagrams for AC input and DC input.

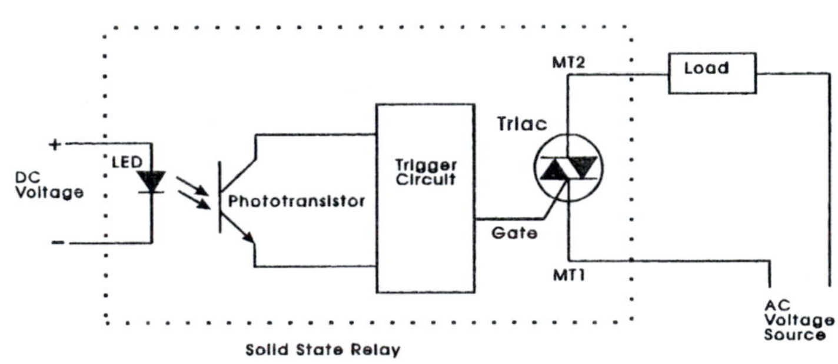

Figure 19–35 An electronic diagram of a solid-state relay that has a DC input signal and 120 output signal.

electromechanical relay in that it is energized by a small voltage, which controls a larger voltage and current.

Figure 19–34 shows a picture of a typical solid-state relay and Figure 19–35 shows four electrical diagrams of the solid-state relay. From the diagrams you can see that the internal circuits of an SSR can use a light-emitting diode or it can use a bridge rectifier to detect the DC input voltage. The output stage of the SSR can be controlled with a triac or two diodes that are connected inverse parallel to allow AC voltage to flow through them. Figure 19–35a shows an LED for the input part of the circuit, and a zero-voltage detection circuit that is used to turn on a triac, which controls the AC voltage in the output circuit. Figure 19–35b shows an SSR that uses a bridge rectifier so its input voltage can be AC or DC. If AC voltage is applied to the input, the voltage passes through the bridge rectifier where it is converted to DC voltage and sent to the LED, where it triggers the zero-voltage detection circuit and finally causes the triac to go into conduction and pass AC current to the load. If the input voltage is DC, it will pass directly through the bridge rectifier and trigger the LED. The zero-voltage detection circuit contains a photo detecting component such as a photo transistor. When these components are encapsulated together it is called an *optoisolation device* or an *optocoupler*. The optoisolation component is manufactured in an integrated circuit (IC) and incorporated into the zero-voltage detection circuit. When current flows through the LED, its light causes the optoisolation device to go into conduction and send a signal to the triac to turn it on and allow it to go into conduction and pass AC current to the load. The AC load is controlled directly by the triac. Figure 19–35c shows an SSR with an LED to accept a DC input voltage, and the dual SCRs connected in an inverse parallel configuration to control the AC output. Figure 19–35d shows an SSR that has a bridge rectifier for its input and has the dual SCRs connected in an inverse parallel configuration to control the AC output. Notice that the power source for the load must be isolated from the input signal.

Questions

1. Explain P-type and N-type material.
2. Explain the operation of a diode (PN) junction and show the input AC waveform and the output DC waveform.
3. Identify the terminals of a transistor and explain its operation.
4. Explain the operation of the SCR and explain the type of circuit where you would find one.
5. Explain the function of the unijunction transistor when it is used with an SCR.

True and False

1. The function of the triac is to provide switching similar to an SCR except the triac operates in an AC circuit.
2. The light-emitting diode (LED) is similar to a DC light bulb in that it has a very tiny filament.
3. The main function of diodes and SCRs is to convert AC voltage to DC voltage.
4. The full-wave bridge rectifier uses one diode.
5. The UJT and diac are important solid-state devices that provide pulse signals to other components to use as a firing signal.

Multiple Choice

1. A PN junction is forward biased when ____
   a. positive voltage is applied to the N material and negative voltage is applied to the P material.
   b. negative voltage is applied to the N material and negative voltage is applied to the P material.
   c. its junction has high resistance.
2. The rectifier _____
   a. converts DC voltage to AC voltage.
   b. converts AC voltage to DC voltage.
   c. can be used as a seven-segment display for numbers.
3. A circuit that has one SCR in it can _____
   a. control AC voltage.
   b. control DC voltage.
   c. control both AC and DC voltage.
4. The transistor operates similar to a relay in that _____
   a. its emitter is like the coil and its base and collector are like a set of contacts.
   b. its collector is like the coil and its base and emitter are like a set of contacts.
   c. its base is like a coil and its emitter and collector are like a set of contacts.
5. The solid-state relay _____
   a. uses an LED to receive a signal and a transistor to switch current.
   b. uses a capacitor to receive a signal and an IC to switch current.
   c. uses a transistor to receive a signal and an IC to switch current.

Problems

1. Draw the symbol for a diode and identify the anode and cathode.
2. Draw the symbol for an NPN and a PNP transistor and identify the emitter, collector, and base.
3. Draw the symbol for a silicon-controlled rectifier (SCR) and identify the anode, cathode, and gate.
4. Draw the electrical symbol for a unijunction transistor and identify base 1, base 2, and emitter.
5. Draw the electrical symbol for a triac and identify MT1, MT2, and gate.

◀ Chapter 20 ▶

Troubleshooting

Objectives

After reading this chapter you will be able to:

1. Explain the fundamentals of troubleshooting.
2. Explain the difference between a symptom and problem.
3. Develop a wiring diagram from a ladder diagram.
4. Develop a ladder diagram from a wiring diagram.
5. Develop a procedure for locating a loss of voltage in a circuit.

As a technician or maintenance person, you will be requested to troubleshoot equipment that has stopped running or is not operating correctly. Many people fear this task because it looks so complex. This chapter will help you gain the skills that make troubleshooting much easier and more predictive. Technicians who have been in the field for many years have developed tests to verify the condition of every component that is found in an electrical control or load circuit. You must believe in the value of these tests and strictly follow their procedures. You can review these tests and their procedures in the previous 19 chapters of this book. After reading this chapter you will have the skills to find problems in any machine that you are requested to troubleshoot. You will also learn troubleshooting processes that can be used every time you troubleshoot something and you will be able to verify the condition of the components you suspect are working correctly or the components you suspect are faulty.

Troubleshooting can be broken into multiple scenarios. You may be asked to troubleshoot a machine that is already in operation and has a wiring diagram, and a ladder diagram that explains the sequence of operation of the machine. In this case, you would use all of the diagrams to test the operation of the machine and determine which parts are inoperative.

In some cases you may be asked to try and troubleshoot a machine that has a wiring diagram but does not have a ladder diagram. In this case, you must convert the wiring diagram to ladder diagram to determine the sequence of operation for the machine. If the machine has only a ladder diagram but does not have a wiring diagram, you must convert the ladder diagram into a wiring diagram so that you can determine the location of the components. In some cases you may be able to open the cabinet door and identify where the components are located.

This chapter explains how to troubleshoot a machine when you have both a wiring diagram and a ladder diagram. It will also explain how to convert a wiring diagram into ladder diagram, and how to convert a ladder diagram into wiring diagram. Both of these skills are very useful when you're working on the job to troubleshoot machines.

Working Safely

The most important point to remember when you are troubleshooting is to work safely and follow all safety procedures provided in the previous chapters in this text. You should strictly follow lockout and tag-out procedures to ensure that the electrical, hydraulic, pneumatic, and mechanical energy sources are safely turned off and secured when you are working on the machine. Some tests require that these systems be fully energized to complete the test. In these cases be sure that you are following all safety rules and are fully aware of the potential for injury when you are working around live circuits. You should also work as though each circuit and power system is live and active any time you are working around a machine.

20.1 What Is Troubleshooting?

Many technicians believe that troubleshooting is a process to determine what is broken on a machine. A little known secret that the best troubleshooting technicians have learned is that it is actually easier to identify what is working correctly on a machine and, through a process of elimination, determine what is broken. This section explains the basic things you should do when you are trying to troubleshoot a machine. You should develop the steps into a procedure that you use each time that you troubleshoot a machine. Many technicians do not use a consistent procedure each time. Instead, they try different things each time they troubleshoot a different machine, and sometimes they try to remember what was broken on the machine the last time and work from there as though the problem is simply repeating itself. You will learn to develop a procedure that you can use every time you troubleshoot a machine.

20.2 The Difference Between Symptoms and Problems

When you are first learning to troubleshoot a machine, you may find what appears to be a problem, but instead is only a symptom of the problem. For example, you may find a blown fuse on a machine, and think you have found the problem. In fact, if you replace the fuse, the machine will begin to operate again. What is hard to understand is that the machine has a motor that is overheating and overworking because it has a bad bearing, and this is what is causing the fuse to blow. Each time you replace the fuse, the motor will start and run again, until it draws enough overcurrent and blows the fuse again. If you continue to replace fuses, the motor will eventually overheat to the point at which it is totally destroyed. In this example, the blown fuse is only a symptom, the bad bearing is the problem. You will learn in this chapter how to determine what are problems and what are symptoms. You may need to revisit some of the early chapters in this text to determine how components, such as relays, transformers, and motors, operate.

20.3 Starting the Troubleshooting Procedure

The first step in troubleshooting a machine should be to determine what the machine was doing prior to stopping. Sometimes this can be done by simply talking to the machine operator or a worker who is assigned to the machine. Sometimes this is the person who was loading and unloading the machine, or it may be a person who was setting up and operating the machine. You can ask a series of questions to determine if the machine was operating correctly prior to this malfunction. For example, you may ask the operator to try and operate different parts of the machine in manual mode. Sometimes you will find that the machine can operate partially or that one part of the machine can operate. If you have not observed this machine operating before, you may need to look around in the general area to see if there are other machines that are similar to the machine that is malfunctioning. You may be able to determine the operation of the machine by observing a similar machine that is close by. If you cannot observe a machine, you may have to ask the operator the basic sequence of operation for the machine.

20.3.1 Observing the Machine and Using All of Your Senses

When your first begin to troubleshoot a machine, it looks like a very large, complex job. As you work with more experienced technicians, you'll find that they use a number of methods to determine what is wrong with the machine. For example, they may spend a few minutes just looking at the machine and trying to get an idea of what components are running or turned on. They may also try to identify what power systems the machine has, such as the type of voltage, if there is hydraulics, or if there is a pneumatic system. You should also try to identify the moving parts on the machine and determine how they operate—Do they use gravity? Do they use belts? Do they use hydraulic cylinders? Another important part of this process is to identify what types of motors the machine has and how many there are. Do the motors have variable frequency drives (VFDs) or other types of acceleration and speed controls or are they connected directly to a motor starter? Finally, you may take a minute to determine whether the machine uses regular hardwired controls or a programmable logic controller (PLC).

A second to step in the process of troubleshooting includes using your other senses beyond vision. For example, you may want to listen for problems, such as motors that are under extreme load, hydraulic pumps that are fully loaded, or belts and pulleys that are squeaking, which indicates they are running improperly and causing friction. These clues will help you later when you're trying to determine what is wrong with the machine. Another sense that is usually overlooked is the sense of smell. In some cases you will be able to smell a motor that is overheating before you're able to determine from some other test that it has a problem. You may also be able to smell a machine that has overheated hydraulic oil, or a machine that has rubber seals that are going. After years of experience, you will learn what a machine smells like when it is operating correctly, and what it smells like when it is faulty.

Some technicians have become very good at using their hands or a screwdriver in their hands to determine if there is any abnormal vibration on the machine. Vibrations that are extreme can be felt and will help to determine if the machine is malfunctioning. There is also sensitive test equipment that can determine if a machine has abnormal vibrations.

20.3.2 Swapping Parts

When you have tested your machine and have identified a faulty component, you may want to swap a part to ensure that the part you suspect is faulty is actually bad. This procedure has several drawbacks. First, you are putting a second machine out of commission to remove the part to do the test. This may create additional problems and loss of more production. The second problem this may create is that some technicians utilize the component swap method too soon, prior to thoroughly testing the part, and basically guess which part is bad and waste a lot of time changing parts. The advantage of swapping a part is that if the part is expensive or has a two- or three-day shipping time, you can ensure that you have actually located the faulty part before you order a new one and wait for it to be shipped.

The basic rule of thumb to follow when swapping parts is to ensure that you have completely tested the part while power is applied and you have completed an alternative test with power off. If you have completed these tests and are sure the part is faulty, you can safely change or swap the part to verify your suspicions.

20.3.3 When to Use a Voltmeter to Troubleshoot

The voltmeter is the meter of choice for most technicians. The voltmeter allows the technician to make tests in the circuit with power applied. It also gives the advantage of being able to make a large number of preliminary tests without disconnecting any wiring or components. If you are called to troubleshoot a machine that has stopped completely, the chances of it having an open circuit, where voltage is interrupted, is very large. The first test that you should make with the voltmeter is to measure the source voltage. You should always set the meter to the highest voltage range until you have determined the actual voltage. After measuring source voltage, you will need to use the ladder diagram for the machine and measure voltage throughout the circuit wherever you suspect the voltage has stopped. When you find the loss of voltage, you can turn power off to the machine, lock it out, disconnect the wire or component that you suspect is faulty, and test it with an ohmmeter.

Some problems that you might encounter when measuring voltage will provide you with inaccurate information. The first problem is voltage feeding backward through a circuit. This usually occurs in three-phase circuits that have one blown fuse. If you have one blown fuse in a three-phase circuit, it is possible for voltage from one of the remaining supply lines that has a good fuse to feed through a transformer winding or motor winding and flow backward up to the bottom of the blown fuse. When you test the output side of each fuse, your meter will measure voltage and it will appear that all of the fuses are good. If you are aware this can happen, you can use a fuse puller to pull all three of the fuses from their holders and test each fuse for continuity with an ohmmeter.

The other problem that is beginning to appear to troubleshooters is odd values of voltage in a circuit. If a circuit has a variable frequency drive, the drive can produce voltages that are above or below 60 Hz. Since most digital and analog voltmeters are designed for 60 Hz, the voltage reading on the meter will be slightly more or less than it actually is, and the technician may feel that the circuit is faulty, when in fact it is operating correctly. You can use a true rms reading voltmeter to get an accurate reading in these types of circuits.

20.3.4 When to Use an Ohmmeter to Troubleshoot

It is important to understand when to use an ohmmeter. All power to a circuit must be turned off before you use an ohmmeter to take measurements, since the ohmmeter presents very low impedance to the circuit. When impedance is low, it will create a short circuit path if you place the meter leads across the terminals that have voltage present. This will cause the large current to create sparks and possibly cause molten metal to explode into your face and it will certainly damage the meter. In most cases, you will use the ohmmeter to make a continuity test with power off, after you have used a voltmeter to locate a faulty switch or wire. When you are making the continuity test, you should zero the meter and use the lowest ohm setting. This setting will detect the slightest amount of resistance in the switch or wire. If the resistance is very high, it is called infinity, and it indicates that an open circuit is present in the contacts or the wire. If you are testing a wire, this test indicates the wire is faulty and must be replaced. If you are testing a switch, be sure you are on the correct terminals that make a contact set, and then change the switch to its other positions. If the meter reading remains at infinity, you have determined the switch contacts are faulty, and the switch must be replaced.

Another reason to use an ohmmeter is to test the actual number of ohms in the windings of a motor, transformer, or coil. When you are testing a single-phase motor, you should measure the resistance in each winding accurately so that you can determine that the winding with the highest resistance reading is the start winding and the winding with the lowest resistance reading is the run winding.

If you are troubleshooting electronic circuits, you will also need to measure the amount of resistance accurately. For example, when you are trying to identify the terminals of a transistor, SCR, or triac, you will need to measure the resistance and the polarity of the meter probes. Be sure to switch the meter to the setting that will provide the most accurate resistance reading.

20.3.5 When to Use a Clamp-on Ammeter to Troubleshoot

A clamp-on ammeter is a valuable troubleshooting instrument that can indicate the presence of current in a circuit. For example, if you suspect that a motor with three-phase supply voltage has one blown fuse, you can quickly determine which leg of the three phases has the blown fuse by measuring the current in each of the three legs with the clamp-on meter. The leg that has the blown fuse will indicate 0 A, and the two legs where the fuses are good will indicate current that is higher than rated for the circuit, because the motor will be drawing locked-rotor current since it will not be able to start. This test is more effective than the voltage drop test, which is subject to feedback voltages.

The clamp-on ammeter can also be used to evaluate the current load of any circuit wire. This is especially helpful in determining how close a circuit is to overloading the circuit wire, which will cause a fuse to blow. You can also tell if a motor is beginning to wear out and have bad bearings or lack of lubrication. If the motor current is higher than its rating, you can begin to suspect that it is overloaded. If the load is within tolerance, you can suspect that the motor needs lubrication or the bearing may be bad.

The clamp-on ammeter can also be used to adjust belts on belt drive systems. You can apply the clamp-on ammeter and adjust the belt tension while watching the meter. If the current increases above the motor rating, the belts are too tight. If the current rating is too low, the belt is not tight enough.

20.3.6 Dividing the Machine Electrical System into the Load Circuit and the Control Circuit to Make Troubleshooting Easier

The part of the electrical system that has the motors in it is called the load circuit. Other electrical loads may include lights and electric heaters. The control circuit includes all of the switches and devices that control voltage to the coil of motor starters, relays, or solenoids. When you are troubleshooting a circuit that has pushbutton switches controlling a motor starter coil, you can quickly eliminate parts of the circuit that are functioning and the parts that are not by seeing if the coil of a motor starter is energized. If the coil of the motor starter is energized, it indicates that all of the switches and the coil are operating correctly, and you should shift your test to the load circuit where the motor is connected to the motor starter.

Another way to understand this concept is to think about troubleshooting a machine the way you would try to locate a specific card in a deck of playing cards. Suppose I pull one card from the deck and then ask you to guess what the card is. As you know, there are 52 cards in a standard deck of cards that has hearts and diamonds that are red cards, and clubs and spades that are black cards. So if you just start guessing, without a method, it may take you 52 guesses to get the correct one. This section shows you a method that you can use later when you are troubleshooting a machine.

For this example, let's say the card I have pulled from the deck is the 9 of hearts. Using this process, the first question would be to ask if the card is a black card. The answer is no, so we can deduce that the card must be red. This one question eliminates 50 percent of the deck. We now know the card must be a heart or diamond and we can stop thinking about all of the spades and clubs in the deck.

The next question will be, is the card a heart? (It does not matter whether I ask if the card is a heart or if I ask if it is a diamond, because the answer will indicate which one it is.) In this case the answer is yes, the card is a heart, so I can stop thinking about all of the diamonds. With two questions, we have eliminated 75 percent of the cards.

Now that I know the card is a heart, I will continue to ask questions that eliminate the remaining cards 50 percent at a time. I know there are 13 hearts and they range from a high of ace and a low of 2, so I will next ask if the card is lower than an 8. The answer is no, so I have eliminated the 2, 3, 4, 5, 6, and 7. The remaining cards are 8, 9, 10, jack, queen, king, and ace. Since the queen or jack is the midpoint, I would next ask if the card is below a queen. The answer is yes, so now I know the card must be a jack, 10, or 9. Then the next question is, is the card lower than a 10? Since the answer is yes, I now know the card is the 9 of hearts.

You may be wondering what this card example has to do with troubleshooting. The four suits of cards—diamonds, spades, hearts, and clubs—represent the four power systems on most machines: electrical, hydraulic, pneumatic, and mechanical. When you come upon a broken machine, your first questions should determine which of these systems are working and which are inoperative. You need to develop questions and tests that will eliminate the most possibilities. Let's say that you have found that the air pressure and the mechanical parts of the system operate correctly, but the hydraulic and electrical systems will not energize. The next test would be to see if you can get the electrical system energized. You can see that this process cuts down the number of parts of the machine that could possibly be faulty.

When you are testing the electrical system, you would check the load and control system that controls the hydraulic pump. The load circuit consists of the circuit that shows the three-phase supply voltage connected to the motor starter and the hydraulic pump connected to the motor starter. The control circuit includes the master stop pushbutton, master start pushbutton, and the coil of the motor starter for the hydraulic pump. The process of selecting tests that will eliminate large portions of the machine will logically help you find the problem on any machine every time you troubleshoot.

20.3.7 Why It Is Important to Test a Circuit with Power Applied

You may wonder why it is important to test a circuit with voltage applied to the circuit. You may even think that this is a dangerous practice. You will find that it is a very safe way to test a circuit and it is also the most reliable. The problem with testing a circuit for continuity with the power off is that it is a very slow process because all of the wires must be disconnected before they can be reliably tested, and it is very easy to have faulty readings from parallel circuits that you are not aware of. If you expect to become a qualified electrical technician, you must be able to repeat the voltage-loss test explained in the previous section without making mistakes. You must also be able to make the test quickly and accurately.

20.3.8 Troubleshooting the Pushbutton Switches in the Control Circuit

Testing the control circuit in a system is a simple procedure. You should remember that the control circuit may contain several pushbutton switches that may be faulty or not operate correctly. Figure 20–1 shows such a circuit. The problem could be any one of several things, such as the loss of the power supply, a loose or broken wire, a faulty pilot switch, or a faulty load (motor starter or relay coil). This means that you must troubleshoot the complete circuit instead of picking on individual components and testing them at random. The test that will be outlined in the following procedure is called the *voltage-loss* test and you should learn to use this test whenever you are trying to locate a faulty component or wire in an inoperable circuit. We will use the control circuit diagram shown in Figure 20–2.

Start the procedure by testing for voltage across the load terminals (the two terminals of the motor starter coil). If voltage is present at the terminals of the motor starter coil, you can be sure that the power supply and all the pilot switches and interconnecting wires are working correctly and the remainder of the troubleshooting test should focus on the coil.

If no voltage is present at the terminals of the coil, the test should focus on the loss of voltage somewhere between the source and where it travels through the pilot switches to the coil. If voltage is not present at the power supply, be sure to check for a blown fuse, open disconnect, or other power supply problems. If voltage is present, it may be 24, 120, or 240 V or higher, and it may also be AC or DC. At this point in the test, all that

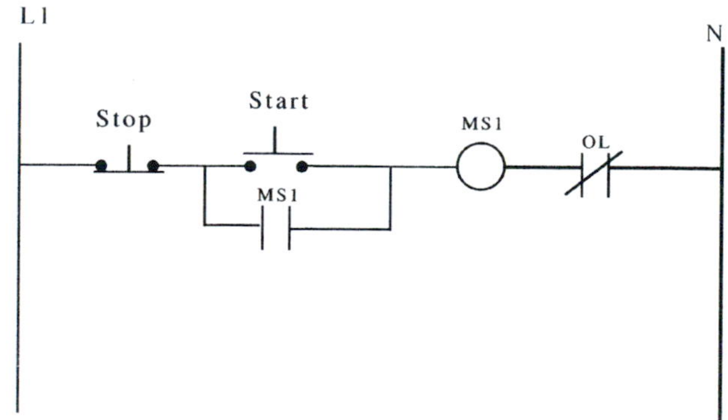

Figure 20–1 Control circuit diagram for troubleshooting.

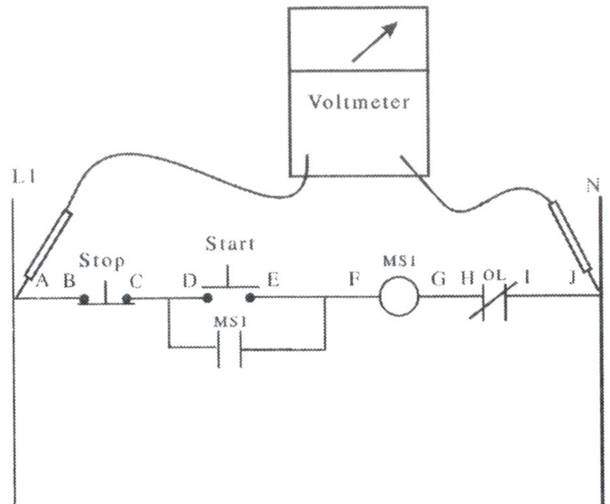

Figure 20–2 Wiring diagram with test points for voltage-loss test.

you are concerned with is that the supply voltage is the same voltage that is specified on the electric control diagram.

After you have determined that the proper amount of supply voltage is available, the remainder of the tests should focus on the pilot switches and interconnecting wires. The fastest and most accurate test for locating an open wire, loose terminal, or faulty pilot switch is to test for a voltage drop (loss). This test requires that power remain applied to the circuits while you touch the meter probes to each test point. *You must be aware of where you are placing your hands, tools, and meter probes at all times because severe electrical shock could result if you come in contact with the electrical circuit.* Even though leaving power applied to a circuit while you test it presents an electrical safety condition, it is necessary because control circuits tend to be complex and you may not be able to find the problem with the power off.

Figure 20–2 also shows the test points where you should place the voltmeter probes to execute the test. This test can be used to locate problems in any circuit, regardless of the number and types of pilot devices. The test points are identified as A through J. The first test should be made with the voltmeter probes touching points A and J, which is across the power supply. Use this test to make sure that your meter is on the right range. From this point, leave one of the voltmeter probes on point J and move the other probe to point B. If no voltage is available at point B, and you had voltage at point A, this test indicates the wire between the stop pushbutton and the power supply is open. You can turn off the power and test the wire for continuity and you should be able to verify that the wire has an open. A continuity test uses the resistance range on the VOM meter. If the wire has an open, the meter will indicate infinite resistance, and if the wire is good it is said to have continuity and the meter will indicate low resistance. The wire must be isolated (one end removed from its terminals) so that you do not measure resistance from a parallel circuit that may be connected to the wire you are testing. If your continuity test indicates that the wire has an open, you should remove and replace the wire to put the circuit back in service.

If you measured voltage when the meter probe is on point B and point J, the test indicates that the circuit is good to point B. You can move the meter terminal probe to point C. If you do not have voltage at point C and you had voltage at point B, the test indicates that the stop pushbutton is open. Remember that the stop pushbutton is a normally closed pushbutton, so you should have voltage at point C. Check to be sure that the stop button is not depressed. Remember that it may be a maintained type of switch and require someone to return it to its normal condition. If the switch is in its normal condition and no power is present at point C, the switch is faulty and must be replaced.

If you are testing for a stop switch that will not deenergize the relay coil, be sure to test point C and ensure that voltage is not present there when the stop switch is depressed. If voltage is still present at point C even when the stop button is depressed, you can assume that its contacts are welded or stuck together and that it must be replaced.

To continue the test for a loss of voltage in this circuit, we will resume from the last point where voltage is present at point C, and continue the test by moving the probe on to point D. If voltage is present at point C and not at point D, your test indicates that there is an open in the wire that connects the stop button and the start button. Since the voltage test indicates a loss of voltage at point D, you can turn off power to the circuit and test this wire for continuity and if this test indicates an open, you should change the wire.

If you have voltage at point D, you have determined that the circuit is operating correctly to this point and you can move to point E. Your test should indicate that you do not have voltage at point E, because the start pushbutton is a normally open switch. You will need to depress the start pushbutton and check for voltage at point E. If you do not have voltage at point E when the switch is depressed, you should turn off the power and test the switch for continuity. If the switch is faulty during the continuity tests, you should replace it.

If you have voltage at point E, your next test should be at point F. If you do not have voltage at point F, you can predict that the wire between the start pushbutton and the coil of the motor starter has an open in it. You can turn off the power and test this wire for continuity.

If you have voltage at point F, you should move the meter lead from point J to point G. Meter probes will now be touching both sides of the motor starter coil. Since the coil is a load, it should have full voltage applied to it if it is supposed to become energized. If you have voltage at points F and G, you can predict that the circuit wire and switches are all operating correctly and providing voltage to the coil. This is perhaps the most important point in troubleshooting. The main function of all of the wire and switches in every circuit is to provide the proper voltage to the load. In this circuit the load is the motor starter coil and if you measure voltage at its terminals and the contacts are not pulled in, you can predict the circuit wires and switches are operating correctly, and you should suspect the motor starter coil is faulty. You can turn off power to the circuit and test the motor starter coil for continuity. If the coil is good, the test should indicate resistance in the range of 20 to 2000 Ω, depending on how much wire is in the coil. If the test indicates infinite resistance, the coil is open and it should be replaced. Remember, do not apply voltage to a coil that is not installed in a relay.

If you had voltage when the meter leads were on points F and J, but did not have voltage at terminals F to G, your test has indicated that the circuit has an open between point G and point J. The final two tests to locate the problem in the circuit are perhaps the most difficult to understand, because it appears as though the circuit is operating correctly when your test determined that voltage is present at point F to J. In reality, your tests have proven that the wire and switches on the left side of the motor starter coil are operating correctly. Since the coil will not energize, you can then predict that the problem is in the wire between points G to J or overload contacts are open.

The best way to locate the fault in this part of the circuit is to move your left meter lead to point F and leave it there for the reminder of the tests, and resume the test by moving the right meter lead to test point J, then to I, and then to H. The point where you lose voltage is the place where there is an open in the circuit. For example, if you have voltage from point F to point J, but you do not have voltage when you move the right meter probe to point I, it indicates that the wire between J and I has an open. Turn off the power and test the wire for continuity. If your test indicates that you have voltage at terminal F to I, you should move the right meter lead to point H. If you do not have voltage at point H to F, it indicates that the overload contacts are open. You can verify this by turning off the power and testing the overload contacts for continuity. If you have voltage from point F to H, you can predict that the last wire to be tested, wire G to H, has an open. You can turn off the power and test the wire for continuity. Remove and replace the wire if it is faulty.

20.4 Developing a Ladder Diagram from a Wiring Diagram

Many times when you troubleshoot a system, a wiring diagram of the electrical circuit will be provided, but a ladder diagram will not be available. You can develop your own ladder diagram by tracing the circuits in the wiring diagram. The material in this section will provide an example of converting a wiring diagram to a ladder diagram. Figure 20–3 shows a wiring diagram of a typical motor control system that is connected to 230 VAC. You can see in this diagram that the major components in this system are the power supply, transformer, start and stop pushbuttons, motor starter, control relays, and hydraulic pump motor. One trick that will help you with this process is to take a colored pen and mark over each wire in the wiring diagram as you add it to the ladder

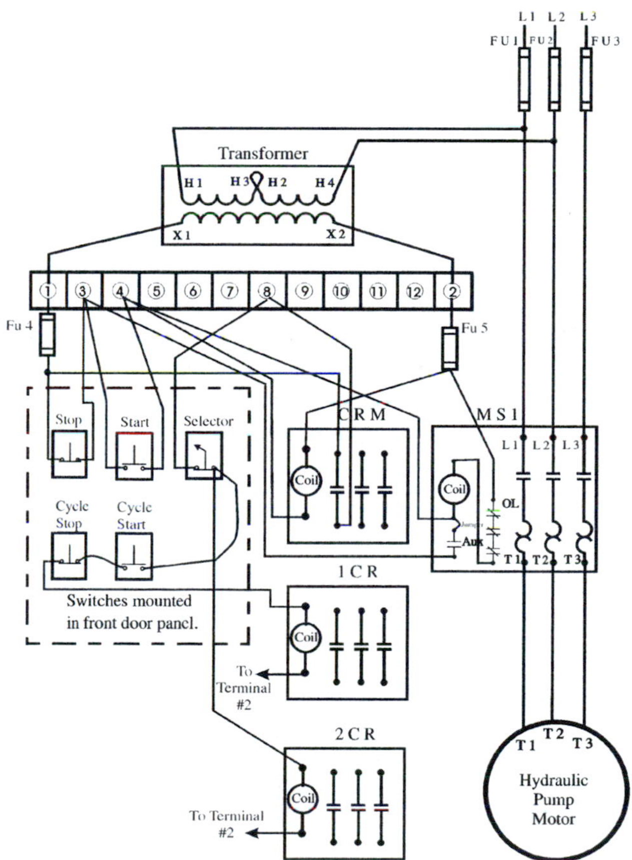

Figure 20–3 Wiring diagram for conversion to ladder.

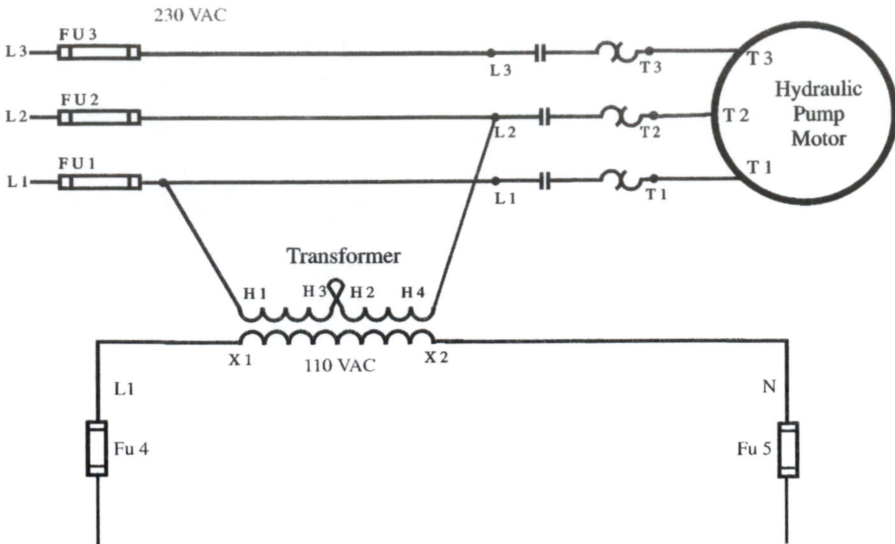

Figure 20–4 Load circuit of the ladder diagram.

diagram. When you have all the wires in the wiring diagram colored, you have completed the ladder diagram.

We will draw the load circuit separate from the control circuit. Since the load circuit has fewer components and is simpler, we will start with it. When you are converting a wiring diagram to a ladder diagram, you can start by locating the loads (motors) and the motor starter controlling each of them. In the example wiring diagram, you can see the motor is at the bottom right, and the motor starter is down the right side. The voltage for the load circuit comes from the three fuses at the top right side of the diagram. When we convert the load part of the wiring diagram to the ladder diagram, you will notice that it looks very similar. Figure 20–4 shows the load circuit across the top of the ladder diagram. The load voltage also provides the primary voltage for the control transformer at terminals H1 and H4, which are connected to L1 and L2. The secondary terminals of the transformer X1 and X2 provide the 110 VAC for the control circuit, so we will also show the transformer and fuses for the control circuit and identify them as L1 and N, which creates the two power rails for the entire control portion of the ladder diagram.

You can always start the ladder diagram of the control circuit with two power rails. The transformer primary winding will be connected to the power supply and the secondary will be connected to L1 and N at the top of the power rails for the ladder diagram. You can start developing the ladder diagram by placing the motor starter coil between the two rails of the ladder diagram. This is shown in Figure 20–5. (You can place the coils of relays and motor starters in any rung of the ladder diagram you are building at this time. The types of switches controlling the coil will determine the actual

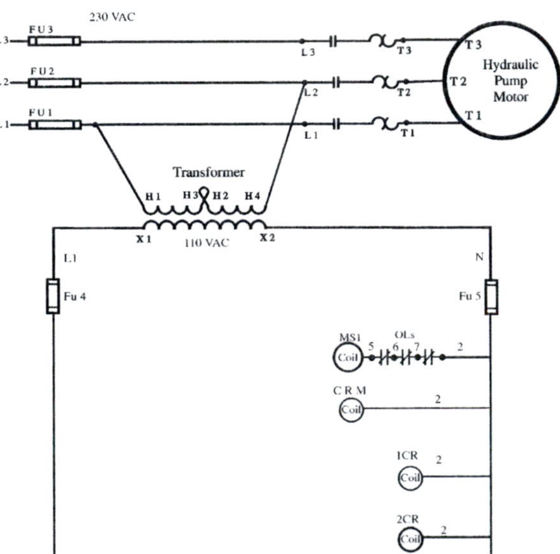

Figure 20–5 Ladder diagram with coils shown in position.

logic of the circuit. You will learn with experience how to locate the rungs in the ladder diagram.)

The next step in the procedure is to find the motor starter and its coil. From the top of the coil you can trace power to the N side of the circuit. From the wiring diagram you can see that power must travel through the three sets of normally closed overloads and then to fuse 5, which is connected to terminal 2, which is connected to N (neutral line of the circuit). The next process is to trace the wires back to L1 (terminal 1). The wire out of the bottom of the motor starter coil runs to a terminal (black dot). From the terminal, one of the wires goes to terminal 4 on the terminal board, and the other wire goes to the top of the auxiliary contacts. To minimize confusion, at this time you should follow only the wire from the motor starter coil to the start pushbutton and disregard the other wires that are connected to the coil and to terminal 4. Be sure to color the wires in the wiring diagram that you have added to the ladder diagram (see the ladder diagram in Figure 20–6).

The next step is to trace the wire from terminal 4 to the right side of the start pushbutton. From the left side of the start pushbutton, the wire runs back to terminal 3 on the terminal board. From terminal 3, a wire runs back to the right side of the stop pushbutton and from the left side of the stop pushbutton the wire runs back to the bottom of fuse 4, which is connected to terminal 1 and L1 (see the ladder diagram in Figure 20–7).

At this point you may get confused by the other wires that are connected to each of the terminals. For example, the auxiliary contacts are shown connected to terminals 3 and 4. You should add the auxiliary contacts and wires connecting the contacts to terminal 3 and terminal 4 at this time.

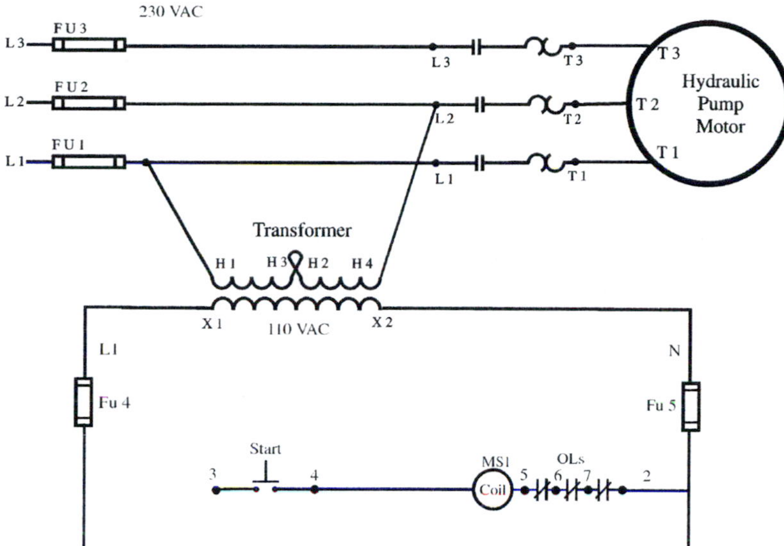

Figure 20–6 Ladder diagram with first line filled in.

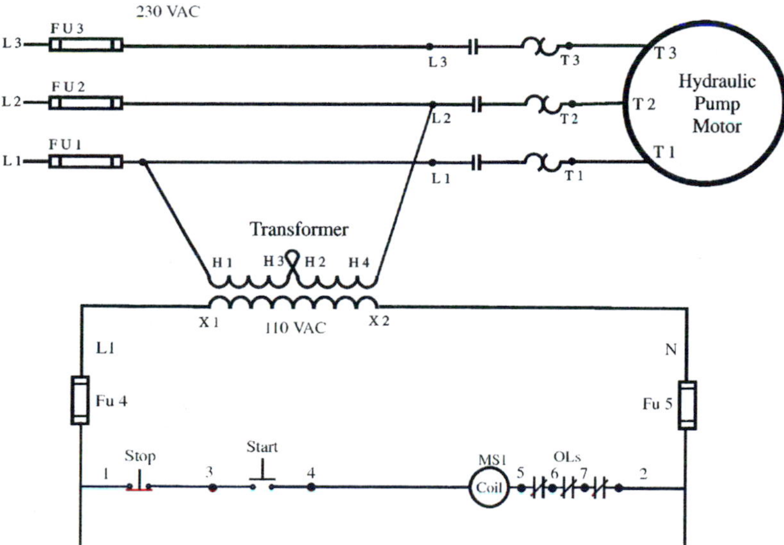

Figure 20–7 Start of ladder diagram with coils in place.

The final component in this circuit is the coil of the control relay master. You have already connected the right side of this coil to terminal 2. The wire from the left side of this coil is connected to terminal 4. When you have added these components and wires to the ladder diagram it will look like Figure 20–8. You should notice that the coil of the CRM is connected in parallel with the coil of MS1. Be sure to color these components and wires on the wiring diagram as you add them to the ladder diagram.

You are now ready to draw the next line of the ladder diagram. You have already drawn the MS coil and the CRM coil, so you are ready to add the next coil, which is the coil of control relay 1CR. The first wire to add to the coil is the wire that runs from the bottom of the 1CR coil to terminal 2 through fuse 5. The wires that run from the top of the 1CR coil connects the coil to the terminal on the left side of the cycle stop switch. From the right side of the cycle stop switch, a wires runs to the cycle start switch. Note that the wire can be a jumper, since the cycle start and cycle stop switches are mounted on the door of the electrical panel and are some distance from the terminal board. In some circuits, the wires would go all the way back to the terminal strip.

The next place the wires run are from the right side of the cycle start pushbutton to the right side of the selector switch. From the left side of the selector switch, the wire runs to terminal 8. From terminal 8 the wire runs back to the bottom of the normally open contacts of control relay CRM. Out of the top of these contacts, the wire runs back to the bottom of fuse 4, which is connected to terminal 1. See Figure 20–9 to see this part of the ladder diagram. Be sure to color the wires in the wiring diagram as you show them in the ladder diagram.

The final part of the ladder diagram is the coil of control relay 2CR. The wire from the right side of the coil is connected to terminal 2, which is connected to N. The wire from the left side of the 2CR coil goes to the left side of the selector switch, which makes its way back to open contacts of control relay CRM and then back to terminal 1 for L1. The coil of 2CR is connected in parallel with the coil of 1CR. This completes the process of converting the wiring diagram to a ladder diagram (Figure 20–10).

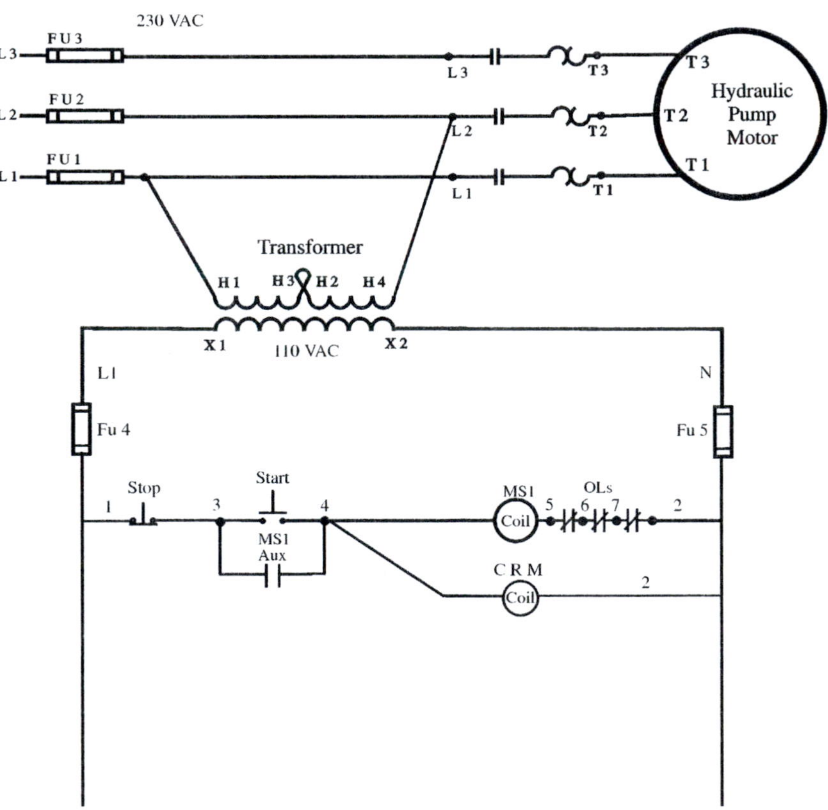

Figure 20–8 Ladder diagram with MS coil and CRM coil.

Figure 20–9 Ladder diagram with 1CR drawn in third rung.

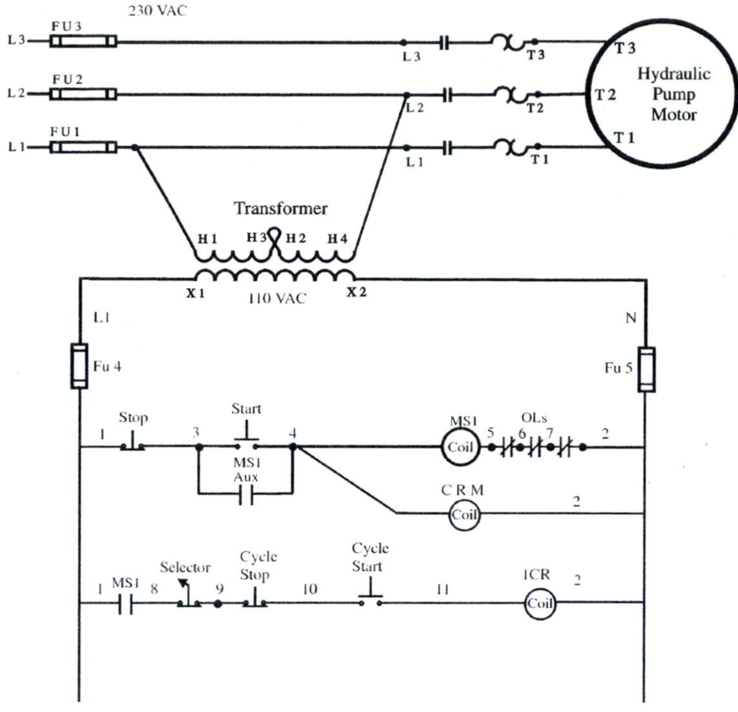

Figure 20–10 Completed ladder diagram.

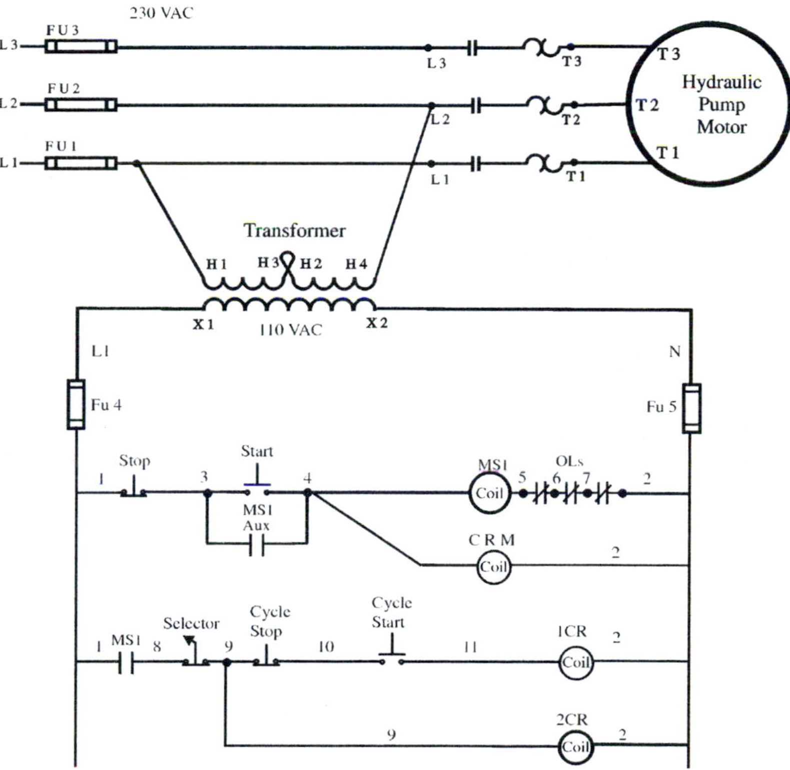

20.5 Converting a Wiring Diagram from a Ladder Diagram

At times you may have a ladder diagram (also called an elementary diagram) that provides the sequence of operation for the machine, but you do not have a wiring diagram that shows information such as the physical location of the terminals and the number of contacts on relays. This usually occurs when a machine is being rewired or when a new circuit is added to the electrical control circuit. When you are wiring the new components into the circuit, you will use many of the same skills that you would use to draw a wiring diagram from the ladder diagram. With practice, you can create a wiring diagram from a ladder diagram in a very short time.

The first step in this process is to make a visual inspection of the electrical cabinet and sketch the components that you see mounted inside the cabinet or components such as pushbutton and selector switches that are mounted in the door of the electrical cabinet. The sketch should be accurate with respect to the location and relative position of each component. It is also important to identify where terminal connections are on a component, such as the terminals where power comes into the device and goes out. The simplest way to identify these contacts is to use identifiers as top of the component and bottom of the component, or left-side terminal and right-side terminal. We will use these identifiers as we develop the diagram.

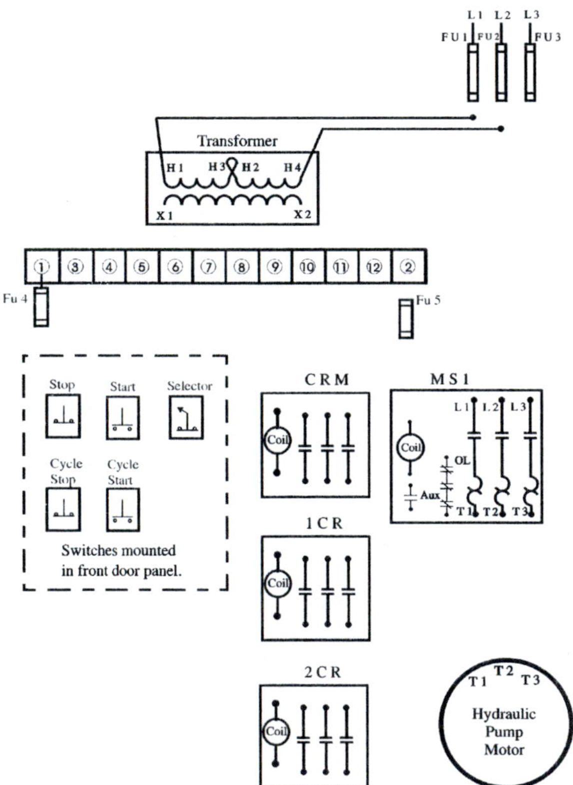

Figure 20–11 Layout of all components in the electrical cabinet to start conversion of the ladder diagram to a wiring diagram.

This section provides an exercise that shows how to develop a wiring diagram from a ladder diagram. In this exercise we will use the picture of the inside of the electrical cabinet and the components mounted in the cabinet door as shown in Figure 20–11. From this picture we will develop the sketch of the components, which will become the skeleton for the wiring diagram. This sketch of the components is also shown in Figure 20–11. We can use the ladder diagram provided in Figure 20–10 to begin making the wiring diagram. We will take advantage of the wire numbers provided in the ladder diagram. We will complete only the first two lines of the ladder diagram for this exercise. In reality, you would need to convert only the part of the ladder diagram where you suspect the problem exists. You would convert the entire diagram only if your company wanted a completed wiring diagram to go with the ladder diagram.

We start the process by copying the wires where voltage enters the ladder diagram. This will include the two wires connected to the primary winding of the transformer, the two wires that run from the secondary terminals at the bottom of the transformer to the terminals at the top of the fuses, the wires that run from the bottom of the fuse on the left that runs to terminal 1, and the wire from the fuse on the right side of the diagram that runs to the #2 terminals on the terminal board. Figure 20–11 also shows this part of the diagram.

Figure 20–12 shows the wires that run from terminal 1 to the terminals on the left side of the master stop pushbutton and from the terminals on the right side of the stop pushbutton to terminal 3 on the terminal board. The diagram in Figure 20–13 shows that the next wire runs from terminal 3 on the terminal board to the terminal on the left side of the master start pushbutton and from the terminal on the right side of the start pushbutton to terminal 4 on the terminal board.

Figure 20–13 also shows the wires that run from terminal 3 to the terminal on the left side of the normally open auxiliary contacts that are mounted on the motor starter, and from the terminal on the right side of the auxiliary contacts back to terminal 4. Figure 20–14 shows the diagram where the next wire connects the overloads of the motor starter to terminal 4 and terminal 7. Note that the terminal on the left side of the first overload is connected to terminal 4 and the

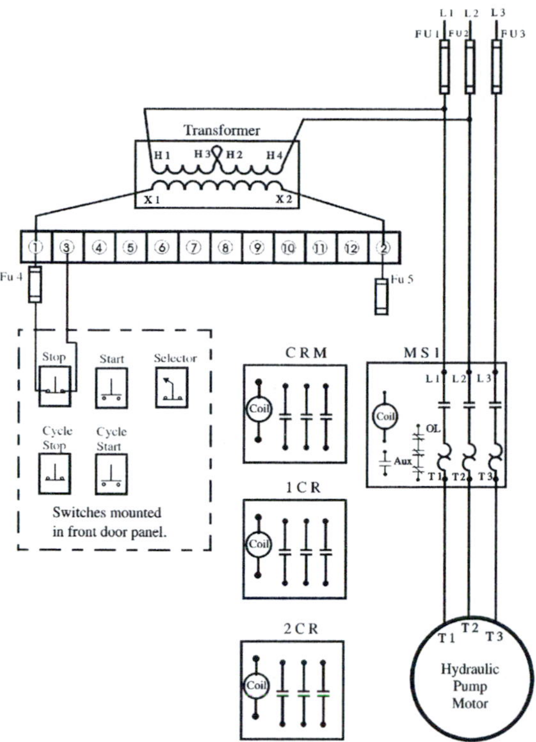

Figure 20–12 Wiring diagram with wires for transformer and motor load drawn in.

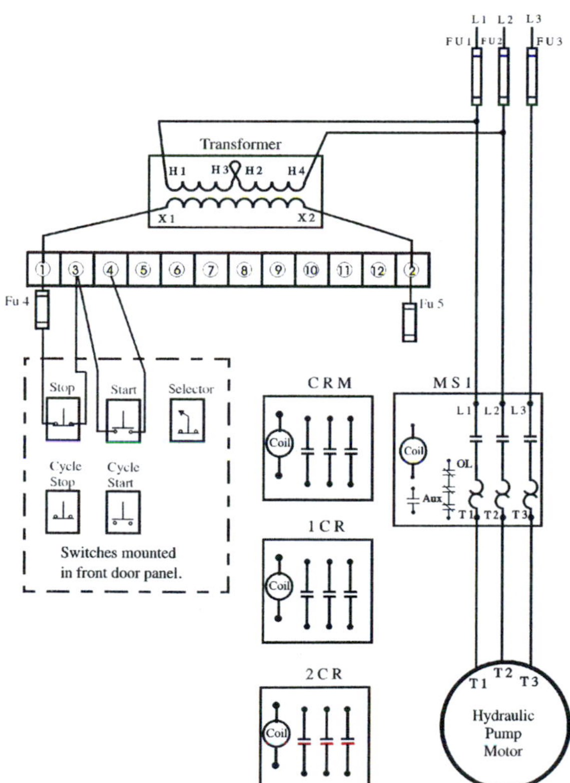

Figure 20–13 Ladder diagram with wires connected between stop and start pushbuttons.

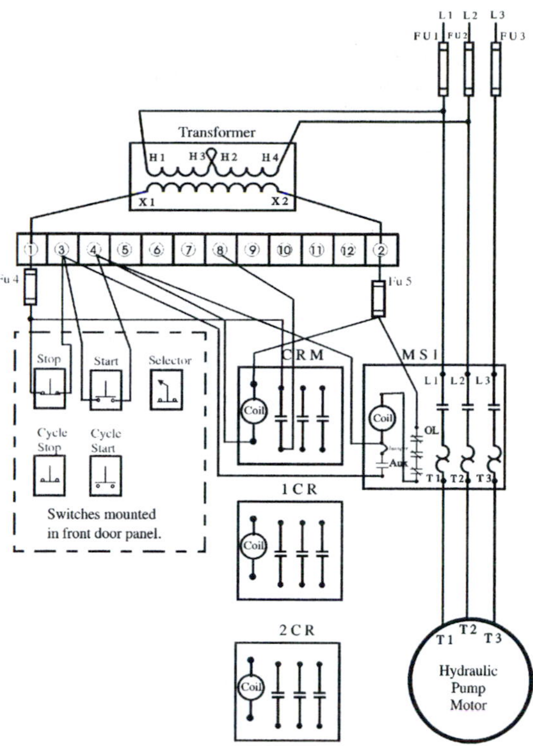

Figure 20–14 Wiring diagram with auxiliary contacts wired in and coils connected to terminal 2.

terminal on the right side of the third overload is connected to terminal 7. The first overload contact is connected directly to the second overload contact, and the second overload contact is connected directly to the third overload contact. (Since the overload contacts and the auxiliary contact are both mounted on the motor starter, some manufacturers may connect the #4 wire of the overloads directly to the auxiliary contacts on the motor starter rather than run the wire all the way back to the terminal board. This practice will save wire, since the distance between the auxiliary contacts and the overloads on the motor starter is relatively short. The #5 wire and the #6 wire that run between the three sets of overload contacts are not sent back to the terminal board. These wires are relatively short since the distance between the overloads is short.)

The final wire, #7, that comes from the right side of the third overload contact connects the last overload to the coil of the motor starter. This wire may also be run directly between the overload and the coil. The coil is also mounted on the motor starter and distance between the coil and the overload contacts is rather short. This practice will save a substantial amount of wire, compared to running the wire back to the terminal board.

The next wire in this part of the diagram runs between the coil of the motor starter and terminal 2. Another wire runs between terminal 7 and the coil of the master control relay (CRM) and from the coil to terminal 2.

The next section of the wiring diagram includes the wires for the second rung of the ladder diagram (Figure 20–15). You can see that the first wire runs between terminal 1 and the top terminal of the first set of normally open contacts that belong to the CRM relay. One end of wire 8 is connected to the bottom terminal of the contacts and the other end is connected to terminal board terminal 8. From terminal 8 on the terminal board the next wire runs to the terminal on the left side of the selector switch. The next wire, #9, is connected to the terminal on the right side of the selector switch on one end and the other end is connected to terminal board terminal 9. From terminal 9 on the terminal board the next wire runs to the terminal on the left side of the cycle stop pushbutton. One end of the #10 wire is connected to the terminal on the right side of the cycle stop pushbutton, and the other end is connected to terminal board terminal 10. From terminal 10 on the terminal board the next wire runs to the left side of the cycle start pushbutton. One end of wire 11 is connected to the right side of the start cycle pushbutton, and the other end is connected to terminal board terminal 11. From terminal 11 on the terminal board the next wire runs to the terminal on the left side of the 1CR coil. The final wire in this rung has one end connected to the left side of the 1CR coil and the other end connected to terminal 2.

Notice in the ladder diagram that a second control relay, 2CR, is wired in parallel to the coil of 1CR. This part of the circuit starts with wire 9, which is connected to terminal 9 in the terminal board on one end and to the terminal on the left side of the 2CR coil. The final wire in this circuit connects the terminal on the right side of the 2CR coil to terminal 2 (Figure 20–16).

20.6 Auto Mode and Manual Mode

When you are troubleshooting a machine, you may find that it can operate in automatic mode and manual mode. A selector switch allows the operator to switch

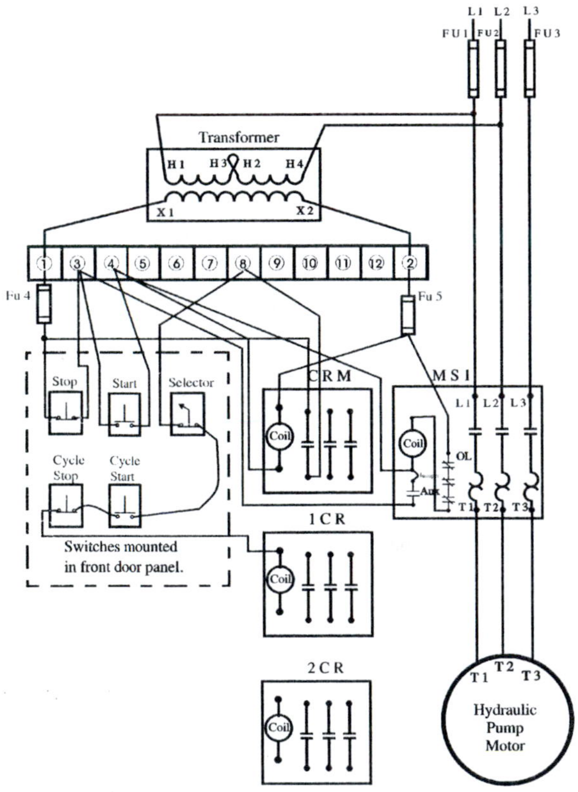

Figure 20–15 Wiring diagram with wires connected to cycle start and cycle stop and to the coil of 1CR.

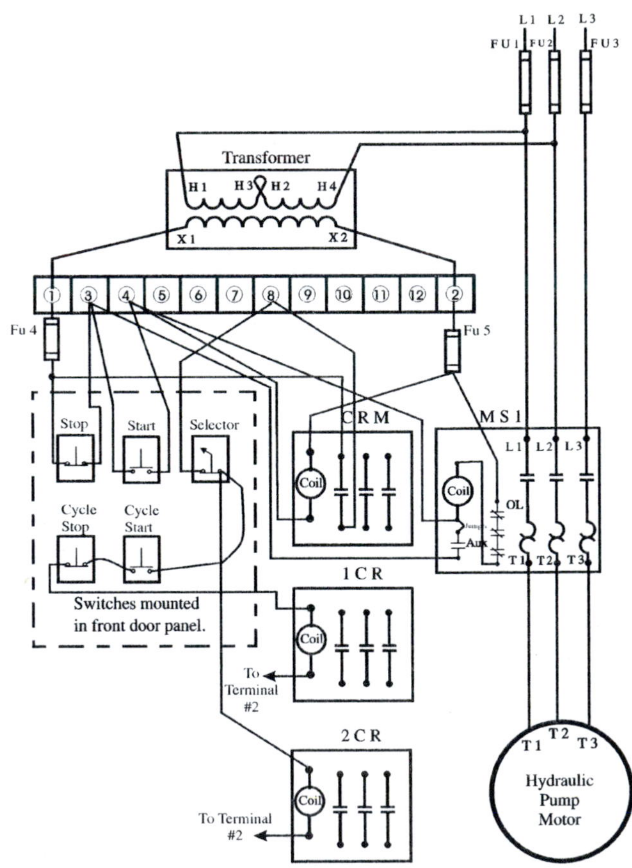

Figure 20–16 Completed wiring diagram.

the machine between automatic mode and manual mode. A good procedure to follow is to try to operate each section and component of the machine in manual mode. If the components operate in manual mode, you know that they are functioning correctly. If the machine does not operate in automatic mode, after testing it in manual mode, you can suspect that the components and wiring in the control circuitry are causing your problems when the system is in automatic mode.

20.7 Review of Troubleshooting

Now that you have seen all of the features of troubleshooting you should remember several important points. First, always work safely and use lockout and tag-out procedures when you are troubleshooting. You should also remember that troubleshooting is the process of determining what systems are working correctly on a machine, so that you can eliminate them from your process and focus on the part of the machine that is malfunctioning. Tests have been presented in this text that will allow you to test any component or control of the electrical system and verify if the part is good or bad. These include in-circuit tests and tests for when a part is removed from a system. If possible, you should test components while they are in circuit. If you determine that a part is faulty, then remove it and test it again when it is out of circuit to verify your previous findings.

When you are troubleshooting, you should use electrical ladder diagrams and wiring diagrams whenever possible. If you have a wiring diagram and need a ladder diagram, you can convert the diagram using the procedures shown in this chapter. You can also convert a ladder diagram to a wiring diagram when you are installing components or need the wiring diagram for other tests.

The final point to remember about troubleshooting is to use all of your senses—vision, hearing, smell, and touch—to evaluate a machine that is malfunctioning. Also be sure to ask personnel who work around the machine how the machine operates under normal condition. Check other similar machines, if they are available, to verify operation. Try to operate all of the sections or components of the machine in manual mode, before you test them in automatic mode.

You must learn to develop a strategy that strictly follows procedures. You should select tests that will eliminate the largest portions of the machine, like the example that showed the best procedure of identifying a card in a deck of cards. Finally, you should keep very accurate and detailed records of your troubleshooting and maintenance functions.

Questions

1. Explain why you should use the voltage loss method to detect an open in a circuit.
2. Explain the difference between a symptom and problem.
3. Explain when it would be appropriate to swap a part that you thought was faulty.
4. Explain why it is important to use the same procedure every time you troubleshoot a machine.
5. Explain when you should use an ammeter to make a test in a circuit.

True and False

1. When you test a component with an ohmmeter, you should always ensure the power is turned off.
2. When you test a component with a voltmeter, you should always ensure the power is turned off.
3. When you are called to troubleshoot a machine, you should use all of your senses to determine what parts of the machine are operating correctly.
4. You could use your "sense of feel" to determine if a machine is vibrating excessively.
5. A clamp-on ammeter could be used to determine which line of the three lines that supply voltage to a motor has a blown fuse.

Multiple Choice

1. A wiring diagram _____
   a. shows the sequence of operation for the circuit.
   b. shows the location of each component in the circuit.
   c. shows how each component operates.
   d. shows how a machine should be troubleshooted.

2. A ladder diagram _____
   a. shows the sequence of operation for the circuit.
   b. shows the location of each component in the circuit.
   c. shows how each component operates.
   d. shows how a machine should be troubleshooted.

3. Troubleshooting may best be described as _____
   a. finding the part of the machine that is broken.
   b. identifying the parts of a machine that are working correctly and through a process of elimination determining what is faulty.
   c. swapping parts until the machine starts running again.
   d. removing parts from a machine and testing each one with an ohmmeter.

4. The control circuit _____
   a. includes the motor and other loads in a circuit.
   b. includes the motor and other controls in a circuit.
   c. includes switches and other controls in a circuit.
   d. All of the above

5. The load circuit _____
   a. includes the motor and other loads in a circuit.
   b. includes the motor and other controls in a circuit.
   c. includes switches and other controls in a circuit.
   d. All of the above

Index

D

E

O

P

R

S